# HARMONIC ANALYSIS AND REPRESENTATIONS OF SEMISIMPLE LIE GROUPS

# MATHEMATICAL PHYSICS AND APPLIED MATHEMATICS

*Editors:*

M. Flato, *Université de Dijon, Dijon, France*

R. Rączka, *Institute of Nuclear Research, Warsaw, Poland*

*with the collaboration of:*

M. Guenin, *Institut de Physique Théorique, Geneva, Switzerland*

D. Sternheimer, *Collège de France, Paris, France*

VOLUME 5

# HARMONIC ANALYSIS AND REPRESENTATIONS OF SEMISIMPLE LIE GROUPS

*Lectures given at the NATO Advanced Study Institute on Representations of Lie Groups and Harmonic Analysis, held at Liège, Belgium, September 5–17, 1977*

*Edited by*

J. A. WOLF, M. CAHEN, AND M. DE WILDE

D. REIDEL PUBLISHING COMPANY

DORDRECHT : HOLLAND/BOSTON : U.S.A./LONDON : ENGLAND

**Library of Congress Cataloging in Publication Data**

CIP

Nato Advanced Study Institute on Representations
of Lie Groups and Harmonic Analysis, Liège 1977.
Harmonic analysis and representations of semisimple Lie groups.

(Mathematical physics and applied mathematics; V. 5)
Includes bibliographies and index.
1. Harmonic analysis—Congresses. 2. Lie groups
Congresses. 3. Representations of groups
Congresses. I. Wolf, Joseph Albert, 1936-
II. Cahen, Michel. III. Wilde, M. De. IV. Nato
Advanced Study Institute. Liège, 1977. V. Title.
VI. Series.
QA403.N37 1977 515′.2433 80-10768
ISBN 90-277-1042-2

---

Published by D. Reidel Publishing Company,
P.O. Box 17, 3300 AA Dordrecht, Holland

Sold and distributed in the U.S.A. and Canada
by Kluwer Boston Inc., Lincoln Building,
160 Old Derby Street, Hingham, MA 02043, U.S.A.

In all other countries, sold and distributed
by Kluwer Academic Publishers Group,
P.O. Box 322, 3300 AH Dordrecht, Holland

D. Reidel Publishing Company is a member of the Kluwer Group



Printed in The Netherlands

# TABLE OF CONTENTS



[A more detailed Table of Contents is given at the beginning of each paper.]

# PREFACE

This book presents the text of the lectures which were given at the NATO Advanced Study Institute on Representations of Lie groups and Harmonic Analysis which was held in Liège from September 5 to September 17, 1977.

The general aim of this Summer School was to give a coordinated introduction to the theory of representations of semisimple Lie groups and to non-commutative harmonic analysis on these groups, together with some glance at physical applications and at the related subject of random walks.

As will appear to the reader, the order of the papers – which follows relatively closely the order of the lectures which were actually given – follows a logical pattern. The two first papers are introductory: the one by R. Blattner describes in a very progressive way a path going from standard Fourier analysis on $\mathbb{R}^n$ to non-commutative harmonic analysis on a locally compact group; the paper by J. Wolf describes the structure of semisimple Lie groups, the finite-dimensional representations of these groups and introduces basic facts about infinite-dimensional unitary representations. Two of the editors want to thank particularly these two lecturers who were very careful to pave the way for the later lectures. Both these chapters give also very useful guidelines to the relevant literature.

In the paper on the role of differential equations in the Plancherel theorem, P. Trombi studies the partial Fourier transform of continuous functions on a reductive group $G$ with values in a complex Frechet space and gives a proof of the Plancherel theorem. V. Varadarajan gives, in a paper on the infinitesimal theory of representation of semisimple Lie groups, a construction of irreducible Harish-Chandra modules which will include in particular fundamental series of representations: the main emphasis in this paper is on the infinitesimal point of view.

In the paper by M. Atiyah and W. Schmid, which is here reproduced from the *Inventiones*, a geometric realization of the discrete series, analogous to the Borel–Weil theorem in the compact case, is used to obtain the main properties of the discrete series of representations. An erratum to this beautiful paper is added at the end. At Liège, W. Schmid gave only two lectures on the regularity of invariant eigendistributions on semisimple Lie groups. A detailed version of these results has appeared in the *Inventiones* and is not reproduced here.

The next paper by M. Flato and D. Sternheimer contains essentially three parts: in the first part (Chapters 1 and 2) one finds an alternative formulation of quantum mechanics on phase space when the associative multiplication of functions is suitably deformed. The second part (Chapters 3 and 4) deals with analytic vectors in group representation. The third part (Chapters 5 and 6) deals with non-linear representations of Lie groups in an infinite-dimensional space and the possibility of linearizing these representations.

The contribution of J. Simon develops the 1-cohomology of representations. It is in particular of fundamental importance for the study of non-linear representations.

The last paper by H. Furstenberg describes various qualitative aspects of the theory of random walks on a Lie group $G$ and in particular boundaries of the group $G$ and $\mu$-harmonic functions ($\mu$=probability measure on $G$).

We are pleased to thank NATO who gave us the basic financial means to organise this Summer School and who gave us useful practical advice in the early stages of the organization.

The University of Liège offered us a warm hospitality and some financial support which was very helpful. The beautiful Sart Tilmant Campus was an ideal location for the lectures. We express here our gratitude.

We thank the Solvay foundation, and in particular its director, I. Prigogine, who gave us some support which was useful to invite mathematicians from non NATO countries.

The firm Faulx et Champagne offered generously to each participant a nice briefcase and we are happy to thank them.

Finally, Springer Verlag gave us the permission necessary to reproduce the Atiyah–Schmid article which appeared in the *Inventiones*. We thank them for their cooperation.

ROBERT J. BLATTNER

# GENERAL BACKGROUND*

## Contents



## 1. Introduction

These lectures will be devoted to the general subject of harmonic analysis on locally compact and Lie groups. We shall begin with the classical problem of decomposing a function of period $2\pi$ on $\mathbb{R}$ into harmonics of the fundamental 'tone': $f(x) \sim \Sigma_{n\in\mathbb{Z}} a_n e^{inx}$. Already in this simple setting we face two questions: (1) What sort of regularity properties should $f$ possess for the decomposition to make any sense at all?; (2) In what sense does the series converge? These questions (or their analogues) will persist throughout our investigations.

After dealing with periodic functions on $\mathbb{R}$, we will take up the situation in $\mathbb{R}^n$. Here harmonics of a fundamental 'tone' are replaced by plane waves, and the summation must be replaced by a multiple integral. The generalization to locally compact Abelian groups is quite simple. In the process of generalizing, we lose those powerful tools (Schwartz distributions) associated with the differentiable structure of $\mathbb{R}^n$, but the generality gained allows application of the theory to situations which could not be touched before, such as the additive or multiplicative groups of a locally compact non-Archimedean field. The main theorems are those of Bessel, Parseval, Plancherel, Herglotz, and Bochner, and the duality theorem of Pontrjagin.

*This work was supported in part by NSF Grant MCS75-17621.



*J. A. Wolf, M. Cahen, and M. De Wilde (eds.), Harmonic Analysis and Representations of Semi-Simple Lie Groups, 1–67.*

To generalize the foregoing to the non-Abelian case requires recasting the problems. One reformulation notes that the plane waves on $\mathbb{R}^n$ (or the unitary characters on a general locally compact Abelian $G$) are eigenfunctions under translation: in the non-Abelian case these translation eigenspaces are replaced by translation invariant subspaces which can be given a complete locally convex topology in such a way as to be topologically irreducible. If these spaces are given invariant Hilbert space structures and if $f \in L_2(G, \mu)$, where $\mu$ is left Haar measure, then the harmonic decomposition of $f$ will result from the decomposition of $L_2(G, \mu)$ under the left regular representation as a direct sum (or direct integral) of irreducible unitary representations. In case $G$ is compact, this is accomplished by the Peter–Weyl theorem. If $G$ is only locally compact, the situation is technically more complicated and will require us to develop direct integral theory. The other reformulation aims to decompose only those $f$ invariant under the inner automorphisms of $G$. Plane waves are replaced by 'characters' of irreducible representations: functions such as $x \mapsto \mathrm{Trace}(\pi(x))$, where $\pi$ is an irreducible representation. Now if $G$ is compact, $\pi$ will be finite-dimensional and so its character will be a well-defined function. But if $G$ is not compact, $\pi$ can be infinite-dimensional so that strictly speaking $\mathrm{Trace}(\pi(x))$ will not be well-defined. In case $G$ is a Lie group, there is hope that $\mathrm{Trace}(\pi(\cdot))$ can be interpreted as a distribution, and if $G$ is semisimple or nilpotent that hope is realized.

With either of the foregoing reformulations there is the problem of finding all irreducible unitary representations of $G$, or at least enough of them to perform the harmonic decomposition of a given $f$. In many cases, this can be done by 'inducing' representations of closed subgroup $H$ of $G$ up to $G$. When $G$ is discrete, induction arises naturally as the (right or left) adjoint of the functor that assigns to every representation of $G$ its restriction to $H$. When $G$ is locally compact, this formulation of induction works in general only when $H$ has finite index in $G$. For arbitrary closed $H$, there is a natural construction due to Mackey and Rieffel that serves well and is natural, although it no longer has a nice functorial interpretation. In case $G$ is a Lie group, it is possible to extend this construction to situations in which one induces from representations of subalgebras of the complexification $\mathfrak{g}_c$ of the Lie algebra $\mathfrak{g}$ of $G$, and this extension can be used to produce important irreducible representations of solvable and semisimple groups.

These notes are organized as follows: In Section 2, we go over harmonic analysis on $\mathbb{R}/2\pi\mathbb{Z}$ and on $\mathbb{R}^n$. These results are extended to locally compact

Abelian $G$ in Section 3. The compact case is covered in Section 4. Section 5 deals with harmonic analysis on general locally compact $G$, including direct integral decomposition theory for unitary representations. Lie groups are dealt with in Section 6, and, finally, in Section 7 we discuss induced representations and their use in producing irreducible representations.

## 2. Harmonic analysis on $\mathbb{R}/2\pi\mathbb{Z}$ and on $\mathbb{R}^n$

Let $f$ be an 'arbitrary' function on $\mathbb{R}$ with values in $\mathbb{C}$ and which has period $2\pi$. Our object is to decompose in some sense $f$ as a linear combination of the harmonic functions $x \mapsto e^{inx}$, $n \in \mathbb{Z}$, so that we will have

$$f(x) \sim \sum_{n \in \mathbb{Z}} a_n e^{inx} \tag{2.1}$$

for some choice of coefficients $\{a_n\}$. We are immediately faced with the problem of giving sense to the sum on the right-hand side of (2.1). One way of doing this is to interpret it as a 'weak sum'; that is, for functions $g$ of period $2\pi$ in a certain class $\mathcal{D}$ we want

$$\int_0^{2\pi} f(x)g(x)\,dx = \sum_{n \in \mathbb{Z}} a_n \int_0^{2\pi} e^{inx} g(x)\,dx, \tag{2.2}$$

where all the integrals are to converge absolutely and where the sum on the right-hand side is to converge unconditionally. The functions in $\mathcal{D}$ are known as *test functions*. For each choice of $\mathcal{D}$ we will have a class of functions $f$ for which (2.2) holds for some choice of $\{a_n\}$ and for all $g \in \mathcal{D}$. Obviously, in order for $\mathcal{D}$ to be a useful class of test functions, the class of decomposable functions $f$ associated to $\mathcal{D}$ must be large; moreover, the coefficients $\{a_n\}$ must be completely determined by $f$.

Suppose $\mathcal{D}$ contains all the harmonic functions $x \mapsto e^{inx}$. Then, if $f \in L_1([0, 2\pi])$ and if (2.2) holds for all $g \in \mathcal{D}$, we must have

$$a_n = \frac{1}{2\pi} \int_0^{2\pi} e^{-inx} f(x)\,dx \quad \text{for} \quad n \in \mathbb{Z}, \tag{2.3}$$

because $\int_0^{2\pi} e^{-inx} e^{imx}\,dx = 2\pi\delta_{mn}$. $\{a_n\}$, regarded as a function on $\mathbb{Z}$, is called the *Fourier Transform* of $f$ and is written $\hat{f}: \hat{f}(n) = a_n$, $n \in \mathbb{Z}$. We henceforth always assume $\mathcal{D}$ contains these harmonics.

One might be tempted to let $\mathscr{D}$ consist of all continuous periodic $g$. Then (2.2) will hold for all $f \in L_2([0, 2\pi])$, as we shall see below (Theorem 2.14). But it will not hold for all $f \in L_1([0, 2\pi])$, which may be seen as follows: Choose ([21], p. 51) a continuous periodic $g$ such that

$$\left\{\sum_{k}{}^{n}_{-n} \int_0^{2\pi} e^{ikx} g(x)\, dx\right\}$$

is unbounded in $n \geqslant 0$. Define the linear functional $\lambda_n$ on $L_1([0, 2\pi])$ by

$$\lambda_n(f) = \sum_{k}{}^{n}_{-n} \lambda_n(f) = \sum_{k-n}^{n} \hat{f}(k) \int_0^{2\pi} e^{ikx} g(x)\, dx.$$

If (2.2) holds for all $f \in L_1([0, 2\pi])$ and this $g$, then the set of functionals $\{\lambda_n\}$ is pointwise bounded. By the Uniform Boundedness Theorem, $\{\lambda_n\}$ is bounded in norm. Now let $\{f_m\}$ be a sequence of periodic functions integrable on $[0, 2\pi]$ such that $f_m \geqslant 0$, $\int_0^{2\pi} f_m(x)\, dx = 1$, and

$$\int_0^{2\pi} f_m(x) h(x)\, dx \to h(0)$$

for all continuous periodic $h$. Then

$$\hat{f}_m(k) \to \frac{1}{2\pi} \quad \text{for all} \quad k \in \mathbb{Z},$$

so that

$$\lambda_n(f_m) \underset{m}{\to} \sum_{k}{}^{n}_{-n} \frac{1}{2\pi} \int_0^{2\pi} e^{ikx} g(x)\, dx,$$

so that $\{\lambda_n(f_m)\}$ is unbounded in $n$ and $m$, which contradicts the supposition that $\|\lambda_n\| \geqslant |\lambda_n(f_m)|$ is supposed to be bounded. Thus for our purposes, letting $\mathscr{D}$ consist of all continuous periodic $g$ is a poor idea.

A much better idea is to let $\mathscr{D}$ consist of the $C^\infty$ periodic functions. Let $g \in \mathscr{D}$. Integration by parts gives

$$(2.4) \qquad \int_0^{2\pi} e^{inx} g'(x)\, dx = -in \int_0^{2\pi} e^{inx} g(x)\, dx.$$

Moreover,

$$(2.5)\qquad \left|\int_0^{2\pi} e^{inx} g(x)\,\mathrm{d}x\right| \leqslant \|g\|_1 \leqslant 2\pi\|g\|_\infty.$$

Repeated application shows that, for any polynomial $P$,

$$(2.6)\qquad \left\{P(n)\int_0^{2\pi} \mathrm{e}^{inx} g(x)\,\mathrm{d}x\right\} \quad \text{is bounded in } n \text{ for all } g\in\mathscr{D}.$$

Then (2.2) holds for all $f\in L_1([0, 2\pi])$. Indeed, we can say much more. For $g\in\mathscr{D}$, set

$$(2.7)\qquad \|g\|_{(k)} = \|g^{(k)}\|_\infty, \qquad k\geqslant 0.$$

Give $\mathscr{D}$ the locally convex topology determined by the seminorms $\|\cdot\|_{(k)}$. With this topology $\mathscr{D}$ will be called the *periodic Schwartz space* of $\mathbb{R}$. Its topological dual $\mathscr{D}'$ is the space of *periodic distributions* on $\mathbb{R}$. If $\mu\in\mathscr{D}'$ and $g\in\mathscr{D}$ we shall write $\int_0^{2\pi} g(x)\,\mathrm{d}\mu(x)$ for $\mu(g)$. Then we have the following theorem ([44], p. 225):

THEOREM 2.8 (Schwartz). *Let* $\mu\in\mathscr{D}'$. *Define* $\hat{\mu}$ *on* $\mathbb{Z}$ *by*

$$\hat{\mu}(n) = \frac{1}{2\pi}\int_0^{2\pi} \mathrm{e}^{-inx}\,\mathrm{d}\mu(x).$$

*Then for each* $g\in\mathscr{D}$ *we have*

$$(2.9)\qquad \int_0^{2\pi} g(x)\,\mathrm{d}\mu(x) = \sum_{n\in\mathbb{Z}} \hat{\mu}(n)\int_0^{2\pi} e^{inx} g(x)\,\mathrm{d}x,$$

*as an unconditional sum. Indeed, we have*

$$(2.10)\qquad \lim_{|n|\to\infty} \hat{\mu}(n)(1+n^2)^{-k} = 0 \text{ for } k \text{ large enough},$$

*and*

$$(2.11)\qquad \mu = \sum_{n\in\mathbb{Z}} \hat{\mu}(n)\,\mathrm{e}^{inx}$$

*as an unconditional sum in* $\mathscr{D}'$. *Here, as always, a periodic function*

$f \in L_1([0, 2\pi])$ *is identified with the distribution*

$$g \mapsto \int_0^{2\pi} g(x) f(x)\, \mathrm{d}x.$$

Equation (2.11) raises another question: Let $f$ belong to some topological function space. Is it true that

$$(2.12) \qquad f(x) = \sum_{n \in \mathbb{Z}} \hat{f}(n)\, \mathrm{e}^{inx}$$

as an unconditional sum in that function space? Here are some answers:

THEOREM 2.13. *For $L_p([0, 2\pi])$, $1 < p < \infty$, the answer is 'Yes'. For $L_1([0, 2\pi])$ and for the space $\mathscr{C}$ of continuous periodic functions under the sup norm, the answer is 'No'.* (Cf. [21], pp. 46–50.)

The case $p = 2$ of (2.13) is of special interest. Then $L_2([0, 2\pi])$ is a Hilbert space, and the validity of (2.12) follows from the fact that

$$\{x \mapsto (2\pi)^{-1/2}\, \mathrm{e}^{inx} : n \in \mathbb{Z}\}$$

is a complete orthonormal set. Of course then, (2.2) holds also with $\mathscr{D} = L_2([0, 2\pi])$, and we have the familiar

THEOREM 2.14 (Parseval). *Let $f, g \in L_2([0, 2\pi])$. Then*

$$\frac{1}{2\pi} \int_0^{2\pi} f(x)\overline{g(x)}\, \mathrm{d}x = \sum_{n \in \mathbb{Z}} \hat{f}(n)\overline{\hat{g}(n)}.$$

(2.14) may be restated as follows: The map $f \mapsto \hat{f}$ is an isometry of $L_2([0, 2\pi], (1/2\pi)\, \mathrm{d}x)$ into $L_2(\mathbb{Z}$, counting measure). That $\{x \mapsto \mathrm{e}^{inx} : n \in \mathbb{Z}\}$ is an orthonormal basis of the former space implies that this isometry is bijective. Thus we have a complete characterization of the Fourier transforms of $L_2$ periodic functions. It is natural to ask if other classes of functions can be characterized by their Fourier transforms. There seems to be no nice characterization of the Fourier transforms of $L_1$ functions.

Finally, we mention

THEOREM 2.15 (Herglotz). *The Fourier transforms of the non-negative periodic measures are precisely the functions* $n \mapsto a_n$, $n \in \mathbb{Z}$, *which are positive definite; i.e. for all choices of functions* $n \mapsto z_n$, $n \in \mathbb{Z}$, *of finite support, we have* $\sum_{n,m \in \mathbb{Z}} a_{n-m} z_n \bar{z}_m \geqslant 0$. (cf. [21], pp. 38–39.)

THEOREM 2.16 (Schwartz). *The Fourier transforms of the* $C^\infty$ *periodic functions are precisely those* $\{a_n\}$ *for which* $\{P(n)a_n\}$ *is bounded for all polynomials P. The Fourier transforms of the periodic distribution are precisely those* $\{a_n\}$ *satisfying the growth condition* (2.10). (cf. [44], pp. 225, 227.)

What about pointwise convergence of (2.12)? There are many classical theorems dealing with this question (cf. [21]). We are content to mention the theorem of Carleson–Hunt ([9, 20]) that proves that (2.12) converges almost everywhere for $f \in L_p$, $p > 1$. We have mentioned that there exist $f \in \mathscr{C}$ for which (2.12) diverges unboundedly for $x = 0$. Finally, Kolmogorov has shown [23] that there exists $f \in L_1$ for which (2.12) diverges *everywhere*.

To generalize the foregoing to (non-periodic) functions on $\mathbb{R}^n$ we must find an analogue for the harmonics. Our analogue is the set of periodic plane waves: $x \mapsto e^{i\xi \cdot x}$. Here $x \in \mathbb{R}^n$, $\xi \in \mathbb{R}^n$, and $\cdot$ is the ordinary 'dot' product. If $f$ is an 'arbitrary' function on $\mathbb{R}^n$, we desire a plane wave decomposition:

$$f(x) \sim \int_{\mathbb{R}^n} a(\xi)\, e^{i\xi \cdot x}\, d\xi. \tag{2.17}$$

In this formula $d\xi$ is Lebesgue measure and $a$ is some function. We have written the decomposition as an integral. If we agree that $a(\xi)\, d\xi$ can be a distribution, we include the possibility that the integral can be interpreted as a sum. Integrating (2.17) weakly, we require

$$\int_{\mathbb{R}^n} f(x) g(x)\, dx = \int_{\mathbb{R}^n} a(\xi) \left( \int_{\mathbb{R}^n} g(x)\, e^{i\xi \cdot x}\, dx \right) d\xi \tag{2.18}$$

for all $g$ in some space $\mathscr{D}$ of test functions.

Let us allow both $f(x)\, dx$ and $a(\xi)\, d\xi$ to be distributions of some sort, say $\mu$ and $\hat{\mu}$. (2.18) then becomes

$$\int_{\mathbb{R}^n} g(x)\, d\mu(x) = \int_{\mathbb{R}^n} \left( \int_{\mathbb{R}^n} g(x)\, e^{i\xi \cdot x}\, dx \right) d\hat{\mu}(\xi). \tag{2.19}$$

The good choice for $\mathscr{D}$ is motivated by the following considerations.

Let $g \in C_c^\infty(\mathbb{R}^n)$, the space of $C^\infty$ functions on $\mathbb{R}^n$ with compact support. Define $\check{g}$ by

$$(2.20) \qquad \check{g}(\xi) = \int_{\mathbb{R}^n} g(x)\, e^{i\xi \cdot x}\, dx, \qquad \xi \in \mathbb{R}^n.$$

Then $\check{g} \in C^\infty(\mathbb{R}^n)$. For $y \in \mathbb{R}^n$, define $\partial_y g$ by

$$(2.21) \qquad (\partial_y g)(x) = (d/dt) g(x + ty)|_{t=0}.$$

If $\eta \in \mathbb{R}^n$, define $p_\eta$ on $\mathbb{R}^n$ by

$$(2.22) \qquad p_\eta(x) = \eta \cdot x, \qquad x \in \mathbb{R}^n.$$

Then differentiation under the integral gives, using the translation invariance of Lebesgue measure,

$$(2.23) \qquad (\partial_y g)\check{} = -ip_y \check{g}$$

and

$$(2.24) \qquad (p_\eta g)\check{} = -i\partial_\eta \check{g}.$$

Thus if D is a linear differential operator with polynomial coefficients, there is another such operator $\check{D}$ such that

$$(2.25) \qquad (Dg)\check{} = \check{D}\check{g} \qquad \text{for all} \qquad g \in C_c^\infty(\mathbb{R}^n).$$

Moreover, the map $D \mapsto \check{D}$ is a bijection of the space of all such operators onto itself. This observation makes reasonable the following definition.

DEFINITION 2.26. The *Schwartz space* $\mathcal{S}(\mathbb{R}^n)$ consists of all $C^\infty$ functions $g$ such that $Dg$ is bounded for each differential operator D with polynomial coefficients. $\mathcal{S}(\mathbb{R}^n)$ is topologized by the seminorms $\|\cdot\|_D$ defined by $\|g\|_D = \|Dg\|_\infty$, $g \in \mathcal{S}(\mathbb{R}^n)$.

THEOREM 2.27 (Schwartz). *$\mathcal{S}(\mathbb{R}^n)$ is a complete locally convex space. The identity injection of $C_c^\infty(\mathbb{R}^n)$ into $\mathcal{S}(\mathbb{R}^n)$ is continuous when $C_c^\infty(\mathbb{R}^n)$ is given the usual inductive limit topology of uniform convergence of functions, together with all derivatives, within compact sets, and the image is dense. For each $\xi \in \mathbb{R}^n$, the right-hand side of* (2.20) *is a continuous function of $g$ for $g \in C_c^\infty(\mathbb{R}^n)$ in the relative topology from $\mathcal{S}(\mathbb{R}^n)$ and so extends by continuity to $g \in \mathcal{S}(\mathbb{R}^n)$. The $\check{g}$ so defined belongs to $\mathcal{S}(\mathbb{R}^n)$. Finally, $g \mapsto \check{g}$ is a topological linear automorphism of $\mathcal{S}(\mathbb{R}^n)$.* (cf. [44], pp. 233–237, 249–250.)

Thus, if $\mu$ belongs to the topological dual $\mathscr{S}'(\mathbb{R}^n)$ of $\mathscr{S}(\mathbb{R}^n)$, there is a unique $\hat{\mu}\in\mathscr{S}'(\mathbb{R}^n)$ such that (2.19) holds. According to (2.27), if $\mu\in\mathscr{S}'(\mathbb{R}^n)$, $\mu|C_c^\infty(\mathbb{R}^n)$ is a distribution which uniquely determines $\mu$. So it makes sense to call $\mu$ a distribution: $\mathscr{S}'(\mathbb{R}^n)$ is the space of *tempered distributions* on $\mathbb{R}^n$.

THEOREM 2.28 (Schwartz). *$\mu\mapsto\hat{\mu}$ is a topological linear automorphism of $\mathscr{S}'(\mathbb{R}^n)$.*

Which distributions on $\mathbb{R}^n$ are tempered? Schwartz (*loc. cit.*) gives several characterizations: for example, a distribution $\mu$ is tempered if and only if it is the restriction to $\mathbb{R}^n$ of a distribution on the one point compactification $S^n$ of $\mathbb{R}^n$, or if and only if it is the result of applying a differential operator with polynomial coefficients to a bounded continuous function. For our purposes it is important to note that $L_p(\mathbb{R}^n$, Lebesgue measure$)\subseteq\mathscr{S}'(\mathbb{R}^n)$; in fact, if $f\in L_p$, $g\in\mathscr{S}(\mathbb{R}^n)$, we have that

$$\int_{\mathbb{R}^n} f(x)g(x)\,\mathrm{d}x=\int_{\mathbb{R}^n} f(x)\frac{1}{(1+x\cdot x)^k}(1+x\cdot x)^k g(x)\,\mathrm{d}x,$$

that $x\mapsto 1/(1+x\cdot x)^k$ belongs to $L_{p/(p-1)}$ for $k$ an integer $\geqslant n(p-1)/2p$, and that $g\mapsto h$ with $h(x)=(1+x\cdot x)^k g(x)$ is continuous from $\mathscr{S}(\mathbb{R}^n)$ to $L_\infty$. Another important fact is that measure $\mu$ is a tempered distribution if $\int_{|x|\leqslant r}|\mathrm{d}\mu|$ has polynomial growth in $r$ and that this condition is necessary for non-negative measures. ([44], p. 242.)

Applying the foregoing to functions in $L_1$, we get

THEOREM 2.29. *If $\mu=f(x)\,\mathrm{d}x$, $f\in L_1$, then $\hat{\mu}=\hat{f}(\xi)\,\mathrm{d}\xi$, where $\hat{f}$ is a continuous function on $\mathbb{R}^n$ vanishing at $\infty$ and given by the formula*

$$\hat{f}(\xi)=\frac{1}{(2\pi)^n}\int_{\mathbb{R}^n} f(x)\,\mathrm{e}^{-i\xi\cdot x}\,\mathrm{d}x.$$

For functions in $L_2$, we have

THEOREM 2.30 (Plancherel). *If $\mu=f(x)\,\mathrm{d}x$, $f\in L_2$, then $\hat{\mu}=\hat{f}(\xi)\,\mathrm{d}\xi$, where $\hat{f}\in L_2$ and is given by*

$$\hat{f}(\xi)=\lim_{r\to\infty}\frac{1}{(2\pi)^n}\int_{|x|\leqslant r} f(x)\,\mathrm{e}^{-i\xi\cdot x}\,\mathrm{d}x,$$

*where the limit is taken in the* norm *of* $L_2$. *Moreover the map* $f \mapsto \hat{f}$ *is a unitary isomorphism of* $L_2(\mathbb{R}^n$, $1/(2\pi)^n$ Lebesgue measure) *with* $L_2(\mathbb{R}^n$, Lebesgue measure). *Moreover,*

$$f(x) = \lim_{r\to\infty} \int_{|x|\leqslant r} \hat{f}(\xi)\, \mathrm{e}^{i\xi\cdot x}\, \mathrm{d}\xi, \text{ where}$$

*the limit is in the* $L_2$ *norm.*

Generalizing (2.15) is

THEOREM 2.31 (Bochner). $\{\hat{\mu} : \mu$ *is a finite positive measure*$\}$ *is precisely the space of continuous functions h such that*

$$\text{(2.32)} \qquad \sum_{ij}{}_1^m z_i \bar{z}_j h(\xi_i - \xi_j) \geqslant 0$$

*for all choices* $z_1, \ldots, z_m \in \mathbb{C}$ *and* $\xi_1, \ldots, \xi_m \in \mathbb{R}^n$, *arbitrary positive integers m.*

Functions satisfying (2.32) are called *positive definite.* Making use of certain algebraic operations on $\mathscr{S}(\mathbb{R}^n)$ or $C_c(\mathbb{R}^n)$ (convolution, involution), it is possible to give an alternate characterization that is generalizable to measures and distributions. We will postpone this until Section 3.

Our final theorem is an important result concerning $L_2$ functions on $\mathbb{R}$. Regard $\mathbb{R}$ as embedded in $\mathbb{C}$.

THEOREM 2.33. *Let* $f \in L_2(\mathbb{R})$. *Then there exists a function F holomorphic in the strip* $\{z \in \mathbb{C} : |\text{Im } z| < a\}$ *such that* $f(x) = F(x)$ *for* $x \in \mathbb{R}$ *and such that* $\int_{\mathbb{R}} |F(x+iy)|^2\, \mathrm{d}x$ *is bounded on* $\{y : |y| < a\}$ *if and only if* $\xi \mapsto \mathrm{e}^{a|\xi|}\hat{f}(\xi)$ *belongs to* $L_2(\mathbb{R})$. (cf. [21], p. 174.)

Theorems of this type were originated by Paley and Wiener [36].

## 3. Locally compact abelian groups

The preceding section has dealt with the harmonic analysis of functions defined on the groups (under addition) $\mathbb{R}^n$, $\mathbb{Z}$, and on $\mathbb{R}/2\pi\mathbb{Z}$, noting that functions on $\mathbb{R}$ of period $2\pi$ can be regarded as functions on $\mathbb{R}/2\pi\mathbb{Z}$. The spaces are locally compact and the operations of addition and $x \mapsto -x$ are continuous. It turns out that one natural domain for the questions of harmonic analysis is that of the class of locally compact Hausdorff topological groups.

DEFINITION 3.1. A *topological group* is a group $G$ which is topologized so that the maps $(x, y) \mapsto xy$ and $x \mapsto x^{-1}$ of $G \times G$ into $G$ and $G$ into $G$, respectively, are continuous.

Simple facts concerning topological groups include the following: (1) $G$ is Hausdorff if and only if it is $T_0$; (2) any open subgroup of $G$ is closed; (3) if $H$ is a subgroup of $G$, then $G/H$ is Hausdorff in the strongest topology making the canonical projection $G \to G/H$ continuous if and only if $H$ is closed; (4) if $H$ is a normal subgroup of $G$, then $G/H$ is a topological group.

In what follows our topological groups will be Hausdorff unless otherwise stated.

Let $G$ be a locally compact group. As examples, think of $\mathbb{R}^n$, $\mathbb{Z}$, GL(2, $\mathbb{R}$) in the topology it inherits from the four-dimensional real space of $2 \times 2$ matrices, and $\mathbb{Z}_p$, the completion of $\mathbb{Z}$ in the $p$-adic norm. A measure on $G$ is a linear functional on $C_c(G)$ continuous when $C_c(G)$ is given the usual inductive limit topology of uniform convergence within compact sets. Our first theorem gives us an analogue of Lebesgue measure on $G$.

THEOREM 3.2 (Haar). *The set of measures $\mu$ on $G$ such that $\mu(\lambda_x f) = \mu(f)$ for all $f \in C_c(G)$, where $(\lambda_x f)(y) = f(x^{-1}y)$, is one-dimensional and contains a positive measure.*

A positive measure given by (3.2) is called a *left Haar measure*. Lebesgue measure and the counting measure are (left) Haar measures on $\mathbb{R}^n$ and $\mathbb{Z}$, respectively. As for $GL(2, \mathbb{R})$, writing its members as $\begin{pmatrix} x_1 & x_2 \\ x_3 & x_4 \end{pmatrix}$, left Haar measure is $|x_1 x_4 - x_2 z_3|^{-2}\, dx_1\, dx_2\, dx_3\, dx_4$. As for (left) Haar measure $\mu$ on $\mathbb{Z}_p$, it is determined by the condition $\mu(p^k \mathbb{Z}_p) = p^{-k}$.

Let $G$ be a locally compact Abelian group. Our aim is to decompose functions, measures, and generalized functions in terms of the analogues of the harmonics ($\mathbb{R}/2\pi\mathbb{Z}$) and the periodic plane waves ($\mathbb{R}^n$). Harmonics and periodic plane waves can be characterized as those bounded continuous functions $f$ on the group $G$ in question such that $f(1) = 1$ and such that $f$ is an eigenfunction for all the operators $\lambda_x$, $x \in G$. These are precisely the *characters* of $G$, that is, the continuous homomorphisms of $G$ into the group $\{z\in\mathbb{C}: |z|=1\}$. Let $\hat{G}$ denote the set of characters of $G$.

LEMMA 3.3. *Let* $\alpha, \beta\in\hat{G}$. *Define* $\alpha\beta$ *and* $\alpha^{-1}$ *on* $G$ *by* $(\alpha\beta)(x)=a(x)\beta(x)$ *and* $\alpha^{-1}(x)=\alpha(x)^{-1}=\overline{\alpha(x)}$ *for* $x\in G$. *Give* $\hat{G}$ *the topology of uniform convergence on compact subsets of* $G$. *Then* $\hat{G}$ *is a topological group.*

$\hat{G}$ is called the *character group* or the *dual group* of $G$. It will turn out that this group is also locally compact, but this fact is most easily seen by getting at $\hat{G}$ another way, which we now do.

DEFINITION 3.4. Let $f, g\in C_c(G)$. Define $f\circ g$ and $f^*$ on $G$ by

$$(f\circ g)(x)=\int_G f(y)g(y^{-1}x)\,\mathrm{d}y$$

and $f^*(x)=\overline{f(x^{-1})}$ for $x\in G$. (Here $\mathrm{d}y$ denotes (left) Haar measure on $G$, chosen once for all.)

LEMMA 3.5. $C_c(G)$ *is closed under* $\circ$ *and* $*$ *and becomes an associative* $*$*-algebra over* $\mathbb{C}$ *when* $\circ$ *is used as multiplication and* $*$ *as the involution (or* $*$*-operation).* $C_c(G)$ *is commutative under* $\circ$.

Using Haar measure, we can form the various $L_p$ spaces $L_p(G)$. We have the inner product $(\cdot,\ \cdot)$ between $L_p(G)$ and $L_{p/(p-1)}(G)$ for $1\leqslant p\leqslant\infty$ given by $(f, g)=\int f(x)\overline{g(x)}\,\mathrm{d}x$. Noting that $\mathrm{d}x=\mathrm{d}(x^{-1})$, we have

LEMMA 3.6. *Let* $f, g, h\in C_c(G)$. *Then* $\|f\circ g\|\leqslant\|f\|_1\|g\|_p$ *and* $\|f^*\|_p=\|f\|_p$ *for* $1\leqslant p\leqslant\infty$. *Moreover* $(f\circ g, h)=(g, f^*\circ h)$.

In view of (3.6), the map $\lambda_f g=f\circ g$, $g\in C_c(G)$, for each $f\in C_c(G)$ is continuous in the $L_2$ norm from $C_c(G)$ into itself and hence extends to a continuous endomorphism of $L_2(G)$, also called $\lambda_f$. $\|\lambda_f\|\leqslant\|f\|_1$. Moreover, the map $f\mapsto\lambda_f$ is a $*$-homomorphism of $C_c(G)$ into the $*$-algebra of all bounded operators on $L_2(G)$. We denote the norm closure of the image of this map by $\mathfrak{A}$. $\mathfrak{A}$ is a commutative $C^*$-algebra.

Now $C_c(G)$ and also $\mathfrak{A}$ has an identity if and only if $G$ is discrete. However, we do have an 'approximate identity' in $C_c(G)$ and $\mathfrak{A}$ as follows:

LEMMA 3.7. *For each neighborhood $V$ of 1 in $G$, choose $u_V \in C_c(G)$ such that $u_V \geq 0$, $\operatorname{Supp}(u_V) \leq V$, and $\int u_V(x)\,dx = 1$. Let $f \in C_c(G)$. For every $1 \leq p \leq \infty$ we have $\lim_V u_V \circ f = f$ in the $L_p$ norm. Moreover, $\lim_V \lambda_{u_V} = I$ in the strong operator topology. (Here our limits are with respect to the directed set of neighborhoods of 1 where $V_1 \geq V_2$ if $V_1 \subseteq V_2$.)*

Let $G^{\#}$ be the space of all continuous $*$-homomorphisms of $\mathfrak{A}$ onto $\mathbb{C}$ under the Gel'fand topology (the topology of simple convergence). We have the following important theorem:

THEOREM 3.8. *Let $\varphi \in G^{\#}$ and suppose that $f \in C_c(G)$ satisfies $\varphi(\lambda_f) \neq 0$. define $\alpha_{\varphi,f}$ on $G$ by*

$$\alpha_{\varphi,f}(x) = \varphi(\lambda_{\lambda_{x^{-1}}f})/\varphi(\lambda_f) \text{ for } x \in G.$$

*Then, if $g \in C_c(G)$ satisfies $\varphi(\lambda_g) \neq 0$, we have $\alpha_{\varphi,f} = \alpha_{\varphi,g}$ so that we may write $\alpha_\varphi$ without ambiguity. $\alpha_\varphi \in \hat{G}$ and the map $\varphi \mapsto \alpha_\varphi$ is a homeomorphism of $G^{\#}$ onto $\hat{G}$. Finally, if $f \in C_c(G)$, we have*

$$\varphi(\lambda_f) = \int f(x)\overline{\alpha_\varphi(x)}\,dx. \tag{3.9}$$

For a proof of this theorem see Loomis ([26], Chapter VII) and the modifications of those arguments due to Williamson [52]. Much of what follows is an adaptation of Williamson's methods suitable to our point of view.

COROLLARY 3.10. *$\hat{G}$ is a locally compact Abelian group.*

As examples of $\hat{G}$, we mention that $\mathbb{R}^n$ can be identified with $\mathbb{R}^{\hat{n}}$ by letting $\xi \in \mathbb{R}^n$ correspond to the character $x \mapsto e^{i\xi \cdot x}$ and $\mathbb{Z}$ can be identified with $(\mathbb{R}/2\pi\mathbb{Z})\hat{}$ by letting $n \in \mathbb{Z}$ correspond to the function $x \mapsto e^{inx}$, which is the pullback to $\mathbb{R}$ of a character of $\mathbb{R}/2\pi\mathbb{Z}$. The topologies and group structures are preserved by these correspondences.

Now formula (3.9) gives our generalization of the Fourier transform to general locally compact Abelian $G$. If $\alpha \in \hat{G}$ and $f \in C_c(G)$, we define $\hat{f}$ on $\hat{G}$ by

$$\hat{f}(\alpha) = \int f(x)\overline{\alpha(x)}\,dx \qquad \text{for} \qquad \alpha \in \hat{G}. \tag{3.11}$$

This formula makes sense if $f \in L_1(G)$. Moreover, since $\|\lambda_f\| \leq \|f\|_1$ for

$f \in C_c(G)$, $\lambda$ extends to a map of $L_1(G)$ into $\mathfrak{A}$; (3.6) tells us that $\circ$ and $*$ extend to $L_1(G)$ which, by (3.5), becomes a Banach$*$-algebra; and $\lambda$ is a continuous $*$-homomorphism of $L_1(G)$ onto a dense subalgebra of $\mathfrak{A}$. By Gel'fand theory, we have the following, the first statement being the *Riemann–Lebesgue lemma*:

THEOREM 3.12. *For $f \in L_1(G)$, $\hat{f} \in C_\infty(\hat{G})$, the space of continuous functions vanishing at $\infty$ on $\hat{G}$. $f \mapsto \hat{f}$ is a $*$-isomorphism of $L_1(G)$ under $\circ$ and $*$ onto a $\|\cdot\|_\infty$-dense $*$-subalgebra of $C_\infty(\hat{G})$ under pointwise multiplication of functions and complex conjugation. Moreover, $\|\hat{f}\|_\infty = \|\lambda_f\| \leq \|f\|_1$.*

In order to prove analogues of the Plancherel and Bochner theorems and the theorem of Pontrjagin that $G$ is canonically isomorphic as a topological group to $\hat{\hat{G}}$, we introduce a space $\mathcal{D}(G)$ of test functions suggested by the paper of Williamson [52].

DEFINITION 3.13. $\|\cdot\|_0$ is defined on $C_c(G)$ by

$$\|f\|_0 = \max(\|f\|_\infty, \|f\|_2, \|\hat{f}\|_\infty).$$

$\mathcal{D}(G)$ is the completion of $C_c(G)$ with respect to $\|\cdot\|_0$.

Let $v_1$, $v_2$, and $v_3$ be defined as the injections of $C_c(G)$ into $C_\infty(G)$, $L_2(G)$, and $C_\infty(\hat{G})$ given by inclusion in the first two cases and the Fourier transform in the third. Giving $C_c(G)$ the $\|\cdot\|_\infty$-norm, $v_1$, $v_2$, and $v_3$ are bounded linear maps. Since $C_\infty(G)$, $L_2(G)$, and $C_\infty(\hat{G})$ are complete we may extend $v_j$ to $\mathcal{D}(G)$, $j=1, 2, 3$.

LEMMA 3.14. *On $\mathcal{D}(G)$ we have $v_1 = v_2$ (i.e. $v_1(g) = v_2(g)$ almost everywhere for $g \in \mathcal{D}(G)$). Moreover, $v_1$ and $v_3$ are injective.*

Thus $\mathcal{D}(G)$ may be regarded as a subspace of $C_\infty(G) \cap L_2(G)$. Moreover, the Fourier transforms can be extended to $\mathcal{D}(G)$ by setting $\hat{g} = v_3(g)$ for $g \in \mathcal{D}(G)$, and we can also define $\lambda_g$ without ambiguity. Note, however, that $\mathcal{D}(G)$ is not in general a subspace of $L_1(G)$ so that $\hat{g}$ is not given by (3.11).

LEMMA 3.15. *If $f, g \in C_c(G)$, then $\|f^*\|_0 = \|f\|_0$ and $\|f \circ g\|_0 \leq \|f\|_0 \|g\|_0$, so that $\circ$ and $*$ extend to $\mathcal{D}(G)$ and the foregoing formulae hold for $f, g \in \mathcal{D}(G)$. $f \mapsto \hat{f}$ is then a $*$-isomorphism of the Banach $*$-algebra $\mathcal{D}(G)$*

*onto a* $\|\cdot\|_\infty$*-dense subalgebra of* $C_\infty(\hat{G})$. *Finally,* $(f, g) = (f \circ g^*)(1)$ *for* $f, g \in \mathcal{D}(G)$.

(This lemma is a simple consequence of (3.6) and the formula $(f \circ g)(x) = (f, \lambda_x g^*)$.)

We introduce a linear functional $\Lambda$ on $\mathcal{D}(G)\hat{}$ by setting $\Lambda(\hat{f}) = f(1)$ for $f \in \mathcal{D}(G)$. That $\Lambda$ is *positive* follows from

LEMMA 3.16. *Let* $\{u_V\}$ *be an approximate identity as in* (3.7). *Let* $f \in \mathcal{D}(G)$. *Then* $f(1) = \lim_V (\lambda_f u_V, u_V)$.

Moreover, simple calculations show that if $\beta \in \hat{G}$, then $\lambda_\beta \hat{f} = (\beta f)\hat{}$ for $f \in C_c(G)$ and hence, by continuity, that $\mathcal{D}(G)$ is closed under multiplication by $\beta$ and that $\lambda_\beta \hat{f} = (\beta f)\hat{}$ for $f \in \mathcal{D}(G)$. Thus $\Lambda(\lambda_\beta \hat{f}) = \Lambda(\hat{f})$ for $f \in \mathcal{D}(G)$. Now this says that if we knew that $\mathcal{D}(G)\hat{} \supseteq C_c(\hat{G})$, we could define a Haar measure $\mu$ on $\hat{G}$ by setting $\mu(\hat{f}) = \Lambda(\hat{f})$ *for* $\hat{f} \in C_c(\hat{G})$. To this end we prove

LEMMA 3.17. *Let* $f, g \in \mathcal{D}(G)$, $H \in C_\infty(\hat{G})$. *Then* $\hat{f}\hat{g}H \in \mathcal{D}(G)\hat{}$.

*Proof.*

$$\|f \circ g \circ h\|_2 \leqslant \|\hat{h}\|_\infty \|\hat{f}\|_\infty \|g\|_2 \leqslant \|\hat{h}\|_\infty \|f\|_0 \|g\|_0,$$

$$|(f \circ g \circ h)(x)| = |(h \circ f, \lambda_x g^*)| \leqslant \|\hat{h}\|_\infty \|f\|_2 \|g^*\|_2 \leqslant \|\hat{h}\|_\infty \|f\|_0 \|g\|_0,$$

and

$$\|\hat{f}\hat{g}\hat{h}\|_\infty \leqslant \|\hat{f}\|_\infty \|\hat{g}\|_\infty \|\hat{h}\|_\infty \leqslant \|\hat{h}\|_\infty \|f\|_0 \|g\|_0$$

for $f, g, h \in C_c(G)$ so that $\|f \circ g \circ h\|_0 \leqslant \|\hat{h}\|_\infty \|f\|_0 \|g\|_0$. Pass to the limit on $f, g$ in $\mathcal{D}(G)$ and on $\hat{h}$ in $C_\infty(\hat{G})$.

PROPOSITION 3.18. $\mathcal{D}(G)\hat{} \supseteq C_c(\hat{G})$.

*Proof.* Let $F \in C_c(\hat{G})$ and set $K = \mathrm{Supp}(F)$. Choose $0 \neq f \in C_c(G)$ so that $f \geqslant 0$. Then $\hat{f}(1) > 0$. Choose $\beta_1, \ldots, \beta_m \in \hat{G}$ so that $\sum_{j\,1}^m \lambda_{\beta_j} |\hat{f}|^2 > 0$ on $K$ and then choose $H \in C_c(\hat{G})$ so that $H \sum_{j\,1}^m \lambda_{\beta_j} |\hat{f}|^2 = F$; that is,

$$F = \sum_{j\,1}^m (\beta_j f)\hat{}\,(\beta_j f)^*\hat{}\, H \in \mathcal{D}(G)\hat{}.$$

THEOREM 3.19. *Haar measure on* $\hat{G}$ *can be normalized so that, for*

*$f \in \mathscr{D}(G)$ such that $\hat{f} \in C_c(\hat{G})$, we have*

$$f(x) = \int \hat{f}(\alpha)\alpha(x)\,d\alpha \qquad \textit{for} \qquad x \in G.$$

*Proof.* $\Lambda_0 = \Lambda | C_c(\hat{G})$ is a Haar integral. Using this Haar measure, we have $f(1) = \int \hat{f}(\alpha)\,d\alpha$. But then

$$f(x) = (\lambda_{x^{-1}} f)(1) = \int (\lambda_{x^{-1}} f)\hat{\ }(\alpha)\,d\alpha = \int \hat{f}(\alpha)\alpha(x)\,d\alpha.$$

We can now attack our main theorems.

THEOREM 3.20 (Plancherel). *If $f \in \mathscr{D}(G)$, then $\hat{f} \in L_2(\hat{G})$. Using the Haar measure of (3.19) on $\hat{G}$, we have $\|\hat{f}\|_2 = \|f\|_2$. The extension of $f \mapsto \hat{f}$ to all of $L_2(G)$ is a unitary isomorphism of $L_2(G)$ onto $L_2(\hat{G})$.*

*Proof.* Let $f \in \mathscr{D}(G)$. Then $\|f\|_2^2 = (f^* \circ f)(1) = \Lambda(|\hat{f}|^2)$. Approximating $|\hat{f}|^2$ from below by functions in $C_c(\hat{G})$, we see that $\|\hat{f}\|_2 \leqslant \|f\|_2$ with equality if $\hat{f} \in C_c(\hat{G})$. It suffices, therefore, to show that $\mathscr{E} = \{f \in \mathscr{D}(G) : \hat{f} \in C_c(G)\}$ is $L_2$-dense in $\mathscr{D}(G)$. This is straightforward using an approximate identity $\{u_V\}$ and then approximating $\hat{u}_V$ in the $\|\cdot\|_\infty$ norm by some $\hat{g}$ with $g \in \mathscr{E}$.

Now for each $x \in G$, we have the function $\tilde{x}$ on $\hat{G}$ defined by $\tilde{x}(\alpha) = \alpha(x)$, $\alpha \in \hat{G}$. From (3.3) we see that $\tilde{x} \in \hat{\hat{G}}$ and that the map $\psi : G \to \hat{\hat{G}}$ defined by $\psi(x) = \tilde{x}$, $x \in G$, is a continuous homomorphism of topological groups.

THEOREM 3.21 (Pontrjagin duality). *$\psi$ is a topological group isomorphism of $G$ onto $\hat{\hat{G}}$.*

*Proof.* We show that $F \mapsto F \circ \psi$ is an $\|\cdot\|_\infty$-norm preserving $*$-isomorphism of $C_\infty(\hat{\hat{G}})$ onto $C_\infty(G)$.

(a) If $F \in C_c(\hat{G})\hat{\ }$, choose $f \in \mathscr{E}$ with $\hat{\hat{f}} = F$. Then $f(x^{-1}) = F(\psi(x))$, $x \in G$, and $F \circ \psi \in C_\infty(G)$. If now $g \in \mathscr{E}$, then $\hat{f} \circ \hat{g} \in C_c(G)$ so that $\hat{f} \circ \hat{g} = \hat{h}$ with $h \in \mathscr{E}$. By (3.19) and the Fubini theorem $h = fg$. Thus, by (3.20), $\|\hat{f} \circ \hat{g}\|_2 = \|fg\|_2 \leqslant \leqslant \|f\|_\infty \|\hat{g}\|_2$. This says that $\|F\|_\infty = \|\lambda_{\hat{f}}\| \leqslant \|f\|_\infty = \|F \circ \psi\|_\infty \leqslant \|F\|_\infty$; i.e. $\|F\|_\infty = \|F \circ \psi\|_\infty$.

(b) Since $C_c(\hat{G})\hat{\ }$ is $\|\cdot\|_\infty$-dense in $C_\infty(\hat{\hat{G}})$, $F \mapsto F \circ \psi$ is a norm preserving $*$-isomorphism of $C_\infty(\hat{\hat{G}})$ onto $C_\infty(G)$.

(c) The theorem follows by showing that $\mathscr{E}$ is $\|\cdot\|_\infty$-dense in $C_\infty(G)$. This is similar to the last part of the proof of (3.20).

COROLLARY 3.22. *$\hat{}$ maps $\mathscr{D}(G)$ isometrically onto $\mathscr{D}(\hat{G})$. $\mathscr{D}(G)$ is a Banach $*$-algebra under pointwise multiplication of functions and complex conjugation. If $f, g \in \mathscr{D}(G)$, then $(f \circ g)\hat{} = \hat{f}\hat{g}$, $(fg)\hat{} = \hat{f} \circ \hat{g}$, $f^*\hat{} = \bar{\hat{f}}$, and $\bar{f}\hat{} = \hat{f}^*$. Also $\hat{\hat{f}}(\hat{x}) = f(x^{-1})$, $x \in G$.*

Taking our cue from (2.19), (2.20), and (2.29), we define $\check{g}(\alpha) = \hat{g}(\alpha^{-1})$ for $g \in \mathscr{D}(G)$, $\alpha \in \hat{G}$. Then if $\mu \in \mathscr{D}'(G)$, the topological dual of $\mathscr{D}(G)$, we can define $\hat{\mu} \in \mathscr{D}'(\hat{G})$ by

$$\hat{\mu}(\check{g}) = \mu(g). \tag{3.23}$$

In this way we define the Fourier transform of members of $\mathscr{D}'(G)$ and obtain an isomorphism of $\mathscr{D}'(G)$ with $\mathscr{D}'(\hat{G})$. Clearly, $\mathscr{D}(G) \subseteq \mathscr{D}'(G)$ *via* the map that makes $f \in \mathscr{D}(G)$ correspond to the functional $g \mapsto \int f(x)g(x)\,\mathrm{d}x$ on $\mathscr{D}(G)$. Then (3.20) and (3.22) imply that (3.23) extends the Fourier transform on $\mathscr{D}(G)$. $\mathscr{D}'(G)$ also contains the finite measures on $G$, since the space $\mathscr{M}_1(G)$ of such is the topological dual of $C_\infty(G)$ under the $\|\cdot\|_\infty$-norm. But $\mathscr{D}'(G)$ contains other measures as well. Indeed, a necessary and sufficient condition that a measure $\mu$ on $G$ belong to $\mathscr{D}'(G)$ is that $|\int g(x)\,\mathrm{d}\mu(x)| \leqslant M\|g\|_0$ for some constant $M$ and for all $g \in C_c(G)$.

LEMMA 3.24. *Let $\mu \in \mathscr{M}_1(G)$. Then $\hat{\mu}$ is a measure on $\hat{G}$ of the form $f(\alpha)\,\mathrm{d}\alpha$, where*

$$f(\alpha) = \int \overline{\alpha(x)}\,\mathrm{d}\mu(x), \qquad \alpha \in \hat{G}.$$

*Proof.* In view of (3.22) and the fact that $\mathscr{E}$ is $\|\cdot\|_0$-dense in $\mathscr{D}(G)$, it suffices to show that

$$\int h(x)\,\mathrm{d}\mu(x) = \int \check{h}(\alpha)\left[\int \overline{\alpha(x)}\,\mathrm{d}\mu(x)\right]\mathrm{d}\alpha$$

for all $h \in \mathscr{E}$. But (3.19) says that

$$h(x) = \int \check{h}(\alpha^{-1})\alpha(x)\,\mathrm{d}\alpha = \int \check{h}(\alpha)\overline{\alpha(x)}\,\mathrm{d}\alpha,$$

and our result follows from the Fubini theorem.

Thus we will write $\hat{\mu}(\alpha) = \int \overline{\alpha(x)}\,\mathrm{d}\mu(x)$. Our last task in this section will be to characterize $\{\hat{\mu} : \mu \in \mathscr{M}_1(G), \mu \geqslant 0\}$.

THEOREM 3.25 (Bochner–Raikov). *$\{\hat{\mu} : \mu \in \mathscr{M}_1(G), \mu \geqslant 0\} = \mathscr{P}(\hat{G})$, the* *space of continuous functions $\varphi$ on $\hat{G}$ such that*

(3.26) $$\sum_{ij\,1}^{m} z_i\bar{z}_j\varphi(\alpha_i\alpha_j^{-1}) \geqslant 0$$

*for all choices $z_1, \ldots, z_m \in \mathbb{C}$, $\alpha_1, \ldots, \alpha_m \in \hat{G}$, arbitrary positive integers m. ($\mathscr{P}(\hat{G})$ is the space of* continuous positive definite *functions on $\hat{G}$.)*

*Proof.* If $\mu \in \mathscr{M}_1(G)$, $\mu \geqslant 0$, then

$$\sum_{ij\,1}^{m} z_i\bar{z}_j \int \overline{(\alpha_i\alpha_j^{-1})(x)}\, \mathrm{d}\mu(x) = \sum_{ij\,1}^{m} z_i\bar{z}_j \int \overline{\alpha_i(x)}\alpha_j(x)\, \mathrm{d}\mu(x)$$
$$= \int \left| \sum_{i\,1}^{m} z_i\overline{\alpha_i(x)} \right|^2 \mathrm{d}\mu(x) \geqslant 0.$$

Moreover, $\hat{\mu}$ is plainly continuous. Thus $\hat{\mu} \in \mathscr{P}(\hat{G})$.

To prove the converse, we must establish some facts about $\mathscr{P}(\hat{G})$. Let $\varphi \in \mathscr{P}(\hat{G})$ and let $\alpha_1 = 1$, $\alpha_2 = \alpha$. Then (3.26) implies that $\begin{bmatrix} \varphi(1) & \varphi(\alpha) \\ \varphi(\alpha^{-1}) & \varphi(1) \end{bmatrix}$ is a positive semidefinite matrix so that $\varphi(\alpha^{-1}) = \overline{\varphi(\alpha)}$, $\varphi(1) \geqslant 0$, and $\varphi(1)^2 - |\varphi(\alpha)|^2 \geqslant 0$. In particular, $\varphi$ is bounded. Next we note that $\int (F \circ F^*)(\alpha)\varphi(\alpha)\, \mathrm{d}\alpha \geqslant 0$ for $F \in C_c(\hat{G})$. This follows by approximating the measure $F(\alpha)\, \mathrm{d}\alpha$ by linear combinations of punctual measures.

Now let $\varphi \in \mathscr{P}(\hat{G})$. Without loss of generality, we suppose that $\varphi(1) = 1$. For $f \in \mathscr{E} \subseteq \mathscr{D}(G)$, as in (3.20), set $\Phi(f) = \int \check{f}(\alpha)\varphi(\alpha)\, \mathrm{d}\alpha$. Now $\mathscr{E}$ is a $*$-subalgebra of $\mathscr{D}(G)$ under pointwise multiplication and complex conjugation. Moreover, we have $\Phi(|f|^2) = \int (f\bar{f})\check{\ }(\alpha)\varphi(\alpha)\, \mathrm{d}\alpha = \int (\check{f} \circ \check{f}^*)(\alpha)\varphi(\alpha)\, \mathrm{d}\alpha \geqslant 0$ and $|\Phi(f)| \leqslant \|\varphi\|_\infty \|\check{f}\|_1 = \|\check{f}\|_1$ for $f \in \mathscr{E}$. The main trick is to prove the following lemma.

LEMMA 3.27. *For $f \in \mathscr{E}$, $|\Phi(f)| \leqslant \Phi(|f|^2)^{1/2}$.*

*Proof.* Choose functions $g_V \in \mathscr{E}$, $V$ a neighborhood of 1 in $\hat{G}$, such that $\{\check{g}_V\}$ is an approximate identity for $\hat{G}$ as in (3.7) with $\check{g}_V^* = \check{g}_V$. Then the Cauchy–Schwarz inequality tells us that $|\Phi(fg_V)|^2 \leqslant \Phi(|f|^2)\Phi(|g_V|^2)$. Clearly $\lim_V \Phi(fg_V) = \Phi(f)$. Moreover, $\{g_V \circ g_V\}$ is an approximate identity for $\hat{G}$, so that $\lim_V \Phi(g_V^2) = \varphi(1)$.

From this lemma, we get that

$$|\Phi(f)| \leqslant \Phi(|f|^{2^n})^{2^{-n}} \leqslant \|(|f|^{2^n})\check{\ }\|_1^{2^{-n}}$$

for all $n$. That is, letting $F=|f|^{\vee}$,

$$|\Phi(f)| \leqslant \|F^{(2^n)}\|_1^{2^{-n}},$$

where $F^{(2^n)}$ is the $2^n$-fold self-convolution of $F$. Passing to the limit on $n$, we get $|\Phi(f)|$ is bounded by the spectral radius of $F$ in $L_1(G)$. But this is $\|\hat{F}\|_\infty$ by (3.12). Thus, by (3.21) and (3.22), $|\Phi(f)| \leqslant \|f\|_\infty \leqslant \|f\|_0$. From this we conclude that $\Phi$ extends to be a member of $\mathscr{D}'(G)$, that $\Phi$ is the restriction to $\mathscr{D}(G)$ of a finite measure on $G$, that the measure $\varphi(\alpha)\,\mathrm{d}\alpha$ on $\hat{G}$ belongs to $\mathscr{D}'(\hat{G})$, and that $\hat{\Phi}=\varphi(\alpha)\,\mathrm{d}\alpha$. Since $\hat{\Phi}(F \circ F^*) \geqslant 0$ for $F \in C_c(\hat{G})$, the same holds for $F \in \mathscr{D}(\hat{G})$ by passing to the limit. Therefore, if $f \in C_c(G)$, $\Phi(|f|^2)=\hat{\Phi}(\check{f} \circ \check{f}^*) \geqslant 0$. Since any $g \in C_c(G)$ such that $g \geqslant 0$ may be written as $|f|^2$, $f \in C_c(G)$, we see that $\Phi$ is a positive measure.

## 4. Compact groups

We would like ultimately to extend as much of the theory developed in the section just completed to non-Abelian groups as can be done. Let $G$ be a locally compact group. What will correspond to the harmonics and periodic plane waves of Section 2 and to the characters of Section 3? If we look for bounded continuous eigenfunctions of all operators $\lambda_x$, $x \in G$, we run the risk of disappointment: indeed, it may be the case that the only one-dimensional space of bounded continuous functions invariant under all $\lambda_x$, $x \in G$, is the space of constant functions. The natural thing to look for then are spaces of bounded continuous functions invariant under all $\lambda_x$, $x \in G$, are which are minimal in some sense. Having found these, we then attempt to decompose functions, measures, generalized functions, etc., into superpositions of functions contained in these subspaces.

This program can be carried out with a minimum of fuss when $G$ is *compact*, which we assume to be the case for the remainder of this section. Choose once for all a left Haar measure $\mathrm{d}x$ on $G$ of total measure 1.

LEMMA 4.1. $\mathrm{d}x$ *is also right invariant; if* $\rho_x$, $x \in G$, *is defined by* $(\rho_x f)(y)=f(yx)$, $y \in G$, *then*

$$\int (\rho_x f)(y)\,\mathrm{d}y = \int f(y)\,\mathrm{d}y, \qquad f \in C(G).$$

*Moreover,* $\int f(y^{-1})\,\mathrm{d}y = \int f(y)\,\mathrm{d}y$ *for* $f \in C(G)$.

*Proof.* $f \mapsto \int(\rho_x f)(y)\,\mathrm{d}y$, $f \in C(G)$, is a left invariant Haar measure of total mass 1 and so equals $dy$. Since Haar measure is right invariant, $f \mapsto \int f(y^{-1})dy$ is a left invariant Haar measure of total mass 1 and so equals $dy$.

We define $\circ$ and $*$ as in (3.4). Then (3.5) holds, except that $C(G)$ is no longer commutative, as does (3.6) and (3.7). Moreover, $\|f\circ g\|_2 \leqslant \|f\|_2\|g\|_1$ for $f, g\in C(G)$, so that we can define the bounded linear operator $\rho_g$ on $L_2(G)$ when $g\in C(G)$ by $\rho_g f = f\circ g$, $f\in C(G)$, and $\|\rho_g\| \leqslant \|g\|_1$. A simple calculation shows that $\rho_g$ commutes with $\lambda_x$, all $x\in G$.

LEMMA 4.2. *Suppose $g\in C(G)$. Then $\rho_g^* = \rho_{g^*}$, $\rho_g$ is Hilbert–Schmidt, and* $\operatorname{Im}\rho_g \subseteq C(G)$.

*Proof.* Indeed, by direct calculation, $(f\circ g, h) = (f, h\circ g^*)$ for $f, g, h\in C(G)$. Moreover,

$$(\rho_g f, h) = \iint_{G\times G} g(x^{-1}y)\overline{f(x)}h(y)\,dx\,dy = (G, \bar{f}\times h)_{G\times G},$$

where $G(x, y) = g(x^{-1}y)$. Therefore the same must hold for $f$, $h\in L_2(G)$ (whence $\bar{f}\times h\in L_2(G\times G)$). Let $\{f_j\}$ be a complete orthonormal set in $L_2(G)$. Then $\{\bar{f}_j\times f_k\}$ is a complete orthonormal set in $L_2(G\times G)$, and we have

$$\sum_{jk} |\rho_g f_j, f_k)|^2 = \sum_{jk} |(G, \bar{f}_j\times f_k)|^2 = \|G\|_2^2 = \|g\|_2^2.$$

Thus $\rho_g$ is Hilbert–Schmidt with Hilbert–Schmidt norm $\|\rho_g\|_{\mathrm{HS}} = \|g\|_2$. Finally, note that $(\rho_g f)(x) = (f, \lambda_x g^*)$ for $f\in C(G)$. Hence, passing to the limit, the same must hold for $f\in L_2(G)$, and $x\mapsto (f, \lambda_x g^*)$ is continuous.

We will call a function $f$ on $G$ a *$G$-finite function* if $\{\lambda_x f : x\in G\}$ spans a finite-dimensional space of functions modulo null functions. We let $\mathcal{R}$ be the space of all continuous $G$-finite functions.

THEOREM 4.3. *$\mathcal{R}$ is a linear subspace dense in $C(G)$ in the $\|\cdot\|_\infty$-norm.*

*Proof.* $\mathcal{R}$ is obviously a linear space. We first show that $\mathcal{R}$ is dense in $L_2(G)$. If $g\in C(G)$, $g = g^*$, then the fact that $\rho_g$ commutes with all $\lambda_x$, $x\in G$, and Lemma 4.2 imply that eigenvectors of $\rho_g$ belonging to non-zero eigenvalues belong to $\mathcal{R}$. Thus, if $f\in L_2(G)$ is orthogonal to $\mathcal{R}$, $f\in\operatorname{Ker}\rho_g$ for all $g\in C(G)$. But $(\rho_g f)(1) = (f, g^*)$ so that such an $f$ is orthogonal to $C(G)$ and hence is zero. Thus $\mathcal{R}$ is dense in $L_2(G)$. Now if $f\in C(G)$, we can find (approximate identity!) $g\in C(G)$ with $\|f - f\circ g\|_\infty < \varepsilon/2$. Choose $h\in\mathcal{R}$

with $\|h-f\|_2<\varepsilon/2\|g\|_2$. Then $\|(h-f)\circ g\|_\infty<\varepsilon/2$ so that $\|f-h\circ g\|_\infty<\varepsilon$. But $h\circ g\in\mathscr{R}$.

This theorem says that $C(G)$ contains lots of finite-dimensional $\lambda$-invariant subspaces and hence lots of minimal ones. We can make this more precise as follows: Defining $\lambda_g$ on $L_2(G)$ for $g\in C(G)$ by $\lambda_g f=g\circ f$ for $f\in C(G)$, we have $(\lambda_g f, h)=\int g(x)(\lambda_x f, h)\,dx$ for $f, h\in C(G)$ and hence for $f, h\in L_2(G)$. Therefore $\lambda_g W\subseteq W$ for every $\lambda$-invariant closed subspace $W$ of $L_2(G)$. Moreover, (4.2) holds for $\lambda_g$ as well as for $\rho_g$. Finally, direct calculation shows that $\lambda_x\lambda_g=\lambda_{\lambda_x g}$ for $x\in G$. Putting these facts together, we see that $\lambda_g W\subseteq\mathscr{R}\cap W$ for all $\lambda$-invariant closed subspaces $W$ of $L_2(G)$ and all $g\in\mathscr{R}$. But $f\in\operatorname{Ker}\lambda_g$ implies that $(g, f^*)=0$ so that $\bigcap\{\operatorname{Ker}\lambda_g\colon g\in\mathscr{R}\}=\{0\}$ by (4.3). Thus we have that $\mathscr{R}\cap W$ is $L_2$-dense in $W$. Now let $\{W_j\}$ be a maximal collection of $L_2$-orthogonal minimal finite-dimensional $\lambda$-invariant subspaces of $C(G)$, and let $W$ be the orthogonal complement of $\Sigma\, W_j$. Since $\{\lambda_x\colon x\in G\}$ is an adjoint closed operator set on $L_2(G)$, we get that $W$ is $\lambda$-invariant, and clearly $W$ is closed in $L_2(G)$. Thus, if $W\neq\{0\}$, it contains $G$-finite continuous functions and hence a minimal finite-dimensional $\lambda$-invariant subspace of $C(G)$, contradicting the maximality of $\{W_j\}$. We have proved

PROPOSITION 4.4. *$L_2(G)$ is a Hilbert space direct sum of minimal finite-dimensional $\lambda$-invariant subspaces of $C(G)$.*

In a sense, this answers the question of $L_2(G)$ harmonic analysis: given the decomposition $\{W_j\}$, we can write any $f\in L_2(G)$ as $\Sigma_j f_j$, where $f\in W_j$ is uniquely determined. The problem is that the decomposition $\{W_j\}$ is unique if and only if $G$ is Abelian. Moreover, we have not yet developed any standard way of getting a hold on the $W_j$. For example, consider the following question: Let $V$ be a finite-dimensional linear space $/\mathbb{C}$ and let $\sigma$ be a continuous homomorphism of $G$ into $\mathrm{GL}(V)$, the topological group of invertible linear operators on $V$. Suppose that $V$ has no proper subspaces invariant under all $\sigma(x)$, $x\in G$. Is $(V, \sigma)$ isomorphic as a vector space with operators to $(W, \lambda)$, where $W$ is a $\lambda$-invariant subspace of $\mathscr{R}$?

The answer to the foregoing question is 'Yes', as may be seen from the following considerations: Let $V'$ be the dual of $V$. For each choice of $u\in V$, $u'\in V'$, let

$$f^\sigma_{u,u'}(x)=\langle u', \sigma(x^{-1})u\rangle,\qquad x\in G,\qquad f^\sigma_{u,u'}\in C(G).$$

Set $(T^\sigma_{u'}u)(x)=f^\sigma_{u,u'}(x)$. Then $T^\sigma_{u'}\in\operatorname{Hom}_{\mathbb{C}}(V, C(G))$. Moreover, $T^\sigma_{u'}\circ\sigma(x)=$

$=\lambda_x \circ T^\sigma_{u'}$. Fix $0 \neq u' \in V'$ and suppose that $T^\sigma_{u'} u = 0$. If $u \neq 0$, then the vector space spanned by $\{\sigma(x^{-1})u : x \in G\}$ is non-zero and $\sigma$-invariant, and hence equals $V$. Thus $\langle u', V \rangle = 0$, contradicting $u' \neq 0$. Thus Ker $T^\sigma_{u'} = \{0\}$, and our assertion is proved. But note what is *not* proved so far: we have *not* shown that $(V, \sigma)$ is isomorphic to one of the $(W_j, \lambda)$. This point is easily settled, however. Let $P_j$ be the orthogonal projection of $L_2(G)$ onto $W_j$. Since $W_j^\perp$ is $\lambda$-invariant, $P_j \circ \lambda_x = \lambda_x \circ P_j$, $x \in G$. Therefore $P_j \circ T^\sigma_{u'}$ is an operator homomorphism of $(V, \sigma)$ into $(W_j, \lambda)$. If it is not an operator isomorphism, either its image is not all of $W_j$ and hence is, since $\lambda$-invariant, zero, or its kernel is not zero and hence is, since $\sigma$-invariant, the whole of $V$. Therefore if $(V, \sigma)$ is not isomorphic to $(W_j, \lambda)$, then Im $T^\sigma_{u'} \subseteq W_j^\perp$. If this holds for all $j$, then Im $R^\sigma_{u'} = \{0\}$, a contradiction. Thus we have proved

PROPOSITION 4.5. *Let $L_2(G) = \bigoplus_j W_j$ as in* (4.4). *Let $(V, \sigma)$ be as above. Then $(V, \sigma)$ is operator isomorphic to $(W_j, \lambda)$ for some $j$.*

The techniques introduced to prove Proposition 4.5 are just what we need to determine the decomposition $\{W_j\}$ completely. We now do that.

DEFINITION 4.6. Let $V$ be a finite-dimensional vector space $/\mathbb{C}$ and let $\sigma$ be a continuous homomorphism into $GL(V)$. $(V, \sigma)$, sometimes written $\sigma$ for short, is called a *finite-dimensional representation* of $G$, abbreviated FDR. If $V$ has no proper subspaces invariant under all $\sigma(x)$, $x \in G$, we say $(V, \sigma)$ is *irreducible*. For any FDR $(V, \sigma)$, $u \in V$, $u' \in V'$, the function $f^\sigma_{u,u'}$ is called a *representative function* of $\sigma$. An FDR $(V, \sigma)$ is called *unitary* if $V$ comes equipped with a Hilbert space structure such that all $\sigma(x)$, $x \in G$, are unitary operators on $V$.

We have seen above that any irreducible FDR $\sigma$ can be found inside $(L_2(G), \lambda)$ and hence can be made unitary. The same is actually true of any FDR.

LEMMA 4.7. *Let $(V, \sigma)$ be an* FDR *of the compact group $G$. Then $V$ carries a Hilbert space structure for which $\sigma$ is unitary.*

*Proof.* Let $[\cdot, \cdot]$ be *any* Hilbert space structure on $V$. Define $(u, v) = \int [\sigma(x^{-1})u, \sigma(x^{-1})v]\, dx$. $(\cdot, \cdot)$ works.

Now it is plain that the set of representative functions of an FDR

$(V, \sigma)$ depends only on the operator isomorphism class of $(V, \sigma)$ and hence so does the linear subspace of $C(G)$ spanned by those functions. Moreover, that subspace is finite dimensional being spanned by $\{f^{\sigma}_{u_i,u'_j} : \{u_i\}$ a basis of $V$, $\{u'_j\}$ a basis of $V'\}$.

Let $(V_1, \sigma_1)$ and $(V_2, \sigma_2)$ be two *irreducible* FDR's of the compact group $G$. Make them unitary. We can then identify $V'_i$ with $V_i$ in a conjugate linear way. If $u_i, v_i \in V_i$, we thereby get, identifying $v_i$ with a member of $V'_i$, a representative function, call it $f^{\sigma_i}_{u_i,v_i}$, given by $f^{\sigma_i}_{u_i,v_i} = (\sigma_i(x^{-1})u_i, v_i)$, $x \in G$. A most fundamental theorem is the following, whose proof we postpone to Section 5, where it will be proved in a more general setting (cf. Theorem 5.30 below).

THEOREM 4.8 (Weyl orthogonality relations). *If $\sigma_1$ and $\sigma_2$ are not operator isomorphic, then $(f^{\sigma_1}_{u_1,v_1}, f^{\sigma_2}_{u_2,v_2}) = 0$ for all $u_1, v_1 \in V_1$ and $u_2, v_2 \in V_2$. If $V_1 = V_2 = V$, $\sigma_1 = \sigma_2 = \sigma$, then*

$$(f^{\sigma}_{u_1,v_1}, f^{\sigma}_{u_2,v_2}) = \frac{1}{n}(u_1, u_2)\overline{(v_1, v_2)}$$

*for all $u_1, u_2, v_1, v_2 \in F$, where $n = \dim V$.*

Let $\hat{G}$ be the set of all operator isomorphism classes of irreducible FDR's of $G$. For each $\alpha \in \hat{G}$, choose a unitary FDR $(V_\alpha, \sigma_\alpha) \in \alpha$. Let $d(\alpha) = \dim V_\alpha$. Choose an orthonormal basis $\{u_{\alpha 1}, \ldots, u_{\alpha d(\alpha)}\}$ of $V_\alpha$. Then (4.8) says that $\{\sqrt{d(\alpha)} f^{\sigma_\alpha}_{u_{\alpha i}, u_{\alpha j}} : \alpha \in \hat{G},\ i, j = 1, \ldots, d(\alpha)\}$ is an orthonormal subset of $L_2(G)$.

PROPOSITION 4.9. *This orthonormal set is complete.*

*Proof.* Let $f \in L_2(G)$ be orthogonal to this orthonormal set. Then $f$ is orthogonal to all representative functions of all irreducible FDR's of $G$. Let $W$ be a minimal finite-dimensional $\lambda$-invariant subspace of $L_2(G)$. Then $(W, \lambda)$ is a unitary irreducible FDR. Let $P$ be the orthogonal projection of $L_2(G)$ on $W$. For every $g \in W$ and $h \in L_2(G)$, $f^{\lambda|W}_{g,Ph}(x) =$ $= (\lambda_{x^{-1}} g, Ph) = (\lambda_{x^{-1}} g, h)$, so that

$$0 = \int (\lambda_{x^{-1}} g, h)\overline{f(x)}\, dx = \iint g(xy)\overline{h(y)}\,\overline{f(x)}\, dy\, dx.$$

Letting $h$ run through $C(G)$, we conclude that

$$y \mapsto \int g(xy)\overline{f(x)}\, dx \text{ vanishes almost everywhere}$$

and therefore identically, since it is continuous. Therefore, setting $y=1$, $f \in W^{\perp}$. By (4.4), $f=0$.

THEOREM 4.10. *For each $\alpha \in \hat{G}$, let $W_\alpha$ be the span of $f^{\sigma_\alpha}_{u,v}$, $u, v \in V_\alpha$ in $C(G)$. Then $L_2(G) = \bigoplus_{\alpha \in \hat{G}} W_\alpha$, a Hilbert space direct sum. Moreover, each $W_\alpha$ is the Hilbert space direct sum of $d(\alpha)$ subspaces $W_{\alpha 1}, \ldots, W_{\alpha d(\alpha)}$ with each $(W_{\alpha j}, \lambda) \in \alpha$.*

This is a corollary of the foregoing and answers the question of how unique the decomposition of (4.4) is: the decomposition of $L_2(G)$ into the $\{W_\alpha\}$ is unique, but the further decomposition of each $W_\alpha$ into the $\{W_{\alpha 1}, \ldots, W_{\alpha d(\alpha)}\}$ is not, corresponding as they do to the images of orthonormal bases of $V_\alpha$ in the projective space of rays of $V_\alpha$.

We know that if $(V, \sigma)$ is an FDR and if $u \in V$, $u' \in V'$, then $f^{\sigma}_{\sigma(x)u,u'} = \lambda_x f^{\sigma}_{u,u'}$. Also $f^{\sigma}_{u,{}^t\sigma(x)u'} = \rho_{x^{-1}} f^{\sigma}_{u,u'}$, $\rho_x$ as in (4.1). For $(x, y) \in G \times G$, set $\tau_{x,y} = \lambda_x \rho_y$. Then $(x, y) \mapsto \tau_{x,y}$ is a homomorphism of $G \times G$ into the group of unitary operators on $L_2(G)$. Clearly each $W_\alpha$ is $\tau$-invariant. The structure of $(W_\alpha, \tau)$ is obtained as follows: The map $(u, u') \mapsto f^{\sigma_\alpha}_{u,u'}$ is a bilinear map of $V_\alpha \times V'_\alpha$ onto $W_\alpha$ and so determines a linear map of $V_\alpha \otimes V'_\alpha$ onto $W_\alpha$. By (4.8) this map, call it $\varphi_\alpha$, is a bijection. Define $\sigma'_\alpha(x)$ on $V'_\alpha$ by $\sigma'_\alpha(x) = {}^t\sigma(x^{-1})$. Then $(V'_\alpha, \sigma'_\alpha)$ is an irreducible FDR. Moreover, we can set, for

$$(x, y) \in G \times G, \ (\sigma_\alpha \times \sigma'_\alpha)(xy) = \sigma_\alpha(x) \otimes \sigma'_\alpha(y)$$

and get the FDR $(V_\alpha \otimes V'_\alpha, \sigma_\alpha \times \sigma'_\alpha)$ of $G \times G$. We have shown:

PROPOSITION 4.11. *$\varphi_\alpha$ is an operator isomorphism of $(V_\alpha \otimes V'_\alpha, \sigma_\alpha \times \sigma'_\alpha)$ with $(W_\alpha, \tau)$.*

By the techniques developed above, one can prove:

PROPOSITION 4.12. *$\mathcal{R}$ is the subspace of $C(G)$ algebraically spanned by the representative functions of irreducible FDR's.*

We must formulate the problem of harmonic analysis of functions, measures, generalized functions, etc., in a manner analogous to (2.1) and (2.2). Let $f$ be a function on $G$. For each $\alpha \in \hat{G}$, we desire a functional $h(\alpha) \in \mathrm{End}_{\mathbb{C}}(V_\alpha)'$, so that

$$(4.13) \qquad f(x) \sim \sum_{\alpha \in \hat{G}} \langle h(\alpha), \sigma_\alpha(x) \rangle, \qquad x \in G,$$

or, in weak form

$$(4.14)\qquad \int f(x)g(x)\,dx = \sum_{\alpha\in\hat{G}} \langle h(\alpha), \int g(x)\sigma_\alpha(x)\,dx\rangle$$

for all $g\in\mathcal{D}$, a space of test functions. We will require, in analogy with Section 2, that $\mathcal{R} \subseteq \mathcal{D} \subseteq C(G)$ and that $\mathcal{D}$ is closed under $\circ$ and $*$. Then for linear functionals $\mu$ on $\mathcal{D}$, that is, for measures and generalized functions, we have

$$(4.15)\qquad \int g(x)\,d\mu(x) = \sum_{\alpha\in\hat{G}} \langle h(\alpha), \int g(x)\sigma_\alpha(x)\,dx\rangle,$$

where the left-hand side uses suggestive notation for $\mu(g)$. There are problems associated with the meaning of convergence of the sum on the right-hand side exactly as there were in Section 2: if $\mathcal{D}=C(G)$ and $\mu$ is a measure, that sum need not be absolute. These problems can be handled by topological methods as was done by introducing $\mathcal{S}(\mathbb{R}^n)$ and $\mathcal{D}(G)$ in Sections 2 and 3, respectively. It will turn out that no problems of this sort arise for $f\in L_2(G)$. In any case, we can determine what $h(\alpha)$ must be.

PROPOSITION 4.16. *The only functionals $h(\alpha)$, $\alpha\in\hat{G}$, that are consistent with* (4.15) *are given by*

$$\langle h(\alpha), T\rangle = d(\alpha)\,\mathrm{Tr}\left(\left[\int \sigma_\alpha(x^{-1})\,dx\right]T\right),$$

*for $T\in\mathrm{End}_{\mathbb{C}}(V_\alpha)$, $\alpha\in\hat{G}$.*

*Proof.* Let $\alpha, \beta\in\hat{G}$. It is trivial to check that

$$\int f^{\sigma_\beta}_{u,u'}(x) f^{\sigma_\alpha}_{v,v'}(x^{-1})\,dx = \begin{cases} 0 & \text{if } \alpha\neq\beta \\ \dfrac{1}{d(\alpha)}\langle u', v\rangle\langle v', u\rangle & \text{if } \alpha=\beta \end{cases},$$

where $u\in V_\beta$, $u'\in V'_\beta$, $v\in V_\alpha$, $v'\in V'_\alpha$, is a Hilbert space free way of stating (4.8). It follows that

$$\int f^\beta_{u,u'}(x)\sigma_\alpha(x)\,dx = \begin{cases} 0 & \text{if } \alpha\neq\beta \\ T^\alpha_{u,u'} & \text{if } \alpha=\beta \end{cases},$$

where $T^\alpha_{u,u'}v = \dfrac{1}{d(\alpha)}\langle u', v\rangle u$. Therefore

$$\langle h(\alpha), T^\alpha_{u,u'}\rangle = \int f^\alpha_{u,u'}(x)\,d\mu(x) = \langle u', \int \sigma_\beta(x^{-1})u\,d\mu(x)\rangle.$$

Since $\{T^{\alpha}_{u,u'} : u \in V_{\alpha}, u' \in V'_{\alpha}\}$ spans $\mathrm{End}_{\mathbb{C}}(V_{\alpha})$, $h(\alpha)$ is uniquely determined. An easy check shows that our formula holds for $T = T^{\alpha}_{u,u'}$.

It is natural to call the operator valued function $\alpha \mapsto d(\alpha) \int \sigma_{\alpha}(x^{-1})\, d\mu(x)$ the *Fourier transform* of $\mu$ and denote it by $\hat{\mu}$. If $\mu = f(x)\,dx$, we will write $\hat{f}$ in place of $\hat{\mu}$. For $g \in \mathscr{D}$, we also define $\check{g}$ by

$$\check{g}(\alpha) = \int g(x)\sigma_{\alpha}(x)\, dx, \qquad \alpha \in \hat{G}.$$

Then (4.15) and (4.16) take the form

$$(4.17) \qquad \mu(g) = \sum_{\alpha \in \hat{G}} \mathrm{Tr}(\hat{\mu}(\alpha)\check{g}(\alpha)).$$

Now $\check{\mathscr{D}} \supseteq \check{\mathscr{R}}$, and $\check{\mathscr{R}}$ consists precisely of the operator valued functions on $\hat{G}$ of finite support, by (4.12) and the calculations in (4.16). Thus $\hat{\mu}$ should be regarded as some sort of measure on $\check{\mathscr{D}}$. We then have the precise meaning of the summation in (4.17) and have, in complete analogy with (2.9) and (3.23),

$$(4.18) \qquad \hat{\mu}(\check{g}) = \mu(g) \quad \text{for} \quad g \in \mathscr{D}.$$

In the following easy result, we assume that each $\sigma_{\alpha}$ is unitary. Moreover, $d$ is the function $\alpha \mapsto d(\alpha)$.

PROPOSITION 4.19. *For $f, g \in \mathscr{D}$, we have $f^{*\vee} = \check{f}^{*}$ and $(f \circ g)^{\vee} = \check{f}\check{g}$. Moreover, for $f, g \in C(G)$, we have $f^{*\wedge} = \hat{f}^{*}$ and $(f \circ g)^{\wedge} = d \cdot \hat{g}\hat{f}$.*

Now let $\mu = \delta_1$, the 'delta function' at 1. Then $\hat{\mu}(\alpha) = d(\alpha) I_{V_{\alpha}}$, $\alpha \in \hat{G}$, so that for $g \in \mathscr{D}$

$$(4.20) \qquad g(1) = \sum_{\alpha \in \hat{G}} d(\alpha)\mathrm{Tr}(\check{g}(\alpha)),$$

where the summation is taken with a grain of salt, as above. Let $f, g \in \mathscr{D}$. Then $(f, g) = (f \circ g^{*})(1)$ and so

$$(4.21) \qquad (f, g) = \sum_{\alpha \in \hat{G}} d(\alpha)\mathrm{Tr}(\check{f}(\alpha)\check{g}^{*}(\alpha)).$$

Now (4.22) is actually true as an absolute summation, as follows from (4.8). This leads to our version of the Plancherel Theorem for compact $G$ as follows: $V_{\alpha}$ being a Hilbert space, $\mathrm{End}_{\mathbb{C}}(V_{\alpha})$ becomes a Hilbert space by setting $(S, T)=\mathrm{Tr}(ST^*)$ for $S, T\in \mathrm{End}_{\mathbb{C}}(V_\alpha)$. We then form the weighted Hilbert space direct sum

$$(4.23)\qquad \mathfrak{H}=\bigoplus_{\alpha\in\hat{G}} d(\alpha)\ \mathrm{End}_{\mathbb{C}}(V_\alpha)$$

consisting of all functions $F$ such that $F(\alpha)\in \mathrm{End}_{\mathbb{C}}(V_\alpha)$ such that

$$(4.24)\qquad \sum_{\alpha\in\hat{G}} d(\alpha)\|F(\alpha)\|^2<\infty.$$

For $F, G\in\mathfrak{H}$, we define

$$(4.25)\qquad (F, G)=\sum_{\alpha\in\hat{G}} d(\alpha)(F(\alpha), G(\alpha)),$$

an absolutely convergent sum $(\cdot,\cdot)$ makes $\mathfrak{H}$ into a Hilbert space. Then we have the following analogue of Plancherel's Theorem:

THEOREM 4.26. *The map $g\mapsto\check{g}$, $g\in\mathscr{D}$, sends $\mathscr{D}$ into $\mathfrak{H}$ and is isometric. It extends to a unitary isomorphism of $L_2(G)$ into $\mathfrak{H}$.*

The weighting $\alpha\mapsto d(\alpha)$ is a measure on $\hat{G}$, known as the *Plancherel measure.* It is basically the Fourier transform of $\delta_1$. Looking at (4.20) another way, we have

$$(4.27)\qquad g(1)=\sum_{\alpha\in\hat{G}} d(\alpha)\int g(x)\chi_\alpha(x)\,\mathrm{d}x,$$

where

$$(4.28)\qquad \chi_\alpha(x)=\mathrm{Tr}(\sigma_\alpha(x)),\qquad x\in G.$$

Now for any FDR $(V, \sigma)$, the function $x\mapsto\mathrm{Tr}(\sigma(x))$ on $G$ is known as the character of $\sigma$. It depends only on the operator isomorphism class of $(V, \sigma)$.

THEOREM 4.29. *The operator isomorphism class of $(V, \sigma)$ is completely determined by its character $\chi_\sigma$.*

*Proof.* Making $\sigma$ unitary, we see that $V$ is a direct sum of minimal $\sigma$-invariant subspaces. Note that for any FDR $(W, \tau)$, we have

$$\chi_\tau(x)=\sum_{i}{}_1^n\, f^\tau_{u_i,u_i'}(x^{-1}),$$

where $\{u_i\}$ and $\{u_i'\}$ are dual bases of $W$ and $W'$, respectively, and $n = \dim W$. Apply (4.8).

Then (4.27) and (4.28) say that the Plancherel measure on $\hat{G}$ arises out of an attempt to expand $\delta_1$ in terms of the characters of irreducible FDR's, or, more properly, in terms of the measure determined by those characters.

We end by noting that if $\mu(\lambda_x \rho_x g) = \mu(g)$ for all $x \in G$, $g \in \mathscr{D}$, then $\hat{\mu}(\alpha)$ commutes with all $\sigma_\alpha(x)$, $x \in G$, and hence is some scalar multiple $\tilde{\mu}(\alpha)$ of $I_{V_\alpha}$ (recall that $\check{\mathscr{R}}(\alpha) = \mathrm{End}_{\mathbf{C}}(V_\alpha)$). Thus we will have

$$(4.30) \qquad \mu(g) = \sum_{\alpha \in \hat{G}} \tilde{\mu}(\alpha) \int g(x) \chi_\alpha(x)\, dx,$$

that is,

$$(4.31) \qquad \mu \sim \sum_{\alpha \in \hat{G}} \tilde{\mu}(\alpha) \chi_\alpha.$$

In this way, the harmonic analysis of such $\mu$, called *central*, may be seen as the expansion of $\mu$ in terms of the characters of irreducible FDR's: these characters have taken over the role of the harmonics of Section 2.

## 5. General locally compact groups

The ideas developed above should convince the reader that representations and their decompositions are the keys to formulating notions of harmonic analysis on locally compact groups. We begin this section with a general discussion of unitary representation theory, following Mackey ([29]).

For any Hilbert space $V$, let $U(V)$ be the group of unitary operators on $V$. Give $U(V)$ the *strong operator topology*, that is, the weakest topology making all the maps $T \mapsto Tv$ of $U(V)$ into $V$, $v \in V$, continuous. This topology coincides with the *weak operator topology* (the weakest topology making all the functions $T \mapsto (Tu, v)$ continuous, $u$, $v \in V$). This topology makes $U(V)$ into a topological group.

DEFINITION 5.1. Let $G$ be a topological group. A (*continuous*) *unitary representation* of $G$ is a continuous homomorphism $\sigma$ of $G$ into $U(V)$ for some Hilbert space $V$. We will use $V(\sigma)$ to denote the Hilbert space for $\sigma$. Two unitary representations $\sigma$ and $\tau$ are *unitarily equivalent*, $\sigma \simeq \tau$, if there is a unitary operator isomorphism of $V(\sigma)$ onto $V(\tau)$. If $W$ is a closed subspace of $V(\sigma)$ invariant under all $\sigma(x)$, $x \in G$, then the restriction of each $\sigma(x)$ to $W$ defines a unitary representation $\sigma_W$ of $G$ into $U(W)$, called a *subrepresentation* of $\sigma$. If $\sigma$ has no non-trivial subrepresentations, it is called *irreducible*. Let $\{\sigma_\iota\}_{\iota\in I}$ be an indexed set of unitary representations of $G$. The operators $\bigoplus_{\iota\in I}\sigma_\iota(x)$, $x\in G$ *on* $\bigoplus_{\iota\in I} V(\sigma_\iota)$ define a unitary representation of $G$ into $U(\bigoplus_{\iota\in I} V(\sigma_\iota))$, written $\bigoplus_{\iota\in I}\sigma_\iota$ and called the *direct sum* of the $\{\sigma_\iota\}$. (Here we are using Hilbert space direct sums, which are completions of algebraic direct sums.)

We have already seen that if $\sigma_W$ is a subrepresentation of $\sigma$, then so is $\sigma_{W^\perp}$ and that $\sigma\simeq\sigma_W\oplus\sigma_{W^\perp}$. It is easy to check that $\lambda$ is a unitary representation of the locally compact group $G$ into $U(L_2(G))$, where the $L_2$ space is taken with respect to left Haar measure on $G$. In analogy with (4.4), we might try to obtain the harmonic analysis of $L_2$ functions on $G$ by decomposing $\lambda$ into a direct sum of irreducible representations. We have already seen in Section 2 that this idea is naive: the irreducible representations of $\mathbb{R}$ are the characters, but $L_2(\mathbb{R})$ contains no eigenspaces for $\lambda$. Indeed what Sections 2 and 3 tell us is that, at a minimum, we must replace direct sums by some sort of continuous direct sum, known as a direct integral. We shall outline the theory of direct integral decompositions below. We already know that for $G$ compact, the direct sum decomposition of $\lambda$ into irreducibles is not unique. What is unique is the set of irreducibles appearing in the decomposition (all of them) and the number of times a given irreducible appears in the decomposition (as many times as the dimension of the irreducible). In general, however, these uniqueness results fail. They hold only for 'good' groups, the type I groups (see below).

Another sort of decomposition of $\lambda$, for $G$ compact, is the subject of the first part of (4.10): the $\{W_\alpha\}$ there have the property that no subrepresentation (except 0) of $\lambda_{W_\alpha}$ is equivalent to any subrepresentation of $\lambda_{W_\beta}$ if $\alpha\neq\beta$, as will follow from (5.7) below. We introduce some terminology.

DEFINITION 5.2. $\sigma$ and $\tau$ are *disjoint* (written $\sigma\downarrow\tau$) if no non-zero subrepresentation of $\sigma$ is equivalent to any subrepresentation of $\tau$. $\sigma$ is *primary* if it cannot be written as a non-trivial direct sum of disjoint subrepresentations.

Now (5.13) below implies that $\lambda_{W_\alpha}$ is primary. In other words, (4.10) decomposes $\lambda$ as a direct sum of disjoint primary subrepresentations. Again, to generalize this result to general locally compact $G$, we must replace direct sums with direct integrals. This time, however, the decomposition will be essentially unique. Now among the primary representations of a given group are direct sums of a number of copies of a given irreducible representation. For type I groups, all primary representations are of this type as a matter of definition; for non type I groups, there are others.

We now develop these notions formally.

DEFINITION 5.3. Let $\sigma$, $\tau$ be unitary representations of the topological group $G$. $R(\sigma, \tau)$ will denote the space of all continuous operator homomorphisms of $V(\sigma)$ into $V(\tau)$. Members of $R(\sigma, \tau)$ are called *intertwining operators.*

Clearly $R(\sigma, \tau)^* = R(\tau, \sigma)$ and $R(\sigma, \tau) \circ R(\pi, \sigma) \subseteq R(\pi, \tau)$ for representations $\pi, \sigma, \tau$. Thus $R(\sigma, \sigma)$ is a$*$-algebra of operators with $I$ on $V(\sigma)$, and $R(\sigma, \sigma)$ is closed in the weak operator topology: $R(\sigma, \sigma)$ is a *von Neumann algebra.* If $E$ is an orthogonal projection in $V(\sigma)$, then $E \in R(\sigma, \sigma)$if and only if Im $E$ is $\sigma$-invariant. The resultant subrepresentation of $\sigma$ will also be denoted by $\sigma_E$. In this way, the study of the subrepresentations of $\sigma$ is reduced to the study of the orthogonal projections in $R(\sigma, \sigma)$.

LEMMA 5.4. *Let $\sigma$ and $\tau$ be representations of $G$ and let $T \in R(\sigma, \tau)$. Then* Ker $T$ and $\overline{\text{Im } T}$ *are invariant under $\sigma$ and $\tau$, respectively. Moreover,* $\sigma_{(\text{Ker } T)^\perp} \simeq \tau_{\overline{\text{Im } T}}$.

*Proof.* Use the polar decomposition of $T$.

PROPOSITION 5.5 (Schur's lemma). *Let $\sigma$ be a representation of $G$. Then* $\dim_{\mathbb{C}} R(\sigma, \sigma) = 1$ *if and only if $\sigma$ is irreducible.*

*Proof.* Use the spectral theorem on self-adjoint operators in $R(\sigma, \sigma)$.

DEFINITION 5.6. Let $\sigma$, $\tau$ be representations of $G$. $\sigma \leqslant \tau$ means $\sigma$ is equivalent to a subrepresentation of $\tau$. $\sigma \prec \tau$ ($\sigma$ is *covered* by $\tau$) means no non-zero subrepresentation of $\sigma$ is disjoint from $\tau$. $\sigma \sim \tau$ ($\sigma$ is *quasi-equivalent* to $\tau$) means $\sigma \prec \tau$ and $\tau \prec \sigma$.

PROPOSITION 5.7. (a) $\sigma \wr \tau$ *if and only if* $R(\sigma, \tau) = \{0\}$.

(b) $\sigma \leqslant \tau$ *implies* $\sigma \prec \tau$.

(c) $\sigma \prec \tau$ *and* $\tau \wr \pi$ *imply* $\sigma \wr \pi$.

(d) $\sigma \wr \tau_\iota, \iota \in I$ *implies* $\sigma \wr \bigoplus_{\iota \in I} \tau_\iota$.

(e) *Given* $\sigma$ *and* $\tau$, *we may write* $\sigma \simeq \sigma_1 \oplus \sigma_2$ *and* $\tau \simeq \tau_1 \oplus \tau_2$ *with* $\sigma_1 \simeq \tau_1$ *and* $\sigma_2 \wr \tau_2$.

(f) *Given* $\sigma$ *and* $\tau$, *we may write* $\sigma \simeq \sigma_1 \oplus \sigma_2$ *with* $\sigma_1 \prec \tau$, $\sigma_2 \wr \tau$, *and* $\sigma_1 \wr \sigma_2$.

(g) $\pi \prec \sigma$ *and* $\sigma \prec \tau$ *imply* $\pi \prec \tau$ (*so* $\sim$ *is an equivalence*).

*Proof.* (a) follows from (5.4) and (d) follows from (a). (b), (c) and (g) are trivial. (e) and (f) are proved by exhaustion.

LEMMA 5.8. *Let E be an orthogonal projection in* $R(\sigma, \sigma)$. *Then* $\sigma_E \wr \sigma_{I-E}$ *if and only if* $E \in CR(\sigma, \sigma)$, *the center of* $R(\sigma, \sigma)$.

*Proof.* $R(\sigma_E, \sigma_{I-E}) = (I-E)R(\sigma, \sigma)|\text{Im}\, E$. Hence by (5.7(a)), $\sigma_E \wr \sigma_{I-E}$ if and only if $(I-E)R(\sigma, \sigma)E = \{0\}$ if and only if $TE = ETE$ for all $T \in R(\sigma, \sigma)$ if and only if (take adjoints!) $ET = ETE$ for all $T \in R(\sigma, \sigma)$ if and only if $E \in CR(\sigma, \sigma)$.

COROLLARY 5.9. $\sigma$ *is primary if and only if* $\dim_{\mathbb{C}} CR(\sigma, \sigma) = 1$.

PROPOSITION 5.10. *Every subrepresentation of a primary representation is primary. If* $\sigma$ *and* $\tau$ *are non-disjoint primary representations, then either* $\sigma \leqslant \tau$ *or* $\tau \leqslant \sigma$. *If* $\sigma \sim \tau$ *and* $\tau$ *is primary, so is* $\sigma$.

According to (5.7(g)), $\sim$ is an equivalence relation and so we can talk of *quasi-equivalence classes.*

PROPOSITION 5.11. *Quasi-equivalence classes are closed under direct summation.*

DEFINITION 5.12. Let $\sigma$ be a primary representation. We say $\sigma$ is *type* I if $\sigma \sim$ an irreducible representation, $\sigma$ is *type* III if $\tau \sim \sigma$ implies $\tau \simeq \sigma$, $\sigma$ is *type* II if it is neither type I nor III.

THEOREM 5.13. *Let* $\sigma$ *be a type I primary representation and let* $\tau$ *and* $\tau_1$ *be irreducible representations quasi-equivalent to* $\sigma$. *Then* $\tau \simeq \tau_1$. *Moreover,* $\sigma$ *is equivalent to the direct sum of a number of copies of* $\tau$ *and that cardinal number, call it n, is well-defined*: *we write* $\sigma \simeq n\tau$. Conversely, *if* $\tau$ *is irreducible and* $\sigma \simeq n\tau$, *then* $\sigma$ *is a type I primary representation.*

COROLLARY 5.14. *The type I primary quasi-equivalence classes are in bijective correspondence with the irreducible equivalence classes.*

DEFINITION 5.15. A topological group is *type* I if all of its primary representations are type I.

Every finite group $G$ is type I because, if $\sigma$ is a representation of $G$ and if $v \in V(\sigma)$, the span of $\sigma(G)v$ is a finite-dimensional invariant subspace and hence contains a minimal invariant subspace. The same is true if, more generally, $G$ is merely compact (cf. Section 6). If $G$ is Abelian and $\sigma$ is a primary representation of $G$, then $\sigma(x) \in CR(\sigma, \sigma)$, $x \in G$, whence (5.9) $\sigma \simeq n\alpha$ for some $\alpha \in \hat{G}$. Hence all Abelian groups are type I. We shall mention other examples below.

Now from the material developed above it should be clear that the problem of decomposing a given representation $\sigma$ into irreducibles is the problem of decomposing $I \in R(\sigma, \sigma)$ into a sum of minimal orthogonal projections in $R(\sigma, \sigma)$, while the problem of decomposing $\sigma$ into disjoint primary representations is the problem of decomposing $I \in CR(\sigma, \sigma)$ into a sum of minimal orthogonal projections in $CR(\sigma, \sigma)$. The generalization of these direct sum decompositions to direct integral decompositions involves replacing the von Neumann algebras generated by these sets of projection by arbitrary maximal Abelian von Neumann algebras in $R(\sigma, \sigma)$ and $CR(\sigma, \sigma)$, respectively (in the latter case getting $CR(\sigma, \sigma)$ itself since $CR(\sigma, \sigma)$ is already Abelian).

To carry out this program, we must establish the relationship between representations of a locally compact group $G$ and representations of the convolution algebra $C_c(G)$. Using left Haar measure on $G$, we define $\circ$ as in (3.4). We can no longer define * as in (3.4), though, because left Haar measure need not be right invariant.

LEMMA 5.16. *There is a unique continuous homomorphism of $G$ into the positive real numbers under multiplication, call it $\Delta_G$, such that for all $f \in C_c(G)$ and all $x \in G$, we have*

$$\int f(yx^{-1})\,dy = \Delta_G(x) \int f(y)\,dy.$$

*Moreover,*

$$\int f(x^{-1})\,dx = \int f(x)\Delta_G(x^{-1})\,dx.$$

On the basis of this, we redefine * be setting

(5.17) $f^*(x) = \overline{f(x^{-1})}\Delta_G(x)^{-1}, \qquad x \in G.$

Then (3.5) holds, except that $C_c(G)$ is *not* commutative, and (3.6) holds, *except that* $\|f^*\|_p = \|f\|$ *only for* $p=1$. (3.7) holds without change.

Now let $\sigma$ be a unitary representation of $G$. For $f \in C_c(G)$, we define the bounded operator $\sigma(f)$ on $V(\sigma)$ by

$$(5.18) \quad (\sigma(f)u, v) = \int f(x)(\sigma(x)u, v)\,dx \qquad \text{for} \qquad u, v \in V(\sigma).$$

THEOREM 5.19. *$\sigma$ is a $*$-homomorphism of $C_c(G)$ into* $\mathrm{End}_{\mathbb{C}}(V(\sigma))$ *(the space of* continuous *endomorphisms of $V(\sigma)$). Moreover, $\sigma$ is* essential, *that is, $\sigma(C_c(G))V(\sigma)$ is dense in $V(\sigma)$. Finally, if $\{u_V\}$ is an approximate identity, $\sigma(u_V)$ converges to $I_{V(\sigma)}$ in the strong operator topology.* Conversely, *if $\tau$ is an essential $*$-homomorphism of $C_c(G)$ into* $\mathrm{End}_{\mathbb{C}}(V)$ *which is continuous when $C_c(G)$ is given the usual inductive limit topology of uniform convergence* within *compact sets and* $\mathrm{End}_{\mathbb{C}}(V(\sigma))$ *is given the weak operator topology, then there is a unique unitary representation $\sigma$ of $G$ such that $\sigma = \tau$ on $C_c(G)$.*

*Proof.* Not difficult. For the converse, note that $\sigma$ must satisfy $\sigma(x)\tau(f) = \tau(\lambda_x f)$ for all $x \in G, f \in C_c(G)$.

Note that $\|\sigma(f)\| \leqslant \|f\|_1$ for all $f \in C_c(G)$. One can show that if $G$ satisfies the second axiom of countability, then $C_c(G)$ is separable (i.e. contains a countable dense subset in the $L_1$ norm and hence $\sigma(C_c(G))$ is separable in the operator norm. Finally, note also that if $\sigma$ is an FDR (not necessarily unitary) one can still define $\sigma(f), f \in C_c(G)$, by (5.18) and that $\sigma$ is a homomorphism (but not necessarily a $*$-homomorphism) which is essential. The rest of (5.19) holds as well, with * and unitary excised and the weak and strong operator topologies replaced by the norm topology, to which they are equivalent for finite dimensional $V$.

We sketch now a method of direct integral decomposition. Let $V$ be a Hilbert space, $\mathfrak{A}$ a $*$-algebra of bounded operators on $V$, $\mathfrak{M}$ a $*$-subalgebra of $\mathfrak{A}'$ such that $\mathfrak{M}'' = \mathfrak{M} \subseteq \mathfrak{M}'$, where $\mathfrak{S}'$ denotes the commutant of $\mathfrak{S}$ for any set $\mathfrak{S}$ of bounded operators on $V$. Suppose that there is a vector $v_0 \in V$ with $\|v_0\| = 1$ and $\{AMv_0 : A \in \mathfrak{A}, M \in \mathfrak{M}\}$ total in $V$. $\mathfrak{M}$ is Abelian, closed in the weak operator (and hence norm) topology, and contains $I$. Let $X$ be the space of continuous $*$-homomorphisms of $\mathfrak{M}$ onto $\mathbb{C}$, and give $X$ the Gel'fand topology. Let $M \mapsto \hat{M}$ denote the Gel'fand isomorphism of $\mathfrak{M}$ with $C(X)$. For each $v, w \in V$ let $\mu_{v,w}$ be the Borel measure on $X$

such that

$$(5.20)\qquad (Mv, w) = \int_X \hat{M}\, \mathrm{d}\mu_{v,w}, \qquad M \in \mathfrak{M}.$$

For each bounded Baire function $f$ on $X$, let $T_f$ be the bounded operator on $V$ defined by

$$(5.21)\qquad (T_f v, w) = \int_X f\, \mathrm{d}\mu_{v,w}, \qquad v, w \in V;$$

i.e., $T_0$ is the natural extension to the $*$-algebra $B(X)$ of bounded Baire functions on $X$ of the inverse of the Gel'fand isomorphism. We have $\|T_f\| \leqslant \sup_{\xi\in X}|f(\xi)|$; moreover, $T_0$ is a $*$-homomorphism of $B(X)$ into $\mathfrak{M}'' = \mathfrak{M}$.

Now set $\mu = \mu_{v_0,v_0}$. We have

PROPOSITION 5.22. *For each $A \in \mathfrak{M}'$, there is a unique $f_A \in C(X)$ such that*

$$(AMv_0, v_0) = \int_X \hat{M} f_A\, \mathrm{d}\mu, \qquad \text{all } M \in \mathfrak{M}.$$

*The map $A \mapsto f_A$ is linear. Moreover, $f_{A^*} = \overline{f_A}$, $\|f_A\|_\infty = \|A\|$, $f_{MA} = \hat{M} f_A$ for all $A \in \mathfrak{M}'$ and $M \in \mathfrak{M}$, and $f_A \geqslant 0$ if $A \geqslant 0$.*

*Proof.* This is all routine except for the existence. For that, note that the map $\hat{M} \to Mv_0$ extends to an isometry $U$ of $L_2(X, \mu)$ into $V$, that $U^*AU$ commutes with all multiplications by $C(X)$ and hence is multiplication by some bounded Baire function $f$, and that $T_f \in \mathfrak{M}$ implies that $T_f = T_{f_A}$ for some $f_A \in C(X)$.

Now for each $\xi \in X$, set $(A, B)_\xi = f_{B^*A}(\xi)$ for all $A, B \in \mathfrak{A}$. $(\cdot\,,\cdot)_\xi$ is a semi-definite Hilbert inner product on $\mathfrak{A}$ according to (5.22). Call the completion of $\mathfrak{A}$ modulo null length vectors $V(\xi)$ and let $A(\xi)$ denote the image of $A$ in $V(\xi)$. Note that $\|A(\xi)\| \leqslant \|A\|$. Attach to each $\xi \in X$ the Hilbert space $V(\xi)$, getting a bundle of Hilbert spaces over the base space $X$, call it $\mathscr{V}$. Let $\mathfrak{F}$ be the linear space of 'vector fields' (i.e. cross-sections of $\mathscr{V}$ over $X$) consisting of the $\xi \mapsto A(\xi)$ for $A \in \mathfrak{A}$.

$\mathfrak{F}$ is a *continuous family of vector fields* in the sense that $\xi \mapsto \|a(\xi)\|$ is continuous for all $a \in \mathfrak{F}$. Moreover, $\{a(\xi) : a \in \mathfrak{F}\}$ is dense in $V(\xi)$. Let $\mathscr{G}$ be the set of all vector fields $b$ such that, for each $\xi \in X$ and $\varepsilon > 0$, there exist $a \in \mathfrak{F}$ and a neighborhood $N$ of $\xi$ such that $\|b(\eta) - a(\eta)\| < \varepsilon$ for all

$\eta \in N$. Then $\mathscr{G}$ is a maximal (indeed, is *the* maximal) continuous *linear* family of vector fields containing $\mathfrak{F}$. Clearly then, $C(X)\mathscr{G} \subseteq \mathscr{G}$.

For any $b \in \mathscr{G}$, we define $\|b\|_2$ by

$$(5.23) \qquad \|b\|_2^2 = \int_X \|b(\xi)\|^2 \, d\mu(\xi),$$

turning $\mathscr{G}$ into a pre-Hilbert space. Its completion can be realized concretely as follows: Let $\mathscr{B}$ be the smallest class of vector fields containing $\mathscr{G}$ and closed under sequential pointwise limits. $\mathscr{B}$ is the space of *Baire* vector fields determined by $\mathscr{G}$. If $b \in \mathscr{B}$, then $\xi \mapsto \|b(\xi)\|$ is an ordinary Baire function; moreover, $\mathscr{B}$ is a linear space and a module over the Baire functions. Defining $\|b\|_2$ by (5.23) for $b \in \mathscr{B}$, $\{b \in \mathscr{B} : \|b\|_2 < \infty\}$ modulo $\{b \in \mathscr{B} \colon \|b\|_2 = 0\}$ is a pre-Hilbert space which is *complete*. Moreover $\mathscr{G}$ is dense in this space. Thus we have every right to call this space an $L_2$ space, and we denote it by $L_2(\mathfrak{F}, \mu)$.

Next, we note that the map

$$(5.24) \qquad \sum_i{}_1^m A_i M_i v_0 \mapsto \sum_i{}_1^m \hat{M}_i(\cdot) A_i(\cdot)$$

is well-defined and gives an isometry of a dense subspace of $V$ onto a dense subspace of $L_2(\mathfrak{F}, \mu)$. Let $T$ be its closure. Then, if $M \in \mathfrak{M}$, $TMT^{-1}$ is the operator multiplying $b \in L_2(\mathfrak{F}, \mu)$ by $\hat{M}$. Moreover, the operators $TAT^{-1}$, $A \in \mathfrak{A}$ have a simple form. Indeed, more is true. Let $\mathfrak{A}_1 = \{B \in \mathrm{End}_{\mathbb{C}}(V) : B\mathfrak{A},\ \mathfrak{A}B \subseteq \mathfrak{A}\}$, a $*$-algebra with 1 containing $\mathfrak{A}$. For each $\xi \in X$ and each $B \in \mathfrak{A}_1$, the equation

$$(5.25) \qquad \tilde{B}(\xi)A(\xi) = (BA)(\xi) \qquad \text{for} \qquad A \in \mathfrak{A}$$

defines an operator in $\mathrm{End}_{\mathbb{C}}(V(\xi))$. To see this, note that $(BA_1, A_2)_\xi = (A_1, B^*A_2)_\xi$ for $A_1, A_2 \in \mathfrak{A}$. Then repeated applications of the Schwarz inequality give

$$\begin{aligned}(BA, BA)_\xi &\leqslant ((B^*B)^{2^n} A, A)_\xi^{2^{-n}} (A, A)_\xi^{1-2^{-n}} \\ &\leqslant \|A^*(B^*B)^{2^n} A\|^{2^{-n}} (A, A)'^{\,1-2^{-n}} \\ &\leqslant \|b\|^2 \|A\|^{2^{-n+1}} (A, A)'^{\,1-2^{-n}}\end{aligned}$$

for all $n$, so that $\|(BA)(\xi)\| \leqslant \|B\| \, \|A(\xi)\|$ for all $A \in \mathfrak{A}$. Now it is easy to see that, for $B \in \mathfrak{A}_1$, $TBT^{-1}$ is the operator sending every $b(\cdot) \in L_2(\mathfrak{F}, \mu)$ into $\tilde{B}(\cdot)b(\cdot) \in L_2(\mathfrak{F}, \mu)$.

Now suppose that there is a countable subset $\mathfrak{F}_0$ of $\mathfrak{F}$ so that $\{a(\xi) : a\in\mathfrak{F}_0\}$ is total in $V(\xi)$ for all $\xi\in X$. Suppose further that $\mathfrak{S}$ is a $*$-subalgebra of $\mathfrak{A}_1$ such that $(\mathfrak{S}\cup\mathfrak{M})''=\mathfrak{M}'$. Then we have

THEOREM 5.26. *$\{\tilde{B}(\xi) : B\in\mathfrak{S}\}$ is an irreducible set of operators on $V(\xi)$ for $\mu$-almost all $\xi\in X$.*

We apply the foregoing to group representations as follows: Let $\sigma$ be a unitary representation of the locally compact group $G$ on $V$. Let $\mathfrak{M}$ be a $*$-subalgebra of $R(\sigma, \sigma)$ such that $\mathfrak{M}''=\mathfrak{M}\subseteq\mathfrak{M}'$. Suppose that there is a vector $v_0\in V$ with $\|v_0\|=1$ and $\{\sigma(x)Mv_0 : x\in G, M\in\mathfrak{M}\}$ total in $V$. Let $\mathfrak{A}=\sigma(C_c(G))$. Then $\{AMv_0 : A\in\mathfrak{A}, M\in\mathfrak{M}\}$ is total in $V$, so that the previous decomposition can be made relative to this $\mathfrak{M}$ and $\mathfrak{A}$. Moreover, $\sigma(x)\in\mathfrak{A}_1$ for all $x\in G$. Set $\sigma_\xi(x)=\tilde{\sigma}(x)(\xi)$ for $\xi\in X$. Then we have

THEOREM 5.27. *$\sigma_\xi$ is a continuous unitary representation of $G$ on $V(\xi)$. Moreover, $\sigma_\xi(f)=\tilde{\sigma}(f)(\xi)$ for all $f\in C_c(G)$ and $\xi\in X$.*

THEOREM 5.28. *If $\mathfrak{M}$ is maximal Abelian in $R(\sigma, \sigma)$ and if $G$ satisfies the second axiom of countability, then $\sigma_\xi$ is irreducible for $\mu$-almost all $\xi\in X$.*

We can obtain a decomposition of $\sigma$ into primary representations by using a slightly different $\mathfrak{A}$. Suppose $G$ satisfies the second axiom of countability and suppose $V$ is separable. Let $\mathfrak{M}=CR(\sigma, \sigma)$. We can find a countable $*$-algebra$/(\mathbb{Q}+i\mathbb{Q})$ which contains $I$ and is weak operator dense in $R(\sigma, \sigma)$, call it $\mathfrak{R}$. Let $\mathfrak{A}$ now be the algebra generated by $\mathfrak{R}$ and $\sigma(C_c(G))$. Theorem (5.26) applies and we obtain

THEOREM 5.29. *Let $G$ be second countable and $V$ be separable. Let $\mathfrak{M}=CR(\sigma, \sigma)$. For each $x\in G$, $\sigma(x)\in\mathfrak{A}_1$, $\sigma_\xi$ is a representation for each $\xi\in X$, and, for $\mu$-almost all $\xi\in X$, $\sigma_\xi$ is primary. Moreover, there is a $\mu$-null set $N\subseteq X$ such that $\sigma_\xi \wr \sigma_\eta$ for $\xi, \eta\in X-N$ with $\xi\neq\eta$.*

If $G$ is not type I, the decomposition of (5.28) can be wildly non-unique. The decomposition (5.29), on the other hand, is canonical and is called the *central* decomposition. If, in addition, $G$ is type I, the primary factors $\sigma_\xi$ correspond to irreducibles (5.14), and the decomposition may be transferred to $\hat{G}$, the set of all unitary equivalence classes of irreducible unitary representation of $G$. If $\sigma$ is the *left regular representation*, that is, the representation $\lambda$ of $G$ on $L_2(G)$, then we obtain from (5.29) a satisfactory analogue of (4.10) for type I $G$ satisfying the second axiom of countability.

If $\{\alpha\}$ is a $\mu$-atom in the central decomposition of $\lambda$, $\alpha\in\hat{G}$ ($G$ type I, second countable), then $\lambda_\alpha$ is contained in $\lambda$ as a subrepresentation; moreover, $\lambda_\alpha \simeq n\sigma_\alpha$ where $\sigma_\alpha\in\alpha$. Thus $\sigma_\alpha$ is contained in $\lambda$. Such an $\alpha$ is called a *discrete* class. We examine such classes more closely.

Let $G$ be unimodular, that is $\Delta_G\equiv 1$. $\alpha\in\hat{G}$ is called *square integrable* if $f^\sigma_{u,v}\in L_2(G)$ for some non-zero $u$, $v\in V(\sigma)$, $\sigma\in\alpha$. (Here $f^\sigma_{u,v}(x)=(\sigma(x)^{-1}u, v)$.)

THEOREM 5.30. *$\alpha\in\hat{G}$ is discrete if and only if it is square integrable. Let $\alpha$, $\beta$ be discrete, and let $\sigma\in\alpha$, $\tau\in\beta$. Then $(f^\sigma_{u,v}, f^\tau_{u',v'})=0$ for all $u$, $v\in V(\sigma)$ and $u'$, $v'\in V(\tau)$ if $\alpha\neq\beta$. Moreover, there is a constant* d($\alpha$), *called the* formal degree *of $\alpha$, such that $0<d(\alpha)<\infty$ and*

$$(f^\sigma_{u,v}, f^\sigma_{u',v'})=(1/\mathrm{d}(\alpha))(u, u')\overline{(v, v')}$$

*for $u$, $v$, $u'$, $v'\in V(\sigma)$. Finally, if $G$ is compact with total Haar measure* 1 *and $V(\sigma)$ is finite dimensional, then* $\mathrm{d}(\alpha)=\dim V(\sigma)$.

*Proof.* Suppose $\alpha$ is discrete. Let $V$ be a $\lambda$-irreducible subspace of $L_2(G)$ such that $\sigma=(V, \lambda)\in\alpha$. Let $P$ be the orthogonal projection of $L_2(G)$ onto $V$. Let $0\neq u\in V$, $0\neq v\in PC_c(G)$, and write $v=Pw$ with $w\in C_c(G)$. Then

$$f^\sigma_{u,v}(x)=(\lambda_{x^{-1}}u, v)=(u, \lambda_x w)=\int u(y)\overline{w(x^{-1}y)}\,\mathrm{d}y=(u\circ w^*)(x)$$

for $x\in G$. But $u\circ w^*\in L_2(G)$ for $u\in L_2(G)$, $w\in C_c(G)$. Therefore $f^\sigma_{u,v}\in L_2(G)$ and $\alpha$ is square integrable.

Conversely, suppose $\alpha$ is square integrable. Choose $u_0$, $v_0$ non-zero in $V(\sigma)$, $\sigma\in\alpha$, so that $f^\sigma_{u_0,v_0}\in L_2(G)$. Let $V_0$ be the algebraic linear span of $\{\sigma(x)u_0 : x\in G\}$, a dense subspace of $V(\sigma)$ since $\sigma$ is irreducible. Since $f^\sigma_{\sigma(x)u_0,v_0}=\lambda_x f^\sigma_{u_0,v_0}$, the map $T$ given on $V_0$ by $Tu=f^\sigma_{u,v_0}$, $u\in V_0$, maps $V_0$ into $L_2(G)$. Moreover, $T\sigma(x)=\lambda_x T$. Let $h\in C_c(G)$. Then

$$(Tu, h)=\int (\sigma(x^{-1})u, v_0)\overline{h(x)}\,\mathrm{d}x$$

so that $|(Tu, h)|\leqslant\|h\|_1\|u\|\,\|v_0\|$ for $u\in V_0$. Therefore $h\in\mathrm{Dom}(T^*)$, whence $T^*$ is closed and densely defined, and $T$ has a closure, namely $T^{**}$. Thus $T^*T^{**}$ is a self adjoint operator in $V(\sigma)$ which commutes with each $\sigma(x)$, $x\in G$. Using the spectral theorem and the irreducibility of $\sigma$, we see that $T^*T^{**}=c\|v_0\|^2 I$ for some $c>0$. Therefore $T^{**}=c^{1/2}\|v_0\|U$, where $U$ is an isometry of $V(\sigma)$ into $L_2(G)$ such that $U\in R(\sigma, \lambda)$.

We have shown that $\|f^\sigma_{u,v_0}\|_2=c^{1/2}\|u\|\,\|v_0\|$ for $u\in V_0$. If $\{u_n\}\subseteq V_0$ with $u_n\to u$ in $V(\sigma)$ and $f^\sigma_{u_n,v_0}\to T^{**}u$, then $f^\sigma_{u_n,v_0}\to f^\sigma_{u,v_0}$ pointwise, so that $T^{**}u=f^\sigma_{u,v_0}$. This shows that $\|f^\sigma_{u,v_0}\|_2=c^{1/2}\|u\|\,\|v_0\|$ for all $u\in V(\sigma)$. Now fix $u\in V(\sigma)$. By considering the span of $\{\sigma(x)v_0\colon x\in G\}$, call it $V_1$, we obtain analogously a constant $0\leqslant a(u)\leqslant\infty$ such that $\|f^\sigma_{u,v}\|_2= =a(u)^{1/2}\|u\|\,\|v\|$ for $v\in V_1$ and finally for all $v\in V(\sigma)$. Setting $v=v_0$, we get that $a(u)=c$ for all $u\in V(\sigma)$. In summary, $\|f_{u,v}\|_2=c^{1/2}\|u\|\,\|v\|$ for all $u, v\in V(\sigma)$. One consequence of this is that, if $v\neq 0$, the map $u\to c^{-1/2}\|v\|^{-1}f^\sigma_{u,v}$ is a unitary equivalence of $\sigma$ with a subrepresentation of $\lambda$: $\alpha$ is discrete.

Now for square integrable $\alpha$, we have

$$(f^\sigma_{u,v}, f^\sigma_{u',v'})=(1/d(\alpha))(u, u')\overline{(v, v')}$$

for $u$, $u'$, $v$, $v'\in V(\sigma)$, $\sigma\in\alpha$ by polarizing $\|f^\sigma_{u,v}\|_2^2=c\|u\|^2\|v\|^2$ and setting $d(\alpha)=c^{-1}$. On the other hand, if $\alpha$, $\beta$ are square integrable with $\alpha\neq\beta$, then $(f^\sigma_{V(\sigma),v}, \lambda)$ and $(f^\tau_{V(\tau),v'}, \lambda)$ are equivalent to $\sigma$ and $\tau$, respectively, where $\sigma\in\alpha$, $\tau\in\beta$, $0\neq v\in V(\sigma)$, $0\neq v'\in V(\tau)$. $\sigma$ and $\tau$ are inequivalent, hence disjoint. Hence $f^\sigma_{V(\sigma)bv}\perp f^\tau_{V(\tau),v'}$.

Finally, suppose $G$ is compact with total Haar measure 1. Let $V(\sigma)$ be finite-dimensional, a condition that is automatically satisfied (cf. the third paragraph following Theorem 6.9), and let $\{u_i\}$ be an orthonormal basis of $V(\sigma)$. Set $n=\dim V(\sigma)$. Then

$$\begin{aligned} n^2 d(\alpha)^{-1} &= \sum_{ij}\int_G (\sigma(x)^{-1}u_i, u_j)\overline{(\sigma(x)^{-1}u_i, u_j)}\,\mathrm{d}x \\ &= \sum_{ij}\int_G (\sigma(x)^{-1}u_i, u_j)(u_j, \sigma(x)^{-1}u_i)\,\mathrm{d}x \\ &= \sum_{i}\int_G (\sigma(x)^{-1}u_i, \sigma(x)^{-1}u_i)\,\mathrm{d}x = n. \end{aligned}$$

Thus $d(\alpha)=n$.

This is the generalization of (4.8) that was promised.

Now let $\alpha\in\hat{G}$ be discrete. Let $W_\alpha$ be a closed $\lambda$-invariant subspace of $L_2(G)$ such that $(W_\alpha, \lambda)\sim\alpha$ and $(W_\alpha^\perp, \lambda)\,\wr\,\alpha$. Such a $W_\alpha$ exists by exhaustion.

THEOREM 5.31. *$W_\alpha$ is $\rho$-invariant. Let $\sigma \in \alpha$. Then $W_\alpha$ is the closed linear span of $\{f^{\sigma}_{u,v} : u, v \in V(\sigma)\}$. Let $\tau_{x,y} = \lambda_x \rho_y$ for $(x, y) \in G \times G$. Then $(W_\alpha, \tau)$ is unitarily equivalent to $\sigma \times \sigma'$.*

The key step in proving (5.31) is

LEMMA 5.32. *Let $V$ by any closed $\lambda$-invariant irreducible subspace of $L_2(G)$. Let $W$ be any closed $\tau$-invariant subspace of $L_2(G)$ with $W \supseteq V$. Then*

$$W \supseteq \{f^{\lambda|V}_{u,v} : u, v \in V\}.$$

*Proof.* Use the techniques of proof in (5.30).

In this way we get a completely satisfactory analogue of (4.10) and (4.11) for the discrete part of the central decomposition of $\lambda$ for unimodular, type I, second countable $G$.

We finish by saying a word on the Plancherel formula for such $G$. For $g \in C_c(G)$, $\alpha \in \hat{G}$, set $\check{g}(\alpha) = \int g(x)\sigma_\alpha(x)\,dx$, where $\sigma_\alpha \in \alpha$ has been picked once for all. Looking at (4.22), we would like it very much if $\check{g}(\alpha)$ were Hilbert-Schmidt, $\alpha \in \hat{G}$, and if there were a measure $\mu$ of some sort on $\hat{G}$ so that

$$(5.33) \qquad (f, g) = \int_{\hat{G}} \mathrm{Tr}(\check{f}(\alpha)\check{g}^*(\alpha))\,d\mu(\alpha)$$

for all $f, g \in C_c(G)$. For this to make sense, we must specify a Borel $\sigma$-field of subsets of $\hat{G}$ and then know that $\alpha \mapsto \mathrm{Tr}(\check{f}(\alpha)\check{g}^*(\alpha))$ is $\mu$-measurable. The Borel field is specified as follows: Give $C_c(G)$ a new norm $\|\cdot\|_*$ defined by

$$(5.34) \qquad \|f\|_* = \sup_\sigma \|\sigma(f)\|,$$

where the supremum is taken over *all* unitary representations $\sigma$ of $G$. The completion $C^*(G)$ of $C_c(G)$ in this norm is a $*$-algebra under the extensions of $\circ$ and $*$ to $C^*(G)$ and is, in fact, a $C^*$-algebra, called the *group $C^*$-algebra* of $G$. Each $\alpha \in \hat{G}$ specified a primitive ideal in $C^*(G)$, viz. the kernel of $\sigma_\alpha$. The primitive ideal space of $C^*(G)$ has the Jacobson or hull-kernel topology: the basic closed sets are those sets of primitive ideals containing a given ideal of $C^*(G)$. Pull this topology back to $\hat{G}$. It is a theorem of Glimm that, for $G$ second countable this topology is $T_0$ if and only if $G$ is type I. For type I $G$, let $\mathscr{F}$ be the Borel field generated by this topology. The measure $\mu$ on $\mathscr{F}$ is called the *Plancherel measure*. Obviously $f(\check{\alpha})$ need be Hilbert-Schmidt only $\mu$-a.e.

We do not take up here the existence of a Plancherel measure. We merely note that if $\alpha$ is discrete, then $\{\alpha\}$ is an atom for $\mu$ and that $\mu(\{\alpha\})=d(\alpha)$; in complete analogy with (4.22). This follows from

THEOREM 5.35. *Let $\alpha$ be discrete and let $f \in C_c(G)$. Then $\sigma_\alpha(f)$ is Hilbert–Schmidt. Let E be the projection of $L_2(G)$ on the $W_\alpha$ of* (5.31). *Then $\|Ef\|_2^2 = d(\alpha)\|\sigma_\alpha(f)\|_{HS}^2$, where $\|\cdot\|_{HS}$ denotes the Hilbert–Schmidt norm.*

## 6. Representations of Lie Groups

Let $G$ be a Lie group and let $\mathfrak{g}$ be its Lie algebra. Let $\sigma$ be a unitary representation of $G$ on the Hilbert space $V$. We would somehow like to associate with $\sigma$ a representation of $\mathfrak{g}$ as formally skew symmetric operators on a dense subspace of $V$. As a first step in this direction, we recall

THEOREM 6.1 (Stone). *Let $\sigma$ be a unitary representation of $\mathbb{R}$ on the Hilbert space V. Then there is a unique closed, densely defined operator S such that $S^* = -S$ and*

$$\sigma(t) = e^{tS} \qquad \textit{for all} \qquad t \in \mathbb{R}.$$

Now let $\sigma$ be a unitary representation of $G$ as above. Let $\xi \in \mathfrak{g}$. Then $t \mapsto \sigma(\exp t\xi)$ is a unitary representation of $\mathbb{R}$ on $V$, so that there is a skew-adjoint operator $\sigma(\xi)$ in $V$ such that

$$\sigma(\exp t\xi) = e^{t\sigma(\xi)}, \qquad t \in \mathbb{R}.$$

Now the map $\xi \mapsto \sigma(\xi)$ would seem to meet our requirements. Unfortunately, the operators $\sigma(\xi)$, $\xi \in \mathfrak{g}$, will in general have different domains. What we need is a common dense domain $\mathscr{D}$ invariant under all of the $\sigma(\xi)$ and such that $\xi \mapsto \sigma(\xi)|\mathscr{D}$ is a representation of $\mathfrak{g}$. We would moreover like $\mathscr{D}$ to be large enough: we want the closure of $\sigma(\xi)|\mathscr{D}$ to be $\sigma(\xi)$.

One such domain is the following:

DEFINITION 6.2. A vector $v \in V$ is called *differentiable* if $x \mapsto \sigma(x)v$ is a $V$-valued $C^\infty$ function on $G$. The space of all differentiable vectors in $V$ will be denoted by $V^\infty$.

By the definition of the domain of $\sigma(\xi)$, $\operatorname{Dom}\sigma(\xi) \supseteq V^\infty$ for all $\xi \in \mathfrak{g}$.

Obviously $\sigma(\xi)V^\infty \subseteq V^\infty$ and $\sigma(x)V^\infty \subseteq V^\infty$ for all $\xi \in \mathfrak{g}$, $x \in G$. What we do not know yet is whether $V^\infty$ is big. This question is answered by the following theorems.

THEOREM 6.3 (Gårding, Nelson–Stinespring). *Let $\mathscr{D}$ be the algebraic linear span of all vectors in $V$ of the form*

$$\int g(x)\sigma(x)v \, \mathrm{d}x,$$

*where $g \in C_c^\infty(G)$, $v \in V$, and the integral is taken (say) in the Riemann sense. Then $\mathscr{D} \subseteq V^\infty$, $\sigma(\xi)\mathscr{D} \subseteq \mathscr{D}$ for $\xi \in \mathfrak{g}$ and $\sigma(x)\mathscr{D} \subseteq \mathscr{D}$ for $x \in G$. Let $\{\xi_1, \ldots, \xi_n\}$ be a basis of $\mathfrak{g}$. Then $\Delta = \sigma(\xi_1)^2 + \cdots + \sigma(\xi_n)^2$ is essentially self-adjoint (i.e., has self-adjoint closure) on $\mathscr{D}$.*

It follows immediately from (6.3) that $\mathscr{D}$, and hence $V^\infty$, is dense in $V$ (let $g$ run through an approximate identity).

THEOREM 6.4 (Segal). *The closure of $\sigma(\xi)|\mathscr{D}$ is $\sigma(\xi)$, all $\xi \in \mathfrak{g}$.*

THEOREM 6.5. *$\xi \mapsto \sigma(\xi)|V^\infty$ is a representation of $\mathfrak{g}$ by essentially skew-adjoint operators on $V^\infty$.*

In this way, we have associated a representation of $\mathfrak{g}$ to $\sigma$, which we denote by $d\sigma$. Unfortunately, this representation may have bad properties. For example, it is possible for a subspace $W$ of $V^\infty$ to be invariant under $d\sigma$ without its closure $\overline{W}$ being invariant under $\sigma$. For this reason, Harish-Chandra [15] introduced a more restricted class of vectors, which he called *well-behaved*, and which are now called *analytic*.

DEFINITION 6.6. A vector $v \in V$ is called *analytic* if $x \mapsto \sigma(x)v$ is a $V$-valued $C^\omega$ function on $G$; i.e. if $\sigma(x)v$ can be expanded as a convergent power series with coefficients in $V$ in terms of local coordinates in the neighborhood of any point $x_0 \in G$. The space of all analytic vectors in $V$ will be denoted by $V^\omega$.

A related notion is the following (Nelson [33]):

DEFINITION 6.7. Let $S$ be an operator in the Hilbert space $V$. A vector $v \in V$ is called an *analytic vector* for $S$ if $v \in \mathrm{Dom}(S^n)$ for all $n \geqslant 0$ and if

$\sum_n{}_0^\infty(\|S^n v\|/n!)t^n < \infty$ for some $t>0$.

Nelson has proved the following important result:

THEOREM 6.8. *Let $\sigma$ be a unitary representation of G on V. Then $V^\omega$ is dense in V. Indeed, let $\{\xi_1, \ldots, \xi_n\}$ be a basis for $\mathfrak{g}$, and let $\Delta = d\sigma(\xi_1)^2 + \cdots + d\sigma(\xi_n)^2$. Then any analytic vector for the closure $\overline{\Delta}$ of $\Delta$ is an analytic vector for $\sigma$ (and there are plenty of these by the spectral theorem).*

Plainly $V^\omega$ is invariant under $\sigma$ and $d\sigma$.

Having associated a representation $d\sigma$ of $\mathfrak{g}$ to any unitary representation $\sigma$ of $G$, it is natural to want to reverse the process, as can be done if $d\sigma$ is any FDR of $\mathfrak{g}$ and if $G$ is simply connected. In the infinite-dimensional case, however, things can go horribly wrong. Nelson has given an example of two skew-symmetric operators $A$ and $B$ on a Hilbert space having a common invariant domain $\mathscr{D}$ on which all operators $aA+bB$, $a$, $b \in \mathbb{R}$, are essentially skew adjoint and on which $A$ and $B$ commute, but such that $\overline{A}$ and $\overline{B}$ have non-commuting spectral resolutions. Thus we have an example of an apparently beautiful representation of the Lie algebra $\mathfrak{g}$ of $\mathbb{R}^2$ which does not come from a unitary representation of $\mathbb{R}^2$. The problem with this example is that $A^2+B^2$ is not essentially self-adjoint on $\mathscr{D}$. In the positive direction, we have the following.

THEOREM 6.9 (Nelson). *Let G be a connected, simply connected Lie group with Lie algebra $\mathfrak{g}$. Let V be a Hilbert space. Let $\tau$ be a representation of $\mathfrak{g}$ by skew-symmetric operators on a common invariant dense domain $\mathscr{D} \subseteq V$. Let $\{\xi_1, \ldots, \xi_n\}$ be a basis for $\mathfrak{g}$ and suppose that $\Delta = \tau(\xi_1)^2 + \cdots + \tau(\xi_n)^2$ is essentially self-adjoint on $\mathscr{D}$. Then there is a unique unitary representation $\sigma$ of G on V such that $\overline{d\sigma(\xi)} = \overline{\tau(\xi)}$ for all $\xi \in \mathfrak{g}$.*

Our next aim is to extend the notion of the *character* of a representation from compact $G$ to Lie groups, not necessarily compact. If $\sigma$ is a unitary representation then the character of $\sigma$ ought to be the function $x \mapsto \mathrm{Tr}(\sigma(x))$. If $\sigma$ is infinite dimensional (and for many groups, the only finite-dimensional irreducible unitary representation is the one-dimensional identity representation), $\mathrm{Tr}(\sigma(x))$ is, strictly speaking, nonsense. What we would like to do is to interpret $\mathrm{Tr}(\sigma(\cdot))$ as a distribution; that is, we would like to restrict our attention to $\sigma$ such that $\sigma(f)$ is trace class for $f \in C_c^\infty(G)$ and such that $f \mapsto \mathrm{Tr}(\sigma(f))$ is continuous on $C_c^\infty(G)$, where $C_c^\infty(G)$ is given the usual topology of uniform convergence of functions and all their derivatives within compact sets.

One case in which this can be done is: $G$ is a connected semisimple matrix group and $\sigma$ is irreducible. We first pause to indicate some facts.

If $K$ is any compact group, $K$ is type I. Indeed, if $\sigma$ is *any* unitary representation of $K$, then $V(\sigma)$ contains an irreducible FDR: by (4.3) and (5.19), $\sigma(\mathscr{R})v \neq \{0\}$ for $0 \neq v \in V(\sigma)$, and so $\sigma(f)v \neq 0$ for some $f \in \mathscr{R}$; therefore $\sigma(G)\sigma(f)v = \sigma(\lambda_G f)v$ spans a finite-dimensional invariant subspace of $V(\sigma)$. It follows that $\sigma$ is a direct sum of irreducibles.

Now let $G$ be a semisimple matrix group and let $\mathscr{R}$ be the space of all $G$-finite continuous functions on $G$. Then $\mathscr{R}$ is an algebra over $\mathbb{C}$ under pointwise multiplication of functions and is closed under complex conjugation. $\mathscr{R}$ also contains all of the representative functions (4.6) of FDR's of $G$. If $\sigma$ is a faithful FDR of $G$ (which exists because $G$ is already given as a group of matrices), the representative functions of $\sigma$ separate the points of $G$. *A fortiori*, $\mathscr{R}$ separates the points of $G$. Now let $f \in C_c(G)$ and let $N$ be the support of $f$. Suppose that $\int f(x)g(x^{-1})\,dx = 0$ for all $g \in \mathscr{R}$. By the Stone–Weierstrass Theorem $\mathscr{R}|N^{-1}$ is $\|\cdot\|_\infty$-norm dense in $C(N^{-1})$. Therefore $\int_N f(x)g(x^{-1})\,dx = 0$ for all $g \in C(N^{-1})$, so that $f = 0$.

Suppose now that $f \in C_c(G)$ and that $\sigma(f) = 0$ for *every irreducible* FDR $\sigma$ of $G$. Since every FDR $\sigma$ of a semisimple group is a direct sum of irreducible FDR's, we see that $\sigma(f) = 0$ for *every* FDR $\sigma$ *whatsoever*. Let $V$ be a FD $\lambda$-invariant subspace of $C(G)$. Then $\int f(x)\langle u', \lambda_x g\rangle\,dx = 0$ for all $g \in V$, $u' \in V'$. If $u' = \delta_1$, $\langle u', \lambda_x g\rangle = g(x^{-1})$ so that $\int f(x)g(x^{-1})\,dx = 0$ for all $g \in V$. That is, $\int f(x)g(x^{-1})\,dx = 0$ for all $g \in \mathscr{R}$, and hence $f = 0$.

From the structure theory of semisimple Lie groups, $G = KS$, where $K$ is a maximal compact subgroup of $G$ and $S$ is a connected solvable Lie group. By Lie's Theorem, every FDR of $S$ has a one-dimensional subrepresentation. Let $\sigma$ be an irreducible FDR of $G$. Then there exists $0 \neq v \in V(\sigma)$ spanning a $\sigma|S$-invariant subspace of $V(\sigma)$: $\sigma(s)v = \chi(s)v$, $s \in S$, where $\chi\colon S \to \mathbb{C}^\times$ is a homomorphism. Then $\sigma(G)v = \sigma(K)\sigma(S)v \subseteq \subseteq \mathbb{C}^\times\sigma(K)v$. Since $\sigma(G)v$ spans $V(\sigma)$, so does $\sigma(K)v$. Make $\sigma|K$ unitary (4.7). Then $u \mapsto f^{\sigma|K}_{u,v}$ belongs to $R(\sigma|K, (\lambda, L_2(K)))$. Since $\sigma(K)v$ spans $V(\sigma)$, this map is injective. Therefore (4.10) implies that $\sigma|K$ contains each $\alpha \in \hat{K}$ at most $d(\alpha)$ times.

We need a way of going from this fact to a similar statement about irreducible unitary representations of $G$, and this is supplied by the theory of *polynomial identities*. For each integer $m \geqslant 0$, let $P_m(x_1, \ldots, x_m)$ be the polynomial in the non-commuting indeterminants $x_1, \ldots, x_m$ given by

$P_m(x_1, \ldots, x_m) = \Sigma\, \mathrm{sgn}(\pi) x_{\pi(1)} \cdots x_{\pi(m)}$, where the summation extends over all permutations $\pi$ of $\{1, \ldots, m\}$. If $\mathfrak{A}$ is an algebra, we say $\mathfrak{A}$ *satisfies* $P_m$ if $P_m(a_1, \ldots, a_m) = 0$ for all $a_1, \ldots, a_m \in \mathfrak{A}$. Let $M_n = \mathrm{End}_{\mathbb{C}}(\mathbb{C}^n)$. Clearly $M_n$ satisfies $P_m$ for all $m \geqslant n^2 + 1$. Let $m(n)$ be the smallest integer such that $M_n$ satisfies $P_m$ for all $m \geqslant m(n)$. Embedding $M_n$ in $M_{n+1}$ by bordering with a last row and column of zeroes, we see that $m(n+1) \geqslant m(n)$. But more is true: $m(n+1) > m(n)$. To see this, note that $b = P_{m(n)-1}(a_1, \ldots, a_{m(n)-1}) \neq 0$ for some $a_1, \ldots, a_{m(n)-1} \in M_n$. Embedding $M_n$ in $M_{n+1}$ as above, choose a matrix unit $e_{j,n+1}$ such that $be_{j,n+1} \neq 0$. Since $e_{j,n+1} a = 0$ for all $a \in M_n$, we have $P_{m(n)}(a_1, \ldots, a_{m(n)-1}, e_{j,n+1}) = be_{j,n+1} \neq 0$. Finally, we say $\mathfrak{A}$ is of *type* $\mathrm{I}_{\leqslant n}$ if $\mathfrak{A}$ satisfies $P_m$ for all $m \geqslant m(n)$.

We shall make use of the following facts about representations of $K$, which are easily deduced from the material of Section 4: For $\alpha \in \hat{K}$, set $e_\alpha = d(\alpha)\overline{\chi_\alpha}$. Then $e_\alpha^* = e_\alpha$, $e_\alpha \circ e_\alpha = e_\alpha$, and $e_\alpha \circ e_\beta = 0$ for $\alpha \neq \beta$. Moreover, if $\tau$ is a representation of $K$, $E_\alpha = \tau(e_\alpha) \in CR(\tau, \tau)$, and $\tau_{E_\alpha} \sim \alpha$. For $\alpha \in \hat{K}$, we let $C_c^\alpha(G)$ consist of all $f \in C_c(G)$ such that

$$\int_K e_\alpha(y) f(y^{-1}x)\, dy = f(x) = \int_K f(xy) e_\alpha(y^{-1})\, dy, \qquad \text{all } x \in G,$$

a $*$-subalgebra of $C_c(G)$. For representations $\sigma$ of $G$,

$$\sigma(C_c^\alpha(G)) = (\sigma|K)(e_\alpha)\sigma(C_c(G))(\sigma|K)(e_\alpha).$$

LEMMA 6.10 (Godement–Stinespring). *Let $G$ be a semisimple matrix group. Then $C_c^\alpha(G)$ is of type $I_{\leqslant d(\alpha)^2}$, $\alpha \in \hat{K}$.*

*Proof.* Let $\sigma$ be an irreducible FDR of $G$. Let $E_\alpha = (\sigma|K)(e_\alpha)$. Since $\sigma|K$ contains $\alpha$ at most $d(\alpha)$ times, $\dim_{\mathbb{C}} E_\alpha V(\sigma) \leqslant d(\alpha)^2$. Since effectively $\sigma(C_c^\alpha(G)) \subseteq \mathrm{End}_{\mathbb{C}}(E_\alpha V(\sigma))$, $\sigma$ provides a homomorphism of $C_c^\alpha(G)$ into an algebra of type $\mathrm{I}_{\leqslant d(\alpha)^2}$. But we have seen that the irreducible FDR's $\sigma$ form a separating set of such homomorphisms of $C_c^\alpha(G)$. Therefore $C_c^\alpha(G)$ is of type $\mathrm{I}_{\leqslant d(\alpha)^2}$.

COROLLARY 6.11. *Let $\sigma$ be an irreducible unitary representation of $G$. Then, for each $\alpha \in \hat{K}$, $\sigma|K$ contains $\alpha$ at most $d(\alpha)$ times.*

*Proof.* For each $\alpha \in \hat{K}$, let $E_\alpha = (\sigma|K)(e_\alpha)$. Let $\mathfrak{A}$ be the von Neumann algebra generated by $\sigma(G)$. $\mathfrak{A}$ is the von Neumann algebra generated by $\sigma(C_c(G))$. Now $E_\alpha \mathfrak{A} E_\alpha$ is the weak operator closure of $E_\alpha \sigma(C_c(G)) E_\alpha =$

$=\sigma(C_c^\alpha(G))$ and so is of type $\mathrm{I}_{\leqslant d(\alpha)^2}$. But $\mathfrak{A}$ is the commutant of $R(\sigma, \sigma)$ and so, by (5.5), is all of $\mathrm{End}_{\mathbb{C}} V(\sigma)$. If $E_\alpha$ contained more than $d(\alpha)^2$ mutually orthogonal projections, then $E_\alpha \mathrm{End}_{\mathbb{C}}(V(\sigma))E_\alpha$ would contain a copy of $M_k$ with $k > d(\alpha)^2$ and this would force $M_k$ to be of type $\mathrm{I}_{<d(\alpha)^2}$ also, a contradiction. Therefore $\dim_{\mathbb{C}} E_\alpha V(\sigma) \leqslant d(\alpha)^2$. But $\sigma|K \simeq \bigoplus_{\alpha\in\hat{K}} (\sigma|K)_{E_\alpha}$. Now apply (5.13) and (5.14).

COROLLARY 6.12. *G is a type I group.*

*Proof.* Let $\sigma$ be a primary representation of $G$. Let $E_\alpha$, $\alpha \in \hat{K}$, and $\mathfrak{A}$ be defined as in (6.11). The argument of (6.11) shows that, for each $\alpha\in\hat{K}$, $E_\alpha \mathfrak{A} E_\alpha$ contains no more than $d(\alpha)^2$ mutually orthogonal projections. Choose $\alpha\in\hat{K}$ so that $E_\alpha \neq 0$. Then $E_\alpha \mathfrak{A} E_\alpha$ contains a minimal projection $E$. By the spectral theorem $\dim_{\mathbb{C}} E\mathfrak{A}E = 1$. Choose $0 \neq v \in EV(\sigma)$ and set $W = \overline{\mathfrak{A}v}$, the closure of $\mathfrak{A}v$. $W$ is a $\sigma$-invariant closed subspace of $V(\sigma)$. Note that $\dim_{\mathbb{C}} EW = 1$. Now let $W' \subseteq W$ be closed and $\sigma$-invariant. Then either $EW' = \{0\}$ or $EW' = EW$. In the former case $v \perp W'$ so that $\mathfrak{A}v \perp W'$ so that $W \perp W'$ and hence $W' = \{0\}$. In the latter case $v \in EW'$. But $W'$ is $\mathfrak{A}$-invariant so that $v \in W'$ and hence $W' = W$. Therefore $\sigma_W$ is irreducible and the result follows from (5.10).

Now we can show that every irreducible unitary $\sigma$ of $G$ has a distribution character. Corollary 6.11 and Theorem 4.10 imply $\sigma|K \leqslant \lambda^K$, the left regular representation of $K$. If $g \in C(K)$, then the proof of (4.2) applied to $\lambda^K$ rather than $\rho^K$ tells us that $\|\lambda^K(g)\|_{\mathrm{HS}} = \|g\|_2$. Therefore $(\sigma|K)(g)$ is Hilbert–Schmidt, and $\|(\sigma|K)(g)\|_{\mathrm{HS}} = \|g\|_2$. Let $f \in C_c(G)$. Then

$$\begin{aligned}
\|\sigma(f)\|_{\mathrm{HS}} &= \left\| \int_G f(x)\sigma(x)\,\mathrm{d}x \right\|_{\mathrm{HS}} \\
&= \left\| \int_G \int_K f(kx)\sigma(kx)\,\mathrm{d}k\,\mathrm{d}x \right\|_{\mathrm{HS}} \\
&\leqslant \int_G \left\| \left( \int_K f(kx)\sigma(k)\,\mathrm{d}k \right)\sigma(x) \right\|_{\mathrm{HS}} \mathrm{d}x \\
&\leqslant \int_G \|(\rho_x f)|K\|_2\,\mathrm{d}x \leqslant \|f\|_\infty \mu(K\,\mathrm{Supp}(f)),
\end{aligned}$$

where $\mu$ is Haar measure on $G$. Let $\mathfrak{k}$ be the Lie algebra of $K$, and let $\mathfrak{k}_{\mathbb{C}}$

be its complexification. Look at the representation $d(\sigma|K)$ of $\mathfrak{k}$ on $V(\sigma|K)^\infty$ as in (6.5). Extend $d(\sigma|K)$ to $\mathfrak{U}(\mathfrak{k}_\mathbb{C})$, the universal enveloping algebra of $\mathfrak{k}_\mathbb{C}$. Let $E_\alpha$, $\alpha\in\hat{K}$, be as in (6.11) and set $V_\alpha = E_\alpha V(\sigma)$. Then each $V_\alpha \subseteq V(\sigma|K)^\infty$. Moreover, if $\mathfrak{Z}(\mathfrak{k})$ is the center of $\mathfrak{U}(\mathfrak{k}_\mathbb{C})$ and if $D\in\mathfrak{Z}(\mathfrak{k})$, $d(\sigma|K)(D)$ acts on each $V_\alpha$ as a scalar $c_D(\alpha)$ by (5.5) since $d(\sigma|K)(D)$ commuted with all $\sigma(k)$, $k\in K$, on $V(\sigma|K)^\infty$. Now there is an isomorphism $\psi$ of $\mathfrak{Z}(\mathfrak{k})$ with the $W$-invariants of $S(\mathfrak{h}_\mathbb{C})$, where $\mathfrak{h}$ is a maximal Abelian subalgebra of $\mathfrak{k}$, where $S(\mathfrak{h}_\mathbb{C})$ is the symmetric algebra over $\mathfrak{h}_\mathbb{C}$, and where $W$ (the *Weyl group* of $\mathfrak{k}$ with respect to $\mathfrak{h}$) is a finite group of linear automorphisms of $\mathfrak{h}$ generated by reflections. Let $C$ be a maximal connected subset of points in $(i\mathfrak{h})'$, the $\mathbb{R}$-dual of $i\mathfrak{h}$, fixed by no $s\in W$, $s\neq 1$. Then $\hat{K}$ may be identified with $C\cap\Lambda$, where $\Lambda$ is a lattice in $(i\mathfrak{h})'$, in such a way that $c_D(\alpha)=\psi(D)(\alpha)$, where $\psi(D)$ is regarded as a polynomial on $\mathfrak{h}'_\mathbb{C}$, the $\mathbb{C}$-dual of $\mathfrak{h}_\mathbb{C}$. $\psi(\mathfrak{Z}(\mathfrak{k}))$ is just a free Abelian algebra in $\dim_\mathbb{R}\mathfrak{h}$ generators. There exist, therefore, $D\in\mathfrak{Z}(\mathfrak{k})$ so that $c_D(\alpha)\geqslant 1$, all $\alpha\in\hat{K}$. For such $D$, define the bounded operator $A_D$ on $V(\sigma)$ by $A_D = c_D(\alpha)^{-1}$ on $V_\alpha$, $A\in\hat{K}$. Recall that $\dim V_\alpha \leqslant d(\alpha)^2$. Therefore

$$\|A_D\|^2_{\mathrm{HS}} = \sum_{\alpha\in\hat{K}} \|A_D|V_\alpha\|^2_{\mathrm{HS}} = \sum_{\alpha\in\hat{K}} c_D(\alpha)^{-1} d(\alpha)^2.$$

By the structure theory of $\psi(\mathfrak{Z}(\mathfrak{k}))$ just given, we can choose $D\in\mathfrak{Z}(\mathfrak{k})$ so that $\|A_D\|_{\mathrm{HS}}<\infty$. Now we let $\xi\in\mathfrak{g}$ act on $C_c^\infty(G)$ by the rule:

$$(\xi f)(x) = (\mathrm{d}/\mathrm{d}t) f((\exp -t\xi)x)\big|_{t=0}.$$

Then $\sigma(\xi f) = d\sigma(\xi)\sigma(f)$. So if $f\in C_c^\infty(G)$, $\sigma(f) = A_D d\sigma(D)\sigma(f) = A_D\sigma(Df)$. We finally get

$$\begin{aligned}\|\sigma(f)\|_{\mathrm{Tr}} &\leqslant \|A_D\|_{\mathrm{HS}}\|\sigma(Df)\|_{\mathrm{HS}}\\ &\leqslant \|A_D\|_{\mathrm{HS}}\|Df\|_\infty \mu(K\,\mathrm{Supp}(f)), \qquad f\in C_c^\infty(G),\end{aligned}$$

where $\|\cdot\|_{\mathrm{Tr}}$ is the trace norm defined by

$$\|S\|_{\mathrm{Tr}} = \sup_{\|T\|\leqslant 1} \sum_i |(TSu_i, u_i)|,$$

where $\{u_i\}$ is any orthonormal basis of $V(\sigma)$. If $\|S\|_{\mathrm{Tr}}<\infty$, we define $\mathrm{Tr}(S)=\Sigma_i(Su_i, u_i)$ and note that $\mathrm{Tr}(S)$ is independent of the choice of $\{u_i\}$ and that $S\mapsto \mathrm{Tr}(S)$ is continuous in the trace norm. Thus:

THEOREM 6.13. *Let $G$ be a connected semisimple matrix group and let $\sigma$ be an irreducible unitary representation of $G$. Then $\sigma(f)$ is of trace class,*

$f \in C_c^\infty(G)$, *and* $f \mapsto \mathrm{Tr}(\sigma(f))$ *is a distribution on* $G$, *called the* distribution character *of* $\sigma$ *and denoted by* $\Theta_\sigma$.

The proofs of (6.11) and (6.12) are essentially Stinespring's modification [47] of a proof of Godement [14]. The results essentially hold if we drop the assumption that $G$ be a matrix group ([15]). The proof of (6.13) follows Harish-Chandra [16]. A different proof of this result may be found in an important paper of Nelson and Stinespring [34], which contains many other results of interest. Another important class of groups with distribution characters are the connected nilpotent Lie groups [38].

We end with a further word on the Plancherel Theorem. Suppose that $G$ is a Lie group and let $\hat{G}$ denote the set of unitary equivalence classes of irreducible unitary representations as in Section 5. Suppose that every $\sigma_\alpha$, $\alpha \in \hat{G}$, has a distribution character. Suppose further that $G$ is unimodular and type I. Then we should try, as in (4.27) and (4.28), to expand $\delta_1$ in terms of distribution characters $\Theta_\alpha$, $\alpha \in \hat{G}$. More exactly, we should try to find a measure $\mu$ on $\hat{G}$ so that

$$(6.14) \qquad \delta_1 = \int_{\hat{G}} \Theta_\alpha \, \mathrm{d}\mu(\alpha)$$

in the weak sense that

$$(6.15) \qquad g(1) = \int_{\hat{G}} \Theta_\alpha(g) \, \mathrm{d}\mu(\alpha)$$

for all $g \in C_c^\infty(G)$. Then (5.33) will follow for $f, g \in C_c^\infty(G)$ so that $\mu$ will be the Plancherel measure for $G$.

## 7. Induced representations

Let $G$ be a finite group and let $H$ be a subgroup of $G$. Given any FDR $\sigma$ of $G$, we obtain an FDR $\sigma_H$ of $H$ simply by restricting the domain of $\sigma$ to $H$. It is quite natural to ask for a (left or right) adjoint to this restriction functor. That is, we would like to find functors $\tau \mapsto {}^G\tau$ and $\tau \mapsto \tau^G$ from the category of FDR's of $H$ (with $H$-operator homomorphisms as morphisms) to the category of FDR's of $G$ (with $G$-operator homomorphisms as morphisms) so that the functors

$$(7.1) \qquad (\tau, \sigma) \mapsto \mathrm{Hom}_G({}^G\tau, \sigma) \qquad \text{and} \qquad (\tau, \sigma) \mapsto \mathrm{Hom}_H(\tau, \sigma_H)$$

should be naturally equivalent, and the functors

(7.2) $(\sigma, \tau) \mapsto \mathrm{Hom}_G(\sigma, \tau^G)$ and $(\sigma, \tau) \mapsto \mathrm{Hom}_H(\sigma_H, \tau)$

should be naturally equivalent.

These problems are easily solved. Let $C(G)$ and $C(H)$ be the group algebras of $G$ and $H$, respectively, but using the counting measure on $G$ and $H$ instead of the normalized measure of Section 4. Then $C(G)$ is injected in $C(G)$ so that the identity goes to the identity. Representations of $G$ (or $H$) are then just unital $C(G)$ (or $C(H)$) modules. The situation is then as follows: $A$ is an algebra with 1 and $B$ is a subalgebra of $A$ with the same 1. For any unital $A$-module $V$ we have its restriction $V_B$ to $B$. For unital $B$-modules $W$ we replace (7.1) and (7.2) by

(7.3) $(W, V) \mapsto \mathrm{Hom}_A({}^AW, V)$ and $(W, V) \mapsto \mathrm{Hom}_B(W, V_B)$

and

(7.4) $(V, W) \mapsto \mathrm{Hom}_A(V, W^A)$ and $(V, W) \mapsto \mathrm{Hom}_B(V_B, W)$,

respectively, where $X \mapsto {}^AX$ (resp. $X \mapsto X^A$) is the left (resp. right) adjoint functor to be found.

(In (7.3), natural equivalence means this: For every pair $(W, V)$, where $W$ (resp. $V$) is a unital $B$- (resp. $A$- ) module, there is a bijection

$$\theta_{W,V} : \mathrm{Hom}_B(W, V_B) \to \mathrm{Hom}_A({}^AW, V)$$

such that for every choice of pairs $(W_1, V_1)$ and $(W_2, V_2)$ and of maps $f \in \mathrm{Hom}_B(W_2, W_1)$, $g \in \mathrm{Hom}_A(V_1, V_2)$ and $\varphi \in \mathrm{Hom}_B(W_1, V_{1B})$, we have

(7.5) $\theta_{W_2,V_2}(g_B \circ \varphi \circ f) = g \circ \theta_{W_1,V_1}(\varphi) \circ {}^Af.$

The theory of adjoint functors (see [32], Chapter IV) tells us the following:

THEOREM 7.6. (i) *Let $X \mapsto {}^AX$ be a left adjoint to the restriction functor $Y \mapsto Y_B$. For each $W$ define $\kappa_W \in \mathrm{Hom}_B(W, ({}^AW)_B)$ by $\theta_{W,{}^AW}(\kappa_W) = id_{{}^AW}$. Then for every $\psi \in \mathrm{Hom}_B(W, V_B)$, there is a unique $\tilde{\psi} \in \mathrm{Hom}_A({}^AW, V)$ such that*

(7.7) $\psi = \tilde{\psi}_B \circ \kappa_W$,

*viz., $\tilde{\psi} = \theta_{W,V}(\psi)$. Moreover, for $f \in \mathrm{Hom}_B(W_1, W_2)$,*

(7.8) $({}^Af)_B \circ \kappa_{W_1} = \kappa_{W_2} \circ f.$

(ii) *Conversely, let* $X \mapsto {}^AX$ *be a functor from the category of unital B-modules to the category of unital A-modules. Suppose that for each W there is given* $\kappa_W \in \mathrm{Hom}_B(W, ({}^AW)_B)$ *satisfying* (7.8) *and such that for each* $\psi \in \mathrm{Hom}_B(W, V_B)$ *there is a unique* $\tilde{\psi} \in \mathrm{Hom}_A({}^AW, V)$ *satisfying* (7.7). *Then* $X \mapsto {}^AX$ *is a left adjoint to* $Y \mapsto Y_B$; *indeed, the* $\theta_{W,V}$ *giving the natural equivalence in* (7.3) *are defined by* $\theta_{W,V}(\psi) = \tilde{\psi}$.

COROLLARY 7.9. *Any two left adjoints to* $Y \mapsto Y_B$ *are naturally equivalent.*

Now Theorem 7.6 forces our hand as to how to define the left adjoint $X \mapsto {}^AX$. Suppose we have such a left adjoint. Let $\theta_0 : A \otimes W \mapsto {}^AW$ be defined by $\theta_0(a \otimes w) = a\kappa_W(w)$. Then

$$\theta_0(ab \otimes w) = \theta_0(a \otimes bw) \qquad \text{for all} \qquad b \in B,$$

so that $\theta_0$ defines a map $\theta : A \otimes_B W \mapsto {}^AW$ in the usual way: $\theta(a \otimes_B w) = \theta_0(a \otimes w)$. We remind the reader that $A \otimes_B W$ is just $A \otimes W/R$, where $R$ is the subspace of $A \otimes W$ spanned by elements of the form $ab \otimes w - a \otimes bw$ for $a \in A$, $b \in B$, $w \in W$. Moreover, $A$ acts on $A \otimes W$ from the left by the rule $a_1(a \otimes w) = a_1 a \otimes w$, and this action leaves $R$ invariant, giving rise to an action of $A$ on $A \otimes_B W$. It follows that $\theta \in \mathrm{Hom}_A(A \otimes_B W, {}^AW)$. Also note that $\theta(1 \otimes_B w) = \kappa_W(w)$. This leads to the suspicion that $W \mapsto A \otimes_B W$ may itself give a left adjoint to $Y \mapsto Y_B$. This suspicion is justified.

THEOREM 7.10. *For each unital B-module W, let* ${}^AW = A \otimes_B W$. *For each* $f \in \mathrm{Hom}_B(W_1, W_2)$, *we may define* ${}^Af \in \mathrm{Hom}_A({}^AW_1, {}^AW_2)$ *by*

$${}^Af(a \otimes_B w_1) = a \otimes_B f(w_1) \qquad \textit{for} \qquad a \in A, \qquad w_1 \in W_1.$$

*For each W, we may define* $\kappa_W \in \mathrm{Hom}_B(W, ({}^AW)_B)$ *by*

$$\kappa_W(w) = 1 \otimes_B w \qquad \text{for} \qquad w \in W.$$

*The* $X \mapsto {}^AX$ *and* $W \mapsto \kappa_W$ *satisfy the hypotheses of Theorem* 7.6(ii), *so that* $X \mapsto {}^AX$ *is a left adjoint to the restriction functor* $Y \mapsto Y_B$.

*Proof.* The map $F : A \times W_1 \to A \otimes_B W_2$ given by $F(a, w_1) = a \otimes_B f(w_1)$ is bilinear and satisfies $F(ab, w_1) = F(a, bw_1)$ for all $b \in B$. Hence ${}^Af : A \otimes_B W_1 \to A \otimes_B W_2$ is defined and linear. Again, ${}^Af(a_1(a \otimes_B w_1)) = a_1(a \otimes_B f(w_1))$ so that ${}^Af \in \mathrm{Hom}_A({}^AW_1, {}^AW_2)$. Then $X \mapsto {}^AX$ is a functor from unital $B$-modules to unital $A$-modules. Again, $1\otimes_B bw = b\otimes_B w = b(1\otimes_B w)$ for all $b\in B$, $w\in W$, so that $\kappa_W\in \mathrm{Hom}_B(W, ({}^A W)_B)$.

To check (7.8), let $w_1\in W_1$. Then $({}^A f)_B(\kappa_{W_1}(w_1)) = ({}^A f)_B(1\otimes_B w_1) = 1\otimes_B f(w_1) = \kappa_{W_2}(f(w_1))$.

Finally, we check (7.7). Let $\psi\in \mathrm{Hom}_B(W, V_B)$. Then the map $\Psi: A\times W\to V$ given by $\Psi(a, w) = a\psi(w)$ is bilinear and satisfies $\Psi(ab, w) = \Psi(a, bw)$ for all $b\in B$. Hence we may define $\tilde{\psi}: {}^A W\to V$ by $\tilde{\psi}(a\otimes_B w) = a\tilde{\psi}(w)$. We have $\tilde{\psi}(a_1(a\otimes_B w)) = a\tilde{\psi}(a\otimes_B w)$, so that $\tilde{\psi}\in \mathrm{Hom}_A({}^A W, V)$. Moreover,

$$\tilde{\psi}_B(\kappa_W(w)) = \tilde{\psi}(1\otimes_B w) = \psi(w)$$

so that (7.7) holds. As for uniqueness, suppose $\varphi\in \mathrm{Hom}_A({}^A W, V)$ satisfies $\psi = \varphi_B\circ\kappa_W$. Then $\varphi(a\otimes_B w) = \varphi(a(1\otimes_B w)) = a\varphi(\kappa_W(w)) = a\psi(w)$, so that $\varphi = \tilde{\psi}$.

In an entirely similar way, we can reduce the problem of finding a right adjoint $X\mapsto X^A$ to the restriction functor $Y\mapsto Y_B$ to the problem of constructing a functor $X\mapsto X^A$ and a map $W\mapsto \varepsilon_W\in \mathrm{Hom}_B(W_B^A, W)$ such that, for $f\in \mathrm{Hom}_B(W_1, W_2)$,

$$\text{(7.11)} \qquad \varepsilon_{W_2}\circ f_B^A = f\circ\varepsilon_{W_1},$$

and such that, for every $\psi\in \mathrm{Hom}_B(V_B, W)$, there is a unique $\tilde{\psi}\in \mathrm{Hom}_A(V, W^A)$ such that

$$\text{(7.12)} \qquad \psi = \varepsilon_W\circ\tilde{\psi}_B.$$

From this reduction, we are led to define $W^A$ to be $\mathrm{Hom}_B(A, W)$, where $A$ is regarded as a left unital $B$-module and where the $A$-action on $\mathrm{Hom}_B(A, W)$ is given by $(a_1\varphi)(a) = \varphi(aa_1)$ for all $a, a_1\in A$ and $\varphi\in W^A$. If $f\in \mathrm{Hom}_B(W_1, W_2)$, $f^A$ is defined by $(f^A(\varphi))(a) = f(\varphi(a))$ for $\varphi\in W_1^A$, $a\in A$. For a unital $B$-module $W$, $\varepsilon_W$ is defined by $\varepsilon_W(\varphi) = \varphi(1)$ for $\varphi\in W^A$. Finally, if $\psi\in \mathrm{Hom}_B(V_B, W)$, the $\tilde{\psi}$ of (7.12) is given by $(\tilde{\psi}(v))(a) = \psi(\mathrm{av})$ for $v\in V$, $a\in A$.

DEFINITION 7.13. ${}^A W$ (resp $W^A$) is called the $A$-module *induced* (resp. *coinduced*) by the $B$-module $W$.

Thus the problems of finding a left and a right adjoint to the functor that assigns to every representation of a group $G$ the representation of a subgroup $H$ obtained by restriction of operators has a solution for finite $G$. It is natural to ask whether the same questions have positive solutions for a general locally compact group $G$ and closed subgroup $H$. Rieffel [39] has answered these questions in the case of isometric strongly continuous representations on Banach spaces. Briefly, there always exists a right adjoint; however, a left adjoint only exists for $H$ open in $G$.

In more detail, let $\tau$ be an isometric strongly continuous representation of $H$ on the Banach space $W$. Rieffel shows that we may let $W^G = \mathrm{Hom}_{L_1(H)}(L_1(G), W)_e$, which is defined as follows: $L_1(H)$ operates on $L_1(G)$ on the left *via* the formula

$$(7.14) \qquad (f \circ g)(x) = \int_H f(a)g(a^{-1}x)\,\mathrm{d}a$$

for $f \in L_1(H)$, $g \in L_1(G)$, and for almost all $x \in G$, so that $L_1(G)$ becomes an essential left $L_1(H)$-module. Now $W$ is also an essential left $L_1(H)$-module:

PROPOSITION 7.15. *Let $G$ be a locally compact group. Then every isometric strongly continuous representation $\sigma$ of $G$ on a Banach space $V$ turns $V$ into an essential left $L_1(H)$-module such that $\|fv\| \leqslant \|f\|_1 \|v\|$ for $f \in L_1(H)$, $v \in V$ by defining $fv = \int_G f(x)\sigma(x)v\,\mathrm{d}x$ (strong integral). Conversely, every essential left $L_1(H)$-module such that $\|fv\| \leqslant \|f\|_1 \|v\|$ for $f \in L_1(H)$, $v \in V$ comes from such a $\sigma$ in this manner.*

$U = \mathrm{Hom}_{L_1(H)}(L_1(G), W)$ consists of the $L_1(H)$-equivariant bounded linear maps from $L_1(G)$ into $W$ and is a Banach space under the usual operator norm. $U$ is an $L_1(G)$-module under the definition $(g\varphi)(h) = \varphi(h \circ g)$ for $\varphi \in U$, $g, h \in L_1(G)$. Moreover, $\|g\varphi\| \leqslant \|g\|_1 \|\varphi\|$. It is not in general essential, however. So we define $U_e$, the *essential part* of $U$, to be the closure of $L_1(G)U$. $U_e$ is then essential and so corresponds to a representation $\tau^G$ of $G$; hence we set $W^G = U_e$. $\tau \mapsto \tau^G$ satisfies (7.2).

Now $L_1(H)$ also operates on $L_1(G)$ on the right *via* the formula

$$(7.16) \qquad (g \circ f)(x) = \int_H g(xa^{-1}) f(a) \Delta_G(a)^{-1}\,\mathrm{d}a$$

for $f \in L_1(H)$, $g \in L_1(G)$, and for almost all $x \in G$. Hence we may form $L_1(G) \otimes_{L_1(H)} W$, upon which we can put the Schatten semi-norm: if $u \in L_1(G) \otimes_{L_1(H)} W$, we set $\|u\| = \inf\{\Sigma \|g_j\|_1 \|w_j\| : u = \Sigma g_j \otimes_{L_1(H)} w_j\}$. Quotienting out $\{u : \|u\| = 0\}$ and completing, we obtain $L_1(G) \overline{\otimes}_{L_1(H)} W$, which we denote by ${}^G W$. The standard left-action of $L_1(G)$ on $L_1(G) \overline{\otimes}_{L_1(H)} W$ induces an action of $L_1(G)$ on ${}^G W$, turning it into an essential left $L_1(G)$-module such that $\|gu\| \leqslant \|g\|_1 \|u\|$ for $g \in L_1(G)$, $u \in {}^G W$ and hence giving us an isometric strongly continuous Banach representation ${}^G\tau$ of $G$ on ${}^G W$. As stated above, $\tau \mapsto {}^G\tau$ does not in general satisfy (7.1) for all $\sigma$. However, it comes pretty close. Rieffel shows that (7.1) *is* satisfied for all $\sigma$ such that $V(\sigma)$ is reflexive. Note that we do not thereby get a left adjoint to the restriction functor for isometric strongly continuous representations on reflexive Banach spaces: ${}^G W$ is not reflexive in general even if $W$ is reflexive.

For the category of continuous unitary representations (7.1) and (7.2) coincide. It may be shown by slight modification of Rieffel's arguments that if an adjoint to restriction is to exist in this category, $H$ must be open in $G$. However, this condition is not sufficient: Let $G$ be $\mathbb{Z}$ under addition and let $H = \{0\}$. Let $\tau$ be the one-dimensional trivial representation of $H$. Suppose that ${}^G\tau$ is a continuous unitary representation of $G$ satisfying (7.1) for all continuous unitary representations $\sigma$ of $G$. Now Theorem 7.6 is a category-theoretic result. Therefore there exists $\kappa \in R(\tau, {}^G\tau|H)$, such that, for every continuous unitary representation $\sigma$ of $G$ and every $T \in R(\tau, \sigma|H)$, there is a unique $\tilde{T} \in R({}^G\tau, \sigma)$ such that $T = \tilde{T} \circ \kappa$. Now $V(\tau) = \mathbb{C}$. Let $v = \kappa 1 \in V({}^G\tau)$ and $T1 = w$. Then (7.6) becomes: for every continuous unitary representation $\sigma$ of $G$ and $w \in V(\sigma)$, there is a unique $\tilde{T} \in R({}^G\tau, \sigma)$ such that $\tilde{T}v = w$. Obviously, $v \neq 0$. Moreover, $v$ is *G-cyclic* in $V({}^G\tau)$; i.e. ${}^G\tau(G)v$ spans $V({}^G\tau)$. Hence the Hilbert space $V({}^G\tau)$ is *separable*. Let $\chi \in \hat{G}$ and let $\sigma_\chi$ be the representation of $G$ on $\mathbb{C}$ given by $\chi$. Let $w_\chi = 1$ and let $\tilde{T}_\chi \in R({}^G\tau, \sigma_\chi)$ satisfy $\tilde{T}_\chi v = w_\chi$. By (5.4), $\sigma_\chi$ is equivalent to a subrepresentation $\tilde{\sigma}_\chi$ of ${}^G\tau$. Since distinct irreducible representations are disjoint, Proposition 5.7(a) implies that $V(\tilde{\sigma}_{\chi_1}) \perp V(\tilde{\sigma}_{\chi_2})$ if $\chi_1 \neq \chi_2$. But $\hat{G}$ is uncountable, so that $V({}^G\tau)$ cannot be separable, a contradiction.

As the foregoing example indicates, a satisfactory category-theoretic formulation of unitary induction seems possible only when $H$ is open and of finite index in $G$, a much too severe restriction. Nevertheless, we can proceed as follows: Let $\tau$ be a strongly continuous Banach representation of $H$ on the Hilbert space $W$. Then $W$ becomes an essential $C_c(H)$-module. Since $C_c(H)$ operates on $C_c(G)$ on the right *via* (7.16), we can form $C_c(G) \bigotimes_{C_c(H)} W$, a left $C_c(G)$-module. We would like to put a natural (possibly degenerate) Hilbert inner product on $C_c(G) \bigotimes_{C_c(H)} W$ so that its completion, after quotienting out null-vectors, is an essential $C_c(G)$ $*$-module. Such an inner product would define a linear map of $V = C_c(G) \bigotimes_{C_c(H)} W$ into the algebraic dual $\overline{V}'$ of the complex conjugate $\overline{V}$ of $V$. Here $\overline{V}$ has the same underlying Abelian group as does $V$, but scalar multiplication of a vector in $\overline{V}$ by $\lambda \in \mathbb{C}$ corresponds to multiplication by $\bar{\lambda}$ in $V$: denoting the member of $\overline{V}$ corresponding to $v \in V$ by $\bar{v}$, scalar multiplication is defined by $\lambda\bar{v} = \overline{\bar{\lambda}v}$. Now $\overline{V}'$ consists of the conjugate linear maps $\varphi$ from $C_c(G)$ to $\overline{W}'$ such that

$$(7.17) \qquad \varphi(g \circ h) = \tau(h)^* \varphi(g)$$

for $g \in C_c(G)$, $h \in C_c(H)$, where * means conjugate transpose (in the algebraic sense).

LEMMA 7.18. *Let $f$ be a continuous $W$-valued function on $G$ such that*

$$(7.19) \qquad f(xa) = \tau(a)^* f(x)$$

*for $x \in G$, $a \in H$. Define $\varphi_f : C_c(G) \to W$ by*

$$(7.20) \qquad \varphi_f(g) = \int_G \overline{g(x)} f(x)\, dx.$$

*Then, regarded as a map from $C_c(G)$ to $\overline{W}'$, $\varphi_f \in \overline{V}'$.*

We shall denote the space of all such $\varphi_f$ by $V^0$. Then we define a linear map $\varepsilon$: $C_c(G) \bigotimes_{\mathbb{C}} W \to V^0$ by $\varepsilon(g \otimes v) = \varphi_f$, where

$$(7.21) \qquad f(x) = \int_H g(xa)\tau(a)^{*-1} v\, da, \qquad x \in G,$$

for $g \in C_c(G)$, $v \in W$. We shall often write $\varepsilon(g \otimes v)(x) = f(x)$, $x \in G$.

Now $\varepsilon$ is about the simplest map one can manufacture for producing members of $V^0$. But under certain conditions it does much more.

LEMMA 7.22. *$\varepsilon$ defines a map from $V \to V^0$ if and only if $\sigma = \Delta_G^{1/2} \Delta_H^{-1/2} \tau$ is a unitary representation of $H$ on $W$.*

*Proof.* Let $g \in C_c(G)$, $h \in C_c(H)$, $v \in W$, $x \in G$. Then

$$\varepsilon((g \circ h) \otimes v)(x) = \int_H (g \circ h)(xa)\tau(a)^{*-1} v\, da$$

$$= \int_H \int_H g(xab^{-1}) h(b) \Delta_G(b)^{-1} \tau(b)^{*-1} v\, db\, da$$

$$= \int_H \int_H g(xa)h(b)\Delta_G(b)^{-1}\Delta_H(b)\tau(a)^{*-1}\tau(b)^{*-1}v \, db \, da,$$

and

$$\varepsilon(g \otimes \tau(h)v)(x) = \int_H g(xa)\tau(a)^{*-1}\tau(h)v \, da$$

$$= \int_H \int_H g(xa)h(b)\tau(a)^{*-1}\tau(b)v \, db \, da.$$

Therefore we get a map from $V \to V^0$ if and only if

$$\tau(b) = \Delta_G(b)^{-1}\Delta_H(b)\tau(b)^{*-1} \qquad \text{for all} \qquad b \in H,$$

which says that $\Delta_G^{1/2}\Delta_H^{-1/2}\tau$ is unitary.

We shall denote this map from $V \to V^0$ by $\varepsilon$ also.

Suppose now that $\Delta_G^{1/2}\Delta_H^{-1/2}\tau$ is unitary, so that $\tau^{*-1} = \Delta_G\Delta_H^{-1}\tau$. We can then define a sesquilinear form on $V$ by $(u_1, u_2) = \langle \varepsilon(u_1), \bar{u}_2 \rangle$.

LEMMA 7.23. *$(\cdot, \cdot)$ is positive semidefinite on $V$.*

*Proof.* Let $u = \sum_1^n g_i \otimes_{C_c(H)} v_i \in V$. Then

$$(u, u) = \sum_{ij} \langle \varepsilon(g_i \otimes_{C_c(H)} v_i), \overline{g_j \otimes_{C_c(H)} v_j} \rangle$$
$$= \sum_{ij} \int_G \int_H g_i(xa)\overline{g_j(x)}(\tau(a)^{*-1}v_i, v_j) \, da \, dx.$$

Choose $h \in C_c(G)$ such that $\int_H h(xb) \, db = 1$ for all $x \in \mathrm{Supp}(g_i)$ for $i = 1, \ldots, n$. Then

$$(u, u) = \sum_{ij} \int_G \int_H \int_H h(xb)g_i(xa)\overline{g_j(x)}(\tau(a)^{*-1}v_i, v_j) \, db \, da \, dx$$

$$= \sum_{ij} \int_G \int_H \int_H h(x)g_i(xb^{-1}a)\overline{g_j(xb^{-1})}\Delta_G(b)^{-1}(\tau(a)^{*-1}v_i, v_j) \, db \, da \, dx$$

$$=\sum_{ij}\int_G\int_H\int_H h(x)g_i(xa)\overline{g_j(xb)}\Delta_G(b)\Delta_H(b)^{-1}(\tau(b^{-1}a)^{*-1}v_i, v_j)\,\mathrm{d}b\,\mathrm{d}a\,\mathrm{d}x$$

$$=\sum_{ij}\int_G\int_H\int_H h(x)g_i(xa)\overline{g_j(xb)}(\tau(a)^{*-1}v_i, \tau(b)^{*-1}v_j)\,\mathrm{d}b\,\mathrm{d}a\,\mathrm{d}x$$

$$=\int_G h(x)(f(x), f(x))\,\mathrm{d}x\geqslant 0,$$

where $f(x)=\varepsilon(u)(x)$.

In the case that $\sigma=\Delta_G^{1/2}\Delta_H^{-1/2}\tau$ is unitary, we obtain thus a Hilbert space, the completion (after quotienting out null-vectors) of $V$ under $(\cdot,\cdot)$. It is not difficult to check that the natural action of $C_c(G)$ on $V$ gives rise to an essential $*$-representation of $C_c(G)$ on this Hilbert space and hence, by (5.19), a unitary representation $\mathrm{Ind}(\sigma; G)$ called the *unitary representation of G induced by* $\sigma$.

The members of $V(\mathrm{Ind}(\sigma; G))$ can be realized concretely as follows: Let $\mathfrak{F}_\sigma$ be the space of functions $f: G\to V(\sigma)$ such that

(1) $f$ is Bourbaki measurable and locally integrable (see [7], p. 169);

(2) $f(xa)=\Delta_G(a)^{-1/2}\Delta_H(a)^{1/2}\sigma(a)^{-1}f(x)$, $x\in G$, $a\in H$;

(3) $f$ is, in a suitable sense, 'square integrable modulo $H$'. The precise meaning of (3) is inspired by the calculation in (7.23): For $h\in C_c(G)$, define $\theta h\in C_c(G/H)$ by $(\theta h)(\pi x)=\int_H h(xb)\,\mathrm{d}b$, $x\in G$, where $\pi$ is the canonical projection of $G$ onto $G/H$. $\theta$ is a positive surjection, continuous with respect to the usual inductive limit topologies, and hence ${}^t\theta$ is a positive continuous injection of $\mathcal{M}(G/H)$, the space of Radon measures on $G/H$, into $\mathcal{M}(G)$. Now $\theta\circ\rho_a=\Delta_H(a)^{-1}\theta$ for $a\in H$, where $\rho$ is as in (4.1), from which it easily follows that $\mathrm{Im}\,{}^t\theta=\{\mu\in\mathcal{M}(G): {}^t\rho_a\mu=\Delta_H(a)^{-1}\mu,\ a\in H\}$. Hence conditions (1) and (2) imply that $\mu_f=\|f(x)\|^2\,\mathrm{d}x\in\mathrm{Im}\,{}^t\theta$, so that $\mu_f={}^t\theta\nu_f$ for a unique positive $\nu_f\in\mathcal{M}(G/H)$. Condition (3) can now be stated precisely: $\nu_f(G/H)<\infty$.

By definition $\nu_f(\theta h)=\int_G h(x)\|f(x)\|^2\,\mathrm{d}x$. Thus, if $u\in V$, if $f(x)=\varepsilon(u)(x)$, and if $h\in C_c(G)$ as in (7.23), then $(\theta h)\circ\pi=1$ on $\mathrm{Supp}(f)$, $f$ satisfies (1) and (2), $\theta h=1$ on $\mathrm{Supp}(\nu_f)$, and hence $\nu_f(G/H)=\nu_f(\theta h)=\|u\|^2$. This leads to the following definition:

(7.24) $\quad \|f\|=\nu_f(G/H)^{1/2} \quad$ for $\quad f\in\mathfrak{F}_\sigma$.

PROPOSITION 7.25. *$\mathscr{H}=\mathfrak{F}_\sigma/\{f\in\mathfrak{F}_\sigma : \|f\|=0\}$ is a Hilbert space under $\|\cdot\|$. The map $\varepsilon : V\to\mathfrak{F}_\sigma$ defines an isometry of $V\to\mathscr{H}$ with dense range and hence induces a unitary map $\varepsilon$ of $V(\mathrm{Ind}(\sigma, G))$ onto $\mathscr{H}$. Moreover,*

$$\varepsilon\circ \mathrm{Ind}(\sigma; G)(x)=\lambda_x\circ\varepsilon, \qquad x\in G,$$

*where $\lambda$ is as in* (3.2).

We shall almost always use this realization of $\mathrm{Ind}(\sigma; G)$.

When is a unitary representation $\tau$ of $G$ of the form $\mathrm{Ind}(\sigma; G)$ for some unitary representation $\sigma$ of some closed subgroup $H$ of $G$? One answer is provided by the Imprimitivity Theorem. Unitarily induced representations possess an additional structure known as a *system of imprimitivity*. Let $H$, $\pi$, $\sigma$, and $\mathfrak{F}_\sigma$ be as above, and set $\sigma^G=\mathrm{Ind}(\sigma; G)$ regarded as acting on $\mathscr{H}$. Note that this use of '$\sigma^G$' is *not* the same as Rieffel's; it is the notation we shall use henceforth for the unitarily induced representation $\mathrm{Ind}(\sigma, G)$. Let $k\in C_\infty(G/H)$. Then $(k\circ\pi)f$ satisfies (1) and (2) for $f\in\mathfrak{F}_\sigma$. Moreover, $\theta(|k\circ\pi|^2 h)=|k|^2\theta h$ for all $h\in C_c(G)$, from which it follows that $\nu_{(k\cdot\pi)f}=|k|^2\nu_f$ so that $(k\circ\pi f\in\mathfrak{F}_\sigma$ and $\|(k\circ\pi)f\|\leqslant\|k\|_\infty\|f\|$. Defining $Pk$: $\mathscr{H}\to\mathscr{H}$ by $(Pk)f=(k\circ\pi)f$ for $f\in\mathfrak{F}_\sigma$, we see that $P$ is a $*$-representation of $C_\infty(G/H)$ on $\mathscr{H}$ of norm 1. We have

$$P(\lambda_x k)=\sigma^G(x)(Pk)\sigma^G(x)^{-1} \tag{7.26}$$

for $x\in G$, $k\in C_\infty(G/H)$, where $(\lambda_x k)(\pi y)=k(\pi(x^{-1}y))$. Moreover, $P$ is essential.

DEFINITION 7.27. Let $\tau$ be any unitary representation of $G$ and let $H$ be a closed subgroup of $G$. A *system of imprimitivity for $\tau$ based on $G/H$* is an essential $*$-representation $P$ of $C_\infty(G/H)$ on $V(\tau)$ of norm 1 satisfying (7.26).

This notion is due to Mackey [27] in the modification of [4]. So every induced representation has a system of imprimitivity. The *Mackey Imprimitivity Theorem* states that the converse is true:

THEOREM 7.28. *Let $\tau$ be a unitary representation of $G$ and let $P$ be a system of imprimitivity for $\tau$ based on $G/H$, $H$ a closed subgroup of $G$. Then there exist a unitary representation $\sigma$ of $H$ and a unitary equivalence $T\in R(\tau, \sigma^G)$ such that $T\circ(Pk)=(P_1k)\circ T$ for all $k\in C_\infty(G/H)$, where $P_1$ is the system of imprimitivity associated with $\sigma^G$.*

We sketch here a beautiful new proof of this theorem due to Bent Ørsted; the details can be found in [35]. For $u, v \in V(\tau)$, there exists $\lambda_{u,v} \in \mathcal{M}(G \times G)$ such that

$$(P(\theta h)\tau(f)u, v) = \int_{G \times G} h(x) f(y)\, d\lambda_{u,v}(x, y)$$

for $h \in C_c(G)$. For $a \in H$ and $x, y \in G$, we have

$$d\lambda_{u,v}(xa, y) = \Delta_H(a)\, d\lambda_{u,v}(x, y). \tag{7.29}$$

Let $\mathscr{D} = \tau(C_c(G))V(\tau)$. For $u, v \in \mathscr{D}$, there exists $p_{u,v} \in C(G)$ such that

$$(P(\theta h)u, v) = \int_G h(z) p_{u,v}(z)\, dz \tag{7.30}$$

for $h \in C_c(G)$: indeed, for $f, g \in C_c(G)$ and $u, v \in V(\tau)$,

$$p_{\tau(f)u, \tau(g)v}(z) = \int_{G \times G} \Delta_G(x)^{-1} f(zx^{-1}y)\overline{g(zx^{-1})}\, d\lambda_{u,v}(x, y).$$

Clearly, for $x, z \in G$

$$p_{u,v}(x^{-1}z) = p_{\tau(x)u, \tau(x)v}(z); \tag{7.31}$$

and

$$p_{u,v}(za) = \Delta_G(a)^{-1}\Delta_H(a) p_{u,v}(z) \tag{7.32}$$

for $a \in H$ by (7.29). Moreover, $p_{u,u} \geqslant 0$.

Now define $(u, v)_1 = p_{u,v}(1)$ for $u, v \in \mathscr{D}$, a positive semi-definite sesquilinear form on $\mathscr{D}$, and let $W$ be the completion of $\mathscr{D}/\{u \in \mathscr{D} : (u, u)_1 = 0\}$, a Hilbert space. Let $\zeta$ be the canonical map of $\mathscr{D}$ into $W$. By (7.31) and (7.32),

$$\Delta_G(a)^{-1}\Delta_H(a)(\zeta\tau(a)u, \zeta\tau(a)v)_1 = (\zeta u, \zeta v)_1 \quad \text{for} \quad a \in H,$$

so that there is a unitary operator $\sigma(a)$ on $W$ given by

$$\sigma(a)\zeta u = \Delta_G(a)^{-1/2}\Delta_H(a)^{1/2}\zeta(\tau(a)u) \tag{7.33}$$

for $u \in \mathscr{D}$. $\sigma$ is a unitary representation of $H$ on $W$. For $u \in \mathscr{D}$, define $\tilde{u} : G \to W$ by $\tilde{u}(x) = \zeta(\tau(x)^{-1}u)$. By (7.33), $\tilde{u}$ satisfies (1) and (2) in the definition of induced representation. Moreover, $(P_1(\theta h)\tilde{u}, \tilde{v}) = (P(\theta h)u, v)$ for $u, v \in \mathscr{D}$

and $h \in C_c(G)$, and $\lambda_z \tilde{u} = \widetilde{\tau(z)u}$ for $z \in G$. Thus $\tilde{u} \in \mathfrak{F}_f$, and $u \mapsto \tilde{u}$ extends to a unitary isomorphism $T: V(\tau) \to V(\sigma^G)$ satisfying (7.28).

Let $\sigma_i$ be unitary representations of the closed subgroup $H$ of $G$, $i=1, 2$. Let $T \in R(\sigma_1, \sigma_2)$. Let $\mathfrak{F}_{\sigma_i}$ be defined as above for $\sigma_i$. If $f \in \mathfrak{F}_{\sigma_i}$, then $T \circ f \in \mathfrak{F}_{\sigma_2}$, and $f \mapsto T \circ f$ defines an operator $\tilde{T} \in R(\sigma_1^G, \sigma_2^G)$. Moreover, if $P_i$ is the system of imprimitivity for $\sigma_i^G$,

(7.34) $\quad (P_2 k) \circ \tilde{T} = \tilde{T} \circ (P_1 k)$

for $k \in C_\infty(G/H)$. Conversely, (see [4]),

PROPOSITION 7.35. *Let $S \in R(\sigma_1^G, \sigma_2^G)$ satisfy (7.34) (with $S$ replacing $\tilde{T}$). Then $S = \tilde{T}$ for some (unique) $T \in R(\sigma_1, \sigma_2)$.*

We illustrate the power of (7.28) and (7.35) by describing Mackey's analysis [28] of $\hat{G}$, where $G$ is second countable, $G = NH$, $H$ and $N$ closed in $G$, $N$ Abelian and normal in $G$, and $N \cap H = \{1\}$. Let $\tau$ be an irreducible unitary representation of $G$. Let $\tau_N$ (resp. $\tau_H$) be the restriction of $\tau$ to $N$ (resp. $H$). By 3.12, $\tau_N$ gives an essential $*$-representation $P$ of $C_\infty(\hat{N})$ such that $P\tilde{f} = \tau_N(f), f \in C_c(N)$. $H$ operates on $\hat{N}$ on the left *via* $(h\alpha)(n) = \alpha(h^{-1}nh)$ for $h \in H, n \in N, \alpha \in \hat{N}$. Hence we can define $\lambda_h$ acting on $C_\infty(\hat{N})$ by $(\lambda_h \varphi)(\alpha) = \varphi(h^{-1}\alpha)$ for $\varphi \in C_\infty(\hat{N})$, $\alpha \in \hat{N}$. One easily checks that

(7.36) $\quad P(\lambda_h \varphi) = \tau_H(h)(P\varphi)\tau_H(h)^{-1}$.

Moreover, $\tau$ being irreducible is equivalent to $(\tau_H, P)$ being irreducible. Suppose now that the action of $H$ on $\hat{N}$ is 'nice': there exists a countable collection of $H$-invariant Borel subsets of $\hat{N}$ which separate the $H$-orbits in $\hat{N}$. If $P$ is extended to the algebra of bounded Borel functions on $\hat{N}$, (7.36) continues to hold for bounded Borel $\varphi$. If $\varphi$ is the characteristic function of a Borel set, $P\varphi$ is a projection; and if $\varphi$ is $H$-invariant, $P\varphi$ commutes with $P(C_\infty(\hat{N}))$ and $\tau_H(H)$. It follows that $P$ vanishes off some unique $H$-orbit $\mathcal{O}$ in $\hat{N}$. Let $\alpha \in \mathcal{O}$, and identify $\mathcal{O}$ with $H/H_\alpha$, where $H_\alpha$ is the closed subgroup $\{h \in H : h\alpha = \alpha\}$ of $H$, *via* the map $h \mapsto h\alpha$, $h \in H$. By (7.36), $P$ defines a system of imprimitivity for $\tau_H$ based on $H/H_\alpha$. By (7.28), $(\tau_H, P)$ is unitarily equivalent to $(\sigma^G, P_1)$ for some unitary representation $\sigma$ of $H_\alpha$, where $P_1$ is the system of imprimitivity associated to $\mathrm{Ind}(\sigma, H)$. By (7.35), $\sigma$ is irreducible. We may reverse this process: every $\alpha \in \hat{N}$ and irreducible representation $\sigma$ of $H_\alpha$ gives rise to an irreducible representation of $G$. Moreover, it may be shown that $G_\alpha = NH_\alpha$ is a closed subgroup of $G$, that $nh \mapsto \alpha(n)\sigma(h)$ defines an irreducible unitary representation $\alpha\sigma$ of $G_\alpha$, and that $\tau$ is unitarily equivalent to $(\alpha\sigma)^G$.

The foregoing construction is called the Mackey 'Little Group Method'. It may be used to prove the Stone–von Neumann theorem classifying $\hat{G}$ if $G$ is the Heisenberg group [49] and the Wigner theorem [51] classifying $\hat{G}$ if $G$ is the Poincaré group. Mackey has vastly extended the Little Group Method to cover the situation where $G$ is no longer a semi-direct product, where $N$ is merely a closed normal subgroup of $G$ such that $G/N$ acts nicely on $\hat{N}$. For this, see [30].

Now (7.35) tells us how to compute $R((\sigma_1^G, P_1), (\sigma_2^G, P_2))$ for $\sigma_i$ a representation of $H$, and $P_i$ the system of imprimitivity associated to $\sigma_i^G$. We would now like to compute $R(\sigma_1^G, \sigma_2^G)$ for $\sigma_i$ a representation of $H_i$, $i=1, 2$. We can handle this in case $G$ is a Lie group.

Let $G$ be a Lie group, $H$ a closed subgroup, $\sigma$ a unitary representation of $H$. Form $\mathfrak{F}_\sigma$ and $\sigma^G$ as above. It is easy to see that ${}^0\mathfrak{F}_\sigma^\infty$, the set of $f \in \mathfrak{F}_\sigma$ which are $C^\infty$ and have compact support modulo $H$, is contained in $V(\sigma^G)^\infty$. Let $\mathfrak{F}_\sigma^\infty$ be the set of $C^\infty$ functions in $\mathfrak{F}_\sigma$. By (6.5) we have a representation $d\sigma^G$ of the Lie algebra $\mathfrak{g}$ of $G$ on $V(\sigma^G)^\infty$. Extend $d\sigma^G$ to $\mathfrak{U}(\mathfrak{g}_\mathbb{C})$, the universal enveloping algebra of the complexification $\mathfrak{g}_\mathbb{C}$ of $\mathfrak{g}$.

PROPOSITION 7.37. *Suppose* $\dim V(\sigma) < \infty$. *Then* $V(\sigma^G)^\infty \subseteq \mathfrak{F}_\sigma^\infty$. *Suppose, moreover, that* $D \in \mathfrak{U}(\mathfrak{g}_\mathbb{C})$ *is an elliptic left-invariant differential operator on G of order* $> \frac{1}{2} \dim (G/H)$. *Then for every compact subset K of G there exist* $c_K > 0$ *such that*

$$\|f(x)\| \leqslant c_K(\|d\sigma^G(D)f\| + \|f\|)$$

*for all* $f \in V(\sigma^G)^\infty$ *and all* $x \in K$. (See [2], Section 4.)

Now let $\sigma_i$ be a unitary representation of $H_i \subseteq G$, $i=1, 2$. Suppose that $\dim V(\sigma_2) < \infty$. Let $T \in R(\sigma_1^G, \sigma_2^G)$. Clearly $TV(\sigma_1^G)^\infty \subseteq V(\sigma_2^G)^\infty$. Let $\mathfrak{F}_{\sigma i}$ and $\varepsilon_i$ be defined by $\sigma_i$ (see 7.21–7.25). Then $\varepsilon_1(g \otimes v) \in {}^0\mathfrak{F}_{\sigma_i}^\infty$ for $g \in C_c^\infty(G)$, $v \in V(\sigma_1)$. Therefore $T\varepsilon_1(g \otimes v) \in \mathfrak{F}_{\sigma_2}^\infty$ by (7.37), so that we may define $r_T : C_c^\infty(G) \to \mathrm{Hom}_\mathbb{C}(V(\sigma_1), V(\sigma_2))$ by

$$r_T(g)v = (T\varepsilon_1(g \otimes v))(1) \tag{7.38}$$

for $g \in C_c^\infty(G)$, $v \in V(\sigma_1)$.

It follows from (7.21) that $\varepsilon_1((\lambda_x g) \otimes v) = \lambda_x \varepsilon_1(g \otimes v)$ for $x \in G$, so that

$$(T\varepsilon_1(g \otimes v))(x^{-1}) = r_T(\lambda_x g)v; \tag{7.39}$$

it also follows that

$$\varepsilon_1((\rho_a g) \otimes v) = \Delta_G(a)^{-1/2} \Delta_{H_1}(a)^{-1/2} \varepsilon_1(g \otimes \sigma_1(a)^{-1} v) \tag{7.40}$$

for $a \in H_1$.

Then (7.39) implies

(7.41) $\quad T \mapsto r_T$ is injective,

and

(7.42) $\quad r_T(\lambda_b g) = \Delta_G(b)^{1/2}\Delta_{H_2}(b)^{-1/2}\sigma_2(b)r_T(g)$

for $b \in H_2$; and (7.40) implies

(7.43) $\quad r_T(\rho_a g) = \Delta_G(a)^{-1/2}\Delta_{H_1}(a)^{-1/2}r_T(g)\sigma_1(a)^{-1}$

for $a \in H_1$. Now (7.23) implies that for every compact subset $K$ of $G$, there exists $C_K > 0$ such that $\|\varepsilon_1(g \otimes v)\| \leqslant C_K \|g\|_\infty \|v\|$ for $\mathrm{Supp}(g) \subseteq K$. Let $D$ be an elliptic member of $\mathfrak{U}(\mathfrak{g}_{\mathbb{C}})$ of order $> \frac{1}{2}\dim(G/H_2)$. Then (7.37) implies

(7.44) $\quad \|r_T(g)\| \leqslant c_{\{1\}} C_K \{\|D^\square g\|_\infty + \|g\|_\infty\} \|T\|$

for all $g$ with $\mathrm{Supp}(g) \subseteq K$, where $D^\square g = (Dg')'$ and $g'(x) = g(x^{-1})$. Thus

THEOREM 7.45. *$T \mapsto r_T$ is an injection of $R(\sigma_1^G, \sigma_2^G)$ into $\mathcal{M}$, the space of $\mathrm{Hom}_{\mathbb{C}}(V(\sigma_1), V(\sigma_2))$-valued distributions on G of order $2\gamma(\frac{1}{4}\dim(G/H_2))$ which satisfy* (7.42) *and* (7.43). *Here $\gamma$ is the 'least integer strictly greater than' function.*

*Remarks.* (1) The details of the proof of (7.45) may be found in [2]. Bruhat's version of this theorem, which antedates [2] and is more general, is contained in [8].

(2) If $H_2$ is open in $G$, (7.37) is not needed so that $\mathcal{M}$ may be taken to consist of measures and $\dim V(\sigma_2)$ need not be restricted. The same holds if $G$ has arbitrarily small neighborhoods of 1 invariant under the adjoint action of $H_2$ (see [5]), for example if $H_2$ is compact.

(3) Theorem 7.45 is most useful when $\sigma_1 = \sigma_2 = \sigma$ and it can be shown that $\dim \mathcal{M} = 1$, for then $\sigma$ must be irreducible; or when $\sigma_1 \neq \sigma_2$ and it can be shown that $\dim \mathcal{M} = 0$, for then $\sigma_1$ and $\sigma_2$ must be disjoint.

To determine $\mathcal{M}$ requires a knowledge of the $H_2 : H_1$ double coset structure of $G$. This is most easily done if there is only a countable number of such double cosets, for then $G$ is a disjoint union $\bigcup_1^\infty G_j$ where $G_i$ is a union of double cosets each open in $\bigcup_j^\infty G_j$, which itself is a closed sub manifold of $G$. For example, let $H_2 x H_1$ be an open double coset and let $\mu \in \mathcal{M}$. Then $\mu | H_2 x H_1$ may be shown to be a function $h \in C^\omega(H_2 x H_1,$

$\mathrm{Hom}_{\mathbb{C}}(V(\sigma_1), V(\sigma_2)))$, that is

$$\mu(g)=\int_G h(y)g(y)\,dy$$

for $g\in C_c^\infty(H_2xH_1)$, and then $h$ must satisfy

$$h(by)=\Delta_G(b)^{1/2}\Delta_{H_2}(b)^{-1/2}\sigma_2(b)h(y) \tag{7.46}$$

and

$$h(ya)=\Delta_G(a)^{-1/2}\Delta_{H_1}(a)^{1/2}h(y)\sigma_1(a) \tag{7.47}$$

for $y\in H_2xH_1$, $a\in H_1$, $b\in H_2$, according to (7.42) and (7.43), respectively. Set $\tau_i=\Delta_G^{-1/2}\Delta_H^{1/2}\sigma_i$. For $y\in G$, define the representation $\tau_{iy}$ of $yH_iy^{-1}$ by $\tau_{iy}(z)=\tau_i(y^{-1}zy)$ for $z\in yH_iy^{-1}$. Set $H_0=H_2\cap(xH_1x^{-1})$. Then (7.46) and (7.47) imply that

$$h(x)\in R(\tau_{1x}|H_0, (\tau_2|H_0)^{*-1}). \tag{7.48}$$

Conversely, any $S\in R(\tau_{1x}|H_0, (\tau_2|H_0)^{*-1})$ defines a unique $h$ satisfying (7.46) and (7.47) and such that $h(x)=S$ by means of

$$h(bxa)=\tau_2(b)^{*-1}S\tau_1(a) \tag{7.49}$$

for $b\in H_2$, $a\in H_1$. Of course, it is by no means obvious that any such $S$ defines an $h$ which is the restriction to $H_2xH_1$ of some distribution in $\mathcal{M}$, much less some $r_T$. To determine when this happens requires a lot of delicate analysis. However, it follows immediately that, if

$$R(\tau_{1x}|H_0, (\tau_2|H_0)^{*-1})=\{0\},$$

then $r_T|H_2xH_1=0$ also. One can get similar results concerning the contribution to $\mathcal{M}$ of double cosets further down the stratification $\{G_i\}$; there is the complication here that, if $\mu\in\mathcal{M}$ is supported by $\bigcup_j{}_1^\infty G$, then its restriction to some double coset in $G_i$ need not be a function on $G_i$: there may be transverse derivatives.

Bruhat [8] has shown how to handle such problems. Using this mechanism, he proved the following theorem:

THEOREM 7.50. *Let $G$ be a connected semisimple Lie group and let $G=KAN$ be its Iwasawa decomposition. Let $M$ (resp. $\tilde{M}$) be the centralizer (resp. normalizer) of $A$ in $K$ and set $P=MAN$, a minimal parabolic subgroup of $G$. $W_r=\tilde{M}/M$ is the restricted Weyl group of $G$. $\tilde{M}$ normalizes $MA$ so* *that $W_r$ operates on $(MA)\hat{}$. N is normal in P and every finite-dimensional irreducible unitary representation of P comes from one of $P/N \simeq MA$; thus $W_r$ operates on the set $P^0$ of such. If $\sigma \in P^0$ and $s \in W_r$, $\sigma_s$ will denote the transform of $\sigma$ by s.*

(i) *If $\sigma \in P^0$ and $\{s \in W_r : \sigma_s = \sigma\} = \{1\}$, then $\sigma^G$ is irreducible.*

(ii) *If $\sigma_1, \sigma_2 \in P^0$, then $\sigma_1^G \simeq \sigma_2^G$ if and only if $\sigma_{1s} \simeq \sigma_2$ for some $s \in W_r$.*

A key point in Bruhat's application of his version of (7.45) to this situation is the so-called *Bruhat decomposition*:

LEMMA 7.51. *Each $P : P$ double coset meets $\tilde{M}$ in exactly one M-coset of $\tilde{M}$. Thus if $x_1, \ldots, x_n$ are distinct representatives in $\tilde{M}$ for $W_r$, where $W_r$ has order n, then $Px_1P, \ldots, Px_nP$ is a complete set of distinct $P : P$ double cosets of G.*

The representations of $G$ constructed in (7.50) form the (*continuous*) *principal series* of representations of $G$; they are associated to the minimal parabolic subgroup $P$ of $G$. Theorem 7.50 states that almost all of them are irreducible. More precise results on the irreducibility of these representations have been obtained by Parthasarathy, Ranga Rao, and Varadarajan [37], Kostant [24], Wallach [50], and Knapp and Stein [22].

In their work on representations of $SL(n, \mathbb{R})$, Gel'fand and Graev [12] introduced a generalization of the notion of induced representation that included the multiplier representations of $SL(2, \mathbb{R})$ on spaces of functions holomorphic on $\{z \in \mathbb{C} : |z| < 1\}$ which had been defined by Bargmann [1]. We end these notes by describing these representations in the general setting given in [3] and noting some applications.

Let $G$ be a Lie group and $H$ a closed subgroup of $G$ with Lie algebras $\mathfrak{g}$ and $\mathfrak{h}$, respectively. Let $\mathfrak{g}_{\mathbb{C}}$ (resp. $\mathfrak{h}_{\mathbb{C}}$) be the complexification of $\mathfrak{g}$ (resp. $\mathfrak{h}$). Let $\sigma$ be a finite-dimensional unitary representation of $H$ and form the space $\mathfrak{F}_\sigma$ of functions from $G$ to $V(\sigma)$ which realize $V(\sigma^G)$ as above. Set $\tau = \Delta_G^{1/2}\Delta_H^{-1/2}\sigma$ and, for $\xi \in \mathfrak{h}$, define $d\tau(\xi) = (d/dt)\tau(\exp t\xi)|_{t=0}$. Then $d\tau$ is a representation of $\mathfrak{h}$ on $V(\sigma)$ such that $d\tau((\text{ad } a)\xi) = \tau(a)\, d\tau(\xi)\tau(a)^{-1}$ for $a \in H$, $\xi \in \mathfrak{h}$. Moreover, if $f \in \mathfrak{F}_\sigma$, then $\xi f = -d\tau(\xi) f$ for $\xi \in \mathfrak{h}$ because of condition (2) in the definition of $\mathfrak{F}_\sigma$; here $\xi$ is regarded as a left invariant differential operator on $G$. Plainly this equation holds for $\xi \in \mathfrak{h}_{\mathbb{C}}$ as well, where $d\tau$ is extended complex linearly to $\mathfrak{h}_{\mathbb{C}}$.

Our generalization, which may be termed 'infinitesimal induction', assumes that, in addition to the $G$, $H$, and $\sigma$ above, we are given a subalgebra $\mathfrak{m}$ of $\mathfrak{g}_{\mathbb{C}}$ and a representation $\rho$ of $\mathfrak{m}$ on $V(\sigma)$ such that:

(1) $\mathfrak{m}\cap\bar{\mathfrak{m}}=\mathfrak{h}_{\mathbb{C}}$,

(2) $(\text{ad } H)\mathfrak{m}=\mathfrak{m}$,

(3) $\mathfrak{m}+\bar{\mathfrak{m}}$ is a subalgebra of $\mathfrak{g}_{\mathbb{C}}$,

(4) $\rho|\mathfrak{h}_{\mathbb{C}}=d\tau$, and

(5) $\rho((\text{ad } a)\eta)=\tau(a)\rho(\eta)\tau(a)^{-1}$ for $a\in H$, $\eta\in\mathfrak{m}$.

We let $\mathfrak{F}_{\sigma,\rho}$ consist of those $f\in\mathfrak{F}_\sigma$ such that:

(i) $f$ is $C^\omega$ on the leaves of the foliation $\mathfrak{n}=(\mathfrak{m}+\bar{\mathfrak{m}})\cap\mathfrak{g}$,

(ii) $(\eta+\rho(\eta))f=0$ for $\eta\in\mathfrak{m}$.

It may then be proved that $\mathfrak{F}_{\sigma,\rho}$ defines a closed $\lambda$-invariant subspace of $\mathfrak{F}_\sigma$, determining a subrepresentation $\text{Ind}(\sigma, \rho; G)$ of $\sigma^G$. This subrepresentation may very well equal 0.

Now an $f$ in $\mathfrak{F}_\sigma$ defines a measurable, locally integrable section $s_f$ of the vector bundle $E$ over $G/H$ associated to $\tau$: indeed, $E=(G\times V(\sigma))/H$, where $H$ acts on $G\times V(\sigma)$ on the right by $(x, v)a=(xa, \tau(a)^{-1}v)$ for $a\in H$, so that the map $x\mapsto(x, f(x))$ of $G\to G\times V(\sigma)$ is $H$-equivariant and so defines a map $s_f: G/H\to E$. Conversely, every measurable, locally integrable section is $s_f$ for some $f$ satisfying (1) and (2) in the definition of $\mathfrak{F}_\sigma$. What condition on $s_f$ corresponds to (i) and (ii) above?

Let $\pi$: $G\to G/H$ be the canonical map. Then $\pi_*(\xi_{xa})=\pi_*([\text{ad } a)\xi]_x)$ for $\xi\in\mathfrak{g}_{\mathbb{C}}$, $x\in G$, $a\in H$. Therefore $\mathfrak{c}_{\pi x}=\pi_*(\mathfrak{m}_x)$, $x\in G$, is a well-defined subspace of $T_{\pi x}(G/H)_{\mathbb{C}}$, as is $\mathfrak{d}_{\pi x}=\pi_*(\mathfrak{n})$ of $T_{\pi x}(G/H)$. The fact that $\mathfrak{m}$ and $\mathfrak{n}$ are Lie algebras implies

(6) $\mathfrak{c}$ and $\mathfrak{d}$ are involutive distributions,

while (1) and the definition of $\mathfrak{n}$ imply

(7) $\mathfrak{c}\cap\bar{\mathfrak{c}}=\{0\}$ and

(8) $\mathfrak{c}+\bar{\mathfrak{c}}=\mathfrak{d}_{\mathbb{C}}$.

So $\mathfrak{d}$ is a foliation, the leaves of which are endowed with a complex structure by $\mathfrak{c}$ (Cauchy–Riemann equations).

We define a similar structure on $E$ as follows: If $w\in V(\sigma)$, define the vector field $\partial_w$ on $V(\sigma)$ as in (2.21). Then if $(x, v)\in G\times V(\sigma)$, set

$$\hat{\mathfrak{m}}_{(x,v)}=\{(\eta_x, -(\partial_{\rho(\eta)v})_v : \eta\in\mathfrak{m}\}+\mathfrak{r}_{(x,v)},$$

where $\mathfrak{r}_{x,v}$ consists of all complex tangent vectors to $G\times V(\sigma)$ at $(x, v)$ which are tangent to $\{x\}\times V(\sigma)$ and which annihilate all functions $\varphi$ on $G\times V(\sigma)$ of the form $\varphi(y, w)=f(w)$, where $f\in V(\sigma)'$, the complex dual of $V(\sigma)$. Also set $\hat{\mathfrak{h}}_{(x,v)}=\{(\xi_x, -(\partial_{\rho(\xi)v})_v : \xi\in\mathfrak{h}\}$. Then we have that $\hat{\mathfrak{m}}$ and $\hat{\mathfrak{h}}$ are involutive distributions on $G\times V(\sigma)$ and

(1′) $\hat{\mathfrak{m}}\cap\bar{\hat{\mathfrak{m}}}=\hat{\mathfrak{h}}_{\mathbb{C}}$,

(2′) $\hat{\mathfrak{m}}$ is invariant under the right action of $H$ on $G \times V(\sigma)$ and $\hat{\mathfrak{h}}$ is the foliation tangent to the $H$-orbits in $G \times V(\sigma)$, and

(3′) $\hat{\mathfrak{m}} + \bar{\hat{\mathfrak{m}}}$ is an involutive distribution.

Thus $\hat{\mathfrak{m}}$ and $(\hat{\mathfrak{m}} + \bar{\hat{\mathfrak{m}}}) \cap T(G \times V(\sigma))$ determine involutive distributions $\hat{\mathfrak{c}}$ and $\hat{\mathfrak{d}}$ on $E$, and we have $\hat{\mathfrak{c}} \cap \bar{\hat{\mathfrak{c}}} = \{0\}$ and $\hat{\mathfrak{c}} + \bar{\hat{\mathfrak{c}}} = \hat{\mathfrak{d}}_{\mathbb{C}}$.

LEMMA 7.52. *Let $f \in \mathfrak{F}_\sigma$. Then $f \in \mathfrak{F}_{\sigma,\rho}$ if and only if $s_f$ is $C^\infty$ on the leaves of $\mathfrak{d}$ and $(s_f)_* \mathfrak{c}_m \subseteq \hat{\mathfrak{c}}_{sf(m)}$ for all $m \in G/H$.*

Now $\mathfrak{c}$ (resp. $\hat{\mathfrak{c}}$) determines what may be termed a *partially complex structure* on $G/H$ (resp. $E$), so that $E$ becomes a partially holomorphic vector bundle over $G/H$. The sections determined by $\mathfrak{F}_{\sigma,\rho}$ are partially holomorphic sections. If $\mathfrak{m} + \bar{\mathfrak{m}} = \mathfrak{g}_{\mathbb{C}}$, then $G/H$ and $E$ are complex manifolds, $E$ is a holomorphic vector bundle over $G/H$, and $\mathfrak{F}_{\sigma,\rho}$ consists of holomorphic sections.

One can prove an intertwining operator theorem for these representations along the lines of Theorem 7.45. In case $\mathfrak{m} + \bar{\mathfrak{m}} = \mathfrak{g}_{\mathbb{C}}$, one can prove more:

PROPOSITION 7.53. *If $\sigma$ is irreducible, then* $\mathrm{Ind}(\sigma, \rho; G)$ *is either* 0 *or irreducible. If $\sigma_1 \downarrow \sigma_2$, then* $\mathrm{Ind}(\sigma_1, \rho_1; G) \downarrow \mathrm{Ind}(\sigma_2, \rho_2; G)$.

Gel'fand and Graev had in mind constructing representations of a connected semisimple $G$ as follows: Suppose $G$ has finite center and let $K$ be a maximal compact subgroup of $G$ with Lie algebra $\mathfrak{k}$. Let $\mathfrak{g} = \mathfrak{k} + \mathfrak{s}$ be the Cartan decomposition of $\mathfrak{g}$ using $\mathfrak{k}$, and let $\theta$ be the corresponding Cartan involution of $\mathfrak{g}$. Extend $\theta$ to $G$. Let $B$ be a $\theta$-stable Cartan subgroup of $G$ with Lie algebra $\mathfrak{b}$. Then $\mathfrak{b} = (\mathfrak{b} \cap \mathfrak{k}) + (\mathfrak{b} \cap \mathfrak{s})$. Let $(b_1, \ldots, b_k)$ (resp. $(b_{k+1}, \ldots, b_r)$) be an ordered $\mathbb{R}$-basis of $\mathfrak{b} \cap \mathfrak{s}$ (resp. $\mathfrak{b} \cap \mathfrak{k}$). Lexicographically order the set of roots $\Phi$ of $(\mathfrak{g}_{\mathbb{C}}, \mathfrak{b}_{\mathbb{C}})$ with respect to $(b_1, \ldots, b_r)$ and let $\Phi^+$ be the subset of positive roots. Let $\mathfrak{m} = \mathfrak{b}_{\mathbb{C}} + \Sigma_{\varphi \in \Phi^+} \{\xi \in \mathfrak{g}_{\mathbb{C}} : [\eta, \xi] = = \varphi(\eta)\xi,\ \eta \in \mathfrak{b}_{\mathbb{C}}\}$, let $(\mathfrak{m} \cap \bar{\mathfrak{m}}) \cap \mathfrak{g} = \mathfrak{h}$. Let $H_0$ be the analytic subgroup of $G$ determined by $\mathfrak{h}$ and let $H = BH_0$. Now let $\chi$ be a unitary character of $B$, and let $d\chi$ be the corresponding representation of $\mathfrak{b}_{\mathbb{C}}$. Extend $d\chi$ to be a representation $\rho$ of $\mathfrak{m}$ by making it vanish on $[\mathfrak{m}, \mathfrak{m}]$. There is a unique character $\sigma$ of $H$ such that $\sigma | B = \chi$ and $d(\sigma | H_0) = \rho | \mathfrak{h}$. The Gel'fand-Graev representation corresponding to $\chi$ is $\mathrm{Ind}(\sigma, \rho; G)$.

If $B$ is a compact Cartan subgroup, then $\mathfrak{h} = \mathfrak{b}$, $H = B$, and $\mathfrak{m} + \bar{\mathfrak{m}} = \mathfrak{g}_{\mathbb{C}}$. Moreover $E$ is a holomorphic complex line bundle over the complex manifold $G/B$. The representations Ind($\sigma$, $\rho$; $G$) which are not zero form the *holomorphic discrete series*. They are always irreducible (7.53). One does not obtain all the discrete series of $G$ in this way: Harish-Chandra has shown that $G$ has discrete series representations if and only if it has a compact Cartan subgroup $B$ [18]. However, if $G$ is simple, it has holomorphic discrete series if and only if, in addition, $G/K$ has a $G$-invariant complex structure [17]. Moreover, one obtains all the discrete series for such a $G$ if and only if the Weyl group of $(\mathfrak{g}, \mathfrak{b})$ has at most index 2 in the Weyl group of $(\mathfrak{g}_{\mathbb{C}}, \mathfrak{b}_{\mathbb{C}})$.

Taking his cue from the Bott–Borel–Weil Theorem [6], Langlands conjectured that *all* discrete series representations of a connected semisimple $G$ with compact Cartan subgroup $B$ could be realized using the complex line bundle $E$ constructed above by looking at the action of $G$ on some sort of $L_2$ sheaf cohomology arising from the sheaf of germs of holomorphic cross sections of $E$ over $B$. The precise conjecture is stated in ([25], p. 256). This conjecture has been proved by Wilfried Schmid [42, 43].

*University of California, Los Angeles*

## References

[1] Bargmann, V., 'Irreducible unitary representations of the Lorentz group', *Ann. of Math.* (2) **48** (1947), 568–640.
[2] Blattner R. J., 'On induced representations', *Amer. J. Math* **83** (1961), 79–98.
[3] Blattner, R. J., 'On induced representations II: infinitesimal induction', *Amer. J. Math.* **83** (1961), 499–512.
[4] Blattner, R. J., 'On a theorem of G. W. Mackey', *Bull. Amer. Math. Soc.* **68** (1962), 585–587.
[5] Blattner, R. J., 'A theorem on induced representations', *Proc. Amer. Math. Soc.* **13** (1962), 881–884.
[6] Bott, R., 'Homogeneous vector bundles', *Ann. of Math.* (2) **66** (1957), 203–248.
[7] Bourbaki, N., *Intégration* (2ᵉ édition), Chapters I–IV, Hermann, Paris, 1965.
[8] Bruhat, F., 'Sur les représentations induites des groupes de Lie', *Bull. Soc. Math. France* **84** (1956), 97–205.
[9] Carleson, L., 'On convergence and growth of partial sums of Fourier series', *Acta Math.* **116** (1966), 135–157.
[10] Chevalley, C., *Theory of Lie Groups I*, Princeton University Press, Princeton, 1946.
[11] Gårding, L., 'Note on continuous representations of Lie groups', *Proc. Nat. Acad. Sci. USA* **33** (1947), 331–332.
[12] Gel'fand I. M., and Graev, M. I., 'Unitary representations of the real unimodular group (principal nondegenerate series)', *Izv. Akad. Nauk. SSSR Ser. Mat.* 17 (1953), 189–248.

[13] Glimm, J., 'Type I C* algebras', *Ann. of Math.* (2) **73** (1961), 572–612.
[14] Godement, R., 'A theory of spherical functions I', *Trans. Amer. Math. Soc.* **73** (1952), 496–556.
[15] Harish-Chandra, 'Representations of a semisimple Lie group on a Banach space I', *Trans. Amer. Math. Soc.* **75** (1953), 185–243.
[16] Harish-Chandra, 'Representations of semisimple Lie groups III', *Trans. Amer. Math. Soc.* **76** (1954), 234–253.
[17] Harish-Chandra, 'Representations of semi-simple Lie groups IV', *Amer. J. Math.* **77** (1955), 743–777.
[18] Harish-Chandra, 'Discrete series for semisimple Lie groups II', *Acta. Math.* **166** (1966), 1–111.
[19] Hochschild, G., *The structure of Lie groups*, Holden-Day, San Francisco, 1965.
[20] Hunt, R. A., 'On the convergence of Fourier series', *Orthogonal Expansions and their Continuous Analogues*, Southern Illinois University Press, Carbondale, 1968, pp. 234–255.
[21] Katznelson, Y., *An Introduction to Harmonic Analysis*, Wiley, New York, 1968.
[22] Knapp A. W., and Stein, E. M., 'Intertwining operators for semisimple groups', *Ann. of Math.* (2) **93** (1971), 489-578.
[23] Kolmogorov, A. N., 'Une série de Fourier-Lebesgue divergente presque partout', *Fund. Math.* **4** (1923), 324–328.
[24] Kostant, B., 'On the existence and irreducibility of certain series of representations', *Bull. Amer. Math. Soc.* **75** (1969), 627–642.
[25] Langlands, R. P., 'Dimension of spaces of automorphic forms', *Proceedings of Symposia in Pure Mathematics, Vol.* 9, American Mathematical Society, Providence, 1966, pp. 253–257.
[26] Loomis, L., *An Introduction to Abstract Harmonic Analysis*, Van Nostrand, New York, 1953.
[27] Mackey, G. W., 'Imprimitivity for representations of locally compact groups I', *Proc. Nat. Acad. Sci. USA* **35** (1949), 537–545.
[28] Mackey, G. W., 'Induced representations of locally compact groups I', *Ann. of Math* (2) **55** (1952), 101–139.
[29] Mackey, G. W., *The Theory of Unitary Group Representations*, University of Chicago Press, Chicago, 1976.
[30] Mackey, G. W., 'Unitary representations of group existensions I', *Acta. Math.* **99** (1958), 265–311.
[31] Mackey, G. W., 'Infinite-dimensional group representations', *Bull. Amer. Math. Soc.* **69** (1963), 628–686.
[32] MacLane, S., *Categories for the Working Mathematician*, Springer-Verlag, New York, 1972.
[33] Nelson, E., 'Analytic vectors', *Ann. of Math.* (2) **70** (1959), 572–615.
[34] Nelson, E., and Stinespring, W. F., 'Representations of elliptic operators in an enveloping algebra', *Amer. J. Math.* **81** (1959), 547–560.
[35] Ørsted, B., 'Induced representations and a new proof of the imprimitivity theorem', *J. Functional Analysis* 31 (1979), 355–359.
[36] Paley, R. E. A. C., and Wiener, N., *Fourier Transforms in the Complex Domain*, American Mathematical Society, Providence, 1934.

[37] Parathasarathy, K. R., Ranga Rao, R., and Varadarajan, V. S., 'Representations of complex semisimple Lie groups and Lie algebras', *Ann. of Math.* (2) 85 (1967), 383–429.
[38] Pukanszky, L., *Leçons sur les représentations des groupes*, Dunod, Paris, 1967.
[39] Rieffel, M., 'Induced Banach representations of Banach algebras and locally compact groups', *J. Functional Analysis* **1** (1967), 443–491.
[40] Rieffel, M., 'Induced representations of $C^*$-algebras', *Advances in Math.* **13** (1974), 176–257.
[41] Riesz, F., and Sz.-Nagy, B., *Functional Analysis*, Ungar, New York, 1955.
[42] Schmid, W., 'On a conjecture of Langlands', *Ann. of Math.* (2) **93** (1971), 1–42.
[43] Schmid, W., '$L^2$-cohomology and the discrete series', *Ann. of Math.* (2) **103** (1976), 375–394.
[44] Schwartz, L., *Théorie des distributions* (nouvelle édition), Hermann, Paris, 1966.
[45] Segal, I. E., 'A class of operator algebras which are determined by groups', *Duke Math. J.* **18** (1951), 221–265.
[46] Segal, I. E., 'An extension of Plancherel's formula to separable unimodular groups', *Ann. of Math.* (2) 52 (1950), 272–292.
[47] Stinespring, W. F., 'A semisimple matrix group is of type I', *Proc. Amer. Math. Soc.* **9** (1958), 965–967.
[48] Stone, M. H., 'Linear transformations in Hilbert space III: Operational methods and group theory', *Proc. Nat. Acad. Sci. USA* **16** (1930), 172–175.
[49] von Neumann, J., 'Die Eindeutigkeit der Schrödingerschen Operatoren', *Math. Ann.* **104** (1931), 570–578.
[50] Wallach, N., 'Cyclic vectors and irreducibility for principal series representations', *Trans. Amer. Math. Soc.* **158** (1971), 107–113.
[51] Wigner, E., 'On unitary representations of the inhomogeneous Lorentz group', *Ann. of Math.* (2) **40** (1930), 149–204.
[52] Williamson, J. H., 'Remarks on the Plancherel and Pontryagin theorems', *Topology* **1** (1962), 73–80.

JOSEPH A. WOLF

# FOUNDATIONS OF REPRESENTATION THEORY FOR SEMISIMPLE LIE GROUPS

## CONTENTS





*J. A. Wolf, M. Cahen, and M. De Wilde (eds.), Harmonic Analysis and Representations of Semi-Simple Lie Groups, 69–130.*

## 0. Introduction

In this paper I hope to communicate some of the basic facts on unitary representations of semisimple Lie groups, staring with the material on general harmonic analysis in R. Blattner's paper, and leading into the more advanced subjects in the papers of M. Atiyah and W. Schmid (the nature of distribution characters; analytic construction of the discrete series), P. Trombi (Plancherel formula for semisimple Lie groups) and V. S. Varadarajan (the infinitesimal approach, including algebraic construction of the discrete series).

The material to be covered is reasonably described by the Table of Contents and the three chapter introductions, so I will only add a word on references.

Chapter I. A good general reference is V. S. Varadarajan, *Lie Groups, Lie Algebras and Their Representations*, Prentice-Hall, 1974. Also good: J. E. Humphreys, *Introduction to Lie Algebras and Representation Theory*, Springer-Verlag GTM #9, 1972, as well as the excellent introductions by Hochschild, Mostow, Samelson and others. This material is absolutely standard.

Chapter II. Besides the books of Varadarajan and Humphreys, one should add Chapter 7 of J. Dixmier, *Algèbres Enveloppantes*, Gauthier-Villars, 1974, and the last chapter of S. Helgason, *Differential Geometry and Symmetric Spaces*, Academic Press, 1972.

Chapter III. Here the most comprehensive reference is G. Warner, *Harmonic Analysis on Semisimple Lie Groups, I*, Springer-Verlag, 1972. Also, my 'Unitary Representations on Partially Holomorphic Cohomology Spaces,' AMS Memoir #138, 1974, may be useful. And of course there are various papers of Harish-Chandra, Schmid, and others.

## Chapter I

# STRUCTURE OF SEMISIMPLE LIE GROUPS

Chapter I presents a brief resume, with occasional indications of proofs, of the theory of semisimple Lie groups up to (but not including) Cartan's highest weight theory for finite-dimensional representations and the theory of parabolic subgroups. We start with some basic notions of linear algebra (Section 1) and do the representations of $\mathfrak{sl}(2)$ which have a highest or lowest weight vector (Section 2). We then discuss some basic facts about semisimple (Section 3) and reductive (Section 4) groups. After that, we look at root systems and the Weyl group (Section 5), Weyl bases and real forms (Section 6), Dynkin diagrams and classification (Section 7), and have a brief glimpse of the structure of real semisimple groups (Section 8). All this can be viewed as a sort of study guide to Varadarajan's book, through Chapter 4, Section 5.

## 1. Preliminaries

Let $V$ be a vector space and $S$ a set of linear transformations of $V$. Then $S$ is called

*irreducible* if $V$ has no proper $S$-invariant subspace,
*semisimple* if every $S$-invariant subspace has an $S$-invariant complement,
*nilpotent* if $\dim V < \infty$ and $V$ has a basis in which every element of $S$ has matrix of the form $\begin{pmatrix} 0 & & * \\ & \ddots & \\ 0 & & 0 \end{pmatrix}$,
*unipotent* if $\{s-1 : s \in S\}$ is nilpotent.

A single linear transformation $s$ of $V$ is called semisimple, nilpotent, unipotent, if the set $\{s\}$ has that property. So $s$ is nilpotent if and only if some power $s^n$ is nilpotent, and $s$ is semisimple if and only if it is diagonable over the algebraic closure of the base field.

Let $G$ be a group (resp. $\mathfrak{g}$ a Lie algebra). A *representation* $\pi$ of $G$ (resp. $\mathfrak{g}$) on $V$ is a homomorphism $\pi: G \to GL(V)$ of $G$ to the group of invertible linear transformations of $V$ (resp. Lie algebra homomorphism $\pi: \mathfrak{g} \to \mathfrak{gl}(V)$ to the set of linear transformations of $V$ with Lie product $[a, b] = ab - ba$). The representation $\pi$ is called irreducible, semisimple, unipotent (in the group case) or nilpotent (in the algebra case) if its image has that property.

From now on 'representation' means 'continuous representation' for Lie groups.

Now suppose that $V$ is a real or complex vector space and that $G$ is a Lie group with Lie algebra $\mathfrak{g}$, and for the moment suppose further that $\dim V < \infty$. If $\pi$ is a representation of $G$ on $V$, then it is real analytic as a map $G \to GL(V)$, and it induces a representation (denoted $d\pi$ or $\dot{\pi}$ or $\pi$) of $\mathfrak{g}$ on $V$ by

$$d\pi(\xi) \colon v \mapsto \left.\frac{\mathrm{d}}{\mathrm{d}t}\right|_{t=0} \pi(\exp_G(t\xi))v.$$

One recovers $\pi$ from $d\pi$ – at least on the identity component $G_0$ of $G$ – by $\pi(\exp_G(t\xi)) = \mathrm{e}^{t \cdot d\pi(\xi)}$. On the other hand, if $\psi$ is a representation of $\mathfrak{g}$ on $V$, and if $\tilde{G}$ is the connected simply connected Lie group with Lie algebra $\mathfrak{g}$, then $\tilde{\pi}(\exp_{\tilde{G}}(\xi)) = \mathrm{e}^{\psi(\xi)}$ defines a representation of $\tilde{G}$ on $V$ with $d\tilde{\pi} = \psi$. Here $\tilde{G}$ is the universal covering group of $G_0$, and $\tilde{\pi}$ may or may not factor through $G_0$, and even then it need not extend to $G$.

Now suppose that $G$ is connected and that $\pi$ is a representation of $G$ on $V$. The relation between $\pi$ and $d\pi$ shows: $\pi$ is irreducible if and only if $d\pi$ is irreducible, $\pi$ is semisimple if and only if $d\pi$ is semisimple, and $\pi$ is unipotent if and only if $d\pi$ is nilpotent.

Let $\mathscr{U}(\mathfrak{g})$ denote the universal enveloping algebra of $\mathfrak{g}_{\mathbf{C}}$. It is the tensor algebra mod the ideal generated by the $\xi \otimes \eta - \eta \otimes \xi - [\xi, \eta]$. If we view $\mathfrak{g}_{\mathbf{C}}$ as the Lie algebra of left invariant vector fields on $G$ then $\mathscr{U}(\mathfrak{g})$ is the associative algebra of all left invariant differential operators on $G$. If $\psi$ is a representation of $\mathfrak{g}$ on $V$ then it 'extends' uniquely to an associative algebra homomorphism $\mathscr{U}(\mathfrak{g}) \to \mathfrak{gl}(V)$, also denoted $\psi$, which is irreducible or semisimple if and only if $\psi \colon \mathfrak{g} \to \mathfrak{gl}(V)$ is. Further, $\psi \colon \mathfrak{g} \to \mathfrak{gl}(V)$ is semisimple or nilpotent if and only if the associative algebra $\psi(\mathscr{U}(\mathfrak{g}))$ has that property in the category of associative algebras. Finally, if $\psi$ is irreducible and $0 \neq v \in V$ then $V = \psi(\mathscr{U}(\mathfrak{g})) \cdot v$.

## 2. Representations of $\mathfrak{sl}(2)$

Let $\mathfrak{g} = \mathfrak{sl}(2)$, linear Lie algebra with basis

$$h = \begin{pmatrix} 1 & 0 \\ 0 & -1 \end{pmatrix}, \quad e = \begin{pmatrix} 0 & 1 \\ 0 & 0 \end{pmatrix}, \quad f = \begin{pmatrix} 0 & 0 \\ 1 & 0 \end{pmatrix}$$

and multiplication table

$$[h,e]=2e, \qquad [h,f]=-2f, \qquad [e,f]=h.$$

Let $\psi$ be an *algebraically* irreducible representation of $\mathfrak{g}$ on some vector space $V$, and suppose that $h$ has an eigenvector, i.e.

$$V_\lambda=\{v\in V: \psi(h)v=\lambda v\}, \qquad \lambda\in\mathbb{C},$$

is nonzero for some value $\lambda=\lambda_0$. Writing $\psi(h)\psi(e)=\psi[h,e]+\psi(e)\psi(h)$ and $\psi(h)\psi(f)=\psi[h,f]+\psi(f)\psi(h)$ we see

$$\psi(e)V_\lambda\subset V_{\lambda+2} \qquad \text{and} \qquad \psi(f)V_\lambda\subset V_{\lambda-2}.$$

By irreducibility, now,

$$V=\sum_{n=-\infty}^{\infty} V_{\lambda_0+2n} \qquad \text{algebraic direct sum.}$$

In particular, if $\dim V<\infty$ then $\psi(h)$ is semisimple and $\psi(e)$ and $\psi(f)$ are nilpotent.

Now suppose that $V$ has a 'highest weight vector,' that is a nonzero vector $v$ in some $V_\lambda$ such that $\psi(e)v=0$. Of course this is automatic if $\dim V<\infty$. Then we take $\lambda_0=\lambda$, $v_0=v$, and define $v_n=\psi(f)^n v_0\in V_{\lambda_0-2n}$. Evidently the algebraic span of the $v_n$ is stable under $\psi(f)$ and $\psi(h)$. I claim that it also is stable under $\psi(e)$. For one has

$$ef^n=f^n e+nf^{n-1}(h-n+1) \qquad \text{in } \mathscr{U}(\mathfrak{g}) \text{ for } n=1,2,\ldots$$

(easy induction), so

$$\psi(e)v_n=\psi(f)^n\psi(e)v_0+n(\lambda_0-n+1)v_{n-1}=n(\lambda_0-n+1)v_{n-1}.$$

Now, again by irreducibility,

$$V \text{ has basis } \{v_0, v_1,\ldots\} \text{ and } \mathfrak{g} \text{ acts by } \psi(f)v_n=v_{n+1},$$
$$\psi(h)v_n=(\lambda_0-2n)v_n, \qquad \text{and } \psi(e)v_n=n(\lambda_0-n+1)v_{n-1}.$$

*Case 1.* $\lambda_0$ is an integer $\geqslant 0$. Then $\psi(e)v_n=0$ for $n=\lambda_0+1>0$. If $v_n\neq 0$ then $\{v_n, v_{n+1},\ldots\}$ spans an invariant subspace, contradicting irreducibility. Thus $V$ has finite dimension $\lambda_0+1$, basis $\{v_0, v_1,\ldots,v_{\lambda_0}\}$, with $\psi$ as above. And for any integer $d\geqslant 0$ these formulae define an irreducible representation of degree $d+1$ of $\mathfrak{g}$.

*Case 2.* $\lambda_0$ is not an integer $\geqslant 0$. Then $\psi(e)v_n\neq 0$ for all $n\geqslant 1$ (easy induction), so dim $V=\infty$. Later we will see that $V$ is the space of $SO(2)$-finite vectors in an irreducible unitary representation $\pi_{\lambda_0}$ of $SL(2, \mathbb{R})$ just when $\lambda_0$ is a negative integer; for $\lambda_0 < -1$ it will be an 'antiholomorphic discrete series' representation, and for $\lambda_0 = -1$ it will be a summand of the unique reducible 'principal series' representation of $SL(2; \mathbb{R})$.

If $V$ has a 'lowest weight vector' $v_0 \in V_{\lambda_0}$, $v_0 \neq 0$ but $\psi(f)v_0 = 0$, then as above

$$V \text{ has basis } \{v_n = \psi(e)^n v_0\}_{n=0,1,\ldots,} \text{ and } \mathfrak{g} \text{ acts by}$$

$$\psi(e)v_n = v_{n+1}, \qquad \psi(h)v_n = (\lambda_0 + 2n)v_n, \qquad \psi(f)v_n = n(\lambda_0 + n - 1)v_{n-1}.$$

If $\lambda_0$ is an integer $\leqslant 0$ then, as above, $v_{-\lambda_0+1} = 0$ and we have the irreducible $\mathfrak{g}$-module of dimension $|\lambda_0| + 1$. Also,

*Case 3.* $\lambda_0$ is not an integer $\leqslant 0$. Then $\psi(f)v_n \neq 0$ for all $n \geqslant 1$ (easy induction), dim $V = \infty$, and it will turn out that $V$ is the space of $SO(2)$-finite vectors in an irreducible unitary representation $\pi_{\lambda_0}$ of $SL(2; \mathbb{R})$ just when $\lambda_0$ is a positive integer; for $\lambda_0 > 1$ it will be a 'holomorphic discrete series' representation, and for $\lambda_0 = 1$ it will be the other summand of the unique reducible 'principal series' representation of $SL(2; \mathbb{R})$.

If $V$ has neither highest weight vector nor lowest weight vector, one needs some analytic tools to see dim $V_\lambda \leqslant 1$; when we do that, we will have the spaces of $SO(2)$-finite vectors in the unitary 'principal' and 'complementary' series representations of $SL(2; \mathbb{R})$.

Finally, we note that the irreducible finite-dimensional representations of $\mathfrak{sl}(2; \mathbb{C})$ and $SL(2; \mathbb{C})$ are given by the formulae in Case 1, and those of $\mathfrak{su}(2)$ and $SU(2)$ follow by restriction.

## 3. Cartan's criterion for semisimplicity

Fix a Lie algebra $\mathfrak{g}$. If $\mathfrak{h}$ and $\mathfrak{k}$ are subalgebras then $[\mathfrak{h}, \mathfrak{k}]$ denotes the subalgebra spanned by the $[\xi, \eta]$, $\xi \in \mathfrak{h}$ and $\eta \in \mathfrak{k}$. $\mathfrak{g}$ is *solvable* if the derived series $\mathfrak{g} = \mathfrak{g}^{(0)} \supset \mathfrak{g}^{(1)} \supset \ldots$, $\mathfrak{g}^{(i+1)} = [\mathfrak{g}^{(i)}, \mathfrak{g}^{(i)}]$, terminates at 0, *nilpotent* if the series $\mathfrak{g} = \mathfrak{g}^{[0]} \supset \mathfrak{g}^{[1]} \supset \ldots$, $\mathfrak{g}^{[i+1]} = [\mathfrak{g}, \mathfrak{g}^{[i]}]$, terminates at 0. Note that $\mathfrak{g}$ is nilpotent just when $\mathrm{ad}_{\mathfrak{g}}$ is nilpotent.

$\mathfrak{g}$ is *semisimple* if it has no nonzero solvable ideal.

A sum of solvable (resp. nilpotent) ideals in $\mathfrak{g}$ again is a solvable (resp.

nilpotent) ideal. Thus $\mathfrak{g}$ has a *radical* (= solvable radical) and a *nilradical* (= nilpotent radical) defined by

rad $\mathfrak{g}$: maximal solvable ideal in $\mathfrak{g}$

nilrad $\mathfrak{g}$: maximal nilpotent ideal in $\mathfrak{g}$

and $\mathfrak{g}/\text{rad}\,\mathfrak{g}$ is semisimple.

THE LEVI–WHITEHEAD THEOREM. *$\mathfrak{g}$ has a (necessarily semisimple) subalgebra $\mathfrak{s}$ such that the projection $\mathfrak{g} \to \mathfrak{g}/\text{rad}\,\mathfrak{g}$ restricts to an isomorphism of $\mathfrak{s}$ onto $\mathfrak{g}/\text{rad}\,\mathfrak{g}$.*

In other words $\mathfrak{g}$ is the semidirect sum $\text{rad}\,\mathfrak{g} + \mathfrak{s}$. Any such algebra $\mathfrak{s}$ is called a *Levi subalgebra* or Levi factor of $\mathfrak{g}$.

THE MAL'CEV–HARISH-CHANDRA THEOREM. *Any two Levi subalgebras of $\mathfrak{g}$ are conjugate by an automorphism* $e^{\text{ad}(\xi)}$, $\xi \in [\text{rad}\,\mathfrak{g}, \mathfrak{g}]$.

These results can be stated thus: Every semisimple subalgebra of $\mathfrak{g}$ is contained in a maximal one, any two maximal ones are $\exp(\text{ad}[\text{rad}\,\mathfrak{g}, \mathfrak{g}])$-conjugate, and if $\mathfrak{s}$ is a maximal one then $\mathfrak{g} = \text{rad}\,\mathfrak{g} + \mathfrak{s}$ semidirect sum. These theorems are not easy; they use cohomology of $\mathfrak{g}/\text{rad}\,\mathfrak{g}$ for a certain representation.

ENGEL'S THEOREM. *If $\pi$ is a finite-dimensional representation of $\mathfrak{g}$ such that each $\pi(\xi)$ is nilpotent, then $\pi$ is nilpotent, so in particular $\mathfrak{g}/\ker(\pi)$ is nilpotent.*

If $G$ has nilpotent Lie algebra $\mathfrak{g}$ and we apply Engel's Theorem to the adjoint representation then, from the Campbell–Hausdorff Theorem, there is a polynomial map $p$: $\mathfrak{g} \times \mathfrak{g} \to \mathfrak{g}$ such that $\exp_G(\xi) \cdot \exp_G(\eta) = \exp_G(p(\xi, \eta))$ for $\xi, \eta \in \mathfrak{g}$. It follows that $\exp_G\colon \mathfrak{g} \to G$ is a covering space for $\mathfrak{g}$ nilpotent and $G$ connected. Thus, in a simply connected nilpotent Lie group, every analytic subgroup is closed and exp(center of $\mathfrak{g}$) = (center of $G$).

LIE'S THEOREM. *Let $\pi$ be a representation of a solvable Lie algebra $\mathfrak{g}$ on a finite-dimensional vector space $V$, both over the same algebraically closed field. Then there is a finite set $\{\lambda_1, \ldots, \lambda_r\}$ of linear functionals on $\mathfrak{g}$ that vanish on $[\mathfrak{g}, \mathfrak{g}]$, and there is a composition series $V = V_1 \supsetneqq \cdots \supsetneqq V_r \supsetneqq V_{r+1} = 0$ of $V$ under $\pi(\mathfrak{g})$, such that the action $\pi_i$ of $\mathfrak{g}$ on $V_i/V_{i+1}$ is* *given, in an appropriate basis, by*

$$\pi_i(\xi)=\begin{pmatrix}\lambda_i(\xi) & & * \\ & \ddots & \\ 0 & & \lambda_i(\xi)\end{pmatrix}, \qquad \xi\in\mathfrak{g}.$$

Some consequences: (1) *if* $\pi$ *is irreducible then* dim $V=1$, (2) *a Lie algebra* $\mathfrak{g}$ *is solvable if and only if* $[\mathfrak{g}, \mathfrak{g}]$ *is nilpotent*, (3) *if* $\xi$, $\eta$, $\zeta\in\mathfrak{g}$ *then* trace $(\pi([\xi, \eta]) \cdot \pi(\zeta))=0$.

Further, the *Cartan–Killing form* on $\mathfrak{g}$ is the bilinear form $\langle\xi, \eta\rangle=\text{trace}\,(\text{ad}(\xi)\cdot\text{ad}(\eta))$. It is symmetric, and is invariant in the sense $\langle[\zeta, \xi], \eta\rangle+\langle\xi, [\zeta, \eta]\rangle=0$.

CARTAN'S CRITERION. (i) $\mathfrak{g}$ *is solvable if and only if* $\langle\mathfrak{g}, [\mathfrak{g}, \mathfrak{g}]\rangle=0$, (ii) $\mathfrak{g}$ *is semisimple if and only if* $\langle\,,\rangle$ *is nondegenerate* ($\langle\xi, \mathfrak{g}\rangle=0\Rightarrow\xi=0$).

If $\mathfrak{h}$ is an ideal in $\mathfrak{g}$, so is $\mathfrak{h}^{\perp}=\{\xi\in\mathfrak{g}: \langle\xi, \mathfrak{h}\rangle=0\}$, and thus also $\mathfrak{h}\cap\mathfrak{h}^{\perp}$. Whenever $\mathfrak{k}$ is an ideal the Cartan–Killing forms satisfy $\langle,\rangle_{\mathfrak{k}}=\langle,\rangle_{\mathfrak{g}}|_{\mathfrak{k}\times\mathfrak{k}}$, so $\mathfrak{h}\cap\mathfrak{h}^{\perp}$ is solvable by Cartan's Criterion. If $\mathfrak{g}$ is semisimple this says $\mathfrak{h}\cap\mathfrak{h}^{\perp}=0$ so, again by Cartan's Criterion, $\mathfrak{g}=\mathfrak{h}\oplus\mathfrak{h}^{\perp}$. A Lie algebra is *simple* if it is nonAbelian and has no proper ideal, i.e. if the adjoint representation is nontrivial and irreducible. Now, if $\mathfrak{g}$ is semisimple then it is a direct sum of simple ideals, and the only ideals are the partial sums.

WEYL'S THEOREM. *Every finite-dimensional representation of a semisimple Lie group is semisimple.*

So, for example, every finite-dimensional representation of $\mathfrak{sl}(2)$ is a direct sum of irreducible representations listed in Section 2. Weyl's original proof: a finite-dimensional representation of a compact group $K$ is semisimple (easy), so the same holds for the Lie algebra $\mathfrak{k}$ of a compact Lie group, thus also for $\mathfrak{k}_{\mathbb{C}}$ and then for any real Lie algebra $\mathfrak{g}$ with $\mathfrak{k}_{\mathbb{C}}\simeq\mathfrak{g}_{\mathbb{C}}$; and (this is the hard part) if $\mathfrak{g}$ is real semisimple then there is a compact Lie group $K$ with $\mathfrak{k}_{\mathbb{C}}\simeq\mathfrak{g}_{\mathbb{C}}$. A more algebraic proof uses cohomology.

Now we transcribe these results over to Lie groups. If $H$ is a normal (resp. Abelian, resp. nilpotent, resp. solvable) subgroup of a Lie group $G$, then its closure has the same property, so we have closed subgroups

rad $G$: maximal connected normal solvable subgroup of $G$,

nilrad $G$: maximal connected normal nilpotent subgroup of $G$, whose respective Lie algebras are rad $\mathfrak{g}$ and nilrad $\mathfrak{g}$.

*G is semisimple* if rad $G$ is trivial, i.e. if rad $\mathfrak{g}=0$. In any case $G/\text{rad } G$ is semisimple. If $G$ is connected then the Levi decomposition $\mathfrak{g}=\text{rad }\mathfrak{g}+\mathfrak{s}$ goes over to $G=(\text{rad } G)\cdot S$, $S$ analytic subgroup for $\mathfrak{s}$ and $S\cap\text{rad } G$ discrete; and here one has a semidirect product when $G$ is simply connected.

Weyl's Theorem goes over directly to connected semisimple Lie groups and then to semisimple Lie groups $G$ with $G/G_0$ finite: *for such groups, every finite-dimensional real or complex representation is semisimple.*

## 4. Reductive Groups and Trace Forms

We say that a Lie group $G$ (resp. Lie algebra $\mathfrak{g}$) is *reductive* if it has a finite-dimensional semisimple representation with discrete kernel (resp. kernel zero). For example let $\mathfrak{g}=\mathfrak{z}\oplus\mathfrak{g}'$ where $\mathfrak{z}$ is Abelian and $\mathfrak{g}'$ is semisimple, choose a basis $\{\zeta_1, \ldots, \zeta_s\}$ of $\mathfrak{z}$, and consider

$$\pi: \sum x_i\zeta_i+\eta\to\begin{pmatrix} x_1 & & \\ & \ddots & \\ & & x_s\end{pmatrix}\oplus \text{ad}_{\mathfrak{g}'}(\eta), \qquad \eta\in\mathfrak{g}';$$

then $\pi$ is faithful and semisimple, so $\mathfrak{g}$ is reductive. In fact we are about to see that this is the only example.

If $\pi$ is a finite-dimensional representation of $\mathfrak{g}$, then the associated *trace form* is the symmetric bilinear form on $\mathfrak{g}$ given by

$$\langle\xi, \eta\rangle_\pi=\text{trace}(\pi(\xi)\cdot\pi(\eta)).$$

It is invariant in the sense $\langle[\zeta, \xi], \eta\rangle_\pi+\langle\xi, [\zeta, \eta]\rangle_\pi=0$. The Cartan-Killing form of $\mathfrak{g}$ is the trace form of the adjoint representation.

THEOREM. *The following conditions are equivalent*: (i) $\mathfrak{g}$ *is reductive,* (ii) $\mathfrak{g}$ *has a nondegenerate* (qua bilinear form) *trace form,* (iii) $\mathfrak{g}=\mathfrak{z}\oplus\mathfrak{g}'$ *where* $\mathfrak{z}$ *is its center and* $\mathfrak{g}'=[\mathfrak{g}, \mathfrak{g}]$ *is a semisimple ideal* (the 'semisimple part'), (iv) *the adjoint representation* $\text{ad}_{\mathfrak{g}}$ *is semisimple.*

*Indication of proof.* If $\text{ad}_{\mathfrak{g}}$ is semisimple it is of the form $\pi_0\oplus\pi_1\oplus\ldots\oplus\pi_r$ where $\pi_0$ represents by zero and the other $\pi_i$ are irreducible and nontrivial. Now $\mathfrak{g}=\mathfrak{g}_0\oplus\mathfrak{g}_1\oplus\ldots\oplus\mathfrak{g}_r$ where $\mathfrak{g}_i$ is the representation space of $\pi_i$, so $\mathfrak{g}_0$ is the center and the other $\mathfrak{g}_i$ are simple. Thus (iv) implies (iii) with $\mathfrak{g}_0=\mathfrak{z}$ and $\mathfrak{g}_1\oplus\ldots\oplus\mathfrak{g}_r=\mathfrak{g}'$. The converse (iii)$\Rightarrow$(iv) is obvious. Given (iii), the representation described just after the definition of 'reductive' is faithful, semisimple, and has nondegenerate trace form; so (iii) implies (i) and (ii). From Schur's Lemma and Engel's Theorem, an irreducible linear Lie algebra is either semisimple or of the form (some scalars) $\oplus$(semisimple) and thus (i) implies (iii). Given (ii), the method used to decompose a semisimple algebra as a direct sum of simple ideals, can be modified to prove (iii).

We also need the relative concept. A closed subgroup $H \subset G$ (resp. subalgebra $\mathfrak{h} \subset \mathfrak{g}$) is *reductive in* $G$ (resp. $\mathfrak{g}$) if $\mathrm{Ad}_{G|H}$ (resp. $\mathrm{ad}_{\mathfrak{g}|\mathfrak{h}}$) is semisimple. In that case it is reductive.

THEOREM. *Let* $\mathfrak{g}$ *be reductive,* $\mathfrak{g} = \mathfrak{z} + \mathfrak{g}'$ *as in* (iii) *above, and let* $\mathfrak{h}$ *be a subalgebra of* $\mathfrak{g}$. *Then these are equivalent*: (i′) $\mathfrak{h}$ *is reductive in* $\mathfrak{g}$, (ii′) *some trace form on* $\mathfrak{g}$ *is nondegenerate on* $\mathfrak{h}$, (iii′) *the Cartan–Killing form of* $\mathfrak{g}'$ *is nondegenerate on the image of* $\mathfrak{h}$ *under the projection* $\mathfrak{g} \to \mathfrak{g}'$ *with kernel* $\mathfrak{z}$.

COROLLARY. *If* $\mathfrak{g}$ *is reductive, and* $\Gamma$ *is a group of automorphisms whose action on* $\mathfrak{g}$ *is semisimple, then the fixed point set* $\mathfrak{g}^{\Gamma}$ *is reductive in* $\mathfrak{g}$.

## 5. Root systems – basic properties

By *Cartan subalgebra* of a Lie algebra $\mathfrak{g}$, we mean a nilpotent subalgebra $\mathfrak{h} \subset \mathfrak{g}$ which is its own normalizer, i.e.

$$\mathfrak{h} = \{\xi \in \mathfrak{g} : [\xi, \mathfrak{h}\,] \subset \mathfrak{h}\,\}.$$

If the base field is algebraically closed, say $\mathbb{C}$, we apply Lie's Theorem to $\mathrm{ad}_{\mathfrak{g}|\mathfrak{h}}$, and nilpotence of $\mathfrak{h}$ gives a decomposition $\mathfrak{g} = \Sigma_{\lambda \in \mathfrak{h}^*} \mathfrak{g}_\lambda$ where $\mathfrak{h}^*$ is the linear dual space and

$$\mathfrak{g}_\lambda = \{\eta \in \mathfrak{g} : (\mathrm{ad}_{\mathfrak{g}}(\xi) - \lambda(\xi))^m \eta = 0 \qquad \text{for} \qquad m \gg 0\},$$

Now the fact that $\mathfrak{h}$ is nilpotent and self-normalizing says: $\mathfrak{g}_0 = \mathfrak{h}$. Let $\Delta = \Delta(\mathfrak{g}, \mathfrak{h}) = \{\lambda \in \mathfrak{h}^* : \lambda \neq 0 \text{ and } \mathfrak{g}_\lambda \neq 0\}$; it is a finite subset of $\mathfrak{h}^*$ called the *root system* of $\mathfrak{g}$ relative to $\mathfrak{h}$, and we have

$$\mathfrak{g} = \mathfrak{h} + \sum_{\lambda \in \Delta} \mathfrak{g}_\lambda, \qquad \textit{root space decomposition.}$$

If the base field is not algebraically closed, say $\mathbb{R}$, then we pass to the algebraic closure by scalar extension; $\mathfrak{h}^{\mathbb{C}}$ is a Cartan subalgebra of $\mathfrak{g}^{\mathbb{C}}$ and we have $\mathfrak{g}^{\mathbb{C}} = \mathfrak{h}^{\mathbb{C}} + \Sigma_{\lambda \in \Delta} \mathfrak{g}_\lambda^{\mathbb{C}}$.

Define polynomial functions $p_i$ on $\mathfrak{g}$ by

$$\det(\tau \cdot 1 - \mathrm{ad}(\xi)) = \sum (-1)^{m-i} p_i(\xi)\tau^i$$

where $\tau$ is an indeterminate and $m = \dim \mathfrak{g}$. The smallest integer $r \geqslant 0$ with $p_r$ not identically zero is called the *rank* of $\mathfrak{g}$. It is not changed by scalar extension of the base field, and rank $\mathfrak{g} = \dim \mathfrak{g}$ only when $\mathfrak{g}$ is nilpotent. If $l = \mathrm{rank}\ \mathfrak{g}$, then an element $\xi \in \mathfrak{g}$ is called *regular* if $p_l(\xi) \neq 0$, *singular* otherwise. The regular elements are dense.

THEOREM. *The Cartan subalgebras of* $\mathfrak{g}$ *are just the*

$$\mathfrak{h}_\xi = \{\eta \in \mathfrak{g}: \mathrm{ad}(\xi)^m \eta = 0 \quad \textit{for} \quad m \gg 0\}$$

*where* $\xi$ *is a regular element of* $\mathfrak{g}$. *In particular they all have the same dimension*, rank $\mathfrak{g}$.

CHEVALLEY'S THEOREM. *Let G be a connected Lie group with Lie algebra* $\mathfrak{g}$. *If* $\mathfrak{g}$ *is a complex Lie algebra then any two Cartan subalgebras are* Ad($G$)-*conjugate. If* $\mathfrak{g}$ *is a real Lie algebra then there are only finitely many* Ad($G$)-*conjugacy classes of Cartan subalgebras.*

Now we specialize to the case where $\mathfrak{g}$ is semisimple. An element $\xi \in \mathfrak{g}$ is called *semisimple* if ad($\xi$) is semisimple, *nilpotent* if ad($\xi$) is nilpotent. Looking at the Jordan normal form of ad($\xi$) on $\mathfrak{g}$ or $\mathfrak{g}^{\mathbb{C}}$, we see $\xi = \xi_s + \xi_n$ with $\xi_s$ semisimple, $\xi_n$ nilpotent, and both ad($\xi_s$) and ad($\xi_n$) polynomials in ad($\xi$). If $\xi$ is regular it follows that $\xi$ is semisimple. But if $\xi$ is semisimple then the Cartan–Killing form is nondegenerate on the centralizer $\mathfrak{g}^\xi = \{\eta \in \mathfrak{g}: [\xi, \eta] = 0\}$, which thus is reductive in $\mathfrak{g}$.

CONCLUSIONS. (i) *an element* $\xi \in \mathfrak{g}$ *is semisimple if and only if it is contained in a Cartan subalgebra,* (ii) *a subalgebra* $\mathfrak{h} \subset \mathfrak{g}$ *is Cartan if and only if it is a maximal Abelian subalgebra and consists of semisimple elements.*

Now fix a complex semisimple Lie algebra $\mathfrak{g}$ and a Cartan subalgebra $\mathfrak{h}$, and consider the root space decomposition $\mathfrak{g} = \mathfrak{h} + \Sigma_{\lambda \in \Delta} \mathfrak{g}_\lambda$. The Cartan–Killing form $\langle\,,\rangle$ is nondegenerate on $\mathfrak{h}$, so to each $\nu \in \mathfrak{h}^*$ we have $h_\nu \in \mathfrak{h}$ defined by $\nu(\xi) = \langle h_\nu, \xi\rangle$. We transfer $\langle\,,\rangle$ to $\mathfrak{h}^*$ by $\langle \nu, \mu\rangle = \langle h_\nu, h_\mu\rangle$. The elementary basic facts are

(i) if $\lambda \in \Delta$ then $\dim \mathfrak{g}_\lambda = 1$, and $-\lambda \in \Delta$,

(ii) if $\lambda, \mu \in \Delta$ with $\lambda + \mu \neq 0$ then $[\mathfrak{g}_\lambda, \mathfrak{g}_\mu] = \mathfrak{g}_{\lambda+\mu}$ and $\mathfrak{g}_\lambda \perp \mathfrak{g}_\mu$,

(iii) if $\xi \in \mathfrak{g}_\lambda$ and $\eta \in \mathfrak{g}_{-\lambda}$ then $[\xi, \eta] = \langle \xi, \eta \rangle h_\lambda$,

(iv) $\mathfrak{g}$ is the $\langle\,,\rangle$-orthogonal sum of $\mathfrak{h}$ and the $\mathfrak{g}_\lambda + \mathfrak{g}_{-\lambda} (\lambda \in \Delta)$.

Looking at the representation of $\mathfrak{g}[\alpha] = \mathfrak{g}_{-\alpha} + h_\alpha \cdot \mathbb{C} + \mathfrak{g}_\alpha$ on $\mathfrak{g}$, described in Section 2, one sees

(v) Let $\alpha, \lambda \in \Delta$ with $\alpha \neq \pm\lambda$. Then there are integers $p, q \geqslant 0$ such that $2\langle\lambda, \alpha\rangle / \langle\alpha, \alpha\rangle = q - p$ and $\lambda + k\alpha \in \Delta$ precisely when $k$ is an integer with $-q \leqslant k \leqslant p$.

(vi) If $\lambda \in \Delta$, $c \in \mathbb{C}$, $c\lambda \in \Delta$ then $c = \pm 1$.

The $h_\lambda$, $\lambda \in \Delta$, span a real form $\mathfrak{h}_{\mathbb{R}}$ of $\mathfrak{h}$ on which the Cartan–Killing form is positive definite. Thus also $\Delta$ spans a real form $\mathfrak{h}^*_{\mathbb{R}}$ of $\mathfrak{h}^*$ which is positive definite. If $\alpha \in \Delta$ then the reflections in the hyperplanes $\alpha = 0$ and $\alpha^\perp$ are

$$s_\alpha \colon \mathfrak{h} \to \mathfrak{h} \qquad \text{by} \qquad s_\alpha(\xi) = \xi - \left(\frac{2\alpha(\xi)}{\langle \alpha, \alpha \rangle}\right) h_\alpha,$$

and

$$s_\alpha \colon \mathfrak{h}^* \to \mathfrak{h}^* \qquad \text{by} \qquad s_\alpha(\lambda) = \lambda - \left(\frac{2\langle \alpha, \lambda \rangle}{\langle \alpha, \alpha \rangle}\right) \alpha.$$

From (v), the $s_\alpha$ permute $\Delta$, hence generate a finite group $W = W(\mathfrak{g}, \mathfrak{h})$, the *Weyl group.* If $w \in W$ then $w(\Delta) = \Delta$.

A subset $\Delta^+ \subset \Delta$ is a *positive root system* if (a) $\Delta$ is disjoint union of $\Delta^+$ and $-\Delta^+$ and (b) whenever $\lambda, \mu \in \Delta^+$ with $\lambda + \mu \in \Delta$ one has $\lambda + \mu \in \Delta^+$.

Each root $\lambda \in \Delta$ defines a hyperplane $(\lambda = 0)$ in $\mathfrak{h}_{\mathbb{R}}$, and $\mathfrak{h}_{\mathbb{R}} \setminus \bigcup_{\lambda \in \Delta} (\lambda = 0)$ is a disjoint union of convex open cones which are its topological components. Each such cone is called a *Weyl chamber.*

A Weyl chamber $\mathscr{C} \subset \mathfrak{h}_{\mathbb{R}}$ defines a positive root system $\Delta^+ = \{\lambda \in \Delta : \lambda > 0$ on $\mathscr{C}\}$. Conversely a positive root system $\Delta^+$ defines a Weyl chamber $\mathscr{C} = \{\xi \in \mathfrak{h}_{\mathbb{R}} : \lambda(\xi) > 0$ for all $\lambda \in \Delta^+\}$.

A positive root is called *simple* if it is not the sum of two positive roots. Let $S = S(\Delta^+)$ denote the system of simple roots for $\Delta^+$, and enumerate $S = \{\alpha_1, \ldots, \alpha_l\}$. Then $S$ is a basis of $\mathfrak{h}^*_{\mathbb{R}}$, so $l = \operatorname{rank} \mathfrak{g}$ and the $\alpha_i$ are linearly independent. Every $\alpha \in \Delta$ has expression $\alpha = \Sigma n_i \alpha_i$ where the $n_i$ are integers, all $\geqslant 0$ if $\alpha \in \Delta^+$, all $\leqslant 0$ if $-\alpha \in \Delta^+$, so in particular $S$ determines $\Delta^+$. Also, $s_{\alpha_i}(\Delta^+) = (\Delta^+ \setminus \{\alpha_i\}) \cup \{-\alpha_i\}$. Now the Weyl chamber $\mathscr{C} = \mathscr{C}(\Delta^+)$ determines $S$ directly as $\{\alpha \in \Delta : \alpha > 0$ on $\mathscr{C}$ and $\overline{\mathscr{C}} \cap (\alpha = 0)$ is open in $(\alpha = 0)\}$. Conversely, of course, $\mathscr{C} = \{\xi \in \mathfrak{h}_{\mathbb{R}} :$ each $\alpha_i(\xi) > 0\}$.

The Weyl group $W$ acts in a simply transitive manner on the set of all Weyl chambers (resp. set of all positive root systems, resp. set of all simple root systems). If $S$ is a simple root system then $\{s_\alpha\colon \alpha\in S\}$ is a minimal generating set for $W$. Also, if $\alpha\in\Delta$ then $\overline{\mathscr{C}}\cap(\alpha=0)$ is open in $(\alpha=0)$ for some chamber, so $\Delta=W(S)$.

## 6. Isomorphism; Compact and Split Real Forms

Select elements $e_\lambda\in\mathfrak{g}_\lambda$, $\lambda\in\Delta$, such that $\langle e_\lambda, e_{-\lambda}\rangle=-1$. So $[e_\lambda, e_{-\lambda}]=-h_\lambda$ by (iii). If $\lambda$, $\mu\in\Delta$ with $\lambda+\mu\neq 0$, define numbers $n_{\lambda,\mu}$ by

$$[e_\lambda, e_\mu]=n_{\lambda,\mu}e_{\lambda+\mu}\quad \text{if } \lambda+\mu\in\Delta,\quad n_{\lambda,\mu}=0\quad \text{if } \lambda+\notin\Delta.$$

Then one has $n_{\lambda,\mu}+n_{\mu,\lambda}=0$, and one can prove

(vii) if $\lambda$, $\mu$, $\nu\in\Delta$ and $\lambda+\mu+\nu=0$ then $n_{\lambda,\mu}=n_{\mu,\nu}=n_{\nu,\lambda}$

(viii) if $\kappa$, $\lambda$, $\mu$, $\nu\in\Delta$, none the negative of any other, and if $\kappa+\lambda+\mu+\nu=0$, then $n_{\kappa,\lambda}n_{\mu,\nu}+n_{\lambda,\mu}n_{\kappa,\nu}+n_{\mu,\kappa}n_{\lambda,\nu}=0$

(ix) If $\alpha$, $\lambda\in\Delta$, $\alpha\neq\pm\lambda$, and $p,q\geqslant 0$ as in (v), then $n_{\alpha,\lambda}n_{-\alpha,-\lambda}=\frac{1}{2}\langle\alpha,\alpha\rangle p(q+1)$.

Suppose that $\mathfrak{g}_1$ is another complex semisimple Lie algebra, $\mathfrak{h}_1$ a Cartan subalgebra of $\mathfrak{g}_1$, and $\Delta_1=\Delta(\mathfrak{g}_1, \mathfrak{h}_1)$ the root system. Let $f\colon \mathfrak{h}\to\mathfrak{h}_1$ be a linear isomorphism such that $f^*\Delta_1=\Delta$.

THEOREM. *$f$ extends to an isomorphism of $\mathfrak{g}$ onto $\mathfrak{g}_1$.*

First, $f$ is an isometry relative to the Cartan–Killing form; for

$$\langle f\xi, f\xi'\rangle_1=\sum_{\Delta_1}\gamma(f\xi)\gamma(f\xi')=\sum_{\Delta_1}(f^*\gamma)(\xi)\cdot(f^*\gamma)(\xi')$$

$$=\sum_{\Delta}\lambda(\xi)\lambda(\xi')=\langle\xi,\xi'\rangle.$$

Choose a positive root system $\Delta_1^+$ for $(\mathfrak{g}_1, \mathfrak{h}_1)$, let $S_1$ be the corresponding simple root system, and set $\Delta^+=f^*\Delta_1^+$ and $S=f^*S_1$. Then $\Delta^+$ is a positive root system, and $S$ the corresponding simple root system, for $(\mathfrak{g}, \mathfrak{h})$. Enumerate $S_1=\{\beta_i\}$ and $S=\{\alpha_i\}$ where $\alpha_i=f^*\beta_i$. We say that a root $\lambda\in\Delta^+$ (resp. $\gamma\in\Delta_1^+$) has *level* $\Sigma n_i$ if $\lambda=\Sigma n_i\alpha_i$ (resp. $\gamma=\Sigma n_i\beta_i$). As above, select $e_\lambda\in\mathfrak{g}_\lambda(\lambda\in\Delta)$ with $\langle e_\lambda, e_{-\lambda}\rangle=-1$ and define $n_{\lambda,\mu}$ by $[e_\lambda, e_\mu]=n_{\lambda,\mu}e_{\lambda+\mu}$ if $\lambda+\mu\in\Delta$, $n_{\lambda,\mu}=0$ if $\lambda+\mu\notin\Delta$. We want elements $f_\gamma\in(\mathfrak{g}_1)_\gamma$ such that $\langle f_\gamma, f_{-\gamma}\rangle_1=-1$ and $[f_\gamma, f_\delta]=n_{f^*\gamma,f^*\delta}f_{\gamma+\delta}$ if $\gamma+\delta\in\Delta_1$; then $f(e_{f^*\gamma})=f_\gamma$ will give an isomorphism of $\mathfrak{g}$ onto $\mathfrak{g}_1$. Suppose that $k>0$ is an integer and we have such $f_\gamma$ for $|\text{level }\gamma|<k$. Let $\varepsilon\in\Delta_1^+$ of level $k$. If there do not exist $\gamma$, $\delta\in\Delta_1$ with $|\text{level }\gamma|$, $|\text{level }\delta|<k$ then let $f_\varepsilon\in(\mathfrak{g}_1)_\varepsilon$ be any nonzero element. Otherwise choose $\gamma$, $\delta\in\Delta_1$, $|\text{level }\gamma|<k$, $|\text{level }\delta|<k$ with $\gamma+\delta=\varepsilon$ and define

$f_\varepsilon$ by $[f_\gamma, f_\delta]=\eta_{\gamma,\delta}f_\varepsilon$, define $f_{-\varepsilon}\in(\mathfrak{g}_1)_{-\varepsilon}$ by $\langle f_\varepsilon, f_{-\varepsilon}\rangle_1=-1$. Using (vii), (viii) and (ix) one shows that $f_{\pm\varepsilon}$ are independent of choice of $\gamma$, $\delta$. Thus $f:\mathfrak{h}\to\mathfrak{h}_1$ extends to $f:\mathfrak{g}\cong\mathfrak{g}_1$.

Suppose, in the above theorem, that $\mathfrak{g}=\mathfrak{g}_1$, $\mathfrak{h}=\mathfrak{h}_1$ and $f(\xi)=-\xi$ for all $\xi\in\mathfrak{h}$. Then in the extension $f:\mathfrak{g}\cong\mathfrak{g}$, we have $f(e_\lambda)=c_\lambda e_{-\lambda}$, so $f^2(e_\lambda)==c_\lambda c_{-\lambda}e_\lambda$. But $\langle e_\lambda, e_{-\lambda}\rangle=\langle f(e_\lambda), f(e_{-\lambda})\rangle$ shows $c_\lambda c_{-\lambda}=1$. So we have

COROLLARY. *$\mathfrak{g}$ has an automorphism $f$ which is $-1$ on $\mathfrak{h}$, and any such automorphism is involutive.*

A *Weyl basis* of $\mathfrak{g}$ is a basis of the form $\{h_i\}\cup\{e_\lambda\}$ where $\{h_i\}$ is a real basis of $\mathfrak{h}_{\mathbb{R}}$, $e_\lambda\in\mathfrak{g}_\lambda$, $\langle e_\lambda, e_{-\lambda}\rangle=-1$, and the $n_{\lambda,\mu}=n_{-\lambda,-\mu}$. To construct a Weyl basis: start with an automorphism $f$ such that $f|_{\mathfrak{h}}=-1$, choose a positive root system $\Delta^+$, and for each $\lambda\in\Delta^+$ select $e_\lambda\in\mathfrak{g}_\lambda$ such that $\langle e_\lambda, f(e_\lambda)\rangle=-1$. Set $e_{-\lambda}=f(e_\lambda)$. Then the $n_{\lambda,\mu}=n_{-\lambda,-\mu}$.

A real subalgebra $\mathfrak{g}_0\subset\mathfrak{g}$ is called a *real form* if $(\mathfrak{g}_0)_{\mathbb{C}}=\mathfrak{g}$. A real form $\mathfrak{g}_0$ is *split* if it has a Cartan subalgebra $\mathfrak{h}_0$ on which all the roots are real-valued. This means, setting $\mathfrak{h}=(\mathfrak{h}_0)_{\mathbb{C}}$, that the root space decomposition $\mathfrak{g}=\mathfrak{h}+\Sigma_\Delta\mathfrak{g}_\lambda$ intersects $\mathfrak{g}_0$ to give a real root space decomposition $\mathfrak{g}_0=\mathfrak{h}_0+\Sigma_\Delta(\mathfrak{g}_0)_\lambda$. A split real form is also called a normal real form. If $\{h_i\}\cup\{e_\lambda\}$ is a Weyl basis of $\mathfrak{g}$, then (ix) gives us

$$n_{\alpha,\lambda}^2=\tfrac{1}{2}\langle\alpha,\alpha\rangle p(q+1)\geqslant 0,$$

so the $n_{\alpha,\lambda}$ are real. As the roots take real values on the real span $\mathfrak{h}_{\mathbb{R}}$ of $\{h_i\}$, now *the real span of that Weyl basis is a split real form of* $\mathfrak{g}$.

A real form $\mathfrak{g}_u\subset\mathfrak{g}$ is called *compact* if the Cartan–Killing form is negative definite on it. An equivalent formulation: *if $\xi\in\mathfrak{g}_u$ then every eigenvalue of* ad$(\xi)$ *is pure imaginary.* If $G$ has Lie algebra $\mathfrak{g}$, then the analytic subgroup for a real form $\mathfrak{g}_u$ is a compact group if and only if $\mathfrak{g}_u$ is a compact real form of $\mathfrak{g}$. Also, if $\mathfrak{g}_u$ is a compact real form of $\mathfrak{g}$ and $\mathfrak{h}_u$ is a Cartan subalgebra, then the roots of $\mathfrak{g}$ relative to $\mathfrak{h}=(\mathfrak{h}_u)_{\mathbb{C}}$ take pure imaginary values on $\mathfrak{h}_u$, so $\mathfrak{h}_u=i\mathfrak{h}_{\mathbb{R}}$.

THEOREM. *Let $\mathfrak{g}$ be a complex semisimple Lie algebra and $G$ any Lie group with Lie algebra $\mathfrak{g}$. Then $\mathfrak{g}$ has compact real forms, and any two are* Ad$(G)$*-conjugate.*

To see this, let $\mathfrak{h}$ be a Cartan subalgebra of $\mathfrak{g}$, $\{h_i\}\cup\{e_\lambda\}$ a Weyl basis, and $f$ an automorphism of $\mathfrak{g}$ with $f|_{\mathfrak{h}}=-1$ and $f(e_\lambda)=e_{-\lambda}$. Let $\mathfrak{g}_0$ be the split real form of $\mathfrak{g}$ spanned over $\mathbb{R}$ by $\{h_i\}\cup\{e_\lambda\}$ and let $\sigma_0$ denote complex conjugation of $\mathfrak{g}$ over $\mathfrak{g}_0$. Then $\sigma_u = f\circ\sigma_0$ is a real algebra automorphism of $\mathfrak{g}$ which is complex conjugation over some vector space real form $\mathfrak{g}_u$. Now $\mathfrak{g}_u$ is a Lie algebra real form. If $\xi = h + \Sigma c_\lambda e_\lambda \in \mathfrak{g}$, $h\in\mathfrak{h}$, then $\sigma_u(\xi) = \sigma_u(h) + \Sigma\bar{c}_\lambda e_{-\lambda}$, so $\xi\in\mathfrak{g}_u$ if and only if $h\in i\mathfrak{h}_{\mathbb{R}}$ and $\bar{c}_\lambda = c_{-\lambda}$. In that case, since $i\mathfrak{h}_{\mathbb{R}}$ is negative definite, $\langle\xi,\xi\rangle = \langle h,h\rangle + \Sigma|c_\lambda|^2 \leqslant 0$ with equality just when $\xi = 0$. So $\mathfrak{g}_u$ is a compact real form. Let $\mathfrak{g}_u'$ be another compact real form, $\mathfrak{h}_u'$ a Cartan subalgebra. Since $\mathfrak{h}' = (\mathfrak{h}_u')_{\mathbb{C}}$ is Ad($G$)-conjugate to $\mathfrak{h}$ we may suppose that $\mathfrak{h}_u' = i\mathfrak{h}_{\mathbb{R}}$. Let $\sigma$, $\sigma'$ be conjugations of $\mathfrak{g}$ over $\mathfrak{g}_u$, $\mathfrak{g}_u'$. Then $\tau = \sigma\sigma'$ is an automorphism of $\mathfrak{g}$ which is the identity on $\mathfrak{h}$, so $\tau$ preserves each $\mathfrak{g}_\lambda$. Now we have nonzero $b_\lambda$ with $\sigma'(e_\lambda) = b_\lambda e_{-\lambda}$, and $\langle e_\lambda, \sigma' e_\lambda\rangle < 0$ so $b_\lambda$ is real and positive. Conjugating by an element of $\exp(\mathfrak{h}_{\mathbb{R}})$ we may assume $b_\lambda = 1$ for $\lambda$ in some simple root system, and then we have $\mathfrak{g}_u = \mathfrak{g}_u'$.

## 7. Dynkin diagrams and classification

Let $\mathfrak{g}$ be a complex semisimple Lie algebra, $\mathfrak{h}$ a Cartan subalgebra, and $S$ a simple $\mathfrak{h}$-root system for $\mathfrak{g}$. Enumerate $S = \{\alpha_1, \ldots, \alpha_l\}$. The *Cartan matrix* $A_S = (a_{ij})$ is defined by $a_{ij} = 2\langle\alpha_i, \alpha_j\rangle/\langle\alpha_i, \alpha_i\rangle$. Here $a_{ii} = 2$, and for $i \neq j$ we have $a_{ij} = 0, -1, -2$ or $-3$.

THEOREM. $\mathfrak{g}$ *determines the Cartan matrix* $A_S$ *up to permutation-equivalence, i.e. conjugation by a permutation matrix.* $A_S$ *determines* $\mathfrak{g}$ *up to isomorphism.*

First, if $G$ has Lie algebra $\mathfrak{g}$, then any two Cartan subalgebras are Ad($G$)-equivalent, and given $\mathfrak{h}$ any two simple root systems are equivalent under the Weyl group. Second, if $S'$ is a simple root system for $(\mathfrak{g}', \mathfrak{h}')$ with $A_S = A_{S'}$, then we have an isometry of $\mathfrak{h}$ onto $\mathfrak{h}'$ which pulls $S'$ back to $S$, hence pulls the root system $\Delta' = \Delta(\mathfrak{g}', \mathfrak{h}')$ back to $\Delta$, and thus extends to an isomorphism of $\mathfrak{g}$ onto $\mathfrak{g}'$.

The information carried by the Cartan matrix $A_S$ can be summarized in a *Dynkin diagram* as follows. For each $\alpha\in S = \{\alpha_1, \ldots, \alpha_l\}$ we have a vertex, represented by a small circle. Vertices $\alpha_i$, $\alpha_j$ are jointed by $a_{ij}a_{ji}$ lines if $i\neq j$. If $0 \neq a_{ij} \neq -1$, so $a_{ij}$ is $-2$ or $-3$, then one has $a_{ji} = -1$, so the number of lines joining two vertices is 0, 1, 2 or 3. If the number is 2 or 3, say $a_{ij} = -2$ or $-3$ and $a_{ji} = -1$, then we insert an arrow toward the vertex representing the shorter root $\alpha_i$. (Another convention: darken the shorter root. But this causes problems later with Satake diagrams.)

Every such diagram is a disjoint union of copies of the irreducible diagrams, which give the isomorphism classes of complex simple Lie algebras

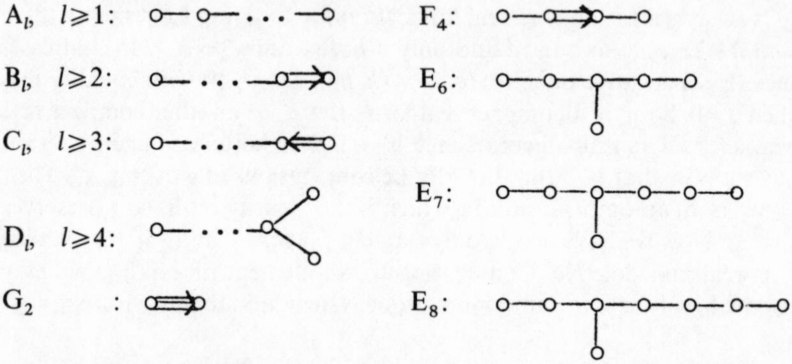


A splitting of the Dynkin diagram of $A_S$ as disjoint union of subdiagrams, corresponds to decomposition of $S$ as disjoint union of mutually orthogonal sets, which in turn corresponds to splitting of as direct sum of ideals. So $\mathfrak{g}$ is simple just when its Dynkin diagram is connected.

*Type* A$_l$. Here $\mathfrak{g}=\mathfrak{sl}(l+1;\mathbb{C})$, complex $(l+1)\times(l+1)$ matrices of trace 0, and we can choose

$$\mathfrak{h}=\left\{\begin{pmatrix} a_1 & & \\ & \ddots & \\ & & a_{l+1}\end{pmatrix} : \sum a_i=0\right\} \qquad \text{with}$$

$$\alpha_i\begin{pmatrix} a_1 & & \\ & \ddots & \\ & & a_{l+1}\end{pmatrix}=a_i-a_{i+1}.$$

The simply connected group is the complex *special linear group*

$$G=SL(l+1;\mathbb{C})=\{g\in GL(l+1;\mathbb{C}): \det g=1\}.$$

We also have

$$\begin{array}{lll} \text{split:} & \mathfrak{g}_0=\mathfrak{sl}(l+1;\mathbb{R}) \quad \text{and} & G_0=SL(l+1;\mathbb{R}) \\ \text{compact:} & \mathfrak{g}_u=\mathfrak{su}(l+1) \quad \text{and} & G_u=SU(l+1) \end{array}$$

*Type* B$_l$. Here $\mathfrak{g}=\mathfrak{o}(2l+1;\mathbb{C})$, antisymmetric complex $(2l+1)\times(2l+1)$ matrices. The simply connected group is a 2-sheeted cover of the complex special orthogonal group $SO(2l+1;\mathbb{C})$. The split real form is the identity component $SO(l, l+1)$ of the real orthogonal group of

$$(x, y)=\sum_{1}^{l} x_i y_i - \sum_{l+1}^{2l+1} x_j y_j,$$

and the compact real form is the ordinary special orthogonal group $SO(2l+1)$.

*Type* $C_l$. Here $\mathfrak{g}=\mathfrak{sp}(l;\mathbb{C})$, complex $2l\times 2l$ matrices that annihilate the antisymmetric form

$$A\colon \mathbb{C}^{2l}\times\mathbb{C}^{2l}\to\mathbb{C} \quad\text{by}\quad A((x, x'), (y, y'))={}^t x\cdot y' - {}^t x'\cdot y$$

The simply connected group is $G=Sp(l;\mathbb{C})$, complex symplectic group, and one has the real forms

$$\text{split: } \mathfrak{g}_0=\mathfrak{sp}(l;\mathbb{R}) \text{ and } G_0=Sp(l;\mathbb{R}), \text{ real matrices in } \mathfrak{g} \text{ and } G$$

$$\text{compact: } \mathfrak{g}_u=\mathfrak{sp}(l;\mathbb{C})\cap\mathfrak{u}(2l)=\mathfrak{sp}(l),\quad G_u=Sp(l;\mathbb{C})\cap U(2l)=Sp(l).$$

*Type* $D_l$. Here $\mathfrak{g}=\mathfrak{o}(2l;\mathbb{C})$, $G=SO(2l;\mathbb{C})$, $G_0=SO(l, l)$ (split), $G_u=SO(2l)$ (compact).

*Type* $G_2$. Here $\mathfrak{g}$ is the algebra of derivations of the complex Cayley algebra $\mathscr{C}$ and $G$ is the automorphism group. The split real forms $\mathfrak{g}_0$, $G_0$ are the derivation algebra and automorphism group of the real form of $\mathscr{C}$ that is not a division algebra, and the compact real forms $\mathfrak{g}_u$ and $G_u$ correspond to the real Cayley division algebra.

Type $F_4$ can be described in terms of the 27-dimensional exceptional Jordan algebra, but this sort of thing is not so useful for $E_6$, $E_7$, $E_8$.

Beside the classification, Dynkin diagrams allow an easy description of the automorphism group $\mathrm{Aut}(\mathfrak{g})$ of a complex semisimple Lie algebra $\mathfrak{g}$. First, let $\mathrm{Int}(\mathfrak{g})$ denote the group of *inner automorphisms*, that is, automorphisms of the form $\mathrm{Ad}(g)$ where $g$ runs through a connected Lie group with Lie algebra $\mathfrak{g}$. $\mathrm{Int}(\mathfrak{g})$ is generated by the *elementary automorphisms* $\exp(\mathrm{ad}(\xi))$, $\xi$ nilpotent in $\mathfrak{g}$. Second, note that $\mathrm{Int}(\mathfrak{g})$ is transitive on the pairs $(\mathfrak{h}, \Delta^+)$ where $\mathfrak{h}$ is a Cartan subalgebra and $\Delta^+$ is a positive $\mathfrak{h}$-root system for $\mathfrak{g}$, and if $\gamma\in\mathrm{Int}(\mathfrak{g})$ with $\gamma(\mathfrak{h})=\mathfrak{h}$ and $\gamma^*\Delta^+=\Delta^+$ then (*simple transitivity of the Weyl group*) $\gamma|_{\mathfrak{h}}=1$. Third, note that every symmetry of the Dynkin diagram extends to an automorphism of $\mathfrak{g}$. Assembling these facts,

$$\mathrm{Int}(\mathfrak{g}) \text{ is a normal subgroup of finite index in } \mathrm{Aut}(\mathfrak{g}),$$
$$\mathrm{Aut}(\mathfrak{g})/\mathrm{Int}(\mathfrak{g}) \text{ is the symmetry group of the Dynkin diagram.}$$

If $\mathfrak{g}$ is simple of type $A_1$, $B_l$, $C_l$, $G_2$, $F_4$, $E_7$ or $E_8$, the Dynkin diagram has no nontrivial symmetries, so $\mathrm{Aut}(\mathfrak{g})=\mathrm{Int}(\mathfrak{g})$. If $\mathfrak{g}$ is simple of type $A_l$ $(l>1)$, $D_l$ $(l>4)$ or $E_6$, the Dynkin diagram has symmetry group $\mathbb{Z}_2$ given by

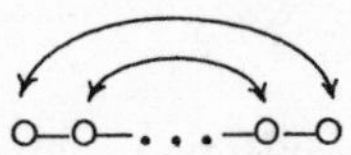
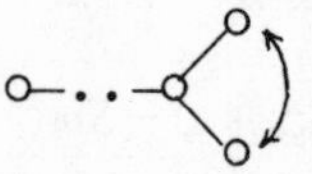
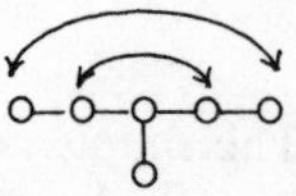

So $\mathrm{Aut}(\mathfrak{g})/\mathrm{Int}(\mathfrak{g})\cong\mathbb{Z}_2$. And if $\mathfrak{g}$ is simple of type $D_4$ then the symmetry group of the diagram is the permutation group $\mathfrak{S}_3$ on the extremal vertices, so $\mathrm{Aut}(\mathfrak{g})/\mathrm{Int}(\mathfrak{g})$ has order 6. If $\mathfrak{g}$ is not simple, one also has permutations of isomorphic simple ideals.

## 8. Real semisimple groups

We say that a group $G$ is of *Harish-Chandra class* if (i) $G$ is a reductive Lie group, (ii) $G$ has only finitely many topological components, (iii) the derived group $[G, G]$ has finite center, and (iv) if $g\in G$ then $\mathrm{Ad}(g)\in\mathrm{Int}(\mathfrak{g}_{\mathbb{C}})$. This is the class that contains connected semisimple Lie groups with finite center and is closed under passage to certain reductive subgroups that we will encounter when we study parabolic subgroups. From now on, $G$ is of Harish-Chandra class

Let $G^0$ denote the identity component. Simply because $G$ is locally compact with $G/G^0$ finite, every compact subgroup of $G$ is contained in a maximal compact subgroup, and any two maximal compact subgroups are $\mathrm{Ad}(G^0)$-conjugate. If $K$ is a maximal compact subgroup of $G$, then $K\cap G^0=K^0$, $K$ meets every component of $G$, and $G/K$ has the structure of riemannian symmetric space of curvature $\leqslant 0$. The symmetry gives an involutive automorphism $\theta$ of $G$, called a *Cartan involution*, with fixed point set $G^\theta=K$. Some examples:

| $G$ | $\theta$ | $K$ |
|---|---|---|
| $GL(n;\mathbb{R})$ | $\theta(g)={}^t g^{-1}$ | $O(n)$ |
| $U(p, q)$ | $\theta(g)=\begin{pmatrix} -I_p & \\ & I_q \end{pmatrix} g \begin{pmatrix} -I_p & \\ & I_q \end{pmatrix}$ | $U(p)\times U(q)$ |
| connected complex semisimple Lie group | conjugation over $G_u$ | $G_u$ |

THEOREM. *A closed subgroup $H\subset G$ is reductive in $G$ if, and only if, $\theta(H)=H$ for some Cartan involution $\theta$ of $G$.*

If $\mathfrak{h}$ is a Cartan subalgebra of $\mathfrak{g}$ the corresponding *Cartan subgroup* is $H=\{g\in G\colon \mathrm{Ad}(g)\xi=\xi$ for all $\xi\in\mathfrak{h}\}$, closed subgroup with identity component $H^0=\exp(\mathfrak{h})$. Since $\mathfrak{h}$ is reductive in $\mathfrak{g}$, it is stable under a Cartan involution $\theta$, and evidently $\theta(H)=H$. So we have

COROLLARY. *Fix a Cartan involution $\theta$ of $G$; then every Cartan subgroup of $G$ is conjugate to a $\theta$-stable Cartan subgroup of $G$.*

Fix a Cartan involution $\theta$, thus also a maximal compact subgroup $K=G^\theta$; this decomposes $\mathfrak{g}$ into $(\pm 1)$-eigenspaces,

$$\mathfrak{g}=\mathfrak{k}+\mathfrak{s}, \qquad \textit{Cartan decomposition,}$$

where $\mathfrak{k}$ is the Lie algebra of $K$. $[\mathfrak{k}, \mathfrak{k}]\subset\mathfrak{k}$, $[\mathfrak{k}, \mathfrak{s}]\subset\mathfrak{s}$, and $[\mathfrak{s}, \mathfrak{s}]\subset\mathfrak{k}$. The Killing form is negative on $\mathfrak{k}$, positive on $\mathfrak{s}$, and $\langle\mathfrak{k}, \mathfrak{s}\rangle=0$. So a subalgebra $\mathfrak{l}\subset\mathfrak{g}$ is reductive in $\mathfrak{g}$ if and only if $\mathfrak{l}=\mathfrak{l}\cap\mathfrak{k}+\mathfrak{l}\cap\mathfrak{s}$ for some Cartan involution $\theta$.

A subspace $\mathfrak{a}\subset\mathfrak{g}$ is called a *Cartan subalgebra of* $(\mathfrak{g}, \mathfrak{k})$ if it is maximal among the subspaces such that

$$\mathfrak{a}\subset\mathfrak{s} \qquad \text{and} \qquad [\mathfrak{a}, \mathfrak{a}]=0.$$

THEOREM. *Any two Cartan subalgebras of* $(\mathfrak{g}, \mathfrak{k})$ *are* $\mathrm{Ad}_G(K)$*-conjugate, so in particular they have the same dimension.*

This dimension is called the *rank* of the symmetric space $G/K$ and the *real rank* of the group $G$. Note that $G$ is a split real form of a complex group $G_{\mathbb{C}}$ if and only if $\text{rank}_{\mathbb{R}} G = \text{rank } G$. Also note that $\mathfrak{g}_u = \mathfrak{k} + \sqrt{-1}\mathfrak{s}$ is a compact real form $\mathfrak{g}_{\mathbb{C}}$.

Fix a Cartan subalgebra $\mathfrak{a}$ of $(\mathfrak{g}, \mathfrak{k})$. Every $\xi \in \mathfrak{s}$ belongs to some Cartan subalgebra of $(\mathfrak{g}, \mathfrak{k})$. It follows that

$$\mathfrak{s} = \bigcup_{k \in K} \text{Ad}(k) \cdot \mathfrak{a} \text{ and so } \exp_G(\mathfrak{s}) = \bigcup_{k \in K} \text{Ad}(k) \cdot \exp_G(\mathfrak{a}).$$

Writing $A = \exp_G(\mathfrak{a})$ and using $G = K \cdot \exp_G(\mathfrak{s})$, this gives a decomposition $G = KAK$. In a moment we will refine that decomposition.

The centralizer $Z_G(A)$ of $A$ in $G$ is $\theta$-stable and intersects $\exp_G(\mathfrak{s})$ in $A$ only. So we have

$$Z_G(A) = M \times A \qquad \text{where} \qquad M = Z_G(A) \cap K = Z_K(A).$$

Let $\mathfrak{t}$ (resp. $T$) be a Cartan subalgebra of $\mathfrak{m}$ (resp. Cartan subgroup of $M$). Then $\mathfrak{h} = \mathfrak{t} + \mathfrak{a}$ (resp. $H = T \times A$) is a Cartan subalgebra of $\mathfrak{g}$ (resp. Cartan subgroup of $G$) stable under $\theta$.

Every $\mathfrak{h}_{\mathbb{C}}$-root of $\mathfrak{g}_{\mathbb{C}}$ takes real values on $\mathfrak{a}$ and pure imaginary values on $\mathfrak{t}$. From the former, we have an $\mathfrak{a}$-root decomposition

$$\mathfrak{g} = (\mathfrak{m} + \mathfrak{a}) + \sum_{\nu \in \Delta_{\mathfrak{a}}} \mathfrak{g}_\nu$$

Here $\Delta_{\mathfrak{a}} \subset \mathfrak{a}^* \backslash \{0\}$ is the set of linear functionals $\nu$ on $\mathfrak{a}$ such that $\mathfrak{g}_\nu = \{\eta \in \mathfrak{g}: [\xi, \eta] = \nu(\xi)\eta \text{ for all } \xi \in \mathfrak{a}\} \neq 0$; $\Delta_{\mathfrak{a}}$ is the $\mathfrak{a}$-*root system* of $\mathfrak{g}$. The elements of $\Delta_{\mathfrak{a}}$ are called $\mathfrak{a}$-*roots* or *restricted roots*. They are the $\lambda|_{\mathfrak{a}}$ where $\lambda \in \Delta(\mathfrak{g}_{\mathbb{C}}, \mathfrak{h}_{\mathbb{C}})$ with $\lambda|_{\mathfrak{a}} \neq 0$. As in the case of ordinary roots, we have positive root systems $\Delta_{\mathfrak{a}}^+$ corresponding to Weyl chambers

$$\mathscr{C}\text{: topological component of } \mathfrak{a} \backslash \bigcup_{\nu \in \Delta_{\mathfrak{a}}} (\text{hyperplane } \nu = 0)$$

by $\Delta_{\mathfrak{a}}^+ = \{\nu \in \Delta_{\mathfrak{a}}: \nu > 0 \text{ on } \mathscr{C}\}$ and $\mathscr{C} = \{\xi \in \mathfrak{a}: \nu(\xi) > 0 \text{ for all } \xi \in \Delta_{\mathfrak{a}}^+\}$. There is a Weyl group $W = W(\mathfrak{g}, \mathfrak{a})$ as before, sometimes called the "baby Weyl group" of $G$. It is simply transitive on the set of positive $\mathfrak{a}$-root systems.

Fix a positive $\mathfrak{a}$-root system $\Delta_{\mathfrak{a}}^+$. Then we have nilpotent subalgebras of $\mathfrak{g}$

$$\mathfrak{n} = \sum_{\nu \in \Delta_{\mathfrak{a}}^+} \mathfrak{g}_\nu \qquad \text{and} \qquad \mathfrak{n}^- = \theta(\mathfrak{n}) = \sum_{\nu \in \Delta_{\mathfrak{a}}^+} \mathfrak{g}_{-\nu},$$

and connected simply connected nilpotent subgroups of $G$

$$N = \exp_G(\mathfrak{n}) \qquad \text{and} \qquad N^- = \theta(N) = \exp_G(\mathfrak{n}^-).$$

Their respective $G$-normalizers are the *minimal parabolic subgroups*

$$B=MAN \qquad \text{and} \qquad B^-=\theta(B)=MAN^-.$$

Note that $B$ and $B^-$ are conjugate – by any element of $K$ that normalizes $\mathfrak{a}$ and gives the Weyl group element sending $\Delta_{\mathfrak{a}}^+$ to its negative. More generally, any two minimal parabolic subgroups of $G$ are conjugate.

THE IWASAWA DECOMPOSITION. $G=KAN$; *more precisely, the map* $K\times A\times N\to G$, *by* $(k, a, n)\mapsto kan$, *is a diffeomorphism.*

The space $G/B\cong K/M$ is called a *maximal boundary* of $G/K$. It appears both in the construction of 'principal series' respresentations and in the theory of bounded harmonic functions.

Let $\mathfrak{a}^+$ denote the positive Weyl chamber for $\Delta_{\mathfrak{a}}^+$, $A^+=\exp_G(\mathfrak{a}^+)$, and $\overline{A^+}$ the closure of $A^+$ in $G$. The promised refinement of $G=KAK$ is: $G=K\cdot\overline{A^+}\cdot K$.

Let $\Delta$ be the $\mathfrak{h}_{\mathbb{C}}$-root system of $\mathfrak{g}_{\mathbb{C}}$, so $\Delta_{\mathfrak{a}}=\{\lambda|_{\mathfrak{a}}\colon \lambda\in\Delta,\ \lambda|_{\mathfrak{a}}\neq 0\}$. We say that choices $\Delta^+$, $\Delta_{\mathfrak{a}}^+$ of positive root systems are *consistent* if $\lambda\in\Delta^+$, $\lambda|_{\mathfrak{a}}\neq 0$ implies $\lambda|_{\mathfrak{a}}\in\Delta_{\mathfrak{a}}^+$. Given $\Delta_{\mathfrak{a}}^+$, the consistent choices of $\Delta^+$ are in one-one correspondence with the positive $\mathfrak{t}_{\mathbb{C}}$-root systems on $\mathfrak{m}_{\mathbb{C}}$; so consistent choices exist.

Now fix consistent choices $\Delta^+$, $\Delta_{\mathfrak{a}}^+$ of positive $\mathfrak{h}_{\mathbb{C}}$-root and $\mathfrak{a}$-root systems. Let $S$ and $S_{\mathfrak{a}}$ denote the corresponding simple systems. The *Satake diagram* of $G$ is obtained as follows. In the Dynkin diagram $\mathscr{D}_S$ of $S$, we darken the vertices corresponding to simple roots $\alpha\in S$ with $\alpha|_{\mathfrak{a}}=0$, and we join (by 2-headed arrows) vertices corresponding to simple roots $\alpha$, $\alpha'\in S$ with $0\neq\alpha|_{\mathfrak{a}}=\alpha'|_{\mathfrak{a}}$. Following Araki, we list all real simple $\mathfrak{g}$ except those where $\mathfrak{g}$ is compact or complex. Here $\lambda_i=\alpha_i|_{\mathfrak{a}}$, $m(\lambda)=\dim\mathfrak{g}_\lambda$, $\mathbb{H}$ denotes the quaternions, $l=\operatorname{rank}\mathfrak{g}$, and we identify the real exceptional Lie algebras by listing the type of a maximal compact subalgebra $\mathfrak{k}$ as subscript with $T_1$ meaning 1-dimensional Abelian.

| | $\mathfrak{g}$ | $\mathscr{D}_S$ | $\mathscr{D}_{S_a}$ | $m(\lambda_i)$ | $m(2\lambda_i)$ |
|---|---|---|---|---|---|
| AI | $\mathfrak{sl}(l+1;\mathbb{R})$ |  |  | 1 | 0 |
| AII | $\mathfrak{sl}(l+1;\mathbb{H})$ | 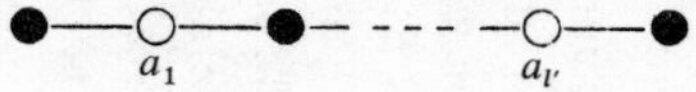 |  $(l=2l'+1,\ l'\geqslant 1)$ | 4 | 0 |
| AIII | $\mathfrak{su}(p,\ l+1-p)$ | 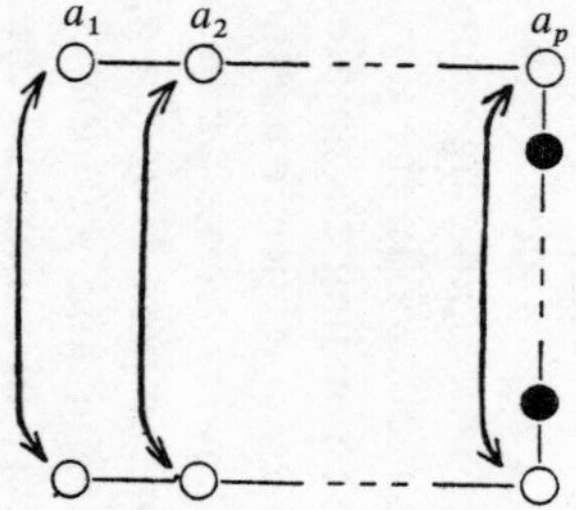 | 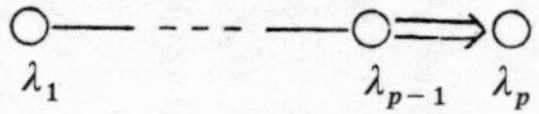 $\left(2\leqslant p\leqslant \frac{l}{2}\right)$ | 2 (for $i<p$)<br>$2(l-2p+1)$ (for $i=p$) | 0<br>1 |
| | $\mathfrak{su}(l'+1,\ l'+1)$ | 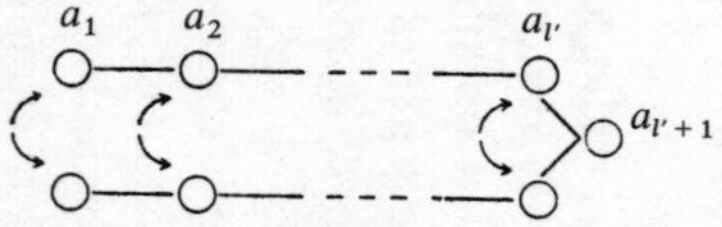 |  $(l=2l'+1,\ l'\geqslant 1)$ | 2 (for $i\leqslant l'$)<br>1 ($i=l'+1$) | 0<br>0 |
| | $\mathfrak{su}(1,\ l)$ | 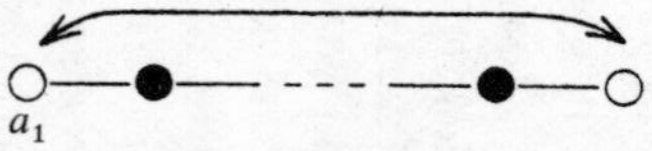 | 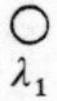 | $2(l-1)$ | 1 |
| BI | $\mathfrak{so}(p,\ 2l+1-p)$ |  |  $(l\geqslant 2,\ 2\leqslant p\leqslant l)$ | 1 ($i<p$)<br>$2(l-p)+1$ ($i=p$) | 0<br>0 |

| | | | | | |
|---|---|---|---|---|---|
| BII | $\mathfrak{so}(1, 2l)$ |   |   | $2l-1$ | 0 |
| CI | $\mathfrak{sp}(L; \mathbb{R})$ |   | 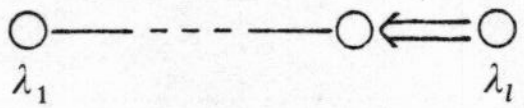  | 1 | 0 |
| CII | $\mathfrak{sp}(p, l-p)$ | 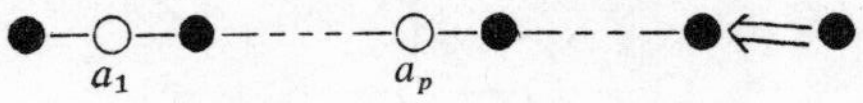  |   $\left(l \geqq 3, 1 \leqq p \leqq \frac{l-1}{2}\right)$ | 4 $(i<p)$ <br> $4(l-2p)$ $(i=p)$ | 0 <br> 3 |
| | $\mathfrak{sp}(l', l')$ | 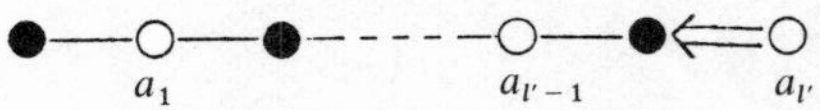  |   $(l=2l', l' \geqq 2)$ | 4 $(i<l')$ <br> 3 $(i=l')$ | 0 <br> 0 |
| DI | $\mathfrak{so}(p, 2l-p)$ | 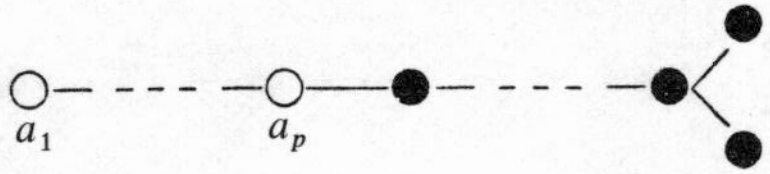  |   $(l \geqq 4, 2 \leqq p \leqq l-2)$ | 1 $(i<p)$ <br> $2(l-p)$ $(i=p)$ | 0 |
| | $\mathfrak{so}(l-1, l+1)$ | 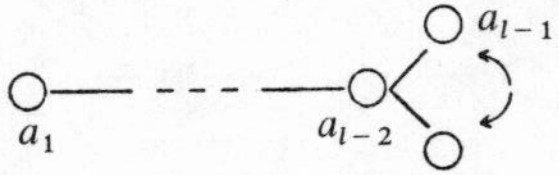  |   | 1 $(i<l-1)$ <br> 2 $(i=l-1)$ | 0 |
| | $\mathfrak{so}(l, l)$ | 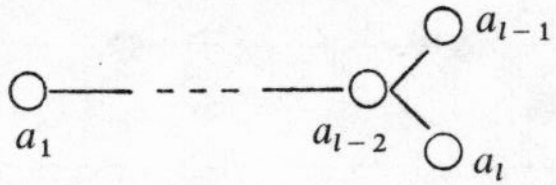  | 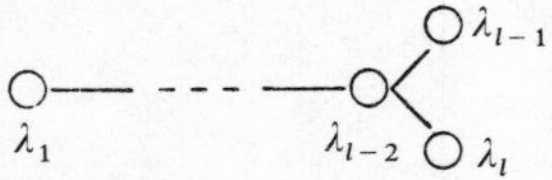  | 1 | 0 |

| | | | | | |
|---|---|---|---|---|---|
| | $\mathfrak{so}(1, 2l-1)$ |  | 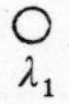 | $2(l-1)$ | 0 |
| DII | $\mathfrak{so}^*(4l')$ | 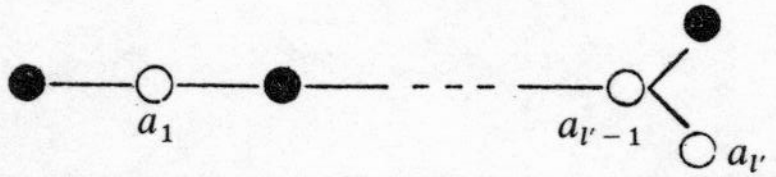 |  $(l=2l', l'\geqq 2)$ | 4 $(i<l')$<br>1 $(i=l')$ | 0<br>0 |
| | $\mathfrak{so}^*(4l'+2)$ | 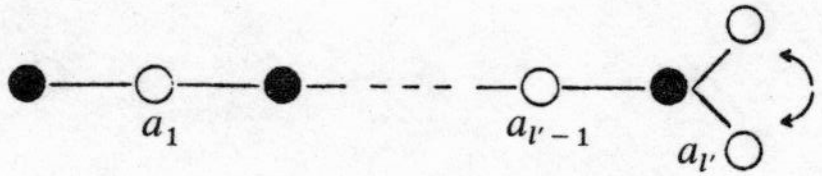 |  $(l=2l'+1, l'\geqq 2)$ | 4 $(i<l')$<br>4 $(i=l')$ | 0<br>1 |
| EI | $\mathfrak{e}_{6,C_4}$ | 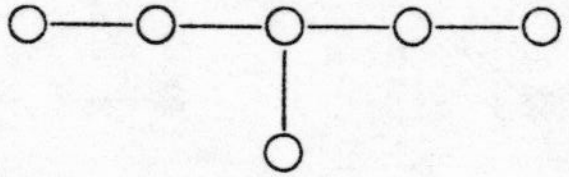 | 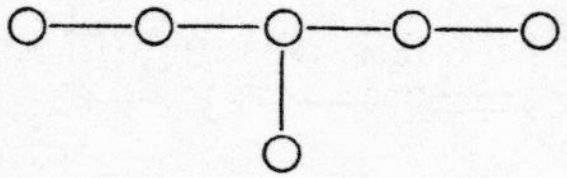 | 1 | 0 |
| EII | $\mathfrak{e}_{6,A_5A_1}$ | 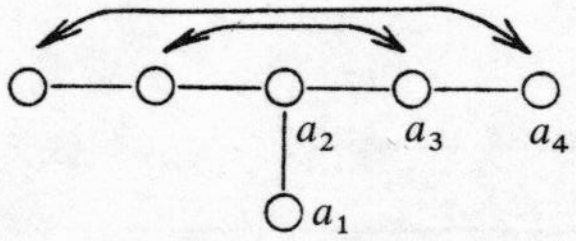 | 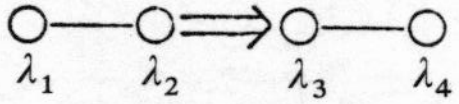 | 1 $(i=1, 2)$<br>2 $(i=3, 4)$ | 0<br>0 |
| EIII | $\mathfrak{e}_{6,D_5T_1}$ | 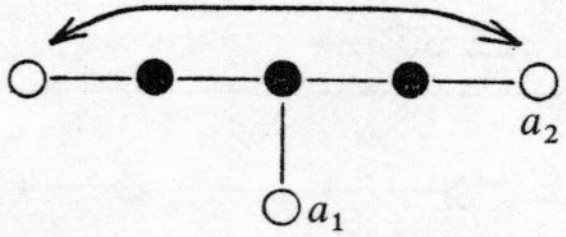 | 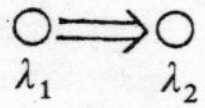 | 6 $(i=1)$<br>8 $(i=2)$ | 0<br>1 |

| | | | | | |
|---|---|---|---|---|---|
| EIV | $\mathfrak{e}_{6,F_4}$ | 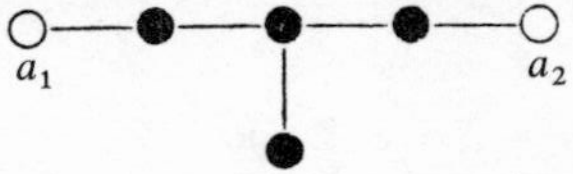 | 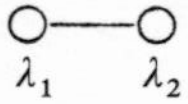 | 8 | 0 |
| EV | $\mathfrak{e}_{7,A_7}$ | 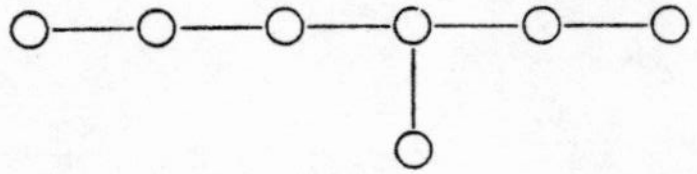 | 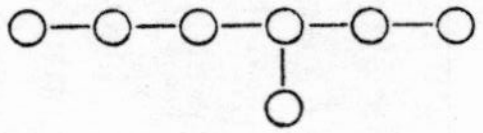 | 1 | 0 |
| EVI | $\mathfrak{e}_{7,D_6A_1}$ | 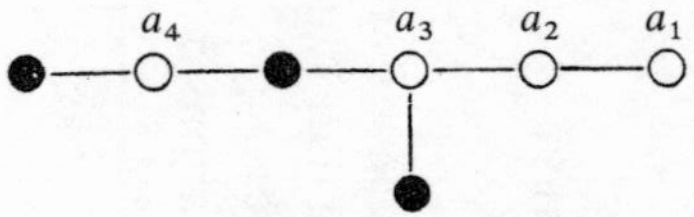 | 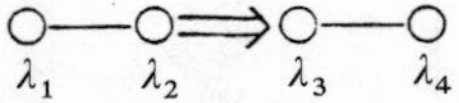 | 1 $(i=1, 2)$<br>4 $(i=3, 4)$ | 0<br>0 |
| EVII | $\mathfrak{e}_{7,E_6T_1}$ | 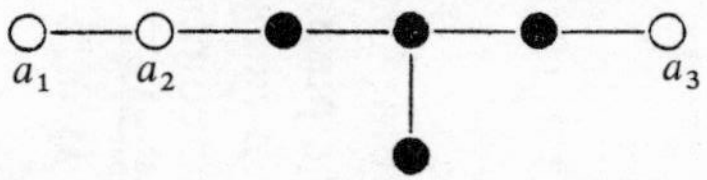 | 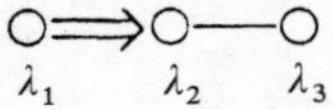 | 1 $(i=1)$<br>8 $(i=2, 3)$ | 0<br>0 |
| EVIII | $\mathfrak{e}_{8,D_8}$ | 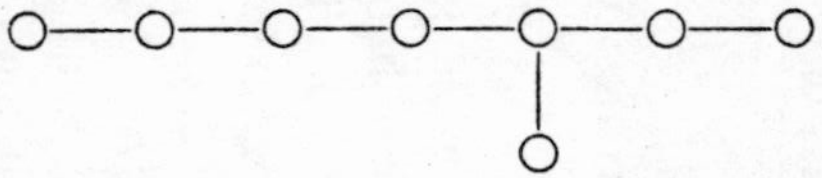 | 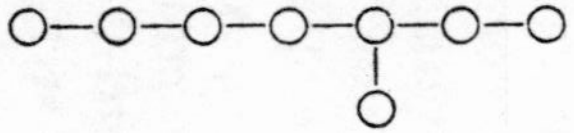 | 1 | 0 |
| EIX | $\mathfrak{e}_{8,E_7A_1}$ |  | 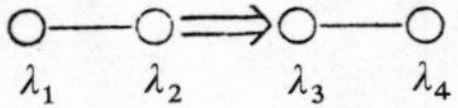 | 1 $(i=1, 2)$<br>8 $(i=3, 4)$ | 0<br>0 |
| FI | $\mathfrak{f}_{4,C_3A_1}$ |  |  | 1 | 0 |
| FII | $\mathfrak{f}_{4,B_4}$ | 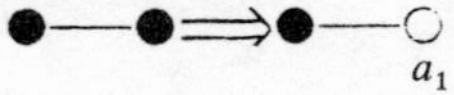 |  | 8 | 7 |
| G | $\mathfrak{g}_{2,A_1A_1}$ |  |  | 1 | 0 |

CHAPTER II

# FINITE-DIMENSIONAL REPRESENTATION THEORY

This chapter is intended to present a short tour of the finite-dimensional representation theory for reductive and semisimple Lie groups. We start by looking at Verma modules and representations with highest weights (Section 9) and specializing (in Section 10) to E. Cartan's highest weight theory for finite-dimensional irreducible representations. Then we recall some results on invariants (Section 11), define the Harish-Chandra homomorphism (Section 12), and prove the Weyl Character Formula (Section 13). In Section 14 we try to fix these ideas by specializing to $\mathfrak{sl}(2)$. We then illustrate to some extent how these results can be applied to homogeneous vector bundles, proving Frobenius' Reciprocity Theorem for compact groups and the Borel-Weil Theorem for compact semisimple groups in Section 15. Finally, in Section 16, we specialize to the decomposition of the $L_2$ space of a compact symmetric space and give Cartan's highest weight theory for class one representations.

## 9. REPRESENTATIONS WITH HIGHEST WEIGHT

Let $\mathfrak{g}$ be a complex reductive Lie algebra. Choose a Cartan subalgebra $\mathfrak{h}$, let $\Delta$ denote the root system, choose a positive root system $\Delta^+$, and let $S=\{\alpha_1,\ldots,\alpha_l\}$ be the simple roots. Then we have

*root lattice* $\Lambda_{\mathrm{rt}}=\mathbb{Z}[S]=$ integral linear combinations of roots,

*weight lattice* $\Lambda_{\mathrm{wt}}=\{\lambda\in\mathfrak{h}^*: 2\langle\lambda,\alpha\rangle/\langle\alpha,\alpha\rangle\in\mathbb{Z}$ for all $\alpha\in S\}$,

*dominant weights* $\Lambda^+_{\mathrm{wt}}=\{\lambda\in\Lambda_{\mathrm{wt}}: 2\langle\lambda,\alpha\rangle/\langle\alpha,\alpha\rangle\geqslant 0$ for all $\alpha\in S\}$,

and, for convenience,

$\Lambda^+_{\mathrm{rt}}=$ non-negative integral linear combinations from $S$.

The latter gives a partial order on $\mathfrak{h}^*$: $\lambda\geqslant\mu$ if $\lambda-\mu\in\Lambda^+_{\mathrm{rt}}$. We will also have use for the notation $\rho=\frac{1}{2}\Sigma_{\Delta^+}\lambda$ and for

$$\mathfrak{n}=\sum_{\lambda\in\Delta^+}\mathfrak{g}_\lambda \qquad\text{and}\qquad \mathfrak{b}=\mathfrak{h}+\mathfrak{n},$$

$$\mathfrak{n}^-=\sum_{\lambda\in\Delta^+}\mathfrak{g}_{-\lambda} \qquad\text{and}\qquad \mathfrak{b}^-=\mathfrak{h}+\mathfrak{n}^-,$$

Finally, we will have use for the *partition function* $P=P_{\mathfrak{g}}$ on $\mathfrak{h}^*$ given by

$$P(\lambda)=\text{number of distinct expressions } \lambda=\sum n_i\alpha_i,$$

where the $n_i$ are integers $\geqslant 0$ and the $\alpha_i$ run over $\Delta^+$. Here note $P(\lambda)>0$ if and only if $\lambda\in\Lambda^+_{\mathrm{rt}}$.

Of course $\Lambda_{\mathrm{rt}}$ is a lattice only in $(\mathfrak{h}\cap[\mathfrak{g},\mathfrak{g}])_{\mathbb{R}}$.

Let $\psi$ represent $\mathfrak{g}$ on a vector space $V$. If $\lambda\in\mathfrak{h}^*$, the corresponding *weight space* $V_\lambda=\{v\in V\colon \psi(\xi)v=\lambda(\xi)v \text{ for } \xi\in\mathfrak{h}\}$. We call $\dim V_\lambda$ the *multiplicity* of the weight $\lambda$ for $\psi$. If $V_\lambda\neq 0$ we say that $\lambda$ is a *weight* of $\psi$ and that any $v\in V_\lambda$ is a *$\lambda$-weight vector*. The algebraic sum $\Sigma_{\lambda\in\mathfrak{h}^*}V_\lambda$ is an invariant subspace because $\psi(\mathfrak{g}_\mu)V_\lambda\subset V_{\lambda+\mu}$. Here we will be concerned with the case where $V=\Sigma_{\lambda\in\mathfrak{h}^*}V_\lambda$; that will be automatic when $\dim V<\infty$. And as in Section 2 it is automatic if $\psi$ is irreducible and has a nonzero weight vector.

Fix $\lambda\in\mathfrak{h}^*$. We define the *Verma module* $M(\lambda)$ to be $\mathscr{U}(\mathfrak{g})\otimes_{\mathscr{U}(\mathfrak{b})}\mathbb{C}$ where $\mathscr{U}(\mathfrak{b})$ acts on $\mathbb{C}$ by

$$\tau_{\lambda-\rho}\colon h+n\mapsto(\lambda-\rho)(h) \qquad \text{for} \qquad h\in\mathfrak{h}, \qquad n\in\mathfrak{n}.$$

In other words, $M(\lambda)=\mathscr{U}(\mathfrak{g})/\{\mathscr{U}(\mathfrak{g})\mathfrak{n}+\Sigma_{\xi\in\mathfrak{h}}\mathscr{U}(\mathfrak{g})(\xi-(\lambda-\rho)(\xi))\}$. Note that $u\mapsto u\otimes 1$ is an $\mathfrak{n}^-$-module isomorphism of $\mathscr{U}(\mathfrak{n}^-)$ onto $M(\lambda)$. If we enumerate

$$\Delta^+=\{\alpha_1,\ldots,\alpha_n\}$$

and choose $0\neq e_\alpha\in\mathfrak{g}_\alpha$ then the Poincaré–Birkhoff–Witt Theorem gives

$$M(\lambda)_\mu=\sum_{\substack{p_i\in\mathbb{Z},\,p_i\geqslant 0,\\ \lambda-\rho-\Sigma p_i\alpha_i=\mu}}(e^{p_1}_{-\alpha_1}\cdots e^{p_n}_{-\alpha_n})\otimes\mathbb{C}$$

In particular, $\dim M(\lambda)_\mu=P(\lambda-\rho-\mu)$ is finite, and is zero unless $\lambda-\rho-\mu\in\Lambda^+_{\mathrm{rt}}$, that is, $\lambda-\rho\geqslant\mu$. In consequence, $M(\lambda)=\Sigma_{\nu\in\Lambda^+_{\mathrm{rt}}}M(\lambda)_{\lambda-\rho-\nu}$, sum of weight spaces; and $M(\lambda)$ has highest weight space $M(\lambda)_{\lambda-\rho}=1\otimes\mathbb{C}$, which is annihilated by $\mathfrak{n}$ and which generates $M(\lambda)$ under $\mathscr{U}(\mathfrak{n}^-)$. Now denote $v_{\lambda-\rho}=1\otimes 1\in M(\lambda)_{\lambda-\rho}$.

THEOREM. *Let $\psi$ represent $\mathfrak{g}$ on a vector space $V$ and suppose there exists $v\in V_{\lambda-\rho}$ such that $\psi(\mathfrak{n})v=0$ and $\psi(\mathscr{U}(\mathfrak{g}))v=V$. Then* (i) *there is a unique $\mathfrak{g}$-module homomorphism $M(\lambda)\to V$ sending $v_{\lambda-\rho}$ to $v$; it is surjective, and is injective if and only if $\psi(u)$ is one-to-one for $0\neq u\in\mathscr{U}(\mathfrak{n}^-)$*, (ii) *$V$ is a sum of weight spaces, each weight is in $\lambda-\rho-\Lambda^+_{\mathrm{rt}}$, each weight space has* *finite dimension, and the highest weight space* $V_{\lambda-\rho}$ *has dimension* 1. *Every* $\mathfrak{g}$*-endomorphism of* $V$ *is scalar, so* $\psi|_{\text{center of } \mathcal{U}(\mathfrak{g})}$ *is an algebra homomorphism* $\mathcal{Z}(\mathfrak{g})\to\mathbb{C}$.

For existence and uniqueness of the $\mathfrak{g}$-map $M(\lambda)\to V$ sending $v_{\lambda-\rho}$ to $v$ we note that relations in $M(\lambda)$ are generated by $\mathfrak{n}\cdot v_{\lambda-\rho}=0$ and $v_{\lambda-\rho}\in M(\lambda)_{\lambda-\rho}$ only. It is surjective because $\mathcal{U}(\mathfrak{n}^-)\cdot v_{\lambda-\rho}=M(\lambda)$ and $\mathcal{U}(\mathfrak{n}^-)\cdot v=V$. The assertions on weights of $V$ follow from the facts on weights of $M(\lambda)$. If $T$ is a $\mathfrak{g}$-endomorphism of $V$ then $\xi\in\mathfrak{h}$ gives $\psi(\xi)Tv= = T\psi(\xi)v=(\lambda-\rho)(\xi)\cdot Tv$ so $Tv\in V_{\lambda-\rho}$, say $Tv=tv$ where $t\in\mathbb{C}$, and now if $u\in\mathcal{U}(\mathfrak{g})$ then $T\psi(u)v=\psi(u)tv=t\psi(u)v$ shows $T=c\cdot 1$. Finally, if $M(\lambda)\to V$ is injective then $\psi(u)$ is injective for $0\neq u\in\mathcal{U}(\mathfrak{n}^-)$ from the corresponding fact on $M(\lambda)$; and if $M(\lambda)\to V$ is not injective then some $u\cdot v_{\lambda-\rho}$, $0\neq u\in\mathcal{U}(\mathfrak{n}^-)$, goes to zero, so $\psi(u)$ is not injective.

Following Harish-Chandra's notation, we write $\chi_\lambda$: $\mathcal{Z}(\mathfrak{g})\mapsto\mathbb{C}$ for the action of the center of $\mathcal{U}(\mathfrak{g})$ on $M(\lambda)$. Since we generally look from the viewpoint of a group representation, we call $\chi_\lambda$ the *infinitesimal character* of a $\mathfrak{g}$-module generated by a highest weight vector of weight $\lambda-\rho$.

Every proper $\mathfrak{g}$-submodule of $M(\lambda)$ is contained in $\Sigma_{\mu\neq\lambda-\rho}M(\lambda)_\mu$, so there is a maximal such submodule. Denote

$$L(\lambda)=M(\lambda)/\text{maximal proper } \mathfrak{g}\text{-submodule}.$$

Then the representation of $\mathfrak{g}$ on $L(\lambda)$ is an irreducible representation with highest weight $\lambda-\rho$.

THEOREM. *Let* $V$ *be an irreducible* $\mathfrak{g}$*-module with highest weight* $\lambda-\rho$; *then* $V\cong L(\lambda)$.

For $V$ is a quotient of $M(\lambda)$.

## 10. Finite-dimensional representations

Suppose that $\mathfrak{g}$ is semisimple and $\psi$ is a finite dimensional representation, say on $V$. Then $V=\Sigma_{\lambda\in\mathfrak{h}^*}V_\lambda$ because $\psi$ is sum of irreducibles, and, in each irreducible, Lie's Theorem applied to $\mathfrak{h}$ gives a nonzero weight vector. For each root $\alpha\in\Delta$ denote

$$\mathfrak{g}[\alpha]=\mathfrak{g}_{-\alpha}+h_\alpha\mathbb{C}+\mathfrak{g}_\alpha\cong\mathfrak{sl}(2;\mathbb{C})$$

and consider $\psi|_{\mathfrak{g}[\alpha]}$ on a weight vector $v_\lambda$. From Section 2,

$$2\langle\lambda, \alpha\rangle/\langle\alpha, \alpha\rangle \text{ is an integer, and is } \geqslant 0 \text{ in case } \psi(\mathfrak{g}_\alpha)v_\lambda = 0.$$

In particular *every weight* $\lambda$ *of* $\psi$ *satisfies* $\lambda \in \Lambda_{\text{wt}}$, *and if* $\lambda$ *is a maximal weight in the sense* $\psi(\mathfrak{n})V_\lambda = 0$ *then* $\lambda \in \Lambda^+_{\text{wt}}$. Further *if* $w \in W$ *and* $\lambda$ *is a weight, then* $w(\lambda)$ *is a weight of the same multiplicity*. For if $\alpha \in \Delta$ and $n = 2\langle\lambda, \alpha\rangle/\langle\alpha, \alpha\rangle$ then, looking at $\mathfrak{g}[\alpha]$, $\psi(e_{-\alpha})^n$ injects $V_\lambda$ into $V_{\lambda - n\alpha}$, so $s_\alpha(\lambda) = \lambda - n\alpha$ is a weight of multiplicity $\geqslant$ (mult of $\lambda$).

Now suppose further that $\psi$ is irreducible. Then $\psi$ *has a highest weight* $\lambda \in \Lambda^+_{\text{wt}}$ because $\dim V < \infty$, $V \cong L(\lambda + \rho)$ in consequence, and so $\lambda$ is unique and has multiplicity 1. If $\mu$ is any weight of $\psi$ then $\mu \leqslant \lambda$, and since $W(\mu)$ meets $\Lambda^+_{\text{wt}}$ one can show $\langle\mu, \mu\rangle \leqslant \langle\lambda, \lambda\rangle$.

These considerations also hold for reductive $\mathfrak{g} = \mathfrak{z} \oplus [\mathfrak{g}, \mathfrak{g}]$ provided that $\psi|_{\mathfrak{z}}$ is semisimple, and that condition is automatic in case $\psi$ is irreducible.

É. Cartan's 'highest weight theory' for finite-dimensional representations:

THEOREM. *If* $\mathfrak{g}$ *is a complex reductive Lie algebra, then the map* $\lambda \mapsto L(\lambda + \rho)$ *is a bijection of* $\Lambda^+_{\text{wt}}$ *onto the set of equivalence classes of irreducible finite-dimensional* $\mathfrak{g}$*-modules.*

To see this, first consider a representation $\psi$ of $\mathfrak{g}$ on $V$ where $V$ is generated by a highest weight vector $v \in V_\lambda$ and, for every simple root $\alpha$, $\psi(e_{-\alpha})^n \cdot v = 0$ for $n$ sufficiently large; we claim that $\psi$ is irreducible and finite dimensional. For if $\dim V > 1$ then $\psi(e_{-\alpha}) \cdot v \neq 0$ for some simple root $\alpha$, so $v$ is in a finite-dimensional nontrivial irreducible $\mathfrak{g}[\alpha]$-submodule $V^0$ of $V$; then $\mathfrak{g} \otimes V^0 \to V$ by $(\xi, w) \mapsto \psi(\xi)w$ is a surjective $\mathfrak{g}[\alpha]$-module map, and $\mathfrak{g} \otimes V^0$ is a finite sum of finite-dimensional $\mathfrak{g}[\alpha]$-modules, so $V$ is a finite sum of finite dimensional $\mathfrak{g}[\alpha]$-modules; now $\dim V < \infty$, and irreducibility comes from $\dim V_\lambda = 1$. Second, let $\lambda \in \Lambda^+_{\text{wt}}$, let $J$ be the maximal proper $\mathfrak{g}$-submodule of $M(\lambda + \rho)$ and let $v_\lambda = 1 \otimes 1 \in M(\lambda + \rho)_\lambda$; for every simple root $\alpha \in S$ let $m_\alpha = 1 + 2\langle\lambda, \alpha\rangle/\langle\alpha, \alpha\rangle$; we claim $J = \Sigma_{\alpha \in S} \mathcal{U}(\mathfrak{g}) e^{m_\alpha}_{-\alpha} \cdot v_\lambda$ and that $J$ has finite codimension in $M(\lambda + \rho)$. For if $\alpha \in S$ then $m_\alpha = 2\langle\lambda + \rho, \alpha\rangle/\langle\alpha, \alpha\rangle$ is a positive integer and one can check that the $\mathfrak{g}$-submodule (say $Y_\alpha$) of $M(\lambda + \rho)$ generated by $e^{m_\alpha}_{-\alpha} \cdot v_\lambda$ is

$$\mathcal{U}(\mathfrak{n}^-) e^{m_\alpha}_{-\alpha} \cdot v_\lambda \subsetneqq M(\lambda + \rho);$$

so $\Sigma_{\alpha \in S} Y_\alpha \subset J$; and our first consideration shows that $M(\lambda + \rho)/\Sigma_{\alpha \in S} Y_\alpha$ is a finite-dimensional simple $\mathfrak{g}$-module. The theorem is proved.

An addendum: Let $w_0 \in W$ be the element that sends $\Delta^+$ to $-\Delta^+$. If $\mathfrak{g}$ is simple then $w_0 = -1$ except when $\mathfrak{g}$ is of type $A_l\,(l>1)$, $D_{2n+1}$ or $E_6$, in which cases $-w_0$ acts on $\Delta^+$ by

$$\overset{\frown}{\underset{\smile}{\circ\!-\!\circ\ \cdots\ \circ\!-\!\circ}} \quad \text{or} \quad \circ\!-\!\cdots\!-\!\circ\!<\!\!\!\overset{\circ}{\underset{\circ}{\updownarrow}} \quad \text{or} \quad \overset{\frown}{\underset{\smile}{\circ\!-\!\circ\!-\!\underset{\displaystyle\circ}{\circ}\!-\!\circ\!-\!\circ}} \quad .$$

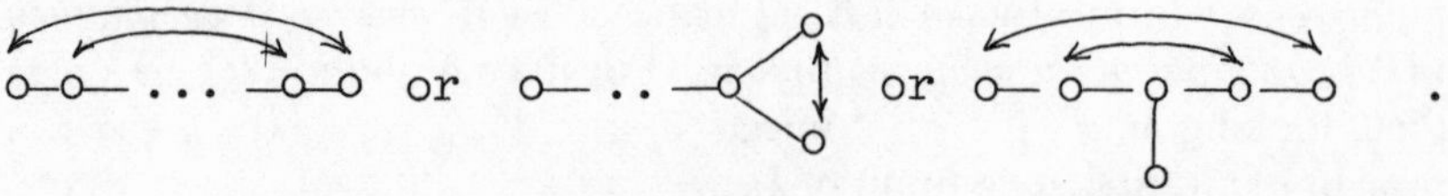

Now suppose that $V$ is a finite-dimensional irreducible $\mathfrak{g}$-module with highest weight $\lambda$. Then $V$ has lowest weight $w_0(\lambda)$, and $V^*$ is the finite-dimensional irreducible $\mathfrak{g}$-module with highest weight $-w_0(\lambda)$.

Let us write $\psi_\lambda$ for the finite-dimensional irreducible representation of highest weight $\lambda \in \Lambda^+_{\text{wt}}$. If $\mathfrak{g}$ is semisimple we can 'describe' $\psi_\lambda$ on the Dynkin diagram as follows. Whenever the non-negative integer $2\langle\lambda, \alpha\rangle / \langle\alpha, \alpha\rangle$ is positive, we write it next to the vertex that corresponds to the simple root $\alpha$. Thus, for the classical groups, the ordinary ('vector') representation is 'described' as

$$\overset{1}{\circ}\!-\!\circ\!-\!\cdots\!-\!\circ, \qquad \overset{1}{\circ}\!-\!\circ\!-\!\cdots\!-\!\circ\!<\!\!\!\overset{\circ}{\underset{\circ}{}},$$

$$\overset{1}{\circ}\!-\!\circ\!-\!\cdots\!-\!\circ\!\!\Rightarrow\!\!\circ \quad \text{or} \quad \overset{1}{\circ}\!-\!\circ\!-\!\cdots\!-\!\circ\!\!\Leftarrow\!\!\circ .$$

The adjoint representations are

$A_1$: $\overset{2}{\circ}$

$A_l, l>1$: $\overset{1}{\circ}\!-\!\circ\!-\!\cdots\!-\!\circ\!-\!\overset{1}{\circ}$

$B_2$: $\circ\!\!\Rightarrow\!\!\overset{2}{\circ}$

$B_l, l>2$: $\circ\!-\!\overset{1}{\circ}\!-\!\cdots\!-\!\circ\!\!\Leftarrow\!\!\circ$

$C_l, l\geqslant 3$: $\overset{2}{\circ}\!-\!\circ\!-\!\cdots\!-\!\circ\!\!\Leftarrow\!\!\circ$

$D_l, l\geqslant 4$: $\circ\!-\!\overset{1}{\circ}\!-\!\cdots\!-\!\circ\!<\!\!\!\overset{\circ}{\underset{\circ}{}}$

$G_2$: $\overset{1}{\circ}\!\!\Rrightarrow\!\!\circ$

$F_4$: $\overset{1}{\circ}\!-\!\circ\!\!\Rightarrow\!\!\circ\!-\!\circ$

$E_6$: $\circ\!-\!\circ\!-\!\underset{\underset{\displaystyle\circ\ 1}{|}}{\circ}\!-\!\circ\!-\!\circ$

$E_7$: $\circ\!-\!\circ\!-\!\underset{\underset{\displaystyle\circ}{|}}{\circ}\!-\!\circ\!-\!\circ\!-\!\circ$ $\quad 1$

$E_8$: $\circ\!-\!\circ\!-\!\underset{\underset{\displaystyle\circ}{|}}{\circ}\!-\!\circ\!-\!\circ\!-\!\circ\!-\!\circ$ $\quad 1$

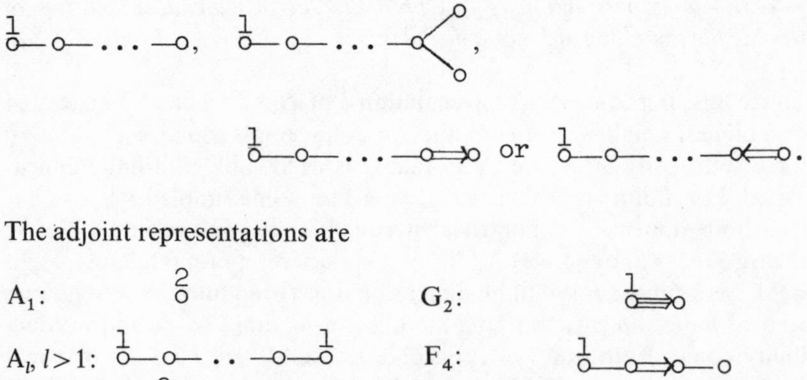

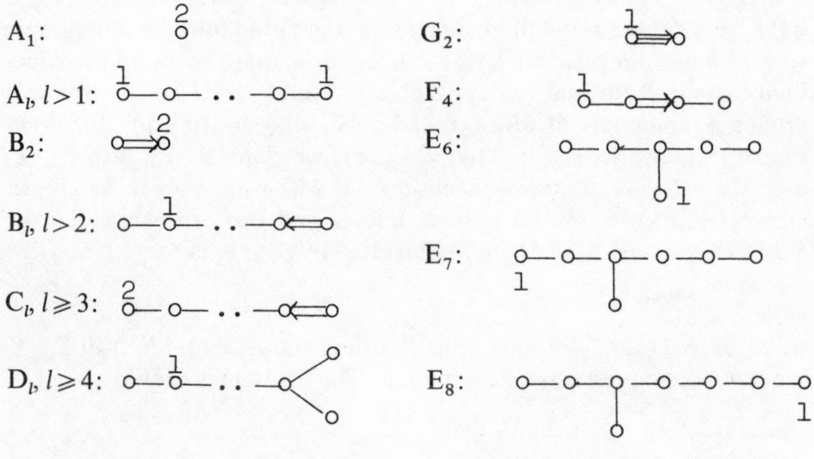

If $\mathfrak{g}$ is simple, then $\psi_1$ is equivalent to its dual $\psi_{-w_0(\lambda)}$ if and only if its Dynkin diagram is stable under

$$\underset{\alpha_1\ \alpha_2\ \alpha_3\quad\ \ \alpha_{l-2}\ \alpha_{l-1}\ \alpha_l}{\circ\!-\!\circ\!-\!\circ\!-\cdots-\!\circ\!-\!\circ\!-\!\circ}\qquad \text{for } A_l\ (l\geqslant 2),$$

$$\text{for } D_{2n+1},$$

$$\text{for } E_6.$$

One also has some handy computational tricks, such as

$$\Lambda^k\left(\underset{\alpha_1\ \alpha_2\qquad\quad \alpha_l}{\overset{1}{\circ}\!-\!\circ\!-\cdots-\!\circ}\right)=\underset{\alpha_1\ \alpha_2\qquad\ \alpha_k\qquad\quad \alpha_l}{\circ\!-\!\circ\!-\cdots-\!\overset{1}{\circ}\!-\cdots-\!\circ}$$

$$\Lambda^k\left(\underset{\alpha_1\qquad\qquad \alpha_l}{\overset{1}{\circ}\!-\!\circ\!-\cdots-\!\circ\!\Leftarrow\!\circ}\right)=\underset{\alpha_1\qquad\ \ \alpha_k\qquad\qquad \alpha_l}{\circ\!-\cdots-\!\overset{1}{\circ}\!-\cdots-\!\circ\!\Leftarrow\!\circ}$$

$$\Lambda^k\left(\underset{\alpha_1\qquad\qquad \alpha_l}{\overset{1}{\circ}\!-\!\circ\!-\cdots-\!\circ\!\Rightarrow\!\circ}\right)=\underset{\alpha_1\qquad\ \ \alpha_k\qquad\qquad \alpha_l}{\circ\!-\cdots-\!\overset{1}{\circ}\!-\cdots-\!\circ\!\Rightarrow\!\circ}$$

$$\Lambda^k\left(\overset{1}{\circ}\!-\!\circ\!-\cdots-\!\circ\!\!<\!\!\begin{smallmatrix}\circ\\\circ\end{smallmatrix}\right)=\underset{\alpha_1\qquad\quad \alpha_k}{\circ\!-\cdots-\!\overset{1}{\circ}\!-\cdots-\!\circ\!\!<\!\!\begin{smallmatrix}\circ\\\circ\end{smallmatrix}}$$

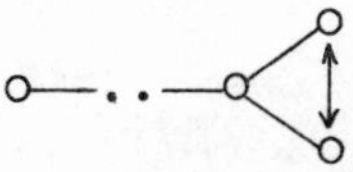

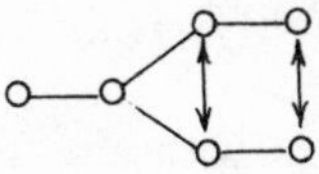

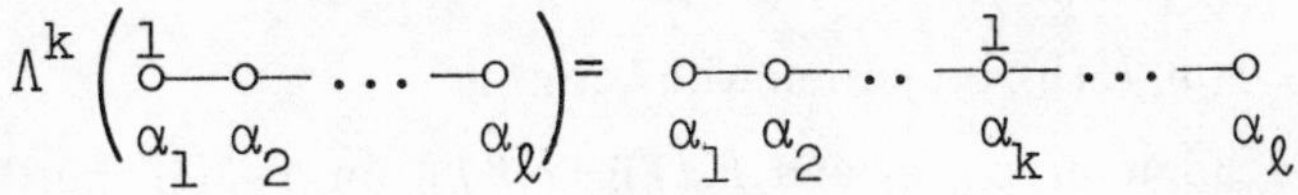

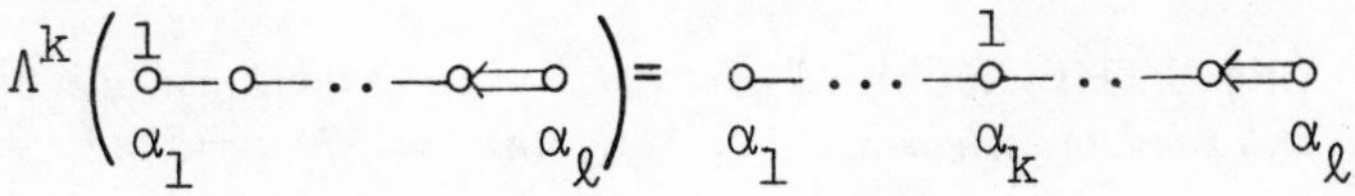

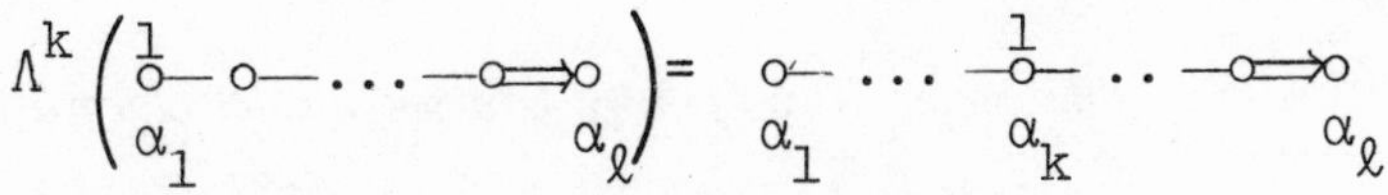

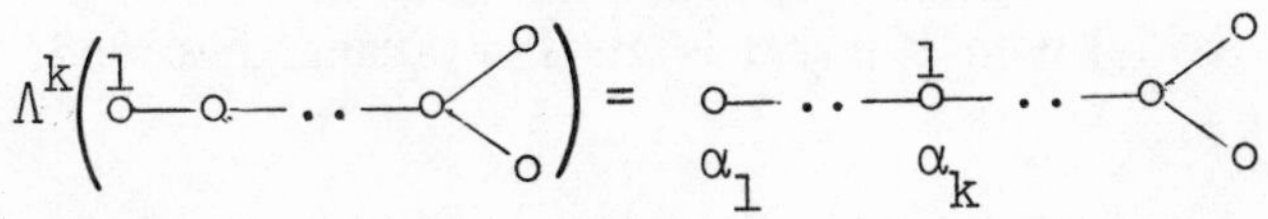

## 11. INVARIANT POLYNOMIALS

Let $f$ be a polynomial function on $\mathfrak{g}$, i.e. an element of the symmetric algebra $S(\mathfrak{g}^*)$. We say that $f$ is *invariant* if it is annihilated by every $\xi \in \mathfrak{g}$, i.e. preserved by every $\gamma \in \mathrm{Int}(\mathfrak{g})$. Denote the space of all such polynomial invariants by $S(\mathfrak{g}^*)^{\mathfrak{g}}$. Similarly one has the space $S(\mathfrak{h}^*)^W$ of $W$-invariant polynomials on $\mathfrak{h}$. Evidently the restriction map

$$i: S(\mathfrak{g}^*) \to S(\mathfrak{h}^*) \qquad \text{by} \qquad i(f) = f|_{\mathfrak{h}}$$

is an algebra homomorphism that maps $S(\mathfrak{g}^*)^{\mathfrak{g}}$ into $S(\mathfrak{h}^*)^W$.

THEOREM. (i) *For each integer* $n \geqslant 0$, *the space* $S^n(\mathfrak{g}^*)^{\mathfrak{g}}$ *of invariants homogeneous of degree n is spanned by the* $\xi \mapsto \mathrm{trace}(\psi(\xi)^n)$ *as* $\psi$ *runs over the semisimple finite-dimensional representations of* $\mathfrak{g}$, (ii) *similarly* $S^n(\mathfrak{h}^*)^W$ *is spanned by the* $\xi \mapsto \mathrm{trace}(\psi)^n)$, (iii) $i$: $S(\mathfrak{g}^*)^{\mathfrak{g}} \to S(\mathfrak{h}^*)^W$ *is an isomorphism.*

The Cartan–Killing form on $[\mathfrak{g}, \mathfrak{g}]$, direct sum with any nondegenerate symmetric bilinear form on the center $\mathfrak{z}$ of $\mathfrak{g}$, defines isomorphisms $a$: $S(\mathfrak{g}) \mapsto S(\mathfrak{g}^*)$ and $b$: $S(\mathfrak{h}) \to S(\mathfrak{h}^*)$, the latter $W$-equivariant. If $J$ denotes the ideal in $S(\mathfrak{g})$ generated by $\mathfrak{n} \cup \mathfrak{n}^-$ then $S(\mathfrak{g}) = S(\mathfrak{h}) \oplus J$. Now let

$$j: S(\mathfrak{g}) \to S(\mathfrak{h}), \text{ projection with kernel } J.$$

Then $j$ is a homomorphism and $i(a(f)) = b(j(f))$ for $f \in S(\mathfrak{g})$. Further, $j|_{S(\mathfrak{g})^{\mathfrak{g}}}$ *is an isomorphism of* $S(\mathfrak{g})^{\mathfrak{g}}$ *onto* $S(\mathfrak{h})^W$.

CHEVALLEY'S THEOREM. *There are* $l = \mathrm{rank}\,\mathfrak{g}$ *homogeneous, algebraically independent, elements* $f_1, \dots, f_l \in S(\mathfrak{g})^{\mathfrak{g}}$ *which generate* $S(\mathfrak{g})^{\mathfrak{g}}$. *Their degrees* $v_1, \dots, v_l$ *are independent (except for a permutation) of choice of the* $f_i$, *and* $\sum v_i = \frac{1}{2}(l + \dim \mathfrak{g})$.

COROLLARY. *The center* $\mathscr{Z}(\mathfrak{g})$ *of* $\mathscr{U}(\mathfrak{g})$ *is a polynomial algebra in* $l$ *indeterminates.*

For the Poincaré–Birkhoff–Witt Theorem gives a vector space isomorphism of $\mathscr{U}(\mathfrak{g})$ onto $S(\mathfrak{g})$ that is $\mathrm{Ad}(G)$-invariant, thus sends $\mathscr{Z}(\mathfrak{g})$ onto $S(\mathfrak{g})^{\mathfrak{g}}$.

## 12. THE HARISH-CHANDRA HOMOMORPHISM

Enumerate $\Delta^+ = \{\alpha_1, \dots, \alpha_n\}$ and choose a basis $\{h_1, \dots, h_l\}$ of $\mathfrak{h}$. Then $\mathscr{U}(\mathfrak{g})$ has vector space basis consisting of the

$$u(\bar{q}, \bar{m}, \bar{p}) = e_{-\alpha_1}^{q_1} \dots e_{-\alpha_n}^{q_n} h_1^{m_1} \dots h_l^{m_l} e_{\alpha_1}^{p_1} \dots e_{\alpha_n}^{p_n},$$

where $\bar{q}$, $\bar{m}$, $\bar{p}$ are multi-indices. If $\xi \in \mathfrak{h}$ then

$$[\xi, u(\bar{q}, \bar{m}, \bar{p})] = \left(\sum_1^n (p_i - q_i)\alpha_i\right)(h)u(\bar{q}, \bar{m}, \bar{p}),$$

and this gives the decomposition

$$\mathscr{U}(\mathfrak{g}) = \sum_{\lambda \in \Lambda \mathrm{rt}} \mathscr{U}(\mathfrak{g})_\lambda \qquad \text{under } \mathfrak{h}.$$

As ad($\xi$) is a derivation, $\mathscr{U}(\mathfrak{g})_\lambda \mathscr{U}(\mathfrak{g})_\mu \subset \mathscr{U}(\mathfrak{g})_{\lambda+\mu}$, and in particular $\mathscr{U}(\mathfrak{g})_0 = \mathscr{U}(\mathfrak{g})^{\mathfrak{h}}$ is a subalgebra of $\mathscr{U}(\mathfrak{g})$. As $\mathscr{U}(\mathfrak{g})_0$ is spanned by the $u(\bar{q}, \bar{m}, \bar{p})$, $\Sigma p_i\alpha_i = \Sigma q_i\alpha_i$,

$$\mathscr{U}(\mathfrak{g})\mathfrak{n} \cap \mathscr{U}(\mathfrak{g})_0 = \mathfrak{n}^- \mathscr{U}(\mathfrak{g}) \cap \mathscr{U}(\mathfrak{g}_0): \qquad \text{call it } L.$$

So *$L$ is a 2-sided ideal in $\mathscr{U}(\mathfrak{g})_0$, and* one checks that $\mathscr{U}(\mathfrak{g})_0 = \mathscr{U}(\mathfrak{h}) \oplus L$. Now consider the projection

$$\phi: \mathscr{U}(\mathfrak{g})_0 \to \mathscr{U}(\mathfrak{h}), \textit{ Harish-Chandra homomorphism.}$$

THEOREM. *Let $V$ be a $\mathfrak{g}$-module generated by a highest weight vector $v \in V_\lambda$, and let $\chi$: $\mathscr{Z}(\mathfrak{g}) \to \mathbb{C}$ its infinitesimal character. If $z \in \mathscr{Z}(\mathfrak{g})$ then $\chi(z) = \phi(z)(\lambda)$.*

For $z = \phi(z) + \Sigma \mu_i \eta_i$ with $\mu_i \in \mathscr{U}(\mathfrak{g})$ and $\eta_i \in \mathfrak{n}$, so

$$\chi(z)v = z \cdot v = \phi(z) \cdot v + \sum \mu_i \eta_i \cdot v = \phi(z) \cdot v = \phi(z)(\lambda)v.$$

Now let $\gamma$ denote the 'shift by $\rho$' automorphism of $S(\mathfrak{h})$, $\gamma(p)(\lambda) = p(\lambda - \rho)$.

THEOREM. *$\gamma \circ \phi|_{\mathscr{Z}(\mathfrak{g})}$ is an isomorphism of $\mathscr{Z}(\mathfrak{g})$ onto $S(\mathfrak{h})^W$, independent of the choice $\Delta^+$ of positive root system.* We call $\gamma \circ \phi|_{\mathscr{Z}(\mathfrak{g})}$ the *Harish-Chandra isomorphism* of $\mathscr{Z}(\mathfrak{g})$ onto $S(\mathfrak{h})^W$.

Toward the end of Section 9 we agreed to write $\chi_\lambda$ for the infinitesimal character of a $\mathfrak{g}$-module generated by a highest weight vector of weight $\lambda - \rho$. Now observe

$$\chi_\lambda(z) = \phi(z)(\lambda - \rho) = (\gamma \circ \phi)(z)(\lambda).$$

Furthermore, one can check (i) *if $\chi$: $\mathscr{Z}(\mathfrak{g}) \to \mathbb{C}$ is any homomorphism then there exists $\lambda \in \mathfrak{h}^*$ such that $\chi = \chi_\lambda$*, (ii) *if $\lambda, \lambda' \in \mathfrak{h}^*$ then $\chi_\lambda = \chi_{\lambda'}$ if and only if $\lambda' \in W(\lambda)$, and* (iii) *if $w_0 \in W$ is the element that sends $\Delta^+$ to $-\Delta^+$, and if $u \to u^T$ is the anti-automorphism of $\mathscr{U}(\mathfrak{g})$ generated by $\mathfrak{g} \ni \xi \mapsto -\xi$, then for all $\lambda \in \mathfrak{h}^*$ and all $z \in \mathscr{Z}(\mathfrak{g})$ one has $\chi_{\lambda+\rho}(z) = \chi_{-w_0(\lambda)+\rho}(z^T)$.*

## 13. Formal characters

Let $\mathbb{Z}\langle \mathfrak{h}^*\rangle$ denote the set of all measures $f$ on $\mathfrak{h}^*$ such that (i) if $\lambda \in \mathfrak{h}^*$ then $f(\lambda) \in \mathbb{Z}$ and (ii) for some $\mu \in \mathfrak{h}^*, f$ has support in $\mu - \Lambda_{\mathrm{rt}}^+$. Then $\mathbb{Z}\langle \mathfrak{h}^*\rangle$ is a ring, with ordinary addition and with convolution product. For convenience with the product, let $e^\lambda$ denote the measure supported in $\{\lambda\}$ with value 1 at $\lambda$; then the multiplication is

$$\left(\sum_{\lambda\in\mathfrak{h}^*} c'_\lambda e^\lambda\right)\left(\sum_{\mu\in\mathfrak{h}^*} c''_\mu e^\mu\right) = \sum_{\nu\in\mathfrak{h}^*}\left(\sum_{\lambda+\mu=\nu} c'_\lambda c''_\mu\right) e^\nu,$$

in other words $e^\lambda e^\mu = e^{\lambda+\mu}$; the $\Sigma_{\lambda+\mu=\nu} c'_\lambda c''_\mu$ are finite sums because of the condition on supports, and if $\operatorname{supp}(f) \subset \alpha - \Lambda_{\mathrm{rt}}^+$ and $\operatorname{supp}(f') \subset \beta - \Lambda_{\mathrm{rt}}^+$ then $\operatorname{supp}(ff') \subset (\alpha+\beta) - \Lambda_{\mathrm{rt}}^+$.

The Weyl group $W$ preserves the set of all finitely supported measures in $\mathbb{Z}\langle \mathfrak{h}^*\rangle$, but does not preserve $\mathbb{Z}\langle \mathfrak{h}^*\rangle$.

We say that a $\mathfrak{g}$-module $V$ *has a formal character* if $V = \Sigma_{\lambda\in\mathfrak{h}^*} V_\lambda$ and each $\dim V_\lambda < \infty$. Then the *formal character* of $V$ is $\mathrm{ch}(V) = \Sigma_{\lambda\in\mathfrak{h}^*} (\dim V_\lambda) e^\lambda$. If $V'$ is submodule then also $V'$ and $V/V'$ have formal characters, and $\mathrm{ch}(V) = \mathrm{ch}(V') + \mathrm{ch}(V/V')$. If $V_1$, $V_2$ are $\mathfrak{g}$-modules that have formal characters, so does $V_1 \otimes V_2$, and $\mathrm{ch}(V_1 \otimes V_2) = \mathrm{ch}(V_1)\mathrm{ch}(V_2)$.

*Example*: $\mathrm{ch}\, M(\lambda) = (\Sigma_{w\in}\varepsilon(w)e^{w\rho})^{-1} e^\lambda \in \mathbb{Z}\langle\mathfrak{h}^*\rangle$, where $\varepsilon(w) = \det_{\mathfrak{h}}(w) = \pm 1$. This is seen as follows. Set

$$d = \left(\sum_{w\in W} \varepsilon(w) e^{w\rho}\right) = e^{-\rho} \prod_{\alpha\in\Delta^+} (e^\alpha - 1)$$

$$= e^\rho \prod_{\alpha\in\Delta^+} (1 - e^{-\alpha}) = \prod_{\alpha\in\Delta^+} (e^{\alpha/2} - e^{-\alpha/2})$$

and

$$k = \sum_{\gamma\in\Lambda_{\mathrm{rt}}^+} P(\gamma) e^{-\gamma} = \prod_{\alpha\in\Delta^+} (1 + e^{-\alpha} + e^{-2\alpha} + \ldots).$$

Then $d$, $k \in \mathbb{Z}\langle \mathfrak{h}^*\rangle$ and $e^{-\rho} d = \Pi_{\alpha\in\Delta^+}(1 - e^{-\alpha})$, so $ke^{-\rho}d = 1$, i.e. $d$ has inverse $e^{-\rho}k$ in $\mathbb{Z}\langle \mathfrak{h}^*\rangle$. But $\mathrm{ch}\, M(\lambda) = e^{\lambda-\rho}\Sigma_{\gamma\in\Lambda_{\mathrm{rt}}^+} P(\gamma) e^{-\gamma} = e^{-\rho} k e^\lambda$.

We now express a fairly general formal character in terms of characters of Verma modules. More precisely, *let $V$ be a $\mathfrak{g}$-module with infinitesimal character $\chi_{\lambda_0}$ and with formal character* $\mathrm{ch}(V) \in \mathbb{Z}\langle\mathfrak{h}^*\rangle$, *let* $D_V = \{\lambda \in W(\lambda_0)$:

$\lambda-\rho+\Lambda_{\mathrm{rt}}^{+}$ *meets* supp ch$(V)\}$; *then* ch$(V)$ *is a* $\mathbb{Z}$*-linear combination of* $\{\mathrm{ch}M(\lambda)\colon \lambda\in D_V\}$. First, if $V\neq 0$ then $D_V\neq 0$. For if $\mu-\rho$ is a maximal element of supp ch$(V)$ and $m=\dim V_{\mu-\rho}$ we have a $\mathfrak{g}$-homomorphism

$$\phi\colon M(\mu)\otimes\mathbb{C}^m\to V$$

that maps $M(\mu)_{\mu-\rho}\otimes\mathbb{C}^m$ isomorphically onto $V_{\mu-\rho}$. So $M(\mu)$ has infinitesimal character $\chi_\mu=\chi_{\lambda_0}$, i.e. $\mu\in W(\lambda_0)$, so $\mu\in D_V$. Second, let $L$ and $N$ be the kernel and cokernel of $\phi$,

$$0\to L\to M(\mu)\otimes\mathbb{C}^m\xrightarrow{\phi} V\to N\to 0.$$

They have infinitesimal character $\chi_{\lambda_0}$ and formal character in $\mathbb{Z}\langle\mathfrak{h}^*\rangle$. Evidently $D_N\subset D_V$; but $\mu\notin D_N$, so the cardinality $|D_N|<|D_V|$. By induction, ch$(N)$ is a $\mathbb{Z}$-linear combination from $\{\mathrm{ch}(\lambda)\colon \lambda\in D_N\}$. Similarly $D_L\subsetneqq D_V$ so ch$(L)$ is a $\mathbb{Z}$-linear combination from $\{\mathrm{ch}(\lambda)\colon \lambda\in D_L\}$. Now the assertion follows because $\mathrm{ch}(V)=-\mathrm{ch}(L)+m\cdot\mathrm{ch}\,M(\mu)+\mathrm{ch}(N)$.

WEYL CHARACTER FORMULA. *Let* $V$ *be a finite-dimensional irreducible* $\mathfrak{g}$*-module with highest weight* $\lambda$. *Then*

$$\mathrm{ch}(V)=\left(\sum_{w\in W}\varepsilon(w)e^{w\rho}\right)^{-1}\left(\sum_{w\in W}\varepsilon(w)e^{w(\lambda+\rho)}\right).$$

For $V\cong L(\lambda+\rho)$, hence has infinitesimal character $\chi_{\lambda+\rho}$, so $d\cdot\mathrm{ch}(V)$ is a $\mathbb{Z}$-linear combination of the $e^{w(\lambda+\rho)}$, $w\in W$. If $w\in W$ then $w(d)=\varepsilon(w)d$ and $w(\mathrm{ch}(V))=\mathrm{ch}(V)$, so now $d\cdot\mathrm{ch}(V)=n\Sigma_{w\in W}\,\varepsilon(w)e^{w(\lambda+\rho)}$ for some integer $n$. But $\dim V_\lambda=1$, so $n=1$.

KOSTANT MULTIPLICITY FORMULA. *Let* $\mu$ *be a weight of a finite-dimensional irreducible* $\mathfrak{g}$*-module of highest weight* $\lambda$. *Then* $\mu$ *has multiplicity* $\Sigma_{w\in W}\varepsilon(w)P(w(\lambda+\rho)-(\mu+\rho))$.

For in our earlier notation,

$$\mathrm{ch}(V)=(ke^{-\rho})(d\cdot\mathrm{ch}(V))=\left(\sum_{\gamma\in\Lambda_{\mathrm{rt}}^{+}}P(\gamma)e^{-\gamma-\rho}\right)\left(\sum_{w\in W}\varepsilon(w)e^{w(\lambda+\rho)}\right),$$

so the coefficient of $e^\mu$ is

$$\sum_{w(\gamma+\rho)-(\lambda+\rho)=\mu}P(\gamma)\varepsilon(w)=\sum_{w\in W}\varepsilon(w)P(w(\lambda+\rho)-(\mu+\rho)).$$

WEYL DEGREE FORMULA. *The finite-dimensional irreducible $\mathfrak{g}$-module of highest weight $\lambda$ has dimension*

$$\dim L(\lambda+\rho)=\prod_{\alpha\in\Delta^+}\frac{\langle\lambda+\rho,\alpha\rangle}{\langle\rho,\alpha\rangle}.$$

For this, consider the function $\deg(\Sigma c_\nu e^\nu)=\Sigma c_\nu$ on the sub-ring $\mathbb{Z}[\mathfrak{h}^*]$ of finitely supported elements in $\mathbb{Z}\langle\mathfrak{h}^*\rangle$. Set $d_\nu=\Sigma_{w\in W}\varepsilon(w)e^{w(\nu)}$, so $\mathrm{ch}(V)=$ $=d_\rho^{-1}\cdot d_{\lambda+\rho}$. Given a root $\alpha$, $e^\lambda\mapsto\langle\lambda,\alpha\rangle e^\lambda$ extends to a derivation $\partial_\alpha$ of $\mathbb{Z}[\mathfrak{h}^*]$. Now apply $\partial=\Pi_{\alpha\in\Delta^+}\partial_\alpha$ to $d_\rho\mathrm{ch}(V)=d_{\lambda+\rho}$ using Liebnitz' Rule for the $\partial_\alpha$:

$$\deg\partial(d_\rho)\cdot\deg\mathrm{ch}(V)=\deg\partial(d_\rho\,\mathrm{ch}(V))=\deg\partial(d_{\lambda+\rho}).$$

As $\deg\partial(e^\nu)=\Pi_{\alpha\in\Delta^+}\langle\nu,\alpha\rangle$ now

$$\deg\partial(d_\nu)=\sum_{w\in W}\varepsilon(w)\prod_{\alpha\in\Delta^+}\langle w\nu,\alpha\rangle=|W|\cdot\prod_{\alpha\in\Delta^+}\langle\nu,\alpha\rangle.$$

So

$$\left\{|W|\cdot\prod_{\alpha\in\Delta^+}\langle\rho,\alpha\rangle\right\}\deg\mathrm{ch}(V)=|W|\cdot\prod_{\alpha\in\Delta^+}\langle\lambda+\rho,\alpha\rangle$$

which gives the formula for the dimension of $V=L(\lambda+\rho)$.

A Weyl character formula for the infinite-dimensional irreducible modules $L(\lambda+\rho)$ would have to rely either on a new idea or on a knowledge of the Jordan–Holder series (it exists) of $M(\lambda+\rho)$. Here one knows (i) every irreducible subquotient of $M(\lambda)$ is isomorphic to $L(\mu)$ for some $\mu\in W(\lambda)\cap(\lambda-\Lambda_{\mathrm{rt}}^+)$, and (ii) a certain root-chain condition says which $\mu$ actually occur. But the multiplicities cause problems.

## 14. Case of $\mathfrak{sl}(2)$ and $\mathfrak{su}(2)$

Let $\mathfrak{g}=\mathfrak{sl}(2;\mathbb{C})$ and $\mathfrak{h}=\left\{\begin{pmatrix}t&0\\0&-t\end{pmatrix}: t\in\mathbb{C}\right\}$. We use the root order $\Delta^+=\{\alpha\}$ where $\alpha\begin{pmatrix}t&0\\0&-t\end{pmatrix}=2t$, so $e_\alpha=\begin{pmatrix}0&1\\0&0\end{pmatrix}$ and $e_{-\alpha}=\begin{pmatrix}0&0\\1&0\end{pmatrix}$. Then $\Lambda_{\mathrm{rt}}=$ $=\{n\alpha: n\in\mathbb{Z}\}$ with $\Lambda_{\mathrm{rt}}^+$ given by $n\geqslant 0$, $\Lambda_{\mathrm{wt}}=\{\frac{1}{2}n\alpha: n\in\mathbb{Z}\}$ with $\Lambda_{\mathrm{wt}}^+$ given by $n\geqslant 0$. The irreducible $\mathfrak{g}$-module of highest weight $\frac{1}{2}(n-1)\alpha$ is the one that has dimension $n$, and its formal character is

$$\mathrm{ch}\, L(\tfrac{1}{2}n) = \frac{e^{n\alpha/2} - e^{-n\alpha/2}}{e^{\alpha/2} - e^{-\alpha/2}}.$$

Let's compare this with the representation on the group level: there, writing $\pi_n$ for the irreducible of degree $n$, highest weight $\frac{1}{2}(n-1)\alpha$

$$\mathrm{trace}\, \pi_n \begin{pmatrix} e^t & 0 \\ 0 & e^{-t} \end{pmatrix} = \sum_{1 \leq j \leq n} e^{(n+1-2j)t} = \frac{e^{nt} - e^{-nt}}{e^t - e^{-t}}$$

The same holds by restriction for $\mathfrak{g}_0 = \mathfrak{sl}(2;\mathbb{R})$ and $SL(2;\mathbb{R})$, and $\mathfrak{g}_u = \mathfrak{su}(2)$ and $SU(2)$.

## 15. Homogeneous vector bundles

Let $G$ be a compact group and $K$ a closed subgroup, and consider the homogeneous space $X = G/K$. The 'left regular' representation of $G$ on $L_2(X)$ comes out of the Peter–Weyl theorem as follows. Write $E(\pi)$ for the space of a representation $\pi \in \hat{G}$, so $L_2(G) = \bigoplus_{\pi \in \hat{G}} E(\pi) \otimes E(\pi^*)$. Then, if $l$ and $r$ denote the left and right regular representations of $G$,

$$\begin{aligned} L_2(X) = L_2(G)^{r(K)} &= \Big\{ \bigoplus_{\pi \in \hat{G}} E(\pi) \otimes E(\pi^*) \Big\}^{r(K)} \\ &= \bigoplus_{\pi \in \hat{G}} E(\pi) \otimes E(\pi^*)^{\pi^*(K)} \\ &= \bigoplus_{\pi \in \hat{G}} m(1_K, \pi|_K) E(\pi) \end{aligned}$$

as unitary $l(G)$-module. This technique is due to Hermann Weyl.

More generally consider a homogeneous vector bundle $\mathscr{V} \to X$, say with typical fibre $V = V(\kappa)$ for some $\kappa \in \hat{K}$. Here $\mathscr{V} \to X$ is associated to the principal $K$-bundle $G \to X$ by the action $\kappa$ of $K$ on $V$. We recall the construction: $\mathscr{V} = G \times_K V$, set of equivalence classes from $G \times V$ under $(gk, v) \sim \sim (g, \kappa(k)v)$. A section $\sigma: X \to \mathscr{V}$ is of the form $gK \mapsto [g, f_\sigma(g)]$ with $f_\sigma: G \to V$ such that $f_\sigma(gk) = \kappa(k)^{-1} f_\sigma(g)$, and every such function $f$ determines a section $\sigma_f$. So now the space of $L_2$ sections of $\mathscr{V} \to X$, as unitary $G$-module, is

$$\begin{aligned} L_2(X; \mathscr{V}) = \{L_2(G) \otimes V\}^{(r \otimes \kappa)(K)} &= \bigoplus_{\pi \in \hat{G}} E(\pi) \otimes \{E(\pi^*) \otimes V\}^{(\pi^* \otimes \kappa)(K)} \\ &= \bigoplus_{\pi \in \hat{G}} m(1_K, \pi^*|_K \otimes \kappa) E(\pi) = \bigoplus_{\pi \in \hat{G}} m(\kappa, \pi) E(\pi). \end{aligned}$$

Glancing back to the definition of induced representation this gives us the following theorem.

FROBENIUS RECIPROCITY THEOREM. *If* $\pi \in \hat{G}$ *and* $\kappa \in \hat{K}$ *then* $m(\pi, \operatorname{Ind}_K^G(K)) = m(\kappa, \pi)$, *whenever G is a compact group and K is a closed subgroup.*

*An application.* Let $G$ be a compact simply connected Lie group and $T$ a Cartan subgroup (maximal torus). Relative to a positive $\mathfrak{t}_{\mathbb{C}}$-root system $\Delta^+$ for $\mathfrak{g}_{\mathbb{C}}$ we have the Borel subalgebra

$$\mathfrak{b} = \mathfrak{t}_{\mathbb{C}} + \sum_{\Delta^+} (\mathfrak{g}_{\mathbb{C}})_{-\alpha} = \mathfrak{t}_{\mathbb{C}} + \mathfrak{n}^-$$

and the Borel subgroup $B = HN^-$, $H = \exp(\mathfrak{t}_{\mathbb{C}})$ and $N^- = \exp(\mathfrak{n}^-)$, in $\mathfrak{g}_{\mathbb{C}}$ and $G_{\mathbb{C}}$. Given $\lambda \in \Lambda_{\mathrm{wt}}^+$ we have well defined holomorphic homomorphism

$$e^\lambda : B \to \mathbb{C} \backslash \{0\} \quad \text{by} \quad e^\lambda(\exp_{G_{\mathbb{C}}}(\xi + \eta)) = e^{\lambda(\xi)} \quad \text{for} \quad \xi \in \mathfrak{t}_{\mathbb{C}}, \quad \eta \in \mathfrak{n}^-.$$

It is a representation of $B$ on $\mathbb{C}$ so we have the associated

$$\mathscr{L}_\lambda \to G_{\mathbb{C}}/B \text{ holomorphic homogeneous line bundle}$$

A section $\sigma: G_{\mathbb{C}}/B \to \mathscr{L}_\lambda$ is holomorphic if and only if it is a continuously differentiable and satisfies

$$\eta(f_\sigma) = 0 \quad \text{for all} \quad \eta \in \mathfrak{n}^-, \text{ where } \eta(f)(x) = (\mathrm{d}/\mathrm{d}t)|_{t=0} f(\exp(t\eta)x).$$

We denote the space of holomorphic sections by, say, $\mathscr{H}^0(\mathscr{L}_\lambda)$.

One more often is interested in $\mathscr{L}_\lambda \to G/T$. Here note $G \cap B = T$, so $G/T \subset G_{\mathbb{C}}/B$, where it is open by dimension and closed because $G$ is compact. This gives the complex structure on $G/T$ (one for each choice of $\Delta^+$) and, from the viewpoint of $G/T$, the complex structure on $\mathscr{L}_\lambda$.

BOREL–WEIL THEOREM. $G_{\mathbb{C}}$ *acts on* $\mathscr{H}^0(\mathscr{L}_\lambda)$ *by the irreducible representation of highest weight* $\lambda$. *In particular*

$$\dim \mathscr{H}^0(\mathscr{L}_\lambda) = \prod_{\alpha \in \Delta^+} \frac{\langle \lambda + \rho, \alpha \rangle}{\langle \rho, \alpha \rangle}.$$

This specializes to various formulae in algebraic geometry, such as the formulae for dimensions of linear systems of divisors on complex Grassmannians.

A short proof of Borel–Weil. Let $V$ be the irreducible $G_{\mathbb{C}}$-module of highest weight $\lambda$, $\pi$ the action of $G_{\mathbb{C}}$ on $V$ and $V^*$ and $\pi^*$ their dual. Denote the pairing by $(v, v^*)$. The lowest weight of $V$ is $-w_0(\lambda)$; choose a weight vector $0 \neq v \in V_{-w_0(\lambda)}$. The lowest weight of $V^*$ is $-\lambda$; choose a weight vector $0 \neq v^* \in V^*_{-\lambda}$. Now define

$$f: G_{\mathbb{C}} \to V \qquad \text{by} \qquad f(g) = (v, \pi^*(g)v^*).$$

If $\eta \in \mathfrak{n}^-$ then $d\pi(\eta)v = 0$ so

$$\begin{aligned} \eta(f)(g) &= (d/dt)|_{t=0}(\pi(\exp(t\eta))^{-1}v, \pi^*(g)v^*) \\ &= -(d\pi(\eta)v, \pi^*(g)v^*) = 0; \end{aligned}$$

so $\eta(f) = 0$. And similarly $f(g \cdot \exp(\eta)) = f(g)$. If $\xi \in \mathfrak{t}_{\mathbb{C}}$ then

$$\begin{aligned} f(g \cdot \exp(\xi)) &= (v, \pi^*(g)\pi^*(\exp(\xi))v^*) \\ &= (v, \pi^*(g)e^{-\lambda(\xi)}v^*) = e^{-\lambda(\xi)}f(g) \end{aligned}$$

because $v^* \in V^*_{-\lambda}$. Now

$$0 \neq f \in \mathscr{H}^0(\mathscr{L}_\lambda).$$

Let

$$W = \text{span}\,\{g(f): g \in G_{\mathbb{C}}\} \subset \mathscr{H}^0(\mathscr{L}_\lambda).$$

Here note that

$$[g(f)](g') = f(g^{-1}g') = (\pi(g)v, \pi^*(g')v^*),$$

so $W$ consists of

$$f_u: g \to (u, \pi^*(g)v^*), \qquad u \in V,$$

with $g(f_u) = f_{\pi(g)u}$. Now apply the Peter–Weyl Theorem: the $L_2$ sections of $\mathscr{L}_\lambda \to G/T$ constitute the space $\bigoplus_{\nu \in \Lambda^+_{wt}} E(\nu) \otimes E(\nu)^*_{-\lambda}$. The condition that an $L_2$ section $\sigma$ be holomorphic: decompose $\eta \in \mathfrak{n}^-$ as $\eta_1 + i\eta_2$ with $\eta_j \in \mathfrak{g}$, then $\eta_1(f_\sigma) + i\eta_2(f_\sigma) = 0$, which just says that the corresponding $E(\nu)^*_{-\lambda}$ is a lowest weight space. Now, if $E(\nu) \otimes E(\nu)^*_{-\lambda}$ contributes to $\mathscr{H}^0(\mathscr{L}_\lambda)$ then $E(\nu)^*$ has lowest weight $-\lambda$, i.e. $\nu = \lambda$. Now $W = \mathscr{H}^0(\mathscr{L}_\lambda) = E(\lambda)$.

## 16. Functions on symmetric spaces

We consider an important case of the decomposition

$$L_2(G/K) = \bigoplus_{\pi \in \hat{G}} m(1_K, \pi|_K)E(\pi)$$

of the first paragraph of Section 15. Write $\pi_\nu$ for the class in $\hat{G}$ with highest weight $\nu\in\Lambda^+_{\mathrm{wt}}$ relative to a positive system $\Delta^+$ of $\mathfrak{h}_{\mathbf{C}}$-roots on $\mathfrak{g}_{\mathbf{C}}$.

THEOREM. *If $G/K$ is a symmetric space, i.e. if $K$ has finite index in the fixed point set $G^\theta$ of an involutive automorphism $\theta$, then $m(1_K, \pi|_K)\leqslant 1$ for every $\pi\in\hat{G}$, so $L_2(G/K)=\bigoplus_{m(1_K,\pi)\neq 0} E(\pi)$.*

*Example*: the 2-sphere $S^2=SU(2)/U(1)$, and $m(1_{U(1)}, \pi_{\text{degree } n})=0$ for $n$ even, $=1$ for $n$ odd. So for each integer $k\geqslant 0$ one has the subspace $E(k\alpha)$ of dimension $2k+1$ in $L_2(S^2)$ generated by the $SU(2)$-translates of the spherical harmonics of degree $k$, and $L_2(S^2)=\bigoplus_{0\leqslant k<\infty} E(k\alpha)$.

Fix a symmetric space $G/K$ say $K$ finite index in $G^\theta$ where $\theta$ is an involutive automorphism, and decompose $\mathfrak{g}=\mathfrak{k}+\mathfrak{s}$ into the $(\pm 1)$-eigenspaces of $\theta$. Extending by complex linearity and restricting, $\theta$ is a Cartan involution of $\mathfrak{g}_0=\mathfrak{k}+\mathfrak{s}_0$, $\mathfrak{s}_0=i\mathfrak{s}$. Let $\mathfrak{a}_0\subset\mathfrak{s}_0$ be a CSA of $(\mathfrak{g}_0, \mathfrak{k})$, $\mathfrak{t}$ a CSA in the $\mathfrak{k}$-centralizer $\mathfrak{m}$ of $\mathfrak{a}_0$, and $\mathfrak{h}=\mathfrak{t}+\mathfrak{a}$ where $\mathfrak{a}=i\mathfrak{a}_0$. Then $\mathfrak{h}$ is a CSA of $\mathfrak{g}$ and we use that choice of $\mathfrak{h}$. We also use a choice of $\Delta^+$ compatible with some system $\Delta^+_{\mathfrak{a}_0}$ of positive $\mathfrak{a}_0$-roots on $\mathfrak{g}_0$.

THEOREM. *Let $\nu\in\Lambda^+_{\mathrm{wt}}$. Then $m(1_{K_0}, \pi_\nu|_{K_0})=1$ if and only if* (i) $\nu|_{\mathfrak{t}}=0$, *so in effect $\nu\in\mathfrak{a}_0^*$, and* (ii) *if $\alpha$ is a simple $\mathfrak{a}_0$-root then $\langle\nu, \alpha\rangle/\langle\alpha, \alpha\rangle$ is an integer $\geqslant 0$.*

This is Cartan's highest weight theory for the *class* 1 *representations* of $G$ relative to $K$, which thus describes $L_2$ of a compact symmetric space.

## CHAPTER III

# INFINITE-DIMENSIONAL REPRESENTATIONS

This third chapter presents a brief and somewhat sketchy introduction to the theory of unitary representations of reductive and semisimple Lie groups $G$. The basic fact for an irreducible unitary representation $\pi$ of $G$ on a Hilbert space $\mathscr{H}$, is that every irreducible representation $\kappa$ of a maximal compact subgroup $K \subset G$ has multiplicity $m(\kappa, \pi|_K) \leqslant \dim \kappa$. This yields up the infinitesimal character $\chi_\pi: \mathscr{Z}(\mathfrak{g}) \to \mathbb{C}$ and the distribution character $\theta_\pi: C_c^\infty(G) \to \mathbb{C}$, and consequently the differential equations

$$z(\theta_\pi) = \chi_\pi(z)\theta_\pi \qquad \text{for} \qquad z \in \mathscr{Z}(\mathfrak{g})$$

which are the starting point for serious harmonic analysis on $G$.

Unfortunately the bound $m(\kappa, \pi|_K) \leqslant \dim \kappa$ has not been proved purely within the context of unitary representations, so we must consider representations of $G$ on a Banach space. The basic facts on Banach representations are in Section 17. Then we look at $C^\infty$ vectors (Section 18), analytic and $K$-finite vectors (Section 19), and finally the $K$-multiplicities (Section 20). We then establish the existence of the distribution character and a few of its properties in Section 21.

The representations that enter into harmonic analysis on a reductive or semisimple group $G$, come in several 'series,' one for each conjugacy class of Cartan subgroup $H$. The series for a compact CSG is fundamental in that it is the basic building block for the other series. More precisely, $H$ defines a certain reductive subgroup $M \subset G$ such that $T = H \cap M$ is a compact CSG in $M$, and the '$H$-series' for $G$ is constructed from the '$T$-series' for $M$ by elementary methods. In Section 22 we state the basic facts on the 'discrete series,' which is the series for a compact Cartan subgroup. Then in Section 23 we define the 'cuspidal parabolic' subgroup $P = MAN$ associated to the Cartan $H = T \times A$, and in Section 24 we describe the $H$-series and state the Plancherel Theorem for $G$.

## 17. BANACH REPRESENTATIONS

Let $G$ be a locally compact group countable at infinity. If $\mathscr{B}$ is a Banach space let $\mathrm{Aut}(\mathscr{B})$ denote the group of all topological (bounded, bounded inverse) automorphisms of $\mathscr{B}$. A homomorphism $\pi$: $G \to \mathrm{Aut}(\mathscr{B})$ is a *Banach representation* of $G$ on $\mathscr{B}$ if it satisfies the following equivalent continuity conditions

(i) $G \times \mathscr{B} \to \mathscr{B}$, by $(g, v) \mapsto \pi(g)v$, is continuous

(ii) if $v \in \mathscr{B}$ then $G \to \mathscr{B}$, by $g \mapsto \pi(g)v$, is continuous

(iii) if $v \in \mathscr{B}$ and $v^* \in \mathscr{B}^*$ then $g \mapsto \langle \pi(g)v, v^* \rangle$ is continuous.

We say that $\pi$ is *topologically irreducible* (TI) if $\mathscr{B}$ has no proper closed $\pi(G)$-invariant subspace. That notion usually is not the right one. If $f \in C_c(G)$ then $\|\pi(x)\|$ is bounded on supp$(f)$ so we have a bounded operator $\pi(f) = \int_G f(x)\pi(x)\, dx$. Suppose, given $T: \mathscr{B} \to \mathscr{B}$ bounded, $n \geqslant 1$, $\{v_1, \ldots, v_n\} \subset \mathscr{B}$ and $\varepsilon > 0$ there exists $f \in C_c(G)$ with $\|(\pi(f) - T)v_i\| < \varepsilon$ for $i = 1, \ldots, n$. Then $\pi$ is *topologically completely irreducible* (TCI). Equivalent formulation: use the algebra $M_c(G)$ of compactly supported Radon measures in place of $C_c(G)$; here $\langle \pi(\mu)v, v^* \rangle = \int_G \langle \pi(x)v, v^* \rangle\, d\mu(x)$ for $v \in \mathscr{B}$, $v^* \in \mathscr{B}^*$.

If $\pi$ is TCI it is TI; for if $0 \neq v \in \mathscr{B}$ then $\pi(C_c(G))v$ is dense in $\mathscr{B}$. The converse holds if $\pi$ is 'finite-dimensionally spanned' (FDS), which means: the linear span of the ranges of $\{\pi(\mu): \mu \in M_c(G)$ and $\pi(\mu)$ has finite rank$\}$ is dense in $\mathscr{B}$.

SCHUR'S LEMMA. *If $\pi$ is TCI, then every bounded operator on $\mathscr{B}$ that commutes with all the $\pi(x)$, $x \in G$, is scalar.*

For this, let $0 \neq v \in \mathscr{B}$ and let $T$ be the operator. If $v$ and $Tv$ are linearly independent we have a net $\{f_\alpha\} \subset C_c(G)$ with $\{\pi(f_\alpha)v\} \to v$ and $\{\pi(f_\alpha)Tv\} \to v$. But $\pi(f_\alpha)T = T\pi(f_\alpha)$ so $v = \lim \pi(f_\alpha)Tv = T \cdot \lim \pi(f_\alpha)v = Tv$, contradiction. Now $Tv = c(v)v$, $c(v) \in \mathbb{C}$, for all $v \in V$. If $u \neq 0 \neq v$ take a net $\{f_\alpha\} \subset C_c(G)$, $\{\pi(f_\alpha)u\} \to v$, and apply $T$ to see $c(u) = c(v)$.

Let $Z$ be the center of $G$. If $\pi$ is TCI now $\pi|_Z$ is a homomorphism $\zeta_\pi: Z \to \mathbb{C} \backslash \{0\}$, called the *central character of* $\pi$.

If $\pi$ is a unitary representation of $G$ on a Hilbert space, then $\pi$ is TI if and only if every bounded operator that commutes with all the $\pi(x)$, $x \in G$, is scalar. This comes right out of the spectral theorem.

THEOREM. *If $\pi$ is unitary and TI then $\pi$ is TCI.*

To see this, let $\mathfrak{A}$ denote the von Neumann algebra of all bounded $T$ on the Hilbert space $\mathscr{H}$ such that if $\{v_1, \ldots, v_n\} \subset \mathscr{H}$ and $\varepsilon > 0$ then, for some $f \in M_c(G)$, each $\|(\pi(f) - T)v_i\| < \varepsilon$. As $\pi(G) \subset \pi(M_c(G)) \subset \mathfrak{A}$, the commutant $\mathfrak{A}' = \mathbb{C}$, so $\mathfrak{A} = (\mathfrak{A}')'$ consists of all bounded operators.

## 18. SMOOTH VECTORS

Now $G$ is a Lie group countable at infinity. Fix a Banach representation $\pi$ of $G$ on $\mathcal{B}$. A vector $v \in \mathcal{B}$ is *differentiable* ($=C^\infty$) if $g \mapsto \pi(g)v$ is a $C^\infty$ map $G \to \mathcal{B}$, that is, if $g \mapsto \langle \pi(g)v, v^* \rangle$ is a $C^\infty$ function on $G$ for every $v^* \in \mathcal{B}^*$. Write $\mathcal{B}_\infty$ for the space of $C^\infty$ vectors in $\mathcal{B}$. If $v \in \mathcal{B}$ and $f \in C_c^\infty(G)$ then $\pi(f)v \in \mathcal{B}_\infty$; so $\pi(C_c^\infty(G))\mathcal{B}$ is a dense subspace of $\mathcal{B}$ contained in $\mathcal{B}_\infty$; in particular $\mathcal{B}_\infty$ is dense in $\mathcal{B}$. The *differentiable representation* $\pi_\infty$ of $G$ associated to $\pi$ is the representation $g \mapsto \pi(g)|_{\mathcal{B}\infty}$, where $\mathcal{B}_\infty$ carries the subspace topology from $\mathcal{B}$.

$\pi_\infty$ lifts to a representation of $\mathcal{U}(\mathfrak{g})$ on $\mathcal{B}_\infty$, by

$$\pi_\infty(\xi)v = (\mathrm{d}/\mathrm{d}t)\pi(\exp(t\xi))v|_{t=0}$$

for $\xi \in \mathfrak{g}$ and $v \in \mathcal{B}_\infty$. Here $\pi(C_c^\infty(G))\mathcal{B}$ is $\pi_\infty(\mathcal{U}(\mathfrak{g}))$-stable; in fact if $D \in \mathcal{U}(\mathfrak{g})$, $f \in C_c^\infty(G)$ and $v \in \mathcal{B}$ then $\pi_\infty(D)\pi(f)v = \pi(Df)v$.

THEOREM. *Suppose that every* $\mathrm{Ad}(g) \in \mathrm{Int}\,(\mathfrak{g}_C)$, *so that every* $\mathrm{Ad}(g)$ *is trivial on the center* $\mathcal{Z}(\mathfrak{g})$ *of* $\mathcal{U}(\mathfrak{g})$. *If* $\pi$ *is TCI then* $\mathcal{Z}(\mathfrak{g})$ *is represented by scalars on* $\mathcal{B}_\infty$, *that is*, $\pi_\infty$ *has an infinitesimal character.*

This is a little bit tricky. One views $\mathcal{U}(\mathfrak{g})$ as the distributions on $G$ supported at 1, so it sits in the convolution algebra $\mathcal{D}_c(G)$ of compactly supported distributions, and $\mathcal{Z}(g)$ sits in the center of $\mathcal{D}_c(G)$. One lifts $\pi_\infty$ to a representation of $\mathcal{D}_c(G)$ on $\mathcal{B}_{(\infty)} = \pi(C_c^\infty(G))\mathcal{B}$ and carries it over to a representation of $\mathcal{D}_c(G)$ on $\mathcal{B}^*_{(\infty)} = \pi^*(C_c^\infty(G))\mathcal{B}^*$, which is weakly dense in $\mathcal{B}^*$. Then one argues as in Schur's Lemma to show that $\pi_\infty(Z)$, $Z$ central in $\mathcal{D}_c(G)$, is a scalar operator $\chi_\pi(Z) \cdot 1$ on $\mathcal{B}_{(\infty)}$, and a weak continuity argument shows $\pi_\infty(Z) = \chi_\pi(Z) \cdot 1$ on $\mathcal{B}_\infty$.

COROLLARY. *Unitary TI representations of connected Lie groups have infinitesimal characters.*

Let $K$ be a compact Lie group, eventually a maximal compact subgroup of $G$. As usual, $\hat{K}$ denotes the set of equivalence classes of finite dimensional irreducible representations of $K$. (*Every TI Banach representation of a compact group is finite dimensional.*) Given $\kappa \in \hat{K}$ we have the *normalized character* $\tau_\kappa(k) = (\dim \kappa)\text{trace } \kappa(k)$. In $L_2(K) = \bigoplus_{\kappa \in \hat{K}} V(\kappa) \otimes V(\kappa)^*$, left or right convolution by $\overline{\tau_\kappa}$ is orthogonal projection to $V(\kappa) \otimes V(\kappa)^*$.

Let $\pi$ be a Banach representation of $K$ on $\mathscr{B}$. If $\kappa \in \hat{K}$ then $\pi(\overline{\tau_\kappa})$ is a continuous projection of $\mathscr{B}$ onto the $K$-isotypic subspace $\mathscr{B}(\kappa)$ of type $\kappa$. In other words, $\mathscr{B}(\kappa)$ consists of all $v \in \mathscr{B}$ such that span$(\pi(K)v)$ is finite dimensional and that $K$ acts on it by a multiple of $\kappa$.

THEOREM. *If $v \in \mathscr{B}_\infty$ then the Fourier series $\Sigma_{\kappa \in \hat{K}}(\overline{\tau_\kappa})v$ converges absolutely ($\Sigma \|\pi(\overline{\tau_\kappa})v\| < \infty$) to $v$.*

The proof runs as follows. Fix a positive definite Ad$(K)$-invariant bilinear form $\langle\, ,\rangle$ on $\mathfrak{k}$, let $\{\xi_1, \ldots, \xi_n\}$ be an orthonormal basis, and consider the 'Casimir' $\Omega = -\Sigma \xi_i^2$. Then $\Omega \in \mathscr{Z}(\mathfrak{k})$, so by Schur's Lemma and skew adjointness of $\kappa(\xi_i)$, if $\kappa \in \hat{K}$ then $\kappa(1+\Omega) = c_\kappa \geqslant 1$ and $(1+\Omega)\tau_\kappa = c_\kappa \tau_\kappa$. If $v \in \mathscr{B}_\infty$ now $\pi(\overline{\tau_\kappa})\pi_\infty(1+\Omega)v = c_\kappa \pi(\overline{\tau_\kappa})v$. Since $K$ is compact we have $a \geqslant 1$ such that $\|\pi(k)\| \leqslant a$ for all $k \in K$, so $\|\pi(\overline{\tau_\kappa})v\| \leqslant (\dim \kappa)^2 a \|v\|$. For each integer $m \geqslant 0$ now $\pi(\overline{\tau_\kappa})v = c_\kappa^{-m} \pi(\overline{\tau_\kappa})\pi_\infty((1+\Omega)^m)v$ gives us

$$\|\pi(\overline{\tau_\kappa})v\| \leqslant c_\kappa^{-m} (\dim \kappa)^2 a \|\pi_\infty((1+\Omega)^m)v\|.$$

To use this, we need

(*) if $m$ is large enough then $\sum_{\kappa \in \hat{K}} (\dim \kappa)^2 c_\kappa^{-m} < \infty$.

Frobenius Reciprocity reduces the proof of (*) to the case where $K$ is connected. Now with $K$ connected, choose a maximal torus $T$ and consider a positive system $\Delta^+$ of $\mathfrak{t}_{\mathbb{C}}$-roots of $\mathfrak{k}_{\mathbb{C}}$. Then $\hat{K}$ is parameterized by $\Lambda_{wt}^+$, intersection of the weight lattice $\Lambda_{wt}$ with a cone in $\sqrt{-1}\mathfrak{t}^*$, say by $\kappa_\lambda \leftrightarrow \lambda$. Here $\dim \kappa_\lambda = \Pi_{\alpha \in \Delta^+} \langle \lambda + \rho, \alpha \rangle / \langle \rho, \alpha \rangle$, which is a polynomial $p(\lambda)$, and $c_{\kappa_\lambda} = 1 + \|\lambda + \rho\|^2 - \|\rho\|^2$. Now, outside of a finite set $F \subset \hat{K}$ one has $\|\lambda + \rho\|^2 - \|\rho\|^2 \geqslant \frac{1}{2}\|\lambda\|^2$, so

$$\sum_{\kappa \in \hat{K} \setminus F} (\dim \kappa)^2 c_\kappa^{-m} \leqslant 2^m \sum_{\lambda \in \Lambda_{wt}^+} (1 + \|\lambda\|^2)^{-m} |p(\lambda)|^2,$$

which is finite for m $\gg 0$. That proves (*). Now

$$\sum_{\kappa \in \hat{K}} \|\pi(\overline{\tau_\kappa})v\| \leqslant \left\{ \sum_{\kappa \in \hat{K}} (\dim \kappa)^2 c_\kappa^{-m} \right\} a \|\pi_\infty((1+\Omega^m)v\|$$

shows absolute convergence of the Fourier series $\Sigma \pi(\overline{\tau_\kappa})v$ for $v \in \mathscr{B}_\infty$. To complete the proof, we must show that

$$v - v_0 = 0 \qquad \text{where } v_0 = \sum_{\kappa \in \hat{K}} \pi(\overline{\tau_\kappa})v.$$

First, every $\pi(\overline{\tau_\kappa})(v - v_0) = 0$. Let $\{f_n\}$ be a $C^\infty$ approximate identity in $L_1(K)$. For each $n$ we have a finite $F_n \subset \hat{K}$ and a function

$$h_n \in \sum_{\kappa \in F_n} V(\kappa) \otimes V(\kappa)^*$$

with $\sup|h_n - f_n| < 2^{-n}$. Now

$$v - v_0 = \lim_{n \to \infty} \pi(f_n)(v - v_0) = \lim_{n \to \infty} \pi(h_n)(v - v_0),$$

but

$$\pi(h_n)(v - v_0) = \pi\left(h_n * \sum_{F_n} \overline{\tau_\kappa}\right)(v - v_0) = \sum_{\kappa \in F_n} \pi(h_n)\pi(\overline{\tau_\kappa})(v - v_0) = 0.$$

Now again, $G$ is a Lie group countable at infinity, $K$ is a compact subgroup, and $\pi$ is a Banach representation of $G$ on $\mathscr{B}$.

THEOREM. *The space* $\Sigma_{\kappa \in \hat{K}} \mathscr{B}_\infty \cap \mathscr{B}(\kappa)$ *is dense in* $\mathscr{B}$.

First, as in the last theorem, if $f \in C^\infty(G)$ (resp. $f \in C_c^\infty(G)$) then the series $\Sigma_{\kappa \in \hat{K}} \overline{\tau_\kappa} * f$ and $\Sigma_{\kappa \in \hat{K}} f * \overline{\tau_\kappa}$ converge absolutely to $f$ in $C^\infty(G)$ (resp. in $C_c^\infty(G)$). Second, let $v \in \mathscr{B}$ and $\varepsilon > 0$, choose $f \in C_c^\infty(G)$ such that $\|\pi(f)v - v\| < \varepsilon$, and choose a compact set $F = KF$ in which $f$ is supported. The $L_1$ norm is a continuous seminorm on $C_c^\infty(G)$. If $A \subset \hat{K}$ is finite set $\overline{\tau_A} = \Sigma_{\kappa \in A} \overline{\tau_\kappa}$, so $f - \overline{\tau_A} * f$ has support in $F$ and $\|\pi(f - \overline{\tau_A} * f)v\| \leqslant c\|f - \overline{\tau_A} * f\|_1$ where $c = \sup_{x \in F}\|\pi(x)v\|$. Choose $A$ so that $\|f - \overline{\tau_A} * f\|_1 < \varepsilon/c$, then

$$\|\pi(\overline{\tau_A} * f)v - v\| \leqslant \|\pi(f - \overline{\tau_A} * f)v\| + \|\pi(f)v - v\| < 2\varepsilon.$$

COROLLARY. $\mathscr{B}_\infty \cap \mathscr{B}(\kappa)$ *is dense in* $\mathscr{B}(\kappa)$.

## 19. Harish-Chandra Module

$G$ is a Lie group countable at infinity, $K$ is a compact subgroup, and $\pi$ is a Banach representation of $G$ on $\mathscr{B}$. A vector $v \in \mathscr{B}$ is *analytic* if $x \mapsto \pi(x)v$ is an analytic map $G \to \mathscr{B}$. Equivalent: if $v^* \in \mathscr{B}$ then $x \mapsto \langle \pi(x)v, v^* \rangle$ is an analytic function on $G$. We write $\mathscr{B}_\omega$ for the space of all analytic vectors in

$\mathscr{B}$. It is $\pi(G)$-stable, so we have the *analytic representation* $\pi_\omega$ of $G$ on $\mathscr{B}_\omega$. Note that $\mathscr{B}_\omega \subset \mathscr{B}_\infty$, and if $u \in \mathscr{U}(\mathfrak{g})$ then $\pi_\infty(u)\mathscr{B}_\omega \subset \mathscr{B}_\omega$. The representation of $\mathscr{U}(\mathfrak{g})$ on $\mathscr{B}_\omega$ is also denoted $\pi_\omega$. The term is justified by a glance at some Taylor series, showing the following proposition.

PROPOSITION. *If $v \in \mathscr{B}_\omega$ then there is a neighborhood $\mathcal{O}$ of 0 in $\mathfrak{g}$ such that $\Sigma_{m=0}^{\infty}(1/m!)\pi_\omega(\xi)^m v$ converges to $\pi(\exp(\xi))v$ for all $\xi \in \mathcal{O}$.*

One studies analytic vectors because of the following corollary.

COROLLARY. *If $\mathscr{B}_0$ is a $\pi_\omega(\mathscr{U}(\mathfrak{g}))$-stable subspace of $\mathscr{B}_\omega$ then its closure is a $\pi(G)$-stable subspace of $\mathscr{B}$.*

We are going to get analogs of the results of Section 18 for analytic vectors, and in the process obtain information needed to define global characters of representations. Of course $\mathscr{B}_\omega$ is only useful because of the following theorem.

NELSON'S THEOREM. *$\mathscr{B}_\omega$ is dense in $\mathscr{B}$.*

If $\kappa \in \hat{K}$ denote $\mathscr{B}_\omega(\kappa) = \mathscr{B}_\omega \cap \mathscr{B}(\kappa)$ and $\mathscr{B}_K = \Sigma_{\kappa \in \hat{K}} \mathscr{B}_\omega(\kappa)$. Then $\mathscr{B}_K = = \mathscr{B}_\omega \cap \Sigma_{\kappa \in \hat{K}} \mathscr{B}(\kappa)$ and we have $\mathscr{B}_K \subset \mathscr{B}_\omega \subset \mathscr{B}_\infty$.

THEOREM. *$\mathscr{B}_K$ is dense in $\mathscr{B}$.*

For let $u \in \mathscr{B}$ and $\varepsilon > 0$. As $\mathscr{B}_\omega$ is dense in $\mathscr{B}$ we have $v \in \mathscr{B}_\omega$ with $\|u - v\| < \varepsilon/2$. As $v \in \mathscr{B}_\infty$, $\Sigma_{\kappa \in \hat{K}} \pi(\overline{\tau_\kappa})v$ converges absolutely to $v$, so some partial sum $w = \Sigma_{\kappa \in F} \pi(\overline{\tau_\kappa})v$, $F \subset \hat{K}$ finite, has $\|v - w\| < \varepsilon/2$. If $f$: $G \to \mathscr{B}$ is analytic and $h$: $K \to \mathbb{C}$ is analytic then $x \mapsto \int_K f(xk)h(k)\,dk$ is an analytic map $G \to \mathscr{B}$; so $w \in \mathscr{B}_\omega$ and $\|u - w\| < \varepsilon$.

COROLLARY. *If $\kappa \in \hat{K}$ and* $\dim \mathscr{B}(\kappa) < \infty$ *then $\mathscr{B}(\kappa) \subset \mathscr{B}_\omega$.*

For $\pi(\overline{\tau_\kappa})\mathscr{B}_\omega$ is a dense subspace of $\mathscr{B}(\kappa)$ contained in $\mathscr{B}_\omega$.

COROLLARY. *$\mathscr{B}_K$ is $\pi_\omega(\mathscr{U}(\mathfrak{g}))$-invariant.*

For if $D \in \mathscr{U}(\mathfrak{g})$ and $k \in K$ then $\pi(k)\pi_\omega(D)v = \pi_\omega(\mathrm{ad}(k)D)\pi(k)v$ for $v \in \mathscr{B}_K$. The $\mathrm{Ad}(k)D$, $k \in K$, lie in a finite-dimensional subspace of $\mathscr{U}(\mathfrak{g})$ and the $\pi(k)v$,

$k \in K$, are in a finite-dimensional subspace of $\mathscr{B}$. So we have a representation $\pi_K$ of $\mathscr{U}(\mathfrak{g})$ on $\mathscr{B}_K$.

THEOREM. *Let G be a connected semisimple Lie group with finite center and K a maximal compact subgroup of G. Let π be a TI Banach representation of G on* $\mathscr{B}$. *Then π has infinitesimal character if and only if it is TCI. In that case, π is K-finite (each* $\mathscr{B}(\kappa)$ *has finite dimension), so each* $\mathscr{B}(\kappa) \subset \mathscr{B}_\omega$ *and* $\mathscr{B}_K = \Sigma_{\kappa \in \hat{K}} \mathscr{B}(\kappa)$.

If $\pi$ is TCI it has infinitesimal character.

Let $\pi$ have infinitesimal character. Choose $0 \neq v \in \mathscr{B}_K$. Then $\mathscr{U}(\mathfrak{g})v = \pi_\omega(\mathscr{U}(g))v$ has $\pi(G)$-stable closure, hence is dense in $\mathscr{B}$. As $v \in \mathscr{B}_K$, $\mathscr{U}(\mathfrak{g})v \subset \mathscr{B}_K$, so $\mathscr{U}(\mathfrak{g})v = \Sigma \mathscr{U}(\mathfrak{g})v \cap \mathscr{B}_\omega(\kappa)$. Let $\mathfrak{J}$ be the annihilator of $v$ in $\mathscr{U}(\mathfrak{k})$. It is a left ideal of finite codimension, $\mathscr{U}(\mathfrak{k})$ acts semisimply on $\mathscr{U}(\mathfrak{k})/\mathfrak{J}$, and a calculation shows that $\overline{\mathscr{U}} = \mathscr{U}(\mathfrak{g})/\mathscr{U}(\mathfrak{g})\mathfrak{J}$ is of the form $\Sigma_{n \in \hat{K}} \overline{\mathscr{U}}(\kappa)$ where each $\overline{\mathscr{U}}(\kappa)$ has finite rank as a $\mathscr{Z}(\mathfrak{g})$-module. Since $\overline{\mathscr{U}} \to \mathscr{U}(\mathfrak{g})v$, by $D + \mathscr{U}(\mathfrak{g})\mathfrak{J} \to \pi_\omega(D)v$, is a $\mathscr{U}(\mathfrak{g})$-module map onto $\mathscr{U}(\mathfrak{g})v$, and since $\pi$ has infinitesimal character, each $\mathscr{U}(\mathfrak{g})v \cap \mathscr{B}_\omega(\kappa)$ has finite dimension. By density of $\mathscr{U}(\mathfrak{g})v$ now each $\dim \mathscr{B}_\omega(\kappa) < \infty$. By density of $\mathscr{B}_K$ now each $\mathscr{B}_\omega(\kappa) = \mathscr{B}(\kappa)$. So $\pi$ is $K$-finite, hence FDS, and thus is TCI.

COROLLARY. *Let G be a connected semisimple Lie group with finite center. If π is an irreducible unitary representation of G and if* $f \in L_1(G)$ *then* $\pi(f)$ *is a completely continuous operator. In other words, G is* CCR *('liminaire'), and so G is a group of type* I *('postliminaire').*

For $\pi$ is TCI, hence $K$-finite, so the $\pi(\overline{\tau_\kappa})\pi(f) = \pi(\overline{\tau_\kappa} * f)$ are operators of finite rank. $\Sigma_{\kappa \in \hat{K}} \overline{\tau_\kappa} * f$ converges $L_1$ to $f$ so $\Sigma_{\kappa \in \hat{K}} \pi(\overline{\tau_\kappa})\pi(f)$ converges strongly to $\pi(f)$. Now $\pi(f)$ is a strong limit of finite rank operators.

Note that this theorem and its corollary hold whenever $G$ is reductive such that (i) every $\mathrm{Ad}(g) \in \mathrm{Int}(\mathfrak{g}_{\mathbb{C}})$ and (ii) $Z$ has a closed normal Abelian subgroup $Z \subset Z_G(G_0)$ with $ZG_0$ of finite index in $G$ and $Z \cap G_0$ co-compact in the center of $G_0$. Then $K = \mathrm{Ad}_G^{-1}$ (maximal compact subgroup of $\mathrm{Ad}(G)$) and $\hat{K}$ consists of finite-dimensional representations only.

A representation $\psi$ of an associative algebra $\mathscr{A}$ on a vector space $V$ is *algebraically irreducible* if $V$ has no proper $\psi(\mathscr{A})$-invariant subspace. It is *algebraically completely irreducible* if, given $n \geqslant 0$ and $\{v_1, \ldots, v_N\}$, $\{u_1, \ldots, u_n\} \subset V$ with $\{u_i\}$ linearly independent, there is an $x \in \mathscr{A}$ with $\psi(x)u_i = v_i$ for $1 \leqslant i \leqslant n$.

THEOREM. *Let $G$ be a connected unimodular Lie group, $K$ a compact subgroup, and $\pi$ a $K$-finite Banach representation of $G$ on $\mathscr{B}$. Then the following are equivalent*: (i) *$\pi$ is topologically irreducible,* (ii) *$\pi$ is TCI,* (iii) *$\pi_K$ (of $\mathscr{U}(\mathfrak{g})$ on $\mathscr{B}_K$) is algebraically irreducible,* (iv) *$\pi_K$ is algebraically completely irreducible.*

Now combine the last two theorems. So *$G$ is a connected semisimple Lie group with finite center, $K$ is a maximal compact subgroup, and $\pi$ is a TCI Banach representation of $G$* on $\mathscr{B}$. Then *$\mathscr{U}(\mathfrak{g})$ acts on $\mathscr{B}_K=\Sigma_{\kappa\in\hat{K}}\mathscr{B}(\kappa)$* by *an algebraically completely irreducible representation $\pi_K$, $K$ acts on $\mathscr{B}_K$ by the finite-multiplicity representation $\pi|_K$, and these are compatible in the sense $\pi|_K(k)\circ\pi_K(D)\circ\pi|_K(k)^{-1}=\pi_K(\mathrm{Ad}_G(k)D)$ and $d(\pi|_K)=\pi_K|_{\mathscr{U}(\mathfrak{k})}$.* The space $\mathscr{B}_K$, with the 'compatible representation' $(\pi_K, \pi|_K)$ of the pair $(\mathscr{U}(\mathfrak{g}), K)$, is the *Harish-Chandra module* associated to $\pi$. These modules will be the subject of Varadarajan's paper.

## 20. The $K$-multiplicities

THEOREM. *Let $G$ be a connected semisimple Lie group with finite center and $K$ a maximal compact subgroup of $G$. Let $\pi$ be a TCI Banach representation of $G$ and $\kappa$ an irreducible representation of $K$. Then the multiplicity $m(\kappa, \pi|_K)\leqslant\dim\kappa$.*

The proof depends on the following algebraic theorem.

THEOREM. *Let $\mathscr{A}$ be an associative algebra and $\mathscr{R}$ a set of representations of $\mathscr{A}$ such that* (i) *$\mathscr{R}$ is complete,* i.e. *if $0\neq x\in\mathscr{A}$ then $\psi(x)\neq 0$ for some $\psi\in\mathscr{R}$, and* (ii) $\dim\psi\leqslant n$ *for every $\psi\in\mathscr{R}$ and some fixed integer $n$. Then every TCI Banach representation of $\mathscr{A}$ has dimension $\leqslant n$.*

To see this, let $r(n)$ be the least integer $r$ such that $\Sigma(\text{sign }\sigma)\xi_{\sigma(1)}\cdots\xi_{\sigma(r)}=0$ for all $\{\xi_1,\ldots,\xi_r\}\subset\mathfrak{gl}(n;\mathbb{C})$, where the sum runs over the permutations of $\{1,\ldots,r\}$. Then $r(n)\leqslant n^2+1$ because $\Lambda^r(\mathfrak{gl}(n;\mathbb{C}))=0$ for $r>n^2=\dim\mathfrak{gl}(n;\mathbb{C})$. A combinatorial argument shows $r(n)\geqslant r(n-1)+2$. Now let $\phi$ be a TCI Banach representation of $\mathscr{A}$ on $\mathscr{B}$ with $\dim\mathscr{B}>n$, and let $\mathscr{B}_0$ be a subspace of dimension $n+1$ in $\mathscr{B}$. As $r(n+1)>r(n)$ we have $T_1,\ldots,T_{r(n)}\in\mathrm{Hom}(\mathscr{B}_0,\mathscr{B}_0)$ such that $[T_1,\ldots,T_{r(n)}]=\Sigma(\text{sign }\sigma)T_{\sigma(1)}T_{\sigma(2)}\cdots T_{\sigma(r(n))}\neq 0$. Extend $T_i$ to a bounded linear operator $\tilde{T}_i$ on $\mathscr{B}$. As $\phi$ is TCI. There is a net $\{x_\alpha\}\subset\mathscr{A}$ with $\{\phi(x_\alpha)\}\to\tilde{T}_1$ in the strong topology. So $\lim\,[\phi(x_\alpha), \tilde{T}_2,\ldots,\tilde{T}_{r(n)}]=$

$=[\tilde{T}_1,\ldots,\tilde{T}_{r(n)}]\neq 0$, and thus for some $y_1\in\mathscr{A}$ we have $[\phi(y_1),\tilde{T}_2,\ldots,\tilde{T}_{r(n)}]\neq$ $\neq 0$. Iterating, we have $\{y_1,\ldots,y_{r(n)}\}\subset\mathscr{A}$ with $[\phi(y_1),\ldots,\phi(y_{r(n)})]=$ $=\phi[y_1,\ldots,y_{r(n)}]\neq 0$. So $[y_1,\ldots,y_{r(n)}]\neq 0$, contradicting $\psi[y_1,\ldots,y_{r(n)}]=0$ for all $\psi\in\mathscr{R}$.

Given $\kappa\in\hat{K}$ set $C_{c,\kappa}(G)=\bar{\tau}_\kappa * C_c(G) * \tau_\kappa$. It is an associative algebra under convolution, and its $\pi$-image preserves $\mathscr{B}(\kappa)$, defining a representation $\pi_\kappa$ of $C_{c,\kappa}(G)$ on $\mathscr{B}(\kappa)$. As $\pi$ is TCI, *each* $\pi_\kappa$ *is* TCI. For if $T$ is a bounded linear operator on $\mathscr{B}(\kappa)$ we extend it to $\tilde{T}=T\cdot\pi(\bar{\tau}_\kappa)$ on $\mathscr{B}$, and if $\{f_\alpha\}\subset C_c(G)$ with $\{\pi(f_\alpha)\}\to\tilde{T}$ then $\{\pi_\kappa(\bar{\tau}_\kappa * f_\alpha * \tau_\kappa)\}\to T$. Now the algebraic theorem just proved, applies to $G$ as follows.

THEOREM. *Suppose that $G$ has a complete* (for $C_c(G)$) *set $\mathscr{R}$ of Banach representations such that, for a fixed $\kappa\in\hat{K}$ and a fixed integer $n$, every $\psi\in\mathscr{R}$ satisfies $m(\kappa,\psi|_K)\leqslant n$. Then, if $\pi$ is a TCI Banach representation of $G$ on $\mathscr{B}$, $m(\kappa,\pi|_K)\leqslant n$.*

For the $\psi_\kappa$, $\psi\in\mathscr{R}$, form a complete set of representations of $C_{c,\kappa}(G)$, each of dimension $\leqslant n\dim\kappa$, so $m(\kappa,\pi|_K)\leqslant n$.

Now the theorem $m(\kappa,\pi|_K)\leqslant\dim\kappa$ is reduced to the finding of a complete set $\mathscr{R}$ of Banach representations of $G$ that satisfy $m(\kappa,\psi|_K)\leqslant\dim\kappa$ for all $\kappa\in\hat{K}$.

This is easy for linear groups; there the finite-dimensional irreducible representations form a complete set $\mathscr{R}$ of Banach representations.

THEOREM. *Every finite-dimensional irreducible representation $\psi$ of $G$ is equivalent to a subrepresentation of some $\pi_\nu=\mathrm{Ind}_{AN}^G(e^\nu)$, $\nu\in(\mathfrak{a}_{\mathbf{C}})^*$.*

Here $G=KAN$ is the Iwasawa decomposition, and $e^\nu$ is a quasi-character on $A$ extended to $AN$ by $e^\nu(an)=e^\nu(a)$. Frobenius Reciprocity gives $m(\kappa,\pi_\nu|_K)=m(\kappa,\mathrm{Ind}_{\{1\}}^K\{1\})=\dim\kappa$, so the theorem forces $m(\kappa,\psi|_K)\leqslant\dim\kappa$. The proof: $\psi$ represents $G$ on $V$ and we apply Lie's Theorem to $\psi^*$ and obtain

$$0\neq v^*\in V^* \quad\text{and}\quad \nu\in(\mathfrak{a}_{\mathbf{C}})^* \quad\text{with}\quad \psi^*(an)v^*=e^{-(\nu+\rho)}(a)v^*.$$

If $v\in V$ then $\phi_v\colon x\mapsto\langle\psi(x)^{-1}v,v^*\rangle$ satisfies $\phi_v(xan)=e^{-(\nu+\rho)}(a)\phi_v(x)$, so $v\mapsto\phi_v$ intertwines $V$ with a subrepresentation of $\pi_\nu$. That subrepresentation is equivalent to $\psi$ because $\psi$ is irreducible and $\phi_v(1_G)\neq 0$ for $\langle v,v^*\rangle\neq 0$.

For non-linear groups it is much more difficult to find the set $\mathscr{R}$. But the general idea is the same. The set $\hat{G}$ of irreducible unitary representations of $G$ is a complete set of Banach representations. If $B=MAN$ is a minimal parabolic subgroup of $G$, then we have representations $\pi_{\mu,\nu}$ of $G$ defined by $\mu\in\hat{M}$ and $\nu\in(\mathfrak{a}_{\mathbb{C}})^*$ as follows. Let $E(\mu)$ be the space of $\mu$, so $\mathrm{Ind}_M^K(\mu)$ represents $K$ on

$$V(\mu)=\{f\in L_2(K)\otimes E(\mu)\colon (r\otimes\mu)(m)f=f \qquad \text{for all} \qquad m\in M\}$$

where $r$ is the right regular representation. Then $\pi_{\mu,\nu}=\mathrm{Ind}_B^G(\mu\otimes e^\nu)$ is a Banach representation of $G$ on $V(\mu)$, and Frobenius Reciprocity gives

$$m(\kappa, \pi_{\mu,\nu}|_K)=m(\kappa, \mathrm{Ind}_M^K(\mu))=m(\mu, \kappa|_M)\leqslant \dim \kappa.$$

So the problem is to put every $\pi\in\hat{G}$ into some $\pi_{\mu,\nu}$. Of course this follows from the following theorem of Harish-Chandra.

SUBQUOTIENT THEOREM. *Every $\pi\in\hat{G}$ is equivalent on the K-finite level* (i.e. *as concerns the representation $\pi_K$* of $\mathcal{U}(\mathfrak{g})$ *on $\mathcal{H}_K$* – *'infinitesimal equivalence'*) *to a subquotient of some $\pi_{\mu,\nu}$.*

Originally that was a deep analytic theorem based on a good knowledge of $\mathcal{U}(\mathfrak{g})$ and its representation $\pi_K$. A few years ago, Lepowski gave an enveloping algebra proof of the subquotient theorem, from which the multiplicity theorem comes out fairly early in the argument. Recently Jacquet* (in a conversation with Casselman) gave a short argument based on asymptotics of matrix coefficients, showing more – that every smooth irreducible $K$-finite ('irreducible admissible') representation of $G$ which has an infinitesimal character is infinitesimally equivalent to a subrepresentation of some $\pi_{\mu,\nu}$.

## 21. THE GLOBAL CHARACTER

THEOREM. *Let G be a connected semisimple Lie group with finite center and $\pi$ a TCI Banach representation of G on a Hilbert space $\mathcal{H}$. If $f\in C_c^\infty(G)$, then $\pi(f)$ is a trace class operator on $\mathcal{H}$. Furthermore $\theta_\pi\colon C_c^\infty(G)\to\mathbb{C}$, by $\theta_\pi(f)=\mathrm{trace}\,\pi(f)$, is a distribution on G, and if $\chi_\pi$ is the infinitesimal character of $\pi$ then $z(\theta_\pi)=\chi_\pi(z)\theta_\pi$ for all $z\in\mathcal{Z}(\mathfrak{g})$. Finally, if $\pi$ and $\pi'$ are K-finite* (*in particular if they are irreducible*) *unitary representations of G, then they are unitarily equivalent if and only if $\theta_\pi=\theta_{\pi'}$.*

---

*I am indebted to W. Schmid for this information.

$\theta_\pi$ is called the *global character* or *distribution character* of $\pi$.

This theorem holds for the larger class of reductive groups mentioned in Section 19.

Note that if $\pi$ is a finite-dimensional representation of $G$, then formally

$$\operatorname{trace} \pi(f) = \int_G f(x) \operatorname{trace} \pi(x) \, dx$$

so the global character $\theta_\pi$ exists and, as distribution, is just integration against the classical character trace $\pi(x)$.

We start with a separable locally compact unimodular group $G$ and a compact subgroup $K$. Let $\pi$ be a Banach representation of $G$ on a Hilbert space $\mathscr{H}$ such that, for some fixed $m_\pi$, each $\dim \mathscr{H}(\kappa) \leqslant m_\pi (\dim \kappa)^2$.

THEOREM. *If $f \in L_2(G)$ and $f$ is compactly supported then $\pi(f)$ is a Hilbert-Schmidt operator on $\mathscr{H}$.*

First, if $T$ is a bounded linear operator on $\mathscr{H}$ with bounded inverse then

$$\|\pi(f)\|_{HS} = \|T^{-1} \cdot \pi(f) \cdot T\|_{HS} \leqslant \|T^{-1}\| \cdot \|\pi(f)\|_{HS} \cdot \|T\|.$$

We use this with a $T$ such that $k \to T \cdot \pi(k) \cdot T^{-1}$ is unitary. So we can assume $\pi|_K$ unitary thus the various $\mathscr{H}(\kappa)$ mutually orthogonal. Second, let $F$ be a compact set in $G$ whose interior contains supp($f$), let $\phi \in C_c^+(G)$ with $\phi = 1$ on $KF$, and choose $\{f_n\}$ continuous on $G$ vanishing outside $F$ with $\{f_n\} \to f$ in $L_2(G)$. Then $\{f_n\} \to f$ in $L_1(G)$ so $\|\pi(f_n) - \pi(f)\| \to 0$. In a moment we will see that $\{\pi(f_n)\}$ is Cauchy HS. Then we will have an HS operator $T$ with $\|\pi(f_n) - T\|_{HS} \to 0$. But $\|\pi(f_n) - T\| \leqslant \|\pi(f_n) - T\|_{HS}$. So then $\pi(f) = T$ is HS. To see $\{\pi(f_n)\}$ Cauchy HS we need an estimate: *if $h \in C_c(G)$ vanishes outside $F$ there exists $N = N(F, \pi) > 0$ such that* $\|\pi(h)\|_{HS} \leqslant N \|h\|_{L_2(G)}$. For that, use $\int_G h(x)\pi(x) \, dx = \int_K (\int_G h(kx)\pi(kx) \, dx) \, dk$ to bound

$$\|\pi(h)\|_{HS} \leqslant \int_G \left\| \int_K h(kx)\pi(kx) \, dk \right\|_{HS} dx$$

$$\leqslant \int_G \left\| \int_K h(kx)\pi(k) \, dk \right\|_{HS} \|\pi(x)\| \, dx.$$

As $\kappa \in \hat{K}$ appears at most $m_\pi \dim \kappa$ times in $\pi|_K$ and appears exactly $\dim \kappa$

times in the left regular representation $L_K$ of $K$, and as $\pi|_K$ is unitary,

$$\left\|\int_K h(kx)\pi(k)\,\mathrm{d}k\right\|_{\mathrm{HS}}^2 \quad m_\pi \left\|\int_K h(kx)L_K(k)\,\mathrm{d}k\right\|_{\mathrm{HS}}^2$$

$$= \quad \text{(Peter–Weyl)} \quad m_\pi \int_K |h(kx)|^2\,\mathrm{d}k.$$

If $M = \sup_{KF}\|\pi(x)\|$ now

$$\|\pi(k)\|_{\mathrm{HS}}\, M m_\pi^{1/2} \int_G \left(\int_K |h(kx)|^2\,\mathrm{d}k\right)^{1/2} \mathrm{d}x$$

$$= M m_\pi^{1/2} \int_G \phi(x)\left(\int_K |h(kx)|^2\,\mathrm{d}k\right)^{1/2} \mathrm{d}x$$

$$\leqslant M m_\pi^{1/2}\|\phi\|_{L_2(G)}\|h\|_{L_2(G)}.$$

That proves the estimate, and the Cauchy sequence assertion follows directly.

Now suppose further that $G$ is a Lie group with $K$ and $\pi$ as above.

THEOREM. *If $f \in C_c^\infty(G)$ then $\pi(f)$ is a trace class operator on $\mathscr{H}$.*

As with the Hilbert-Schmidt assertion we may assume $\pi|_K$ unitary. Let $\mathscr{A}$ be the closure of $\{\pi(f) \colon f \in C_c^\infty(G)\}$ in the Banach space of bounded linear operators on $\mathscr{H}$. If $L_G$ and $R_G$ are the left and right regular representations of $G$ then $\pi(x)\pi(f) = \pi(L_G(x)f)$ and $\pi(f)\pi(x) = \pi(R_G(x)f)$. This defines Banach representations

$$l(x)\colon A \mapsto \pi(x)A \qquad \text{and} \qquad r(x)\colon A \to A\pi(x)^{-1}$$

of $G$ on $\mathscr{A}$. If $f \in C_c^\infty(G)$ then $\pi(f)$ is a $C^\infty$ vector for $l$ and $r$. Now let $\psi = l \otimes r$, representation of $G \times G$ on $\mathscr{A}$ by $\psi(x, y)(A) = \pi(x)A\pi(y)^{-1}$. We may assume $K$ connected so, as in Section 18, there exists $D_0 \in \mathscr{Z}([\mathfrak{k}, \mathfrak{k}])$ with $\kappa(D_0) = (\dim\kappa)^2\kappa(1)$ for all $\kappa \in \hat{K}$. If $f \in C_c^\infty(G)$ then $\pi(f)$ is a $C^\infty$ vector for $\psi$. Now $\psi(L_G(D_0)R_G(D_0)f)$ also is $C^\infty$ for $\psi$, and as in Section 18

$$\sum_{\kappa_1, \kappa_2 \in \hat{K}} \|\psi(\overline{\tau_{\kappa_1 \otimes \kappa_2}})\pi(L_G(D_0)R_G(D_0)f)\| < \infty.$$

But

$$\psi(\overline{\tau_{\kappa_1 \otimes \kappa_2}})\pi(L_G(D_0)R_G(D_0)f) = l(\overline{\tau_{\kappa_1}})r(\overline{\tau_{\kappa_2}})l_\infty(D_0)r_\infty(D_0)\pi(f)$$
$$= l_\infty(D_0)l(\overline{\tau_{\kappa_1}})r_\infty(D_0)r(\overline{\tau_{\kappa_2}})\pi(f)$$
$$= (\dim \kappa_1)^2(\dim \kappa_2)^2 l(\overline{\tau_{\kappa_1}})r(\overline{\tau_{\kappa_2}})\pi(f)$$

so

$$(*) \qquad \sum_{\kappa_1,\kappa_2 \in \hat{K}} (\dim \kappa_1)^2(\dim \kappa_2)^2 \|\pi(\overline{\tau_{\kappa_1}})\pi(f)\pi(\overline{\tau_{\kappa_2}})\| < \infty.$$

Now let $\{v_i\colon i \in b_\kappa\}$ be an orthonormal basis of $\mathscr{H}(\kappa)$, $\kappa \in \hat{K}$. The $\mathscr{H}(\kappa)$ are mutually orthogonal because $\pi|_\kappa$ is unitary, so $\{v_i\colon i \in b = \bigcup b_\kappa\}$ is an orthonormal basis of $\mathscr{H}(\kappa)$. Calculate

$$\sum_{i,j \in b} |\langle \pi(f)v_i, v_j\rangle| = \sum_{\kappa_1,\kappa_2 \in \hat{K}, i \in b\kappa_1, j \in b\kappa_2} |\langle \pi(f)v_i, v_j\rangle|$$
$$\leqslant \sum_{\kappa_1,\kappa_2 \in \hat{K}} (\dim \mathscr{H}(\kappa_1))(\dim \mathscr{H}(\kappa_2)) \|\pi(\overline{\tau_{\kappa_1}})\pi(f)\pi(\overline{\tau_{\kappa_2}})\|$$
$$\leqslant m_\pi^2 \sum_{\kappa_1,\kappa_1 \in \hat{K}} (\dim \kappa_1)^2(\dim \kappa_2)^2 \|\pi(\overline{\tau_{\kappa_1}})\pi(f)\pi(\overline{\tau_{\kappa_2}})\| < \infty.$$

With $G$, $K$ and $\pi$ as above, we now prove the following theorem.

THEOREM. *The map* $C_c^\infty(G) \ni f \to \theta_\pi(f) = \text{trace } \pi(f)$ *is a distribution on* $G$.

As before we may assume $\pi|_K$ unitary. We have $D_0$ as above. We start by showing the existence of $D \in \mathscr{U}(\mathfrak{k})$ such that $|\theta_\pi(f)| \leqslant m_\pi \|\pi(L_G(DD_0)f)\|$ for all $f \in C_c^\infty(G)$. First,

$$|\theta_\pi(f)| \leqslant \sum_{\kappa \in \hat{K}} |\text{trace}(\pi(\overline{\tau_\kappa})\pi(f)\pi(\overline{\tau_\kappa}))|$$
$$\leqslant \sum_{\psi \in \hat{K}} (\dim \mathscr{H}(\kappa)) \|\pi(\overline{\tau_\kappa})\pi(f)\|$$
$$\leqslant m_\pi \sum_{\kappa \in \hat{K}} (\dim \kappa)^2 \|\pi(\overline{\tau_\kappa})\pi(f)\|$$
$$= m_\pi \sum_{\kappa \in \hat{K}} \|\pi(\overline{\tau_\kappa})\pi(L_G(D_0)f)\|.$$

As $\pi(L_G(D_0)f)$ is a $C^\infty$ vector for $l$ we have, as in Section 18, some $D \in \mathscr{U}(\mathfrak{k})$

independent of $f$ with

$$\sum_{\kappa\in\hat{K}} \|\pi(\overline{\tau_\kappa})\pi(L_G(D_0)f)\| = \sum_{\kappa\in\hat{K}} \|l(\overline{\tau_\kappa})\pi(L_G(D_0))f\|$$
$$\leqslant \|l(D)\pi(L_G(D_0)f)\| = \|\pi(L_G(DD_0)f)\|$$

yielding $D$ as required. Now let $F \subset G$ be compact and $\{f_n\} \subset C_c^\infty(G)$ such that each $\mathrm{supp}(f_n) \subset F$ and $L_G(D)f_n \to 0$ uniformly on $F$. Then

$$\|\pi(L_G(D')f_n)\| \leqslant \int_G |L_G(D')f_n(x)|\ \|\pi(x)\|\,\mathrm{d}x \to 0 \quad \text{for all} \quad D' \in \mathscr{U}(\mathfrak{g}).$$

So our estimate with $D$ gives

$$|\theta_\pi(f_n)| \leqslant m_\pi \|\pi(L_G(DD_0)f_n\| \to 0$$

which shows that $\theta_\pi$ is a distribution.

For the theorem on semisimple $G$ it remains only to show that equality of $K$-finite unitary characters $\theta_\pi$, $\theta_{\pi'}$ implies unitary equivalence $\pi \cong \pi'$. That involves a certain amount of machinery comparing Naimark equivalence, infinitesimal equivalence. Banach equivalence and unitary equivalence, and we will not go into it.

Let $\pi$ be a TCI representation of $G$ and $\theta_\pi$ its distribution character. The distribution $\theta_\pi$ is *invariant* (under conjugation by elements of $G$) because

$$\theta_\pi(f \cdot \mathrm{Ad}(x)) = \mathrm{trace} \int_G f(xgx^{-1})\pi(g)\,\mathrm{d}g \qquad \text{(definition)}$$
$$= \mathrm{trace} \int_G f(g)\pi(x^{-1}gx)\,\mathrm{d}g \qquad (G \text{ unimodular})$$
$$= \mathrm{trace}\,(\pi(x^{-1})\pi(f)\pi(x)) = \mathrm{trace}\,\pi(f) = \theta_\pi(f)$$

$\theta_\pi$ is an *eigendistribution* (for the center $\mathscr{Z}(\mathfrak{g})$ of $\mathscr{U}(\mathfrak{g})$) because

$$(z\theta_\pi)(f) = \theta_\pi(zf) = \mathrm{trace}(\pi(zf)) = \mathrm{trace}(\pi_\infty(z) \cdot \pi(f))$$
$$= \mathrm{trace}(\chi_\pi(z)\pi(f)) = \chi_\pi(z)\ \mathrm{trace}\ \pi(f) = \chi_\pi(z)\theta_\pi(f)$$

where $\chi_\pi$ is the infinitesimal character of $\pi$. This system of differential equations

$$z\theta_\pi = \chi_\pi(z)\theta_\pi, \qquad z \in \mathscr{Z}(\mathfrak{g}),$$

together with invariance, has a strong influence on $\theta_\pi$. To describe it, we define the *regular set*

$$G' = \{x \in G \colon \mathfrak{g}^{\mathrm{Ad}(x)} \text{ is a Cartan subalgebra of } \mathfrak{g}\}$$

THEOREM. *Let $\theta$ be an invariant eigendistribution on G. Then $\theta$ is a locally $L_1$ function analytic on $G'$.*

This marvellous result of Harish-Chandra has recently been considerably simplified by Schmid and Atiyah, and is the subject of their paper.

## 22. Discrete series

$G$ is a unimodular locally compact group countable at infinity. We write $\hat{G}$ for the set of unitary equivalence classes of irreducible unitary representations. It has a distinguished subset,

$$\hat{G}_{\mathrm{disc}} = \{[\pi] \in \hat{G} \colon [\pi] \subset L_G, \text{ the left regular representation}\}.$$

$\hat{G}_{\mathrm{disc}}$ is called the *discrete series* of $G$. Roughly speaking, it is the part of $\hat{G}$ that occurs discretely in the decomposition $L_2(G) = \int_{\hat{G}} \mathscr{H}_\pi \otimes \mathscr{H}_\pi^* \, d\pi$ relative to 'Plancherel measure' on $\hat{G}$. The Peter–Weyl Theorem $L_2(K) = \bigoplus_{\kappa \in \hat{K}} V(\kappa) \otimes V(\kappa)^*$ for a compact group $K$, says $\hat{K} = \hat{K}_{\mathrm{disc}}$.

The discrete series plays a fundamental role in harmonic analysis on semisimple and reductive groups. There, the Plancherel measure is concentrated on several 'series' of representations, which are constructed from the discrete series of certain subgroups.

Let $G$ be a reductive Lie group with only finite many components, with compact center, such that every $\mathrm{Ad}(\mathfrak{g}) \in \mathrm{Int}(\mathfrak{g}_{\mathbb{C}})$, and such that the derived group $[G, G]$ has finite center. Then Harish-Chandra has given an explicit description of $\hat{G}$, which we now recall.

First, $\hat{G}_{\mathrm{disc}}$ *is non-empty if and only if G has a compact Cartan subgroup.*

Now suppose that $G$ has a compact Cartan subgroup $T$. One can interpose a maximal compact subgroup, say $T \subset K \subset G$. Choose a positive $\mathfrak{t}_{\mathbb{C}}$-root system $\Delta^+$ for $\mathfrak{g}_{\mathbb{C}}$ and define

$$\rho = \tfrac{1}{2} \sum_{\Delta^+} \alpha, \qquad \tilde{\omega}(\cdot) = \prod_{\Delta^+} \langle \cdot, \alpha \rangle \quad \text{and} \quad \Delta_{G,T} = \prod_{\Delta^+} (e^{\alpha/2} - e^{-\alpha/2})$$

Consider the shifted lattice

$$L = \{\lambda \in i\mathfrak{t}^* \colon e^{\lambda - \rho} \text{ is well defined in } T_0\}$$

which parameterizes $\hat{T}_0$ under $\lambda \leftrightarrow e^{\lambda-\rho}$. It has 'regular set' $L' = \{\lambda \in L: \tilde{\omega}(\lambda) \neq 0\}$. Given $\lambda \in L'$ set

$$q(\lambda) = |\{\alpha \in \Delta^+ : \mathfrak{g}_{\mathbb{C}}^{\alpha} \subset \mathfrak{k}_{\mathbb{C}} \quad \text{and} \quad \langle \lambda, \rangle < 0\}| + \\ + |\{\alpha \in \Delta^+ : \mathfrak{g}_{\mathbb{C}}^{\alpha} \not\subset \mathfrak{k}_{\mathbb{C}} \quad \text{and} \quad \langle \lambda, \alpha \rangle > 0\}|$$

so $(-1)^{q(\lambda)} = (-1)^q \operatorname{sign} \tilde{\omega}(\lambda)$ where $q = \frac{1}{2} \dim G/K$.

THEOREM. *If $\lambda \in L'$ there is a unique class $[\pi_\lambda] \in (G_0)\hat{}_{\text{disc}}$ whose distribution character satisfies*

$$\theta_{\pi\lambda}|_{T_0 \cap G'} = (-1)^{q(\lambda)} \frac{1}{\Delta_{G,T}} \sum_{w \in W(G_0, T_0)} \varepsilon(w) e^{w\lambda}$$

*where $W(G_0, T_0)$ is the Weyl group* (normalizer mod centralizer) *of $T_0$ in $G_0$. Every class in $(G_0)\hat{}_{\text{disc}}$ is one of these $[\pi_\lambda]$. Classes $[\pi_\lambda] = [\pi_{\lambda'}]$ if and only if $\lambda' \in W(G_0, T_0)(\lambda)$. In the appropriate normalization of Haar measure, $[\pi_\lambda]$ has formal degree $|\tilde{\omega}(\lambda)|$. Finally, $[\pi_\lambda]$ has central character $e^{\lambda-\rho}|_{ZG_0}$ and infinitesimal character $\chi_\lambda$.*

Set $G\dagger = Z_G(G_0) \cdot G_0$. Then one can check that

$$(G\dagger)\hat{} = \{[\psi \otimes \pi] : [\psi] \in Z_G(G_0)\hat{}, [\pi] \in \hat{G}_0, \psi|_{ZG_0} \otimes \pi(1) = \psi(1) \otimes \pi|_{ZG_0}\}$$

Further, $[\psi \otimes \pi]$ has the same infinitesimal character $\chi_\pi$ as $[\pi]$ and has $L_1^{\text{loc}}$ distribution character

$$\theta_{\psi\otimes\pi}(zg) = (\operatorname{trace} \psi(z))\theta_\pi(g) \quad \text{for} \quad z \in Z_G(G_0),\ g \in G_0.$$

Finally, $[\psi \otimes \pi] \in (G\dagger)\hat{}_{\text{disc}}$ if and only if $[\pi] \in (G_0)\hat{}_{\text{disc}}$.

Let $\gamma = \operatorname{Ind}_{G\dagger}^G(\psi \otimes \pi)$ where $[\psi \otimes \pi] \in (G\dagger)\hat{}$. Its infinitesimal character $\chi_\gamma = \chi_\pi$, its distribution character $\theta_\gamma \in L_1^{\text{loc}}(G)$ with support in $G\dagger$ and formula

$$\theta_\gamma(zg) = \sum_{G/G\dagger} \operatorname{trace} \psi(x^{-1}zx)\theta_\pi(x^{-1}gx),$$

and $[\gamma] \in \hat{G}_{\text{disc}}$ just when $[\pi] \in (G_0)\hat{}_{\text{disc}}$. Now set

$$\pi_{\psi,\lambda} = \operatorname{Ind}_{G\dagger}^G(\psi \otimes \pi_\lambda) \quad \text{for} \quad \lambda \in L',\ \psi \in Z_G(G_0)\hat{},\quad \psi|_{ZG_0} = e^{\lambda-\rho}.$$

THEOREM. *$\hat{G}_{\text{disc}}$ consists precisely of the classes $[\pi_{\psi,\lambda}]$. Classes $[\pi_{\psi,\lambda}] = [\pi_{\psi',\lambda'}]$ just when $([\psi'], [\lambda']) \in W_{G,T}([\psi], [\lambda])$. The class $[\pi_{\psi,\lambda}]$*

*is the unique one in $\hat{G}$ with distribution character such that*

$$\theta_{\pi_{\psi,\lambda}}(zh) = \sum_{1 \leqslant i \leqslant r} (-1)^{q(w_j\lambda)} \mathrm{tr}\,\psi(x_j^{-1} z x_j) \frac{1}{\Delta_{G,T}} \sum_{w \in W_{G_0}} \det(ww_j) e^{ww_j(\lambda)}(h)$$

*where* $G = \bigcup_{1 \leqslant i \leqslant r} x_j G^\dagger$, $x_j$ *normalizing* $T_0$, *and* $w_j \in W_{G,T}$ *is the element represented by* $x_j$. *Further,* $\chi_{\pi_{\psi,\lambda}} = \chi_\lambda$ *and* $[\pi_{\psi,\lambda}{}^*] = [\pi_{\psi^*,-\lambda}]$, *and in the appropriate normalization of Haar measures* $\deg(\pi_{\psi,\lambda}) = r \cdot \dim(\psi) \cdot |\tilde{\omega}(\lambda)|$.

These results hold for the larger class of reductive groups described in Section 19, as do the various analogs of the Borel–Weil Theorem: As in the compact case, $G/T$ is open in $G_C/B^-$ and has complex structure. If $e^\nu \in \hat{T}_0$ and $[\psi] \in Z_G(G_0)\hat{}$ with $\psi|_{ZG_0} = e^\nu|_{ZG_0}$ then $[\psi \otimes e^\nu] \in (Z_G(G_0) \cdot T_0)\hat{} = \hat{T}$ defines a homogeneous holomorphic line bundle $\mathscr{L}_{\psi,\nu} \to G/T$ with $G$-invariant Hermitian metric. One can form the spaces $\mathscr{H}^{0,q}(G/T; \mathscr{L}_{\psi,\nu})$ of $\mathscr{L}_{\psi,\nu}$-valued square integrable harmonic $(0, q)$-forms on $G/T$.

THEOREM. *If* $\nu + \rho \notin L'$ *then every* $\mathscr{H}^{0,q}(G/T; \mathscr{L}_{\psi,\nu}) = 0$. *If* $\nu + \rho \in L'$ *then* (i) $\mathscr{H}^{0,q}(G/T; \mathscr{L}_{\psi,\nu}) = 0$ *for* $q \neq q(\nu + \rho)$, *and* $G$ *acts irreducibly on* $\mathscr{H}^{0,q(\nu+\rho)}(G/T, \mathscr{L}_{\psi,\nu})$ *by the discrete series representation* $[\pi_{\psi,\nu+\rho}]$.

But this does not 'construct' the discrete series; just as in the compact case one needs the decomposition $L_2(G) = \int_{\hat{G}} \mathscr{H}(\pi) \otimes \mathscr{H}(\pi^*)\, d\pi$ to prove it, and in fact one needs a lot of information about the $\pi_{\psi,\lambda}|_K$.

The case of $SL(2; \mathbb{R})$ goes as follows. $G_{\mathbb{C}}/B^-$ is the Riemann sphere $\mathbb{C} \cup \{\infty\}$ with $G_{\mathbb{C}} = SL(2; \mathbb{C})$ acting by linear fractional transformations

$$\begin{pmatrix} a & b \\ c & d \end{pmatrix} : z \mapsto \frac{az + b}{cz + d}$$

and $B^- = \left\{ \begin{pmatrix} a & 0 \\ c & a^{-1} \end{pmatrix} \right\}$ the subgroup fixing 0. For convenience we replace $SL(2; \mathbb{R})$ by its conjugate

$$G = \phi^{-1} \cdot SL(2; \mathbb{R}) \cdot \phi = \left\{ \begin{pmatrix} a & b \\ \bar{b} & \bar{a} \end{pmatrix} : |a|^2 - |b|^2 = 1 \right\},$$

which is $SU(1, 1)$, where

$$\phi = \sqrt{\frac{i}{2}} \begin{pmatrix} 1 & -1 \\ -i & -i \end{pmatrix}.$$

Then

$$T=K=G\cap B^{-}=\left\{k_\theta=\begin{pmatrix}e^{i\theta} & 0\\ 0 & e^{-i\theta}\end{pmatrix}: \theta \text{ real}\right\}$$

and so

$$G/T\cong G(0)=\{z\in\mathbb{C}: |z|<1\}.$$

The positive $\mathfrak{t}_{\mathbb{C}}$-root $\alpha$ is given by $\alpha\begin{pmatrix}t & 0\\ 0 & -t\end{pmatrix}=2t$, and $\rho=\frac{1}{2}\alpha$, so $L=\{h\alpha\colon 2h\in\mathbb{Z}\}$ with $L'$ given by $h\neq 0$, and $\hat{T}$ consists of the $e^{h\alpha}\colon k_\theta\mapsto e^{2ih\theta}$ with $2h\in\mathbb{Z}$. Let $\mathscr{L}_h\to G/T$ denote the holomorphic Hermitian line bundle corresponding to $e^{h\alpha}\in\hat{T}$. It has a holomorphic trivialization under which the $L_2$ holomorphic sections are identified with the holomorphic functions $f$ on the disc $D=\{z\in\mathbb{C}: |z|<1\}$ such that $\int_D|f(z)|^2(1-|z|^2)^{-2h-2}\,dx\,dy<\infty$. This Hilbert space is nonzero just when $h<-\frac{1}{2}$, and then it has orthonormal basis consisting of the functions

$$z\mapsto\left\{\frac{\pi}{-2h-1}\,\frac{\Gamma(-2h+n)}{\Gamma(-2h)\Gamma(n+1)}\right\}^{1/2} z^n, \qquad n \text{ integer} \geqslant 0.$$

Thus, for $h=-1, -3/2, -2, -5/2,\ldots$ we have $\mathscr{H}^{0,0}(G/T;\mathscr{L}_h)\neq 0$, and $G$ acts on that space by the discrete series representation $\pi_{(-h+1/2)\alpha}$. Those representations form the 'holomorphic discrete series' of $G$. For $h>-\frac{1}{2}$, one can see that $\mathscr{H}^{0,1}(G/T;\mathscr{L}_h)\neq 0$, and obtain the 'antiholomorphic discrete series' representations $\pi_{(-h+1/2)\alpha}=\pi^*_{(h-1/2)\alpha}$. These representation exhaust the discrete series of $G\cong SL(2;\mathbb{R})$.

## 23. Cuspidal parabolic subgroups

$G$ is a reductive Lie group of the class studied in Section 22. We fix a Cartan involution $\theta$ or, equivalently, the maximal compact subgroup $K=G^\theta$. Every Cartan subgroup of $G$ is conjugate to a $\theta$-stable one; let $H$ be a $\theta$-stable CSG in $G$. One checks $H=T\times A$ where $T=H\cap K$ and $A=\exp_G(\mathfrak{a})$, $\mathfrak{a}=\{\xi\in\mathfrak{h}:\ \theta(\xi)=-\xi\}$. Further, the $G$-centralizer of $A$ is $\theta$-stable and decomposes

$$Z_G(A)=M\times A, \qquad M=\theta(M), \qquad T \text{ a CSG in } M.$$

Here $M$ is in general noncompact; it is compact if and only if $\mathfrak{a}$ is a CSA of $(\mathfrak{g}, \mathfrak{k})$ – in general $\mathfrak{a}$ is smaller. But $M$ is a reductive Lie group of the class studied in Section 22, and $\hat{M}_{\text{disc}} \neq \emptyset$ because $T$ is a compact CSG of $M$.

Just as in the case of CSA's of $(\mathfrak{g}, \mathfrak{k})$, we have an $\mathfrak{a}$-root space decomposition $\mathfrak{g} = (\mathfrak{m} + \mathfrak{a}) + \Sigma_{\nu \in \Delta_\mathfrak{a}} \mathfrak{g}_\nu$ where $\mathfrak{g}_\nu = \{\eta \in \mathfrak{g}\colon [\xi, \eta] = \nu(\xi)\eta$ for all $\xi \in \mathfrak{a}\}$ and $\Delta_\mathfrak{a} = \{\nu \in \mathfrak{a}^*\colon \nu \neq 0$ and $\mathfrak{g}_\nu \neq \{0\}\}$. Any choice $\Delta_\mathfrak{a}^+$ of positive $\mathfrak{a}$-root system on $\mathfrak{g}$ comes from some choice $\Delta^+$ of positive $\mathfrak{h}_C$-root system on $\mathfrak{g}_C$ by $\Delta_\mathfrak{a}^+ = \{\alpha|_\mathfrak{a}\colon \alpha \in \Delta^+$ and $\alpha|_\mathfrak{a} \neq 0\}$. Denote

$$\mathfrak{n} = \sum_{\nu \in \Delta_\mathfrak{a}^+} \mathfrak{g}_{-\nu} \qquad \text{and} \qquad N = \exp_G(\mathfrak{n})$$

and

$$\mathfrak{p} = \mathfrak{m} + \mathfrak{a} + \mathfrak{n} \qquad \text{and} \qquad P = MAN.$$

Then $P$ is a 'parabolic subgroup' of $G$, and it is called *cuspidal* because $\hat{M}_{\text{disc}} \neq \emptyset$. Every cuspidal parabolic subgroup of $G$ is conjugate to one of the $P = P(\mathfrak{a}, \Delta_\mathfrak{a}^+)$ just described.

In the case of a minimal parabolic subgroup, i.e. the case where $\mathfrak{a}$ is a CSA of $(\mathfrak{g}, \mathfrak{k})$, the Weyl group $W(G, A) = $(normalizer of $A$)/(centralizer of $A$) is transitive on the set of all positive $\mathfrak{a}$-root systems, so $\mathfrak{h}$, or equivalently $\mathfrak{a}$, determines the conjugacy class of $P$. But that is not the case in general. So we say that two cuspidal parabolic subgroups $P_1, P_2 \subset G$ are *associated* if their maximally split CSA are $G$-conjugate, in other words if they are respectively conjugate to groups $P(\mathfrak{a}, \Delta_\mathfrak{a}^+)$ for the same $\mathfrak{a}$ but possibly different choices of $\Delta_\mathfrak{a}^+$.

A few words on parabolic subgroups in general. Fix a CSA of $(\mathfrak{g}, \mathfrak{k})$, say $\mathfrak{a}_0$, and extend it to a $\theta$-stable CSA in $\mathfrak{g}$, say $\mathfrak{h}_0 = \mathfrak{t}_0 + \mathfrak{a}_0$ where $\mathfrak{t}_0$ is any CSA in the $\mathfrak{k}$-centralizer, $\mathfrak{m}_0$, of $\mathfrak{a}_0$. Now fix a positive system $\Delta_{\mathfrak{a}_0}^+$ of $\mathfrak{a}_0$-roots and let $S_{\mathfrak{a}_0}$ be its simple subsystem. To every subset $T \subset S_{\mathfrak{a}_0}$ we associate an algebra $\mathfrak{p}_T = \mathfrak{m}_T + \mathfrak{a}_T + \mathfrak{n}_T$ as follows. Let $\langle T \rangle = \{\nu \in \Delta_{\mathfrak{a}_0}\colon \nu$ is a linear combination of the roots in $T\}$ and set

$$\mathfrak{a}_T = \{\xi \in \mathfrak{a}\colon \nu(\xi) = 0 \qquad \text{for all } \nu \in T\}$$

$$\mathfrak{m}_T = \{\xi \in \mathfrak{a}\colon \langle \xi, \mathfrak{a}_T \rangle = 0\} + \sum_{\nu \in \langle T \rangle} \mathfrak{g}_\nu$$

$$\mathfrak{n}_T = \sum_{\nu \in \Delta_{\mathfrak{a}_0}^+, \nu \notin \langle T \rangle} \mathfrak{g}_{-\nu}.$$

For $T$ empty we get the minimal parabolic $\mathfrak{b}^-$, and for $T = S_{\mathfrak{a}_0}$ we have $\mathfrak{p}_T = \mathfrak{g}$. Every subalgebra of $\mathfrak{g}$ that contains a minimal parabolic subalgebra is $G_0$-conjugate to exactly one of the $\mathfrak{p}_T$ and is called a *parabolic subalgebra* of $\mathfrak{g}$: the $\mathfrak{p}_T$ are called *standard* (rel $\mathfrak{a}_0$) parabolic subalgebras. On the group level

$$N_T=\exp_G(\mathfrak{n}_T),\quad A_T=\exp_G(\mathfrak{a}_T)\quad\text{and}\quad M_T\times A_T=Z_G(A_T),$$

and $P_T=M_TA_TN_T$ is the corresponding (standard) parabolic subgroup of $G$. Note that $P_T$ is cuspidal just when $\mathfrak{a}_T$ is the $(\theta=-1)$-intersection of some CSA of $\mathfrak{g}$. This can be expressed in terms of $\Delta_{\mathfrak{a}_0}$.

## 24. SERIES OF REPRESENTATIONS AND THE PLANCHEREL FORMULA

Let $P=MAN$, cuspidal parabolic subgroup of $G$ constructed from $H=T\times A$ and $\Delta_{\mathfrak{a}}^+$. If $[\eta]\in\hat{M}$ and $\sigma\in\mathfrak{a}^*$, we extend $[\eta\otimes e^{i\sigma}]\in(MA)^\wedge$ to $P$ by triviality on $N$ and get $\pi_{\eta,\sigma}=\mathrm{Ind}_P^G(\eta\otimes e^{i\sigma})$, unitary representation of $G$. If $\eta$ has infinitesimal character $\chi_\nu$ relative to $\mathfrak{t}$ and distribution character $\Psi_\eta$, then $\pi_{\eta,\sigma}$ has infinitesimal character $\chi_{\nu+i\sigma}$ relative to $\mathfrak{h}$ and has distribution character $\Theta_{\pi_{\eta,\sigma}}$ that is supported in the closure of $\bigcup_{g\in G}\mathrm{Ad}(g)(MA)$ and that is independent of choice of $P$ within its association class. Thus $\pi_{\eta,\sigma}$ depends only on $(H,\eta,\sigma)$.

Now the conjugacy class of the Cartan subgroup $H$ specifies a series of unitary representations of $G$,

$$\{[\pi_{\eta,\sigma}]:\quad \eta\in\hat{M}_{\mathrm{disc}}\quad\text{and}\quad\sigma\in\mathfrak{a}^*\},$$

which we will call the *H-series*. Other people use other names, such as 'principal *P*-series' or 'cuspidal principal series' or, and this is misleading, 'degenerate principal series.' Anyway, at the two extremes the names are standard; for $H$ maximally split (i.e. $\mathfrak{a}$ a CSA of $(\mathfrak{g},\mathfrak{k})$), principal series, and for $H$ compact, discrete series.

Now we can characterize and parameterize the $H$-series as follows. Denote the positive $\mathfrak{t}_C$-root system on $\mathfrak{m}_C$ by $\Delta_{\mathfrak{t}}^+$, so $\Delta_{\mathfrak{t}}^+=\{\alpha|_{\mathfrak{t}}:\ \alpha\in\Delta^+$ and $\alpha|_{\mathfrak{a}}=0\}$. Set

$$\rho_{\mathfrak{t}}=\tfrac{1}{2}\sum_{\Delta^+}\nu,\quad \tilde{\omega}_{\mathfrak{t}}(\cdot)=\prod_{\Delta\mathfrak{t}^+}\langle\cdot,\nu\rangle\quad\text{and}\quad \Delta_{M,T}=\prod_{\Delta\mathfrak{t}^+}(e^{\nu/2}-e^{-\nu/2})$$

and consider

$$L_{\mathfrak{t}}'=\{\nu\in i\mathfrak{t}^*:\ e^{\nu-\rho_{\mathfrak{t}}}\in\hat{T}_0\quad\text{and}\quad\tilde{\omega}_{\mathfrak{t}}(\nu)\neq 0\}.$$

Set $M^\dagger=Z_M(M_0)M_0$ as was done for $G$ in Section 22. Thus $\hat{M}_{\mathrm{disc}}$ consists of the classes of the

$$\eta_{\psi,\nu} = \mathrm{Ind}_{M^\dagger}^{M}(\psi \otimes \eta_\nu) \quad \text{for} \quad \nu \in L''_{\mathfrak{t}} \quad \text{and} \quad \psi \in Z_M(M_0)\hat{}$$

where

$$\eta_\nu \in (M_0)\hat{}_{\mathrm{disc}} \text{ for parameter } \nu \text{ and } \psi|_{ZM_0} = \mathrm{e}^{\nu - \rho\mathfrak{t}}|_{ZM_0}.$$

So the $H$-series $\hat{G}_H$ of $G$ consists of the classes of the

$$\pi_{\psi,\nu,\sigma} = \mathrm{Ind}_P^G(\eta_{\psi,\nu} \otimes \mathrm{e}^{i\sigma}); \qquad \psi \in Z_M(M_0), \qquad \nu \in L''_{\mathfrak{t}}, \qquad \sigma \in \mathfrak{a}^*$$

with the consistency condition between $\psi$ and $\nu$. Specializing the calculation of induced characters to which we alluded, $\pi_{\psi,\nu,\sigma}$ has infinitesimal character $\chi_{\nu+i\sigma}$ and it has distribution character $\theta_{\psi,\nu,\sigma}$ that satisfies

$$\theta_{\psi,\nu,\sigma}(ta) = \left|\frac{\Delta_{M,T}(t)}{\Delta_{G,H}(ta)}\right| \sum_{w(ta)\in NG(H)(ta)} \frac{1}{|N_M(T)(wt)|} \Psi_{\eta\chi,\nu}(wt)\, \mathrm{e}^{i\sigma}(a)$$

where $t \in T$, $a \in A$ and $ta \in G'$. The discrete series character $\Psi_{\eta_{\psi,\nu}}$ is specified in Section 22.

Every $H$-series representation is a finite sum of irreducibles, and if $\sigma$ is regular relative to $(\mathfrak{g}, \mathfrak{a})$ then $\pi_{\psi,\nu,\sigma}$ is irreducible. Also, if $H_1$ and $H_2$ are non-conjugate CSG then every $H_1$-series representation of $G$ is disjoint from every $H_2$-series representation.

Harish-Chandra's Plancherel Formula goes roughly as follows. Choose a complete system $\{H_1, \ldots, H_r\}$ of conjugacy classes of $\theta$-stable CSG in $G$, split $H_j = T_j \times A_j$, and consider cuspidal parabolic subgroups $P_j = M_j A_j N_j$. Set $L''_j = L''_{\mathfrak{t}_j}$. Given $\nu \in L''_j$ the set

$$S(\nu) = \{\psi \in Z_{M_j}((M_j)_0)\hat{}: \psi = \mathrm{e}^{\nu - \rho_{\mathfrak{t}_j}} \text{ on } Z_{(M_j)_0}\}$$

is finite, so one has finite sums

$$\pi_{j,\nu+i\sigma} = \sum_{S(\nu)} (\dim \psi)\pi_{\psi,\nu,\sigma} \quad \text{and} \quad \theta_{j,\nu+i\sigma} = \sum_{S(\nu)} (\dim \psi)\theta_{\psi,\nu,\sigma}.$$

THEOREM. *There are analytic $W(G, H_j)$-invariant functions $m_{j,\nu}$ on $\mathfrak{a}_j^*$, $1 \leqslant j \leqslant r$, with the following property. Let $f \in C_c^\infty(G)$, and for $x \in G$ let $r_x f$ denote right translate $g \mapsto f(gx)$. Then*

$$\sum_{1\leqslant j\leqslant r} \sum_{\nu \in L''_j} |\tilde{\omega}_{\mathfrak{t}_j}(\nu)| \int_{\mathfrak{a}_j^*} |\theta_{j,\nu+i\sigma}(r_x f) m_{j,\nu}(\sigma)|\, \mathrm{d}\sigma < \infty$$

*and*

$$f(x)=\sum_{1\leqslant j\leqslant r}\ \sum_{\nu\in L_j''}|\tilde{\omega}_{t_j}(\nu)|\int_{\mathfrak{a}_j^*}\theta_{j,\nu+i\sigma}(r_x f)m_{j,\nu}(\sigma)\,\mathrm{d}\sigma$$

This is the Plancherel formula for $G$, Fourier inversion for the Fourier transform $f\mapsto\hat{f}$ where $\hat{f}:\hat{G}\to\mathbb{C}$ by $\hat{f}(\pi)=\theta_\pi(f)$. It expresses the Plancherel measure

$$\mathrm{d}\pi_{\psi,\nu,\sigma}=(\dim\psi)|\tilde{\omega}_{t_j}(\nu)|m_{j,\nu}(\sigma)\,\mathrm{d}\sigma$$

as a very smooth measure on the union of the various $H$-series. Peter Trombi's paper will be concerned with the proof and the precise nature of the Plancherel measure.

*University of California, Berkeley*

V. S. VARADARAJAN

# INFINITESIMAL THEORY OF REPRESENTATIONS OF SEMISIMPLE LIE GROUPS[1]

*In Memory of my Mother*

## Contents



## 1. Introduction

**1.1.** For any locally compact topological group $G$ satisfying the second axiom of countability, let $\mathscr{E}(G)$ be the set of all equivalence classes of irreducible unitary representations of $G$. Among the central goals of representation theory have been, first of all, to get a good understanding of the structure of $\mathscr{E}(G)$; and secondly, once this is done, to do harmonic analysis on $G$ by setting up isomorphisms between suitable spaces of functions and 'distributions' on $G$ with the spaces of their 'Fourier transforms' defined on $\mathscr{E}(G)$. In this generality we are very far from any definitive solution to these problems. However, when $G$ is a *reductive group defined over a local field*, great progress has been made.

In this paper we shall restrict ourselves to the case of a reductive *real* Lie group, and study the problem of giving an *infinitesimal description* of the most important irreducible representations of such a group.

[1]Research partially supported by National Science Foundation Grant MCS 76-05962.



*J. A. Wolf, M. Cahen and M. De Wilde (eds.), Harmonic Analysis and Representations of Semisimple Lie Groups*, 131–255.

**1.2.** For simplicity of exposition we shall assume that $G$ is connected, semisimple, and has finite center. $K$ is a maximal compact subgroup of $G$. $\mathfrak{g}_0$ (resp. $\mathfrak{k}_0$) is the Lie algebra of $G$ (resp. $K$), and $\mathfrak{g}$ (resp. $\mathfrak{k}$) is the complexification of $\mathfrak{g}_0$ (resp. $\mathfrak{k}_0$). For any Lie algebra $\mathfrak{a}$ over a field $k$ we denote by $U(\mathfrak{a})$ the universal enveloping algebra of $\mathfrak{a}$. We shall always suppose $\mathfrak{a} \subset U(\mathfrak{a})$; and if $\mathfrak{b} \subset \mathfrak{a}$ is a Lie subalgebra of $\mathfrak{a}$, we shall canonically identify $U(\mathfrak{b})$ with the subalgebra of $U(\mathfrak{a})$ generated by 1 and $\mathfrak{b}$. Elements of $U(\mathfrak{g})$ are interpreted as usual as left invariant analytic differential operators on $G$. We recall that $K$ is connected.

Let $\pi$ be a representation of $G$ in a complete locally convex (Hausdorff) topological vector space $V$, i.e., a homomorphism $\pi(x \mapsto \pi(x))$ of $G$ into $\mathrm{Aut}(V)$ such that

$$(x, v) \mapsto \pi(x, v) = \pi(x)v \qquad (x \in G, v \in V)$$

is a continuous map of $G \times V$ into $V$. Let $V^\infty$ be the space of differentiable vectors in $V$, namely, the set of all $v \in V$ such that $\pi(\cdot : v)$ is a $C^\infty$ map of $G$ into $V$. It is then well known that there is a unique representation $\pi$ of $U(\mathfrak{g})$ in the vector space $V^\infty$ such that, for $v \in V^\infty$, and $X_1, \ldots, X_r \in \mathfrak{g}_0$,

$$\pi(X_1 X_2 \ldots X_r)v = (\partial^r/\partial t_1 \ldots \partial t_r)_0 \pi(\exp t_1 X_1 \ldots \exp t_r X_r : v)$$

where the suffix 0 indicates that the derivative is calculated at $t_1 = \cdots = t_r = 0$. $V^\infty$ is $G$-stable and one has

$$\pi(x)\pi(a)\pi(x)^{-1} = \pi({}^x a), \qquad (x \in G, a \in U(\mathfrak{g}))$$

where ${}^x a = \mathrm{Ad}(x) \cdot a$, Ad being the adjoint representation of $G$ in $U(\mathfrak{g})$.

A vector $v \in V$ is called *K-finite* if it lies in a finite-dimensional subspace of $V$ stable under $K$ (or $\mathfrak{k}$). For any $\vartheta \in \mathscr{E}(K)$, let $V_\vartheta$ (resp. $V_\vartheta^\infty$) be the linear span of all subspaces $L \subset V$ (resp. $L \subset V^\infty$) with the following property: $L$ is finite-dimensional, $K$-stable, and the representation of $K$ defined by $L$ is irreducible and belongs to $\vartheta$. Then $V_\vartheta^\infty = V^\infty \cap V_\vartheta$ and $V_\vartheta^\infty$ is dense in $V_\vartheta$, the latter being closed in $V$. We put

$$V^0 = \sum_{\vartheta \in \mathscr{E}(K)} V_\vartheta^\infty, \qquad \text{(algebraic sum).}$$

It is well known that $V^0$ is dense in $V$.

In general $V^0$ will not be stable under $G$. However, it is stable under $U(\mathfrak{g})$ and so defines a $U(\mathfrak{g})$-module. This is an immediate consequence of the following simple but enormously useful lemma.

LEMMA 1.2.1. *Let* $\mathfrak{a}$ *be a finite-dimensional Lie algebra over a field k, and* $\mathfrak{b} \subset \mathfrak{a}$ *a Lie subalgebra. Let L be any* $\mathfrak{a}$*-module* (i.e., *a vector space over k carrying a representation of* $\mathfrak{a}$). *Denote by* $L^0$ *the sum of all finite-dimensional sub* $\mathfrak{b}$*-modules of L* (*considered as a* $\mathfrak{b}$*-module*). *Then*:
(i) *if* $v \in L$, *then* $v \in L^0$ *if and only if* $\dim_k(U(\mathfrak{b})v) < \infty$.
(ii) $L^0$ *is a sub* $\mathfrak{a}$*-module of L.*

(i) is trivial. For (ii), note that if $v \in L^0$ and $L' = U(\mathfrak{b})v$, then for $X \in \mathfrak{a}$, $Xv \in \mathfrak{a} \cdot L'$, $\mathfrak{a} \cdot L'$ is $\mathfrak{b}$-stable and finite-dimensional.

Elements of $L^0$ are called $\mathfrak{b}$-*finite*. If $L^0 = L$, we say that $L$ is $\mathfrak{b}$-*finite*. The assignment

$$L \mapsto L^0$$

is a covariant functor from the category of $\mathfrak{a}$-modules to the category of $\mathfrak{b}$-finite $\mathfrak{a}$-modules. We have

LEMMA 1.2.2. *If* $A, B, C$ *are* $\mathfrak{a}$*-modules and*

$$0 \to A \to B \to C$$

*is exact, then*

$$0 \to A^0 \to B^0 \to C^0$$

*is also exact.*

This is trivial to check. It is important to note that the functor $L \mapsto L^0$ is *not* exact.

**1.3.** For arbitrary $V$, the relation between the $G$-module $V$ and the $\mathfrak{g}$-module $V^0$ is not an incisive one. However, it was discovered by Harish-Chandra that when $V$ is suitably restricted, it is essentially determined by $V^0$. Harish-Chandra's pioneering work on representations of $G$ was based on this correspondence $V \mapsto V^0$, which reduced the study of extensive families of $G$-modules to the study of a natural category of $U(\mathfrak{g})$-modules. In order to introduce these modules we begin with a description of his famous finiteness theorem.

A representation $\pi$ of $G$ in a complete locally convex space $V$ is said to be *irreducible* if 0 and $V$ are the only closed $G$-stable subspaces; $\pi$ is said to have an *infinitesimal character* if there is a homomorphism

$$X_\pi: \mathfrak{Z} \to \mathbb{C}, \qquad (\mathfrak{Z} = \text{center of } U(\mathfrak{g})),$$

called the *infinitesimal character* of $\pi$, such that

$$\pi(z)v = X_\pi(z)v, \qquad (z \in \mathfrak{Z}, v \in V^\infty).$$

$\pi$ is said to be *simple* if it is irreducible and has an infinitesimal character.

THEOREM 1.3.1. *Let $\pi$ be a simple representation of G in a complete locally convex space V. Then*

(i) *All $V_\vartheta (\vartheta \in \mathscr{E}(K))$ are finite-dimensional.*

(ii) *For all* $\vartheta \in \mathscr{E}(K)$, $\dim(V_\vartheta) \leqslant \dim(\vartheta)^2$.

(iii) $V^0 = \Sigma_{\vartheta \in \mathscr{E}(K)} V_\vartheta$ *is an irreducible* $U(\mathfrak{g})$*-module.*

I do not know whether all irreducible representations are simple. We however have

THEOREM 1.3.2. *Irreducible unitary representations are simple and hence possess the properties* (i)–(iii) *of Theorem* 1.3.1.

This theorem suggests that it is very useful to introduce the following definitions.

DEFINITION 1.3.3. A Harish-Chandra $G$-module is a complete locally convex space $V$ together with a representation of $G$ in $V$ such that $\dim(V_\vartheta) < \infty$, $\forall \vartheta \in \mathscr{E}(K)$.

Let $\mathscr{E}(\mathfrak{k})$ be the set of equivalence classes of irreducible $\mathfrak{k}$-modules.

DEFINITION 1.3.4. A Harish-Chandra $\mathfrak{g}$-module is a $U(\mathfrak{g})$-module $V$ such that

(a) $V$ is the algebraic direct sum of finite-dimensional irreducible $\mathfrak{k}$-modules.

(b) If $V_\vartheta$ is the isotypical subspace of $V$ corresponding to $\vartheta \in \mathscr{E}(\mathfrak{k})$, then $\dim(V_\vartheta) < \infty \ \forall \ \vartheta \in \mathscr{E}(\mathfrak{k})$.

THEOREM 1.3.5. (i) *Let V be a Harish-Chandra G-module. Then $V^0$ is a Harish-Chandra* $\mathfrak{g}$*-module.*

(ii) *Let* $\mathscr{L}_G$ (resp. $\mathscr{L}_\mathfrak{g}$) *be the set of all sub G-modules* (*resp. sub* $\mathfrak{g}$*-modules*) *of V* (*resp.* $V^0$), *partially ordered by inclusion. Then, for any* $L \in \mathscr{L}_G$,

$$L^0 = L \cap V^0, \qquad L = \mathrm{Cl}(L^0)$$

(Cl = *Closure*). *Moreover*

$$L \mapsto L^0 = L \cap V^0$$

*is an order-preserving bijection of* $\mathscr{L}_G$ *with* $\mathscr{L}_{\mathfrak{g}}$.

(iii) *V irreducible* $\Leftrightarrow$ *V simple* $\Leftrightarrow$ $V^0$ *is irreducible.*

In the last statement, irreducibility of $V^0$ means the usual algebraic irreducibility, i.e., nonexistence of any $U(\mathfrak{g})$-stable linear subspaces of $V^0$ other than 0 and $V^0$.

It is natural to ask which Harish-Chandra $\mathfrak{g}$-modules $U$ are 'integrable', i.e., of the form $V^0$ for some Harish-Chandra $G$-module $V$. Some care is necessary here since the simply connected covering group $\tilde{G}$ of $G$ need not have finite center. Harish-Chandra proved that this is so, if $U$ is irreducible, that is, there is a $\tilde{G}$-module $V$ (whose underlying space is even a Hilbert space) such that $V^0 \simeq U$ [1a, 1b], see also Lepowsky [1]. Recent work of Casselman [1] has shown that this theorem is actually true *for arbitrary Harish-Chandra* $\mathfrak{g}$*-modules*. Casselman's theorem thus reduces the study of $\tilde{G}$-modules with finite $\tilde{K}$-multiplicities to the study of Harish-Chandra modules for $\mathfrak{g}$. In order that $U$ actually defines a $G$-module, i.e., that there is a $G$-module $V$ with $V^0 \simeq U$, it is then clearly necessary and sufficient that $U$ is of *type* $K$ in the following sense: $U_\vartheta \neq 0$ only for $\vartheta \in \mathscr{E}(K)$. If $U(\mathfrak{g}) \cdot U_{\vartheta_0} = U$ for some $\vartheta_0 \in \mathscr{E}(K)$, then $U$ is of type $K$. In particular, for *irreducible* $U$, in order that $U$ be of type $K$ it suffices to require only that for some $\vartheta_0 \in \mathscr{E}(K)$, $U_{\vartheta_0} \neq 0$.

DEFINITION 1.3.6. Two Harish-Chandra $G$-modules $V_i$ $(i = 1, 2)$ are infinitesimally equivalent if the $\mathfrak{g}$-modules $V_1^0$ and $V_2^0$ are equivalent.

DEFINITION 1.3.7. (i) A Harish-Chandra $G$-module $V$ is said to be unitary if $V$ is a Hilbert space and the representation of $G$ in $V$ is unitary.

(ii) A Harish-Chandra $\mathfrak{g}$-module $V$ is said to be unitary if there is a scalar product $\langle \cdot, \cdot \rangle$ converting $V$ into a preHilbert space such that

$$\langle Xu, v \rangle + \langle u, Xv \rangle = 0$$

for all $u, v \in V$, $X \in \mathfrak{g}_0$.

THEOREM 1.3.8. (i) *Two unitary Harish-Chandra G-modules are in-*

*finitesimally equivalent if and only if they are unitarily equivalent in the usual sense.*

(ii) *If U is a unitary Harish-Chandra* $\mathfrak{g}$*-module of type K, there is a unitary Harish-Chandra G-module V such that* $U \simeq V^0$ *via an isomorphism that preserves scalar products; moreover, V is determined up to unitary equivalence.*

Finally we have the problem of an effective determination of the equivalence class of an irreducible Harish-Chandra $\mathfrak{g}$-module. Harish-Chandra obtained a criterion which has turned out to be extremely useful. To formulate it, let

$$\mathfrak{O} = U(\mathfrak{g})^K$$

i.e., $\mathfrak{O}$ is the centralizer of Ad($K$) in $U(\mathfrak{g})$. For any Harish-Chandra $\mathfrak{g}$-module $V$, each $V_\vartheta$ ($\vartheta \in \mathscr{E}(\mathfrak{k})$) is a module for the algebra $\mathfrak{O}U(\mathfrak{k}) = U(\mathfrak{k})\mathfrak{O}$ (or $\mathfrak{O} \otimes U(\mathfrak{k})$).

THEOREM 1.3.9. (i) *Let V be an irreducible Harish-Chandra module and* $\vartheta \in \mathscr{E}(\mathfrak{k})$. *Then* $V_\vartheta$ *is an irreducible* $\mathfrak{O}U(\mathfrak{k})$*-module.*

(ii) *Suppose* $V_i$ ($i = 1, 2$) *are irreducible Harish-Chandra* $\mathfrak{g}$*-modules. Let* $\vartheta \in \mathscr{E}(\mathfrak{k})$ *be such that* $V_{1,\vartheta} \neq 0$, $V_{2,\vartheta} \neq 0$ *and that the* $\mathfrak{O}U(\mathfrak{k})$*-modules* $V_{1,\vartheta}$ and $V_{2,\vartheta}$ *are equivalent. Then* $V_1$ *and* $V_2$ *are equivalent.*

Although the foregoing is only a bare sketch of the early work of Harish-Chandra (with recent refinements due to others), I hope it is clear that the category of Harish-Chandra modules is a natural object of study in the infinitesimal theory. This was the point of departure, not only of Harish-Chandra's own work [1], but of Parthasarathy, Ranga Rao and Varadarajan [1], Kostant [1] and many others. For details I refer the reader to the book of Dixmier [1] and the early papers of Harish-Chandra [1].

It is also worth mentioning that Harish-Chandra's version of the finite multiplicity theorem is much more general than that stated here. He proved in fact that if $V$ is a cyclic $\mathfrak{g}$-module which is $\mathfrak{k}$-finite and on which $\mathfrak{k}$ acts semisimply, then each $V_\vartheta$ ($\vartheta \in \mathscr{E}(\mathfrak{k})$) is a finite $\mathfrak{Z}$-module, $\mathfrak{Z}$ being the center of $U(\mathfrak{g})$; when $\mathfrak{Z}$ acts through a homomorphism on $V$, the dimensions of the $V_\vartheta$ become finite. For the general theorem, see his papers [1a]. It is obvious that this result is true for all finitely generated $V$.

For additional information and more details, see also Varadarajan [1], see 7 of Part II, and Warner [1]; see also Borel [1].

**1.4.** It is clear from this that the problem of determining $\mathscr{E}(G)$ may be pursued in two different ways. One is to start with $G$ and try to construct its irreducible unitary representations by analytic means. The other is to start with $\mathfrak{g}$ and construct, by algebraic methods, all of the irreducible Harish-Chandra modules for $\mathfrak{g}$, and then determine which of them are unitary. My own feeling is that it is not wise to insist on such strict separation of techniques for this problem. Both ways of looking at representations are valuable, and each complements the other. So in what follows I may move back and forth between the analytic and algebraic frameworks, whenever it is convenient to do this.

If $\pi$ is a simple representation of $G$ in a Hilbert space, it has a *character*. This character, denoted by $\Theta_\pi$, is a distribution on $G$ in the sense of L. Schwartz, and is given for $f \in C_c^\infty(G)$ by

$$\Theta_\pi(f) = \mathrm{tr}(\pi(f))$$

here $C_c^\infty(G)$ is the space of compactly supported smooth functions on $G$, and $\pi(f)$ is the operator (which can be proved to be of trace class)

$$\pi(f) = \int_G f(x)\pi(x)\,\mathrm{d}x,$$

$\mathrm{d}x$ being a Haar measure on $G$. $\Theta_\pi$ is an *invariant eigen distribution*, i.e., it has the following two properties:

(i) $\Theta_\pi$ is invariant under all inner automorphisms of $G$, i.e.,

$$\Theta_\pi(f) = \Theta_\pi({}^a f)\,\forall f \in C_c^\infty(G), \qquad a \in G,$$

where $({}^a f)(x) = f(a^{-1}xa)$ $(x \in G)$.

(ii) $\Theta_\pi$ is an eigendistribution for each $z \in \mathfrak{Z}$.

Using $\mathrm{d}x$ we can define a formal adjoint operation $D \mapsto {}^tD$ on the algebra of smooth differential operators on $G$ by requiring that for all $f_1, f_2 \in C_c^\infty(G)$

$$\int_G (Df_1) f_2\,\mathrm{d}x = \int_G f_1({}^tDf_2)\,\mathrm{d}x.$$

The map $D \mapsto {}^tD$ is an anti-automorphism of period 2. If we regard the elements of $U(\mathfrak{g})$ as left invariant differential operators on $G$, then ${}^t(U(\mathfrak{g})) = U(\mathfrak{g})$, and on $U(\mathfrak{g})$ this is the unique anti-automorphism such that

${}^tX = -X$ for $X \in \mathfrak{g}$. We can now allow the differential operator $D$ to act on distributions $T$ by

$$(DT)(f) = T({}^tDf).$$

If $\varphi \in C^\infty(G)$, and we identify $\varphi$ with the distribution

$$\underline{\varphi}: f \mapsto \int_G f\varphi \, \mathrm{dx},$$

then $D\varphi$ is the distribution

$$D\underline{\varphi}: f \mapsto \int_G f D\varphi \, \mathrm{dx}.$$

So there is no need to distinguish between $\varphi$ and $\underline{\varphi}$ and we shall not do this. With these preliminaries, (ii) above means that

$$z\Theta_\pi = \chi_\pi(z)\Theta_\pi, \qquad (z \in \mathfrak{Z}),$$

thus expressing the fact that $\Theta_\pi$ is an eigen distribution for each element $z$ of $\mathfrak{Z}$, the algebra of all bi-invariant differential operators on $G$, the eigenvalue being $\chi_\pi(z)$; here $\chi_\pi$ is the infinitesimal character of $\pi$.

The distribution $\Theta_\pi$ depends only on the infinitesimal equivalence class of $\pi$; and even more importantly, it *determines* the infinitesimal equivalence class of $\pi$ completely; in particular, for unitary $\pi$, it determines the unitary equivalence class of $\pi$. This property makes it a fundamental tool in the analytical approach to representation theory.

By a profound study of the differential equations satisfied by the characters, Harish-Chandra [3] succeeded in proving that the characters are locally integrable functions. More precisely, he proved the following.

THEOREM 1.4.1. *Let $\Theta$ be a distribution on $G$ such that*

(i) *$\Theta$ is invariant (under all inner automorphisms)*

(ii) *$\Theta$ is $\mathfrak{Z}$-finite*, i.e., *the distributions $z\Theta$ $(z \in \mathfrak{Z})$, form a finite-dimensional space. Then there is a locally integrable invariant function $\theta$ on $G$, analytic on the set $G'$ of regular points of $G$, such that $\Theta = \theta\,\mathrm{dx}$*, i.e.,

$$\Theta(f) = \int_G \theta(x) f(x)\,\mathrm{dx}, \qquad (f \in C_c^\infty(G)).$$

We remark that if $\Theta = \Theta_\pi$, the function $\theta = \theta_\pi$ does not depend on dx (if both sides use the same Haar measure). We also recall the definition of $G'$. Let $l$ = rank of $G$, $n = \dim(G)$. Then $D$ is the invariant analytic function on $G$ that is not $\equiv 0$ such that

$$\det(t+\mathrm{Ad}(x)-1)=t^l D(x)+\cdots+t^n$$

($x \in G$, $t$ an indeterminate). A point $x \in G$ is called *regular* if $D(x) \neq 0$. The set $G'$ of regular points is open, dense in $G$, and $G \backslash G'$ is of measure zero. So, as the function $\theta$ in the theorem is determined uniquely on $G'$, there are no essential ambiguities in the correspondence $\Theta \rightleftarrows \theta$.

The distribution characters of simple representations are therefore invariant class functions that are locally integrable, and are determined once they are known at the regular points of a representative set of Cartan subgroups (CSG). It was substantially by this method that Harish-Chandra determined the discrete series of representations, using which he then went on to determine the part of $\mathscr{E}(G)$ that enters the Plancherel formula, namely the *tempered* equivalence classes.

Let me first define the discrete series. For any separable unimodular locally compact group $H$ we define $\mathscr{E}_2(H) \subset \mathscr{E}(H)$ to be the set of all $\omega \in \mathscr{E}(H)$ with the following property: if $\pi \in \omega$ and $\pi$ acts on the Hilbert space $V$, all the 'matrix entries'

$$f_{\pi;\varphi,\psi}: x \mapsto \langle \pi(x)\varphi, \psi \rangle, \qquad (\varphi, \psi \in V),$$

belong to $L_2(H)$. $\mathscr{E}_2(H)$ is called the *discrete series* for $H$; its members, the *discrete classes*; the reason for so naming is the following equivalent characterization: if $\omega \in \mathscr{E}(H)$, then $\omega \in \mathscr{E}_2(H) \Leftrightarrow$ there is a direct summand of the regular representation of $H$ which is irreducible and whose class is $\omega$.

Not all groups possess a discrete series. If $H$ is a group of class $\mathscr{H}$ (see Varadarajan [1], Part II, Section 1) then, $\mathscr{E}_2(H) \neq \emptyset$ if and only if $H$ has a compact CSG. This is one of the major signposts of the theory and was proved by Harish-Chandra. Assuming that $H$ has a compact CSG, say $T$, Harish-Chandra [4] proceeded to determine completely the characters of the discrete classes of $H$.

Once the discrete series was determined, Harish-Chandra used them to construct all the *tempered* classes in $\mathscr{E}(H)$ for any group $H$ of class $\mathscr{H}$. The importance of temperedness is due to the fact that only the tempered classes enter the decomposition of the regular representation of $H$. I shall now briefly describe these results.

**1.5.** First the discrete series. I refer you to Harish-Chandra's papers [4, 5a] or my Springer Lecture Notes [1] for the precise statements in the case of an arbitrary group of class $\mathscr{H}$. I shall consider here only the case of a connected semisimple $G$, which is the $\mathbb{R}$-analytic group defined by $\mathfrak{g}_0$ in the simply connected complex analytic group $G_c$ (corresponding to $\mathfrak{g}$). We assume that $G$ has a compact CSG $T$. It is possible to arrange matters so that $T \subset K$, $K = U \cap G$ where $U$ is a compact form of $G_c$.

Let $\mathfrak{h}_0 \subset \mathfrak{g}_0$ be the Lie algebra of $T$ and $\mathfrak{h} = \mathbb{C} \cdot \mathfrak{h}_0$. The character group $\hat{T}$ of $T$ is canonically isomorphic to a lattice $\mathbb{L}$ in $(-1)^{1/2}\mathfrak{h}_0^*$ which is precisely the lattice of integral elements of $\mathfrak{h}^*$, i.e., those $\lambda \in \mathfrak{h}^*$ such that $2\langle\lambda, \alpha\rangle/\langle\alpha, \alpha\rangle \in \mathbb{Z}$ for all roots $\alpha$ of $(\mathfrak{g}, \mathfrak{h})$. Let $\mathfrak{w}(\mathfrak{g})$ denote the Weyl group of $(\mathfrak{g}, \mathfrak{h})$; $\mathfrak{w}(\mathfrak{g})$ operates on $\mathbb{L}$; and those $\lambda \in \mathbb{L}$ which are fixed only by the identity element of $\mathfrak{w}(\mathfrak{g})$ are said to be *regular*; we denote by $\mathbb{L}'$ the set of all regular elements in $\mathbb{L}$. $\mathfrak{w}(\mathfrak{k})$ is the subgroup of $\mathfrak{w}(\mathfrak{g})$ generated by those Weyl reflexions $s_\alpha$ for which $\alpha$ is a root of $(\mathfrak{k}, \mathfrak{h})$; $\mathfrak{w}(\mathfrak{k})$ is canonically isomorphic to the Weyl group of $(G, T)$, i.e., to $N_G(T)/T$ where $N_G(T)$ is the normalizer of $T$ in $G$.

The discrete classes in $\mathscr{E}(G)$ are in bijective correspondence with the points of $\mathfrak{w}(\mathfrak{k})\backslash\mathbb{L}'$. To describe them explicitly we fix a positive system of roots for $(\mathfrak{k}, \mathfrak{h})$, say $P_\mathfrak{k}$. Let $\mathbb{L}'(P_\mathfrak{k})$ be the subset of $\mathbb{L}'$ of all $\lambda \in \mathbb{L}'$ such that

$$\langle\lambda, \alpha\rangle > 0 \,\forall \alpha \in P_\mathfrak{k}$$

(if $\mu \in \mathbb{L}$, then $\mu \in \mathbb{L}' \Leftrightarrow \langle\mu, \beta\rangle \neq 0 \;\forall$ roots $\beta$ of $(\mathfrak{g}, \mathfrak{h})$). It is easy to see that $\mathbb{L}'(P_\mathfrak{k})$ is a cross-section for $\mathfrak{w}(\mathfrak{k})\backslash\mathbb{L}'$. For any positive system $P$ of roots of $(\mathfrak{g}, \mathfrak{h})$ we define likewise $\mathbb{L}'(P)$ to be the subset of all $\lambda \in \mathbb{L}'$ such that

$$\langle\lambda, \beta\rangle > 0 \;\forall \beta \in P.$$

Clearly

$$\mathbb{L}'(P_\mathfrak{k}) = \bigcup_{P \supset P_\mathfrak{k}} \mathbb{L}'(P), \qquad \text{(disjoint union)}.$$

For any $P \supset P_\mathfrak{k}$ we define

$$\Delta_P(b) = \xi_{\delta_P}(b) \prod_{\alpha \in P} (1 - \xi_{-\alpha}(b)), \qquad (b \in T),$$

where

$$\delta_P = \tfrac{1}{2} \sum_{\alpha \in P} \alpha$$

and $\xi_\mu$ is the character of $T$ corresponding to $\mu \in \mathbb{L}$; of course $\delta_P \in \mathbb{L}'$.

There is a bijection $\Lambda \mapsto \omega(\Lambda)$ of $\mathbb{L}'(P_\mathfrak{t})$ with $\mathscr{E}_2(G)$. Given $\Lambda \in \mathbb{L}'(P_\mathfrak{t})$, $\omega(\Lambda)$ is characterized by the following properties:

(i) The character $\Theta_\Lambda$ of $\omega(\Lambda)$ is tempered.

(ii) On $T' = T \cap G'$ (recall that $G'$ is the set of regular points of $G$)

$$\Theta_\Lambda = (-1)^q \sum_{s \in \mathfrak{w}(\mathfrak{k})} \varepsilon(s) \xi_{s\Lambda} / \Delta_P$$

where $q = \frac{1}{2}\dim(G/K)$ ($q$ is an integer), $\varepsilon(s) = \det(s)$, and $P$ is the positive system of roots of $(\mathfrak{g}, \mathfrak{h})$ such that $\mathbb{L}'(P)$ contains $\Lambda$.

Actually $\Theta_\Lambda$ is the only invariant eigen distribution on $G$ satisfying (i) and (ii). Moreover, we can replace (i) by

(i′) $\qquad |\Theta_\Lambda(x)| \leq C|D(x)|^{-1/2}, \qquad \forall x \in G',$

where $C > 0$ is a constant.

The infinitesimal character $\chi_\Lambda$ of $\omega(\Lambda)$ is also easy to describe. First we introduce a homomorphism

$$\boldsymbol{\beta}_Q \colon U(\mathfrak{g}) \to U(\mathfrak{h})$$

corresponding to any positive system $Q$ for $(\mathfrak{g}, \mathfrak{h})$ – the so-called Harish-Chandra homomorphism – as follows. For any element $a \in U(\mathfrak{g})$ such that $a$ is centralized by $\mathrm{Ad}(T)$, i.e.,

$$[H, a] = 0 \ \forall H \in \mathfrak{h},$$

$\boldsymbol{\beta}_Q(a)$ is the unique element of $U(\mathfrak{h})$ such that ($\mathfrak{g}_\alpha$ being the root spaces)

$$a \equiv \boldsymbol{\beta}_Q(a), \qquad \left(\bmod \sum_{\alpha \in Q} U(\mathfrak{g})\mathfrak{g}_\alpha\right).$$

That $\boldsymbol{\beta}_Q$ is a homomorphism follows from the fact that

$$\mathfrak{g}_\alpha U(\mathfrak{h}) \subset U(\mathfrak{h})\mathfrak{g}_\alpha$$

for all roots $\alpha$ of $(\mathfrak{g}, \mathfrak{h})$. Then, for all $z \in \mathfrak{Z}$,

$$\chi_\Lambda(z) = \boldsymbol{\beta}_P(z)(\Lambda - \delta_P), \qquad (\Lambda \in \mathbb{L}'(P)).$$

In this formula we are interpreting elements of $U(\mathfrak{h})$ canonically as polynomials on $\mathfrak{h}^*$. It is known that if we define

$$\gamma(z)(\mu) = \boldsymbol{\beta}_P(z)(\mu - \delta_P), \qquad (\mu \in \mathfrak{h}^*, z \in \mathfrak{Z}),$$

then $\gamma(z) \in U(\mathfrak{h})$, is independent of the choice of $P$, and

$$\gamma: \mathfrak{Z} \to U(\mathfrak{h})$$

is actually an *isomorphism* of $\mathfrak{Z}$ with the Weyl group invariants in $U(\mathfrak{h})$:

$$\gamma: \mathfrak{Z} \simeq U(\mathfrak{h})^{\mathfrak{w}(\mathfrak{g})}.$$

Thus,

$$\chi_\Lambda(z) = \gamma(z)(\Lambda), \qquad (z \in \mathfrak{Z}, \Lambda \in \mathbb{L}'),$$

and further that the discrete classes $\omega(\Lambda)$ and $\omega(\Lambda')$ ($\Lambda$, $\Lambda' \in \mathbb{L}'$) have the same infinitesimal character if and only if $\Lambda' \in \mathfrak{w}(\mathfrak{g}) \cdot \Lambda$ (cf. Wolf's paper in this volume).

**1.6.** Let us now give up the assumption that $\mathrm{rk}(G) = \mathrm{rk}(K)$. $G$ no longer need have any discrete series. We then look at the parabolic subgroups (psgrps) of $G$. Any parabolic subgroup $Q$ of $G$ has what is known as a Langlands decomposition:

$$Q = MAN;$$

here $N$ is the unipotent radical of $Q$ (hence a closed normal subgroup of $Q$), and $Q$ is the normalizer of $N$ in $G$; $A$ is a vector group; $M$ is a group of class $\mathscr{H}$; $MA \simeq M \times A$ and $Q \simeq MA \times N$. We consider only *cuspidal* psgrps $Q$, namely those for which $M$ has a compact CSG, or, by virtue of Harish-Chandra's theorem, $\mathscr{E}_2(M) \neq \emptyset$. For instance, as $M$ is compact for a *minimal psgrp $Q$, minimal psgrps are always cuspidal.* Also, if $G$ itself has a compact CSG, then $G$ is itself a cuspidal psgrp; $A = \{1\}$, $N = \{1\}$ in this case.

Given a cuspidal $Q = MAN$, we can define a family of unitary representations of $G$ as follows. Select $\eta_M \in \mathscr{E}_2(M)$, and $\xi \in \hat{A}$. We choose a representation $\sigma_M \in \eta_M$ and define $\pi(\sigma_M, \xi)$ to be the representation of $G$, unitarily induced from $Q$ by the unitary representation

$$man \mapsto \xi(a)\sigma_M(m), \qquad (m \in M, a \in A, n \in N).$$

In case $Q = G$ (so that $G$ has a compact CSG), $\pi(\sigma_G) = \sigma_G$ so that we obtain $\mathscr{E}_2(G)$.

Harish-Chandra has proved that these representations $\pi(\sigma_M, \xi)$ always decompose as a direct sum of finitely many mutually inequivalent subrepresentations, and furthermore, that when $\xi$ is in 'general position' (this can be made precise), $\pi(\sigma_M, \xi)$ is even irreducible. His work also showed that the series of irreducible representations obtained this way are always tempered and depend only on the conjugacy class of $A$; and that by varying $A$ over a set of representatives, *all* tempered irreducible unitary representations are obtained. Finally, any tempered *simple* representation $V$ of $G$ can be proved to be infinitesimally equivalent to a tempered *irreducible unitary* representation. The Plancherel formula for $G$ involves only tempered representations (Harish-Chandra [5c]).

Even when $G$ does not have a compact CSG, one can always look at cuspidal psgrps $Q=MAN$ for which dim($A$) is as small as possible. Let us call them *fundamental*. It is known that for any two such, say $Q_i=M_iA_iN_i$ ($i=1, 2$), $A_1$ and $A_2$ are always conjugate in $G$. So we have an associated series of representations $\pi(\sigma_M, \xi)$ of $G$, called the *fundamental series* of $G$. Harish-Chandra [5c] has proved that *these are always irreducible*. For instance, if $G$ has a single conjugacy class of CSG's, the fundamental series is the principal series, and it exhausts the set of tempered classes in $\mathscr{E}(G)$.

Due to lack of space and time I have not been able to discuss in any detail the various other important contributions to this subject, especially those of Knapp–Zuckermann [1], Langlands [1], Trombi [1], and so on. In addition to all of this I would like to refer to the very interesting work of Atiyah and Schmid [1, 2] in which they treat some of the basic questions of character theory and discrete series from a point of view that is often different from Harish-Chandra's. The reader is referred to the lectures of Atiyah and Schmid in this volume for an account of this work.

It is clear from this discussion that an attempt to understand the infinitesimal structure of the irreducible representations of $G$ should begin with the study of the discrete series, and perhaps, the fundamental series. This will also provide some additional insight into these representations, because, the analytical construction of these representations, coming down as it does ultimately to the discrete series, is, to some extent, indirect. In these lectures, I shall attempt to sketch a method of construction of irreducible Harish-Chandra $\mathfrak{g}$-modules that will prove to be supple enough to include the fundamental series and actually much more. The basic idea for this construction goes back to my paper with Enright [1]; it has since then been developed, very substantially, by Enright in a series of remarkable papers, first with Wallach [1, 2], and then by himself [1–3], till it now appears that his methods are capable of giving a complete view of all the irreducible Harish-Chandra $\mathfrak{g}$-modules.

## 2. Holomorphic discrete series. Verma modules

**2.1.** The representations of the discrete series that are most immediately accessible to the infinitesimal method are the so-called *holomorphic discrete series* representations. These were also the first to be discovered historically; for, the simplest semisimple group with discrete series is $SL(2, \mathbb{R})$; and, for this group, the discrete series representations are all realizable in suitable Hilbert spaces of holomorphic functions (hence the name) on the disc or the upper half plane. They go back to Bargmann [1].

$G=SL(2, \mathbb{R})$, $K=SO(2, \mathbb{R})$. We select a *standard basis* $\{H, X, Y\}$ for $\mathfrak{g}$:

$$\mathbb{C}\cdot H=\mathfrak{h}=\mathfrak{k}$$

$$[H, X]=2X, \qquad [H, Y]=-2Y, \qquad [X, Y]=H.$$

For any $\lambda \in \mathbb{C}$ we define the modules $V_{\pm,\lambda}$ as follows: $V_{+,\lambda}$ is spanned by the vectors

$$v_\lambda^+, v_{\lambda-2}^+, v_{\lambda-4}^+, \ldots,$$

which form a basis; and $H, X, Y$ act on $V_{+,\lambda}$ by

$$Hv_{\lambda-2r}^+=(\lambda-2r)v_{\lambda-2r}^+, \qquad (r\geqslant 0),$$
$$Yv_{\lambda-2r}^+=v_{\lambda-2(r+1)}^+, \qquad (r\geqslant 0),$$
$$Xv_\lambda^+=0, \; Xv_{\lambda-2r}^+=r(\lambda-r+1)v_{\lambda-2(r-1)}, \qquad (r\geqslant 1).$$

Similarly, $V_{-,\lambda}$ is spanned by vectors

$$v_\lambda^-, v_{\lambda+2}^-, v_{\lambda+4}^-, \ldots$$

which form a basis; $H, X, Y$ act on $V_{-,\lambda}$ by

$$Hv_{\lambda+2r}^-=(\lambda+2r)v_{\lambda+2r}^-, \qquad (r\geqslant 0),$$
$$Xv_{\lambda+2r}^-=v_{\lambda+2r+2}^-, \qquad (r\geqslant 0),$$
$$Yv_\lambda^-=0, \qquad Yv_{\lambda+2r}^-=r(-\lambda-r+1)v_{\lambda+2r-2}, \qquad (r\geqslant 1).$$

Suppose now $n$ is an integer $\neq 0$. Define

$$D_n=\begin{cases} V_{-,n+1}, & n>0,\\ V_{+,n-1}, & n<0.\end{cases}$$

Then the $D_n$ ($n \in \mathbb{Z}$, $n\neq 0$) are the $\mathfrak{g}$-modules defined by the discrete series representations of $G$. The characters $\Theta_n$ of $D_n$ are the unique tempered invariant eigendistributions on $G$ whose values at

$$u_\theta = \begin{pmatrix} \cos\theta & \sin\theta \\ -\sin\theta & \cos\theta \end{pmatrix}$$

are given by

$$\Theta_n(u_\theta) = -\operatorname{sign}(n)\frac{e^{in\theta}}{e^{i\theta} - e^{-i\theta}}, \qquad (\theta \neq 0, \pm\pi, \pm 2\pi, \ldots).$$

For all of this, see Bargmann [1], Harish-Chandra [1e], and Lang [1].

The algebra $\mathfrak{Z}$ is generated by the element

$$\omega = H^2 + 2H + 1 + 4YX = H^2 - 2H + 1 + 4XY.$$

Since

$$\omega \cdot v_\lambda^+ = (\lambda+1)^2 v_\lambda^+, \qquad \omega \cdot v_\lambda^- = (\lambda-1)^2 v_\lambda^-$$

and since $v_\lambda^\pm$ generates $V_{\pm,\lambda}$, we get

$$\omega|_{D_n} = n^2 \cdot \text{identity}.$$

So $D_{\pm n}$ have the same infinitesimal character. To distinguish between them we look at the action of $\mathfrak{D} = U(\mathfrak{g})^K$. We note that $\mathfrak{D}$ is freely generated by $H$ and $\omega$ (which commute since $\omega$ is in $\mathfrak{Z}$ = center $U(\mathfrak{g})$). We have

$$H \cdot v_{n+1}^- = (n+1)v_{n+1}^-, \quad H \cdot v_{-n-1}^+ = -(n+1)v_{-n-1}^+, \quad n > 0.$$

Let $P^+$ (resp. $P^-$) be the positive system of roots of $(\mathfrak{g}, \mathfrak{h})$ for which $X$ (resp. $Y$) is the positive root vector. Then, for $u \in \mathfrak{D}$,

$$u \cdot v_{n+1}^- = \boldsymbol{\beta}_{P^-}(u)(n+1)v_{n+1}^-,$$
$$u \cdot v_{-n-1}^+ = \boldsymbol{\beta}_{P^+}(u)(-n-1)v_{-n-1}^+, \qquad (n > 0).$$

All these calculations remain meaningful when $n = 0$. The module $V_{+,-1}$ (resp. $V_{-,1}$) is then denoted by $D_{0,-}$ (resp. $D_{0,+}$); and $D_{0,\pm}$ are the so-called mock discrete series representations.

**2.2.** We shall now show how the modules $D_n$ constructed above may be generalized in a far reaching manner. This was first done by Harish-Chandra in [2a]; we shall essentially follow his treatment. We find it convenient to work in the general context of *Verma modules*. Since the Verma modules will play a crucial role in almost everything of what I am going to discuss, I shall begin with a brief description of these modules and their properties. For details, see Dixmier's book [1]; see also Verma [1], [2], Bernstein, Gel'fand and Gel'fand [1], [2], and Varadarajan [2].

Let $\mathfrak{m}$ be a reductive Lie algebra over $\mathbb{C}$ and $\mathfrak{h} \subset \mathfrak{m}$ a CSA. An $\mathfrak{m}$-module

$M$ is called a *weight module* if

$$M = \bigoplus_{\lambda \in \mathfrak{h}^*} M[\lambda]$$

where, for any $\lambda \in \mathfrak{h}^*$,

$$M[\lambda] = \{v : v \in M, H \cdot v = \lambda(H)v \ \forall H \in \mathfrak{h}\}.$$

The $M[\lambda]$ which are $\neq 0$ are called the *weight spaces*; the corresponding $\lambda$, the *weights*. Suppose $Q$ is a positive system of roots of $(\mathfrak{m}, \mathfrak{h})$. Then $M$ is called a *Q-extreme module* if

(i) $M$ is a weight module

(ii) $\exists$ a weight $\lambda$ and a nonzero $v \in M[\lambda]$ such that

(a) $M = U(\mathfrak{m}) \cdot v$

(b) $\mathfrak{m}_\alpha \cdot v = 0 \ \forall \alpha \in Q$, $\mathfrak{m}_\alpha$ being the root spaces of $\mathfrak{m}$.

It is then known that $M[\lambda] = \mathbb{C} \cdot v$ (i.e., dim $M[\lambda] = 1$) and that the weights of $M$ are of the form $\lambda - \sigma$ where $\sigma$ is either 0 or a sum of elements of $Q$; moreover, all weight spaces are finite dimensional. $\lambda$ is called the *highest weight* (*relative to* $Q$) of $M$. Let $\mathfrak{Z}(\mathfrak{m})$ be the center of $U(\mathfrak{m})$. The elements of $\mathfrak{Z}(\mathfrak{m})$ act as scalars on $M$, i.e., $M$ has an infinitesimal character. If

$$\delta_Q = \tfrac{1}{2} \sum_{\alpha \in Q} \alpha,$$

if $U(\mathfrak{m})^{\mathfrak{h}}$ is the centralizer of $\mathfrak{h}$ in $U(\mathfrak{m})$, and if

$$\beta_{\mathfrak{m},Q} : U(\mathfrak{m})^{\mathfrak{h}} \to U(\mathfrak{m}), \qquad \gamma_{\mathfrak{m}} : \mathfrak{Z}(\mathfrak{m}) \to U(\mathfrak{h}),$$

are the Harish-Chandra homomorphisms defined by

$$\beta_{\mathfrak{m},Q}(u) \equiv u \left( \mathrm{mod} \sum_{\alpha \in Q} U(\mathfrak{m})\mathfrak{m}_\alpha \right), \qquad (u \in U(\mathfrak{m})^{\mathfrak{h}}),$$

$$\gamma_{\mathfrak{m}}(z)(\mu) = \beta_{\mathfrak{m},Q}(z)(\mu - \delta_Q), \qquad (z \in \mathfrak{Z}(\mathfrak{m}), \mu \in \mathfrak{h}^*),$$

then, for all $z \in \mathfrak{Z}(\mathfrak{m})$, $v \in M$, we have

$$z \cdot v = \chi_{\mathfrak{m}, \lambda + \delta_Q}(z) v, \qquad (\chi_{\mathfrak{m},\mu}(z) = \gamma_{\mathfrak{m}}(z)(\mu)).$$

For each root $\alpha$ of $(\mathfrak{m}, \mathfrak{h})$, $X_\alpha$ will denote always a nonzero element in $\mathfrak{m}_\alpha$; $\bar{H}_\alpha$ will always denote the unique element in $[\mathfrak{m}_\alpha, \mathfrak{m}_{-\alpha}]$ $(\subset \mathfrak{h})$ such that $\alpha(\bar{H}_\alpha) = 2$. $\mathfrak{w}(\mathfrak{m})$ is the Weyl group of $(\mathfrak{m}, \mathfrak{h})$. Finally, $\prec_Q$ denotes the usual partial order in $\mathfrak{h}^*$ induced by $Q$: for $\lambda, \lambda' \in \mathfrak{h}^*$, $\lambda' \prec_Q \lambda \Leftrightarrow \lambda - \lambda'$ is either 0 or a sum of elements of $Q$.

LEMMA 2.2.1. *Let $M$ be a $Q$-extreme module with highest weight $\lambda$ and let $u \neq 0$ be in $M[\lambda]$. Let $S \subset Q$ be the set of simple roots in $Q$. Then:*

(i) *Any subquotient module of $M$ contains a $Q$-extreme sub $\mathfrak{m}$-module.*

(ii) *$M$ is irreducible* $\Leftrightarrow \dim(M^{\mathfrak{n}}) = 1$* *where* $\mathfrak{n} = \Sigma_{\alpha \in Q} \mathfrak{m}_\alpha$.

(iii) *If $M'$ is a $Q$-extreme module of highest weight $\lambda'$ and $M'$ is a subquotient of $M$, then $\lambda' \prec_Q \lambda$ and $\exists s \in \mathfrak{w}(\mathfrak{m})$* such that $s(\lambda + \delta_Q) = \lambda' + \delta_Q$.

(iv) *If $\alpha \in S$ and $\lambda_\alpha = \lambda(\bar{H}_\alpha)$ is an integer $\geqslant 0$, and if $u' = X_{-\alpha}^{\lambda_\alpha + 1} \cdot u$, then $X_\gamma \cdot u' = 0 \; \forall \gamma \in Q$, and $u' \in M[\lambda']$ where*

$$\lambda' = s_\alpha(\lambda + \delta_Q) - \delta_Q;$$

*in particular, if $u' \neq 0$, $U(\mathfrak{m}) \cdot u'$ is a $Q$-extreme module of highest weight $\lambda'$ and highest weight vector $u'$.*

(v) *If $\mathfrak{l} \supset \mathfrak{h}$ is a reductive subalgebra of $\mathfrak{m}$, and $Q_{\mathfrak{l}}$ is the set of roots in $Q$ that are roots of $(\mathfrak{l}, \mathfrak{h})$, then $M$ is $\mathfrak{l}$-finite if (and only if) $X^r_{-\alpha} u = 0$ for $r \gg 0$ for each $\alpha \in Q_{\mathfrak{l}}$; in this case,*

$$\lambda(\bar{H}_\alpha) \text{ is an integer } \geqslant 0 \; \forall \alpha \in Q_{\mathfrak{l}}.$$

For (iii) we use Harish-Chandra's theorem that $\gamma_{\mathfrak{m}}$ is an isomorphism of $\mathfrak{Z}(\mathfrak{m})$ with $U(\mathfrak{h})^{\mathfrak{w}(\mathfrak{m})}$, so that,

$$\chi_{\mathfrak{m},\mu} = \chi_{\mathfrak{m},\mu'} \Leftrightarrow \mu' \in \mathfrak{w}(\mathfrak{m}) \cdot \mu, \qquad (\mu, \mu' \in \mathfrak{h}^*).$$

For (iv) it is enough to show that $X_\gamma v' = 0 \; \forall$ *simple* $\gamma$. For $\gamma \neq \alpha$ this is clear since $X_\gamma$ and $X_{-\alpha}$ commute; for $\gamma = \alpha$, it is an $\mathfrak{sl}(2)$ calculation, and follows from the relation

$$X_\alpha X_{-\alpha}^{\lambda_\alpha + 1} = X_{-\alpha}^{\lambda_\alpha + 1} X_\alpha + (\lambda_\alpha + 1) X_{-\alpha}^{\lambda_\alpha} (\bar{H}_\alpha - \lambda_\alpha).$$

Finally, for (v), we use Lemma 1.2.1. to observe that $M$ is $\mathfrak{l}$-finite if and only if $U(\mathfrak{l}) \cdot u$ is $\mathfrak{l}$-finite; from this point onwards, it is a standard result (see Varadarajan [2]).

For any $\lambda \in \mathfrak{h}^*$, the *Verma module* with highest weight $\lambda$ relative to $Q$, denoted by $V_{\mathfrak{m},Q,\lambda}$, may be defined by the following *universal property*:

(i) $V_{\mathfrak{m},Q,\lambda}$ is $Q$-extreme with highest weight $\lambda$.

(ii) If $M$ is a $Q$-extreme $\mathfrak{m}$-module of highest weight $\lambda$ and if $v_\lambda$ (resp. $v_{M,\lambda}$) is a nonzero vector in $V_{\mathfrak{m},Q,\lambda}[\lambda]$ (resp. $M[\lambda]$), there is a unique $\varphi \in \mathrm{Hom}_{\mathfrak{m}}(V_{\mathfrak{m},Q,\lambda}, M)$ such that $\varphi(v_\lambda) = v_{M,\lambda}$.

Such a $\varphi$ necessarily maps $V_{\mathfrak{m},Q,\lambda}$ *onto* $M$. $V_{\mathfrak{m},Q,\lambda}$ is determined uniquely

---

*For any $\mathfrak{m}$-module $N$ and any Lie algebra $\mathfrak{m}' \subset \mathfrak{m}$, $N^{\mathfrak{m}'}$ is the set of all $v \in N$ such that $X \cdot v = 0 \; \forall X \in \mathfrak{m}'$.

up to isomorphism. To prove its existence we need only take

$$V_{\mathfrak{m},Q,\lambda}=U(\mathfrak{m})/J(\mathfrak{m}, Q, \lambda)$$

where $J(\mathfrak{m}, Q, \lambda)$ is the left ideal in $U(\mathfrak{m})$ generated by the $\mathfrak{m}_\alpha$ $(\alpha \in Q)$ and the $H-\lambda(H)1$ $(H \in \mathfrak{h})$. $U(\mathfrak{m})$ acts by natural action and the image of 1 is a vector of weight $\lambda$. Alternatively, $V_{\mathfrak{m},Q,\lambda}$ can be defined as follows: let $\mathbb{C}_\lambda$ denote the one dimensional $\mathfrak{b}$-module ($\mathfrak{b}$ being the Borel subalgebra $\mathfrak{h}+\Sigma_{\alpha \in Q}\mathfrak{m}_\alpha$) with underlying vector space $\mathbb{C}$, on which $H \in \mathfrak{h}$ acts as the scalar $\lambda(H)$ and the $\mathfrak{m}_\alpha$ act trivially; let

$$V_{\mathfrak{m},Q,\lambda}=U(\mathfrak{m}) \bigotimes_{U(\mathfrak{b})} \mathbb{C}_\lambda;$$

here $U(\mathfrak{m})$ is regarded as a right $U(\mathfrak{b})$-module and $U(\mathfrak{m})$ acts on $V_{\mathfrak{m},Q,\lambda}$ from left; $1 \otimes 1$ is then the vector of highest weight $\lambda$. The universal property is a consequence of the universal mapping property of tensor products.

Being $Q$-extreme modules, the $V_{\mathfrak{m},Q,\lambda}$ possess all the properties enunciated in Lemma 2.2.1. We now formulate the lemmas that describe the properties special to the Verma modules.

LEMMA 2.2.2. *Let* $\mathfrak{n}^-=\Sigma_{\alpha \in Q}\mathfrak{m}_{-\alpha}$. *Then, a Q-extreme module M is a Verma module if and only if, as a* $U(\mathfrak{n}^-)$*-module, it is free or even torsion free*; it is then a free* $U(\mathfrak{n}^-)$*-module with one generator.*

LEMMA 2.2.3. *Fix* $\lambda \in \mathfrak{h}^*$. *Then, up to equivalence, there is a unique Q-extreme module of highest weight* $\lambda$ *which is irreducible. If we denote it by* $I_{\mathfrak{m},Q,\lambda}$, *and if* $\varphi$: $V_{\mathfrak{m},Q,\lambda} \to I_{\mathfrak{m},Q,\lambda}$ *is an* $\mathfrak{m}$*-module homomorphism that is surjective, then* $\ker(\varphi)=K_{\mathfrak{m},Q,\lambda}$ *is the unique largest sub* $\mathfrak{m}$*-module of* $V_{\mathfrak{m},Q,\lambda}$ *not containing* $V_{\mathfrak{m},Q,\lambda}[\lambda]$ (then $K_{\mathfrak{m},Q,\lambda} \cap V_{\mathfrak{m},Q,\lambda}[\lambda]=0$).

The Verma modules are special cases of a whole family of modules for $\mathfrak{m}$ obtained as follows. Let $L$ be any $\mathfrak{b}$-module where $\mathfrak{b}$ is the Borel subalgebra $\mathfrak{b}=\mathfrak{h}+\Sigma_{\alpha \in Q}\mathfrak{m}_\alpha$; let

$$V_{\mathfrak{m},Q,L}=U(\mathfrak{m}) \bigotimes_{U(\mathfrak{b})} L;$$

here we regard $U(\mathfrak{m})$ as a right-$U(\mathfrak{b})$-module, and view $V_{\mathfrak{m},Q,L}$ as a $U(\mathfrak{m})$-

*Let $R$ be a ring, *not necessarily commutative*, with unit. Let $L$ be an $R$-module. For each $v \in L$, let $R_v$ be the left ideal of $R$ given by $R_v=\{a\colon a \in R, a \cdot v=0\}$. $L$ is said to be *torsion-free* if $R_v=0$ $\forall v \in L$.

module from the left. The assignment

$$L \mapsto V_{\mathfrak{m},Q,L}$$

is a covariant functor; as $U(\mathfrak{m})$ is a free right $U(\mathfrak{b})$-module, this functor is exact. If $L$ is the one-dimensional module $\mathbb{C}_\lambda$, $V_{\mathfrak{m},Q,L}$ is $V_{\mathfrak{m},Q,\lambda}$.

LEMMA 2.2.4. *Let F be a finite-dimensional* $\mathfrak{m}$*-module and let*

$$F_1 = F \supset F_2 \supset \cdots \supset F_r = 0$$

*be a flag of sub* $\mathfrak{b}$*-modules of F such that*

$$F_i/F_{i+1} \simeq \mathbb{C}_{\nu_i}, \qquad 1 \leqslant i \leqslant r-1$$

*for suitable* $\nu_i \in \mathfrak{h}^*$. *Let*

$$\lambda \in \mathfrak{h}^*, \qquad V = V_{\mathfrak{m},Q,\lambda}, \quad \textit{and} \quad V_i < V_{\mathfrak{m},Q,\mathbb{C}_\lambda \otimes F_i} \qquad (1 \leqslant i \leqslant r).$$

*Then* $V_1 \simeq V \otimes F$ *canonically, as* $\mathfrak{m}$*-modules, and*

$$V_1 \supset V_2 \supset \cdots \supset V_r = 0$$

*is a flag of sub* $\mathfrak{m}$*-modules of* $V \otimes F$ *such that*

$$V_i/V_{i+1} \simeq V_{\mathfrak{m},Q,\nu_i+\lambda}.$$

This lemma is an immediate consequence of the following general principle (see Enright–Wallach [1], Lemma 4.1 where it is formulated as a lemma of Garland–Lepowsky): suppose $\mathfrak{l}$ is a Lie algebra (over some field $k$), $\mathfrak{a} \subset \mathfrak{l}$ a subalgebra, $A$ (resp. $L$) a $U(\mathfrak{a})$-module (resp. $U(\mathfrak{l})$-module) and $L_\mathfrak{a}$ the $U(\mathfrak{a})$-module obtained by regarding $L$ as an $U(\mathfrak{a})$-module; then we have a canonical isomorphism of $U(\mathfrak{l})$-modules

$$U(\mathfrak{l}) \bigotimes_{U(\mathfrak{a})} (A \otimes L_\mathfrak{a}) \simeq \left( U(\mathfrak{l}) \bigotimes_{U(\mathfrak{a})} A \right) \otimes L,$$

under which

$$1 \otimes (\alpha \otimes \lambda) \mapsto (1 \otimes \alpha) \otimes \lambda, \qquad (\alpha \in A, \lambda \in L).$$

Applying this principle, we get

$$V_1 = U(\mathfrak{m}) \bigotimes_{U(\mathfrak{b})} (\mathbb{C}_\lambda \otimes F_\mathfrak{b}) \simeq \left( U(\mathfrak{m}) \bigotimes_{U(\mathfrak{b})} \mathbb{C}_\lambda \right) \otimes F \simeq V_{\mathfrak{m},Q,\lambda} \otimes F.$$

The last statement is clear from the relation

$$V_i/V_{i+1} \simeq U(\mathfrak{m}) \bigotimes_{U(\mathfrak{b})} (\mathbb{C}_\lambda \otimes F_i/\mathbb{C}_\lambda \otimes F_{i+1}) \simeq U(\mathfrak{m}) \bigotimes_{U(\mathfrak{b})} \mathbb{C}_{\lambda+\nu_i}.$$

Verma modules are not always irreducible. But we have

LEMMA 2.2.5. *Let $\lambda \in \mathfrak{h}^*$, $v \in V_{\mathfrak{m},Q,\lambda}[\lambda]$ a nonzero vector. Then:*

(i) *If $\alpha \in Q$ is a simple root and $\lambda(\bar{H}_\alpha)$ is an integer $\geqslant 0$, then $v' = X_{-\alpha}^{\lambda\alpha+1} \cdot v$ is $\neq 0$ and of weight $\lambda' = s_\alpha(\lambda + \delta_Q) - \delta_Q$; moreover, $X_\gamma \cdot v' = 0 \ \forall \gamma \in Q$, so that $U(\mathfrak{m}) \cdot v' \simeq V_{\mathfrak{m},Q,\lambda'}$*

(ii) *$V_{\mathfrak{m},Q,\lambda}$ is irreducible if and only if for each $\alpha \in Q$*

$$(\lambda + \delta_Q)(\bar{H}_\alpha) \notin \{1, 2, 3, \dots\}.$$

For convenience we shall write

$$s' \cdot \mu = s(\mu + \delta_Q) - \delta_Q, \qquad (s \in \mathfrak{w}(\mathfrak{m}), \mu \in \mathfrak{h}^*).$$

Note that by (i) above, if $(\lambda + \delta)(\bar{H}_\alpha)$ is an integer $> 0$ for some $\alpha \in S$, or equivalently, if $\lambda(\bar{H}_\alpha)$ is an integer $\geqslant 0$, we have

$$V_{\mathfrak{m},Q,s'_\alpha \cdot \lambda} \hookrightarrow V_{\mathfrak{m},Q,\lambda}$$

canonically. This result can be generalized quite a bit.

LEMMA 2.2.6. (i) *Let $\lambda \in \mathfrak{h}^*$. Then $V_{\mathfrak{m},Q,\lambda}$ has a Jordan–Hölder series of finite length whose constituents are of the form $I_{\mathfrak{m},Q,\mu}$ with $\mu \in \mathfrak{w}(\mathfrak{m})' \cdot \lambda$* and $\mu \prec_Q \lambda$.

(ii) *The following conditions on a pair of Verma modules $V_{\mathfrak{m},Q,\lambda}$, $V_{\mathfrak{m},Q,\mu}$ are equivalent ($\lambda, \mu \in \mathfrak{h}^*$ are arbitrary):*

(a) $V_{\mathfrak{m},Q,\mu} \hookrightarrow V_{\mathfrak{m},Q,\lambda}$

(b) $I_{\mathfrak{m},Q,\mu}$ *is in the Jordan–Hölder series of* $V_{\mathfrak{m},Q,\lambda}$

(c) $\exists \gamma_1, \gamma_2, \dots, \gamma_k \in Q$ *such that*

$$\lambda \succ_Q s'_{\gamma_1} \lambda \succ_Q s'_{\gamma_2} s'_{\gamma_1} \lambda \succ \cdots \succ_Q s'_{\gamma k} s'_{\gamma k-1} \cdots s'_{\gamma_1} \lambda = \mu$$

(iii) *for any $\lambda, \mu \in \mathfrak{h}^*$, the complex vector space* $\mathrm{Hom}_\mathfrak{m}(V_{\mathfrak{m},Q,\lambda}, V_{\mathfrak{m},Q,\mu})$ *has dimension $\leqslant 1$; and all nonzero elements of it are injections.*

We note that condition (c) of (ii) is equivalent to the following: let $\lambda' = \lambda + \delta_Q$, $\mu' = \mu + \delta_Q$; then $\mu' = s \cdot \lambda'$ where $s \in \mathfrak{w}(\mathfrak{m})$ can be written as $s = s_{\gamma_k} s_{\gamma_{k-1}} \cdots s_{\gamma_1}$ with

$$\lambda'(\bar{H}_{\gamma_1}),\ (s_{\gamma_1}\lambda')(\bar{H}_{\gamma_2}),\ (s_{\gamma_2}s_{\gamma_1}\lambda')(\bar{H}_{\gamma_3}), \ldots, (s_{\gamma_{k-1}} \cdots s_{\gamma_1}\lambda')(\bar{H}_{\gamma_k})$$

being all integers $\geqslant 0$.

LEMMA 2.2.7. *Let $\lambda \in \mathfrak{h}^*$ be such that $\lambda(\bar{H}_\alpha)$ is an integer $\geqslant 0$ for all $\alpha \in Q$. Then:*

(i) $V_{\mathfrak{m},Q,s'\cdot\lambda} \hookrightarrow V_{\mathfrak{m},Q,\lambda}\ \forall s \in \mathfrak{w}(\mathfrak{m})$

(ii) *fix $s_1, s_2 \in \mathfrak{w}(\mathfrak{m})$; then*

$$V_{\mathfrak{m},Q,s_2'\cdot\lambda} \hookrightarrow V_{\mathfrak{m},Q,s_1'\cdot\lambda}$$

*if and only if we can write*

$$s_2 = s_{\gamma_k} s_{\gamma_{k-1}} \cdots s_{\gamma_1} \cdot s_1,$$

*with $\gamma_1, \ldots, \gamma_k \in Q$ and*

$$l(s_{\gamma_1}s_1) = l(s_1) + 1, \quad l(s_{\gamma_2}s_{\gamma_1}s_1) = l(s_{\gamma_1}s_1) + 1, \ldots,$$
$$\ldots, l(s_2) = l(s_{\gamma_{k-1}} \cdots s_{\gamma_1}s_1) + 1$$

*where $l$ is the length function on $\mathfrak{w}(\mathfrak{m})$ defined by $Q$ (or $S$).*

It follows from (ii) above that when $\lambda$ is dominant integral with respect to $Q$, the inclusion

$$V_{\mathfrak{m},Q,s_2'\cdot\lambda} \hookrightarrow V_{\mathfrak{m},Q,s_1'\cdot\lambda}$$

*depends only on the mutual relationship between $s_1$* and $s_2$ and not on $\lambda$, i.e., we have a partial ordering $\leqslant$ on $\mathfrak{w}(\mathfrak{m})$, depending on $Q$, such that, for $s_1, s_2 \in \mathfrak{w}(\mathfrak{m})$,

$$s_2 \leqslant s_1 \Leftrightarrow V_{\mathfrak{m},Q,s_2'\cdot\lambda} \hookrightarrow V_{\mathfrak{m},Q,s_1'\cdot\lambda}$$
$$\forall \lambda \in \mathfrak{h}^* \text{ dominant integral with respect to } Q.$$

For the proofs see, for instance, Bernstein–Gel'fand–Gel'fand [2]. The proof of (i) is however very simple. If $s \in \mathfrak{w}(\mathfrak{m})$ is $\neq 1$,

$$s = s_{\beta_m} s_{\beta_{m-1}} \cdots s_{\beta_1}, \qquad l(s) = m$$

where $\beta_1, \ldots, \beta_m$ are now *simple roots*, i.e., $\in S$. It is then known that

$$\beta_1,\ s_{\beta_1} \cdot \beta_2,\ s_{\beta_2}s_{\beta_1} \cdot \beta_3, \ldots, s_{\beta_{m-1}} \cdots s_{\beta_1} \cdot \beta_m$$

are distinct in $Q$, and are precisely those roots $\gamma$ in $Q$ such that $s\gamma \in -Q$. In particular,

$$\lambda(\bar{H}_{\beta_1}), (s'_{\beta_1} \cdot \lambda)(\bar{H}_{\beta_2}), (s'_{\beta_2} \cdot s'_{\beta_1} \cdot \lambda)(\bar{H}_{\beta_3}), \ldots,$$
$$\ldots, (s'_{\beta_{m-1}} \ldots s'_{\beta_1} \cdot \lambda)(\bar{H}_{\beta_m})$$

are now all integers $\geqslant 0$. Successive applications of (i) of Lemma 2.2.5 now yield

$$V_{\mathfrak{m},Q,\lambda} \hookleftarrow V_{\mathfrak{m},Q,s'_{\beta_1} \cdot \lambda} \hookleftarrow V_{\mathfrak{m},Q,s'_{\beta_2} s'_{\beta_1} \cdot \lambda} \hookleftarrow \cdots \hookleftarrow V_{\mathfrak{m},Q,s' \cdot \lambda}.$$

Moreover, if we write

$$k_1 = \lambda(\bar{H}_{\beta_1}), k_2 = (s'_{\beta_1} \cdot \lambda)(\bar{H}_{\beta_2}), \ldots, k_m = (s'_{\beta_{m-1}} \ldots s'_{\beta_1} \cdot \lambda)(\bar{H}_{\beta_m}),$$

and if $v_\lambda$ is a highest weight vector of $V_{\mathfrak{m},Q,\lambda}$, then

$$X_{-\beta_j}^{k_j+1} X_{-\beta_{j-1}}^{k_{j-1}+1} \ldots X_{-\beta_1}^{k_1+1} \cdot v_\lambda = v_{s'_j \cdot \lambda}$$

is a highest weight vector of $V_{\mathfrak{m},Q,s'_j \cdot \lambda}$, for $1 \leqslant j \leqslant m$.

COROLLARY 2.2.8. *Suppose $\lambda$ is dominant integral with respect to $Q$. Let $t_0 \in \mathfrak{w}(\mathfrak{m})$ be the element maximum length,* i.e., *$t_0 \cdot Q = -Q$. Then $V_{\mathfrak{m},Q,t'_0 \cdot \lambda}$ is irreducible.*

LEMMA 2.2.9. *Suppose $\lambda, \mu \in \mathfrak{h}^*$ and $V_{\mathfrak{m},Q,\mu} \hookrightarrow V_{\mathfrak{m},Q,\lambda}$. Then $\exists$ a nonzero element $y(\mu, \lambda) \in U(\mathfrak{n}^-)$, which is unique up to multiplication by a nonzero complex number, such that $y(\mu, \lambda)$ maps the highest weight space of $V_{\mathfrak{m},Q,\lambda}$ onto the highest weight space of $V_{\mathfrak{m},Q,\mu}$. If $\nu \in \mathfrak{h}^*$ is such that $V_{\mathfrak{m},Q,\nu} \hookrightarrow V_{\mathfrak{m},Q,\mu}$, then $\exists c \in \mathbb{C}^*$, $c$ depending on $\lambda, \mu, \nu$, such that*

$$y(\nu, \mu) y(\mu, \lambda) = c y(\nu, \lambda).$$

*If $\lambda$ is dominant integral with respect to $Q$ and $\mu = s' \cdot \lambda$, then we can take (in the notation above)*

$$y(s' \cdot \lambda, \lambda) = X_{-\beta_m}^{k_m+1} X_{-\beta_{m-1}}^{k_{m-1}+1} \ldots X_{-\beta_1}^{k_1+1}.$$

We shall write

$$d_{s,1}(\lambda) = y(s' \cdot \lambda, \lambda), \qquad (\lambda \text{ dominant integral}).$$

More generally, if $s, t \in \mathfrak{w}(\mathfrak{m})$ are such that

$$V_{\mathfrak{m},Q,t' \cdot \lambda} \hookrightarrow V_{\mathfrak{m},Q,s' \cdot \lambda}, \qquad (\lambda \text{ dominant integral}),$$

we write

$$d_{t,s}(\lambda)=y(t'\cdot\lambda, s'\cdot\lambda).$$

Finally we come to the celebrated Bernstein–Gel'fand–Gel'fand (B-G-G) resolution. Let $\lambda$ be dominant integral with respect to $Q$. Then $I_{\mathfrak{m},Q,\lambda}$, the unique irreducible $\mathfrak{m}$-module with highest weight $\lambda$ (relative to $Q$), is *finite dimensional.* We have

$$I_{\mathfrak{m},Q,\lambda}\simeq V_{\mathfrak{m},Q,\lambda} \Big/ \sum_{s\neq 1} V_{\mathfrak{m},Q,s'\cdot\lambda}$$

and the B–G–G resolution is in fact a resolution of $I_{\mathfrak{m},Q,\lambda}$ in terms of sums of Verma modules. We have (see Bernstein–Gel'fand–Gel'fand [2]).

LEMMA 2.2.10. *Let $\lambda\in\mathfrak{h}^*$ be dominant integral with respect to $Q$. Let $r=\max_{s\in\mathfrak{w}(\mathfrak{m})}l(s)$ and let $t_0\in\mathfrak{w}(\mathfrak{m})$ be such that $l(t_0)=r$. Denote by $I_{\mathfrak{m},Q,\lambda}$ the finite-dimensional irreducible $\mathfrak{m}$-module of highest weight $\lambda$ relative to $Q$. Write*

$$B^{(i)}_{\mathfrak{m},Q,\lambda}=\bigoplus_{\substack{s\in\mathfrak{w}(\mathfrak{m})\\ l(s)=i}} V_{\mathfrak{m},Q,s'\cdot\lambda}, \qquad (1\leqslant i\leqslant r),$$

$$B^{(0)}_{\mathfrak{m},Q,\lambda}=V_{\mathfrak{m},Q,\lambda}.$$

*Then we have an exact resolution*

$$0\longrightarrow B^{(r)}_{\mathfrak{m},Q,\lambda}\xrightarrow{d_r}B^{(r-1)}_{\mathfrak{m},Q,\lambda}\xrightarrow{d_{r-1}}\cdots\longrightarrow B^{(1)}_{\mathfrak{m},Q,\lambda}\xrightarrow{d_1}V_{\mathfrak{m},Q,\lambda}\xrightarrow{\varepsilon}I_{\mathfrak{m},Q,\lambda}\longrightarrow 0.$$

*Here the $d_j$ and $\varepsilon$ are $\mathfrak{m}$-module maps; and $\varepsilon$ the natural surjection.*

It is possible to be more precise on the dependence of the $d_j$ on $\lambda$. For each $s\in\mathfrak{w}(\mathfrak{m})$ we identify $V_{\mathfrak{m},Q,s'\cdot\lambda}$ with the unique submodule of $V_{\mathfrak{m},Q,\lambda}$ isomorphic to it. In particular,

$$V_{\mathfrak{m},Q,s_1'\cdot\lambda}\subset V_{\mathfrak{m},Q,s_2'\cdot\lambda}, \qquad (s_1, s_2\in\mathfrak{w}(\mathfrak{m}), s_1\leqslant s_2).$$

By (iii) of Lemma 2.2.6, any $\mathfrak{m}$-module map $V_{\mathfrak{m},Q,s_1'\cdot\lambda}\rightarrow V_{\mathfrak{m},Q,s_2'\cdot\lambda}$ is a multiple of the above inclusion. So, for $1\leqslant j\leqslant r$, the map

$$d_j\colon B^{(j)}_{\mathfrak{m},Q,\lambda}\rightarrow B^{(j-1)}_{\mathfrak{m},Q,\lambda}$$

can be decribed by a matrix

$$D_j=(d_j(t, s))_{l(t)=j-1,\, l(s)=j}$$

of *complex numbers.* The Bernstein–Gel'fand–Gel'fand resolution is then *independent of* $\lambda$ in the following sense (cf. [2] loc. cit):

LEMMA 2.2.11. *The matrices* $D_j$ *are independent of* $\lambda$. *For given* $j$, $d_j(t, s)=0$ *if* $s \not\leqslant t$ *and* $d_j(t, s)=\pm 1$ *if* $s \leqslant t$. *For* $j=1$, $d_1(1, s)=1\ \forall s$ *with* $l(s)=1$.

Given a Verma module $V_{\mathfrak{m},Q,\lambda}$, the $U(\mathfrak{h}\oplus\mathfrak{n}^-)$-module isomorphism

$$V_{\mathfrak{m},Q,\lambda} \simeq U(\mathfrak{n}^-), \qquad \left(\mathfrak{n}^- = \sum_{-\alpha\in Q} \mathfrak{m}_\alpha\right),$$

enables us to calculate the multiplicities of the weights occurring in $V_{\mathfrak{m},Q,\lambda}$. To this end we introduce the partition function $\mathbb{P}_Q$. Let $\mathbb{Z}^+$ be the set of integers $\geqslant 0$. Then, for any nonempty subset $R \subset Q$, the *partition function* $\mathbb{P}_R$ is a function from $\mathfrak{h}^*$ to $\mathbb{Z}^+$; for any $\mu \in \mathfrak{h}^*$, $\mathbb{P}_R(\mu)$ is the number of functions

$$\nu\colon R \to \mathbb{Z}^+$$

such that $\mu = \Sigma_{\alpha\in R}\nu(\alpha)$, $\alpha$, i.e.,

$$\mathbb{P}_R(\mu) = \# \{\nu\colon \nu \text{ a function from } R \text{ to } \mathbb{Z}^+, \Sigma_{\alpha\in R}\nu(\alpha)\alpha = \mu\}.$$

Clearly, $\mathbb{P}_R(\mu)=0$ unless $\mu$ is either 0 or a sum of elements of $R$; it is always finite. As $V_{\mathfrak{m},Q,\lambda}$ is a free $U(\mathfrak{n}^-)$-module,

$$\dim(V_{\mathfrak{m},Q,\lambda}[\mu]) = \mathbb{P}_Q(\lambda-\mu).$$

Finally, let

$$\mathfrak{n} = \sum_{\alpha\in Q} \mathfrak{m}_\alpha.$$

For any $\mathfrak{m}$-module $M$, let

$$M^{\mathfrak{n}} = \{v\colon v \in M, \mathfrak{n}\cdot v = 0\}.$$

If $M$ is a weight module, let

$$M[\mu]^{\mathfrak{n}} = M^{\mathfrak{n}} \cap M[\mu]$$

for any $\mu \in \mathfrak{h}^*$; in this case, $M^{\mathfrak{n}}$ is $\mathfrak{h}$-stable and

$$M^{\mathfrak{n}} = \bigoplus_{\mu} M[\mu]^{\mathfrak{n}}.$$

LEMMA 2.2.12. *The functors* $M \mapsto M^{\mathfrak{n}}$, $M \mapsto M[\mu]^{\mathfrak{n}}$ ($\mu$ *fixed*), *from the category of weight modules for* $\mathfrak{m}$ *to the category of vector spaces over* $\mathbb{C}$, *are covariant and left exact. Let* $\mathscr{B}_{\mathfrak{m},Q}$ *be the category of all* $\mathfrak{m}$-*modules* $M$ *such that* (a) $M$ *is a weight module* (b) *for each* $\alpha \in Q$, $X_\alpha$ *is locally nilpotent*[1] *on* $M$. *Then, if* $\mu \in \mathfrak{h}^*$ *is* $Q$-*dominant* (i.e., $\mu(\bar{H}_\alpha) \geqslant 0 \; \forall \alpha \in Q$), *the functor* $M \mapsto M[\mu]^{\mathfrak{n}}$ *is exact on* $\mathscr{B}_{\mathfrak{m},Q}$.

This is just Lemma 7 of Enright and Varadarajan [1].

LEMMA 2.2.13. *Let* $F$ *be a finite-dimensional* $\mathfrak{m}$-*module and* $V = V_{\mathfrak{m},Q,\mu}$ *where* $\mu \in \mathfrak{h}^*$. *Let* $M = F \otimes V$. *We then have the following.*

(i) *If* $V$ *is irreducible, we have, for any* $\nu \in \mathfrak{h}^*$,

$$\dim(M[\nu]^{\mathfrak{n}}) \leqslant \dim(F[\nu-\mu]).$$

*Moreover, if in addition* $\nu$ *is* $Q$-*dominant integral,*

$$\dim(M[\nu]^{\mathfrak{n}}) = \dim(F[\nu-\mu])$$

(ii) *Suppose* $\mu = t_0' \cdot \lambda$ *where* $\lambda$ *is* $Q$-*dominant integral. Let*

$$M_s = F \otimes V_{\mathfrak{m},Q,s' \cdot \lambda},$$

*so that* $M = M_{t_0}$. *Then, for any* $s \in \mathfrak{w}(\mathfrak{m})$, *and any* $\nu \in \mathfrak{h}^*$ *which is* $Q$-*dominant integral,*

$$\dim(M_s[\nu]^{\mathfrak{n}}) = \dim(F[s^{-1}(\nu+\delta_Q) - (\lambda+\delta_Q)]).$$

See Lemmas 5.1 and 5.2 of Enright–Wallach [1]. If $W$ is any Verma module and $W = W^{\mathfrak{n}} \oplus W'$ is a $\mathfrak{h}$-module direct sum, the projection

$$E: F \otimes W \to F \otimes W^{\mathfrak{n}}$$

parallel to $F \otimes W'$, gives rise to an injection (as is easily verified)

$$(F \otimes W)^{\mathfrak{n}} \hookrightarrow F \otimes W^{\mathfrak{n}}$$

when $E$ is restricted to $F \otimes W$. In the special case $W = V$, $\dim(V^{\mathfrak{n}}) = 1$ and we get injections

$$(F \otimes V)^{\mathfrak{n}} \hookrightarrow F, \qquad (F \otimes V)[\nu]^{\mathfrak{n}} \hookrightarrow F[\nu-\mu].$$

---

[1] An endomorphism $L$ of a vector space $V$ is called *locally nilpotent* if for each $v \in V$, $L^m v = 0$ for some integer $m = m(v) \geqslant 1$. $B_{\mathfrak{m},Q}$ is the category of weight modules which are $\mathfrak{b}$-finite, $\mathfrak{b}$ being the Borel subalgebra $\mathfrak{h} \oplus \mathfrak{n}$.

For $\nu$ dominant integral, Lemmas 2.2.4 and 2.2.12 give

$$\dim((F \otimes W)[\nu]^{\mathfrak{n}}) = \dim(F[\nu - \mu_W])$$

where $W$ is any Verma module with highest weight $\mu_W$ relative to $Q$. This proves (i) and gives the formula

$$\dim(M_s[\nu]^{\mathfrak{n}}) = \dim(F[s^{-1}(\nu + \delta_Q) - (\lambda + \delta_Q)]).$$

It can be shown that

$$\dim(M_s[\nu]^{\mathfrak{n}}) = \dim(M_{t_0}[(t_0 s^{-1})^{\cdot} \cdot \nu]^{\mathfrak{n}})$$

as a consequence of Proposition 4.7.2 to be discussed in Section 4. Since $V_{\mathfrak{m},Q,t_0' \cdot \lambda}$ is irreducible, we of course have

$$\dim(M_{t_0}[(t_0 s^{-1})^{\cdot} \cdot \nu]^{\mathfrak{n}}) \leqslant \dim(F[s^{-1}(\nu + \delta_Q) - (\lambda + \delta_Q)])$$

by (i).

LEMMA 2.2.14. *Let $M$ be a module in the category $\mathscr{B}_{\mathfrak{m},Q}$ of Lemma 2.2.12.*

(a) $M = \bigoplus_\chi M_\chi$ *where the sum is over the homomorphisms $\chi$ of $\mathfrak{Z}(\mathfrak{m})$ into $\mathbb{C}$, $\mathfrak{Z}(\mathfrak{m})$ being the center of $U(\mathfrak{m})$; and for each $\chi$, $M_\chi$ is the largest submodule of $M$ on which $z - \chi(z)$ is locally nilpotent for each $z \in \mathfrak{Z}(\mathfrak{m})$.*

(b) *If $M$ has a flag of submodules*

$$M = M_1 \supset M_2 \supset \cdots \supset M_r \supset 0 = M_{r+1}$$

*such that for suitable $\nu_1, \ldots, \nu_r \in \mathfrak{h}^*$,*

$$M_i/M_{i+1} \simeq V_{\mathfrak{m},Q,\nu_i}, \qquad (1 \leqslant i \leqslant r),$$

*and if all the $\nu_i + \delta_Q$ are dominant with respect to some positive system $Q'$ of roots of $(\mathfrak{m}, \mathfrak{h})$ ($Q'$ may not be the same as $Q$), then*

$$M \simeq \bigoplus_{1 \leqslant i \leqslant r} V_{\mathfrak{m},Q,\nu_i}.$$

For proving (b), we may assume $M = M_\chi$ so that all the $\nu_i + \delta_Q$ are in the same $\mathfrak{w}(\mathfrak{m})$-orbit; then, the fact that *all* of them are $Q'$-dominant implies that all of them are identical. So $\nu_1 = \cdots = \nu_r = \nu$. Then $\nu$ is the highest weight of $M$, implying that $X_\gamma(M[\nu]) = 0 \ \forall \gamma \in Q$. If $v_i \in M[\nu]$ such that $v_i$ goes over to the highest weight vector of $M_i/M_{i+1} \simeq V_{\mathfrak{m},Q,\nu}$, it is now clear that for $1 \leqslant i \leqslant r$

$$M_i \simeq (U(\mathfrak{m}) \cdot v_i) \oplus M_{i+1}, \qquad U(\mathfrak{m}) \cdot v_i \simeq V_{\mathfrak{m},Q,v}.$$

This gives (b).

**2.3.** We return to $G$, $K$, $\mathfrak{g}_0$, $\mathfrak{k}_0$, $\mathfrak{g}$, $\mathfrak{k}$. *We assume*[1] *that* $\mathrm{rk}(G)=\mathrm{rk}(K)$. Let $\mathfrak{g}_0=\mathfrak{k}_0 \oplus \mathfrak{p}_0$ be a Cartan decomposition of $\mathfrak{g}_0$ and $\mathfrak{p}=\mathbb{C} \cdot \mathfrak{p}_0$. *We also assume that* $[\mathfrak{p}, \mathfrak{p}]=\mathfrak{k}$ (which is equivalent to assuming that $\mathfrak{g}_0$ has no ideals $\subset \mathfrak{k}_0$). Let $\mathfrak{h}_0 \subset \mathfrak{k}_0$ be a CSA of $\mathfrak{g}_0$ and $\mathfrak{h}=\mathbb{C} \cdot \mathfrak{h}_0$. Given a positive system $P$ of roots of $(\mathfrak{g}, \mathfrak{h})$, we are interested in constructing $P$-extreme modules for $\mathfrak{g}$ which are Harish-Chandra modules; since the $P$-extreme modules have finite-dimensional weight spaces, this comes down to requiring that their highest weight vectors be $\mathfrak{k}$-finite.

Let $P_{\mathfrak{k}}$ (resp. $P_n$) be the set of compact (resp. noncompact) roots; here, as usual, we call a root $\alpha$ *compact* or *noncompact* according as $\mathfrak{g}_\alpha \subset \mathfrak{k}$ or $\subset \mathfrak{p}$. A root $\alpha \in P_n$ is said to be *totally positive* (t.p.) if $\exists$ a set $P_\alpha \subset P_n$ such that $\alpha \in P_\alpha$ and $\Sigma_{\beta \in P_\alpha} \mathfrak{g}_\beta$ is stable under $\mathrm{ad}(\mathfrak{k})$. We can always write $\mathfrak{g}_0=\mathfrak{g}_0^{(1)} \oplus \mathfrak{g}_0^{(2)}$ where the $\mathfrak{g}_0^{(i)}$ are ideals[1] of $\mathfrak{g}_0$ having the following property[2]: if $P^{(i)}$ is the set of roots of $P$ belonging to $\mathfrak{g}^{(i)}$, then $P^{(1)}$ has no t.p. root while all roots in $P_n^{(2)}$ are t.p.; it is possible that $\mathfrak{g}_0^{(1)}$ or $\mathfrak{g}_0^{(2)}$ could be 0. This decomposition depends on $P$ but otherwise canonical. It now turns out that if $M$ is a $P$-extreme module for $\mathfrak{g}$ which is $\mathfrak{k}$-finite, then $M$ is a quotient of $M^{(1)} \otimes M^{(2)}$ where $M^{(1)}$ is a finite-dimensional $\mathfrak{g}^{(1)}$-module and $M^{(2)}$ is a $P^{(2)}$-extreme $\mathfrak{k}^{(2)}$-finite $\mathfrak{g}^{(2)}$-module; for irreducible $M$ we even have $M \simeq M^{(1)} \otimes M^{(2)}$. Since we are primarily interested in unitary modules, we may assume $M^{(1)}$ is trivial. So we may as well assume that $\mathfrak{g}_0^{(1)}=0$, i.e., each root in $P_n$ is t.p.; $P$ or $(\mathfrak{g}, \mathfrak{h}, P)$ is then called *admissible.*

From now on we shall assume that $(\mathfrak{g}, \mathfrak{h})$ has admissible positive systems and that $P$ is one of these. A crucial property of $P$ is that

$$\beta \in P_{\mathfrak{k}}, \qquad \beta \text{ simple in } P_{\mathfrak{k}} \Rightarrow \beta \text{ simple in } P.$$

More generally, $\mathfrak{c}_0=\mathrm{center}(\mathfrak{k}_0) \neq 0$; $\dim(\mathfrak{c}_0)=m$ is the number of simple ideals of $\mathfrak{g}_0$; and if $S$ is the set of simple roots in $P$, then

$$S=\{\alpha_1, \ldots, \alpha_k, \beta_1, \ldots, \beta_m\}$$

---

[1] As $\mathrm{rk}(\mathfrak{g}_0)=\mathrm{rk}(\mathfrak{k}_0)$, all ideals of $\mathfrak{g}$ are stable under both the Cartan involution and complex conjugation (of $\mathfrak{g}$ with respect to $\mathfrak{g}_0$).

[2] This follows from the fact that if $\alpha$, $\alpha'$ are noncompact roots with $\alpha$ t.p., then $\alpha+\alpha'$ is not a root unless $-\alpha'$ is t.p.; in this case $\alpha+\alpha'$ is compact.

where (a) $\{\alpha_1, \ldots, \alpha_k\}$ is the set of simple roots of the positive system $P_{\mathfrak{k}}$ of $(\mathfrak{k}, \mathfrak{h})$ (b) the $\beta_j$ are in $P_n$ and $\beta_j|\mathfrak{c}_0 \neq 0 \; \forall j$ (c) if $\gamma \in P_n$, there is a unique $j$ with $1 \leqslant j \leqslant m$ and integers $c_i \geqslant 0 \; (1 \leqslant i \leqslant k)$ such that

$$\gamma = \beta_j + c_1 \alpha_1 + \cdots + c_k \alpha_k.$$

We note that for $\lambda \in \mathfrak{h}^*$,

$$\lambda(\overline{H}_\alpha) \text{ is an integer} \geqslant 0 \; \forall \alpha \in P_{\mathfrak{k}}$$
$$\Leftrightarrow \lambda(\overline{H}_{\alpha_i}) \text{ is an integer} \geqslant 0 \text{ for } 1 \leqslant i \leqslant k.$$

We have two $\delta$'s:

$$\delta_P = \tfrac{1}{2} \sum_{\alpha \in P} \alpha \qquad \delta_{P_{\mathfrak{k}}} = \tfrac{1}{2} \sum_{\alpha \in P_{\mathfrak{k}}} \alpha.$$

But, *since $P_{\mathfrak{k}}$-simple roots are $P$-simple,* $\delta_P(\overline{H}_\alpha) = \delta_{P_{\mathfrak{k}}}(\overline{H}_\alpha) = 1$ for $P_{\mathfrak{k}}$-simple $\alpha \in P_{\mathfrak{k}}$. The two affine actions of $\mathfrak{w}(\mathfrak{k})$, one with respect to $\delta_P$, and the other with respect to $\delta_{P_{\mathfrak{k}}}$, coincide; we write, without ambiguity,

$$s' \cdot \lambda = s(\lambda + \delta_{P_{\mathfrak{k}}}) - \delta_{P_{\mathfrak{k}}} = s(\lambda + \delta_P) - \delta_P \qquad (s \in \mathfrak{w}(\mathfrak{l}), \lambda \in \mathfrak{h}^*).$$

PROPOSITION 2.3.1.   *Let $(\mathfrak{g}, \mathfrak{h}, P)$ be admissible, $\lambda \in \mathfrak{h}^*$. If there is a P-extreme Harish-Chandra module of highest weight $\lambda$, then $\lambda(\overline{H}_\alpha)$ is an integer $\geqslant 0 \; \forall \alpha \in P_{\mathfrak{k}}$. Conversely, if $\lambda \in \mathfrak{h}^*$ is such that $\lambda(\overline{H}_\alpha)$ is an integer $\geqslant 0 \; \forall \alpha \in P_{\mathfrak{k}}$, the irreducible P-extreme module $I_{\mathfrak{g},P,\lambda}$ is necessarily a Harish-Chandra module. Any $\lambda \in \mathfrak{h}^*$ such that $\lambda(\overline{H}_{\alpha_i})$ is an integer $\geqslant 0$ for $1 \leqslant i \leqslant k$ and $\lambda(\overline{H}_{\beta_j}) \in \mathbb{C}$ for $1 \leqslant j \leqslant m$, has this property.*

We now construct, for such $\lambda$, the *universal P-extreme Harish-Chandra module,* i.e., the P-extreme Harish-Chandra module of highest weight $\lambda$ having every Harish-Chandra module of highest weight $\lambda$ as its quotient.

PROPOSITION 2.3.2.   *Let P be an admissible positive system for $(\mathfrak{g}, \mathfrak{h})$ and $\lambda \in \mathfrak{h}^*$ such that $\lambda(\overline{H}_\alpha)$ is an integer $\geqslant 0 \; \forall \alpha \in P_{\mathfrak{k}}$. Then, there is an imbedding*

$$V_{\mathfrak{g},P,s' \cdot \lambda} \hookrightarrow V_{\mathfrak{g},P,\lambda}, \qquad (s \in \mathfrak{w}(\mathfrak{k})).$$

*Moreover,*

$$J_{\mathfrak{g},P,\lambda} = V_{\mathfrak{g},P,\lambda} \Big/ \sum_{\substack{s \in \mathfrak{w}(\mathfrak{l}) \\ s \neq 1}} V_{\mathfrak{g},P,s' \cdot \lambda}$$

*is nonzero and is the universal P-extreme Harish-Chandra module of highest weight* $\lambda$.

The two affine actions of $\mathfrak{w}(\mathfrak{k})$ coincide, as we have already remarked. This gives

$$V_{\mathfrak{g},P,s'\cdot\lambda} \hookrightarrow V_{\mathfrak{g},P,\lambda},$$

so that the $V_{\mathfrak{g},P,s'\cdot\lambda}$ may be identified with their images in $V_{\mathfrak{g},P,\lambda}$. Indeed, if $s = s_{\beta_r} \ldots s_{\beta_1}$ is a reduced expression for $s \in \mathfrak{w}(\mathfrak{k})$ (i.e., with $l(s) = r$ where $l$ is the length function on $\mathfrak{w}(\mathfrak{k})$ relative to $P_{\mathfrak{k}}$), and if $s_j = s_{\beta_j} \cdot s_{\beta_{j-1}} \ldots s_{\beta_1}$, then $\beta_1, s_{\beta_1} \cdot \beta_2, s_{\beta_2} \cdot \beta_3, \ldots$ are all in $P_{\mathfrak{k}}$, so that

$$\lambda \succ_P s_1' \cdot \lambda \succ_P s_2' s_1' \cdot \lambda \ldots \succ_P s' \cdot \lambda.$$

This implies

$$V_{\mathfrak{g},P,s'\cdot\lambda} \hookrightarrow V_{\mathfrak{g},P,s'_{r-1}\cdot\lambda} \hookrightarrow \cdots \hookrightarrow V_{\mathfrak{g},P,s_1'\cdot\lambda} \hookrightarrow V_{\mathfrak{g},P,\lambda}.$$

$$\sum_{1 \neq s \in \mathfrak{w}(\mathfrak{k})} V_{\mathfrak{g},P,s'\cdot\lambda} = \sum_{1 \leqslant i \leqslant k} V_{\mathfrak{g},P,s'_{\alpha_i}\cdot\lambda}.$$

Of course it often happens that $J_{\mathfrak{g},P,\lambda} \simeq I_{\mathfrak{g},P,\lambda}$. We in fact have

PROPOSITION 2.3.3. *Let P be admissible and* $\lambda \in \mathfrak{h}^*$ *such that* $\lambda(\bar{H}_\alpha)$ *is an integer* $\geqslant 0$ $\forall \alpha \in P_{\mathfrak{k}}$. *Write* $\delta_P = \frac{1}{2}\Sigma_{\alpha \in P}\alpha$. *Let* $t_0 \in \mathfrak{w}(\mathfrak{k})$ *be such that* $t_0 \cdot P_{\mathfrak{k}} = -P_{\mathfrak{k}}$. *Then we have the following*

(i) $V_{\mathfrak{g},P,t_0'\cdot\lambda}$ *is irreducible* $\Leftrightarrow (\lambda + \delta_P)(\bar{H}_\alpha \notin \{1, 2, \ldots\}$ $\forall \alpha \in P_n$.

(ii) *If* $V_{\mathfrak{g},P,t_0'\cdot\lambda}$ *is irreducible, then* $J_{\mathfrak{g},P,\lambda}$ *is irreducible.*

For all of this, see Harish-Chandra [2a]. In [2a] he showed that

$$I_{\mathfrak{g},P,\lambda} \text{ unitary} \Rightarrow \lambda \text{ is real on } (-1)^{1/2}\mathfrak{h}_0 \text{ and } \lambda(\bar{H}_\alpha) \leqslant 0 \ \forall \alpha \in P_n$$

(note that $(-1)^{1/2}\mathfrak{h}_0$ is spanned by the $\bar{H}_\alpha$ for $\alpha \in P$); and ([2b, 2c]) that

$$(\lambda + \delta_P)(\bar{H}_\alpha) \leqslant 0 \ \forall \alpha \in P_n \Rightarrow I_{\mathfrak{g},P,\lambda} (\simeq J_{\mathfrak{g},P,\lambda}) \text{ is unitary}$$

(that $I_{\mathfrak{g},P,\lambda} \simeq J_{\mathfrak{g},P,\lambda}$ follows from Proposition 2.3.3). It is not difficult to use these results together with the irreducibility criterion of Proposition 2.3.3 to conclude that

$$(U) \qquad (\lambda + \delta_P)(\bar{H}_\alpha) \leqslant 1 \ \forall \alpha \in P_n \Rightarrow I_{\mathfrak{g},P,\lambda} \text{ is unitary.}$$

[The determination of the $\lambda$ for which $I_{\mathfrak{g},P,\lambda}$ is unitary has attracted some attention (see Wallach [2], Rossi-Vergne [1]). The most complete results appear to have been obtained by R. Parthasarathy (personal communication). Let me give a brief description of Parthasarathy's result. We assume that the group $G$ is the real form of the simply connected complex group corresponding to $\mathfrak{g}$.

We fix a Cartan subalgebra $\mathfrak{h}_0$ of $\mathfrak{g}_0$ such that $\mathfrak{h}_0 \subseteq \mathfrak{k}_0$. $P$ is the positive system of roots of $(\mathfrak{g}, \mathfrak{h})$ adapted to the complex structure of $G/K$ (i.e. regular integral linear forms on $\mathfrak{h}$ dominant with respect to $P$ parameterize, in the sense of Harish-Chandra, the holomorphic discrete series of $G$). $P_\mathfrak{k}$ is the set of compact roots in $P$. The Borel subalgebra of $\mathfrak{g}$-determined by $P$ is denoted by $\mathfrak{b}$. For any parabolic subalgebra $\mathfrak{q}$ of $\mathfrak{g}$ containing $\mathfrak{b}$, we define a linear form $\delta_n(\mathfrak{q})$ on $\mathfrak{h}$ as follows:

Let $\mathfrak{u}$ be the nilradical of $\mathfrak{q}$. Then, with respect to the Cartan decomposition of $\mathfrak{g}$, $\mathfrak{u}$ splits as

$$\mathfrak{u} = \mathfrak{u}_\mathfrak{k} \oplus \mathfrak{u}_\mathfrak{p}, \qquad (\mathfrak{u}_\mathfrak{k} = \mathfrak{u} \cap \mathfrak{k},\ \mathfrak{u}_\mathfrak{p} = \mathfrak{u} \cap \mathfrak{p}).$$

We put

$$\delta_n(\mathfrak{q})(H) = \tfrac{1}{2}\mathrm{tr}\,(\mathrm{ad}\, H)_{\mathfrak{u}_\mathfrak{p}}.$$

Let now $\Lambda$ be an integral linear form on $\mathfrak{h}$ which is dominant with respect to $P$ and let $\langle \Lambda, \alpha \rangle = 0$ for all $\alpha \in P$ whose root space is not contained in $\mathfrak{u}$. Then

THEOREM (PARTHASARATHY). *If* $\mu = \Lambda + 2\delta_n(\mathfrak{q})$, *the corresponding irreducible highest weight representation is unitarizable.*]

The Harish-Chandra module $I_{\mathfrak{g},P,\lambda}$ can now be characterized using the criterion of Theorem 1.3.9.

PROPOSITION 2.3.4. *Let* $(\mathfrak{g}, \mathfrak{h}, P)$ *be admissible,* $\lambda \in \mathfrak{h}^*$ *and let* $\lambda(\bar{H}_\alpha)$ *be an integer* $\geq 0\ \forall \alpha \in P_\mathfrak{k}$. *Then any P-extreme Harish-Chandra module M with highest weight* $\lambda$ *has the following properties*:

(i) *If* $\vartheta(\lambda) \in \mathscr{E}(\mathfrak{k})$ *is the class with* $P_\mathfrak{k}$*-highest weight* $\lambda$, *then* $[M: \vartheta(\lambda)] = 1$, i.e., $\vartheta(\lambda)$ *occurs in M with multiplicity* 1.

(ii) *Let* $\mathfrak{D} = U(\mathfrak{g})^\mathfrak{k}$ *and let* $t_0 \in \mathfrak{w}(\mathfrak{k})$ *be the element such that* $t_0 \cdot P_\mathfrak{k} = -P_\mathfrak{k}$; *then* $\mathfrak{D}$ *acts on* $M_{\vartheta(\lambda)}$ *through the homomorphism*[1]

[1]One can argue as in Theorem 4.10.3 of Varadarajan [2] that for $u \in \mathfrak{D}$, $\boldsymbol{\beta}_P(u)$ is a $\mathfrak{w}(\mathfrak{k})'$-invariant element of $U(\mathfrak{h})$, using the fact that $P_\mathfrak{k}$-simple roots of $P_\mathfrak{k}$ are $P$-simple.

$$u \mapsto \boldsymbol{\beta}_P(u)(t'_0 \cdot \lambda) = \boldsymbol{\beta}_P(u)(\lambda), \qquad (u \in \mathfrak{D}).$$

*In particular,* $J_{\mathfrak{g},P,\lambda}$ *has these properties, while the Harish-Chandra module* $I_{\mathfrak{g},P,\lambda}$ *is the unique irreducible one with properties* (i) *and* (ii).

Concerning the $G$-modules corresponding to the $J_{\mathfrak{g},P,\lambda}$ we have the following (see Harish-Chandra [2b, 2c]):

PROPOSITION 2.3.5. *Let* $\mathrm{rk}(\mathfrak{g}) = \mathrm{rk}(\mathfrak{k})$ *and* $[\mathfrak{p}, \mathfrak{p}] = \mathfrak{k}$. *Let* $\tilde{G}$ *be the simply connected group corresponding to* $\mathfrak{g}_0$. *Let* $P$ *be an admissible positive system of roots of* $(\mathfrak{g}, \mathfrak{h})$. *Suppose* $\lambda \in \mathfrak{h}^*$ *is such that*

(i) $(\lambda + \delta_P)(\bar{H}_\alpha)$ *is an integer* $> 0\ \forall \alpha \in P_{\mathfrak{k}}$,

(ii) $(\lambda + \delta_P)(\bar{H}_\alpha) \leqslant 0\ \forall \alpha \in P_n$.

*Then the Harish-Chandra module* $J_{\mathfrak{g},P,\lambda}$ *lifts to an irreducible unitary representation of* $\tilde{G}$. *If, in addition,* $\Lambda = \lambda + \delta_P$ *is regular, i.e., if the set of conditions* (ii) *above is sharpened to*

(iii) $(\lambda + \delta_P)(\bar{H}_\alpha) < 0\ \forall \alpha \in P_n$,

*then the matrix coefficients of the corresponding representation of* $\tilde{G}$ *are square integrable modulo the center of* $\tilde{G}$; *and in case* $\lambda$ *lies in the lattice of characters of the CSG of K corresponding to* $\mathfrak{h}_0$, *this representation is already defined on G and belongs to* $\mathscr{E}_2(G)$. *In particular, if G is the real analytic group defined by* $\mathfrak{g}_0$ *in the simply connected complex group corresponding to* $\mathfrak{g}$, *and if* $\lambda$ *is integral (i.e.,* $(\lambda + \delta_P)(\bar{H}_\alpha)$ *is an integer* $\forall \alpha \in P$, *which is* $>0$ *or* $<0$ *according as* $\alpha \in P_{\mathfrak{k}}$ *or* $\alpha \in P_n$), *this representation lies in the discrete class* $\omega(\lambda + \delta_P)$ *of* $\mathscr{E}_2(G)$.

**2.4.** For $\mathfrak{g}, \mathfrak{k}$ as above, not all positive systems are admissible; moreover, we can have $\mathrm{rk}(\mathfrak{g}) = \mathrm{rk}(\mathfrak{k})$ without $(\mathfrak{g}, \mathfrak{h})$ having any admissible positive system. *On the other hand, most of the assertions of Propositions* 2.3.4 *and* 2.3.5 *are perfectly meaningful under the sole assumption that* $\mathrm{rk}(\mathfrak{g}) = \mathrm{rk}(\mathfrak{k})$. So the following questions arise naturally:

(1) Given $\lambda \in \mathfrak{h}^*$ such that $\lambda(\bar{H}_\alpha)$ is an integer $\geqslant 0\ \forall \alpha \in P_{\mathfrak{k}}$, does an irreducible Harish-Chandra module with properties (i) and (ii) of Proposition 2.3.4 exist?

(2) Suppose such modules exist; calling them $I_{\mathfrak{g},P,\lambda}$, is it true that $I_{\mathfrak{g},P,\lambda}$ comes from the discrete class $\omega(\lambda + \delta_P)$, when $\lambda + \delta_P$ is regular?

In the paper of Enright–Varadarajan [1], the existence of the Harish-Chandra modules with the required properties was established in full generality; the identification with the discrete series was carried out when $\lambda$ is 'sufficiently regular'; for *all* regular $\lambda+\delta_P$ this was done by Wallach [1]. The construction, by purely infinitesimal methods, of such Harish-Chandra $\mathfrak{g}$-modules is my next theme. Before taking it up however, I want to look at the modules $J_{\mathfrak{g},P,\lambda}$ a little more closely so that their structure as $\mathfrak{k}$-modules may be determined.

**2.5.** Consider for example the case $G=SU(2, 1)$, $K=S(U(2)\times U(1))$. Here $\mathfrak{k}$ has one-dimensional center and $[\mathfrak{k}, \mathfrak{k}]\approx\mathfrak{sl}(2, \mathbb{C})$. Fix a CSA $\mathfrak{h}\subset\mathfrak{k}$ and choose one of the two roots of $(\mathfrak{k}, \mathfrak{h})$ to be positive, say $\alpha$. Then the positive systems of $(\mathfrak{g}, \mathfrak{h})$ that contain $\alpha$ are of the form

$$(\alpha, \beta, \alpha+\beta), \quad (\alpha, -\beta, -\alpha-\beta), \quad (\alpha, -\beta, \alpha+\beta).$$

The first two are admissible but the third one is not. There will be three distinct families of discrete classes, one corresponding to each of these positive systems. *Only the families coming from the first two positive systems can be accounted for by highest weight modules.*

Returning to the set-up of Proposition 2.3.5, we observe that if $(\lambda+\delta_P)(\bar{H}_\alpha)$ is allowed to vanish for some $\alpha\in P_n$, $J_{\mathfrak{g},P,\lambda}$ is no longer a discrete class. For obvious reasons they are called *limits of discrete classes* or, as we prefer to do so, following Lang [1], *mock discrete classes*.

**2.6.** Let us return to the context of Proposition 2.3.2. We wish to examine the structure of $J_{\mathfrak{g},P,\lambda}$ as a $\mathfrak{k}$-module. To do this it is more convenient to describe the $J_{\mathfrak{g},P,\lambda}$ in a slightly different manner. Let

$$\mathfrak{p}^+=\sum_{\alpha\in P_n}\mathfrak{g}_\alpha, \qquad \mathfrak{p}^-=\sum_{\alpha\in -P_n}\mathfrak{g}_\alpha.$$

Then $\mathfrak{g}=\mathfrak{p}^-+\mathfrak{k}+\mathfrak{p}^+$ is a direct sum; $\mathfrak{p}^\pm$ are ad($\mathfrak{k}$)-stable *Abelian* subalgebras; and $\mathfrak{p}^\pm$ are ideals in $\mathfrak{k}+\mathfrak{p}^\pm$ respectively. Write $\mathfrak{q}^\pm=\mathfrak{k}+\mathfrak{p}^\pm$.

Suppose $m_\lambda$ is the highest weight vector (of weight $\lambda$) in $J_{\mathfrak{g},P,\lambda}$ and let $M_\lambda^+=U(\mathfrak{q}^+)\cdot m_\lambda$. Using the relations

$$U(\mathfrak{g})=U(\mathfrak{p}^-)U(\mathfrak{k})U(\mathfrak{p}^+),$$
$$U(\mathfrak{q}^+)=U(\mathfrak{k})U(\mathfrak{p}^+)=U(\mathfrak{p}^+)U(\mathfrak{k}),$$

and the fact that $\lambda-\theta+\beta$ is not a weight of $J_{\mathfrak{g},P,\lambda}$ if $\beta\in P_n$ and $\theta$ is a sum of positive *compact* roots, one finds that $M_\lambda^+$ is stable under $\mathfrak{q}^+$, that $\mathfrak{p}^+$ acts trivially on it, and that it is irreducible as a $\mathfrak{k}$-module with class

$\vartheta(\lambda)$; and furthermore, that $J_{\mathfrak{g},P,\lambda}=U(\mathfrak{p}^-)\cdot M_\lambda^+$. Remembering that $J_{\mathfrak{g},P,\lambda}$ is universal, we get:

PROPOSITION 2.6.1. *Let* $\mathfrak{g}$, $P$, $\lambda$ *be as in Proposition 2.3.2. Let*[1] $I_{\mathfrak{q}^+,P_{\mathfrak{k}},\lambda}$ *be the* $\mathfrak{q}^+$*-module on which* $\mathfrak{p}^+$ *acts trivially and which is, as a* $\mathfrak{k}$*-module, isomorphic to* $I_{\mathfrak{k},P_{\mathfrak{k}},\lambda}$. *Then we have a* $\mathfrak{g}$*-module isomorphism*

$$J_{\mathfrak{g},P,\lambda}\simeq U(\mathfrak{g})\bigotimes_{U(\mathfrak{q}^+)} I_{\mathfrak{q}^+,P_{\mathfrak{k}},\lambda}.$$

COROLLARY 2.6.2. *Let* $I_{\mathfrak{q}^-,P_{\mathfrak{k}},\lambda}$ *be the* $\mathfrak{q}^-$*-module on which* $\mathfrak{p}^-$ *acts trivially and which is, as a* $\mathfrak{k}$*-module, isomorphic to* $I_{\mathfrak{k},P_{\mathfrak{k}},\lambda}$. *Let us regard* $U(\mathfrak{p}^-)$ *as a* $U(\mathfrak{q}^-)$*-module, with* $\mathfrak{k}$ *acting through the adjoint representation, and elements of* $U(\mathfrak{p}^-)$ *through left multiplication. Then, we have a* $U(\mathfrak{q}^-)$*-module isomorphism*

$$J_{\mathfrak{g},P,\lambda}\simeq U(\mathfrak{p}^-)\otimes I_{\mathfrak{q}^-,P_{\mathfrak{k}},\lambda}.$$

We can obtain several important consequences from these results. First we have the following result on the weight spaces.

COROLLARY 2.6.3. *Let* $\Xi_\lambda$ *be the set of weights of* $I_{\mathfrak{k},P_{\mathfrak{k}},\lambda}$. *Then,*

$$J_{\mathfrak{g},P,\lambda}[\mu]\simeq \bigoplus_{\nu\in\Xi_\lambda} U(\mathfrak{p}^-)[\mu-\nu]\otimes I_{\mathfrak{k},P_{\mathfrak{k}},\lambda}[\nu].$$

To calculate dimensions of weight spaces we need to introduce the partition functions. Let $\mathbb{P}_{P_n}$ be the function from $\mathfrak{h}^*$ to $\mathbb{Z}^+$ (set of integers $\geqslant 0$) such that, for any $\mu\in\mathfrak{h}^*$,

$$\mathbb{P}_{P_n}(\mu)=\#\left\{\text{functions } \nu\colon P_n\to\mathbb{Z}^+ \text{ such that } \mu=\sum_{\alpha\in P_n}\nu(\alpha)\alpha\right\}.$$

Then, expressing $\dim(U(\mathfrak{p}^-)[\mu-\nu])$ as $\mathbb{P}_{P_n}(\nu-\mu)$, we get

$$\dim J_{\mathfrak{g},P,\lambda}[\mu]=\sum_{\nu\in\Xi_\lambda}\dim(I_{\mathfrak{k},P_{\mathfrak{k}},\lambda}[\nu])\cdot\mathbb{P}_{P_n}(\nu-\mu).$$

One can use this formula to compute, as Harish-Chandra did in [2], the character of the representation of $G$ corresponding to $J_{\mathfrak{g},P,\lambda}$ on the regular

[1] Recall that $\mathfrak{p}^+$ is an ideal in $\mathfrak{q}^+$ and $\mathfrak{q}^+/\mathfrak{p}^+\simeq\mathfrak{k}$ so that any $\mathfrak{k}$-module may be regarded as a $\mathfrak{q}^+$-module with $\mathfrak{p}^+$ acting trivially.

points of the CSG of $G$ defined by $\mathfrak{h}_0$. Such a calculation would prove that for $\lambda+\delta_P$ as in Proposition 2.3.5 and regular, $J_{\mathfrak{g},P,\lambda}$ comes from the discrete class $\omega(\lambda+\delta_P)$.

COROLLARY 2.6.4. *$J_{\mathfrak{g},P,\lambda}$ is a free $U(\mathfrak{p}^-)$-module of rank equal to* $\dim(I_{\mathfrak{k},P\mathfrak{k},\lambda})$.

For $I_{\mathfrak{k},P\mathfrak{k},\lambda}$ we have the B–G–G resolution (Lemma 2.2.10)

$$0\to B^{(r)}_{\mathfrak{k},P\mathfrak{k},\lambda}\to B^{(r-1)}_{\mathfrak{k},P\mathfrak{k},\lambda}\to\cdots\to B^{(1)}_{\mathfrak{k},P\mathfrak{k},\lambda}\to V_{\mathfrak{k},P\mathfrak{k},\lambda}\to I_{\mathfrak{k},P\mathfrak{k},\lambda}\to 0.$$

We regard the $\mathfrak{k}$-modules occurring here as $\mathfrak{q}^+$-modules in the usual way and so the above is a resolution in the category of $\mathfrak{q}^+$-modules. Since $U(\mathfrak{g})$ is a *free* right $U(\mathfrak{q}^+)$-module, we can tensor the above resolution with $U(\mathfrak{g})$ over $U(\mathfrak{q}^+)$ to get a resolution of $J_{\mathfrak{g},P,\lambda}$.

PROPOSITION 2.6.5. *Let $\mathfrak{g}$, $P$, $\lambda$ be as in Proposition 2.3.2. Then we have an exact resolution*

$$0\to E^{(r)}_{\mathfrak{g},P,\lambda}\to E^{(r-1)}_{\mathfrak{g},P,\lambda}\to\cdots\to E^{(1)}_{\mathfrak{g},P,\lambda}\to E^{(0)}_{\mathfrak{g},P,\lambda}\to J_{\mathfrak{g},P,\lambda}\to 0$$

*where $r$ is the length of the longest element in $\mathfrak{w}(\mathfrak{k})$, and the $E^{(i)}_{\mathfrak{g},P,\lambda}$ are $\mathfrak{g}$-modules given by*

$$E^{(0)}_{\mathfrak{g},P,\lambda}=U(\mathfrak{g})\underset{U(\mathfrak{q}^+)}{\otimes} V_{\mathfrak{q}^+,P\mathfrak{k},\lambda},$$

$$E^{(i)}_{\mathfrak{g},P,\lambda}=U(\mathfrak{g})\underset{U(\mathfrak{q}^+)}{\otimes} B^{(i)}_{\mathfrak{q}^+,P\mathfrak{k},\lambda},\qquad (i>0)$$

*(here $V_{\mathfrak{q}^+,P\mathfrak{k},\lambda}$, $B^{(i)}_{\mathfrak{q}^+,P\mathfrak{k},\lambda}$ are the corresponding modules $V_{\mathfrak{k},P\mathfrak{k},\lambda}$, $B^{(i)}_{\mathfrak{k},P\mathfrak{k},\lambda}$, regarded (trivially) as $\mathfrak{q}^+$-modules).*

We can use this resolution to determine the multiplicities with which the irreducible classes of $\mathfrak{k}$ occur in $J_{\mathfrak{g},P,\lambda}$. For brevity, write $E^{(i)}=E^{(i)}_{\mathfrak{g},P,\lambda}$, $J=J_{\mathfrak{g},P,\lambda}$. For any $\mathfrak{k}$-module $M$, write $M^{\mathfrak{n}_\mathfrak{k}^+}$ for the subspace

$$M^{\mathfrak{n}_\mathfrak{k}^+}=\{v\colon v\in M,\ \mathfrak{n}_\mathfrak{k}^+\cdot v=0\}\qquad \mathfrak{n}_\mathfrak{k}^+=\sum_{\alpha\in P\mathfrak{k}}\mathfrak{k}_\alpha;$$

if $M$ is a weight module for $\mathfrak{h}$, $M^{\mathfrak{n}_\mathfrak{k}^+}$ is $\mathfrak{h}$-stable; we then write, for any $\mu\in\mathfrak{h}^*$,

$$M[\mu]^{\mathfrak{n}_\mathfrak{k}^+}=M^{\mathfrak{n}_\mathfrak{k}^+}\cap M[\mu].$$

Then, we can make use of Lemma 2.2.12 to obtain, for any $P_{\mathfrak{k}}$*-dominant integral* $\nu \in \mathfrak{h}^*$, the following *exact* resolution:

$$0 \to E^{(r)}[\nu]^{\mathfrak{n}\mathfrak{k}^+} \to E^{(r-1)}[\nu]^{\mathfrak{n}\mathfrak{k}^+} \to \cdots \to E^{(0)}[\nu]^{\mathfrak{n}\mathfrak{k}^+} \to J[\nu]^{\mathfrak{n}\mathfrak{k}^+} \to 0$$

Let us write, for $s \in \mathfrak{w}(\mathfrak{k})$

$$E_s = U(\mathfrak{g}) \bigotimes_{U(\mathfrak{q}^+)} V_{\mathfrak{q}^+, P_{\mathfrak{k}}, s' \cdot \nu}.$$

Then the exact resolution obtained above gives the formula

$$(^{**}) \qquad \dim(J[\nu]^{\mathfrak{n}\mathfrak{k}^+}) = \sum_{s \in \mathfrak{w}(\mathfrak{k})} \varepsilon(s) \dim(E_s[\nu]^{\mathfrak{n}\mathfrak{k}^+}),$$

assuming of course that these dimensions are all finite. To calculate the dimensions of the spaces $E_s[\nu]^{\mathfrak{n}\mathfrak{k}^+}$ we need to appeal to Lemma 2.2.13. In fact, by that lemma,

$$\dim((F \otimes V_{\mathfrak{k}, P_{\mathfrak{k}}, s' \cdot \lambda})[\nu]^{\mathfrak{n}\mathfrak{k}^+}) = \dim(F[s^{-1}(\nu + \delta_{P_{\mathfrak{k}}}) - (\lambda + \delta_{P_{\mathfrak{k}}})])$$

for any finite-dimensional $\mathfrak{k}$-module. Hence

$$\dim(E_s[\nu]^{\mathfrak{n}\mathfrak{k}^+}) = \dim(U(\mathfrak{p}^-)[s^{-1}(\nu + \delta_{P_{\mathfrak{k}}}) - (\lambda + \delta_{P_{\mathfrak{k}}})]).$$

In deriving this we must observe that for any $s \in \mathfrak{w}(\mathfrak{k})$

$$E_s \simeq U(\mathfrak{p}^-) \otimes V_{\mathfrak{k}, P_{\mathfrak{k}}, s' \cdot \lambda}, \qquad \text{(as } \mathfrak{k}\text{-modules).}$$

Now, for any $\theta \in \mathfrak{h}^*$,

$$\dim(U(\mathfrak{p}^-)[\theta]) = \mathbb{P}_{P_n}(-\theta).$$

Hence all dimensions on the right side of (**) are finite, and

$$\dim(E_s[\nu]^{\mathfrak{n}\mathfrak{k}^+}) = \mathbb{P}_{P_n}((\lambda + \delta_{P_{\mathfrak{k}}}) - s^{-1}(\nu + \delta_{P_{\mathfrak{k}}})).$$

This gives:

PROPOSITION 2.6.6. *Let* $\mathfrak{g}$, $P$, $\lambda$ *be as in Proposition* 2.3.2. *Then for any* $P_{\mathfrak{k}}$*-dominant integral* $\nu \in \mathfrak{h}^*$,

$$[J_{\mathfrak{g},P,\lambda} : \vartheta(\nu)] = \sum_{s \in \mathfrak{w}(\mathfrak{k})} \varepsilon(s) \mathbb{P}_{P_n}((\lambda + \delta_{P_{\mathfrak{k}}}) - s(\nu + \delta_{P_{\mathfrak{k}}})),$$

*here,* $\vartheta(\nu)$ *is the irreducible class of* $\mathfrak{k}$ *of* $P_{\mathfrak{k}}$*-highest weight* $\nu$, *and* $[J_{\mathfrak{g},P,\lambda} : \vartheta(\nu)]$ *is the multiplicity of the occurrence of* $\vartheta(\nu)$ *in* $J_{\mathfrak{g},P,\lambda}$.

If $\lambda$ is such that

$$(\lambda+\delta_P)(\bar{H}_\alpha)<0 \ \forall \alpha \in P_n, \qquad J_{\mathfrak{g},P,\lambda}=I_{\mathfrak{g},P,\lambda}.$$

So the above formula gives the $K$-multiplicities of the holomorphic discrete series. This formula is of course meaningful under the sole assumption that $\mathrm{rk}(\mathfrak{g})=\mathrm{rk}(\mathfrak{k})$. *That it remains true and gives the K-multiplicities of the discrete classes of G in all cases was the conjecture of Blattner.* It was settled completely and affirmatively by the work of Wilfried Schmid [1] and Henryk Hecht and Wilfried Schmid [1] (for partial results, see R. Parthasarathy [1]). It has also been since completely proved by Enright [1, 2] using his theory. The methods of Enright however go much further and give a very general multiplicity formula for a whole family of $\mathfrak{g}$-modules which include the discrete series, mock discrete series, and the fundamental series. I shall treat this question later on.

Finally, we remark that the pairs $(\mathfrak{g}, \mathfrak{k})$ that we have treated here, namely, those for which $(\mathfrak{g}, \mathfrak{h})$ has an admissible positive system $P$ of roots, are precisely those for which $G/K$ has a $G$-invariant *complex structure*; if $\mathfrak{p}^+=\Sigma_{\beta\in P_n}\mathfrak{g}_\beta$, there exists a unique such structure for which $\mathfrak{p}^+$ is the space of holomorphic tangent vectors at the point of $G/K$ below $1 \in G$, and all such structures are obtained bijectively as $P$ varies over the admissible positive systems of $(\mathfrak{g}, \mathfrak{h})$. The representations corresponding to the $J_{\mathfrak{g},P,\lambda}$ can then be realized in spaces of holomorphic sections of some vector bundles on $G/K$. This accounts for the name 'holomorphic discrete series'. For the treatment, from this point of view, of the discrete series, see Harish-Chandra [2] and Schmid [1, 3].

## 3. THE MODULES $D_{\mathfrak{g},P,\mu}$

**3.1.** We shall assume that $\mathrm{rk}(\mathfrak{g})=\mathrm{rk}(\mathfrak{k})$; $\mathfrak{h}\subset\mathfrak{k}$ is a CSA, $P_\mathfrak{k}$ a positive system of roots of $(\mathfrak{k}, \mathfrak{h})$. Let $\mathfrak{w}(\mathfrak{k})$ be the Weyl group of $(\mathfrak{k}, \mathfrak{h})$, and $t_0 \in \mathfrak{w}(\mathfrak{k})$ such that $t_0 \cdot P_\mathfrak{k}=-P_\mathfrak{k}$, Define

$$\delta_{P_\mathfrak{k}}=\tfrac{1}{2}\sum_{\alpha\in P_\mathfrak{k}}\alpha, \quad s' \cdot \lambda=s(\lambda+\delta_{P_\mathfrak{k}})-\delta_{P_\mathfrak{k}} \quad (s\in\mathfrak{w}(\mathfrak{k}), \lambda\in\mathfrak{h}^*).$$

For reasons which will become clear presently, we change our parametrization slightly and consider a $\mu\in\mathfrak{h}^*$ which is $P_\mathfrak{k}$-integral such that

$$(t'_0 \cdot \mu)(\bar{H}_\alpha) \text{ is an integer } \geqslant 0 \ \forall \alpha \in P_{\mathfrak{k}}.$$

We wish to construct for each such $\mu$ an irreducible Harish-Chandra module $D_{\mathfrak{g},P,\mu}$ with the following properties:

(1) the irreducible class $\vartheta(t'_0 \cdot \mu)$ of $\mathfrak{k}$ occurs exactly one in $D_{\mathfrak{g},P,\mu}$

(2) on the $\vartheta(t'_0 \cdot \mu)$-isotypical subspace of $D_{\mathfrak{g},P,\mu}$, the centralizer $\mathfrak{D}$ of $\mathfrak{k}$ in $U(\mathfrak{g})$ acts via the homomorphism

$$\boldsymbol{\beta}_{P,\mu}: u \mapsto \boldsymbol{\beta}_P(u)(\mu), \qquad (u \in \mathfrak{D}).$$

We also wish to examine the relationship that these modules have (once they are constructed), with the discrete series of the corresponding group $G$.

This was done first by Enright and Varadarajan [1]. I shall outline the main idea behind the construction of $D_{\mathfrak{g},P,\mu}$. Although the main theorem (Theorem 3.3.10) of this section will include a statement identifying $D_{\mathfrak{g},-t_0\cdot P,\mu}$ with the $U(\mathfrak{g})$-module defined by the discrete class $\omega(\Lambda)$ when $\mu = t'_0 \cdot (\Lambda - \delta_{P_{\mathfrak{k}}} + \delta_{P_n})$ and $\Lambda$ is 'sufficiently regular', we shall not seriously discuss the identification question in this section. It will be taken up later on.

We being with the remark that it follows from Harish-Chandra's criterion (Theorem 1.3.9) that $D_{\mathfrak{g},P,\mu}$ is determined uniquely, up to isomorphism. We shall now sketch briefly the main lines of the construction of $D_{\mathfrak{g},P,\mu}$.

**3.2.** In order to motivate the construction we look at the special case when $\mathfrak{g}_0 = \mathfrak{su}(2, 1)$. Then $P_{\mathfrak{k}}$ consists of a single root, say $\alpha$, and there are three positive systems $P$ containing $P_{\mathfrak{k}}$; moreover, $t_0 = s_\alpha$. Let $\lambda = s'_\alpha \cdot \mu$. If $P$ is admissible, we take $D_{\mathfrak{g},P,\mu} = J_{\mathfrak{g},P,\lambda}$ where

$$J_{\mathfrak{g},P,\lambda} = V_{\mathfrak{g},P,\lambda} / V_{\mathfrak{g},P,s'_\alpha\lambda}.$$

However, if $P$ is not admissible, $\alpha$ is not simple in $P$, so that

$$s'_\alpha \cdot \lambda = s_\alpha(\lambda + \delta_{P_{\mathfrak{k}}}) - \delta_{P_{\mathfrak{k}}} \notin \mathfrak{w}(\mathfrak{g})' \cdot \lambda,$$

and we no longer have $V_{\mathfrak{g},P,s'_\alpha\cdot\lambda} \hookrightarrow V_{\mathfrak{g},P,\lambda}$. So we look for a $\mathfrak{g}$-module $W$ with the following properties:

(a) $W$ is a weight module on which $X_\alpha$ acts locally nilpotently while $X_{-\alpha}$ acts injectively;

(b) $V_{\mathfrak{g},P,s'_\alpha\cdot\lambda} \subset W$;

(c) $\exists$ a $v_\lambda \neq 0$ in $W[\lambda]^{\mathfrak{n}\mathfrak{k}^+}$ (i.e., $v_\lambda$ is of weight $\lambda$ and $X_\alpha v_\lambda = 0$) such that $W = U(\mathfrak{g})v_\lambda$ and

$$v_{s'_\alpha \cdot \lambda} = X_{-\alpha}^{\lambda(\bar{H}_\alpha)+1} \cdot v_\lambda$$

is the highest weight vector in $V_{\mathfrak{g},P,s'_\alpha \cdot \lambda}$; moreover, $v_\lambda \notin V_{\mathfrak{g},P,s'_\alpha \cdot \lambda}$.
As we shall see presently, arguments of a very general nature leads to the existence of a $\mathfrak{g}$-module $W$ with the above properties. Let $\bar{V} = W/V_{\mathfrak{g},P,s'_\alpha \cdot \lambda}$. Then $\bar{V} = U(\mathfrak{g}) \cdot \bar{v}_\lambda$, $\bar{v}_\lambda$ being the image of $v_\lambda$. Since $X_\alpha \cdot \bar{v}_\lambda = 0$ and

$$X_{-\alpha}^{\lambda(\bar{H}_\alpha)+1} \cdot \bar{v}_\lambda = 0,$$

$\bar{v}_\lambda$ is $\mathfrak{k}$-finite, hence $\bar{V}$ is a Harish-Chandra module. If $\bar{v} \in \bar{V}[\lambda]^{\mathfrak{n}\mathfrak{k}^+}$, $\bar{v}$ is the image of $v \in W[\lambda]^{\mathfrak{n}\mathfrak{k}^+}$ by Lemma 2.2.12; but then, by the $\mathfrak{k}$-finiteness of $\bar{v}$, $X_{-\alpha}^{\lambda(\bar{H}_\alpha)+1}\, \bar{v} = 0$, giving

$$X_{-\alpha}^{\lambda(\bar{H}_\alpha)+1}\, v \in V_{\mathfrak{g},P,s'_\alpha \cdot \lambda}[s'_\alpha \lambda]^{\mathfrak{n}\mathfrak{k}^+}$$

i.e.,

$$X_{-\alpha}^{\lambda(\bar{H}_\alpha)+1} v \in \mathbb{C} \cdot v_{s'_\alpha \cdot \lambda}.$$

As $X_{-\alpha}$ is injective, $v$ is a multiple of $v_\lambda$. So $\bar{v} \in \mathbb{C} \cdot \bar{v}_\lambda$. This shows that the class $\vartheta(\lambda)$ *occurs exactly once* in $\bar{V}$. Moreover, if $u \in \mathfrak{D}$,

$$X_{-\alpha}^{\lambda(\bar{H}_\alpha)+1} u \cdot v_\lambda = u \cdot v_{s'_\alpha \cdot \lambda} = \boldsymbol{\beta}_P(u)(\mu) v_{s'_\alpha \cdot \lambda} = X_{-\alpha}^{\lambda(\bar{H}_\alpha)+1} (\boldsymbol{\beta}_P(u)(\mu) v_\lambda).$$

Since $X_{-\alpha}$ acts injectively on $W$, this implies that

$$u \cdot v_\lambda = \boldsymbol{\beta}_P(u)(\mu) \cdot v_\lambda, \Rightarrow u \cdot \bar{v}_\lambda = \boldsymbol{\beta}_P(u)(\mu) \bar{v}_\lambda.$$

At the same time,

$$W[\lambda]^{\mathfrak{n}\mathfrak{k}^+} = \mathbb{C} \cdot v_\lambda \Rightarrow \bar{V}[\lambda]^{\mathfrak{n}\mathfrak{k}^+} = \mathbb{C} \cdot \bar{v}_\lambda \Rightarrow [\bar{V} : \vartheta(\lambda)] = 1.$$

So $\bar{V}$ has a *unique* irreducible quotient which is $\neq 0$. We can take it to be $D_{\mathfrak{g},P,\mu}$.

So the main point is the construction of $W$. Clearly it is a question of 'enlarging' $V_{\mathfrak{g},P,\mu}$ into a $\mathfrak{g}$-module so that, at the same time, $V_{\mathfrak{k},P\mathfrak{k},\mu}$ gets 'enlarged' to $V_{\mathfrak{k},P\mathfrak{k},\lambda}$. To do this we simply write $V_{\mathfrak{g},P,\mu}$ as a quotient $(U(\mathfrak{g}) \otimes_{U(\mathfrak{k})} V_{\mathfrak{k},P\mathfrak{k},\mu})/C$; we then obtain the 'enlargement'

$$V_{\mathfrak{g},P,\mu} \hookrightarrow (U(\mathfrak{g}) \bigotimes_{U(\mathfrak{k})} V_{\mathfrak{k},P\mathfrak{k},\lambda})/C = W'$$

with $v_\lambda$ as image of $1 \otimes v_{\mathfrak{k},\lambda}$, $v_{\mathfrak{k},\lambda}$ being the highest weight vector of $V_{\mathfrak{k},P\mathfrak{k},\lambda}$. $W'$ has all the required properties except perhaps that of the injectivity of $X_{-\alpha}$; *it may have torsion as a module over* $X_{-\alpha}$. But if $W'_T$ is the torsion submodule, $C \cap W'_T = 0(!)$ and so we can replace $W'$ by $W = W'/W'_T$.

It should be noted that the starting point of the construction was the Verma $\mathfrak{g}$-module $V_{\mathfrak{g},P,s'_\alpha \cdot \lambda} = V_{\mathfrak{g},P,\mu}$. *It is for this reason that* $\mu$ *was chosen as the parameter for* $D_{\mathfrak{g},P,\mu}$, *rather than* $\lambda$.

**3.3.** The construction of the $D_{\mathfrak{g},P,\mu}$ in the general case proceeds along similar lines. $\mathfrak{g}$ is now only assumed to satisfy $\mathrm{rk}(\mathfrak{g}) = \mathrm{rk}(\mathfrak{k})$, i.e., there may not be any admissible positive system of roots of $(\mathfrak{g}, \mathfrak{h})$. $P$ is a positive system of roots of $(\mathfrak{g}, \mathfrak{h})$. $\mu \in \mathfrak{h}^*$ is such that

$$(t'_0 \cdot \mu)(\bar{H}_\alpha) \text{ is an integer } \geqslant 0 \ \forall \alpha \in P_{\mathfrak{k}}.$$

We write

$$\lambda = t'_0 \cdot \mu, \qquad \mu = t'_0 \cdot \lambda.$$

Of course, for $s \in \mathfrak{w}(\mathfrak{k})$, $v \in \mathfrak{h}^*$,

$$s' \cdot v = s(v + \delta_{P\mathfrak{k}}) - \delta_{P\mathfrak{k}}, \qquad \left(\delta_{P\mathfrak{k}} = \tfrac{1}{2} \sum_{\alpha \in P\mathfrak{k}} \alpha\right).$$

*From now on* $\mathfrak{g}$, $P$, $\lambda$, $\mu$ *will satisfy these conditions.* Write

$$\mathfrak{n}_\mathfrak{k}^+ = \sum_{\alpha \in P\mathfrak{k}} \mathfrak{k}_\alpha, \qquad \mathfrak{n}_\mathfrak{k}^- = \sum_{\alpha \in P\mathfrak{k}} \mathfrak{k}_{-\alpha}.$$

First, some remarks on modules on which $U(\mathfrak{n}_\mathfrak{k}^-)$ acts without torsion. Let $\mathfrak{a}$ be any finite-dimensional Lie algebra and $E$ any $\mathfrak{a}$-module. The $\mathfrak{a}$-*torsion part* of $E$, $E_{T,\mathfrak{a}}$ in symbols, is defined by

$$E_{T,\mathfrak{a}} = \{v : v \in E,\ x \cdot v = 0 \text{ for some } x \neq 0 \text{ in } U(\mathfrak{a})\}.$$

We have the following lemma (see Enright and Wallach [1]) ('torsion lemma'):

LEMMA 3.3.1. *Let* $r = \dim(\mathfrak{a})$. *For any integer* $N \geqslant 0$, *let* $U(\mathfrak{a})^{(N)}$ *be the subspace of* $U(\mathfrak{a})$ *of elements of degree* $\leqslant N$. *Let* $E$ *be any* $\mathfrak{a}$-*module.*

(i) *For* $u \in E$, $u \in E_{T,\mathfrak{a}} \Leftrightarrow \dim(U(\mathfrak{a})^{(N)} \cdot u) = O(N^{r-1})$ *as* $N \to \infty$.

(ii) $E_{T,\mathfrak{a}}$ *is a sub* $\mathfrak{a}$-*module of* $E$; *if* $\mathfrak{a}$ *is a subalgebra of* $\mathfrak{b}$ *and* $E$ *is a* $\mathfrak{b}$-*module, then* $E_{T,\mathfrak{a}}$ *is a sub* $\mathfrak{b}$-*module.*

(iii) $E/E_{T,\mathfrak{a}}$ *is torsion-free, i.e.,* $(E/E_{T,\mathfrak{a}})_{T,\mathfrak{a}} = 0$.

We shall call $E_{T,\mathfrak{a}}$ the *$\mathfrak{a}$-torsion sub-module of $E$.*

The next lemma is the 'enlargement lemma'.

LEMMA 3.3.2. *Let $\mathfrak{b}$ be a Lie algebra of finite dimension over $\mathbb{C}$; $\mathfrak{a}$, $\mathfrak{a}_1$, $\mathfrak{a}_2$, subalgebras of $\mathfrak{b}$, with $\mathfrak{a}_1, \mathfrak{a}_2 \subset \mathfrak{a} \subset \mathfrak{b}$. Assume that* ad $X$ *is nilpotent on $\mathfrak{b}$ $\forall X \in \mathfrak{a}_2$. Let $A$, $A'$ be $\mathfrak{a}$-modules, $B$ a $\mathfrak{b}$-module, and*

$$i: A \hookrightarrow A', \qquad j: A \hookrightarrow B$$

*be $\mathfrak{a}$-module injections. Suppose $B$ and $A'$ are torsion-free for $U(\mathfrak{a}_1)$ and that each element of $U(\mathfrak{a}_2)^+ = U(\mathfrak{a}_2)\mathfrak{a}_2$ acts locally nilpotently on $B$ and $A'$. Then there exists a $\mathfrak{b}$-module $B'$, an $\mathfrak{a}$-module injection $j': A' \hookrightarrow B'$, and a $\mathfrak{b}$-module injection $i': B \hookrightarrow B'$ such that*

(a) *$B'$ is torsion-free for $U(\mathfrak{a}_1)$ and each element of $U(\mathfrak{a}_2)^+$ acts locally nilpotently[1] on $B'$.*

(b) *The diagram*

$$\begin{array}{ccc} A & \xrightarrow{\;i\;} & A' \\ {\scriptstyle j}\downarrow & & \downarrow{\scriptstyle j'} \\ B & \xrightarrow[\;i'\;]{} & B' \end{array}$$

*commutes.*

*Further, if $B = U(\mathfrak{b}) \cdot j(A)$, we can arrange matters so that $B' = U(\mathfrak{b}) \cdot j'(A')$.*

$\exists$ natural maps $\alpha: U(\mathfrak{b}) \bigotimes_{U(\mathfrak{a})} A \to U(\mathfrak{b}) \bigotimes_{U(\mathfrak{a})} A'$, $\beta: U(\mathfrak{b}) \bigotimes_{U(\mathfrak{a})} A \to B_1$,

where $B_1 = U(\mathfrak{b}) \cdot A$, $\alpha$ being a $\mathfrak{b}$-module injection and $\beta$ a $\mathfrak{b}$-module

---

[1] This is equivalent to saying that if $X_1, \ldots, X_m$ is a basis for $\mathfrak{a}_2$, then each $X_i$ is locally nilpotent on $B'$. Indeed, for any *nilpotent* Lie algebra $\mathfrak{c}$ of finite dimension and any $\mathfrak{c}$-module $L$, the following are equivalent:

(a) each element of $U(\mathfrak{c})^+ = U(\mathfrak{c})\mathfrak{c}$ is locally nilpotent on $L$

(b) $\exists$ a basis $X_1, \ldots, X_m$ for $\mathfrak{c}$ such that each $X_i$ is locally nilpotent on $L$.

$L$ is then necessarily $U(\mathfrak{c})$-finite, i.e., $\dim(U(\mathfrak{c})v) < \infty$ $\forall v \in L$. If $\mathfrak{m}$ is a reductive Lie algebra with CSA $\mathfrak{h}$ and positive system of roots $Q$, and if $\mathfrak{c} = \Sigma_{\alpha \in Q} \mathfrak{m}_\alpha$, then any weight module $M$ for $\mathfrak{m}$ that has these properties possesses the following additional ones:

(c) $M$ is $U(\mathfrak{h} \oplus \mathfrak{c})$-finite.

(d) Each $v \in M$ lies in a sub $\mathfrak{m}$-module $M' \subset M$ such that $M'$ is a weight module and $\dim(M'[\lambda]) < \infty$ $\forall \lambda \in \mathfrak{h}^*$.

surjection, with $\alpha(1\otimes a)=1\otimes i(a)$ and $\beta(1\otimes a)=j(a)$. Let $C$ be the kernel of $\beta$ and $C_1=\alpha(C)$. Take $B_1''=(U(\mathfrak{b})\otimes_{U(\mathfrak{a})}A')/C_1$. We then have a diagram such as in (b), above, with $B_1, B_1''$ in place of $B, B'$. Now we have imbeddings $B_1 \hookrightarrow B, B_1 \hookrightarrow B_1''$ and so we can take $B''=B+_{B_1}B_1''$ where $+_{B_1}$ means images of $B_1$ are identified. Elements of $U(\mathfrak{a}_2)^+$ will act locally nilpotently on $B''$, but $B''_{T,\mathfrak{a}_1}$ may not be 0; we use the torsion lemma and replace $B''$ by $B''/B''_{T,\mathfrak{a}_1}=B'$; $B'$ is torsion-free for $U(\mathfrak{a}_1)$.

We now apply this lemma to the following situation. Let $\theta\in\mathfrak{h}^*$ and let $\beta\in P_{\mathfrak{k}}$ be a simple root such that $\theta(\bar{H}_\beta)$ is an integer $\geqslant 0$. We then have the $\mathfrak{k}$-Verma modules $V_{\mathfrak{k},P\mathfrak{k},\theta}$ and $V_{\mathfrak{k},P\mathfrak{k},s_\beta'\cdot\theta}$ with highest weight vectors $v_\theta$ and $v_{s_\beta'\cdot\theta}$ such that

$$V_{\mathfrak{k},P\mathfrak{k},s_\beta'\cdot\theta}\subset V_{\mathfrak{k},P\mathfrak{k},\theta},$$

$$v_{s_\beta'\cdot\theta}=X_{-\beta}^{\theta(\bar{H}_\beta)+1}\cdot v_\theta.$$

Let $W'$ be a $\mathfrak{g}$-module such that

$$V_{\mathfrak{k},P\mathfrak{k},s_\beta'\cdot\theta}\subset W', \qquad W'=U(\mathfrak{g})\cdot v_{s_\beta'\cdot\theta}.$$

$W'$ is then a weight module. We then have the following consequence of Lemma 3.3.2 (with $\mathfrak{b}=\mathfrak{g}$, $\mathfrak{a}=\mathfrak{k}$, $\mathfrak{a}_1=\mathfrak{n}_{\mathfrak{k}}^-$, $\mathfrak{a}_2=\mathfrak{n}_{\mathfrak{k}}^+$).

LEMMA 3.3.3. *Suppose $W'$ is torsion-free for $U(\mathfrak{n}_{\mathfrak{k}}^-)$ and that elements of $U(\mathfrak{n}_{\mathfrak{k}}^+)\mathfrak{n}_{\mathfrak{k}}^+$ act locally nilpotently on $W'$. Then $\exists$ a $\mathfrak{g}$-module $W$ having these two properties and such that $W'\subset W$, $V_{\mathfrak{k},P\mathfrak{k},\theta}\subset W$, $W=U(\mathfrak{g})\cdot v_\theta$, the inclusions*

$$\begin{array}{ccc} W' & \subset & W \\ \cup & & \cup \\ V_{\mathfrak{k},P\mathfrak{k},s_\beta'\cdot\theta} & \subset & V_{\mathfrak{k},P\mathfrak{k},\theta} \end{array}$$

*being compatible. Fix such a $W$. Then one has the following:*

(a) *$W$ is a weight module,*

(b) *$\bar{W}=W/W'$ is $\mathfrak{k}(\beta)$-finite where $\mathfrak{k}(\beta)=\mathfrak{h}\oplus\mathfrak{k}_\beta\oplus\mathfrak{k}_{-\beta}$,*

(c) *If $\nu\in\mathfrak{h}^*$ is $P_{\mathfrak{k}}$-dominant integral, we have the exact sequence*

$$W[\nu]^{\mathfrak{n}_{\mathfrak{k}}^+}\to\bar{W}[\nu]^{\mathfrak{n}_{\mathfrak{k}}^+}\to 0$$

(d) *If $\nu\in\mathfrak{h}^*$ is such that $\nu(\bar{H}_\beta)$ is an integer $\geqslant 0$, we have the injection*

$$X_{-\beta}^{\nu(\bar{H}_\beta)+1}: W[\nu]^{\mathfrak{n}_{\mathfrak{k}}^+}\hookrightarrow W'[s_\beta'\cdot\nu]^{\mathfrak{n}_{\mathfrak{k}}^+}.$$

If $\bar{v}_\theta$ is the image of $v_\theta$ in $\overline{W}$,

$$X_\beta \bar{v}_\theta = 0, \qquad X_{-\beta}^{\theta(\overline{H}_\beta)+1} \cdot \bar{v} = 0.$$

So $\bar{v}_\theta$ is $\mathfrak{k}(\beta)$-finite. This gives (b). For (d), if $v \in W[\nu]^{\mathfrak{n}\mathfrak{k}^+}$, and the image of $v$ in $\overline{W}$, say $\bar{v}$, is not zero, $0 \neq \bar{v} \in \overline{W}[\nu]^{\mathfrak{n}\mathfrak{k}^+}$; and by the $\mathfrak{k}(\beta)$-finiteness of $\overline{W}$, $X_{-\beta}^{\nu(\overline{H}_\beta)+1}\bar{v} = 0$. So

$$X_{-\beta}^{\nu(\overline{H}_\beta)+1} v \in W'[s'_\beta \cdot \nu]^{\mathfrak{n}\mathfrak{k}^+}.$$

COROLLARY 3.3.4. *Suppose* $W'[s'_\beta \cdot \theta]^{\mathfrak{n}\mathfrak{k}^+} = \mathbb{C} \cdot v_{s'_\beta \cdot \theta}$. *Then*

$$W[\theta]^{\mathfrak{n}\mathfrak{k}^+} = \mathbb{C} \cdot v_\theta.$$

Follows from (d) and the injectivity of the action of $X_{-\beta}$ on $W$.

We can now give the basic construction made in the work of Enright and Varadarajan [1]. Let $l$ be the length function of $\mathfrak{w}(\mathfrak{k})$ (relative to $P_k$); $r$, the maximum length; $t_0$, the element of length $r$; and

$$t_0 = s_{\beta_r} s_{\beta_{r-1}} \cdots s_{\beta_1}$$

a reduced expression for $t_0$. Of course $t_0 \cdot P_\mathfrak{k} = -P_\mathfrak{k}$. Let $\mu = t'_0 \cdot \lambda$ where $\lambda \in \mathfrak{h}^*$ is $P_\mathfrak{k}$-dominant integral, i.e., $\lambda(\overline{H}_\alpha)$ is an integer $\geq 0$ $\forall \alpha \in P_\mathfrak{k}$. If we write

$$s_0 = 1, \qquad s_j = s_{\beta_j} s_{\beta_{j-1}} \cdots s_{\beta_1}$$

and denote by $V_{\mathfrak{k},j}$ the Verma module $V_{\mathfrak{k},P\mathfrak{k},s'_j \cdot \lambda}$:

$$V_{\mathfrak{k},j} = V_{\mathfrak{k},P\mathfrak{k},s'_j \cdot \lambda},$$

we then have the canonical inclusions

$$V_{\mathfrak{k},r} \subset V_{\mathfrak{k},r-1} \subset \cdots \subset V_{\mathfrak{k},0}.$$

We know that $r = |P_\mathfrak{k}|$ and that

$$\beta_1, \quad s_{\beta_1} \cdot \beta_2, \quad s_{\beta_2} s_{\beta_1} \cdot \beta_3, \ldots, s_{\beta_{r-1}} \cdots s_{\beta_1} \cdot \beta_r$$

*are precisely all the roots in* $P_\mathfrak{k}$. So, for any $\nu \in \mathfrak{h}^*$ which is $P_\mathfrak{k}$-dominant integral, if we put

$$c_j(\nu) = (s'_j \cdot \nu)(\overline{H}_{\beta_{j+1}}), \qquad (0 \leq j \leq r-1),$$

then

$$c_j(\nu) \text{ is an integer } \geq 0 \ \forall j = 0, 1, \ldots, r-1.$$

Let $v_j$ be a nonzero vector of weight (highest) $s'_j \cdot \lambda$ in $V_{\mathfrak{k},j}$. We may assume that

$$v_{j+1} = X_{\beta_{j+1}}^{c_{j+1}(\lambda)} \cdot v_j, \qquad (j=0, 1, \ldots, r-1).$$

We now begin with the $\mathfrak{g}$-Verma module $V_{\mathfrak{g},P,t'_0\cdot\lambda}$ and the canonical inclusion

$$V_{\mathfrak{k},r} \subset V_{\mathfrak{g},P,t'_0\cdot\lambda}.$$

We write $W_r = V_{\mathfrak{g},P,t'_0\cdot\lambda}$ and note that $W_r$ is a weight module for $\mathfrak{g}$, with

$$W_r = U(\mathfrak{g})\cdot v_r, \qquad W_r[t'_0\cdot\lambda]^{\mathfrak{n}_\mathfrak{k}^+} = \mathbb{C}\cdot v_r.$$

Moreover, $W_r$ is torsion-free for $U(\mathfrak{n}_\mathfrak{k}^-)$ while the elements of $U(\mathfrak{n}_\mathfrak{k}^+)\mathfrak{n}_\mathfrak{k}^+$ act locally nilpotently on $W_r$. Lemma 3.3.3 and Corollary 3.3.4 may now be used repeatedly, to prove the following.

LEMMA 3.3.5. *$\exists$ $\mathfrak{g}$-modules $W_j$ $(0\leqslant j\leqslant r)$ with the following properties (with $\mu = t'_0\cdot\lambda$, $\lambda = t'_0\cdot\mu$, $\lambda$ being $P_\mathfrak{k}$-dominant integral):*

(i) $W_r = V_{\mathfrak{g},P,t'_0\cdot\lambda}$; $W_r \subset W_{r-1} \subset \cdots \subset W_0$.

(ii) *for all j, $W_j$ is a weight module which is torsion-free for $U(\mathfrak{n}_\mathfrak{k}^-)$, and on which elements of $U(\mathfrak{n}_\mathfrak{k}^+)\mathfrak{n}_\mathfrak{k}^+$ act locally nilpotently*

(iii) $$V_{\mathfrak{k},j} \subset W_j,\ W_j = U(\mathfrak{g})\cdot v_j, \qquad (0\leqslant j\leqslant r)$$

(iv) *the inclusions*

$$\begin{array}{ccccccc} V_{\mathfrak{k},r} & \subset & V_{\mathfrak{k},r-1} & \subset \cdots \subset & V_{\mathfrak{k},0} \\ \cap & & \cap & & \cap \\ W_r & \subset & W_{r-1} & \subset \cdots \subset & W_0 \end{array}$$

*are compatible*

(v) $$W_j[s'_j\cdot\lambda]^{\mathfrak{n}_\mathfrak{k}^+} = \mathbb{C}\cdot v_j, \qquad (0\leqslant j\leqslant r),$$
$$v_{j+1} = X_{-\beta_{j+1}}^{c_{j+1}(\lambda)} v_j, \qquad (0\leqslant j\leqslant r-1;\ c_{j+1}(\lambda) = (s'_j\cdot\lambda)\bar{H}_{\beta_{j+1}}))$$

(vi) *$W_j/W_{j+1}$ is $\mathfrak{k}(\beta_{j+1})$-finite $(j=0, 1, \ldots, r-1)$ where $\mathfrak{k}(\beta_{j+1})$ is*

$$\mathfrak{h}\oplus\mathfrak{k}_{\beta_{j+1}}\oplus\mathfrak{k}_{-\beta_{j+1}}$$

(vii) $v_j \notin W_{j+1}$ $(0\leqslant j\leqslant r-1)$; *more generally, if $s\in\mathfrak{w}(\mathfrak{k})$ and*

$$W_j[s'\cdot\lambda]^{\mathfrak{n}_\mathfrak{k}^+} \neq 0,$$

*then $l(s)\geqslant j$, $\qquad (0\leqslant j\leqslant r)$.*

For the assertion (vii) we use induction. Since $W_r$ is a Verma module for $\mathfrak{g}$ with highest weight $t_0' \cdot \lambda$ relative to $P$, $s' \cdot \lambda$ is not even a weight for it if $l(s) < r$. So we are done for $j = r$. Assuming (vii) for $j+1$, take

$$v \in W_j[s' \cdot \lambda]^{\mathfrak{n}_\mathfrak{k}^+}, \qquad v \neq 0.$$

If $v \in W_{j+1}$, $l(s) \geqslant j+1$; *if* $v \notin W_{j+1}$, its image $\bar{v}$ in $W_j/W_{j+1}$ is such that $X_{\beta j+1}\bar{v} = 0$ and so, by (vi), $(s' \cdot \lambda)(\bar{H}_{\beta_{j+1}})$ is an integer $\geqslant 0$; (d) of Lemma 3.3.3 now gives

$$X_{-\beta j+1}^{a+1} v = v' \in W_{j+1}[s'_{\beta j+1} s' \cdot \lambda]^{\mathfrak{n}_\mathfrak{k}^+},$$

$a$ being $(s'\lambda)(\bar{H}_{\beta_{j+1}})$. Since $v' \neq 0$, $l(s_{\beta_{j+1}} s) \geqslant j+1$, so that $l(s) \geqslant j$.

It remains to construct a suitable sub $\mathfrak{g}$-module $\tilde{W}$ such that $v_0 \notin \tilde{W}$ and $W_0/\tilde{W}$ is $\mathfrak{k}$-finite. *We cannot simply take* $W_0 = W_1$. Indeed, the inclusions $W_r \subset \cdots \subset W_0$ imitate the inclusions $V_{\mathfrak{k},r} \subset V_{\mathfrak{k},r-1} \subset \cdots \subset V_{\mathfrak{k},0}$; and it is *not* $V_{\mathfrak{k},0}/V_{\mathfrak{k},1}$ but $V_{\mathfrak{k},0}/\Sigma_{s \neq 1} V_{\mathfrak{k},P\mathfrak{k},s' \cdot \lambda}$ that is the finite-dimensional $\mathfrak{k}$-module with highest weight $\lambda$. In view of the analogy with the $V$'s, this suggests the following procedure for defining $\tilde{W}$. For any $s \in \mathfrak{w}(\mathfrak{k})$, let $R(s)$ be the set of all reduced expressions for $s$. If $U \subset W_0$ is any sub $\mathfrak{g}$-module and if $\gamma_1, \ldots, \gamma_q \in P_\mathfrak{k}$, we write

$$U(\gamma_q, \ldots, \gamma_1) = \{v;\, v \in W_0,\ X_{-\gamma_q}^{a_q} \ldots X_{-\gamma_1}^{a_1} v \in U \text{ for some integers } a_i \geqslant 0\}.$$

We shall see presently that these are sub $\mathfrak{g}$-modules. For any $j$ with $1 \leqslant j \leqslant r$, any $s \in \mathfrak{w}(\mathfrak{k})$ with $l(s) < j$, and any reduced expression $s = s_{\alpha_m} \ldots s_{\alpha_1}$, let us consider $W_j(\alpha_m, \ldots, \alpha_1)$; define $\tilde{W}$ by

$$\tilde{W} = \sum_{1 \leqslant j \leqslant r} \sum_{s : l(s) < j} \sum_{s = s_{\alpha_m}, \ldots, s_{\alpha_1}} W_j(\alpha_m, \ldots, \alpha_1)$$

where the inner sum is over all the set $R(s)$ of all reduced expressions of $s$.

LEMMA 3.3.6. *Let $E$ be a $\mathfrak{g}$-module which is a weight module and on which elements of $U(\mathfrak{n}_\mathfrak{k}^+)\mathfrak{n}_\mathfrak{k}^+$ act locally nilpotently. Let $F \subset E$ be a sub $\mathfrak{g}$-module. For $\gamma_1, \ldots, \gamma_q \in P_\mathfrak{k}$, let*

$$F(\gamma_q, \ldots, \gamma_1) = \{v : v \in E,\ X_{-\gamma_q}^{a_q} \ldots X_{-\gamma_1}^{a_1} v \in F \text{ for some integers } a_j \geqslant 0\}.$$

*Then we have the following*

(i) $F(\gamma_q, \ldots, \gamma_1)$ *is a sub $\mathfrak{g}$-module.*

(ii) *Suppose* $\alpha_1, \ldots, \alpha_m \in P_{\mathfrak{k}}$ *are simple,* $s = s_{\alpha_m} \ldots s_{\alpha_1}$ *is a reduced expression, and* $\nu \in \mathfrak{h}^*$ *such that*

$$b_1 = \nu(\bar{H}_{\alpha_1}),\ b_2 = (s'_{\alpha_1} \cdot \nu)(\bar{H}_{\alpha_2}), \ldots, b_m = (s'_{\alpha_{m-1}} \ldots s'_{\alpha_1} \cdot \nu)(\bar{H}_{\alpha_m})$$

*are all integers* $\geqslant 0$. Write

$$d_{s,1}(\nu) = X^{b_m}_{-\alpha_m} \ldots X^{b_1}_{-\alpha_1}.$$

Then

$$d_{s,1}(\nu)(\bar{F}[\nu]^{\mathfrak{n}_{\mathfrak{k}}^+}) \subset F[s' \cdot \nu]^{\mathfrak{n}_{\mathfrak{k}}^+}, \qquad (\bar{F} = F(\alpha_m, \ldots, \alpha_1)).$$

For (i), observe that $F(\gamma_q, \ldots, \gamma_1)$ is the set of $v \in E$ that are $\mathfrak{k}(\gamma_1)$-finite mod $F(\gamma_q, \ldots, \gamma_2)$, and use Lemma 1.2.1 together with induction on $q$. For (ii) use induction on $m$ to come to the case $m = 1$; here it follows from the $\mathfrak{k}(\alpha_1)$-finiteness of $F(\alpha_1)/F$.

*Remark* 3.3.7. The conditions (ii) on $\nu$ are equivalent to saying that

$$(\nu + \delta_{P_{\mathfrak{k}}})(\bar{H}_\beta) \text{ is an integer } > 0\ \forall \beta \in P_{\mathfrak{k}} \cap s^{-1}(-P_{\mathfrak{k}}).$$

LEMMA 3.3.8. $\tilde{W}$ *has the following properties*:

(i) $\tilde{W}$ is a sub $\mathfrak{g}$-module of $W_0$; and $v_0 \notin \tilde{W}$,

(ii) *For any simple root* $\beta \in P_{\mathfrak{k}}$, $X^{\lambda(\bar{H}_\beta)+1}_{-\beta} \cdot v_0 \in \tilde{W}$.

(iii) *Suppose* $\nu \in \mathfrak{h}^*$ *is such that* $W_0[\nu]^{\mathfrak{n}_{\mathfrak{k}}^+} \not\subset \tilde{W}$; *then* $\nu$ *is* $P_{\mathfrak{k}}$*-dominant integral and we have an injection*

$$d_{t_0,1}(\nu)\colon W_0[\nu]^{\mathfrak{n}_{\mathfrak{k}}^+} \hookrightarrow W_r[t'_0 \cdot \nu]^{\mathfrak{n}_{\mathfrak{k}}^+}.$$

If $v_0 \in W_j(\alpha_m, \ldots, \alpha_1)$ where $s = s_{\alpha_m} \ldots s_{\alpha_1}$ is reduced and $m < j$, $d_{s,1}(\lambda)v_0 \in W_j[s' \cdot \lambda]^{\mathfrak{n}_{\mathfrak{k}}^+}$, contradicting (vii) of Lemma 3.3.5. Now, $W_0[\lambda]^{\mathfrak{n}_{\mathfrak{k}}^+}$ is one-dimensional and spanned by $v_0$, while Lemma 2.2.12 implies that for *any* $P_{\mathfrak{k}}$*-dominant integral* $\nu$,

$$\tilde{W}[\nu]^{\mathfrak{n}_{\mathfrak{k}}^+} = \sum_{1 \leqslant j \leqslant r} \sum_{s: l(s) < j} \sum_{s = s_{\alpha_m} \ldots s_{\alpha_1}} W_j(\alpha_m, \ldots, \alpha_1)[\nu]^{\mathfrak{n}_{\mathfrak{k}}^+}.$$

So, as $v_0 \notin W_j(\alpha_m, \ldots, \alpha_1)$, we see $v_0 \notin \tilde{W}$. Since

$$W_0 = W_r(\beta_r, \beta_{r-1}, \ldots, \beta_1), \qquad t_0 = s_{\beta_r} \ldots s_{\beta_1},$$

the second assertion of (iii) is clear. Suppose $\nu \in \mathfrak{h}^*$ is as in (iii) and $v \in W_0[\nu]^{\mathfrak{n}^+}$ such that $v \notin \tilde{W}$. As image of $v$ in $W_0/W_1$ is nonzero and lies in $(W_0/W_1)[\nu]^{\mathfrak{n}_{\mathfrak{k}}^+}$, $\nu(\bar{H}_{\beta_1}) = b_1$ is an integer $\geqslant 0$ while $X^{b_1}_{-\beta_1}v \in W_1$.

If $v_1 = X^{b_1}_{-\beta_1} v \in W_2$, then $v \in W_2(\beta_1) \subset \tilde{W}$; so $v_1 \notin W_2$. This means $b_2 = (s'_{\alpha_1} \cdot \nu)(\bar{H}_{\beta_2})$ is an integer $\geqslant 0$ while $X^{b_2}_{-\beta_2} v_1 \in W_2$. Again

$$v_2 = X^{b_2}_{-\beta_2} v_1 \notin W_3,$$

as otherwise $v \in W_3(\beta_2, \beta_1) \subset \tilde{W}$. This argument continues till we reach $W_r$. This proves (iii). (ii) follows from (iii) since $X^{\lambda(\bar{H}_\beta)+1}_{-\beta} \cdot v_0 \in W_0[s'_\beta \cdot \lambda]^{\mathfrak{n}\mathfrak{r}^+}$ and $(s'_\beta \cdot \lambda)(\bar{H}_\beta) < 0$.

LEMMA 3.3.9. *$W_0/\tilde{W} = \bar{W}$ is a Harish-Chandra module. If $\bar{v}_0$ is the image of $v_0$ in $\bar{W}$, $\bar{W} = U(\mathfrak{g}) \cdot \bar{v}_0$. The class $\vartheta(\lambda)$ of $\mathscr{E}(\mathfrak{k})$ occurs exactly once in $\bar{W}$; $U(\mathfrak{k}) \cdot \bar{v}_0$ is the corresponding isotypical space, and $\mathfrak{O} = U(\mathfrak{g})^{\mathfrak{k}}$ acts on this space through the homomorphism $\boldsymbol{\beta}_{P,\mu}(\cdot)$.*

Since $W_0[\lambda]^{\mathfrak{n}\mathfrak{r}^+} = \mathbb{C} \cdot v_0$ is $\mathfrak{O}$-stable, $u \cdot v_0 = c \cdot v_0$ for $c \in \mathbb{C}$. Now $d_{t_0,1}(\lambda)$ is injective and centralizes $u$, while $v_r = d_{t_0,1}(\lambda)$ has the property that $u \cdot v_r = \boldsymbol{\beta}_P(u)(t'_0 \cdot \lambda) v_r$, as $W_r = V_{\mathfrak{g},P,t'_0 \cdot \lambda}$. Hence $c = \boldsymbol{\beta}_P(u)(t'_0 \cdot \lambda)$.

We can now formulate the main theorem of Enright and Varadarajan [1]. Assertion (iv) of the theorem below follows from results of Wilfried Schmid [2] that characterize discrete series representations through 'minimal $\mathfrak{k}$-type' properties such as (iii) in the theorem below. We shall discuss this aspect of the theory later on (Section 12).

THEOREM 3.3.10. *Let* rk($\mathfrak{g}$) = rk($\mathfrak{k}$); *$P_{\mathfrak{k}}$, a positive system of roots of $(\mathfrak{k}, \mathfrak{h})$, $\mathfrak{h} \subset \mathfrak{k}$ being a* CSA; *$\mu \in \mathfrak{h}^*$ is such that $\lambda = t'_0 \cdot \mu$ is $P_{\mathfrak{k}}$-dominant integral. Then, for any positive system $P$ of roots of $(\mathfrak{g}, \mathfrak{h})$, with $P_{\mathfrak{k}} \subset P$, $\exists$ a unique (up to isomorphism) irreducible Harish-Chandra $\mathfrak{g}$-module $D_{\mathfrak{g},P,\mu}$ such that:*

(i) *the irreducible class $\vartheta(\lambda)$ of $\mathfrak{k}$ with $P_{\mathfrak{k}}$-highest weight $\lambda$ occurs exactly once in $D_{\mathfrak{g},P,\mu}$,*

(ii) *on the $\vartheta(\lambda)$-isotypical subspace of $D_{\mathfrak{g},P,\mu}$, $\mathfrak{O}$ acts through the homomorphism*

$$\boldsymbol{\beta}_{P,\mu}: u \mapsto \boldsymbol{\beta}_P(u)(\mu), \qquad (\mu \in \mathfrak{O}).$$

*The module $D_{\mathfrak{g},P,\mu}$ has the following further property:*

(iii) *suppose $\nu \in \mathfrak{h}^*$ is $P_{\mathfrak{k}}$-dominant integral such that the class $\vartheta(\nu)$ of $\mathfrak{k}$ occurs in $D_{\mathfrak{g},P,\mu}$; then $\lambda_{-t_0 \cdot P} \prec \nu$. In particular, if $\alpha \in -t_0 \cdot P$ and $\lambda - \alpha$ is $P_{\mathfrak{k}}$-dominant integral, the class $\vartheta(\lambda - \alpha)$ does not occur in $D_{\mathfrak{g},P,\mu}$.*

(iv) *If $\Lambda \in \mathfrak{h}^*$ is integral and sufficiently dominant relative to $P$,* i.e., *if $\Lambda(\bar{H}_\alpha)$ is an integer $\geqslant 0$ $\forall \alpha \in P$, and if $\lambda = \Lambda - \delta_{P_{\mathfrak{k}}} + \delta_{P_n}$, then the $U(\mathfrak{g})$-*

*module arising out of the discrete series representation* $\omega(\Lambda)$ *of* $G$ *is precisely* $D_{\mathfrak{g},-t_0\cdot P,\mu}$ *where* $\mu = t_0' \cdot \lambda$.

**3.4.** Although this is a very substantial analogue of the results of Section 2.3, there are several obvious shortcomings in the method used for the construction of the modules $D_{\mathfrak{g},P,\mu}$. First, it is not clear that $W_0/\tilde{W}$ is the analogue of the universal module $J_{\mathfrak{g},P,\lambda}$. This is clearly related to the fact that the modules $W_j$ are not constructed with universal mapping properties and that the construction uses *one particular reduced expression for* $t_0$. Obvious analogies with the family of $\mathfrak{k}$-Verma modules $V_{\mathfrak{k},P_\mathfrak{k},s'\cdot\lambda}$ suggest that one should attempt to construct modules $W_s$ indexed by $s \in \mathfrak{w}(\mathfrak{k})$, and satisfying certain natural inclusions, and that the $W_s$ must be constructed in a canonical manner. A second shortcoming is the predominant role played by $\lambda$. If multiplicity formulae of Blattner type are to be proved, then one should treat all the classes $\vartheta(\nu)$ on the same footing as $\vartheta(\lambda)$.

In order to overcome these difficulties, Enright [1] introduced the concept of *completion of a* $\mathfrak{g}$*-module with respect to a* $P_\mathfrak{k}$*-simple root*. Using this he showed that if one starts with a $\mathfrak{g}$-module $A$ belonging to a fairly large category $\mathscr{I}$, successive completions of $A$ with respect to the simple roots of $P_\mathfrak{k}$ will yield what he calls a lattice $\{A_s\}_{s\in\mathfrak{w}(\mathfrak{k})}$ of $\mathfrak{g}$-modules above $A$. One thus obtains two very general covariant functors

$$A \mapsto A_1, \qquad A \mapsto \tau(A) = A_1 / \sum_{s \neq 1} A_s.$$

The basic theorem of the theory is now the fact that $\tau(A)$ is a $\mathfrak{k}$-finite $\mathfrak{g}$-module; thus

$$A \mapsto \tau(A)$$

goes from the category $\mathscr{I}$ to the category of $\mathfrak{k}$-finite $\mathfrak{g}$-modules. The constructions of the discrete series or the fundamental series may now be viewed as restricting the above functor to suitable subcategories of $\mathscr{I}$. In [1] he used further the structure theory of the lattices $\{A_s\}$ of modules to obtain multiplicity formulae for the $\mathfrak{k}$-classes occurring in $\tau(A)$ when $A$ is suitably restricted. From now on my aim is to try to describe the main features of these ideas of Enright and the results that flow from them.

## 4. Completions of Modules. Lattices of Modules

4.1. We denote by $\mathfrak{a}$ the three-dimensional Lie algebra over $\mathbb{C}$ with basis $\{H, X, Y\}$ satisfying

$$[H, X]=2X, \qquad [H, Y]=-2Y, \qquad [X, Y]=H.$$

For any $\mathfrak{a}$-module $M$ we write

$$M[c]=\{v: v \in M, H \cdot v=cv\}, \qquad (c \in \mathbb{C}),$$
$$M^X=\{v: v \in M, X \cdot v=0\}, \qquad M[c]^X=M^X \cap M[c], \qquad (c \in \mathbb{C}).$$

We regard $\mathfrak{h}=\mathbb{C} \cdot H$ as a CSA of $\mathfrak{a}$, and the element $\alpha \in \mathfrak{h}^*$ with $\alpha(H)=2$ as the positive root of $(\mathfrak{a}, \mathfrak{h})$. This enables us to speak of weight modules for $\mathfrak{a}$ (namely those modules $M$ which are, as $\mathfrak{h}$-modules, direct sums of the $M[c]$), and Verma modules $V_{\mathfrak{a},\{\alpha\},\lambda}=V_{+,\lambda}$ of Section 2.1 ($\lambda \in \mathbb{C}$); for brevity we denote these by $V_\lambda$.

DEFINITION 4.1.1. (a) A weight module $M$ for $\mathfrak{a}$ is said to be complete (with respect to $H$, $X$, $Y$) if for each integer $n \geqslant 0$, $Y^{n+1}$ induces a bijection of $M[n]^X$ with $M[-n-2]^X$:

$$Y^{n+1}: M[n]^X \simeq M[-n-2]^X.$$

(b) Let $M$ be a weight module for $\mathfrak{a}$. By a completion of $M$ (with respect to $H$, $X$, $Y$) we mean a weight module $M'$ for $\mathfrak{a}$ together with an $\mathfrak{a}$-module injection

$$i: M \hookrightarrow M'$$

such that

(i) $M'/i(M)$ is $U(\mathfrak{a})$-finite,
(ii) $M'$ is complete.

We shall presently introduce the category of $\mathfrak{a}$-modules in which the above definitions have reasonable consequences. Before that we make some remarks.

First, when we consider completions of a module $M$, it is usually convenient to regard $M$ as a submodule of the completion. Secondly, if $M'(\supseteq M)$ is a completion of $M$, the $U(\mathfrak{a})$-finiteness of $M'/M$ implies that for any integer $n \geqslant 0$,

$$M[-n-2]^X=M'[-n-2]^X.$$

Finally, let $M$ be any weight module for $\mathfrak{a}$ on which $Y$ acts injectively. If $c \in \mathbb{C}$ and $v \in M[c]^X$ is nonzero,

$$U(\mathfrak{a}) \cdot v \simeq V_c.$$

Now, the only mappings between the $V_c$ are the imbeddings

$$V_{-n-2} \hookrightarrow V_n, \qquad (n \geqslant 0 \text{ integer}).$$

So, to say that $M$ is complete is to say that given any integer $n \geqslant 0$ and an imbedding $V_{-n-2} \hookrightarrow M$, we have an imbedding $V_n \hookrightarrow M$ such that the following diagram is commutative:

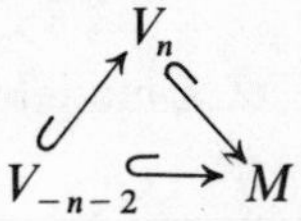

Let us now introduce the category $\mathscr{I}(\mathfrak{a})$ (or $\mathscr{I}(\mathfrak{a}, \mathfrak{h}, \{\mathfrak{a}\})$ to be more precise). A module $M$ for $\mathfrak{a}$ is said to belong to $\mathscr{I}(\mathfrak{a})$ if

(i) $M$ is a weight module,
(ii) $Y$ acts injectively on $M$,
(iii) $X$ acts locally nilpotently on $M$.

We can replace (ii) by (ii′) saying that, as a module for $\mathbb{C}[Y]$, $M$ is free (= torsion free, as $\mathbb{C}[Y]$ is a principal domain).

PROPOSITION 4.1.2. *Let $M \in \mathscr{I}(\mathfrak{a})$. Then we have the following.*

(i) *$M$ has a completion; all completions of $M$ are in $\mathscr{I}(\mathfrak{a})$.*

(ii) *If $M'$, $M''$ are two completions of $M$, with respective imbeddings $i'$, $i''$ of $M$ into $M'$, $M''$, there is a unique $\mathfrak{a}$-module isomorphism $M \xrightarrow{\sim} M''$ such that the following diagram is commutative:*

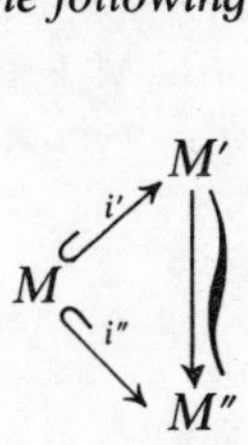

(iii) *If $M_i$ $(i=1, 2)$ are in $\mathscr{I}(\mathfrak{a})$, $M_i'$ the completion of $M_i$ $(M_i \subseteq M_i')$, and if*

$$L: M_1 \to M_2$$

*is any $\mathfrak{a}$-module map, there exists a unique $\mathfrak{a}$-module map*

$$L': M_1' \to M_2'$$

*such that the diagram*

$$\begin{array}{ccc} M_1 & \subset & M'_1 \\ {\scriptstyle L}\downarrow & & \downarrow{\scriptstyle L'} \\ M_2 & \subset & M'_2 \end{array}$$

*is commutative.*

We shall always write $C(M)$ for the completion of a module $M \in \mathscr{I}(\mathfrak{a})$. Thus

$$M \mapsto C(M), \qquad (M \in \mathscr{I}(\mathfrak{a})),$$

is a covariant functor, mapping $\mathscr{I}(\mathfrak{a})$ into itself. If

$$T: M_1 \to M_2, \qquad (M_i \in \mathscr{I}(\mathfrak{a})),$$

is any $\mathfrak{a}$-module map, we denote by $C(T)$ the extension of $T$ such that

$$C(T): C(M_1) \to C(M_2)$$

is an $\mathfrak{a}$-module map.

In order to make full use of the concept of completions, it is necessary to consider the situation where $\mathfrak{a}$ is a subalgebra of a finite-dimensional Lie algebra $\mathfrak{m}$. For any $\mathfrak{m}$-module $M$ let $M_{\mathfrak{a}}$ denote the underlying $\mathfrak{a}$-module. Let $\mathscr{I}_{\mathfrak{m}}(\mathfrak{a})$ be the category of all $\mathfrak{m}$-modules $M$ such that $M_{\mathfrak{a}} \in \mathscr{I}(\mathfrak{a})$. We then have the following sharpening of the preceding proposition.

PROPOSITION 4.1.3. (i) *Let $M$ be an $\mathfrak{m}$-module belonging to $\mathscr{I}_{\mathfrak{m}}(\mathfrak{a})$ Then there exists exactly one $\mathfrak{m}$-module structure on $C(M_{\mathfrak{a}})$ such that $M$ becomes a sub $\mathfrak{m}$-module of it. We write $C(M)$ for this $\mathfrak{m}$-module, so that $C(M)_{\mathfrak{a}} = C(M_{\mathfrak{a}})$.*

(ii) *Let $M_i (i=1, 2)$ be $\mathfrak{m}$-modules in $\mathscr{I}_{\mathfrak{m}}(\mathfrak{a})$. Suppose*

$$T: M_1 \to M_2$$

*is an $\mathfrak{m}$-module map and let*

$$C(T): C(M_1)_{\mathfrak{a}} \to C(M_2)_{\mathfrak{a}}$$

*be the extension of $T$ as an $\mathfrak{a}$-module map. Then $C(T)$ is an $\mathfrak{m}$-module map of $C(M_1)$ into $C(M_2)$:*

$$C(T): C(M_1) \to C(M_2).$$

COROLLARY 4.1.4. *Let $M$ be an $\mathfrak{m}$-module lying in $\mathscr{I}_{\mathfrak{m}}(\mathfrak{a})$, and $M_i$ $(i=1, 2)$ two completions of $M$. Then $\exists$ a unique $\mathfrak{m}$-module isomorphism of $M_1$ with $M_2$ such that*

$$\begin{array}{ccc} & & M_1 \\ & \nearrow & \\ M & & \Big\downarrow\wr \\ & \searrow & \\ & & M_2 \end{array}$$

*is commutative.*

In view of the fundamental importance of the concept of completions of modules for everything that we do later, I shall give a brief sketch of the proof that for any module in $\mathscr{I}_{\mathfrak{m}}(\mathfrak{a})$, its completion exists, is in $\mathscr{I}_{\mathfrak{m}}(\mathfrak{a})$, and that the process of completion is functorial.

*Existence of completion.* Let $M$ be any $\mathfrak{m}$-module such that $M_{\mathfrak{a}} \in \mathscr{I}(\mathfrak{a})$. For any $z \neq 0$ in $M[-n-2]^X$ ($n$ integer $\geqslant 0$) we have a surjective $\mathfrak{m}$-module map

$$j_z\colon U(\mathfrak{m}) \bigotimes_{U(\mathfrak{a})} V_{-n-2} \to M_z = U(\mathfrak{m}) \cdot z$$

with

$$j_z(1 \otimes v_{-n-2}) = z$$

($v_\lambda$ is the highest weight vector of the Verma module $V_\lambda$). On the other hand, we have the imbedding

$$V_{-n-2} \hookrightarrow V_n \quad (v_{-n-2} \mapsto Y^{n+1} v_n)$$

that extends to an $\mathfrak{m}$-module imbedding

$$L_{-n-2} \hookrightarrow L_n \qquad (1 \otimes v_{-n-2} \mapsto 1 \otimes Y^{n+1} v_n)$$

where of course we are writing $L_\lambda = U(\mathfrak{m}) \otimes_{U(\mathfrak{a})} V_\lambda$. We let $z$ vary over a basis of $M[-n-2]^X$ for each integer $n \geqslant 0$ and regard $M$ as a quotient of $L'$,

$$L' = M \oplus \bigoplus_z L_{-n-2}, \qquad j\colon L' \to M,$$

where $j$ = the identity of $M$ and $= j_z$ on the summand $L_{-n-2}$ corresponding to $z$. Further there is now a natural imbedding $L' \hookrightarrow L$ where

$$L = M \oplus \bigoplus_z L_n.$$

Let $J'$ be the kernel of $j$ and $J_0$ its image in $L$; $L/J_0$ may have $Y$-torsion and so we introduce $J$ as the set of all $v \in L$ such that $Y^m v \in J_0$ for some $m \geqslant 0$. $J$ is a sub $\mathfrak{m}$-module; and by Lemma 3.3.1, $\bar{M} = L/J$ is torsion free for $Y$; moreover we have

$$M \hookrightarrow \bar{M}$$

As $V_n/V_{-n-2}$ is $U(\mathfrak{a})$-finite, $L_n/L_{-n-2}$ is also so. So $\bar{M}/M$ (we regard $M$ as a sub $\mathfrak{m}$-module of $\bar{M}$) is $U(\mathfrak{a})$-finite. Hence, for any integer $n \geqslant 0$,

$$\bar{M}[-n-2]^X = M[-n-2]^X.$$

From this it is clear that $\bar{M}$ is a completion of $M$.

*Functoriality.* Let $M_i$ $(i=1, 2)$ be $\mathfrak{m}$-modules such that $(M_i)_{\mathfrak{a}} \in \mathscr{I}(\mathfrak{a})$ and let $T: M_1 \to M_2$ an $\mathfrak{m}$-module map. Let the $\mathfrak{m}$-module $\bar{M}_i$ be a completion of $M$, $M_i \subset \bar{M}_i$, $\Gamma$ the graph of $T$. Then $\Gamma$ is a sub $\mathfrak{m}$-module of $M_1 \oplus M_2$:

$$\Gamma \subset M_1 \oplus M_2 \subset \bar{M}_1 \oplus \bar{M}_2.$$

Denote by $\bar{\Gamma}$ the set of all $v \in \bar{M}_1 \oplus \bar{M}_2$ such that $v$ is $U(\mathfrak{a})$-finite mod $\Gamma$. Then $\bar{\Gamma}$ is a sub $\mathfrak{m}$-module of $\bar{M}_1 \oplus \bar{M}_2$. It is easy to check that the projection

$$\mathrm{pr}_1 : \bar{\Gamma} \to \bar{M}_1$$

is an *isomorphism* (onto). Hence $\bar{\Gamma}$ is the graph of an $\mathfrak{m}$-module map $\bar{T}: \bar{M}_1 \to \bar{M}_2$ that extends $T$. For uniqueness we show that any $\mathfrak{m}$-module map $\bar{M}_1 \to \bar{M}_2$ that vanishes on $M_1$ is identically zero. In proving this as well as in the proof that $\mathrm{pr}_1$ is an isomorphism we use the following lemma (cf. Lemma 2.2.12):

LEMMA 4.1.5. *Let $A_1$, $A_2$ be weight modules for $\mathfrak{a}$ with $A_1 \subset A_2$. Suppose $A_2/A_1$ is $U(\mathfrak{a})$-finite and that $X$ acts locally nilpotently on $A_2$. Then $A_2$ is generated (as an $\mathfrak{a}$-module) by $A_1$ and the various $A_2[n]^X$ ($n$ integer $\geqslant 0$).*

*From now on we shall use completions only for modules in $\mathscr{I}(\mathfrak{a})$.*

PROPOSITION 4.1.6. (i) *Let $M_i \in \mathscr{I}(\mathfrak{a})$ $(i=1, 2)$, with $M_1 \subset M_2$. Let*

$$\mathrm{Cl}_{M_2}(M_1) = \{v : v \in M_2,\ v \text{ is } U(\mathfrak{a})\text{-finite mod } M_1\}$$

(*closure of $M_1$ in $M_2$*). *Then $\mathrm{Cl}_{M_2}(M_1)$ is an $\mathfrak{a}$-module, which is an $\mathfrak{m}$-module if the $M_i$ are $\mathfrak{m}$-modules. If $M_2$ is complete* (resp. *$M_1$ is complete*),

$$\mathrm{Cl}_{M_2}(M_1) \simeq C(M_1) \quad (\text{resp. } \mathrm{Cl}_{M_2}(M_1) = M_1).$$

(ii) *Completions of direct sums are direct sums of completions. Direct summands of complete modules are complete* (all modules in $\mathscr{I}(\mathfrak{a})$ or $\mathscr{I}_{\mathfrak{m}}(\mathfrak{a})$).

**4.2.** Let us look at some examples of complete modules. First of all, the Verma modules $V_\lambda$ ($\lambda \in \mathbb{C} \backslash \mathbb{Z}$) are complete. If $\lambda$ is an integer, then $V_\lambda$ is complete if $\lambda + 1 \geqslant 0$ while

$$C(V_{-\lambda-2}) = V_\lambda, \qquad (\lambda \text{ integer}, \lambda + 1 \geqslant 0).$$

Let $\lambda$ be an integer $\geqslant 0$. A simple argument shows that $H + \lambda + 2$ and $X^{\lambda+2}$ generate a proper left ideal in $U(\mathfrak{a})$, i.e., if

$$\mathfrak{M}'_\lambda = U(\mathfrak{a})(H + \lambda + 2) + U(\mathfrak{a})X^{\lambda+2}$$

then $1 \notin \mathfrak{M}'_\lambda$. Let us write $W$ for the $U(\mathfrak{a})$-module $U(\mathfrak{a})/\mathfrak{M}'_\lambda$. Poincare–Birkhoff–Witt theorem shows that the elements

$$Y^r X^s, \qquad (0 \leqslant s \leqslant \lambda + 1, r \geqslant 0),$$

are linearly independent modulo $\mathfrak{M}'_\lambda$. Hence $W \in \mathscr{I}(\mathfrak{a})$. If

$$\pi \colon U(\mathfrak{a}) \to W$$

is the natural map, the vectors

$$w_{r,s} = \pi(Y^r X^s), \qquad (r \geqslant 0, 0 \leqslant s \leqslant \lambda + 1),$$

form a basis for $W$. Clearly, as $\pi(1) = w_{0,0}$,

$$W = U(\mathfrak{a}) \cdot w_{0,0}, \qquad H \cdot w_{0,0} = -(\lambda + 2) w_{0,0},$$

$$X^s \cdot w_{0,0} = w_{0,s} \qquad (0 \leqslant s \leqslant \lambda + 1),$$

$$H \cdot w_{0,\lambda+1} = \lambda w_{0,\lambda+1}, \qquad X w_{0,\lambda+1} = 0.$$

We now introduce the Casimir element

$$\omega = (H+1)^2 + 4YX$$

and denote by $T_\lambda$ the $U(\mathfrak{a})$-submodule of $W$ which is the largest on which $\omega - (\lambda+1)^2$ is nilpotent. Since

$$(\omega - (\lambda+1)^2) w_{0,\lambda+1} = 0$$

we see that $T_\lambda \neq 0$. From general theory we know that $T_\lambda$ is a direct summand of $W$ and that there is a unique complementary summand. So we can speak of the $U(\mathfrak{a})$-module projection

$$W \to T_\lambda$$

without any ambiguity. Let $v_0$ be the image of $w_{0,0}$ in $T_\lambda$ under this projection. Since $\pi(X^{\lambda+1}) = w_{0,\lambda+1} \in T_\lambda$ and since we also have

$$\pi(X^{\lambda+1}) = X^{\lambda+1} \cdot \pi(1) = X^{\lambda+1} \cdot w_{0,0},$$

we get $X^{\lambda+1} \cdot v_0 = w_{0,\lambda+1}$. At the same time, as $w_{0,0}$ generates $W$, $v_0$ generates $T_\lambda$:

$$T_\lambda = U(\mathfrak{a}) \cdot v_0, \qquad H \cdot v_0 = -(\lambda+2)v_0, \qquad w_{0,\lambda+1} = X^{\lambda+1} \cdot v_0.$$

Furthermore,

$$U(\mathfrak{a}) \cdot w_{0,\lambda+1} \simeq V_\lambda.$$

We thus have an exact sequence of $U(\mathfrak{a})$-modules

$$0 \to V_\lambda \to T_\lambda \to V' \to 0, \qquad (V' = T_\lambda / V_\lambda,\ V_\lambda = U(\mathfrak{a}) \cdot w_{0,\lambda+1}).$$

Now $v_0 \notin V_\lambda$; for, as $Hv_0 = -(\lambda+2)v_0$, $v_0$ would then be in $V_\lambda^X$ so that $Xv_0 = 0$, which will mean $X^{\lambda+1}v_0 = w_{0,\lambda+1} = 0$, a contradiction. So the image $v_0'$ of $v_0$ in $V'$ is $\neq 0$. Hence

$$0 \neq V' = U(\mathfrak{a}) \cdot v_0'.$$

Since $X^{\lambda+1} \cdot v_0 \in V_\lambda$, $X^{\lambda+1} \cdot v_0' = 0$. Let $d \geqslant 0$ be the largest of all integers $t \geqslant 0$ such that $X^t \cdot v_0' \neq 0$; then $0 \leqslant d \leqslant \lambda$. Clearly, as

$$H \cdot (X^d \cdot v_0') = (-\lambda - 2 + 2d) X^d \cdot v_0',$$

it follows that

$$\omega \cdot (X^d \cdot v_0') = (\lambda + 1 - 2d)^2 \cdot (X^d \cdot v_0').$$

So, as $V'$ is a quotient of $T_\lambda$, $(\lambda+1)^2 = (\lambda+1-2d)^2$. This gives, since $0 \leqslant d \leqslant \lambda$, $d = 0$. Hence $X \cdot v_0' = 0$, which, combined with $H \cdot v_0' = -(\lambda+2)v_0'$, gives $V' \simeq V_{-\lambda-2}$. So the exact sequence under study becomes

$$0 \to V_\lambda \to T_\lambda \to V_{-\lambda-2} \to 0.$$

This *does not split*. For, if it does, we would have $T_\lambda[-\lambda-2] \subset T_\lambda^X$, which is a contradiction, since $v_0 \in T_\lambda[-\lambda-2]$ but $X \cdot v_0 \neq 0$. At the same time, $[\omega - (\lambda+1)^2]^2$ is zero on $T_\lambda$, since $\omega - (\lambda+1)^2$ is zero on both $V_\lambda$ and $V_{-\lambda-2}$. In particular,

$$[\omega-(\lambda+1)^2]^2\cdot v_0=0, \qquad [\omega-(\lambda+1)^2]\cdot v_0\neq 0,$$
$$T_\lambda[-\lambda-2]^X=V_\lambda[-\lambda-2].$$

The module $T_\lambda$ is now easily seen to be the module $U(\mathfrak{a})/\mathfrak{M}_\lambda$ where $\mathfrak{M}_\lambda$ is the left ideal generated by $H+\lambda+2$, $X^{\lambda+2}$, $[\omega-(\lambda+1)^2]^2$:

$$T_\lambda\simeq U(\mathfrak{a})/\mathfrak{M}_\lambda, \quad \mathfrak{M}_\lambda=U(\mathfrak{a})\cdot\{H+\lambda+2, X^{\lambda+2}, [\omega-(\lambda+1)^2]^2\}.$$

The fact that $T_\lambda[-\lambda-2]^X=V_\lambda[-\lambda-2]$ makes it clear that $T_\lambda$ is complete. It is also *indecomposable* within the category $\mathscr{I}(\mathfrak{a})$. We now have:

PROPOSITION 4.2.1. (i) *The $V_\lambda$ ($\lambda\in\mathbb{C}$) and the $T_\lambda$ ($\lambda$ integer $\geqslant 0$) are precisely all the indecomposable objects of the category $\mathscr{I}(\mathfrak{a})$. Among these, the $V_\lambda$ with $\lambda\notin\{-2,-3,\dots\}$ and the $T_\lambda$ ($\lambda$ integer $\geqslant 0$), are the complete ones.*

(ii) *Every module in $\mathscr{I}(\mathfrak{a})$ is a direct sum (not necessarily finite) of indecomposable ones.*

(iii) *Fix an integer $\lambda\geqslant 0$. Then $T_\lambda$ is characterized as the unique $\mathfrak{a}$-module $T\in\mathscr{I}(\mathfrak{a})$ containing a vector $u\neq 0$ with*

$$T=U(\mathfrak{a})\cdot u, \quad u\in T[-\lambda-2], \quad X^{\lambda+2}u=0, \quad [\omega-(\lambda+1)^2]^2u=0.$$

*It can also be characterized as the unique $T\in\mathscr{I}(\mathfrak{a})$ such that $\exists$ an exact sequence (necessarily non-split)*

$$0\to V_\lambda\to T\to V_{-\lambda-2}\to 0$$

*and* $\dim T[-\lambda-2]^X=1$.

PROPOSITION 4.2.2. *Let $I_{\lambda+1}$ be the finite-dimensional $\mathfrak{a}$-module of highest weight $\lambda+1$; let $M=I_{\lambda+1}\otimes V_\nu$ ($\nu\in\mathbb{C}$, $\lambda$ an integer $\geqslant 0$).*

(i) *If $\nu\notin\mathbb{Z}$, we have, canonically,*

$$M\simeq\bigoplus_{0\leqslant r\leqslant\lambda+1} V_{\lambda+1+\nu-2r}.$$

(ii) *If $\nu\in\mathbb{Z}$, let $\Xi$ be the set $\{\lambda+1+\nu-2r\}_{0\leqslant r\leqslant\lambda+1}$ and let $\Xi'$ be the set of all integers $\mu'\geqslant 0$ such that $\mu'$ and $-\mu'-2$ are both in $\Xi$; let $\Xi''$ be the set of all $\mu''\in\Xi$ such that $\mu''\notin\Xi'$, $\mu''\notin-\Xi'-2$. Then we have, canonically,*

$$M\simeq\bigoplus_{\mu'\in\Xi'} T_{\mu'}\oplus\bigoplus_{\mu''\in\Xi''} V_{\mu''}.$$

The point is that $M$ has a flag whose successive quotients are the $V_\mu$ ($\mu \in \Xi$) in some order (Lemma 2.2.4). Moreover, it follows from Lemma 2.2.13 (see also Lemmas 5.1, 5.2 and 5.3 of Enright and Wallach [1]) that if $\nu$ is an integer $<0$, $\dim(M[\mu]^X) \leqslant 1\ \forall \mu$ while for $\nu$ integral $\geqslant 0$, $\dim(M[\mu]^X)$ is $\leqslant 1$ ($\mu$ integer $<0$) and $\geqslant 1$ ($\mu \in \Xi$ is integral and $\geqslant 0$). This gives the proposition.

COROLLARY 4.2.3. *We have,*

$$I_{\lambda+1} \otimes V_{-1} = \begin{cases} \bigoplus_{0 \leqslant i \leqslant \lambda/2} T_{2i} & (\lambda \text{ even}), \\ V_{-1} \oplus \bigoplus_{0 \leqslant i \leqslant (\lambda-1)/2} T_{2i+1} & (\lambda \text{ odd}). \end{cases}$$

COROLLARY 4.2.4. *With $\nu$ an integer $\geqslant 0$ we have*

$$I_{\lambda+1} \otimes V_\nu = \begin{cases} \bigoplus_{0 \leqslant i \leqslant \lambda+1} V_{\nu-\lambda-1+2i} & (0 \leqslant \lambda \leqslant \nu), \\ \bigoplus_{0 \leqslant i \leqslant (\lambda-\nu-1)/2} T_{2i} \oplus \bigoplus_{\substack{(\lambda+1-\nu)/2 \leqslant \\ \leqslant j \leqslant (\lambda+1+\nu)/2}} V_{2j} & (\lambda \geqslant \nu+1,\ \lambda-\nu-1 \text{ even}), \\ V_{-1} \oplus \bigoplus_{0 \leqslant i < (\lambda-\nu-2)/2} T_{2i+1} \oplus \bigoplus_{(\lambda-\nu)/2 \leqslant j \leqslant (\lambda+\nu)/2} V_{2j+1} & (\lambda \geqslant \nu+1,\ \lambda-\nu-1 \text{ odd}). \end{cases}$$

From these calculations we get the important

PROPOSITION 4.2.5. *Let $M$ be an $\mathfrak{m}$-module with $M_\mathfrak{a} \in \mathscr{I}(\mathfrak{a})$ and $F$ a finite dimensional $\mathfrak{m}$-module. Then*

$$C(F \otimes M) = F \otimes C(M).$$

*In particular, $F \otimes M$ is complete if $M$ is so.*

As $F \otimes C(M)/F \otimes M$ is $U(\mathfrak{a})$-finite it is enough to prove the second assertion; and for this we may assume $\mathfrak{m} = \mathfrak{a}$. One comes down to the case when $F = I_{\lambda+1}$, $M = V_\nu$ ($\nu \neq -2, -3, \dots$) or $M = T_\mu$ ($\mu$ integer $\geqslant 0$). For $I_{\lambda+1} \otimes V_\nu$ ($\nu \neq -2, -3, \dots$), completeness is clear from the proposition and corollaries above; as $T_\mu$ is a direct summand of $I_{\mu+1} \otimes V_{-1}$, $I_{\lambda+1} \otimes T_\mu$ is a direct summand of $(I_{\lambda+1} \otimes I_{\mu+1}) \otimes V_{-1}$ and so is complete.

As our final examples of complete modules we consider the case when $\mathfrak{l}$ is a reductive Lie algebra over $\mathbb{C}$ with a CSA $\mathfrak{b}$, and a positive system of roots $Q$. Let $\alpha \in Q$ be a *simple* root and let us take $H = \bar{H}_\alpha$, $X = X_\alpha$, $Y = X_{-\alpha}$. So $\mathfrak{a} = \mathbb{C} \cdot \bar{H}_\alpha + \mathbb{C} \cdot X_\alpha + \mathbb{C} \cdot X_{-\alpha}$.

PROPOSITION 4.2.6. *Let* $M = V_{\mathfrak{l},Q,\lambda}$ $(\lambda \in \mathfrak{b}^*)$. *Then M is complete if* $\lambda(\bar{H}_\alpha) \neq -2, -3, \ldots$; if $\lambda(\bar{H}_\alpha) \in \{-2, -3, \ldots\}$, *then the completion of M is* $V_{\mathfrak{l},Q,s'_\alpha \cdot \lambda}$.

Follows from the fact that if $\mathfrak{l}^-(\alpha)$ is the span $\Sigma_{\beta \in Q, \beta \neq \alpha} \mathfrak{l}_{-\beta}$, *then* $\mathfrak{l}^-(\alpha)$ *is* ad($\mathfrak{a}$)-*stable and*

$$M_{\mathfrak{a}} \simeq U(\mathfrak{l}^-(\alpha)) \otimes V_{\lambda(\bar{H}_\alpha)}.$$

**4.3.** The example of $T_\lambda$ ($\lambda$ integer $\geqslant 0$) shows that one can have complete modules $A$, $B$ $(A \subset B)$ for which $B/A$ is not complete. *So* $M \mapsto C(M)$ *is not an exact functor*. We have however the following easily proved consequence of Lemma 2.2.12.

PROPOSITION 4.3.1. *Let* $M_i$ $(i = 1, 2, 3)$ *be* $\mathfrak{a}$-modules in $\mathscr{I}(\mathfrak{a})$. *If*

$$0 \to M_2 \to M_1 \to M_3 \to 0$$

*is exact, then*

$$0 \to C(M_2) \to C(M_1) \to C(M_3)$$

*is exact; and*

$$0 \to C(M_2) \to C(M_1) \to C(M_3) < 0$$

*is exact if and only if*

$$0 \to M_2^X \to M_1^X \to M_3^X \to 0$$

*is exact.*

For long exact sequences we have the following generalization of Proposition 4.3.5.

PROPOSITION 4.3.2. *Suppose* $r \geqslant 3$ *and*

$$0 \to M_1 \to M_2 \to \cdots \to M_r \to 0$$

*is an exact sequence of modules in* $\mathscr{I}(\mathfrak{a})$. *Then*

$$0 \to C(M_1) \to C(M_2) \to \cdots \to C(M_r) \to 0$$

*is exact if and only if the following sequence is exact*:

$$(*) \qquad 0 \to M_1^X \to M_2^X \to \cdots \to M_r^X \to 0.$$

*A sufficient condition for the exactness of* (*) *is that* $X$ *is surjective on* $M_i$ *for* $i \leqslant r-2$, i.e.,

$$X(M_i) = M_i \quad \textit{for} \quad 1 \leqslant i \leqslant r-2.$$

If we consider the segment

$$C(M_{i-1}) \to C(M_i) \to C(M_{i+1})$$

and write $R_i =$ Image of $C(M_{i-1})$ and $R'_i = \ker(C(M_i) \to C(M_{i+1}))$, then $M_i \cap R'_i \subset R_i \subset R'_i$. Since $R'_i / M_i \cap R'_i$ is $U(\mathfrak{a})$-finite, $R'_i / R_i$ is $U(\mathfrak{a})$-finite. So, by Lemma 4.1.5, to prove $R'_i = R_i$ it is enough to check that $R'_i[n]^X \subset R_i$ for any integer $n \geqslant 0$. If $a_i \in R'_i[n]^X$, $Y^{n+1} a_i \in M_i[-n-2]^X$, so that $Y^{n+1} a_i =$ $=$ image of $a'_{i-1} \in M_{i-1}[-n-2]^X$, provided (*) is exact; and $a'_{i-1} =$ $= Y^{n+1} a_{i-1}$ for $a_{i-1} \in C(M_{i-1})$, showing that $a_i =$ Image of $a_{i-1}$. The other assertions are straightforward.

**4.4.** From now on we shall fix a reductive Lie algebra $\mathfrak{m}$ over $\mathbb{C}$ with a CSA $\mathfrak{h}$, and a positive system $Q$ of roots of $(\mathfrak{m}, \mathfrak{h})$. For any *simple* root $\alpha \in Q$ we have the Lie algebra

$$\mathfrak{a}^{(\alpha)} = \mathbb{C} \cdot \bar{H}_\alpha + \mathbb{C} \cdot X_\alpha + \mathbb{C} \cdot X_{-\alpha}$$

and we can consider completions with respect to $\bar{H}_\alpha$, $X_\alpha$, $X_{-\alpha}$.

It is not difficult to see that if $M$ is a weight module for $\mathfrak{h}$ such that $M_{\mathfrak{a}}^{(\alpha)}$ belongs to $\mathscr{I}(\mathfrak{a}^{(\alpha)})$, then the completion of $M$ with respect to $\bar{H}_\alpha$, $X_\alpha$, $X_{-\alpha}$ is also a weight module, and further, that, $M$ is complete if and only if for any $\mu \in \mathfrak{h}^*$ with $\mu(\bar{H}_\alpha)$ an integer $\geqslant 0$, we have a bijection

$$X_{-\alpha}^{\mu(\bar{H}_\alpha)+1} \colon \; M[\mu]^{X_\alpha} \simeq M[s'_\alpha \cdot \mu]^{X_\alpha}.$$

We write $C_\alpha$ for the functor of completion with respect to $\bar{H}_\alpha$, $X_\alpha$, $X_{-\alpha}$, and refer to it as *completion with respect to* $\alpha$. For instance, if $C_\alpha(M) = M$, we say $M$ is *complete with respect to* $\alpha$, or $\alpha$*-complete*. Since we shall be interested in varying $\alpha$, it is convenient to introduce the following definition. $\alpha$ *will always be simple in* $Q$.

DEFINITION 4.4.1. By $\mathscr{I}(\mathfrak{m})$ (or $\mathscr{I}(\mathfrak{m}, \mathfrak{h}, Q)$, if greater precision is required) we mean the category of all $\mathfrak{m}$-modules $M$ with the following properties.

(i) $M$ is a weight module with integral weights, i.e., for any weight $\lambda$ of $M$, $\lambda(\bar{H}_\alpha) \in \mathbb{Z}$ for all roots $\alpha$ of $(\mathfrak{m}, \mathfrak{h})$.

(ii) Let

$$\mathfrak{n}^- = \sum_{\beta \in -Q} \mathfrak{m}_\beta;$$

then $M$ is torsion free for $U(\mathfrak{n}^-)$.

(iii) Let

$$\mathfrak{n}^+ = \sum_{\beta \in Q} \mathfrak{m}_\beta;$$

then, elements of $U(\mathfrak{n}^+)\mathfrak{n}^+$ act locally nilpotently on $M$, or, equivalently, $M$ is $\mathfrak{n}^+$-finite.

PROPOSITION 4.4.2. *Let $M \in \mathscr{I}(\mathfrak{m})$ and $\alpha \in Q$ a simple root. Then $C_\alpha(M) \in \mathscr{I}(\mathfrak{m})$ also.*

To verify that $C_\alpha(M)$ is $U(\mathfrak{n}^+)$-finite we must show that $\dim(U(\mathfrak{n}^+) \cdot a) < \infty$ whenever $a \in C_\alpha(M)[\mu]^{X_\alpha}$ for some $\mu \in \mathfrak{h}^*$ with $\mu(\bar{H}_\alpha)$ an integer $\geqslant 0$ (cf. Lemma 4.1.5). Since $\alpha$ is simple, $\mathfrak{p}_\alpha = \mathfrak{m}_{-\alpha} + \mathfrak{h} + \mathfrak{n}^+$ is a *subalgebra.* Let $L = U(\mathfrak{p}_\alpha)a$, $L_1 = U(\mathfrak{p}_\alpha)a_1$ where $a_1 = X^m_{-\alpha}a$ lies in $M$ (surely the case for $m \gg 0$). Since $M$ is $U(\mathfrak{n}^+)$-finite, $\exists$ a finite set $\Phi \subset \mathfrak{h}^*$ such that all weights of $L_1$ are of the form $\varphi + r\alpha$ with $\varphi \in \Phi$, $r \in \mathbb{Z}$. But $L/L_1$ is $U(\mathfrak{a}^{(\alpha)})$-finite; so weights of $L$ are also of the same form. In particular, this is so for all weights of $U(\mathfrak{n}^+) \cdot a$. Since $X_\alpha \cdot a = 0$, these weights are of the form

$$\mu + \sum_{\beta \in Q \setminus \{\alpha\}} C_\beta \cdot \beta$$

with $C_\beta$ integral and $\geqslant 0$; so, $\exists$ only finitely many of them. This proves that $\dim(U(\mathfrak{n}^+) \cdot a) < \infty$.

PROPOSITION 4.4.3. *Let $M \in \mathscr{I}(\mathfrak{m})$ and suppose that $M$ is complete with respect to $\alpha$. Then, for any $\mu \in \mathfrak{h}^*$ with $\mu(\bar{H}_\alpha)$ integral and $\geqslant 0$, we have a bijection*

$$X_{-\alpha}^{\mu(\bar{H}_\alpha)+1} : M[\mu]^{\mathfrak{n}^+} \simeq M[s'_\alpha \cdot \mu]^{\mathfrak{n}^+}.$$

Follows from the fact that if $\beta \in Q$ is simple and distinct from $\alpha$, $X_\beta$ commutes with $X_{-\alpha}$.

**4.5.** We shall now consider the functors of the form

$$M \mapsto C_{\gamma_1}(C_{\gamma_2}(\ldots(C_{\gamma_r}(M)))\ldots)$$

for modules $M \in \mathscr{I}(\mathfrak{m})$, $\gamma_1, \ldots, \gamma_r$ being simple roots in $Q$.

DEFINITION 4.5.1. $\mathscr{J}(\mathfrak{m})$ (or more precisely $\mathscr{J}(\mathfrak{m}, \mathfrak{h}, Q)$) is the subcategory of all $M \in \mathscr{I}(\mathfrak{m})$ with the following property. Suppose $t \in \mathfrak{w}(\mathfrak{m})$ and that

$$t = s_{\gamma_1} \ldots s_{\gamma_d} = s_{\beta_1} \ldots s_{\beta_d}$$

are two reduced expressions for $t$, $\beta_j$, $\gamma_i$ being simple roots from $Q$. Then there is a unique isomorphism of $\mathfrak{m}$-modules

$$C_{\gamma_1}(\ldots(C_{\gamma_d}(M))\ldots) \simeq C_{\beta_1}(\ldots(C_{\beta_d}(M))\ldots)$$

such that

$$\begin{array}{ccc} & \nearrow & C_{\gamma_1}(\ldots(C_{\gamma_d}(M))\ldots) \\ M & & \downarrow \\ & \searrow & C_{\beta_1}(\ldots(C_{\beta_d}(M))\ldots) \end{array}$$

is commutative.

PROPOSITION 4.5.2. *Let $M \in \mathscr{I}(\mathfrak{m})$ and let*

$$t = s_{\gamma_1} \ldots s_{\gamma_d} = s_{\beta_1} \ldots s_{\beta_d}$$

*be two reduced expressions for the element $t \in \mathfrak{w}(\mathfrak{m})$, $\gamma_i$, $\beta_j$ being simple roots in $Q$. Then $\exists$ at most one isomorphism*

$$C_{\gamma_1}(\ldots(C_{\gamma_d}(M))\ldots) \simeq C_{\beta_1}(\ldots(C_{\beta_d}(M))\ldots)$$

*that makes the diagram*

$$\begin{array}{ccc} & \nearrow & C_{\gamma_1}(\ldots(C_{\gamma_d}(M))\ldots) \\ M & & \downarrow \\ & \searrow & C_{\beta_1}(\ldots(C_{\beta_d}(M))\ldots) \end{array}$$

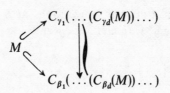

*commutative. In particular, if there is one isomorphism, it is the only one such; and the existence of such an isomorphism for each $t \in \mathfrak{w}(\mathfrak{m})$ and each pair of reduced expressions for $t$ is sufficient to ensure that $M \in \mathscr{I}(\mathfrak{m})$.*

PROPOSITION 4.5.3. *Let $\mathfrak{b}$ be a finite-dimensional Lie algebra with $\mathfrak{m} \subset \mathfrak{b}$. Suppose $M$ is a $\mathfrak{b}$-module such that the underlying $\mathfrak{m}$-module $M_{\mathfrak{m}}$ belongs to $\mathscr{I}(\mathfrak{m})$. Then, for any $t \in \mathfrak{w}(\mathfrak{m})$ and any two reduced expressions*

$$t = s_{\gamma_1} \dots s_{\gamma_d} = s_{\beta_1} \dots s_{\beta_d}, \qquad (\gamma_i, \beta_j \in Q \text{ and simple}),$$

*the modules*

$$C_{\gamma_1}(\dots(C_{\gamma_d}(M_{\mathfrak{m}}))\dots) \quad \text{and} \quad C_{\beta_1}(\dots(C_{\beta_d}(M_{\mathfrak{m}}))\dots)$$

*can be given $\mathfrak{b}$-module structures that are uniquely determined by the requirement that the natural injections of $M$ into them be $\mathfrak{b}$-module injections. Moreover, if $M_{\mathfrak{m}} \in \mathscr{I}(\mathfrak{m})$, the unique $\mathfrak{m}$-module isomorphism*

$$C_{\gamma_1}(\dots(C_{\gamma_d}(M_{\mathfrak{m}}))\dots) \xrightarrow{\sim} C_{\beta_1}(\dots(C_{\beta_d}(M_{\mathfrak{m}}))\dots)$$

*that commutes with the inclusions of $M_{\mathfrak{m}}$ into these two modules is a $\mathfrak{b}$-module isomorphism.*

Both these propositions can be proved by essentially the same argument. First assume that $M \in \mathscr{I}(\mathfrak{m})$ and write

$$L = C_{\gamma_1}(\dots(C_{\gamma_d}(M))\dots), \qquad N = C_{\beta_1}(\dots(C_{\beta_d}(M))\dots).$$

Also write

$$L_{j,j+1,\dots,d} = C_{\gamma_j}(\dots(C_{\gamma_d}(M))\dots), \qquad (1 \leqslant j \leqslant d).$$

Suppose now $T$ is an isomorphism

$$L \xrightarrow{\sim} N$$

such that

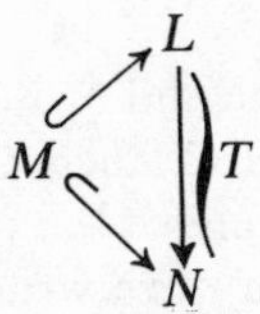


is commutative. It is then clear that if we define

$$N_{j,j+1,\dots,d} = T(L_{j,j+1,\dots,d})$$

then (cf. Proposition 4.1.6)

$$L_d = \mathrm{Cl}_{L,\gamma d}(M),\ L_{j,j+1,\ldots,d} = \mathrm{Cl}_{L,\gamma j}(L_{j+1,\ldots,d}), \qquad (1 \leqslant j < d),$$

where $\mathrm{Cl}_{L,\gamma j}$ means closure in $L$ with respect to the algebra $\mathfrak{a}^{(\gamma j)}$. Hence

$$N_d = \mathrm{Cl}_{N,\gamma d}(M),\ N_{j,j+1,\ldots,d} = \mathrm{Cl}_{N,\gamma j}(N_{j+1,\ldots,d}), \qquad (1 \leqslant j < d).$$

These relations show that the $N_{j,j+1,\ldots,d}$ *are determined within N intrinsically*, and so, if $T'$ is another isomorphism $L \simeq N$ commuting with the $M \hookrightarrow L$, $M \hookrightarrow N$, one has

$$L_{j,j+1,\ldots,d} \underset{T,T'}{\cong} N_{j,j+1,\ldots,d}, \qquad (1 \leqslant j \leqslant d).$$

So Proposition 4.1.2 and an easy induction on $d-j$ shows $T = T'$. Furthermore, if $M$ has a $\mathfrak{b}$-module structure, the $L_{j,j+1,\ldots,d}$ inherit unique $\mathfrak{b}$-module structures. The $N_{j,j+1,\ldots,d}$, being closures of modules, also acquire unique $\mathfrak{b}$-module structures, by Proposition 4.1.6. The isomorphisms

$$T\colon L_{j,j+1,\ldots,d} \cong N_{j,j+1,\ldots,d}$$

then become $\mathfrak{b}$-module isomorphisms (Corollary 4.1.4).

In other words, if $t \in \mathfrak{w}(\mathfrak{m})$ and

$$t = s_{\gamma_1} \ldots s_{\gamma d} = s_{\beta_1} \ldots s_{\beta d}, \qquad (\gamma_i, \beta_j \in Q \text{ and simple}),$$

are two reduced expressions for $t$, the functors

$$C_{\gamma_1}(\ldots(C_{\gamma d}(\cdot))\ldots) \quad \text{and} \quad C_{\beta_1}(\ldots(C_{\beta d}(\cdot))\ldots)$$

are *naturally isomorphic* on $\mathscr{J}(\mathfrak{m})$. We thus permit ourselves to write

$$C_t(\cdot) = C_{\gamma_1}(\ldots(C_{\gamma d}(\cdot))\ldots)$$

and regard $C_t$, for each $t \in \mathfrak{w}(\mathfrak{m})$, as a covariant functor

$$C_t\colon\ \mathscr{J}(\mathfrak{m}) \to \mathscr{I}(\mathfrak{m}).$$

If $\mathfrak{b}$ is a finite dimensional Lie algebra containing $\mathfrak{m}$ and $M$ is a $\mathfrak{b}$-module whose underlying $\mathfrak{m}$-module $M_{\mathfrak{m}}$ belongs to $\mathscr{J}(\mathfrak{m})$, then $C_t(M_{\mathfrak{m}})$ is a $\mathfrak{b}$-module, the $\mathfrak{b}$-module structure being unique under the requirement that $M \hookrightarrow C_t(M_{\mathfrak{m}})$ is a $\mathfrak{b}$-module injection ($t \in \mathfrak{w}(\mathfrak{m})$); we write $C_t(M)$ for $C_t(M_{\mathfrak{m}})$ equipped with this $\mathfrak{b}$-module structure.

DEFINITION 4.5.4. Let $\mathfrak{b}$ be a finite-dimensional Lie algebra containing $\mathfrak{m}$. Then $\mathscr{J}_{\mathfrak{b}}(\mathfrak{m})$ (resp. $\mathscr{I}_{\mathfrak{b}}(\mathfrak{m})$) is the category of $\mathfrak{b}$-modules $M$ such that $M_{\mathfrak{m}} \in \mathscr{J}(\mathfrak{m})$ (resp. $\mathscr{I}(\mathfrak{m})$).

PROPOSITION 4.5.5. *For each* $t \in \mathfrak{w}(\mathfrak{m})$,

$$C_t\colon M \mapsto C_t(M)$$

*is a covariant functor from* $\mathscr{J}_{\mathfrak{b}}(\mathfrak{m})$ *to* $\mathscr{I}_{\mathfrak{b}}(\mathfrak{m})$.

**4.6.** We are now in a position to introduce the notion of a lattice of modules above a module of $\mathscr{J}(\mathfrak{m})$. Recall that $\mathscr{J}(\mathfrak{m})$ is in fact $\mathscr{J}(\mathfrak{m}, \mathfrak{h}, Q)$.

DEFINITION 4.6.1. Let $A$ be any $\mathfrak{m}$-module belonging to the category $\mathscr{I}(\mathfrak{m})$. A lattice of $\mathfrak{m}$-modules above $A$ (or more precisely, a $\mathfrak{w}(\mathfrak{m})$-lattice), is a collection of $\mathfrak{m}$-modules $\{A_s\}_{s \in \mathfrak{w}(\mathfrak{m})}$, having the following properties:

(i) if $t_0 \in \mathfrak{w}(\mathfrak{m})$ is such that $t_0 \cdot Q = -Q$, then

$$A_{t_0} = A.$$

(ii) If $s \in \mathfrak{w}(\mathfrak{m})$ and $\gamma \in Q$ is a simple root such that $l(s_\gamma s) = l(s)+1$ ($l$ being the length function on $\mathfrak{w}(\mathfrak{m})$ relative to $Q$), then $A_s$ contains $A_{s_\gamma s}$ and is isomorphic to $C_\gamma(A_{s_\gamma s})$:

$$A_{s_\gamma s} \subset A_s,\ A_s \simeq C_\gamma(A_{s_\gamma s}), \qquad (l(s_\gamma s) = l(s)+1).$$

Notice first of all that if $A \in \mathscr{I}(\mathfrak{m})$ has a lattice above it, then we have the following properties:

(i) $A \in \mathscr{J}(\mathfrak{m})$, $A_s \in \mathscr{I}(\mathfrak{m})$,

(ii) $A_{t_0} = A \subset A_s \subset A_1$, $(s \in \mathfrak{w}(\mathfrak{m}))$

(iii) $A_s = C_{st_0}(A)$, $(C_1(A) = A)$.

In fact, recall that the element $t_0$ has the property

$$t_0 = s_1 s_2\ (s_1, s_2 \in \mathfrak{w}(\mathfrak{m})) \Rightarrow l(t_0) = l(s_1) + l(s_2).$$

So, if $s \in \mathfrak{w}(\mathfrak{m})$, we can write

$$t_0 = s_{\beta_1} \ldots s_{\beta_d} s_{\gamma_1} \ldots s_{\gamma_q}, \quad t_0 s^{-1} = s_{\beta_1} \ldots s_{\beta_d}, \quad s = s_{\gamma_1} \ldots s_{\gamma_q}$$

where the $\beta_i$'s and $\gamma_j$'s are in $Q$ and simple, and all expressions above are reduced. Then $l(s_{\beta_{j+1}} \ldots s_{\beta_d} s) = q + d - j$ and we have

$$A_s = C_{\beta_d}(A_{s_{\beta_d} s}) = C_{\beta_d}(C_{\beta_{d-1}}(A_{s_{\beta_{d-1}} s_{\beta_d} s})) = \cdots$$

giving

$$A_s = C_{\beta_d}(C_{\beta_{d-1}}(\ldots(C_{\beta_1}(A))\ldots)).$$

If we write $s' = st_0$, then $s' = s_{\beta_d} \ldots s_{\beta_1}$ is a reduced expression for $s'$, and $A_s$ is the result of applying the functor $C_{\beta_d} \ldots C_{\beta_1}$ to $A$. As $A_s$ depends only on $s$ (or $s' = st_0$), we see at once that $A \in \mathscr{J}(\mathfrak{m})$ and that $A_s = C_{st_0}(A)$.

Conversely, let $A \in \mathscr{J}(\mathfrak{m})$ and let

$$A_1 = C_{t_0}(A), \qquad (A \subset C_{t_0}(A)).$$

For any $s \in \mathfrak{w}(\mathfrak{m})$ we define

$$A_s = \mathrm{Cl}_{\beta_d}(\mathrm{Cl}_{\beta_{d-1}}(\ldots(\mathrm{Cl}_{\beta_1}(A)\ldots)$$

where Cl means closure in $A_1$, and

$$st_0 = s_{\beta_d} \ldots s_{\beta_1}, \qquad (\beta_j \text{ simple in } Q),$$

is a reduced expression. It is not difficult to verify that $A_s$ so defined depends only on $s$ and not on the choice of the reduced expression for $st_0$; and that $\{A_s\}_{s \in \mathfrak{w}(\mathfrak{m})}$ is a lattice of $\mathfrak{m}$-modules above $A$. This gives the following theorem (Enright [1]).

THEOREM 4.6.2. *Let $A$ be an $\mathfrak{m}$-module belonging to the category $\mathscr{I}(\mathfrak{m})$ $(= \mathscr{I}(\mathfrak{m}, \mathfrak{h}, Q))$. Then we have the following.*

(i) *$\exists$ a lattice $A_s$ $(s \in \mathfrak{w}(\mathfrak{m}))$ above $A$ if and only if $A \in \mathscr{J}(\mathfrak{m})$.*

(ii) *Suppose $\mathfrak{b}$ is a finite-dimensional Lie algebra containing $\mathfrak{m}$ and that $A \in \mathscr{J}_{\mathfrak{b}}(\mathfrak{m})$. Then, if $A_s$ $(s \in \mathfrak{w}(\mathfrak{m}))$ is a lattice above $A$, each $A_s$ can be made into a $\mathfrak{b}$-module containing $A$ as a sub $\mathfrak{b}$-module; this can be done in only one way; and $A_s \in \mathscr{I}_{\mathfrak{b}}(\mathfrak{m})\ \forall s$.*

(iii) *Suppose $A$ is as in* (ii) *and $\{A_s\}$, $\{A'_s\}$ are two lattices above $A$. Then there is a unique $\mathfrak{b}$-module isomorphism of $A_1 \simeq A'_1$ such that*

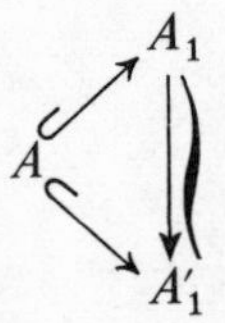


*is commutative. Calling this isomorphism $\psi$, we have moreover,*

$$\psi(A_s) = A'_s, \qquad (s \in \mathfrak{w}(\mathfrak{m})),$$

i.e.,

$$A_s \overset{\psi}{\simeq} A'_{s'}.$$

*Example* 4.6.3. There are two simple classes of examples where one already has a lattice furnished by nature.

(a) $\mathfrak{m}$, $\mathfrak{h}$, $Q$ are as above. We take $\lambda \in \mathfrak{h}^*$ to be dominant integral and define

$$A_s = V_{\mathfrak{m},Q,s'\cdot\lambda}, \qquad (s \in \mathfrak{w}(\mathfrak{m})).$$

Here we are identifying $V_{\mathfrak{m},Q,s'\cdot\lambda}$ as a submodule of $V_{\mathfrak{m},Q,\lambda}$ in the natural way:

$$V_{\mathfrak{m},Q,s'\cdot\lambda} \subset V_{\mathfrak{m},Q,\lambda}, \qquad (s \in \mathfrak{w}(\mathfrak{m})).$$

Proposition 4.2.6 now implies that the $A_s$ form a lattice above $A_{t_0}$; it is only necessary to note that if $s \in \mathfrak{w}(\mathfrak{m})$ and $\gamma \in Q$ is simple such that $l(s_\gamma s) = l(s)+1$, then for any $\lambda$ as above, $(s' \cdot \lambda)(\bar{H}_\gamma)$ is an integer $\geqslant 0$ so that $C_\gamma(V_{\mathfrak{m},Q,(s_\gamma s)'\cdot\lambda}) = V_{\mathfrak{m},Q,s'\cdot\lambda}$. In particular,

$$A_{t_0} = V_{\mathfrak{m},Q,t_0'\cdot\lambda} \in \mathscr{J}(\mathfrak{m}).$$

(b) $\mathfrak{m}$, $\mathfrak{h}$, $Q$ as above; but now $\mathfrak{m} \subset \mathfrak{g}$ where $\mathfrak{g}$ is *s.s.*, has same rank as $\mathfrak{m}$ (so that $\mathfrak{h}$ is a CSA of $\mathfrak{g}$), and $P$ is a positive system of roots of $(\mathfrak{g}, \mathfrak{h})$ such that

(i) $Q \subset P$,

(ii) $Q$-simple roots of $Q$ are $P$-simple.

Let $t_0 \in \mathfrak{w}(\mathfrak{m})$ be such that $t_0 \cdot Q = -Q$. For any $s \in \mathfrak{w}(\mathfrak{m})$ we have, by (ii),

$$s' \cdot \lambda = s(\lambda + \delta_P) - \delta_P = s(\lambda + \delta_Q) - \delta_Q.$$

So, if we take $\lambda \in \mathfrak{h}^*$ such that

$$\lambda(\bar{H}_\alpha) \text{ is an integer } \geqslant 0 \ \forall \alpha \in Q$$

and define

$$A_s = V_{\mathfrak{g},P,s'\cdot\lambda}, \qquad (s \in \mathfrak{w}(\mathfrak{m})),$$

then using (ii) above we argue as in (a) above to conclude that the following are true:

(i) the $A_s$ $(s \in \mathfrak{w}(\mathfrak{m}))$ form a lattice of $\mathfrak{g}$-modules over

$$A_{t_0} = V_{\mathfrak{g},P,t_0'\cdot\lambda}.$$

(ii) $\quad V_{\mathfrak{g},P,t\delta\cdot\lambda}\in\mathscr{I}_{\mathfrak{g}}(\mathfrak{m}),\qquad V_{\mathfrak{g},P,s'\cdot\lambda}\in\mathscr{I}_{\mathfrak{g}}(\mathfrak{m})\ \forall s\in\mathfrak{w}(\mathfrak{m}).$

These conditions are satisfied in the set up of Proposition 2.3.2 with $\mathfrak{m}=\mathfrak{k}$, $P$ = an admissible positive system for $(\mathfrak{g}, \mathfrak{h})$, and $Q=P_k$.

It is clearly desirable to prove that the category $\mathscr{J}(\mathfrak{m})$, and more generally, the category $\mathscr{J}_{\mathfrak{b}}(\mathfrak{m})$, is fairly extensive. The next two propositions suggest that this is indeed so. It would be tempting to conjecture that $\mathscr{I}(\mathfrak{m})=\mathscr{J}(\mathfrak{m})$; but this appears to be difficult to establish.[1]

PROPOSITION 4.6.4. *Let* $\mathfrak{m}$, $\mathfrak{h}$, $Q$ *be as above. Then we have the following.*

(i) *The Verma module* $V_{\mathfrak{m},Q,\mu}\in\mathscr{J}(\mathfrak{m})\ \forall$ *integral* $\mu\in\mathfrak{h}^*$.

(ii) *If* $A\in\mathscr{J}(\mathfrak{m})$ *and* $F$ *is a finite-dimensional* $\mathfrak{m}$*-module, then* $F\otimes A\in\mathscr{J}(\mathfrak{m})$.

(iii) *If* $A\in\mathscr{J}(\mathfrak{m})$, $B\subset A$ *a sub* $\mathfrak{m}$*-module such that* $A/B$ *is* $U(\mathfrak{n}^-)$*-torsion free, then* $B\in\mathscr{J}(\mathfrak{m})$.

(iv) *If* $L$ *is a finite-dimensional* $\mathfrak{b}$*-module with integral weights, then* $U(\mathfrak{m})\otimes_{U(\mathfrak{b})}L\in\mathscr{J}(\mathfrak{m})$.

(ii) follows from Proposition 4.2.5. To prove (iii) we use the same type of argument as was used in proving Propositions 4.5.2 and 4.5.3. If $E=C_t(A)$ $(t\in\mathfrak{w}(\mathfrak{m}))$, with $A\subset E$, it is a question of characterizing

$$\mathrm{Cl}_{\gamma_1}(\mathrm{Cl}_{\gamma_2}(\ldots(\mathrm{Cl}_{\gamma_d}(B))\ldots)=F$$

within $E$, $t=s_{\gamma_1}\ldots s_{\gamma_d}$ being a reduced expression for $t$. This is done by the following lemma.

LEMMA 4.6.5. *Let notation and assumptions be as above, i.e.,* $A\in\mathscr{J}(\mathfrak{m})$, $E=C_t(A)\supset A$, $B\subset A$ *with* $A/B$ *free of* $U(\mathfrak{n}^-)$*-torsion.* Then

$$F=\mathrm{Cl}_{\gamma_1}(\mathrm{Cl}_{\gamma_2}(\ldots(\mathrm{Cl}_{\gamma_d}(B)\ldots)$$

*is given by*

$$F=\{e\in E\colon y\cdot e\in B \textit{ for some nonzero } y\in U(\mathfrak{n}^-)\}$$

i.e., *$F$ is the $U(\mathfrak{n}^-)$-torsion* mod *$B$ of $E$.*

Since

$$0\to B\to A\to A/B\to 0$$

is an exact sequence of modules in $\mathscr{I}(\mathfrak{m})$, the sequence

[1] *Added in proof.* This has been done now by Bouaziz and Deodhar (independently); see, for example, Deodhar's forthcoming paper in *Inventiones Mathematicae*.

$$0 \to C_\beta(B) \to C_\beta(A) \to C_\beta(A/B)$$

is exact, $\beta \in Q$ being simple. In particular,

$$C_\beta(A)/C_\beta(B) \hookrightarrow C_\beta(A/B)$$

showing that $C_\beta(A)$ has no $U(\mathfrak{n}^-)$-torsion mod $C_\beta(B)$. So

$$C_\beta(B) = \{v \in C_\beta(A) : y \cdot v \in B \text{ for some } y \neq 0 \text{ in } U(\mathfrak{n}^-)\}.$$

Moreover, $C_\beta(B) \subset C_\beta(A)$ with $C_\beta(A)/C_\beta(B)$ having no $U(\mathfrak{n}^-)$-torsion. So the argument can be continued with completions with respect to another simple root; and so on.

For proving (i), we know that $V_{\mathfrak{m},Q,t_0'\cdot\lambda} \in \mathscr{J}(\mathfrak{m})$ if $\lambda$ is $Q$-dominant integral. Given any integral $\mu \in \mathfrak{h}^*$, $\exists$ $Q$-dominant integral $\xi$ such that $\mu - \xi + 2\delta_Q$ is dominant integral with respect to $-Q$, so that $t_0'(\mu - \xi)$ is dominant integral with respect to $Q$. So $V_{\mathfrak{m},Q,\mu-\xi} \in \mathscr{J}(\mathfrak{m})$. But then, if $F_\xi$ is the finite dimensional $\mathfrak{m}$-module with highest weight $\xi$, $F_\xi \otimes V_{\mathfrak{m},Q,\mu-\xi} \in \mathscr{J}(\mathfrak{m})$; moreover,

$$V_{\mathfrak{m},Q,\mu} \simeq U(\mathfrak{m}) \cdot (v_\xi \otimes v_{\mu-\xi}) \subseteq F_\xi \otimes V_{\mathfrak{m},Q,\mu-\xi}$$

(the $v$'s are highest weight vectors) while Lemma 2.2.4 shows that $F_\xi \otimes V_{\mathfrak{m},Q,\mu-\xi}$ has a 'composition series' with successive quotients as Verma modules and with smallest submodule as $U(\mathfrak{m}) \cdot (v_\xi \otimes v_{\mu-\xi})$. In particular,

$$(F_\xi \otimes V_{\mathfrak{m},Q,\mu,\xi})/(U(\mathfrak{m}) \cdot (v_\xi \otimes v_{\mu-\xi}))$$

is $U(\mathfrak{n}^-)$-torsion free (even $U(\mathfrak{n}^-)$-free). The conclusion (i) follows from (ii) and (iii). To prove (iv) we use the following lemma.

LEMMA 4.6.6.[1] *Let $L$ be a finite-dimensional $\mathfrak{b}$-module with integral weights. As usual, for any $\nu \in \mathfrak{h}^*$, let $\mathbb{C}_\nu$ be the $\mathfrak{b}$-module which is one-dimensional of weight $\nu$ on which $\mathfrak{n}^+$ is trivial. Then $\exists$ $Q$-dominant integral $\lambda \in \mathfrak{h}^*$ and a finite-dimensional $\mathfrak{m}$-module $F$ such that*

$$L \hookrightarrow F_\mathfrak{b} \otimes \mathbb{C}_{-\mu}$$

---

[1]Let $\lambda_i$ $(1 \leqslant i \leqslant N)$ be the weights of $L$ (with multiplicities). Choose $\mu$ to be $Q$-dominant integral with all $\mu + \lambda_i$ likewise. Then

$$U(\mathfrak{m}) \otimes_{U(\mathfrak{b})} (L \otimes \mathbb{C}_\mu) \simeq \bigoplus_{1 \leqslant i \leqslant N} V_{\mathfrak{m},Q,\mu+\lambda_i}.$$

Observe that $(1 \otimes L \otimes \mathbb{C}_\mu) \cap \oplus_i M_i = 0$ where $M_i$ is the finite codimensional submodule of $V_{\mathfrak{m},Q,\mu+\lambda_i}$. Hence $L \otimes \mathbb{C}_\mu \hookrightarrow F$ for $F = \oplus_i I_{\mathfrak{m},Q,\mu+\lambda_i}$.

*where $F_{\mathfrak{b}}$ is the* $\mathfrak{b}$*-module underlying* $F$.

To prove (iv) we write $M=(F_{\mathfrak{b}}\otimes\mathbb{C}_{-\mu})/L$ and tensor the exact sequence

$$0\to L\to F_{\mathfrak{b}}\otimes\mathbb{C}_{-\mu}\to M\to 0$$

by $U(\mathfrak{m})$ over $U(\mathfrak{b})$ to get

$$0\to U(\mathfrak{m})\bigotimes_{U(\mathfrak{b})} L\to F\otimes V_{\mathfrak{m},Q,-\mu}\to U(\mathfrak{m})\bigotimes_{U(\mathfrak{b})} M\to 0;$$

we now appeal to (i)–(iii).

Proposition 4.6.4 already shows that $\mathscr{J}(\mathfrak{m})$ is fairly extensive. For our needs, what is decisive is the following even more striking result of Enright and Wallach:

PROPOSITION 4.6.7. *Let* $\mathfrak{g}$ *be a s.s. Lie algebra over* $\mathbb{C}$ *containing* $\mathfrak{m}$ *such that* $\mathrm{rk}(\mathfrak{g})=\mathrm{rk}(\mathfrak{m})$. *Let $P$ be a positive system of roots of* $(\mathfrak{g}, \mathfrak{h})$ *such that $Q$ is the subset of roots of* $(\mathfrak{m}, \mathfrak{h})$ *belonging to $P$. Suppose $\mu$ is a $Q$-integral element of* $\mathfrak{h}^*$. *Then the* $\mathfrak{g}$*-Verma module* $V_{\mathfrak{g},P,\mu}$ *belongs to* $\mathscr{J}_{\mathfrak{g}}(\mathfrak{m})$, i.e.,

$$V_{\mathfrak{g},P,\mu}\in\mathscr{J}_{\mathfrak{g}}(\mathfrak{m}).$$

In particular, $\exists$ a $\mathfrak{w}(\mathfrak{m})$*-lattice of* $\mathfrak{g}$*-modules above* $V_{\mathfrak{g},P,\mu}$.

There is an even more general proposition in which $\mathrm{rk}(\mathfrak{g})\neq\mathrm{rk}(\mathfrak{m})$ but $V_{\mathfrak{g},P,\mu}$ is proved to be in $\mathscr{J}(\mathfrak{m})$ provided $(\mathfrak{g}, \mathfrak{m}, \mathfrak{h})$ is a regular triple (Section 6.1). These beautiful results of Enright–Wallach are difficult to prove and depend on the existence of resolutions of $V_{\mathfrak{g},P,\mu}$ in terms of modules in $\mathscr{J}_{\mathfrak{g}}(\mathfrak{m})$ of a special type (see Enright [1], Enright-Wallach [1]). We shall sketch proofs of these results in Section 4.8 (Theorem 4.8.6).

**4.7.** We shall now describe some of the properties of lattices, or what comes to the same thing, of the functors

$$A\mapsto A_s, \qquad (s\in\mathfrak{w}(\mathfrak{m})).$$

Let $\mathfrak{n}^+=\Sigma_{\beta\in Q}\,\mathfrak{m}_\beta$. First we have

PROPOSITION 4.7.1. (i) *If* $s\in\mathfrak{w}(\mathfrak{m})$ *is fixed, then the functor*

$$A\mapsto A_s, \qquad (A\in\mathscr{J}(\mathfrak{m})),$$

*is covariant and commutes with taking direct sums.*

(ii) *If* $s \in \mathfrak{w}(\mathfrak{m})$ *is fixed and*

$$0 \to A \xrightarrow{i} B \xrightarrow{j} C \to 0, \qquad (A, B, C \in \mathscr{J}(\mathfrak{m})),$$

*is exact, then*

$$0 \to A_s \xrightarrow{i} B_s \xrightarrow{j} C_s$$

*is also exact* $\forall s$ *and*

$$i(A_s) = i(A_1) \cap B_s, \qquad j(B_s) \subseteq j(B_1) \cap C_s.$$

(iii) *If* $A, B, C \in \mathscr{J}(\mathfrak{m})$ *and if*

$$0 \to A \to B \to C \to 0,$$

$$0 \to A^{\mathfrak{n}^+} \to B^{\mathfrak{n}^+} \to C^{\mathfrak{n}^+} \to 0,$$

*are both exact, then for each* $s \in \mathfrak{w}(\mathfrak{m})$,

$$0 \to A_s^{\mathfrak{n}^+} \to B_s^{\mathfrak{n}^+} \to C_s^{\mathfrak{n}^+} \to 0$$

*is exact.*

For proving (iii) we use downward induction on $l(s)$ and Proposition 4.3.1.

In a lattice $\{A_s\}_{s \in \mathfrak{w}(\mathfrak{m})}$ of $\mathfrak{m}$-modules, the key feature is revealed by Proposition 4.4.3. It shows that if $\gamma \in Q$ is simple, then one can go from $A_s^{\mathfrak{n}^+}$ to $A_{s_\gamma s}^{\mathfrak{n}^+}$ via $X_{-\gamma}^m$ ($m \geqslant 0$ suitably chosen) provided of course that $l(s_\gamma s) = l(s) + 1$. Repeatedly employing this idea one can transfer many issues involving $A_1$ to corresponding issues concerning the initial module $A$ where they are presumably easier to deal with. The next proposition shows how this technique may be used for calculating dimensions of $\mathfrak{n}^+$-invariants. Here we must recall the definition of the elements $d_{s,1}(\nu) \in U(\mathfrak{n}^-)$ (see Section 2.2): if $\nu \in \mathfrak{h}^*$ is such that

$$(\nu + \delta_Q)(\overline{H}_\beta) \text{ is an integer } > 0 \ \forall \beta \in Q \cap s^{-1}(-Q)$$

and if

$$s = s_{\alpha_m} \ldots s_{\alpha_1}, \qquad (\alpha_j \in Q, \text{ simple}),$$

is a reduced expression, then

$$d_{s,1}(\nu) = X_{-\alpha_m}^{k_m + 1} \ldots X_{-\alpha_1}^{k_1 + 1}$$

where

$$k_1 = \nu(\bar{H}_{\alpha_1}), k_2 = (s'_{\alpha_1} \cdot \nu)(\bar{H}_{\alpha_2}), \ldots, k_m = (s'_{\alpha_{m-1}} \cdots s'_{\alpha_1} \cdot \nu)(\bar{H}_{\alpha_m})$$

(these are all integers $\geqslant 0$ by the assumption on $\nu$). Changing the reduced expression changes $d_{s,1}(\nu)$ only by a nonzero scalar factor because of the following characteristic property of $d_{s,1}(\nu)$: if $\nu$ is as above,

$$V_{\mathfrak{m},Q,s'\cdot\nu} \subset V_{\mathfrak{m},Q,\nu},$$
$$d_{s,1}(\nu)(V_{\mathfrak{m},Q,\nu}[\nu]) = V_{\mathfrak{m},Q,s'\cdot\nu}[s'\nu].$$

PROPOSITION 4.7.2. *Let $A \in \mathscr{J}(\mathfrak{m})$ and $(A_s)$ the lattice above $A$. Suppose $s_1, s_2 \in \mathfrak{w}(\mathfrak{m})$ are such that $l(s_1 s_2) = l(s_1) + l(s_2)$. Then, for all $\nu \in \mathfrak{h}^*$ such that $(\nu + \delta_Q)(\bar{H}_\alpha)$ is an integer $> 0 \,\forall \alpha \in Q \cap s_1^{-1}(-Q)$,*

$$d_{s_1,1}(\nu): A_{s_2}[\nu]^{\mathfrak{n}^+} \to A_{s_1 s_2}[s'_1 \cdot \nu]^{\mathfrak{n}^+}$$

*is a bijection. In particular, $\forall \nu \in \mathfrak{h}^*$ which are $Q$-dominant integral,*

$$d_{t_Q s^{-1},1}(\nu): A_s[\nu]^{\mathfrak{n}^+} \to A[(t_Q s^{-1})' \cdot \nu]^{\mathfrak{n}^+}$$

*is a bijection for each $s \in \mathfrak{w}(\mathfrak{m})$.*

This follows by starting with a reduced expression

$$s_1 = s_{\alpha_m} \cdots s_{\alpha_1}$$

for $s_1$ and successively applying Proposition 4.4.3.

PROPOSITION 4.7.3. *Let $\mathfrak{b}$ be a finite-dimensional Lie algebra containing $\mathfrak{m}$ and let $A$ be any $\mathfrak{b}$-module belonging to $\mathscr{J}_{\mathfrak{b}}(\mathfrak{m})$. Let $\{A_s\}$ be the lattice above $A$. Then, for any finite-dimensional $\mathfrak{b}$-module $F$, $F \otimes A$ lies in $\mathscr{J}_{\mathfrak{b}}(\mathfrak{m})$ and we have a canonical $\mathfrak{b}$-module isomorphism*

$$(F \otimes A)_s \simeq F \otimes A_s.$$

*In particular,*

$$\{F \otimes A_s\}_{s \in \mathfrak{w}(\mathfrak{m})}$$

*is the lattice above $A$.*

PROPOSITION 4.7.4. *Let notation be as above. Let $A \in \mathscr{J}_{\mathfrak{b}}(\mathfrak{m})$. Suppose $x \in U(\mathfrak{b})^{\mathfrak{m}}$. Then $x$ leaves invariant all the subspaces $A_s[\mu]^{\mathfrak{n}^+}$ $(s \in \mathfrak{w}(\mathfrak{m})$, $\mu \in \mathfrak{h}^*)$. If $\nu \in \mathfrak{h}^*$ is $Q$-dominant integral, the diagram*

$$\begin{array}{ccc} A_1[\nu]^{\mathfrak{n}^+} & \xrightarrow{x} & A_1[\nu]^{\mathfrak{n}^+} \\ d_{t_0,1}(\nu) \Big\| & & \Big\| d_{t_0,1}(\nu) \\ A[t_0' \cdot \nu]^{\mathfrak{n}^+} & \xrightarrow{x} & A[t_0' \cdot \nu]^{\mathfrak{n}^+} \end{array}$$

*is commutative.*

COROLLARY 4.7.5. *Suppose $\nu \in \mathfrak{h}^*$ is $Q$-dominant integral and* $\dim(A[t_0' \cdot \nu]^{\mathfrak{n}^+}) = 1$. *Then*

$$\dim(A_1[\nu]^{\mathfrak{n}^+}) = 1.$$

**4.8.** In this section we discuss the beautiful and deep result of Enright and Wallach [1], which asserts that certain $\mathfrak{g}$-Verma modules lie in $\mathscr{J}_{\mathfrak{g}}(\mathfrak{m})$. The main idea behind the proof is to show that such a result is a consequence of the existence of resolutions of these $\mathfrak{g}$-Verma modules in terms of $\mathfrak{g}$-modules whose underlying $\mathfrak{m}$-modules are especially simple.

First we consider the set up $\mathfrak{m}$, $\mathfrak{h}$, $Q$ as in Section 4.7. We introduce a definition.

DEFINITION 4.8.1. An $\mathfrak{m}$-module $M$ is said to be of type (R) if it is a direct sum (of possibly an infinite number) of $\mathfrak{m}$-modules of the form $F \otimes V_{\mathfrak{m},Q,\mu}$, where $F$ is a finite-dimensional $\mathfrak{m}$-module and $t_0' \cdot \mu$ is $Q$-dominant integral.

It is clear that if $M$ is a module of type (R), then $M \in \mathscr{J}(\mathfrak{m})$. The basic result we need is

PROPOSITION 4.8.2. *Let $N$ be a module in $\mathscr{J}(\mathfrak{m})$. Suppose $N$ has a resolution in terms of modules of type* (R), *i.e.,* $\exists$ *a long exact sequence*

$$\text{(LE)}: \quad 0 \to M_m \to M_{m-1} \to \cdots \to M_0 \to N \to 0$$

*such that $M_i$ is of type* (R) *for $0 \leqslant i \leqslant m$. Then $N \in \mathscr{J}(\mathfrak{m})$. Moreover, for any $s \in \mathfrak{w}(\mathfrak{m})$, we have the long exact sequence*

$$(\text{LE}_s): \quad 0 \to M_{m,s} \to M_{m-1,s} \to \cdots \to M_{0,s} \to N_s \to 0.$$

We start with (LE) and apply successively various completions. To ensure that at each stage the long sequences always *stay* exact, we use the following lemma which depends on Proposition 4.3.2.

LEMMA 4.8.3. *Let $M$ be a module of type* (R). *Denote by $\{M_s\}_{s\in\mathfrak{w}(\mathfrak{m})}$ a lattice above $M$. If $s\in\mathfrak{w}(\mathfrak{m})$ and $\gamma\in Q$ is a simple root such that $l(s_\gamma s)=l(s)+1$, then $X_\gamma$ is surjective on $M_{s_\gamma s}$.*

We first come down to the case $M=F\otimes V_{\mathfrak{m},Q,\mu}$ where $F$ is a finite-dimensional $\mathfrak{m}$-module and $t_0'\cdot\mu$ is $Q$-dominant integral. Next we use Lemma 5.8 of Enright-Wallach [1], according to which an endomorphism of a vector space $U_1\otimes U_2$ of the form $L_1\otimes 1+1\otimes L_2$ where $L_1^r=0$ for some $r\geqslant 1$, is surjective as soon as $L_2$ is surjective. So we may assume

$$M=V_{\mathfrak{m},Q,t_0'\cdot\lambda},$$

where $\lambda$ is $Q$-dominant integral. Then we have $M_1=V_{\mathfrak{m},Q,\lambda}$ and

$$M_s=V_{\mathfrak{m},Q,s'\cdot\lambda}\subset M_1\ \forall s\in\mathfrak{w}(\mathfrak{m}).$$

If $s$ and $\gamma$ are as in the lemma, $((s_\gamma s)'\cdot\lambda)(\bar{H}_\gamma)=n$ is an integer $<0$; on the other hand (see the argument for Proposition 4.2.6), the $\mathfrak{a}^{(\gamma)}$-module underlying $M_{s_\gamma s}$ is a direct sum of modules for $\mathfrak{a}^{(\gamma)}$ of the form $F\otimes V_n$ where $F$ is finite-dimensional, and $V_n$ is the Verma-module for $\mathfrak{a}^{(\gamma)}$ of highest weight $n$. Since $n<0$, $X_\gamma$ is surjective on $V_n$; so it is surjective on $M_{s_\gamma s}$ by the above mentioned lemma of Enright-Wallach [1].

Proposition 4.8.2 is now easy to prove. Let $t\in\mathfrak{w}(\mathfrak{m})$ and let

$$t=s_{\gamma_1}\dots s_{\gamma_d}=s_{\beta_1}\dots s_{\beta_d},\qquad (\gamma_i,\beta_j\in Q \text{ and simple}),$$

be two reduced expressions for $t$. Write

$$C_i^+=C_{\gamma_i},\quad C_i^*=C_{\beta_i},\quad C^+=C_1^+\dots C_d^+,\quad C^*=C_1^*\dots C_d^*.$$

The modules $M_r$ $(0\leqslant r\leqslant m)$ have lattices $\{M_{r,s}\}_{s\in\mathfrak{w}(\mathfrak{m})}$ above them, and (cf. Section 4.6)

$$C_q^+\dots C_d^+(M_r)=M_{r,t_q^+t_0},\qquad (t_q^+=s_{\gamma_q}s_{\gamma_{q+1}}\dots s_{\gamma_d}).$$

Since $l(t_{p+1}^+t_0)=l(t_p^+t_0)+1$ for $1\leqslant p\leqslant d$ $(t_{d+1}^+=1)$ and $t_{p+1}^+t_0=s_{\gamma_p}t_p^+t_0$, Lemma 4.8.3 shows that $X_{\gamma_p}$ is surjective on all the $C_{p+1}^+\dots C_d^+(M_r)$. A simple downward induction on $p$ and Proposition 4.3.2 now show that for all $p=1,2,\dots,d$, we have the following long exact sequence:

$$(\mathrm{LE}_p):\quad 0\to C_p^+\dots C_d^+(M_m)\to C_p^+\dots C_d^+(M_{m-1})\to\cdots$$
$$\cdots\to C_p^+\dots C_d^+(M_0)\to C_p^+\dots C_d^+(N)\to 0.$$

For $p=1$, this gives the long exact sequence

$$(\mathrm{LE})^+:\quad 0\to C_t(M_m)\to\cdots\to C_t(M_0)\to C^+(N)\to 0.$$

So we have a canonical isomorphism

$$C^+(N)\simeq C_t(M_0)/\text{Image of } C_t(M_1).$$

Similarly

$$C^*(N)\simeq C_t(M_0)/\text{Image of } C_t(M_1)$$

giving a canonical isomorphism

$$C^+(N)\simeq C^*(N)$$

which is essentially what one had to prove.

We are reduced to establishing resolutions of Verma modules. We assume $\mathfrak{m}\subset\mathfrak{g}$ where $\mathfrak{g}$ is $s\cdot s$, $\mathfrak{m}$ is reductive in $\mathfrak{g}$, and $\bar{\mathfrak{h}}$ is a CSA of $\mathfrak{g}$ such that $(\mathfrak{g}, \mathfrak{m}, \bar{\mathfrak{h}})$ is a regular triple in the sense of Section 6.1, below. This is more general than assuming that $\mathrm{rk}(\mathfrak{g})=\mathrm{rk}(\mathfrak{m})$. Let $P$ be a regular positive system of roots of $(\mathfrak{g}, \bar{\mathfrak{h}})$ and $Q$, the corresponding set of roots of $(\mathfrak{m}, \mathfrak{h})$ where $\mathfrak{h}=\mathfrak{m}\cap\bar{\mathfrak{h}}$. We put

$$\mathfrak{n}=\sum_{\alpha\in P}\mathfrak{g}_\alpha,\quad \mathfrak{n}_\mathfrak{m}=\sum_{\alpha\in Q}\mathfrak{m}_\alpha,\quad \mathfrak{b}=\bar{\mathfrak{h}}+\mathfrak{n},\quad \mathfrak{b}_\mathfrak{m}=\mathfrak{h}+\mathfrak{n}_\mathfrak{m}.$$

Since $P$ is regular, $\mathfrak{b}\supset\mathfrak{b}_\mathfrak{m}$.

The next lemma, which is Lemma 4.2 of Enright–Wallach [1], is the starting point for establishing the required resolutions.

LEMMA 4.8.4. *Fix* $\bar{\lambda}\in\bar{\mathfrak{h}}^*$ *and let* $\mathbb{C}_{\bar{\lambda}}$ (resp. $\mathbb{C}_\lambda$) *denote the one-dimensional* $\mathfrak{b}$*-module* (resp. $\mathfrak{b}_\mathfrak{m}$*-module*) *on which* $\Sigma_{\alpha\in P}\mathfrak{g}_\alpha$ (resp. $\Sigma_{\alpha\in Q}\mathfrak{g}_\alpha$) *acts trivially and having weight* $\bar{\lambda}$ *(resp.* $\lambda=\bar{\lambda}|_\mathfrak{h}$*). Let* $\Lambda^j(\mathfrak{b}/\mathfrak{b}_\mathfrak{m})$ *be the jth exterior power of the* $\mathfrak{b}_\mathfrak{m}$*-module* $\mathfrak{b}/\mathfrak{b}_\mathfrak{m}$*. Write, for* $0\leqslant j\leqslant m=\dim(\mathfrak{b}/\mathfrak{b}_\mathfrak{m})$,

$$L_j=U(\mathfrak{b})\underset{U(\mathfrak{b}_\mathfrak{m})}{\otimes}(\Lambda^j(\mathfrak{b}/\mathfrak{b}_\mathfrak{m})\otimes\mathbb{C}_\lambda)$$

*and regard the* $L_j$ *as* $U(\mathfrak{b})$*-modules in the usual way. Then we have a long exact sequence of* $U(\mathfrak{b})$*-modules*

$$(\mathrm{LE}_\mathfrak{b}):\quad 0\to L_m\to L_{m-1}\to\cdots\to L_0\to\mathbb{C}_{\bar{\lambda}}\to 0.$$

To get the resolution $(\mathrm{LE}_\mathfrak{b})$ one begins with the relative Lie algebra homology resolution (see Bernstein–Gel'fand–Gel'fand [2], Section 9) of

$\mathfrak{b}$-modules:

$$0 \to U(\mathfrak{b}) \underset{U(\mathfrak{b}_\mathfrak{m})}{\otimes} \Lambda^m(\mathfrak{b}/\mathfrak{b}_\mathfrak{m}) \to U(\mathfrak{b}) \underset{U(\mathfrak{b}_\mathfrak{m})}{\otimes} \Lambda^{m-1}(\mathfrak{b}/\mathfrak{b}_\mathfrak{m}) \to \cdots$$
$$\cdots \to U(\mathfrak{b}) \underset{U(\mathfrak{b}_\mathfrak{m})}{\otimes} \Lambda^0(\mathfrak{b}/\mathfrak{b}_\mathfrak{m}) \to \mathbb{C} \to 0.$$

Tensor this by $\mathbb{C}_\lambda$ and note that for any $\mathfrak{b}_\mathfrak{m}$-module $F$, we have a canonical isomorphism of $\mathfrak{b}$-modules

$$(U(\mathfrak{b}) \underset{U(\mathfrak{b}_\mathfrak{m})}{\otimes} F) \otimes \mathbb{C}_{\bar\lambda} \simeq U(\mathfrak{b}) \underset{U(\mathfrak{b}_\mathfrak{m})}{\otimes} (F \otimes \mathbb{C}_\lambda).$$

We then obtain ($\mathrm{LE}_\mathfrak{b}$).

PROPOSITION 4.8.5. *Let the notation be as above and let*

$$E_j = U(\mathfrak{g}) \underset{U(\mathfrak{b}_\mathfrak{m})}{\otimes} (\Lambda^j(\mathfrak{b}/\mathfrak{b}_\mathfrak{m}) \otimes \mathbb{C}_\lambda), \qquad (0 \leqslant j \leqslant m).$$

*Regard the $E_j$ as $\mathfrak{g}$-modules in the usual way. Then*

$$E_j \simeq U(\mathfrak{g}) \underset{U(\mathfrak{m})}{\otimes} A_j$$

*where*

$$A_j = U(\mathfrak{m}) \underset{U(\mathfrak{b}_\mathfrak{m})}{\otimes} (\Lambda^j(\mathfrak{b}/\mathfrak{b}_\mathfrak{m}) \otimes \mathbb{C}_\lambda)$$

*and we have a long exact sequence of $\mathfrak{g}$-modules*

$$(\mathrm{LE}_\mathfrak{g}): \quad 0 \to E_m \to E_{m-1} \to \cdots \to E_0 \to V_{\mathfrak{g},P,\bar\lambda} \to 0.$$

*Moreover, we have*:

(a) $E_j \in \mathscr{J}_\mathfrak{g}(\mathfrak{m})$, $(0 \leqslant j \leqslant m)$.

(b) *Suppose $\bar\lambda$ is such that for any weight $\theta$ of*

$$\Lambda(\mathfrak{b}/\mathfrak{b}_\mathfrak{m}) = \bigoplus_{0 \leqslant r \leqslant m} \Lambda^r(\mathfrak{b}/\mathfrak{b}_\mathfrak{m}), \qquad t'_0 \cdot (\lambda + \theta)$$

*is $Q$-dominant integral; then for any $j = 0, 1, \ldots, m$, the $\mathfrak{m}$-module $(E_j)_\mathfrak{m}$, underlying $E_j$, is of type* (R)[1].

To get ($\mathrm{LE}_\mathfrak{g}$) we tensor ($\mathrm{LE}_\mathfrak{b}$) by $U(\mathfrak{g})$ over $U(\mathfrak{b})$. That $E_j \simeq U(\mathfrak{g}) \otimes_{U(\mathfrak{m})} A_j$

[1]See Definition 4.8.1. This condition (b) on $\bar\lambda$ will be referred to as $\lambda$ (or $\bar\lambda$) being strongly-$Q$-dominant integral.

is straightforward. An easy argument shows that if $\mathfrak{q}$ is any ad($\mathfrak{m}$)-stable subspace of $\mathfrak{g}$ which is complementary to $\mathfrak{m}$, and if $S(\mathfrak{q})$ is the symmetric algebra over $\mathfrak{q}$ regarded as an $\mathfrak{m}$-module,

$$(E_j)_{\mathfrak{m}} \simeq S(\mathfrak{q}) \otimes A_j.$$

The conclusion (a), namely that $(E_j)_{\mathfrak{m}}$ belongs to $\mathscr{I}(\mathfrak{m})$, now follows from (ii) and (iv) of Proposition 4.6.4. If $\bar{\lambda}$ satisfies the condition described in (b), we deduce from Lemmas 2.2.4 and 2.2.14 that

$$A_j \simeq \bigoplus_{1 \leqslant r \leqslant m_j} V_{\mathfrak{m},Q,\lambda+\theta_{j,r}}$$

where $\theta_{j,r}$ $(1 \leqslant r \leqslant m_j)$ are all the weights of $\Lambda^j(\mathfrak{b}/\mathfrak{b}_{\mathfrak{m}})$ (counted with multiplicities). The conclusion (b) is immediate.

THEOREM 4.8.6. *Let* $(\mathfrak{g}, \mathfrak{m}, \bar{\mathfrak{h}})$ *be a regular triple, and* $P$, $Q$ *as above,* $P$ *regular. Then, if* $\lambda = \bar{\lambda}|_{\mathfrak{b}}$ *is* $Q$*-integral,*

$$V_{\mathfrak{g},P,\bar{\lambda}} \in \mathscr{I}_{\mathfrak{g}}(\mathfrak{m}).$$

*In particular,* $\exists$ *a* $\mathfrak{w}(\mathfrak{m})$*-lattice of* $\mathfrak{g}$*-modules above* $V_{\mathfrak{g},P,\bar{\lambda}}$.

If $\bar{\lambda}$ is as in (b) of Proposition 4.8.5, i.e., strongly-$Q$-dominant integral, $V_{\mathfrak{g},P,\bar{\lambda}}$ lies in $\mathscr{I}_{\mathfrak{g}}(\mathfrak{m})$, in view of Proposition 4.8.2. For general $\bar{\lambda}$, we can use a tensoring argument as in (i) of Proposition 4.6.4, since $\bar{\lambda} - r\delta_P$ is strongly-$Q$-dominant integral for $r \gg 0$.

4.9. Fix $\mathfrak{m}$, $\mathfrak{h}$, $Q$ as above. Let $\mathfrak{b}$ be a finite dimensional Lie algebra over $\mathbb{C}$ with $\mathfrak{b} \supset \mathfrak{m}$. Let $\mathscr{I}(\mathfrak{m})$, $\mathscr{I}_{\mathfrak{b}}(\mathfrak{m})$ have their usual meanings. Denote by $\mathscr{F}(\mathfrak{b}, \mathfrak{m})$ the category of all $\mathfrak{m}$-finite $\mathfrak{b}$-modules on which $\mathfrak{m}$ acts semi-simply[1], and by $\mathscr{H}(\mathfrak{b}, \mathfrak{m})$, the subcategory of $\mathscr{F}(\mathfrak{b}, \mathfrak{m})$ consisting of all $A \in \mathscr{F}(\mathfrak{b}, \mathfrak{m})$ such that each irreducible finite dimensional module for $\mathfrak{m}$ occurs in $A$ only finitely many times. If $\mathfrak{b} = \mathfrak{m}$, we write $\mathscr{F}(\mathfrak{m}) = \mathscr{F}(\mathfrak{m}, \mathfrak{m})$, $\mathscr{H}(\mathfrak{m}) = \mathscr{H}(\mathfrak{m}, \mathfrak{m})$. For $\mathfrak{m} = \mathfrak{k}$, $\mathfrak{b} = \mathfrak{g}$, $\mathscr{H}(\mathfrak{g}, \mathfrak{k})$ is simply the category of Harish Chandra modules.

[1] If $A$ is a $\mathfrak{b}$-module, then $A \in \mathscr{F}(\mathfrak{b}, \mathfrak{m})$ if and only if $A_{\mathfrak{m}}$ is the direct sum of finite-dimensional irreducible $\mathfrak{m}$-modules.

Suppose now $A \in \mathscr{J}(\mathfrak{m})$. We then take a lattice $\{A_s\}$ above $A$ and put[1]

$$\tau(A) = A / \sum_{s \neq 1} A_s.$$

The fundamental theorem of Enright [1] may now be formulated as follows.

THEOREM 4.9.1. *The assignment*

$$\tau: A \mapsto \tau(A) = A / \sum_{s \neq 1} A_s, \qquad (A \in \mathscr{J}(\mathfrak{m})),$$

*is a covariant functor from* $\mathscr{J}(\mathfrak{m}, \mathfrak{h}, Q)$ *to* $\mathscr{F}(\mathfrak{m})$. *Its restriction to* $\mathscr{J}_{\mathfrak{b}}(\mathfrak{m})$ *is a covariant functor from* $\mathscr{J}_{\mathfrak{b}}(\mathfrak{m})$ *to* $\mathscr{F}(\mathfrak{b}, \mathfrak{m})$. *If* $\mathscr{J}_f(\mathfrak{m})$ (resp. $\mathscr{J}_f(\mathfrak{b}, \mathfrak{m})$) *is the subcategory of all* $A \in \mathscr{J}(\mathfrak{m})$ (resp. $\mathscr{J}_{\mathfrak{b}}(\mathfrak{m})$) *such that*

$$\dim_{\mathbb{C}} A[t'_0 \cdot \lambda]^{\mathfrak{n}^+} < \infty \ \forall \lambda \in \mathfrak{h}^*$$

which are $Q$-dominant integral, then

$$\tau(A) \in \mathscr{H}(\mathfrak{m}), \qquad (\text{resp. } \tau(A) \in \mathscr{H}(\mathfrak{b}, \mathfrak{m})),$$

$\forall A \in \mathscr{J}_f(\mathfrak{m})$ (resp. $A \in \mathscr{J}_f(\mathfrak{b}, \mathfrak{m})$).

In proving this we may of course work in the setting where $\mathfrak{b} = \mathfrak{m}$; indeed, we have seen that once we start with an $\mathfrak{m}$-module having a $\mathfrak{b}$-module structure, all subsequent $\mathfrak{m}$-modules inherit canonical $\mathfrak{b}$-module structures, and all $\mathfrak{m}$-module maps become $\mathfrak{b}$-module maps.

$\tau(A)$ is $\mathfrak{a}^{(\beta)}$-finite for each simple root $\beta \in Q$ since $A_{s_\beta} \subset \Sigma_{s \neq 1} A_s$. That $\tau(A) \in \mathscr{F}(\mathfrak{m})$ now follows from the easily proved result that any $U(\mathfrak{n}^+)$-finite weight module for $\mathfrak{m}$, which is $\mathfrak{a}^{(\beta)}$-finite $\forall$ simple $\beta \in Q$, is in $\mathscr{F}(\mathfrak{m})$. If $\nu \in \mathfrak{h}^*$ is $Q$-dominant integral,

$$A_1[\nu]^{\mathfrak{n}^+} \to \tau(A)[\nu]^{\mathfrak{n}^+} \to 0$$

is exact (Lemma 2.2.12), while (Proposition 4.7.2)

$$d_{t_0,1}(\nu): A_1[\nu]^{\mathfrak{n}^+} \simeq A[t'_0 \cdot \nu]^{\mathfrak{n}^+}$$

is a bijection, so that

$$\dim(\tau(A)[\nu]^{\mathfrak{n}^+}) \leqslant \dim(A_1[\nu]^{\mathfrak{n}^+}) = \dim(A[t'_0 \cdot \nu]^{\mathfrak{n}^+}).$$

---

[1] $\tau(A)$ could be 0 of course. We note that categories of modules over a $\mathbb{C}$-algebra $R$ are always supposed to include 0, the zero-dimensional vector space regarded as an $R$-module.

*Remark* 4.9.2. Simple examples of the correspondence

$$A \mapsto \tau(A)$$

arise when $A = V_{\mathfrak{m}, Q, t_0' \cdot \lambda}$, $\lambda$ being $Q$-dominant integral; then $\tau(A)$ is the finite-dimensional irreducible $\mathfrak{m}$-module of $Q$-highest weight $\lambda$. When $\mathfrak{m} = \mathfrak{k}$, $\mathfrak{b} = \mathfrak{g}$, $P$ an admissible positive system for $(\mathfrak{g}, \mathfrak{h})$, $Q = P_{\mathfrak{k}}$, and $\lambda$ is $P_{\mathfrak{k}}$-dominant integral,

$$\tau(V_{\mathfrak{g}, P, t_0' \cdot \lambda}) = J_{\mathfrak{g}, P, \lambda}$$

by Proposition 2.3.2. Finally, let us assume only that $\mathrm{rk}(\mathfrak{g}) = \mathrm{rk}(\mathfrak{k})$ and that $P$ is a positive system for $(\mathfrak{g}, \mathfrak{h})$; let $\lambda \in \mathfrak{h}^*$ be $P_{\mathfrak{k}}$-dominant integral. Let

$$A = V_{\mathfrak{g}, P, t_0' \cdot \lambda}.$$

Then we can build the lattice $\{A_s\}$ $(s \in \mathfrak{w}(\mathfrak{k}))$ above $A$ (Theorem 4.8.6). Let

$$t_0 = s_{\beta_r} s_{\beta_{r-1}} \ldots s_{\beta_1}$$

be a reduced expression for $t_0$. Define

$$W_r = A, \qquad v_r = v_{t_0' \cdot \lambda}$$

where $v_{t_0' \cdot \lambda}$ is the highest weight vector of $A$. We can now find vectors

$$v_0, v_1, \ldots, v_r$$

such that

$$\begin{cases} v_j \in A_{s_j}[s_j' \cdot \lambda]^{\mathfrak{n}^+} & (s_j = s_{\beta_j} s_{\beta_{j-1}} \ldots s_{\beta_1} (j \geqslant 1),\ s_0 = 1) \\ X_{-\beta_{j+1}}^{c_{j+1}} v_j = v_{j+1} & (0 \leqslant j \leqslant r-1,\ c_{j+1} = (s_j' \cdot \lambda)(\bar{H}_{\beta_{j+1}}) + 1) \end{cases}$$

and then take $W_j = U(\mathfrak{g}) \cdot v_j$ $(0 \leqslant j \leqslant r)$. Then the $W_j$ have the properties encountered in the construction of Enright-Varadarajan [1] (Lemma 3.3.5).

PROPOSITION 4.9.3. *Let $\mu \in \mathfrak{h}^*$ be $Q$-dominant integral and let $A \in \mathscr{J}_{\mathfrak{b}}(\mathfrak{m})$. Then the $U(\mathfrak{b})^{\mathfrak{m}}$-module $\tau(A)[\mu]^{\mathfrak{n}^+}$ is a quotient of the $U(\mathfrak{b})^{\mathfrak{m}}$-module $A[t_0' \cdot \mu]^{\mathfrak{n}^+}$.*

Clearly, since

$$A_1[\mu]^{\mathfrak{n}^+} \to \tau(A)[\mu]^{\mathfrak{n}^+} \to 0$$

is exact while, by Proposition 4.7.4.

$$A_1[\mu]^{\mathfrak{n}^+} \simeq A[t'_0 \cdot \mu]^{\mathfrak{n}^+}$$

as $U(\mathfrak{b})^{\mathfrak{m}}$-modules.

PROPOSITION 4.9.4. (i) *The functor* $\tau$ *takes direct sums to direct sums.*
(ii) *If* $A \in \mathscr{I}_{\mathfrak{b}}(\mathfrak{m})$ *and* $F$ *is a finite-dimensional* $\mathfrak{b}$-*module*

$$\tau(F \otimes A) = F \otimes \tau(A).$$

PROPOSITION 4.9.5. *Suppose* $A, B, C \in \mathscr{I}(\mathfrak{m})$ *are such that*
(a) $0 \to A \to B \to C \to 0$ *is exact,*
(b) $0 \to A^{\mathfrak{n}^+} \to B^{\mathfrak{n}^+} \to C^{\mathfrak{n}^+} \to 0$ *is exact.*
*Then*

$$\tau(A) \to \tau(B) \to \tau(C) \to 0$$

*is exact.*

The $\mathfrak{m}$-finiteness of $\tau(M)$ for $M \in \mathscr{I}_{\mathfrak{b}}(\mathfrak{m})$ implies that $\tau(M)[\nu]^{\mathfrak{n}^+} = 0$ unless $\nu$ is $Q$-dominant integral. By Lemma 2.2.12 we go from $\tau(M)$ to $M_1$. We then use Proposition 4.7.1.

Finally, here is a simple but useful sufficient condition for $\tau(A)$ to be different from 0.

PROPOSITION 4.9.6. *Let* $A \in \mathscr{I}(\mathfrak{m})$ *and let* $\{A_s\}$ *be the lattice above* $A$. *Suppose there is a* $\nu \in \mathfrak{h}^*$ *with the following properties:*
(a) $\nu$ *is* $Q$-*dominant integral*
(b) $A[t'_0 \cdot \nu]^{\mathfrak{n}^+} \neq 0$ *while* $A[s' \cdot \nu]^{\mathfrak{n}^+} = 0 \ \forall s \neq t_0$.
*Then* $\tau(A)[\nu]^{\mathfrak{n}^+}$ *and* $A[t'_0 \cdot \nu]^{\mathfrak{n}^+}$ *are equivalent* $U(\mathfrak{b})^{\mathfrak{m}}$-*modules. In particular,* $\tau(A) \neq 0$ *and in fact, the class* $\vartheta(\nu)$ *of* $\mathscr{E}(\mathfrak{m})$, *with highest weight* $\nu$ *relative to* $Q$, *occurs in* $\tau(A)$ *with multiplicity* $\dim(A[t'_0 \cdot \nu]^{\mathfrak{n}^+})$.

For, by Proposition 4.7.2,

$$A_1[\nu]^{\mathfrak{n}^+} \neq 0, \qquad A_s[\nu]^{\mathfrak{n}^+} = 0, \qquad (s \neq 1).$$

If $v$ is a nonzero element in $A_1[\nu]^{\mathfrak{n}^+}$ and if $v \in \Sigma_{s \neq 1} A_s$, then Lemma 2.2.12 shows that $v = \Sigma_{s \neq 1} v_s$ where $v_s \in A_s[\nu]^{\mathfrak{n}^+}$; so all $v_s = 0$, giving $v = 0$, a contradiction. Hence image of $v$ in $\tau(A)$ is nonzero. Thus we have a $U(\mathfrak{b})^{\mathfrak{m}}$-module isomorphism

$$A_1[\nu]^{\mathfrak{n}^+} \simeq \tau(A)[\nu]^{\mathfrak{n}^+}.$$

Combining with Proposition 4.7.4 we get the result.

## 5. The modules $W_{\mathfrak{g},P,\lambda}$ when $\mathrm{rk}(\mathfrak{g})=\mathrm{rk}(\mathfrak{k})$

**5.1.** We assume now that $\mathfrak{g}_0$ is a real s.s. Lie algebra, and $\mathfrak{g}_0=\mathfrak{k}_0+\mathfrak{p}_0$ a Cartan decomposition of $\mathfrak{g}_0$; $\mathfrak{g}, \mathfrak{k}, \mathfrak{p}$ are the complexifications of $\mathfrak{g}_0, \mathfrak{k}_0, \mathfrak{p}_0$. *We assume further that* $\mathrm{rk}(\mathfrak{g})=\mathrm{rk}(\mathfrak{k})$ and choose a CSA $\mathfrak{g}\subset\mathfrak{k}$. $P$ denotes a positive system of roots of $(\mathfrak{g}, \mathfrak{h})$; and $P_\mathfrak{k}$, the subset of $P$ of roots of $(\mathfrak{k}, \mathfrak{h})$. We write $\mathfrak{w}(\mathfrak{g})$ and $\mathfrak{w}(\mathfrak{k})$ for the Weyl groups of $(\mathfrak{g}, \mathfrak{h})$ and $(\mathfrak{k}, \mathfrak{h})$ respectively. We define $t_0 \in \mathfrak{w}(\mathfrak{k})$ by $t_0 \cdot P_\mathfrak{k}=-P_\mathfrak{k}$. Let $P_n=P\backslash P_\mathfrak{k}$ and

$$\delta_P=\tfrac{1}{2}\sum_{\alpha\in P}\alpha, \qquad \delta_{P_\mathfrak{k}}=\tfrac{1}{2}\sum_{\alpha\in P_\mathfrak{k}}\alpha, \qquad \delta_{P_n}=\delta_P-\delta_{P_\mathfrak{k}}.$$

Theorem 4.8.6 implies that for any $\lambda\in\mathfrak{h}^*$ which is $P_\mathfrak{k}$-integral, the Verma $\mathfrak{g}$-module $V_{\mathfrak{g},P,\lambda}$ belongs to the category $\mathscr{J}_\mathfrak{g}(\mathfrak{k}, \mathfrak{h}, P_\mathfrak{k})$. We now define, $\forall P_\mathfrak{k}$-integral $\lambda\in\mathfrak{h}^*$, the $\mathfrak{g}$-modules

$$V_{P,\lambda,s}=(V_{\mathfrak{g},P,\lambda})_s, \qquad (s\in\mathfrak{w}(\mathfrak{k})),$$

as the elements of the lattice above $V_{\mathfrak{g},P,\lambda}$. Let

$$W_{\mathfrak{g},P,\lambda}=W_{P,\lambda}=\tau(V_{\mathfrak{g},P,\lambda}).$$

We are now in a position to formulate the main theorems of Enright [1].

THEOREM 5.1.1. *Let $\lambda\in\mathfrak{h}^*$ be $P_\mathfrak{k}$-integral, $\mu=t_0'\cdot\lambda$. Then we have the following.*

(i) *$W_{P,\lambda}$ is a Harish-Chandra module. It is $\neq 0$ whenever $t_0'\cdot\lambda$ is $P_\mathfrak{k}$-dominant integral.*

(ii) *Let $\mu=t_0'\cdot\lambda$ be $P_\mathfrak{k}$-dominant integral. Then $W_{P,\lambda}$ contains the class $\vartheta(\mu)$ of $\mathscr{E}(\mathfrak{k})$ of $P_\mathfrak{k}$-highest weight $\mu$ exactly once; and, on the corresponding isotypical space, $\mathfrak{D}=U(\mathfrak{g})^\mathfrak{k}$ acts through the homomorphism*

$$\boldsymbol{\beta}_{P,\lambda}: u\mapsto\boldsymbol{\beta}_P(u)(\lambda), \qquad (u\in\mathfrak{D}).$$

*Moreover, if $\nu\in\mathfrak{h}^*$ is $P_\mathfrak{k}$-dominant integral and $\vartheta(\nu)$ the class in $\mathscr{E}(\mathfrak{k})$ of $P_\mathfrak{k}$-highest weight equal to $\nu$,*

$$[W_{P,\lambda}:\vartheta(\nu)]\leqslant\dim(V_{\mathfrak{g},P,\lambda}[t_0'\cdot\nu]^{\mathfrak{n}_\mathfrak{k}^+})$$

*where, as usual,*

$$\mathfrak{n}_\mathfrak{k}^+=\sum_{\alpha\in P_\mathfrak{k}}\mathfrak{n}_\alpha.$$

*In particular,* $\vartheta(\nu)$ *can occur in* $W_{P,\lambda}$ *only when*

$$\lambda \underset{-t_0 \cdot P}{\prec} \nu.$$

(iii) *With* $\mu$ *still* $P_{\mathfrak{k}}$*-dominant integral, we have* $\dim(W_{P,\lambda}[\mu]^{\mathfrak{n}_{\mathfrak{k}}^+})=1$; *if* $W^0_{P,\lambda}$ *is the* $U(\mathfrak{g})$*-module generated by this one-dimensional space,* $W^0_{P,\lambda}$ *contains a unique proper largest sub* $\mathfrak{g}$*-module, denoted by* $M_{\mathfrak{g},P,\lambda}$; *and*

$$W^0_{P,\lambda}/M_{\mathfrak{g},P,\lambda} \simeq D_{\mathfrak{g},P,\lambda}.$$

COROLLARY 5.1.2. *Suppose* $\gamma \in -t_0 \cdot P$ *is such that* $\lambda-\gamma$ *is* $P_{\mathfrak{k}}$*-dominant integral. Then the class* $\vartheta(\lambda-\gamma)$ *in* $\mathscr{E}(\mathfrak{k})$ *does not occur in* $W_{P,\lambda}$.

The fundamental question is of course when $W_{P,\lambda}$ is irreducible. The following theorem of Enright [1] gives a powerful criterion for this.

THEOREM 5.1.3. *Let* $\lambda \in \mathfrak{h}^*$, $\mu = t'_0 \cdot \lambda$. *Suppose that* $\lambda$ *is* $P_{\mathfrak{k}}$*-integral and that the* $\mathfrak{g}$*-Verma module* $V_{\mathfrak{g},P,\lambda}$ *is irreducible,* i.e.,

$$(\lambda+\delta_P)(\bar{H}_\alpha) \notin \{1, 2, \ldots\} \text{ for each } \alpha \in P.$$

*Then* $W_{P,\lambda}$ *is a finitely generated Harish Chandra module whose Jordan-Hölder constituents are all equivalent. Moreover, if either* $\mu$ *is* $P_{\mathfrak{k}}$*-dominant integral* or *if* $\lambda+\delta_P$ *is regular, then* $W_{P,\lambda}$ *is nonzero and irreducible.*

THEOREM 5.1.4. *Let* $\lambda \in \mathfrak{h}^*$, $\mu = t'_0 \cdot \lambda$. *Suppose that* $\lambda$ *is* $P_{\mathfrak{k}}$*-integral and that the* $\mathfrak{g}$*-Verma module* $V_{\mathfrak{g},P,\lambda}$ *is irreducible. Then*:

(i) *If* $\langle \lambda+\delta_P, \beta\rangle = 0$ *for some P-simple* $\beta \in P_{\mathfrak{k}}$, *then* $W_{P,\lambda}=0$

(ii) *If* $\lambda+\delta_P$ *is dominant with respect to* $-P$, *then* $W_{P,\lambda}$ *is nonzero and irreducible if and only if* $\langle \lambda+\delta_P, \beta\rangle < 0$ *for each P-simple* $\beta \in P_{\mathfrak{k}}$.

For these results see Section 11. In Section 12 we describe the relation between these modules and the discrete and mock-discrete series of representations of the corresponding group.

## 6. The Fundamental Series

**6.1.** The original construction of Enright and Varadarajan [1] dealt only with the case $\mathrm{rk}(\mathfrak{g}_0)=\mathrm{rk}(\mathfrak{k}_0)$. It was observed by Enright and Wallach [1] that the same method was applicable to the case when $\mathrm{rk}(\mathfrak{g}_0) \neq \mathrm{rk}(\mathfrak{k}_0)$ provided only that the CSA $\mathfrak{h}_0$ of $\mathfrak{g}_0$ is fundamental. This was a far reaching observation; for, the representations of $\mathfrak{g}$ that were obtained by Enright and Wallach [2] are those of the *fundamental series*. The results of Enright [1] prove that these representations are irreducible and describe their structure as $\mathfrak{k}$-modules. The irreducibility is, in the unitary case, due to Harish-Chandra [5]; his treatment was analytical.

We work with a real s.s. Lie algebra $\mathfrak{g}_0$ with a Cartan decomposition $\mathfrak{g}_0 = \mathfrak{k}_0 + \mathfrak{s}_0$ and a CSA $\mathfrak{h}_0$ such that

$$\mathfrak{h}_0 = (\mathfrak{h}_0 \cap \mathfrak{k}_0) + (\mathfrak{h}_0 \cap \mathfrak{s}_0), \ \mathfrak{h}_0 \cap \mathfrak{k}_0 \text{ is a CSA of } \mathfrak{k}_0.$$

Such CSA's are called *fundamental*. They are all conjugate under the adjoint group of $\mathfrak{k}_0$. Let $\mathfrak{g}$, $\mathfrak{k}$, $\mathfrak{h}$ be the complexification of $\mathfrak{g}_0$, $\mathfrak{k}_0$, $\mathfrak{h}_0$ respectively. It is then easy to obtain the following properties of $\mathfrak{g}$, $\mathfrak{k}$, $\mathfrak{h}$:

(i) $\mathfrak{g}$ is s.s. over $\mathbb{C}$; $\mathfrak{k} \subset \mathfrak{g}$ is reductive.

(ii) $\mathfrak{h}$ is a CSA of $\mathfrak{g}$ such that $\mathfrak{h}_1 = \mathfrak{h} \cap \mathfrak{k}$ is a CSA of $\mathfrak{k}$.

(iii) $\mathfrak{h}$ is the centralizer of $\mathfrak{h}_1$ in $\mathfrak{g}$, i.e., $\mathfrak{h}_1$ contains points that are regular in $\mathfrak{g}$.

(iv) if $\mathfrak{h}_{\mathbb{R}}$ is the $\mathbb{R}$-linear span of the $H_\alpha$ ($a$ a root of $(\mathfrak{g}, \mathfrak{h})$), i.e., if $\mathfrak{h}_{\mathbb{R}}$ is the real form of $\mathfrak{h}$ on which all roots of $(\mathfrak{g}, \mathfrak{h})$ take real values, $\mathfrak{h}_{\mathbb{R}} \cap \mathfrak{h}_1$ is a real form of $\mathfrak{h}_1$.

Following Enright and Wallach [1] we call $(\mathfrak{g}, \mathfrak{k}, \mathfrak{h})$ a *regular triple. From now on we shall assume that* $(\mathfrak{g}, \mathfrak{k}, \mathfrak{h})$ *is a regular triple.*

The condition (iii) is equivalent to saying that no root of $(\mathfrak{g}, \mathfrak{h})$ vanishes identically on $\mathfrak{h}_1$ or on $\mathfrak{h}_{1,\mathbb{R}}$ where

$$\mathfrak{h}_{1,\mathbb{R}} = \mathfrak{h}_1 \cap \mathfrak{h}_{\mathbb{R}}.$$

If $\mathfrak{h}'_{\mathbb{R}}$ is the set of points of $\mathfrak{h}_{\mathbb{R}}$ that are regular in $\mathfrak{g}$, then the positive systems of $(\mathfrak{g}, \mathfrak{h})$ are in bijective correspondence $(P \rightleftharpoons C_P)$ with the connected components of $\mathfrak{h}'_{\mathbb{R}}$; $P$ is simply the set of roots of $(\mathfrak{g}, \mathfrak{h})$ that are $>0$ on $C_P$. Since $\mathfrak{h}'_{\mathbb{R}} \cap \mathfrak{h}_1 \neq \emptyset$, some of the $C_P$ will meet $\mathfrak{h}_{1,\mathbb{R}}$; we call such $P$ *regular* and denote by $P_{\mathfrak{k}}$ the set of all roots of $(\mathfrak{k}, \mathfrak{h}_1)$ that are $>0$ on $C_P \cap \mathfrak{h}_{1,\mathbb{R}}$. Evidently

$$P_{\mathfrak{k}} = \{\beta : \beta \text{ a root of } (\mathfrak{k}, \mathfrak{h}), \ \beta = \alpha|\mathfrak{h}_1 \text{ for some } \alpha \in P\}.$$

The map

$$P \mapsto P_{\mathfrak{k}}$$

is surjective from the set of all regular positive systems of $(\mathfrak{g}, \mathfrak{h})$ to the set of *all* positive systems of $(\mathfrak{k}, \mathfrak{h})$. By abuse of notation we often write

$$P \supset P_{\mathfrak{k}};$$

this will always mean that $P$ is regular and $P_{\mathfrak{k}}$ is the set of restrictions to $\mathfrak{h}_1$ of the elements of $P$. The following three lemmas are of crucial importance.

LEMMA 6.1.1. *Let* $\mathfrak{w}(\mathfrak{k})$ (resp. $\mathfrak{w}(\mathfrak{g})$) *be the Weyl group of* $(\mathfrak{k}, \mathfrak{h})$ (resp. $(\mathfrak{g}, \mathfrak{h})$). *Then, every element s of* $\mathfrak{w}(\mathfrak{k})$ *is the restriction to* $\mathfrak{h}_1$ *of a unique element* $\tilde{s}$ *of* $\mathfrak{w}(\mathfrak{g})$ *that stabilizes* $\mathfrak{h}_1$; *and*

$$s \mapsto \tilde{s}$$

*is an injection of the group* $\mathfrak{w}(\mathfrak{k})$ *into the group* $\mathfrak{w}(\mathfrak{g})$.

In view of this result we shall identify $\mathfrak{w}(\mathfrak{k})$ as a subgroup of $\mathfrak{w}(\mathfrak{g})$.

LEMMA 6.1.2. *Let P be a regular positive system of roots of* $(\mathfrak{g}, \mathfrak{h})$. *Let* $\lambda \in \mathfrak{h}^*$ *and let V be the Verma* $\mathfrak{g}$*-module* $V_{\mathfrak{g},P,\lambda}$. *Then we have the following.*

(i) *V is* $\mathfrak{h}_1$*-finite,* i.e., *for any* $\nu_1 \in \mathfrak{h}_1^*$, $V[\nu_1]$ *is finite-dimensional*

(ii) *If* $\lambda_1 = \lambda|\mathfrak{h}_1$, *then* $\dim(V[\lambda_1]) = 1$; *in particular,* $V[\lambda_1] = \mathbb{C} \cdot v_\lambda$

(iii) *If* $\nu_1 \in \mathfrak{h}_1^*$ *and* $V[\nu_1] \neq 0$, *then we can write*

$$\lambda_1 - \nu_1 = \sum_{\alpha \in P} m_\alpha \alpha|_{\mathfrak{h}_1}$$

*where the* $m_\alpha$ *are integers* $\geqslant 0$.

Select $H_0 \in \mathfrak{h}_{1,\mathbb{R}}$ such that if $\alpha$ (resp. $\beta$) is a root of $(\mathfrak{g}, \mathfrak{h})$ (resp. $(\mathfrak{k}, \mathfrak{h})$), $\alpha \in P$ (resp. $\beta \in P_{\mathfrak{k}}$) if and only if $\alpha(H_0) > 0$ (resp. $\beta(H_0) > 0$). Let $c = \min_{\alpha \in P} \alpha(H_0)$. Now,

$$V[\nu_1] = \sum_{\nu \in \mathfrak{h}^*, \nu|\mathfrak{h}_1 = \nu_1} V[\nu].$$

If a $V[\nu]$ is $\neq 0$, $\lambda - \nu = \Sigma m_\alpha \alpha$ where the $m_\alpha$ are integers $\geqslant 0$ and $\alpha$ varies in $P$; and so, $m_\alpha \leqslant c^{-1} (\lambda(H_0) - \nu_1(H_0))$, proving that $\nu$ is restricted to be in a finite set. This proves (i) and (iii). For (ii) note that if $\nu_1 = \lambda_1 = \lambda|\mathfrak{h}_1$, all the $m_\alpha$ are 0. So $\nu = \lambda$.

LEMMA 6.1.3. *Let* $U(\mathfrak{g})^{\mathfrak{h}_1}$ *denote the centralizer of* $\mathfrak{h}_1$ *in* $U(\mathfrak{g})$. *Fix a regular positive system P of roots of* $(\mathfrak{g}, \mathfrak{h})$. *Then, for any* $u \in U(\mathfrak{g})^{\mathfrak{h}_1}$, *there is a unique* $\boldsymbol{\beta}_P(u) \in U(\mathfrak{h})$ *such that*

$$u \equiv \boldsymbol{\beta}_P(u) \bmod \sum_{\alpha \in P} U(\mathfrak{g})\mathfrak{g}_\alpha.$$

*The map*

$$\boldsymbol{\beta}_P: u \mapsto \boldsymbol{\beta}_P(u), \qquad (u \in U(\mathfrak{g})^{\mathfrak{h}_1},$$

*is a homomorphism. If* $\lambda \in \mathfrak{h}^*$ *and* $V = V_{\mathfrak{g},P,\lambda}$, *then* $U(\mathfrak{g})^{\mathfrak{h}_1}$ leaves $V[\lambda]$ *stable and acts on it through the homomorphism*

$$\boldsymbol{\beta}_{P,\lambda}: u \mapsto \boldsymbol{\beta}_P(u)(\lambda), \qquad (u \in U(\mathfrak{g})^{\mathfrak{h}_1}.$$

The point here is that no sum of the form $\Sigma_{\alpha \in P} m_\alpha \alpha$ with all $m_\alpha \geqslant 0$ can vanish identically on $\mathfrak{h}_1$. Hence if $u \in U(\mathfrak{g})^{\mathfrak{h}_1}$, the existence of $\beta_P(u)$, its uniqueness, and the fact that $\boldsymbol{\beta}_P$ is a homomorphism, all follow as in the usual case of $U(\mathfrak{g})^{\mathfrak{h}}$. The last statement is clear since $V[\lambda] = V[\lambda_1]$.

COROLLARY 6.1.4. *Let* $\mathfrak{D} = U(\mathfrak{g})^{\mathfrak{k}}$. *Then*

$$\boldsymbol{\beta}_P: u \mapsto \boldsymbol{\beta}_P(u), \qquad (u \in \mathfrak{D}),$$

*is a homomorphism of* $\mathfrak{D}$ *into* $U(\mathfrak{h})$. *If* $\lambda \in \mathfrak{h}^*$ and $V = V_{\mathfrak{g},P,\lambda}$, *then* $V[\lambda]$ *is stable under* $\mathfrak{D}$ *and* $\mathfrak{D}$ *acts on it via the homomorphism*

$$\boldsymbol{\beta}_{P,\lambda}: u \mapsto \boldsymbol{\beta}_P(u)(\lambda), \qquad (u \in \mathfrak{D}).$$

**6.2.** Fix a regular positive system $P$ for $(\mathfrak{g}, \mathfrak{h})$ and let $P_{\mathfrak{k}}$ be the corresponding positive system for $(\mathfrak{k}, \mathfrak{h})$. Let

$$\delta_P = \tfrac{1}{2} \sum_{\alpha \in P}, \qquad \delta_{P_{\mathfrak{k}}} = \tfrac{1}{2} \sum_{\beta \in P_{\mathfrak{k}}} \beta.$$

For any $s \in \mathfrak{w}(\mathfrak{k})$ we write, as usual,

$$s' \cdot v_1 = s(v_1 + \delta_{P_{\mathfrak{k}}}) - \delta_{P_{\mathfrak{k}}}, \qquad (v_1 \in \mathfrak{h}_1^*).$$

Let $t_0 \in \mathfrak{w}(\mathfrak{k})$ be defined by

$$t_0 \cdot P_{\mathfrak{k}} = -P_{\mathfrak{k}}.$$

If $v \in \mathfrak{h}^*$, we shall call $v$ $P_{\mathfrak{k}}$-integral or $P_{\mathfrak{k}}$-dominant and so on, if $v|\mathfrak{h}_1$ has the corresponding properties.

Let $\lambda \in \mathfrak{h}^*$ be $P_{\mathfrak{k}}$-integral. Consider the Verma $\mathfrak{g}$-module $V = V_{\mathfrak{g},P,\lambda}$. Let

$$\mathfrak{n}_{\mathfrak{k}}^{\pm} = \sum_{\pm\beta \in P_{\mathfrak{k}}} \mathfrak{k}_\beta, \qquad \mathfrak{n}^{\pm} = \sum_{\pm\beta \in P} \mathfrak{g}_\beta.$$

Then it is clear that

$$\mathfrak{n}_{\mathfrak{k}}^{\pm} \subset \mathfrak{n}^{\pm}.$$

Consequently $V \in \mathscr{I}_{\mathfrak{g}}(\mathfrak{k}, \mathfrak{h}_1, P_{\mathfrak{k}})$. Furthermore, by Theorem 4.8.6, $V \in \mathscr{J}_{\mathfrak{g}}$ $(\mathfrak{k}, \mathfrak{h}_1, P_{\mathfrak{k}})$ even. So, by Theorem 4.9.1, $\tau(V)$ is a Harish-Chandra $\mathfrak{g}$-module. We write $W_{\mathfrak{g},P,\lambda} = W_{P,\lambda}$ for this module:

$$W_{\mathfrak{g},P,\lambda} = W_{P,\lambda} = \tau(V_{\mathfrak{g},P,\lambda}).$$

We shall follow Enright [1] and refer to $\{W_{\mathfrak{g},P,\lambda}\}$ *as the fundamental series of* $\mathfrak{g}$*-modules associated to* $(\mathfrak{g}, \mathfrak{k}, \mathfrak{h})$.

**6.3.** I shall now describe the main results of Enright [1] concerning the $W_{\mathfrak{g},P,\lambda}$. The analogy with Section 5 is clear (see also Enright-Wallach [1]).

THEOREM 6.3.1. *Let* $\lambda \in \mathfrak{h}^*$ *be such that* $\lambda_1 = \lambda | \mathfrak{h}_1$ *is* $P_{\mathfrak{k}}$*-integral. Write* $\mu_1 = t_0' \cdot \lambda_1$. *Then we have the following.*

(i) $W_{\mathfrak{g},P,\lambda}$ *is a Harish-Chandra module; it is* $\neq 0$ *whenever* $\mu_1$ *is* $P_{\mathfrak{k}}$*-dominant integral.*

(ii) *Suppose* $\mu_1$ *is* $P_{\mathfrak{k}}$*-dominant integral. Then* $W_{\mathfrak{g},P,\lambda}$ *contains the class* $\vartheta(\mu_1)$ *of* $\mathscr{E}(\mathfrak{k})$, *of* $P_{\mathfrak{k}}$*-highest weight* $\mu_1$, *exactly once; and, on the corresponding isotypical subspace of* $W_{\mathfrak{g},P,\lambda}$, $\mathfrak{D} = U(\mathfrak{g})^{\mathfrak{k}}$ *acts through the homomorphism*

$$\boldsymbol{\beta}_{P,\lambda}: u \mapsto \boldsymbol{\beta}_P(u)(\lambda) \qquad (u \in \mathfrak{D}).$$

*Moreover, if* $\nu_1 \in \mathfrak{h}_1^*$ *is* $P_{\mathfrak{k}}$*-dominant integral and* $\vartheta(\nu_1)$ *is the class of* $\mathscr{E}(\mathfrak{k})$ *of* $P_{\mathfrak{k}}$*-highest weight* $\nu_1$,

$$[W_{\mathfrak{g},P,\lambda} : \vartheta(\nu_1)] \leqslant \dim(V_{\mathfrak{g},P,\lambda}[t_0 \cdot \nu_1]^{\mathfrak{n}_{\mathfrak{k}}^+})$$

*where, as usual,*

$$\mathfrak{n}_{\mathfrak{k}}^+ = \sum_{\beta \in P_{\mathfrak{k}}} \mathfrak{k}_\beta.$$

COROLLARY 6.3.2. *If* $\nu_1 \in \mathfrak{h}_1^*$ *is* $P_{\mathfrak{k}}$*-dominant integral and the class* $\vartheta(\nu_1)$ *of* $\mathscr{E}(\mathfrak{k})$ *of* $P_{\mathfrak{k}}$*-highest weight* $\nu_1$ *occurs in* $W_{\mathfrak{a},P,\lambda}$, *then*[1]

$$\nu_1 = t_0' \cdot \lambda_1 + \sum_{\gamma \in -t_0 \cdot P} m(\gamma)(\gamma | \mathfrak{h}_1)$$

*where the* $m(\gamma)$ *are integers* $\geqslant 0$. *In particular, if* $\gamma \in -t_0 \cdot P$ *is such that* $\nu_1 = t_0' \cdot \lambda_1 - (\gamma | \mathfrak{h}_1)$ *is* $P_{\mathfrak{k}}$*-dominant integral, the class* $\vartheta(\nu_1)$ *does not occur in* $W_{\mathfrak{g},P,\lambda}$.

---

[1]Of course, for any $s \in \mathfrak{w}(\mathfrak{k})$, $s \cdot P$ is also a regular positive system.

THEOREM 6.3.3. *Let $\lambda \in \mathfrak{h}^*$ be such that $\mu_1 = t_0' \cdot \lambda_1$ is $P_{\mathfrak{k}}$-dominant integral, $\lambda_1$ being $\lambda|\mathfrak{h}_1$. Then there exists a unique (up to isomorphism) irreducible Harish Chandra module $D_{\mathfrak{g},P,\lambda}$ with the following properties:*

(i) *the class $\vartheta(\mu_1)$ of $\mathscr{E}(\mathfrak{k})$ of $P_{\mathfrak{k}}$-highest weight $\mu_1$ occurs exactly once in $D_{\mathfrak{g},P,\lambda}$*

(ii) *On the isotypical subspace of $D_{\mathfrak{g},P,\lambda}$ corresponding to this class $\vartheta(\mu_1)$, $\mathfrak{D} = U(\mathfrak{g})^{\mathfrak{k}}$ acts through the homomorphism*

$$\boldsymbol{\beta}_{P,\lambda}: u \mapsto \boldsymbol{\beta}_P(u)(\lambda), \qquad (u \in \mathfrak{D}).$$

*Moreover, $D_{\mathfrak{g},P,\lambda}$ is a subquotient of $W_{\mathfrak{g},P,\lambda}$ described as follows:*

$$D_{\mathfrak{g},P,\lambda} \simeq W^0_{\mathfrak{g},P,\lambda}/M_{\mathfrak{g},P,\lambda}$$

*where $W^0_{\mathfrak{g},P,\lambda}$ is the $U(\mathfrak{g})$-module generated by the one-dimensional space $W_{\mathfrak{g},P,\lambda}[\mu_1]^{\mathfrak{n}_{\mathfrak{k}}^+}$, and $M_{\mathfrak{g},P,\lambda}$ is the unique largest proper sub $\mathfrak{g}$-module of $W^0_{\mathfrak{g},P,\lambda}$.*

These existence and uniqueness theorems are supplemented by the following irreducibility theorem of Enright [1].

THEOREM 6.3.4. *Let $\lambda \in \mathfrak{h}^*$ and $\lambda_1 = \lambda|_{\mathfrak{h}_1}$. Suppose that $V_{\mathfrak{g},P,\lambda}$ is irreducible. Then $W_{\mathfrak{g},P,\lambda}$ is nonzero and irreducible if either $\mu_1 = t_0' \cdot \lambda_1$ is $P_{\mathfrak{k}}$-dominant integral or if $\lambda + \delta_P$ is regular. In either case, $W_{\mathfrak{g},P,\lambda} \simeq D_{\mathfrak{g},P,\lambda}$.*

I shall not describe the relationship between the modules $W_{\mathfrak{g},P,\lambda}$ and the so-called fundamental series of representations of the group $G$ corresponding to $\mathfrak{g}$. For a detailed treatment the reader is referred to Enright and Wallach [2]. In Section 12, I shall discuss the case when $\mathrm{rk}(\mathfrak{g}) = \mathrm{rk}(\mathfrak{k})$. In that situation, the corresponding group representations form the discrete and mock-discrete series.

## 7. THE $\mathfrak{k}$-STRUCTURE OF THE MODULES $W_{\mathfrak{g},P,\lambda}$. THE BLATTNER FORMULAE

**7.1.** I shall now formulate the theorem of Enright [1] that gives the $\mathfrak{k}$-structure of the modules $W_{\mathfrak{g},P,\lambda}$. We work in the framework of Section 6.1, i.e., that $(\mathfrak{g}, \mathfrak{k}, \mathfrak{h})$ is a regular triple and that $P$ is a regular positive system for $(\mathfrak{g}, \mathfrak{h})$; $\mathfrak{h}_1 = \mathfrak{h} \cap \mathfrak{k}$. Enright's theorem then gives an expression for the multiplicity $[W_{\mathfrak{g},P,\lambda}: \vartheta(\nu)]$, when $\nu$ varies over the set of $P_{\mathfrak{k}}$-dominant integral elements of $\mathfrak{h}_1^*$, as an alternating sum of certain partition functions.

This is the formula that was originally conjectured for the discrete series by Blattner, and proved by Schmid and Hecht. Enright's theorem establishes an analogous formula for all the modules $W_{P,\lambda}$ and is thus very much more general.

Let $\lambda \in \mathfrak{h}^*$ be such that $\lambda_1 = \lambda|\mathfrak{h}_1$ is $P_{\mathfrak{k}}$-integral. We write $(V_{P,\lambda,s})$ for the lattice above $V_{\mathfrak{g},P,\lambda}$:

$$V_{P,\lambda,s} = (V_{\mathfrak{g},P,\lambda})_s, \qquad (s \in \mathfrak{w}(\mathfrak{k})).$$

As usual, let

$$W_{P,\lambda} = \tau(V_{\mathfrak{g},P,\lambda}).$$

Let $l(t_0) = r$ and let us define, in analogy with the B–G–G resolutions

$$\mathscr{V}_\lambda^{(0)} = V_{P,\lambda,1},$$

$$\mathscr{V}_\lambda^{(i)} = \bigoplus_{\substack{s \in \mathfrak{w}(\mathfrak{k}) \\ l(s) = i}} V_{P,\lambda\pm}, \qquad (1 \leqslant i \leqslant r).$$

We then have the following theorem (Enright [1]).

THEOREM 7.1.1. *Let $\lambda$ be $P_{\mathfrak{k}}$-integral and let $V_{\mathfrak{g},P,\lambda}$ be irreducible. Then:*

(a) *We have an exact sequence of $\mathfrak{g}$-modules*

$$0 \to V_{\mathfrak{g},P,\lambda} = \mathscr{V}_\lambda^{(r)} \xrightarrow{d_r} \cdots \longrightarrow \mathscr{V}_\lambda^{(1)} \xrightarrow{d_1} V_{P,\lambda,1} \xrightarrow{\varepsilon} W_{P,\lambda} \longrightarrow 0$$

*where $\varepsilon$ is the natural surjection.*

(b) *If $\nu \in \mathfrak{h}_1^*$ is $P_{\mathfrak{k}}$-dominant integral, we have the exact sequence*

$$0 \to V_{\mathfrak{g},P,\lambda}[\nu]^{\mathfrak{n}_{\mathfrak{k}}^+} \to \mathscr{V}_\lambda^{(r-1)}[\nu]^{\mathfrak{n}_{\mathfrak{k}}^+} \to \cdots \to \mathscr{V}_\lambda^{(1)}[\nu]^{\mathfrak{n}_{\mathfrak{k}}^+} \to W_{P,\lambda}[\nu]^{\mathfrak{n}_{\mathfrak{k}}^+} \to 0.$$

It is possible to specify the $d_i$ more precisely. They are essentially defined in the same way as the corresponding maps for the B–G–G resolution. The above theorem is however adequate for determining the $\mathfrak{k}$-structure of the $W_{P,\lambda}$.

The dimensions of the spaces $W_{P,\lambda}[\nu]^{\mathfrak{n}_{\mathfrak{k}}^+}$ can now be computed as alternating sums. Using Proposition 4.7.2 we immediately get

COROLLARY 7.1.2. *We have, $\forall\, \nu \in \mathfrak{h}_1^*$ which are $P_{\mathfrak{k}}$-dominant integral,*

$$\dim(W_{P,\lambda}[\nu]^{\mathfrak{n}_{\mathfrak{k}}^+}) = \sum_{s \in \mathfrak{w}(\mathfrak{k})} \varepsilon(s) \dim(V_{\mathfrak{g},P,\lambda}[(t_0 s^{-1})' \cdot \nu]^{\mathfrak{n}_{\mathfrak{k}}^+}).$$

On the other hand, we have the following lemma.

LEMMA 7.1.3. *Let $\lambda \in \mathfrak{h}^*$, $\lambda_1 = \lambda | \mathfrak{h}_1$ and let $\lambda$ be $P_{\mathfrak{k}}$-integral. Suppose $V_{\mathfrak{g},P,\lambda}$ is irreducible,* i.e.,

$$(\lambda + \delta_P)(\bar{H}_\alpha) \notin \{1, 2, \ldots\}, \qquad \forall \alpha \in P_{\mathfrak{k}}.$$

*Then, for any $\nu \in \mathfrak{h}_1^*$,*

$$\dim(V_{\mathfrak{g},P,\lambda}[\nu]^{\mathfrak{n}_{\mathfrak{k}}^+}) = \mathbb{P}_{-P_n}(\nu - \lambda_1)$$

*where $\mathbb{P}_{-P_n}$ is the usual partition function* (see Definition 11.3.1).

It is now clear that we have the following theorem.

THEOREM 7.1.4. *Let $(\mathfrak{g}, \mathfrak{k}, \mathfrak{h})$ be a regular triple and $P$ a regular positive system. Suppose $\lambda \in \mathfrak{h}^*$ is such that $\lambda_1 = \lambda | \mathfrak{h}_1^*$ is $P_{\mathfrak{k}}$-integral and is such that $V_{\mathfrak{g},P,\lambda}$ is irreducible, i.e.,*

$$(\lambda + \delta_P)(\bar{H}_\alpha) \notin \{1, 2, \ldots\}, \qquad \forall \alpha \in P.$$

*Then, for any $\nu \in \mathfrak{h}_1^*$ which is $P_{\mathfrak{k}}$-dominant integral, the multiplicity $[W_{P,\lambda}: \vartheta(\nu)]$ with which the irreducible class $\vartheta(\nu)$ (in $\mathscr{E}(\mathfrak{k})$ of $P_{\mathfrak{k}}$ highest weight $\nu$) occurs in $W_{P,\lambda}$ is given by the following formula, where we write $\mu_1 = t_0' \cdot \lambda_1$ (so that $\mu_1 + \delta_{P_{\mathfrak{k}}} = t_0(\lambda_1 + \delta_{P_{\mathfrak{k}}})$):*

$$[W_{P,\lambda}: \vartheta(\nu)] = \sum_{s \in \mathfrak{w}(\mathfrak{k})} \varepsilon(s) \mathbb{P}_{-t_0 P_n}(s(\nu + \delta_{P_{\mathfrak{k}}}) - (\mu_1 + \delta_{P_{\mathfrak{k}}})).$$

For these results, see Section 11.

## 8. INVARIANT FORMS

The remainder of this article will be devoted to sketching the main ideas underlying the proofs of the irreducibility theorem and the Blattner formulae for the modules $W_{\mathfrak{g},P,\lambda}$. The basic tool necessary for this is that of the invariant bilinear forms associated with modules for $\mathfrak{g}$. In this section we shall discuss this notion and the main theorems relating to it. In Section 9, we shall indicate how these results can be used to prove the main theorems when $\lambda$ is in "good position'. To make the transition to general $\lambda$ it is necessary to use the technique of tensoring by finite dimensional modules. This technique, due to Fomin-Shapovalov [1] and Zuckerman [1], will be reviewed briefly in Section 10; the applications of it to the theory of the modules $W_{\mathfrak{g},P,\lambda}$ will be treated in Section 11.

**8.1.** Let $\mathfrak{l}$ be a finite-dimensional Lie algebra over $\mathbb{C}$, and $M$ any $\mathfrak{l}$-module. We begin by introducing the notion of an invariant bilinear form for $M$ which is a variant of the classical notion. Let $\sigma$ be an involutive anti-automorphism on $\mathfrak{l}$:

$$\sigma^2 = id, \qquad [X, Y]^\sigma = [Y^\sigma, X^\sigma], \qquad (X, Y \in \mathfrak{l}).$$

We extend $\sigma$ to an involutive anti-automorphism $a \to a^\sigma$ of $U(\mathfrak{l})$. Then, *an invariant form for M (relative to $\sigma$)* is a bilinear map

$$\varphi: M \times M \to \mathbb{C}$$

such that for all $a \in U(\mathfrak{l})$, $m, m' \in M$,

$$\varphi(a \cdot m, m') = \varphi(m, a^\sigma \cdot m').$$

Note that this differs from the classical definition of invariant forms which uses the *canonical anti-automorphism* of $U(\mathfrak{l})$ instead of $\sigma$, namely the one which is $-id$ on $\mathfrak{l}$.

The complex vector space of invariant forms for $M$ is denoted by $\mathrm{Inv}_{\mathfrak{l}}(M)$. Although everything depends on $\sigma$, we shall suppress this in our notations; in any given context we shall work with only one $\sigma$. Moreover, once $\sigma$ is chosen, everything that we do is canonical.

For bilinear forms, the definitions of symmetry, skew-symmetry, and nondegeneracy, are made as usual. Thus, if $\varphi(M \times M \to \mathbb{C})$ is a bilinear form, $\varphi$ is called *symmetric* (resp. *skew-symmetric*) if $\varphi(m, m') = \varphi(m', m)$ (resp. $\varphi(m, m') = -\varphi(m', m)$) $\forall\, m, m' \in M$; $\varphi$ is called *nondegenerate* iff

(i) $\varphi(m, m') = 0\ \forall\, m' \in M \Rightarrow m = 0$

(ii) $\varphi(m, m') = 0\ \forall\, m \in M \Rightarrow m' = 0$

are both satisfied. The set of nondegenerate invariant forms is denoted by $\mathrm{Inv}_{\mathfrak{l}}^{\times}(M)$.

Let $M'$ be the algebraic dual of $M$. We convert $M'$ into a $U(\mathfrak{l})$-module by defining

$$(a \cdot \lambda)(m) = \lambda(a^\sigma \cdot m), \qquad (\lambda \in M', m \in M, a \in U(\mathfrak{l})).$$

Since $M'$ can be very big, the functor $M \mapsto M'$ does not have many reasonable properties. However we shall see presently that, when $\mathfrak{l}$ is reductive, we can modify this functor suitably so that it is very well behaved on a certain naturally defined category of $\mathfrak{l}$-modules.

Consider now the case when $\mathfrak{l} = \mathfrak{m}$ is reductive. As usual $\mathfrak{h} \subset \mathfrak{m}$ is a CSA and $Q$ a positive system of roots of $(\mathfrak{m}, \mathfrak{h})$. We shall assume that $\sigma$ fixes each element of $\mathfrak{h}$. That such a $\sigma$ exists is easy to see.

For, by the theory of real forms for the semisimple algebra $[\mathfrak{m}, \mathfrak{m}]$, $\exists$ an involutive automorphism $\tau$ of $[\mathfrak{m}, \mathfrak{m}]$ such that $\tau = -id$ on the CSA $\mathfrak{h} \cap [\mathfrak{m}, \mathfrak{m}]$ of $[\mathfrak{m}, \mathfrak{m}]$; we then take $\sigma = -\tau$ on $[\mathfrak{m}, \mathfrak{m}]$ and $-id$ on center$(\mathfrak{m})$. Note that

$$\mathfrak{m}_\alpha^\sigma = \mathfrak{m}_{-\alpha}, \qquad U(\mathfrak{n}^+)^\sigma = U(\mathfrak{n}^-)$$

where, as usual,

$$\mathfrak{n}^\pm = \sum_{\pm\alpha \in Q} \mathfrak{m}_\alpha.$$

*Whenever we operate in the context of* $(\mathfrak{m}, \mathfrak{h})$, *we shall always assume that* $\sigma = id$ *on* $\mathfrak{h}$.

We now introduce the category $\mathbb{B}(\mathfrak{m})$ of all $\mathfrak{m}$-modules $M$ such that

(a) $M$ is a weight module for $\mathfrak{h}$.

(b) The weight spaces of $M$ are all finite-dimensional.

For any $M \in \mathbb{B}(\mathfrak{m})$ we denote by $M^\sigma$ the submodule of $M'$ consisting of all $\mathfrak{h}$-finite elements. The following proposition is easy to prove.

PROPOSITION 8.1.1. (a) *$M \mapsto M^\sigma$ is a contravariant functor from $\mathbb{B}(\mathfrak{m})$ to $\mathbb{B}(\mathfrak{m})$ and is essentially involutive in the sense that $(M^\sigma)^\sigma$ and $M$ are naturally isomorphic.*

(b) *For any $M \in \mathbb{B}(\mathfrak{m})$, $M^\sigma$ has the same weights as $M$; if $\mu, \nu$ are weights of $M$, $M[\mu]$ and $M^\sigma[\nu]$ are orthogonal when $\mu \neq \nu$ and are in duality when $\mu = \nu$.*

(c) *$M^\sigma$ has a finite Jordan–Hölder composition series if and only if $M$ has one; the lengths of the composition series are then identical. In particular, $M^\sigma$ is irreducible if and only if $M$ is irreducible.*

If $M \in \mathbb{B}(\mathfrak{m})$ and we write, for any $T \in \mathrm{Hom}_{\mathfrak{m}}(M, M^\sigma)$,

$$\beta_T(m, m') = (Tm)(m'), \qquad (m, m' \in M),$$

it is immediate that $\beta_T \in \mathrm{Inv}_{\mathfrak{m}}(M)$ and that the map

$$T \mapsto \beta_T$$

is a linear isomorphism of $\mathrm{Hom}_{\mathfrak{m}}(M, M^{\sigma})$ onto $\mathrm{Inv}_{\mathfrak{m}}(M)$. Moreover, $T$ is an isomorphism if and only if $\beta_T$ is nondegenerate. Thus $M$ admits nondegenerate invariant forms if and only if $M$ and $M^{\sigma}$ are isomorphic.

If $M \in \mathbb{B}(\mathfrak{m})$ and has finite Jordan–Hölder composition series of length $l$, it is not difficult to show that $\mathrm{Inv}_{\mathfrak{m}}(M)$ is finite-dimensional and in fact that its dimension is $\leqslant l^2$; $\mathrm{Inv}_{\mathfrak{m}}^{\times}(M)$, the subset of nondegenerate elements of $\mathrm{Inv}_{\mathfrak{m}}(M)$, is a Zariski open subset of $\mathrm{Inv}_{\mathfrak{m}}(M)$; it is therefore dense as soon as it is nonempty. For irreducible $M \in \mathbb{B}(\mathfrak{m})$, $\mathrm{Inv}_{\mathfrak{m}}(M)$ is at most one-dimensional; in this case, $\mathrm{Inv}_{\mathfrak{m}}^{\times}(M)$ is precisely the set of nonzero elements.

**8.2.** It follows from Proposition 8.1.1 that if $M$ is a highest weight vector module with highest weight $\lambda$, then $\lambda$ is also the highest weight of $M^{\sigma}$. Hence if $M$ is irreducible, $M \simeq M^{\sigma}$ and so $\dim_{\mathbb{C}}(\mathrm{Inv}_{\mathfrak{m}}(M))=1$. So, by pulling back, if $M$ is assumed only to be a highest weight vector module, $M$ admits nonzero invariant forms. On the other hand, if $\lambda$ is the highest weight of $M$ and $(\cdot, \cdot)$ any invariant form, $(M[\lambda], M[\lambda'])=0$ if $\lambda'$ is a weight $\neq \lambda$; hence, $(M[\lambda], M[\lambda])=0$ implies $(M[\lambda], M)=0$ which gives, as $U(\mathfrak{m}) \cdot M[\lambda]=M$, that $(\cdot, \cdot)=0$. This shows that $\mathrm{Inv}_{\mathfrak{m}}(M)$ is one-dimensional and consists only of *symmetric* forms. It is easy to describe such forms directly. Let $v_{\lambda}$ be a nonzero vector in $M$ of weight $\lambda$ and let $\boldsymbol{\beta}_Q$ be the Harish-Chandra homomorphism $U(\mathfrak{m})^{\mathfrak{h}} \to U(\mathfrak{h})$; we extend $\boldsymbol{\beta}_Q$ linearly to all of $U(\mathfrak{m})$ by making it 0 on all nonzero weight spaces of the adjoint representation of $\mathfrak{h}$ on $U(\mathfrak{m})$, and denote the extension also by $\boldsymbol{\beta}_Q$. It is then easy to see that

$$\boldsymbol{\beta}_Q(x^{\sigma})=\boldsymbol{\beta}_Q(x), \qquad (x \in U(\mathfrak{m})),$$

and that

$$(\cdot, \cdot)\colon (x \cdot v_{\lambda}, y \cdot v_{\lambda}) \mapsto \boldsymbol{\beta}_Q(y^{\sigma}x), \qquad (x, y \in U(\mathfrak{m})),$$

defines an element of $\mathrm{Inv}_{\mathfrak{m}}(M)$ forming a basis of it. These remarks are applicable to the case $M=V_{\mathfrak{m},Q,\lambda}$. If $I_{\mathfrak{m},Q,\lambda}$ is the irreducible module of $Q$-highest weight $\lambda$, it admits a one-dimensional space of invariant forms whose nonzero elements are symmetric, nondegenerate, and whose pull-backs to $V_{\mathfrak{m},Q,\lambda}$ are precisely all the invariant forms on $V_{\mathfrak{m},Q,\lambda}$.

The Verma modules treated above are special cases of modules for

$U(\mathfrak{m})$ which are induced from modules for $\mathfrak{b}$, where, as usual,

$$\mathfrak{b}=\mathfrak{h}\oplus\mathfrak{n}^{+}=\mathfrak{h}\oplus\sum_{\alpha\in Q}\mathfrak{m}_{\alpha}.$$

Let $\mathbb{F}(\mathfrak{b})$ be the category of all finite-dimensional modules for $\mathfrak{b}$ on which $\mathfrak{h}$ acts semisimply, i.e., which are weight modules for $\mathfrak{h}$. For any module $L\in\mathbb{F}(\mathfrak{b})$ we define $U(L)$ as the induced module $U(\mathfrak{m})\otimes_{U(\mathfrak{b})}L$. We know of course that $L\mapsto U(L)$ is an exact functor from the category $\mathbb{F}(\mathfrak{b})$ to the category $\mathbb{B}(\mathfrak{m})$, which is covariant. We shall now examine the question of constructing invariant forms for the modules $U(L)$. The main result is the following Proposition which may be regarded as an infinitesimal analogue of Bruhat's classical determination of intertwining forms for induced representations.

PROPOSITION 8.2.1. *Let $L\in\mathbb{F}(\mathfrak{b})$. For any $\varphi\in\mathrm{Inv}_{\mathfrak{m}}(U(L))$ let $\mathrm{Res}_L\varphi=\underline{\varphi}$ be the bilinear form on $L\times L$ given by*

$$\underline{\varphi}(u_1,u_2)=\varphi(1\otimes u_1,1\otimes u_2),\qquad (u_j\in L).$$

*Then $\underline{\varphi}\in\mathrm{Inv}_{\mathfrak{b}}(L)$, and*

$$\mathrm{Res}_L\colon\varphi\mapsto\underline{\varphi}$$

*is a linear isomorphism of $\mathrm{Inv}_{\mathfrak{m}}(U(L))$ onto $\mathrm{Inv}_{\mathfrak{b}}(L)$.*

The key point in the proof of this result is the fact that the map $(n,b)\to n^{\sigma}b$ $(n\in U(\mathfrak{n}^{+}), b\in U(\mathfrak{b}))$ extends to a linear isomorphism $U(\mathfrak{n}^{+})\otimes U(\mathfrak{b})\simeq U(\mathfrak{m})$. If $\underline{\varphi}=0$,

$$\varphi(n^{\sigma}\otimes v,1\otimes v')=\underline{\varphi}(v,n\cdot v')=0\ \forall v,v'\in L$$

so that $\varphi=0$ on $U(L)\times(1\otimes L)$; this implies $\varphi=0$. For proving the surjectivity of $\mathrm{Res}_L$, let $\psi\in\mathrm{Inv}_{\mathfrak{b}}(L)$ and define $\varphi_1$ as the bilinear form on $(1\otimes L)\times U(L)$ such that

$$\varphi_1(1\otimes v,n^{\sigma}b\otimes v')=\psi(n\cdot v,b\cdot v'),\quad (n\in U(\mathfrak{n}^{+}),b\in U(\mathfrak{b}),v,v'\in L)$$

Using the hypothesis that

$$\psi(a\cdot w,w')=\psi(w,a\cdot w')\ \forall a\in U(\mathfrak{h}),\quad w,w'\in L,$$

it is easily verified that

$$\varphi_1(1\otimes b'\cdot v,x)=\varphi_1(1\otimes v,b'^{\sigma}x)\ \forall v\in L,\quad b'\in U(\mathfrak{b}),\quad x\in U(L).$$

We can then define $\varphi$ as the bilinear form on $U(L)\times U(L)$ such that

$$\varphi(y\otimes v, x)=\varphi_1(1\otimes v, y^\sigma x), \quad \forall\, v\in L,\ y\in U(\mathfrak{m}), \quad x\in U(L).$$

COROLLARY 8.2.2. *Let* $L\in\mathbb{F}(\mathfrak{b})$. *If* $L\neq 0$, *then* $\mathrm{Inv}_{\mathfrak{m}}(U(L))\neq 0$.

Recalling that $\mathrm{Inv}_{\mathfrak{m}}^{\times}(U(L))$ is a dense open subset of $\mathrm{Inv}_{\mathfrak{m}}(U(L))$ as soon as it is nonempty, we get

COROLLARY 8.2.3. *Let* $L, M\in\mathbb{F}(\mathfrak{b})$, *with* $L\subset M$. *Let* $\mathrm{Res}_{U(L)}^{U(M)}\varphi$ *denote, for any* $\varphi\in\mathrm{Inv}_{\mathfrak{m}}(U(M))$, *the restriction of* $\varphi$ *to* $U(L)\times U(L)$. *Then*

(i) $\mathrm{Res}_{U(L)}^{U(M)}: \mathrm{Inv}_{\mathfrak{m}}(U(M))\to\mathrm{Inv}_{\mathfrak{m}}(U(L))$ *is surjective*

(ii) *If* $\mathrm{Inv}_{\mathfrak{m}}^{\times}(U(M))$ *and* $\mathrm{Inv}_{\mathfrak{m}}^{\times}(U(L))$ *are nonempty*, $\exists\ \varphi_M\in\mathrm{Inv}_{\mathfrak{m}}^{\times}(U(M))$ *such that* $\mathrm{Res}_{U(L)}^{U(M)}\varphi_M\in\mathrm{Inv}_{\mathfrak{m}}^{\times}(U(L))$.

If $R_1, R_2\in\mathbb{B}(\mathfrak{m})$ with $R_2\subset R_1$ and if $\varphi_1\in\mathrm{Inv}_{\mathfrak{m}}(R_1)$ is such that

$$\varphi_2=\varphi_1\big|_{R_2\times R_2}$$

is nondegenerate, it is clear that $R_2$ is a direct summand; in fact $R_1=R_2\oplus R_2^{\perp}$ where $R_2^{\perp}=\{u\colon u\in R_1,\ \varphi_1(v, u)=0\ \forall\, v\in R_2\}$. If moreover $\varphi_1$ is also nondegenerate, $\varphi_1\big|_{R_2^{\perp}\times R_2^{\perp}}$ is nondegenerate, so that $R_1/R_2$ admits nondegenerate invariant forms. This observation leads to the following

PROPOSITION 8.2.4. (a) *Let* $L, M\in\mathbb{F}(\mathfrak{b})$ *with* $L\subset M$. *If* $U(L)$ *admits nondegenerate invariant forms,* $U(L)$ *is a direct summand of* $U(M)$. *If, in addition,* $U(M)$ *also admits invariant forms that are nondegenerate, then* $U(M)/U(L)\simeq U(M/L)$ *also admits nondegenerate invariant forms.*

(b) *Let* $L_i\in\mathbb{F}(\mathfrak{b})$, $0\leqslant i\leqslant m$, *and let*

$$0\to L_m\to L_{m-1}\to\cdots\to L_1\to L_0\to 0$$

*be a long exact sequence. Suppose that the* $\mathfrak{m}$*-modules* $U(L_i)$ $(1\leqslant i\leqslant m)$ *admit nondegenerate invariant forms. Then*:

(i) $U(L_0)$ *also admits nondegenerate invariant forms.*

(ii) *The long exact sequence*

$$0\to U(L_m)\to U(L_{m-1})\to\cdots\to U(L_1)\to U(L_0)\to 0$$

*splits at every level,* i.e., $\exists$ $\mathfrak{m}$*-modules* $V_i\subset U(L_i)$ *for* $1\leqslant i\leqslant m-1$ *such that*

$$U(L_i)=V_i\oplus(\textit{Image of } U(L_{i+1})), \qquad (1\leqslant i\leqslant m-1).$$

8.3. I shall now discuss the important technique of transferring an invariant form from a module to its completion. Let

$$\mathfrak{a}=\mathbb{C}\cdot H+\mathbb{C}\cdot X+\mathbb{C}\cdot Y$$

be the three-dimensional simple Lie algebra with

$$[H, X]=2X, \qquad [H, Y]=-2Y, \qquad [X, Y]=H.$$

Invariance will be relative to an involutive antiautomorphism $\sigma$ of $\mathfrak{a}$ such that $H^\sigma=H$; we can modify $X$ and $Y$ so that we have

$$H^\sigma=H, \qquad X^\sigma=Y, \qquad Y^\sigma=X.$$

Let $M$ be any $U(\mathfrak{a})$-module belonging to the category $\mathscr{I}(\mathfrak{a})$; let $C(M)$ be the completion of $M$. Given an invariant form $\varphi_M$ on $M\times M$ its transfer $C(\varphi_M)$ will be an invariant form on $C(M)\times C(M)$. $C(\varphi_M)$ will however *not* be an extension of $\varphi_M$; in fact, $C(\varphi_M)$ will vanish on $M\times M$, and even on $(M\times C(M))\cup(C(M)\times M)$. The typical example is the case when $M=V_{-n-2}$, $C(M)=V_n$ where $n$ is an integer $\geqslant 0$. As $M$ is irreducible, $\varphi_M$ is unique up to a multiplying scalar while $C(\varphi_M)$ has to vanish on

$$(V_{-n-2}\times V_n)\cup(V_n\times V_{-n-2})$$

since it is the pull back of a form on $(V_n/V_{-n-2})\times(V_n/V_{-n-2})$. We can normalize $C(\varphi_M)$ by requiring that

$$C(\varphi_M)(1_n, 1_n)=\xi\cdot\varphi_M(Y^{n+1}\cdot 1_n, Y^{n+1}\cdot 1_n)$$

where $\xi$ is a constant $\neq 0$ and $1_n$ is a cyclic vector of weight $n$ in $V_n$. The next proposition shows that this can be done in general.

PROPOSITION 8.3.1. *Let $\xi_n(n=0, 1, 2, \ldots)$ be a sequence of nonzero constants. Let $\varphi=(\cdot, \cdot)$ be an invariant form on $M\times M$ where $M\in\mathscr{I}(\mathfrak{a})$' Then $\exists$ a unique invariant form $C(\varphi)={}^C(\cdot, \cdot)$ on $C(M)\times C(M)$ such that*

(i) *${}^C(\cdot, \cdot)$ is zero on $(M\times C(M))\cup(C(M)\times M)$*

(ii) *for any integer $n\geqslant 0$ and $a, b\in C(M)[n]^X$,*

$$^C(a, b)=\xi_n(Y^{n+1}a, Y^{n+1}b).$$

Let $L=C(M)/M$. Then $L=\bigoplus_{n\geqslant 0}L_n$ where, for each integer $n\geqslant 0$, $L_n$ is an $\mathfrak{a}$-module of the form $I_n\otimes V$, with $I_n$ as the irreducible $\mathfrak{a}$-module of highest weight $n$, and $\mathfrak{a}$ acting trivially on $V$. Let $1_n$ be a nonzero vector of $I_n$ of weight $n$. Then $L_n[n]^X=1_n\otimes V$; and so, if $\psi_n$ is the invariant form for $I_n$ with $\psi_n(1_n, 1_n)=1$, the map $\psi_V\mapsto\psi_n\otimes\psi_V$ is an isomorphism of the space of all bilinear forms on $V\times V$ with the space of invariant bilinear forms on $L_n\times L_n$. For any integer $n\geqslant 0$ there is a well-defined bilinear form $\theta_n$ on $L_n[n]^X\times L_n[n]^X$ such that

$$\theta_n(\pi(a), \pi(b))=\xi_n(Y^{n+1}a, Y^{n+1}b), \qquad (a, b\in C(M)[n]^X),$$

where $\pi$ is the natural map $C(M) \to L$. By the earlier remark $\theta_n$ extends to an invariant form for $L_n$, and this uniquely. So $\exists$ a unique invariant form for $L$ such that it coincides on each $L_n$ with the above mentioned extension of $\theta_n$; the various $L_n$ will be mutually orthogonal with respect to this form. Its pullback to $C(M) \times C(M)$ is the required form $C(\varphi)$.

The constants $\xi_n$ can be completely arbitrary (but $\neq 0$). It is remarkable however that they are essentially uniquely determined *if one requires the transfer process to commute with tensoring by finite-dimensional modules.*

PROPOSITION 8.3.2. *The following two statements are equivalent*:

(i) $\exists$ *a constant* $\xi \neq 0$ *such that* $\xi_n = \xi / n!(n+1)!$ *for all* $n = 0, 1, 2, 3, \ldots$ $(0 = 1)$.

(ii) *For any* $M \in \mathscr{I}(\mathfrak{a})$, *any finite-dimensional* $\mathfrak{a}$*-module* $F$, *and arbitrary invariant forms* $\varphi_M$ (*for* $M$) *and* $\varphi_F$ (*for* $F$),

$$C(\varphi_F \otimes \varphi_M) = \varphi_F \otimes C(\varphi_M)$$

(*here we identify* $C(F \otimes M)$ *canonically with* $F \otimes C(M)$).

We remark that any finite-dimensional module for $\mathfrak{a}$, being the direct sum of irreducible such modules, always admits invariant forms, even symmetric and nondegenerate ones. So the condition (ii) above is meaningful.

(ii)$\Rightarrow$(i). Take $M = V_{-n-2}$, $C(M) = V_n \supset M$ ($n$ an integer $\geqslant 0$), $F = \mathbb{C}e_1 \oplus \mathbb{C}e_{-1}$ where $He_{\pm 1} = \pm e_{\pm 1}$, $Xe_1 = Ye_{-1} = 0$, $Ye_1 = e_{-1}$, $Xe_{-1} = e_1$, so that $F$ is the two-dimensional irreducible module for $\mathfrak{a}$. Then

$$F \otimes V_n \simeq V_{n-1} \otimes V_{n+1}, \qquad F \otimes V_{-n-2} \simeq V_{-n-1} \otimes V_{-n-3}.$$

We may assume that $\varphi_F(e_1, e_1) = \varphi_M(v_{-n-2}, v_{-n-2}) = 1$ where $v_{-n-2}$ is a nonzero vector of $M[-n-2]$. A simple calculation shows that $V_{n+1} \subset$ $\subset F \otimes V_n$ is generated by $v'_{n+1} = e_1 \otimes v_n$. Equating the expressions obtained for $(\varphi_F \otimes C(\varphi_M))(v'_{n+1}, v'_{n+1})$ and $C(\varphi_F \otimes \varphi_M)(v'_{n+1}, v'_{n+1})$ we get $\xi_n = (n+1)(n+2)\xi_{n+1}$.

(i)$\Rightarrow$(ii). Using the fact that any irreducible finite-dimensional module for $\mathfrak{a}$ of highest weight $m \geqslant 0$ is a canonical direct summand of $I_2 \otimes \cdots \otimes I_2$ (m times), where $I_2$ is the irreducible module of dimension 2, it is easy to come down to the case $F = I_2$. Write

$$\Phi = \varphi_M, \qquad \Psi = \varphi_F \otimes \Phi, \qquad L_n = (F \otimes C(M))[n]^X$$

($n$ integer $\geqslant 0$). It is then a question of verifying that $C(\Psi)=\varphi_F \otimes C(\Phi)$ on $L_n \times L_n$, i.e., that

$$C(\Psi)(a, b)=(\varphi_F \otimes C(\Phi))(a, b), \qquad (a, b \in L_n),$$

where we may additionally assume that $a \in F \otimes C(A)$, $b \in F \otimes C(B)$, $A$ and $B$ being indecomposable submodules of $M$. The only nontrivial case is when $A$ and $B$ are both $\simeq V_{-r-2}$ where $r$ is an integer $\geqslant 0$. In this case the calculation is essentially the same as in the earlier argument.

*From now on we shall use the transfer process only for the choice of constants given by*

$$\xi_n = 1/n!(n+1)!, \qquad (n=0, 1, 2, \ldots, ; 0!=1).$$

**8.4.** It is obvious that the process of transferring will be valuable only if it does not change when we go over the 'relative' situation. Suppose $\mathfrak{l}$ is a finite-dimensional Lie algebra containing $\mathfrak{a}$. Let us assume that $\sigma$ is the restriction to $\mathfrak{a}$ of an involutive anti-automorphism (also denoted by $\sigma$) of $\mathfrak{l}$ that stabilizes $\mathfrak{a}$. If $M \in \mathscr{I}_{\mathfrak{l}}(\mathfrak{a})$, then $C(M)$ is also an $\mathfrak{l}$-module; the question naturally arises whether, for any $\mathfrak{l}$-invariant form $\varphi$ on $M \times M$, the transferred form $C(\varphi)$ (which is a priori only $\mathfrak{a}$-invariant) is actually $\mathfrak{l}$-invariant. It is remarkable that this is indeed so.

THEOREM 8.4.1 (*Invariance theorem*). *Let $\mathfrak{l}$ be a finite-dimensional Lie algebra containing $\mathfrak{a}$. Let $\sigma$ be an involutive anti-automorphism of $\mathfrak{l}$ such that $\mathfrak{a}^\sigma = \mathfrak{a}$ and $H^\sigma = H$. Then, for any $\mathfrak{l}$-module $M$ whose underlying $\mathfrak{a}$-module belongs to $\mathscr{I}(\mathfrak{a})$, and any $\mathfrak{l}$-invariant form $\varphi$ on $M \times M$ (relative to $\sigma$), the transferred form $C(\varphi)$ on $C(M) \times C(M)$ is also $\mathfrak{l}$-invariant.*

For the proof see Enright [1]. His proof is based upon a long and nontrivial calculation. Since the actual details of this proof will not be used in the rest of this article, we do not discuss them here.

The following proposition follows at once from the above theorem and the fact that $\mathrm{Inv}_{\mathfrak{m}}(V_{\mathfrak{m},Q,\lambda})$ is of dimension 1.

PROPOSITION 8.4.2. *Let $\mathfrak{m}$, $\mathfrak{h}$, $Q$ have their usual meanings. For any $\mu \in \mathfrak{h}^*$ let $\varphi_\mu$ be a nonzero invariant form for the Verma module $V_{\mathfrak{m},Q,\mu}$. For each simple root $\gamma \in Q$, let $C_\gamma$ be the functor of completion with respect to $\gamma$. Then:*

(a) *Let $\gamma \in Q$ and let $\lambda \in \mathfrak{h}^*$ be such that $\lambda(\bar{H}_\gamma)$ is an integer $\geqslant 0$ (so that*

$V_{\mathfrak{m},Q,\lambda}=C_\gamma(V_{\mathfrak{m},Q,s'_\gamma\cdot\lambda})$). *Then* $\exists$ *a constant* $K_\gamma\neq 0$ *such that*

$$C_\gamma(\varphi_{s'_\gamma\cdot\lambda})=K_\gamma\cdot\varphi_\lambda.$$

(b) *Let* $F$ *be a finite-dimensional* $\mathfrak{m}$*-module and let* $\varphi_F$ *be an* $\mathfrak{m}$*-invariant form on* $F\times F$. *Let* $\lambda\in\mathfrak{h}^*$ *be* $Q$*-dominant integral. Let* $s\in\mathfrak{w}(\mathfrak{m})$ *and let* $\gamma\in Q$ *be a simple root such that* $l(s_\gamma s)=l(s)+1$. *Then* $\exists$ *a constant* $K_{s,\gamma}\neq 0$ *such that*

$$C_\gamma(\varphi_F\otimes\varphi_{(s_\gamma s)'\cdot\lambda})=K_{s,\gamma}\cdot\varphi_F\otimes\varphi_{s'\cdot\lambda}.$$

For bilinear forms on modules for semisimple Lie algebras see also J. C. Jantżen [1].

## 9. Structure theory for the modules $V_{P,\lambda,s}$ and $W_{P,\lambda}$ when $\lambda$ is in good position

**9.1.** From now on we restrict our attention to the regular triple $(\mathfrak{g},\mathfrak{k},\mathfrak{h})$ associated with a real s.s. Lie algebra $\mathfrak{g}_0$ and a maximal compact subalgebra $\mathfrak{k}_0\subset\mathfrak{g}_0$. In other words, $\mathfrak{g}$ (resp. $\mathfrak{k}$) is the complexification of $\mathfrak{g}_0$ (resp. $\mathfrak{k}_0$), while $\mathfrak{h}$ is the complexification of a fundamental CSA $\mathfrak{h}_0$ of $\mathfrak{g}_0$ stable under the Cartan involution (corresponding to $\mathfrak{k}_0$) of $\mathfrak{g}_0$; if $\mathfrak{g}_0=\mathfrak{k}_0\oplus\mathfrak{p}_0$ is the Cartan decomposition of $\mathfrak{g}_0$ and $\mathfrak{p}=\mathbb{C}\cdot\mathfrak{p}_0$, then $\mathfrak{h}=(\mathfrak{h}\cap\mathfrak{k})\oplus(\mathfrak{h}\cap\mathfrak{p})$, and $\mathfrak{h}_1=\mathfrak{h}\cap\mathfrak{k}$ is a CSA of $\mathfrak{k}$.

Let $P$ be a regular positive system for $(\mathfrak{g},\mathfrak{h})$, and $P_\mathfrak{k}$, the associated positive system for $(\mathfrak{k},\mathfrak{h}_1)$. As usual we put

$$\mathfrak{b}=\mathfrak{h}+\mathfrak{n}^+,\quad \mathfrak{b}_\mathfrak{k}=\mathfrak{h}_1+\mathfrak{n}_\mathfrak{k}^+,\quad \mathfrak{n}^\pm=\sum_{\pm\alpha\in P}\mathfrak{g}_\alpha,\quad \mathfrak{n}_\mathfrak{k}^\pm=\sum_{\pm\beta\in P_\mathfrak{k}}\mathfrak{k}_\beta.$$

We then know that $V_{\mathfrak{g},P,\lambda}$ has a lattice above it for any $\lambda\in\mathfrak{h}^*$ such that $\lambda_1=\lambda|_{\mathfrak{h}_1}$ is $\Delta_\mathfrak{k}$-integral. As in Section 6, we put, for $s\in\mathfrak{w}(\mathfrak{k})$,

$$V_{P,\lambda,s}=(V_{\mathfrak{g},P,\lambda})_s,\quad W_{P,\lambda}=\tau(V_{\mathfrak{g},P,\lambda})=V_{P,\lambda,1}/\sum_{s\neq 1}V_{P,\lambda,s}.$$

We saw in Section 6 that the $W_{P,\lambda}$ are Harish-Chandra modules and constitute the so-called fundamental series for $(\mathfrak{g}_0,\mathfrak{k}_0,\mathfrak{h}_0)$. In this section we shall describe the main results for the $W_{P,\lambda}$ under the assumption that $\lambda_1=\lambda|_{\mathfrak{h}_1}$ is strongly $-P_\mathfrak{k}$-dominant integral. We shall refer to this by saying that $\lambda$ *is in good position* (for the definition of strongly $-P_\mathfrak{k}$-dominant integral $\lambda_1$ see footnote to Proposition 4.8.5).

**9.2.** *The* $\mathfrak{k}$*-splitting theorem for* $V_{\mathfrak{g},P,\lambda}$ *and* $V_{P,\lambda,s}$. *Let* $\lambda\in\mathfrak{h}^*$. Then we have, by Proposition 4.8.5, the exact resolution of $\mathfrak{g}$-modules

$$(\mathscr{R}) \qquad 0 \to E_m \to E_{m-1} \to \cdots \to E_0 \to V_{\mathfrak{g},P,\lambda} \to 0$$

where $m = \dim(\mathfrak{b}/\mathfrak{b}_{\mathfrak{k}})$, and for $0 \leqslant j \leqslant m$,

$$E_j = U(\mathfrak{g}) \bigotimes_{U(\mathfrak{b}\mathfrak{k})} (\Lambda^j(\mathfrak{b}/\mathfrak{b}_{\mathfrak{k}}) \otimes \mathbb{C}_{\lambda_1}).$$

Clearly

$$E_j \simeq U(\mathfrak{g}) \bigotimes_{U(\mathfrak{k})} A_j,$$

$$A_j = U(\mathfrak{k}) \bigotimes_{U(\mathfrak{b}\mathfrak{k})} (\Lambda^j(\mathfrak{b}/\mathfrak{b}_{\mathfrak{k}}) \otimes \mathbb{C}_{\lambda_1}).$$

The $\mathfrak{k}$-splitting theorem of Enright ([1], Theorem 6.17) asserts that the long exact sequence $(\mathscr{R})$ splits at every level provided $\lambda$ is in good position.

THEOREM 9.2.1. *Let $\lambda$ be an element of $\mathfrak{h}^*$ in good position. Then the long exact sequence of $\mathfrak{g}$-modules*

$$0 \to E_m \to E_{m-1} \to \cdots \to E_0 \to V_{\mathfrak{g},P,\lambda} \to 0$$

*splits at each level as $\mathfrak{k}$-modules*, i.e., $\exists$ *sub $\mathfrak{k}$-modules $F_i \subset E_i$* $(0 \leqslant i \leqslant m-1)$

$$E_i = F_i \oplus (\text{Image of } E_{i+1}), \qquad (0 \leqslant i \leqslant m-1).$$

*In particular, the $\mathfrak{k}$-module underlying $V_{\mathfrak{g},P,\lambda}$ is a direct summand of $S(\mathfrak{p}) \otimes V_{\mathfrak{k},P\mathfrak{k},\lambda_1}$, $S(\mathfrak{p})$ being the symmetric algebra over $\mathfrak{p}$, considered as a $\mathfrak{k}$-module.*

The last statement is clear since, as $\mathfrak{k}$-modules

$$E_0 \simeq U(\mathfrak{g}) \bigotimes_{U(\mathfrak{k})} A_0 \simeq U(\mathfrak{g}) \bigotimes_{U(\mathfrak{k})} V_{\mathfrak{k},P\mathfrak{k},\lambda_1} \simeq S(\mathfrak{p}) \otimes V_{\mathfrak{k},P\mathfrak{k},\lambda_1}.$$

Although the proof of this is long and involves many details, it can be divided into three steps.

*Step* 1. One considers a complex analogous to $(\mathscr{R})$ but involving only *symmetric algebras.* Let $\varepsilon(\mathfrak{g} \to \mathfrak{k})$ be the projection corresponding to the direct sum $\mathfrak{g} = \mathfrak{k} \oplus \mathfrak{p}$ and let

$$\mathfrak{p}^+ = \varepsilon(\mathfrak{b}).$$

We then have a complex of $\mathfrak{b}_{\mathfrak{k}}$-modules

$$0 \to S(\mathfrak{p}) \otimes \Lambda^m(\mathfrak{p}^+) \xrightarrow{\partial} S(\mathfrak{p}) \otimes \Lambda^{m-1}(\mathfrak{p}^+) \xrightarrow{\partial} \cdots$$
$$\cdots \xrightarrow{\partial} S(\mathfrak{p}) \otimes \Lambda^1(\mathfrak{p}^+) \xrightarrow{\partial} S(\mathfrak{p}) \xrightarrow{\varepsilon} S(\mathfrak{p}/\mathfrak{p}^+) \to 0$$

where the maps $\partial$ are $\mathfrak{b}_{\mathfrak{k}}$-module maps defined by (cf. Enright-Wallach [1])

$$\partial(a \otimes X_1 \wedge \ldots \wedge X_j) = \sum_{1 \leqslant i \leqslant j} (-1)^{i-1}(aX_i) \otimes X_1 \wedge \ldots \wedge \hat{X}_i \wedge \ldots \wedge X_j.$$

It is known that this complex is exact and that the associated complexes obtained by restricting to elements of given degree are also exact. More precisely, for any $j \geqslant -m$, we have an exact complex

$$0 \to S_j(\mathfrak{p}) \otimes \Lambda^m(\mathfrak{p}^+) \xrightarrow{\partial} S_{j+1}(\mathfrak{p}) \otimes \Lambda^{m-1}(\mathfrak{p}^+) \xrightarrow{\partial} \cdots$$
$$\cdots \xrightarrow{\partial} S_{j+m-1}(\mathfrak{p}) \otimes \Lambda^1(\mathfrak{p}^+) \xrightarrow{\partial} S_{j+m}(\mathfrak{p}) \xrightarrow{\varepsilon} S_{j+m}(\mathfrak{p}/\mathfrak{p}^+) \to 0.$$

Tensoring from the left by $U(\mathfrak{k})$ over $U(\mathfrak{b}_{\mathfrak{k}})$, and from the right by $\mathbb{C}_{\lambda_1}$, we get the exact complex

$$(\dagger) \qquad 0 \to D_{j,m} \to D_{j+1,m-1} \to \cdots \to D_{j+m,0} \to C_{j+m} \to 0$$

where, with $L_{j+r,m-r} = S_{j+r}(\mathfrak{p}) \otimes \Lambda^{m-r}(\mathfrak{p}^+) \otimes \mathbb{C}_{\lambda_1}$

$$D_{j+r,m-r} = U(\mathfrak{k}) \bigotimes_{U(\mathfrak{b}_{\mathfrak{k}})} L_{j+r,m-r}$$
$$\simeq S_{j+r}(\mathfrak{p}) \otimes (U(\mathfrak{k}) \bigotimes_{U(\mathfrak{b}_{\mathfrak{k}})} (\Lambda^{m-r}(\mathfrak{p}^+) \otimes \mathbb{C}_{\lambda_1})).$$

The second set of equivalences above shows (as $\lambda_1$ is strongly $-P_{\mathfrak{k}}$-dominant integral) that $D_{j+r,m-r}$ is of type (R) in the sense of Definition 4.8.1 and hence we see that the $D_{j+r,m-r}$ admit nondegenerate $\mathfrak{k}$-invariant forms. On the other hand, $D_{j+r,m-r} \simeq U(L_{j+r,m-r})$ in the notation of Section 8.2. So Proposition 8.2.4 implies that *the exact sequence* $(\dagger)$ *splits at every level.*

*Step* 2. The exact sequence

$$0 \to E_m \to E_{m-1} \to \cdots \to E_0 \to V_{\mathfrak{g},P,\lambda} \to 0$$

is built up out of its filtered parts *which are also exact* (see Section 5 of Enright–Wallach [1]):

$$(*) \qquad 0 \to E^j_m \to E^{j+1}_{m-1} \to \cdots \to E^{j+m}_0 \to V^{j+m} \to 0$$

where, with $U^l(\mathfrak{g})$ as the space of elements of $U(\mathfrak{g})$ of degree $\leqslant l (=0$ if $l<0)$,

$$E^{j+r}_{m-r} = (U^{j+r}(\mathfrak{g})U(\mathfrak{k})) \underset{U(\mathfrak{k})}{\otimes} (U(\mathfrak{k}) \underset{U(\mathfrak{b}\mathfrak{k})}{\otimes} (\Lambda^{m-r}(\mathfrak{p}^+) \otimes \mathbb{C}_{\lambda_1}))$$

$$V^{j+m} = \text{Image of } E^{j+m}_0 \text{ in } V_{\mathfrak{g},P,\lambda}.$$

We now compare $(*)$ with the exact sequence obtained from (†):

$$(\dagger\dagger) \qquad 0 \to D^j_m \to D^{j+1}_{m-1} \to \cdots \to D^{j+m}_0 \to C^{j+m} \to 0$$

where

$$D^l_q = \bigoplus_{0 \leqslant t \leqslant l} D_{t,q}.$$

The comparison is made by means of natural $\mathfrak{k}$-module isomorphisms

$$\Phi: D^l_q \simeq E^l_q$$

and gives rise to diagrams

$$\begin{array}{ccccccccc} 0 \to & E^j_m & \to & E^{j+1}_{m-1} & \to \cdots \to & E^{j+m}_0 & \to V^{j+m} \to 0 \\ & \uparrow\wr\,\Phi & & \uparrow\wr\,\Phi & & \uparrow\wr\,\Phi & \\ 0 \to & D^j_m & \to & D^{j+1}_{m-1} & \to \cdots \to & D^{j+m}_0 & \to C^{j+m} \to 0. \end{array}$$

*Step* 3. Although these diagrams *are not commutative*, they commute up to terms of highest degree (roughly speaking). This allows one to transfer the splitting property from the lower row to the upper row, using induction on degree.

**9.3.** In order to gain insight into the structure of $W_{P,\lambda} = \tau(V_{\mathfrak{g},P,\lambda})$ it is clearly necessary to know, with $A = V_{\mathfrak{g},P,\lambda}$ and $A_s = V_{P,\lambda,s}$ ($s \in \mathfrak{w}(\mathfrak{k})$), when it is true that an element $u \in A_1[\nu]^{\mathfrak{n}\mathfrak{k}^+}$ does not belong to $\Sigma_{s \neq 1} A_s$, $\nu \in \mathfrak{h}^*$ being $P_{\mathfrak{k}}$-dominant integral; for, the images in $\tau(A)$ of such $u$ span $\tau(A)^{\mathfrak{n}\mathfrak{k}^+}$. A study of examples suggests that this is so if and only if $d_{t_0,1}(\nu)u$ generates a sub $\mathfrak{k}$-module of $A$ which is a direct summand of $A$. We thus introduce, in the framework of $(\mathfrak{m}, \mathfrak{h}, Q)$,

DEFINITION 9.3.1. Let $M$ be a module in $\mathscr{I}(\mathfrak{m})$; $\mu \in \mathfrak{h}^*$; and $z \in M[\mu]^{\mathfrak{n}^+}$. Then $z$ is said to be a split invariant (of weight $\mu$) if the submodule $U(\mathfrak{m}) \cdot z$

(which is $\simeq V_{\mathfrak{m},Q,\mu}$) is a direct summand of $M$.

DEFINITION 9.3.2. Let $M \in \mathscr{I}(\mathfrak{m})$ and $(M_s)$ the lattice above $M$. We say that $M$ has the split invariant property iff for any $Q$-dominant integral $\nu \in \mathfrak{h}^*$ and any $u \in M_1[\nu]^{\mathfrak{n}^+}$,

$$u \notin \sum_{s \neq 1} M_s \Leftrightarrow d_{t_0,1}(\nu)u \quad \text{is a split invariant of } M.$$

We denote by $^*\mathscr{I}(\mathfrak{m})$ (actually $^*\mathscr{I}(\mathfrak{m}, \mathfrak{h}, Q)$) the subcategory of $\mathscr{I}(\mathfrak{m})$ of all modules which possess the split invariant property. If $\mathfrak{l}$ is a finite-dimensional Lie algebra with $\mathfrak{l} \supset \mathfrak{m}$, $^*\mathscr{I}_{\mathfrak{l}}(\mathfrak{m})$ is defined to be the subcategory $\mathscr{I}_{\mathfrak{l}}(\mathfrak{m}) \cap {}^*\mathscr{I}(\mathfrak{m})$.

*Remark* 9.3.3. It appears likely that $^*\mathscr{I}(\mathfrak{m}) = \mathscr{I}(\mathfrak{m})$. This issue can be finessed so far as our main goals are concerned. There is however a simple argument which shows that if $0 \neq u \in M_1[\nu]^{\mathfrak{n}^+}$ for a $Q$-dominant integral $\nu \in \mathfrak{h}^*$ and if $d_{t_0,1}(\nu)u$ is a split invariant of $M$, then $u \notin \Sigma_{s \neq 1} M_s$. In fact, if $u \in \Sigma_{s \neq 1} M_s$, then $u = \Sigma_{\gamma \in Q, \gamma\,\text{simple}} u_\gamma$ where $u_\gamma \in M_{s_\gamma}[\nu]^{\mathfrak{n}^+}$; if $t_\gamma = t_0 s_\gamma$, then $l(t_\gamma) = l(t_0) - 1$ and $t_0 = s_{\gamma'} t_\gamma$, with $\gamma'$ simple. So

$$d_{t_0,1}(\nu)u = \sum_\gamma d_{s_{\gamma'},1}(t'_\gamma \cdot \nu) \cdot (d_{t_\gamma,1}(\nu)u_\gamma),$$

showing that $d_{t_0,1}(\nu)u$ is in the $\mathfrak{m}$-module generated by vectors of weights strictly greater than $t'_0 \cdot \nu$; it cannot therefore be a split invariant.

PROPOSITION 9.3.4. *Let $M_i$ $(i \in I)$ be modules in $\mathscr{I}(\mathfrak{m})$ and let $M = \bigoplus_{i \in I} M_i$. Then*

$$M \in {}^*\mathscr{I}(\mathfrak{m}) \Leftrightarrow M_i \in {}^*\mathscr{I}(\mathfrak{m}) \text{ for all } i \in I.$$

This is straightforward.

The main results concerning split invariants are the following

THEOREM 9.3.5. *Let $F$ be a finite-dimensional $\mathfrak{m}$-module and $\mu \in \mathfrak{h}^*$ a $Q$-dominant integral linear function on $\mathfrak{h}$. Then*

$$F \otimes V_{\mathfrak{m},Q,t'_0 \cdot \mu} \in {}^*\mathscr{I}(\mathfrak{m}).$$

THEOREM 9.3.6. *Let $\mathfrak{g}, \mathfrak{k}, \mathfrak{h}$ be as before and let $\lambda \in \mathfrak{h}^*$ be in good position. Then*

$$V_{\mathfrak{g},P,\lambda} \in {}^*\mathscr{I}_{\mathfrak{g}}(\mathfrak{k}).$$

Theorem 9.3.5 leads at once to Theorem 9.3.6. For, by the $\mathfrak{k}$-splitting theorem (9.2.1), $(V_{\mathfrak{g},P,\lambda})_{\mathfrak{k}}$ is a direct summand of $S(\mathfrak{p}) \otimes V_{\mathfrak{k},P_{\mathfrak{k}},\lambda_1}$ which belongs to ${}^*\mathscr{I}(\mathfrak{k})$ by Theorem 9.3.5 since $t_0' \cdot \lambda_1$ is $P_{\mathfrak{k}}$-dominant integral.

Theorem 9.3.5 is proved using the theory of invariant forms. Let $A = F \otimes V_{\mathfrak{m},Q,t_0' \cdot \mu}$ and let $(A_s)$ be the lattice above $A$. We use Proposition 8.4.2. With notation as in that Proposition it is clear that the form $\varphi_F \otimes \varphi_\mu$ is the pull back to $A_1$ of the form $\varphi_F \otimes \bar{\varphi}_\mu$ on $\tau(A) = F \otimes I_{\mathfrak{m},Q,\mu}$ ($\bar{\varphi}_\mu$ being the invariant nondegenerate form on $I_{\mathfrak{m},Q,\mu}$.) We may assume $\varphi_F$ symmetric so that all forms in sight are symmetric. The nondegeneracy of $\varphi_F \otimes \bar{\varphi}_\mu$ now shows that $\Sigma_{s \neq 1} A_s$ is the radical ($= A_1^\perp$) of $\varphi_F \otimes \varphi_\mu$. So, if $u \in A_1[v]^{\mathfrak{n}^+}$, $v \in \mathfrak{h}^*$ being $Q$-dominant integral, and $u \notin \Sigma_{s \neq 1} A_s$, $\exists\, v \in A_1[v]^{\mathfrak{n}^+}$ such that $(\varphi_F \otimes \varphi_\mu)(u, v) \neq 0$. Exploiting the fact that $\varphi_F \otimes \varphi_\mu$ is the result of transferring $\varphi_F \otimes \varphi_{t_0' \cdot \mu}$ across successive completions, this gives, for

$$u_0 = d_{t_0,1}(v)u, \qquad v_0 = d_{t_0,1}(v)u,$$

the result $(\varphi_F \otimes \varphi_{t_0' \cdot \mu})(u_0, v_0) \neq 0$. In other words, if $B = U(\mathfrak{m}) \cdot u_0$, then $v_0 \notin B^\perp$. So, as $C = U(\mathfrak{m}) \cdot v_0$ is irreducible, $C \cap B^\perp = 0$, giving (as $\varphi_F \otimes \varphi_{t_0' \cdot \mu}$ is nondegenerate) $C^\perp + B = Q$, on taking orthogonal complements. As $B$ is irreducible, it is clear that $A = B \oplus C^\perp$.

Using appropriate tensoring arguments one can remove the restriction on $\lambda$ in Theorem 9.3.5. This will be done in Section 11.

**9.4.** *The irreducibility of* $W_{P,\lambda}$. The results of the preceding $n^0 s$ enable one to prove the irreducibility of $W_{P,\lambda}$, when $\lambda$ is in good position. First we have

LEMMA 9.4.1. *Let $\mathfrak{g}$, $\mathfrak{k}$, $\mathfrak{h}$ be as before. Then $\exists$ an involutive anti-automorphism $\sigma$ of $\mathfrak{g}$ such that $\sigma$ leaves $\mathfrak{k}$ stable and is the identity on $\mathfrak{h}$.*

It is in proving this lemma that we make use of the restrictions made at the beginning of Section 9, namely that $(\mathfrak{g}, \mathfrak{k})$ is the complexification of $(\mathfrak{g}_0, \mathfrak{k}_0)$, $\mathfrak{h}$, the complexification of $\mathfrak{h}_0$, $\mathfrak{h}_0$ being a fundamental CSA of $\mathfrak{g}_0$. For a proof of this lemma under these circumstances, see Enright [1], Lemma 10.1. In fact we could have made the general assumption:

(*) $\exists$ an involutive anti-automorphism $\sigma$ of $\mathfrak{g}$ with $\sigma$ leaving $\mathfrak{k}$ stable such that $\sigma$ is the identity on $\mathfrak{h}$.

I do not know whether (*) is true for any regular triple $(\mathfrak{g}, \mathfrak{k}, \mathfrak{h})$.

PROPOSITION 9.4.2. *Let $\lambda \in \mathfrak{h}^*$ be in good position. Let $\mu_1 = t'_0 \cdot \lambda_1$ where $\lambda_1 = \lambda|_{\mathfrak{h}_1}$. Then:*

(a) *The $\mathfrak{g}$-module $U(\mathfrak{g}) \otimes_{U(\mathfrak{k})} V_{\mathfrak{k},P_{\mathfrak{k}},\lambda_1}$ lies in $\mathscr{J}_{\mathfrak{g}}(\mathfrak{k})$; and*

$$(U(\mathfrak{g}) \bigotimes_{U(\mathfrak{k})} V_{\mathfrak{k},P_{\mathfrak{k}},\lambda_1})_s \simeq U(\mathfrak{g}) \bigotimes_{U(\mathfrak{k})} V_{\mathfrak{k},P_{\mathfrak{k}},s'\cdot\mu_1}$$

*for all $s \in \mathfrak{w}(\mathfrak{k})$.*

(b) $\exists$ *$\mathfrak{g}$-module surjections $\varepsilon_s$ ($s \in \mathfrak{w}(\mathfrak{k})$),*

$$\varepsilon_s \colon U(\mathfrak{g}) \bigotimes_{U(\mathfrak{k})} V_{\mathfrak{k},P_{\mathfrak{k}},s'\cdot\mu_1} \to V_{P,\lambda,s}$$

*such that*

$$\varepsilon_s(\mathbb{C} \cdot 1 \otimes v_{s'\cdot\mu_1}) = V_{P,\lambda,s}[s' \cdot \mu_1]^{\mathfrak{n}_{\mathfrak{k}}^+}.$$

(c) *$W_{P,\lambda}[\mu_1]^{\mathfrak{n}_{\mathfrak{k}}^+}$ and $V_{P,\lambda,s}[s' \cdot \mu_1]^{\mathfrak{n}_{\mathfrak{k}}^+}$ are one-dimensional and equivalent as $U(\mathfrak{g})^{\mathfrak{k}}$-modules; indeed, $U(\mathfrak{g})^{\mathfrak{k}}$ acts on both of them via the homomorphism $\beta_{P,\lambda_1}$.*

(d) *The one-dimensional space $W_{P,\lambda}[\mu_1]^{\mathfrak{n}_{\mathfrak{k}}^+}$ is $U(\mathfrak{g})$-cyclic for $W_{P,\lambda}$; the one-dimensional space $V_{P,\lambda,s}[s' \cdot \mu_1]^{\mathfrak{n}_{\mathfrak{k}}^+}$ is $U(\mathfrak{g})$-cyclic for $V_{P,\lambda,s}$.*

(a) is clear. For (b) we recall that when $\lambda$ is in good position, we can start with the exact resolution

$$0 \to E_m \to E_{m-1} \to \cdots \to E_0 \to V_{\mathfrak{g},P,\lambda} \to 0$$

and apply completion functors (with respect to simple roots in $P_{\mathfrak{k}}$) successively to get, for each $s \in \mathfrak{w}(\mathfrak{k})$, the exact resolution

$$0 \to (E_m)_s \to (E_{m-1})_s \to \cdots \to (E_0)_s \to V_{P,\lambda,s} \to 0.$$

Since $E_0 = U(\mathfrak{g}) \otimes_{U(\mathfrak{k})} V_{\mathfrak{k},P_{\mathfrak{k}},\lambda_1}$, we get

$$(E_0)_s = U(\mathfrak{g}) \bigotimes_{U(\mathfrak{k})} V_{\mathfrak{k},P_{\mathfrak{k}},s'\cdot\mu_1}$$

giving the first assertion in (b) and the result

$$V_{P,\lambda,s} = U(\mathfrak{g}) \cdot \varepsilon_s(1 \otimes v_{s' \cdot \mu_1}).$$

Since $V_{\mathfrak{g},P,\lambda}[\lambda_1]^{\mathfrak{n}_{\mathfrak{t}}^+} = \mathbb{C} \cdot v_{\lambda_1}$ is one-dimensional, and $d_{t_0 s^{-1},1}(s' \cdot \mu_1)$ is a linear bijection of $V_{P,\lambda,s}[s' \cdot \mu_1]^{\mathfrak{n}^+}$ onto $V_{\mathfrak{g},P,\lambda}[\lambda_1]^{\mathfrak{n}^+}$ (Proposition 4.7.2), we see that $V_{P,\lambda,s}[s' \cdot \mu_1]^{\mathfrak{n}^+}$ is one-dimensional, hence spanned by $\varepsilon_s(1 \otimes v_{s' \cdot \mu_1})$. This proves (b) and the second statement in (d). The assertion (c) is clear from Proposition 4.7.4 and Corollary 6.1.4, provided one knows that $W_{P,\lambda}[\mu_1]^{\mathfrak{n}_{\mathfrak{t}}^+}$ is one-dimensional; but this is a consequence of Proposition 4.9.6 and Lemma 6.1.2.

We can now obtain the irreducibility theorem.

THEOREM 9.4.3. *Let $\mathfrak{g}$, $\mathfrak{k}$, $\mathfrak{h}$, $P$ and $P_{\mathfrak{k}}$ be as above. Let $\lambda \in \mathfrak{h}^*$, $\lambda_1 = \lambda|\mathfrak{h}_1$, $\mu_1 = t_0' \cdot \lambda_1$. Assume that:*

(a) *$\lambda$ is in good position.*

(b) *$V_{\mathfrak{g},P,\lambda}$ is irreducible.*

*Then $W_{P,\lambda}$ is $\neq 0$, is irreducible, and admits a nondegenerate symmetric $\mathfrak{g}$-invariant form. Moreover, the class $\vartheta(\mu_1)$ occurs in $W_{P,\lambda}$ with multiplicity one.*

The Harish-Chandra module $W_{P,\lambda}$ is cyclically generated by the isotypical subspace $W_{P,\lambda,\vartheta}$ corresponding to the equivalence class $\vartheta$ of the irreducible $\mathfrak{k}$-module of $P_{\mathfrak{k}}$-highest weight $\mu_1$; moreover the class $\vartheta$ occurs exactly once in $W_{P,\lambda}$ (Proposition 9.4.2). So, if $W_{P,\lambda}$ admits a nondegenerate invariant form, an easy argument shows that any sub $\mathfrak{g}$-module $W' \neq W$ is orthogonal to $W_{P,\lambda,\vartheta}$, hence contained in $W^{\perp}$, hence 0, proving the irreducibility of $W_{P,\lambda}$. So it is a question of proving that $W_{P,\lambda}$ admits a nondegenerate $\mathfrak{g}$-invariant form.

Since $V_{\mathfrak{g},P,\lambda}$ is irreducible, it admits a nondegenerate $\mathfrak{g}$-invariant form, even a symmetric one, say $\langle \cdot, \cdot \rangle_0$. Transferring this across the lattice $(V_{P,\lambda,s})$ we obtain, using the theory of Section 8, a symmetric $\mathfrak{g}$-invariant form $\langle \cdot, \cdot \rangle_1$ for the module $V_{P,\lambda,1}$. The fact that $\langle \cdot, \cdot \rangle_1$ factors through to a nondegenerate (clearly symmetric) form for $W_{P,\lambda}$ is now a consequence of the following lemma:

LEMMA 9.4.4. *$\Sigma_{s \neq 1} V_{P,\lambda,s}$ is the radical of $\langle \cdot, \cdot \rangle_1$*, i.e., *one has the relation*

$$(V_{P,\lambda,1})^{\perp} = \sum_{s \neq 1} B_{P,\lambda,s}.$$

For brevity, write

$$A=V_{\mathfrak{g},P,\lambda}, \qquad A_s=V_{P,\lambda,s},$$
$$B=U(\mathfrak{k})\cdot v_{\lambda_1}\simeq V_{\mathfrak{k},P\mathfrak{k},\lambda_1},$$
$$B_s=U(\mathfrak{k})\cdot v_{s'\cdot\mu_1}\simeq V_{\mathfrak{k},P\mathfrak{k},s'\cdot\mu_1}.$$

It is then easily seen that $\Sigma_{s\neq 1}B_s$ is the radical of $\langle\cdot,\cdot\rangle_1|_{B_1\times B_1}$. In particular, $B_s\perp B_1$ for $s\neq 1$. On the other hand, let us consider $A$ and $A_1$ as $\mathfrak{k}$-modules and decompose them as direct sums

$$A=\bigoplus_{\chi} A_{\chi}, \qquad A_1=\bigoplus_{\chi} A_{1,\chi},$$

where $\chi$ runs through the homomorphisms center$(U(\mathfrak{k}))\rightarrow\mathbb{C}$, and $A_\chi$ (resp. $A_{1,\chi}$) is the generalized eigen subspace of $A$ (resp. $A_1$) corresponding to $\chi$, namely, the largest sub $\mathfrak{k}$-module on which the endomorphisms $z-\chi(z)\cdot 1$ are locally nilpotent $\forall\, z\in$ center$(U(\mathfrak{k}))$. An easy argument shows that if $\chi_1$ is the homomorphism of center$(U(\mathfrak{k}))$ into $\mathbb{C}$ according to which center $(U(\mathfrak{k}))$ acts on $V_{\mathfrak{k},P\mathfrak{k},\lambda_1}$, $B=A_{\chi_1}$; on the other hand, for any $\chi$, $A_{1,\chi}=(A_\chi)_1$. So $B_1=A_{1,\chi_1}$. This shows that for $s\neq 1$, $B_s\perp A_{1,\chi}\ \forall\,\chi$, and hence $B_s\perp A_1$.

As $U(\mathfrak{g})\cdot B_s=A_s$, we find $\Sigma_{s\neq 1}A_s\subset A_1^{\perp}$. Let $R=A_1^{\perp}$. If $R\neq\Sigma_{s\neq 1}A_s$, then $R/\Sigma_{s\neq 1}A_s$ is $\neq 0$ and $\mathfrak{k}$-finite. So $\exists$ a $P_{\mathfrak{k}}$-dominant integral $\xi\in\mathfrak{h}_1^*$ and a $z\in R[\xi]^{\mathfrak{n}\mathfrak{k}^+}$ such that $z\notin\Sigma_{s\neq 1}A_s$. As $A\in {}^*\mathscr{I}_{\mathfrak{g}}(\mathfrak{k})$ (Theorem 9.3.6) $z'=d_{t_0,1}(\xi)z$ is a split invariant of $A$. So, if $C=U(\mathfrak{k})\cdot z'$, then $C\simeq V_{\mathfrak{k},p\mathfrak{k},t_0'\cdot\xi}$ and $A=C\oplus D$ where $D$ is a sub $\mathfrak{k}$-module of $A$. Then $D^{\perp}\cap C^{\perp}=0$ while

$$D^{\perp}\simeq A/D\simeq C\simeq V_{\mathfrak{k},P\mathfrak{k},t_0'\cdot\xi}.$$

This shows that $D^{\perp}[t_0'\cdot\xi]\not\subset C^{\perp}$, i.e., if $w'$ is a nonzero vector in $D^{\perp}[t_0'\cdot\xi]$, then $w'\in A[t_0'\cdot\xi]^{\mathfrak{n}\mathfrak{k}^+}$ and $\langle w',z'\rangle_0\neq 0$. But then, as $\langle\cdot,\cdot\rangle_1$ is obtained from $\langle\cdot,\cdot\rangle_0$ by successive transfers, $\langle w,z\rangle_1\neq 0$ where $w\in A_1[\xi]^{\mathfrak{n}\mathfrak{k}^+}$ is such that $d_{t_0,1}(\xi)w=w'$. This contradicts the assumption that $z\in R$.

**9.5.** *Resolution of $W_{P,\lambda}$ in terms of the* $\mathfrak{g}$*-modules* $V_{P,\lambda,s}$. In Section 2 we have recalled the B–G–G resolution. Thus, if $\mathfrak{m}$ is a reductive Lie algebra over $\mathbb{C}$, with CSA $\mathfrak{h}$ and positive system of roots $Q$, then, for any $Q$-dominant integral $\mu\in\mathfrak{h}^*$, the irreducible finite-dimensional $\mathfrak{m}$-module $I_{\mathfrak{m},Q,\mu}$ (with $Q$-highest weight $\mu$) has an exact resolution

$$\text{(B–G–G)}\quad 0\rightarrow B_\mu^{(r)}\rightarrow B_\mu^{(r-1)}\rightarrow\cdots\rightarrow B_\mu^{(1)}\rightarrow B_\mu^{(0)}\rightarrow I_{\mathfrak{m},Q,\mu}\rightarrow 0$$

where

$$B_\mu^{(j)} = \bigoplus_{l(s)=j} V_{\mathfrak{m},Q,s'\cdot\mu} \qquad (1 \leqslant j \leqslant r = l(t_0)), \qquad B_\mu^{(0)} = V_{\mathfrak{m},Q,\mu}.$$

On the other hand we have on many occasions stressed the analogy of the configuration of modules $(V_{\mathfrak{m},Q,s'\cdot\mu})$ with the configuration of modules $(A_s)$ of a lattice above a module $A$. This suggests the following question: let $A \in \mathscr{J}(\mathfrak{m})$ and let $(A_s)_{s\in\mathfrak{w}(\mathfrak{m})}$ be the lattice above $A$. Define, with $r = l(t_0)$ as usual,

$$\mathscr{B}^{(j)} = \bigoplus_{l(s)=j} A_s \qquad (1 \leqslant j \leqslant r), \qquad \mathscr{B}^{(0)} = A_1.$$

Then, does there exist an exact resolution (of $\mathfrak{m}$-modules)

$$0 \to \mathscr{B}^{(r)} \to \mathscr{B}^{(r-1)} \to \cdots \to \mathscr{B}^1 \to \mathscr{B}^{(0)} \xrightarrow{\varepsilon} \tau(A) \to 0.$$

Here of course $\tau(A) = \mathscr{B}^{(0)}/\mathscr{B}^{(1)} = A_1/\Sigma_{s\neq 1} A_s$, and $\varepsilon$ is the natural surjection. The answer is affirmative (in the context of regular triples $(\mathfrak{g}, \mathfrak{k}, \mathfrak{h})$) when $A = V_{\mathfrak{g},P,\lambda}$, provided $\lambda_1 = \lambda|_{\mathfrak{h}_1}$ is $\Delta_{\mathfrak{k}}$-integral and $V_{\mathfrak{g},P,\lambda}$ is irreducible. We shall see in Section 11 that such a resolution of $W_{P,\lambda} = \tau(V_{\mathfrak{g},P,\lambda})$ leads to explicit multiplicity formulae for the irreducible $\mathfrak{k}$-classes occurring in $W_{P,\lambda}$. I shall now sketch Enright's argument [1] for getting the resolution of $W_{P,\lambda}$ when $\lambda$ is in good position.

To formulate this result with precision we introduce some notation. Let $(\mathfrak{g}, \mathfrak{k}, \mathfrak{h})$ be as above. Let $P$ be a regular positive system of roots of $(\mathfrak{g}, \mathfrak{h})$ and let $P_{\mathfrak{k}}$ be the corresponding positive system of roots of $(\mathfrak{k}, \mathfrak{h}_1)$. We then have a partial ordering $\leqslant$ in $\mathfrak{w}(\mathfrak{k})$ characterized by

$$s_1 \leqslant s_2 \Leftrightarrow V_{\mathfrak{k},P_{\mathfrak{k}},s_1'\cdot\mu} \hookrightarrow V_{\mathfrak{k},P_{\mathfrak{k}},s_2'\cdot\mu} \ \forall\, P_{\mathfrak{k}}\text{-dominant integral } \mu \in \mathfrak{h}_1^*$$

(see Section 2). Hence, for a $P_{\mathfrak{k}}$-dominant integral $\mu \in \mathfrak{h}_1^*$, if we identify all the $V_{\mathfrak{k},P_{\mathfrak{k}},s'\cdot\mu}$ with their images in $V_{\mathfrak{k},P_{\mathfrak{k}},\mu}$, then any $\mathfrak{k}$-module map

$$\delta\colon B_\mu^{(i)} \to B_\mu^{(i-1)} \qquad B_\mu^{(i)} = \bigoplus_{l(s)=i} V_{\mathfrak{k},P_{\mathfrak{k}},s'\cdot\mu}$$

can be represented by a matrix $(\delta_{s_1,s_2})_{l(s_1)=i-1,l(s_2)=i}$ of complex numbers. So, if we write $d_j$ for the maps of the resolution (B–G–G),

$$d_j\colon B_\mu^{(j)} \to B_\mu^{(j-1)}$$

then the matrix, say $\mathbb{D}_j = (d_{j,s_1,s_2})_{l(s_1)=j-1,l(s_2)=j}$, representing $d_j$, is independent of $\mu$: $d_{j,s_1,s_2} \in \{0, +1, -1\}$, and $d_{j,s_1,s_2} = 0$ unless $s_2 \leqslant s_1$ (Lemma 2.2.11).

Let $\lambda \in \mathfrak{h}^*$ and let $(V_{P,\lambda,s})_{s \in \mathfrak{w}(\mathfrak{k})}$ be the lattice of $\mathfrak{g}$-modules above the $\mathfrak{g}$-Verma module $V_{\mathfrak{g},P,\lambda}$. We then have

THEOREM 9.5.1. *Let $\mathfrak{g}, \mathfrak{k}, \mathfrak{h}, P, P_{\mathfrak{k}}$ be as above. Suppose $\lambda \in \mathfrak{h}^*$ and assume that $\lambda$ is in good position. We then have the following*:

(a) *If $s_1, s_2 \in \mathfrak{w}(\mathfrak{k})$ and $s_1 \leqslant s_2$, then $V_{P,\lambda,s_1} \subset V_{P,\lambda,s_2}$.*

(b) *Define the $\mathfrak{g}$-modules $\mathscr{V}_\lambda^{(i)}$ by*

$$\mathscr{V}_\lambda^{(0)} = V_{P,\lambda,1}, \qquad \mathscr{V}_\lambda^{(i)} = \bigoplus_{l(s)=i} V_{P,\lambda,s} \qquad (1 \leqslant i \leqslant r = l(t_0))$$

*and let the $\mathfrak{g}$-module maps*

$$d_i: \mathscr{V}_\lambda^{(i)} \to \mathscr{V}_\lambda^{(i-1)}$$

*be defined by the matrices $\mathbb{D}_i$ occurring in the* B–G–G *resolution. Then*

$$0 \to \mathscr{V}_\lambda^{(r)} \xrightarrow{d_r} \mathscr{V}_\lambda^{(r-1)} \longrightarrow \cdots \xrightarrow{d_1} \mathscr{V}_\lambda^{(0)} \xrightarrow{\varepsilon} W_{P,\lambda} \to 0$$

*is an exact resolution, $\varepsilon$ being the natural surjection.*

For brevity write $A = V_{\mathfrak{g},P,\lambda}$; and for any $\mathfrak{g}$-module $M$ write $M_{\mathfrak{k}}$ for the underlying $\mathfrak{k}$-module. From the $\mathfrak{k}$-splitting theorem of Section 9.2 (Theorem 9.2.1) we know that $A_{\mathfrak{k}}$ is a direct summand of $(E_0)_{\mathfrak{k}}$ where

$$E_0 = U(\mathfrak{g}) \bigotimes_{U(\mathfrak{k})} V_{\mathfrak{k},P_{\mathfrak{k}},\lambda_1}.$$

Thus we may write

$$(E_0)_{\mathfrak{k}} = A_{\mathfrak{k}} \oplus B$$

where $B$ is a sub $\mathfrak{k}$-module of $(E_0)_{\mathfrak{k}}$. On the other hand, we also know that $(E_0)_{\mathfrak{k}}$ is $\simeq S(\mathfrak{p}) \otimes V_{\mathfrak{k},P_{\mathfrak{k}},\lambda_1}$, so that

$$(E_0)_{\mathfrak{k}} \simeq \bigoplus_{i \in I} F_i \otimes V_{\mathfrak{k},P_{\mathfrak{k}},t_0' \cdot \mu_1}, \qquad (\mu_1 = t_0' \cdot \lambda_1),$$

where $F_i$ $(i \in I)$ are finite-dimensional $\mathfrak{k}$-modules and $\mu_1$ is clearly $P_{\mathfrak{k}}$-dominant integral.

It follows from these remarks that $(E_0)_{\mathfrak{k}}$ and $A_{\mathfrak{k}}$ belong to $\mathscr{J}(\mathfrak{k})$. So $B \in \mathscr{J}(\mathfrak{k})$ by Proposition 4.6.4. Obviously, we then have canonical isomorphisms

$$(E_0)_{\mathfrak{k},s} \simeq A_{\mathfrak{k},s} \oplus B_s, \qquad (s \in \mathfrak{w}(\mathfrak{k})).$$

The assertion (a) is now already proved. For, if $s_1 \leqslant s_2$, we have

$$V_{\mathfrak{k},P\mathfrak{k},s_1'\cdot\mu_1} \subset V_{\mathfrak{k},P\mathfrak{k},s_2'\cdot\mu_1}$$

which gives $(E_0)_{\mathfrak{k},s_1} \subset (E_0)_{\mathfrak{k},s_2}$ and hence implies both the inclusions

$$A_{\mathfrak{k},s_1} \subset A_{\mathfrak{k},s_2}, \qquad B_{s_1} \subset B_{s_2}.$$

Moreover, the B–G–G resolution of $I_{\mathfrak{k},P\mathfrak{k},\mu_1}$ gives rise to the exact complex

$$0 \to \mathscr{E}^{(r)} \xrightarrow{\bar{d}_r} \mathscr{E}^{(r-1)} \xrightarrow{\bar{d}_{r-1}} \cdots \xrightarrow{\bar{d}_0} \mathscr{E}^{(0)} \to \tau((E_0)_{\mathfrak{k}}) \to 0$$

where

$$\mathscr{E}^{(0)} = (E_0)_{\mathfrak{k},1}, \qquad \mathscr{E}^{(i)} = \bigoplus_{l(s)=i} (E_0)_{\mathfrak{k},s}, \qquad (1 \leqslant i \leqslant r),$$

$$\bar{d}_i = id \otimes d_i, \qquad \left(\mathscr{E}^{(i)} \simeq S(\mathfrak{p}) \otimes \left(\bigoplus_{l(s)=i} V_{\mathfrak{k},P\mathfrak{k},s'\cdot\mu_1}\right)\right).$$

In view of the assertion (a) we can now observe that the above complex is the direct sum of the complexes

$$0 \to \mathscr{A}^{(r)} \to \mathscr{A}^{(r-1)} \to \cdots \to \mathscr{A}^{(0)} \to \tau(A_{\mathfrak{k}}) \to 0,$$

$$0 \to \mathscr{B}^{(r)} \to \mathscr{B}^{(r-1)} \to \cdots \to \mathscr{B}^{(0)} \to \tau(B) \to 0,$$

where

$$\mathscr{A}^{(0)} = A_{\mathfrak{k},1}, \qquad \mathscr{B}^{(0)} = B_1, \qquad \mathscr{A}^{(i)} = \bigoplus_{l(s)=i} A_{\mathfrak{k},s}, \qquad \mathscr{B}^{(i)} = \bigoplus_{l(s)=i} B_s.$$

The exactness of the complex $(\mathscr{E}^{(i)})_{0 \leqslant i \leqslant r}$ then gives the exactness of both $(\mathscr{A}^{(i)})$ and $(\mathscr{B}^{(i)})$. The assertion (b) is now clear.

## 10. Tensoring by finite-dimensional modules

**10.1.** In order to extend the theorems of Section 9 to the cases where $\lambda$ is not in good position we use the technique of tensoring by finite-dimensional modules. It was first introduced by Fomin and Shapovalov [1]; they used it to settle a question that was open for some time; namely, the coefficients of the exponentials, which appear in the expression for the restriction of an irreducible character (of a s.s. Lie group) to the various Cartan subgroups, are *constants*. This technique was refined and fashioned into a very useful tool in representation theory by Zuckerman (see [1], where he discusses some of the applications of his theory).

Zuckerman as well as Fomin and Shapovalov operate in the category of

Harish-Chandra modules. If $M$ and $M'$ are two such modules which are irreducible and whose infinitesimal characters are parametrized by $\lambda, \lambda' \in \mathfrak{h}^*$, then their technique allows one to go back and forth between $M$ and $M'$ whenever $\lambda$ and $\lambda'$ are in the same chamber. This however is not sufficient for our purposes; we also need to relate $M$ and $M'$ when $\lambda$ and $\lambda'$ lie in different chambers. Enright [1] has developed a variant of the ideas of Zuckerman and Fomin–Shapovalov to handle this case. He has also treated the case when $M$ and $M'$ are Verma modules. We now proceed to describe briefly the results of Zuckerman, Fomin–Shapovalov, and Enright.

**10.2.** *Tensoring of Verma modules.* We work in the context of $(\mathfrak{m}, \mathfrak{h}, Q)$. For brevity we denote the Verma module $V_{\mathfrak{m},Q,\lambda}$ by $V_\lambda$ ($\lambda \in \mathfrak{h}^*$). For any $Q$-integral $\mu \in \mathfrak{h}^*$ let $F^\mu$ be the finite-dimensional irreducible module with $\mu$ as an extreme weight.

Let $\mathfrak{Z}(\mathfrak{m})$ be the center of $U(\mathfrak{m})$. The homomorphisms $\mathfrak{Z}(\mathfrak{m}) \to \mathbb{C}$ are then parametrized by the $\mathfrak{w}(\mathfrak{m})$-orbits in $\mathfrak{h}^*$. For any $\lambda \in \mathfrak{h}^*$ we write $[\lambda]$ to denote the orbit of $\lambda$ as well as the corresponding homomorphism $\mathfrak{Z}(\mathfrak{m}) \to \mathbb{C}$. The parametrization is such that $\mathfrak{Z}(\mathfrak{m})$ acts on $V_\lambda$ through $[\lambda+\delta]$, where $\delta = \delta_Q$.

If $M$ is any $\mathfrak{m}$-module, then, for each $\lambda \in \mathfrak{h}^*$ we denote by $M^{[\lambda]}$ the sub $\mathfrak{m}$-module of all $v \in M$ such that $(z-[\lambda](z)\cdot 1)^m v = 0$ for each $z \in \mathfrak{Z}(\mathfrak{m})$ if $m \gg 0$. The $M^{[\lambda]}$ are linearly independent for distinct $\lambda$; $M = \bigoplus_\lambda M^{[\lambda]}$ if and only if $\dim(\mathfrak{Z}(\mathfrak{m})v) < \infty$ for each $v \in M$; this is the case for example if $M$ is a weight module and all its weight-spaces are finite-dimensional.

DEFINITION 10.2.1. *Let $A$. $B$ be $\mathfrak{m}$-modules; $\mu \in \mathfrak{h}^*$, $\mu$ being $Q$-integral. We say that $A$ is linked to $B$ by $\mu$,*

$$A \xrightarrow[(\mu)]{} B$$

*in symbols, iff for some $\nu \in \mathfrak{h}^*$,*

$$B \simeq (F^\mu \otimes A)^{[\nu]}.$$

Note that given $A$ and $B$, the orbit $[\nu]$ is uniquely determined by the fact that $B = B^{[\nu]}$ necessarily.

We now have the following result (Enright [1]).

THEOREM 10.2.2. *Let $\lambda$, $\xi \in \mathfrak{h}^*$. Assume that* (a) *$\lambda+\delta$ is regular,*

(b) $\lambda - \xi$ *is integral,* (c) $V_\lambda$ and $V_\xi$ *are both irreducible. Then there is an integer* $r \geqslant 1$ *and* $\mu_i, \nu_i \in \mathfrak{h}^*$ ($\mu_i$ *integral*) *such that*

$$(\lambda \to \xi) \colon V_\lambda \xrightarrow[(\mu_1)]{} V_{\nu_1} \xrightarrow[(\mu_2)]{} V_{\nu_2} \to \cdots \to V_{\nu_{r-1}} \xrightarrow[(\mu_r)]{} V_\xi.$$

*If in addition* $\xi + \delta$ *is regular, we can choose* $r$ *and* $\mu_i, \nu_i$ *so that not only* $(\lambda \to \xi)$ *but also the linkages* $(\xi \to \lambda)$ *below are valid:*

$$(\xi \to \lambda) \colon V_\xi \xrightarrow[(-\mu_r)]{} V_{\nu_{r-1}} \xrightarrow[(-\mu_{r-1})]{} \cdots \longrightarrow V_{\nu_1} \xrightarrow[(-\mu_1)]{} V_\lambda.$$

For the proof one may assume $\mathfrak{m}$ to be semisimple. For any positive system $R$ let $C_R$ denote the subset (=chamber) of all $\nu \in \mathfrak{h}^*$ with $\langle \mathrm{Re}(\nu), \alpha \rangle \geqslant 0$ $\forall \alpha \in R$; set $\delta_R = \frac{1}{2}\Sigma_{\alpha \in R}\alpha$.

The basic linkages are, for *arbitrary* $R$,

(*)

$$V_{\eta+\nu} \xrightarrow[(-\nu)]{} V_\eta, \qquad (\eta + \delta \in C_R,\ \nu \in C_R,\ \nu \text{ integral})$$

$$V_\eta \xrightarrow[(\nu)]{} V_{\eta+\nu}, \qquad (\eta + \delta \in C_R,\ \nu \in C_R,\ \nu \text{ integral},\ \eta + \delta \text{ regular}).$$

These are proved by showing, under the conditions on $\eta$ and $\nu$, that if $\nu'$ (resp. $\nu''$) is a weight of $F^{-\nu}$ (resp. $F^\nu$), then $[\eta + \nu + \nu' + \delta] = [\eta + \delta]$ if and only if $\nu' = -\nu$ (resp. $[\eta + \nu'' + \delta] = [\eta + \nu + \delta]$ if and only if $\nu'' = \nu$).

A simple argument shows that it is enough to prove the simultaneous linkages $(\lambda \to \xi)$ and $(\xi \to \lambda)$ under the condition that $\xi + \delta$ is also regular.

*Case I*: $\lambda + \delta$ *and* $\xi + \delta$ *are both in a single* $C_R$. Write $\lambda - \xi = \varphi_1 - \varphi_2$ where $\varphi_1$ and $\varphi_2$ are $R$-dominant integral. Then $\lambda + \varphi_2 = \xi + \varphi_1 = \zeta$ and the result follows quickly from (*).

*Case II*. $\lambda + \delta$ *and* $\xi + \delta$ *are in distinct chambers.* The essential case is when $\lambda + \delta \in C_R$, $\xi + \delta \in C_{R'}$ where $R'$ is *adjacent* to $R$, i.e., $R' = s_\gamma R$ where $\gamma \in R$ is *simple* in $R$. The idea is now to consider a variable $R$-dominant integral $\nu \in \mathfrak{h}^*$ which goes to infinity in such a way that (a) $\langle \nu, \beta \rangle \to \infty$ if $\beta$ is simple in $R$ and $\beta \neq \gamma$, (b) $\langle \nu, \gamma \rangle = 0(1)$. Let $\bar{\lambda} = \lambda + \nu$, $\bar{\xi} = \xi + s_\gamma \nu$. In view of Case I it is enough to prove the linkages

$$C_\xi \xrightarrow[(\bar{\lambda} - \bar{\xi})]{} V_{\bar{\lambda}}, \qquad V_{\bar{\lambda}} \xrightarrow[(\bar{\xi} - \bar{\lambda})]{} V_{\bar{\xi}}.$$

To prove the first of these for example we consider a weight $\theta$ of $F^{\lambda - \xi}$ and an element $s \in \mathfrak{w}(\mathfrak{m})$ such that $s(\theta + \bar{\xi} + \delta) = \bar{\lambda} + \delta$. Since $\langle \nu, \beta \rangle \geqslant 0$ for $\beta$ simple in $R$ and $\beta \neq \gamma$, we find that $R \backslash \{\gamma\}$ is stable under $s^{-1}$, i.e., $s = 1$ or $s = s_\gamma$. We rule out $s = s_\gamma$ by remarking that, as both $V_\lambda$ and $V_\xi$

are irreducible, $(\lambda+\delta)(\bar{H}_\gamma)$ is *not* an integer. The proof of Case II can now be completed easily.

From this theorem we get the following corollary.

COROLLARY 10.2.3. *Let* $\lambda$, $\xi \in \mathfrak{h}^*$. *Assume that* (a) $\lambda+\delta$ *is regular* (b) $\lambda-\xi$ *is integral* (c) *both* $V_\lambda$ *and* $V_\xi$ *are irreducible. Then there is a finite-dimensional* $\mathfrak{m}$*-module* $F$ *such that* $V_\xi$ *is a direct summand of* $F \otimes V_\lambda$.

It is clear that one can develop a more general theory of tensoring for modules of the category $\mathcal{I}(\mathfrak{m})$. Due to limitations of space and time I do not consider this here.

**10.3.** *Tensoring of Harish-Chandra modules.* We go over to the framework involving $(\mathfrak{g}, \mathfrak{k})$. We fix a $\theta$-stable Cartan subalgebra $\mathfrak{l}_0 \subset \mathfrak{g}_0$ and write $\mathfrak{l} = \mathbf{C} \cdot \mathfrak{l}_0$. Let $\mathcal{A}(\mathfrak{g}, \mathfrak{k}) = \mathcal{A}$ be the category of Harish-Chandra modules which are finitely generated as $U(\mathfrak{g})$-modules. We use the $\mathfrak{w}(\mathfrak{g})$-orbits in $\mathfrak{l}^*$ to parametrize the homomorphisms $\mathfrak{Z}(\mathfrak{g}) \to \mathbf{C}$ where $\mathfrak{Z}(\mathfrak{g}) = \text{center } U(\mathfrak{g})$. As in the previous number, we denote, for $\nu \in \mathfrak{l}^*$, by $[\nu]$ the orbit of $\nu$ as well as the corresponding homomorphism of $\mathfrak{Z}(\mathfrak{g})$. Given any $\mathfrak{g}$-module $M$ we have as before the sub $\mathfrak{g}$-modules $M^{[\nu]}$ $(\nu \in \mathfrak{l}^*)$. If $M \in \mathcal{A}$, $M$ is the direct sum of the various $M^{[\nu]}$. We write $\mathcal{A}^{[\nu]}$ for the subcategory of all $M \in \mathcal{A}$ with $M = M^{[\nu]}$. The functor $M \mapsto M^{[\nu]}$ from $\mathcal{A}$ to $\mathcal{A}^{[\nu]}$ is covariant and exact. We denote it by $P_\nu$.

We begin by formulating Zuckerman's result. Fix a positive system $\Delta^+$ of roots of $(\mathfrak{g}, \mathfrak{l})$. Let $\mathfrak{l}_{\mathbb{R}}$ be the real linear span of the root vectors $H_\alpha$; $C_0^+$, the closed Weyl chamber in $\mathfrak{l}_{\mathbb{R}}^*$ defined by $\Delta^+$; for each $\nu \in \mathfrak{l}^*$ such that $\text{Re}(\nu) \in C_0^+$, let $\Delta_\nu^+$ be the set of $\Delta^+$-simple roots $\alpha$ such that $\langle \text{Re}(\nu), \alpha \rangle = 0$. We now define

$$C^+ = \{\nu \in \mathfrak{l}^* \mid \text{Re}(\nu) \in C_0^+,\ \langle \text{Im}(\nu), \alpha \rangle \geqslant 0\ \forall \alpha \in \Delta_\nu^+\}.$$

It is then obvious that $C^+$ is a fundamental domain for the action of $\mathfrak{w}(\mathfrak{g})$ on $\mathfrak{l}^*$. As before we write, for any integral $\mu \in \mathfrak{l}^*$, $F^\mu$ for the finite-dimensional irreducible $\mathfrak{g}$-module of extreme weight $\mu$.

Following Zuckerman [1] we introduce, for any $\lambda \in C^+$ and integral $\mu \in \mathfrak{l}^*$, the functors $\Phi_{\lambda+\mu}^\lambda$ and $\Psi_\lambda^{\lambda+\mu}$ on $\mathcal{A} = \mathcal{A}(\mathfrak{g}, \mathfrak{k})$ defined by

$$\text{(Z)} \qquad \begin{aligned} \Phi_{\lambda+\mu}^\lambda &= p_{\lambda+\mu} \circ F^\mu \otimes (\ ) \circ p_\lambda \\ \Psi_\lambda^{\lambda+\mu} &= p_\lambda \circ F^{-\mu} \otimes (\ ) \circ p_{\lambda+\mu}. \end{aligned}$$

Thus, if $A \in \mathscr{A}^{[\lambda]}$ and $B \in \mathscr{A}^{[\lambda+\mu]}$, we have

$$\Phi^{\lambda}_{\lambda+\mu}(A) = (F^{\mu} \otimes A)^{[\lambda+\mu]}, \qquad \Psi^{\lambda+\mu}_{\lambda}(B) = (F^{-\mu} \otimes B)^{[\lambda]}.$$

Zuckerman's theorems are now as follows.

THEOREM 10.3.1. *Let $\lambda \in C^{+}$, and $\mu$ a $\Delta^{+}$-dominant integral element of* $\mathfrak{l}^{*}$. *Then*:

(i) *If $\alpha \in \mathscr{A}^{[\lambda]}$ and $\Phi^{\lambda}_{\lambda+\mu}(A) = 0$, then $A = 0$*

(ii) *If $\lambda$ and $\lambda+\mu$ have identical stabilizers in $\mathfrak{w}(\mathfrak{g})$, then*

$$\Phi^{\lambda}_{\lambda+\mu}(\mathscr{A}^{[\lambda]} \to \mathscr{A}^{[\lambda+\mu]}) \qquad \textit{and} \qquad \Psi^{\lambda+\mu}_{\lambda}(\mathscr{A}^{[\lambda+\mu]} \to \mathscr{A}^{[\lambda]})$$

*are isomorphisms of categories, and each is a natural inverse of the other.*[1]

DEFINITION 10.3.2. A module $M \in \mathscr{A}$ is said to be primary if it is either 0, or if it is nonzero and if all its composition factors in a Jordan–Hölder series are mutually equivalent.

Recall that $M$ has a finite Jordan–Hölder series (see Dixmier [1]). Zuckerman's second result is now the following:

THEOREM 10.3.3. *Let notation be as above. Then*:

(i) $\Psi^{\lambda+\mu}_{\lambda}$ *maps primary modules to primary modules*

(ii) *If $A \in \mathscr{A}^{[\lambda]}$ is irreducible, every composition factor $B$ of $\Phi^{\lambda}_{\lambda+\mu}(A)$ must map under $\Psi^{\lambda+\mu}_{\lambda}$ to a primary module whose composition factors are all equivalent to $A$; and $\Psi^{\lambda+\mu}_{\lambda}(B) \neq 0$ for at least one $B$.*

(iii) *If $A \in \mathscr{A}^{[\lambda]}$ is irreducible, and $B \in \mathscr{A}^{[\lambda+\mu]}$ is an irreducible module so that $\Psi^{\lambda+\mu}_{\lambda}(B)$ is nonzero and has all composition factors equivalent to $A$, then $B$ is a composition factor of $\Phi^{\lambda}_{\lambda+\mu}(A)$.*

We refer to Zuckerman for the proofs. They use the global character theory of Harish-Chandra. Since, by the subquotient theorem, any irreducible module in $\mathscr{A}$ has a global character, one can define global characters for all the modules of $\mathscr{A}$. If $\theta(M)$ is the global character of $M \in \mathscr{A}$, then Zuckerman's basic lemma is the result that for any $A \in \mathscr{A}^{[\lambda]}$, the length of the module $\Psi^{\lambda+\mu}_{\lambda}(\Phi^{\lambda}_{\lambda+\mu}(A))$ is $m = [\mathfrak{w}(\lambda) : \mathfrak{w}(\lambda+\mu)]$ and that

---

[1] This means that each of the composite functors $\Psi^{\lambda+\mu}_{\lambda} \circ \Phi^{\lambda}_{\lambda+\mu}$ and $\Phi^{\lambda}_{\lambda+\mu} \circ \Psi^{\lambda+\mu}_{\lambda}$ is naturally equivalent to the identity functor. Also note that if $\lambda \in C^{+}$ and $\mu \in \mathfrak{l}^{*}$ is $\Delta^{+}$-dominant integral, the stabilizer of $\lambda+\mu$ in $\mathfrak{w}(\mathfrak{g})$ is contained in the stabilizer of $\lambda$ in $\mathfrak{w}(\mathfrak{g})$.

all composition factors of $\Psi_{\lambda}^{\lambda+\mu}(\Phi_{\lambda+\mu}^{\lambda}(A))$ are equivalent to $A$; here, for any $\nu \in \mathfrak{l}^*$, $\mathfrak{w}(\nu)$ denotes the stabilizer of $\nu$ in $\mathfrak{w}(\mathfrak{g})$.

Unfortunately, these results are not completely adequate for our purposes since, the 'initial' parameter $\lambda$ and the 'final' parameter $\lambda+\mu$, are in the same chamber. It is necessary to consider the case when they lie in different chambers. Of course we can confine ourselves to the situation where they belong to *adjacent* chambers. Enright [1] has formulated one such result.

PROPOSITION 10.3.4. *Let $R$ be a positive system of roots of* $(\mathfrak{g}, \mathfrak{l})$; *let $\gamma \in R$ be a simple root and let $R' = s_\gamma R$. Let $c > 0$ be a constant. Write $C_R$* (resp. $C_{R'}$) *for the chamber associated with $R$* (resp. $R'$), *so that $\lambda \in C_R$ if and only if* $\langle \mathrm{Re}(\lambda), \alpha \rangle \geqslant 0\ \forall \alpha \in R$. Let $\mathfrak{S}_c(R, R')$ *be the subset of* $\mathfrak{l}^* \times \mathfrak{l}^*$ *of all* $(\lambda, \lambda')$ *such that*

(a) $\lambda \in C_R$, $\lambda' \in C_{R'}$
(b) $\lambda$ *and* $\lambda'$ *are both regular*
(c) $\lambda - \lambda'$ *is integral*
(d) $\|\lambda - \lambda'\| \leqslant c$, $|\lambda(\bar{H}_\gamma)| \leqslant c$, $|\lambda'(\bar{H}_\gamma)| \leqslant c$, *and*
(e) $\lambda(\bar{H}_\gamma)$ (*hence also* $\lambda'(\bar{H}_\gamma)$ *by* (c)) *is not an integer.*

*Then we can find a constant $c' > 0$ with the following property. Suppose* $(\lambda, \lambda') \in \mathfrak{S}_c(R, R')$ *and*

$$\langle \mathrm{Re}(\lambda), \beta \rangle > c', \qquad \langle \mathrm{Re}(\lambda'), \beta \rangle > c', \qquad \beta \in R \cap R';$$

*then the functors*

$$\Phi_{\lambda'}^{\lambda} = p_{\lambda'} \circ F^{\lambda' - \lambda} \otimes (\,) \circ p_\lambda,$$
$$\Psi_{\lambda}^{\lambda'} = p_{\lambda} \circ F^{\lambda - \lambda'} \otimes (\,) \circ p_{\lambda'},$$

*are isomorphisms from* $\mathscr{A}^{\lambda]}$ *to* $\mathscr{A}^{\lambda']}$ *and vice versa, and each is a natural inverse of the other.*

The proof is a combination of Zuckerman's ideas based on global character theory and the technique introduced in the proof of Theorem 10.2.2.

## 11. The main theorems

We now have all the results that are needed to complete the proofs of the main theorems. Our notation is the usual one. In particular, $\mathfrak{g}$, $\mathfrak{k}$, $\mathfrak{h}$, $P$, $P_{\mathfrak{k}}$ are as in Section 9.

**11.1.** *The irreducibility of the modules $W_{P,\lambda}$.* The first issue is of course to prove the irreducibility of $W_{P,\lambda}$ under the least restrictive conditions. We begin with

THEOREM 11.1.1. *Let $\lambda \in \mathfrak{h}^*$, $\lambda_1 = \lambda|_{\mathfrak{h}_1}$. Assume that*

(a) *$\lambda_1$ is $\Delta_{\mathfrak{k}}$-integral.*

(b) *$V_{\mathfrak{g},P,\lambda}$ is irreducible.*

*Then*

$$V_{\mathfrak{g},P,\lambda} \in {}^*\mathscr{J}_{\mathfrak{g}}(\mathfrak{g}, \mathfrak{h}_1, P_{\mathfrak{k}}).$$

The point is that for any integer $m \gg 0$, $\lambda' = \lambda - m\delta_P$ is in good position, $V_{\mathfrak{g},P,\lambda'}$ is irreducible, and $\lambda' + \delta_P$ is regular. So, by Theorem 9.3.6, $V_{\mathfrak{g},P,\lambda'}$ lies in ${}^*\mathscr{J}_{\mathfrak{g}}(\mathfrak{k})$. By Corollary 10.2.3, there is a finite-dimensional $\mathfrak{g}$-module $F$ such that $V_{\mathfrak{g},P,\lambda}$ is a direct summand of $F \otimes V_{\mathfrak{g},P,\lambda'}$. On the other hand, $(V_{\mathfrak{g},P,\lambda'})_{\mathfrak{k}}$ is a direct summand of $S(\mathfrak{p}) \otimes V_{\mathfrak{k},P_{\mathfrak{k}},\lambda'_1}$, by Theorem 9.2.1; the suffix $\mathfrak{k}$ denotes as usual the underlying $\mathfrak{k}$-module. Hence $(V_{\mathfrak{g},P,\lambda})_{\mathfrak{k}}$ is a direct summand of $(F_{\mathfrak{k}} \otimes S(\mathfrak{p})) \otimes V_{\mathfrak{k},P_{\mathfrak{k}},\lambda'_1}$. Theorem 9.3.5 now guarantees that $(V_{\mathfrak{g},P,\lambda})_{\mathfrak{k}} \in {}^*\mathscr{J}(\mathfrak{k})$.

For the irreducibility of $W_{P,\lambda}$ we have the following very general result; it is the main theorem in Enright [1].

THEOREM 11.1.2. *Let $\lambda \in \mathfrak{h}^*$, $\lambda = \lambda|_{\mathfrak{h}_1}$. Assume that $\lambda_1$ is $\Delta_{\mathfrak{k}}$-integral, and that $V_{\mathfrak{g},P,\lambda}$ is irreducible. Let $W_{P,\lambda} = \tau(V_{\mathfrak{g},P,\lambda})$. Then*:

(i) *$W_{P,\lambda}$ belongs to $\mathscr{A}^{[\lambda+\delta_P]}$ and is possibly 0 but is always primary* (*see Definition* 10.3.2).

(ii) *If $\lambda + \delta_P$ is regular, $W_{P,\lambda}$ is nonzero and irreducible.*

(iii) *If $t'_0 \cdot \lambda_1$ is $P_{\mathfrak{k}}$-dominant integral, $W_{P,\lambda}$ is nonzero and irreducible. In this case, $W_{P,\lambda}[\mu_1]^{\mathfrak{n}_{\mathfrak{k}}^+}$ is one-dimensional; the irreducible representation of $\mathfrak{k}$ with $P_{\mathfrak{k}}$-highest weight $\mu_1$ occurs exactly once in $W_{P,\lambda}$; and on the corresponding isotypical space, $U(\mathfrak{g})^{\mathfrak{k}}$ acts via the homomorphism $\boldsymbol{\beta}_{P,\lambda_1}$.*

Consider first the context of (iii). The results of Section 6.1 make it clear that $V_\lambda[\lambda_1]^{\mathfrak{n}_{\mathfrak{k}}^+}$ is one-dimensional and $U(\mathfrak{g})^{\mathfrak{k}}$ acts on this one-dimensional space via $\boldsymbol{\beta}_{P,\lambda_1}$ ($V_\lambda = V_{\mathfrak{g},P,\lambda}$). If we now use Proposition 4.9.6, we get the second assertion of (iii). If we now know that all the composition factors of $W_{P,\lambda}$ are equivalent, the irreducibility of $W_{P,\lambda}$ would follow from the relation $\dim(W_{P,\lambda}[\mu_1]^{\mathfrak{n}_{\mathfrak{k}}^+}) = 1$. So it remains to prove (i) and (ii).

Also (i) follows from (ii). To see this, write $\delta = \delta_P$ and let notation be as in Section 10. Since $V_{\mathfrak{g},P,\lambda} = V^{[\lambda+\delta]}_{\mathfrak{g},P,\lambda}$, it is easy to see that $V_{P,\lambda,1} = V^{[\lambda+\delta]}_{P,\lambda,1}$ and so

$W_{P,\lambda}=W_{P,\lambda}^{[\lambda+\delta]}$. For $\lambda$ in good position, $W_{P,\lambda}$ is irreducible by Theorem 9.4.3. In the general case, we can find $\lambda'$ in good position such that $V_{\mathfrak{g},P,\lambda'}$ is irreducible and $V_{\mathfrak{g},P,\lambda}$ is a direct summand of $F\otimes V_{\mathfrak{g},P,\lambda'}$ for a suitable finite-dimensional $\mathfrak{g}$-module (Corollary 10.2.3). So $W_{P,\lambda}$ is a direct summand of $F\otimes W_{P,\lambda'}$, hence finitely generated as a $U(\mathfrak{g})$-module. So $W_{P,\lambda}\in \mathscr{A}^{[\lambda+\delta]}$. Let now $R$ be the positive system of roots determined by the requirement: $\alpha\in R$ if and only if either $(\lambda+\delta)(\bar{H}_\alpha)>0$, or $(\lambda+\delta)(\bar{H}_\alpha)=0$ and $-\alpha\in P$. Let $\lambda'=\lambda+\delta_R$. Then $\lambda'+\delta$ is regular, belongs to $C_R$, and $V_{\lambda'}$ is irreducible. So, by (ii), $W_{P,\lambda}$ is irreducible. However, from the work in Section 10.2 we have

$$(F^{-\delta_R}\otimes V_{\lambda'})^{[\lambda+\delta]}\simeq V_\lambda$$

so that

$$W_{P,\lambda}\simeq(F^{-\delta_R}\otimes W_{P,\lambda'})^{[\lambda+\delta]}\simeq \Psi_{\lambda+\delta}^{\lambda+\delta+\delta_R}(W_{P,\lambda'}).$$

From Zuckerman's result (Theorem 10.3.3) we infer that $W_{P,\lambda}$ is primary.

So it remains to prove (ii). Let $\Lambda$ be the set of all $\lambda\in\mathfrak{h}^*$ having the following properties: (a) $\lambda_1$ is $\Delta_{\mathfrak{k}}$-integral, (b) $\lambda+\delta$ is regular, (c) $V_\lambda$ is irreducible; for any positive system $R$ of roots of $(\mathfrak{g}, \mathfrak{h})$ let $\Lambda(R)$ be the subset of $\lambda\in\Lambda$ with $\lambda+\delta\in C_R$. Write $A(\lambda)$ for the assertion (ii) for $\lambda$. The proof that $A(\lambda)$ is true for all $\lambda\in\Lambda$ is quite analogous to the proof of Theorem 10.2.2.

*Step I*: *$A(\lambda)\Leftrightarrow A(\lambda+\nu)$ if $\lambda\in\Lambda(R)$ and $\nu$ is $R$-dominant integral.* Here $\lambda+\delta$ and $\lambda+\nu+\delta$ are both regular and in $C_R$. So, by (ii) of Theorem 10.3.1, $\Phi_{\lambda+\delta+\nu}^{\lambda+\delta}$ is an isomorphism from the category $\mathscr{A}^{[\lambda+\delta]}$ to the category $\mathscr{A}^{[\lambda+\delta+\nu]}$. But, as

$$W_{P,\lambda+\nu}=\tau(V_{\lambda+\nu})\simeq\tau((F^\nu\otimes V_\lambda)^{[\lambda+\nu+\delta]})\simeq\Phi_{\lambda+\delta+\nu}^{\lambda+\delta}(W_{P,\lambda})$$

it is clear that $A(\lambda+\nu)\Leftrightarrow A(\lambda)$.

*Step II*: *$A(\lambda)\Leftrightarrow A(\lambda')$ if $\lambda, \lambda'\in\Lambda(R)$ and $\lambda-\lambda'$ is integral.* We can write $\lambda-\lambda'=\nu'-\nu$ where $\nu$ and $\nu'$ are $R$-dominant integral. As $\lambda+\nu=\lambda'+\nu'$, $A(\lambda)\Leftrightarrow A(\lambda+\nu)=A(\lambda'+\nu')\Leftrightarrow A(\lambda')$, by Step I.

*Step III*: *$A(\lambda)$ is true $\forall\lambda\in\Lambda(-P)$.* Whatever be $\lambda$, $\lambda-m\delta\in\Lambda(-P)$ if $m$ is an integer $\gg 0$; and, for $m\gg 0$ we also have the fact that $\lambda-m\delta$ is in good position. So $A(\lambda-m\delta)$ is true for integers $m\gg 0$, by Theorem 9.4.3; the validity of $A(\lambda)$ for $\lambda\in\Lambda(-P)$ is now clear from Step II.

*Step IV: $A(\lambda) \Leftrightarrow A(\lambda')$ if $\lambda \in \Lambda(R')$ and $\lambda - \lambda'$ is integral, $R$ and $R'$ being adjacent.* Since $R$ and $R'$ are adjacent, $R' = s_\gamma R$ where $\gamma$ (resp. $-\gamma$) is simple in $R$ (resp. $R'$). Since both $V_\lambda$ and $V_{\lambda'}$ are irreducible it follows that $\lambda(\bar{H}_\gamma)$ and $\lambda'(\bar{H}_\gamma)$ are *not integers*. Write $\lambda = \bar{\lambda} + v, \bar{\lambda}' + s_\gamma v$ where $v$ is integral and varies in such a way that $\langle v, \gamma \rangle = O(1)$ and $\langle v, \beta \rangle \to +\infty$ for each $\beta \in R \cap R'$. Proposition 10.3.4 is now applicable and enables us to conclude the following: provided $\min_{\beta \in R \cap R'} \langle v, \beta \rangle$ is sufficiently large, $\Phi_{\bar{\lambda}'}^{\bar{\lambda}}$ is an isomorphism of the category $\mathscr{A}^{[\bar{\lambda}]}$ onto the category $\mathscr{A}^{[\bar{\lambda}']}$. On the other hand (cf. proof of Theorem 10.2.2),

$$(F^{\bar{\lambda}' - \bar{\lambda}} \otimes V_{\bar{\lambda}})^{[\bar{\lambda}' + \delta]} \simeq V_{\bar{\lambda}'}$$

so that

$$\Phi_{\bar{\lambda}'}^{\bar{\lambda}}(W_{P,\bar{\lambda}}) \simeq W_{P,\bar{\lambda}'}.$$

This shows that $A(\bar{\lambda}) \Leftrightarrow A(\bar{\lambda}')$. So $A(\lambda) \Leftrightarrow A(\lambda')$, since, by Step I, $A(\lambda) \Leftrightarrow A(\bar{\lambda})$ and $A(\lambda') \Leftrightarrow A(\bar{\lambda}')$.

*Step V: $A(\lambda)$ is true $\forall \lambda \in \Lambda$.* This is proved for $\lambda \in \Lambda(R)$ by induction on $|P \cap R|$. For $|P \cap R| = 0$, $R = -P$ and this is Step III. If $|R \cap P| > 0$, $R \neq -P$ and so $\exists$ simple $\gamma \in R$ such that $\lambda(\bar{H}_\gamma)$ is *not* an integer. Let $R' = s_\gamma R$. If the integer $m$ is $\gg 0$, $\lambda' = \lambda + m\delta_{R'}$ lies in $\Lambda(R')$. As $|P \cap R'| < |P \cap R|$, $A(\lambda')$ is true. So $A(\lambda)$ is true by Step IV.

Similar, and in fact simpler arguments prove

THEOREM 11.1.3. *Let $\lambda \in \mathfrak{h}^*$, $\lambda_1 = \lambda|_{\mathfrak{h}_1}$ and assume that*

(a) *$\lambda_1$ is $\Delta_{\mathfrak{l}}$-integral.*
(b) *$V_{\mathfrak{g},P,\lambda}$ is irreducible.*
(c) *$W_{P,\lambda} \neq 0$.*

*Then $W_{P,\lambda}$ admits a symmetric nondegenerate invariant form.*

Theorem 9.4.3 implies this for $\lambda - m\delta$ when $m$ is an integer $\gg 0$. On the other hand it is obvious that if a module $B \in \mathscr{A}$ admits a symmetric nondegenerate form, the same is true for the $B^{[v]}$. Theorem 10.2.2 then enables us to go from $\lambda - m\delta$ to $\lambda$.

**11.2.** *Resolutions of $W_{P,\lambda}$ in terms of $V_{P,\lambda,s}$.* We now turn to the question of constructing resolutions for the modules $W_{P,\lambda}$. Let the notation be as in Section 9.5. We have the following theorem.

THEOREM 11.2.1. *Let $\lambda \in \mathfrak{h}^*$, $\lambda_1 = \lambda|_{\mathfrak{h}_1}$. Assume that* (a) $\lambda_1$ is $\Delta_{\mathfrak{k}}$*-integral and* (b) $V_{\mathfrak{g},P,\lambda}$ *is irreducible. Let* $(V_{P,\lambda,s})$ *be the lattice above* $V_{\mathfrak{g},P,\lambda}$. *Then:*

(i) *If* $s_1, s_2 \in \mathfrak{w}(\mathfrak{k})$ *and* $s_1 \leqslant s_2$, *then* $V_{P,\lambda,s_1} \subset V_{P,\lambda,s_2}$.

(ii) *Define the* $\mathfrak{g}$*-modules* $\mathscr{V}_\lambda^{(i)}$ *by*

$$\mathscr{V}_\lambda^{(0)} = V_{P,\lambda,1}, \quad \mathscr{V}_\lambda^{(i)} = \bigoplus_{l(s)=i} V_{P,\lambda,s}, \qquad (1 \leqslant i \leqslant r),$$

*and let the* $\mathfrak{g}$*-module maps*

$$d_i \colon \mathscr{V}_\lambda^{(i)} \to \mathscr{V}_\lambda^{(i-1)}$$

*be defined by the matrices* $\mathbb{D}_i$ *occurring in the* B–G–G *resolution. Then*

$$0 \to \mathscr{V}_\lambda^{(R)} \xrightarrow{d_r} \mathscr{V}_\lambda^{(r-1)} \to \cdots \xrightarrow{d_1} \mathscr{V}_\lambda^{(0)} \xrightarrow{\varepsilon} W_{P,\lambda} \to 0$$

*is an exact resolution,* $\varepsilon$ *being the natural surjection.*

By Theorem 9.5.1 this is true for $\lambda$ in good position. For integers $m \gg 0$ consider $\lambda' = \lambda - m\delta$. Then the above remarks show that the theorem is true for $\lambda'$. By Corollary 10.2.3 $V_\lambda$ is a direct summand of $F \otimes V_{\lambda'}$ for a suitable finite dimensional $\mathfrak{g}$-module $F$, i.e.,

$$F \otimes V_{\lambda'} = A \oplus V_\lambda$$

where $A$ is a sub $\mathfrak{g}$-module. Clearly $A \in \mathscr{J}_{\mathfrak{g}}(\mathfrak{k})$ and $(F \otimes V_{\lambda'})_s = A_s \oplus V_{P,\lambda,s}$ $\forall s \in \mathfrak{w}(\mathfrak{k})$. This gives

$$A_{s_1} \subset A_{s_2}, \quad V_{P,\lambda,s_1} \subset V_{P,\lambda,s_2}, \qquad (s_1 \leqslant s_2),$$

and shows that the complex associated with $F \otimes V_{\lambda'}$ is the direct sum of the complexes associated with $A$ and $V_\lambda$, respectively; since the former is exact, the two latter ones are also exact complexes.

**11.3.** *The* $\mathfrak{k}$*-multiplicities for the modules* $W_{P,\lambda}$. Let $\mathscr{E}(\mathfrak{k})$ be the set of equivalence classes of irreducible finite-dimensional representations of $\mathfrak{k}$. For any $P_{\mathfrak{k}}$-dominant integral $\nu \in \mathfrak{h}_1^*$ let $\vartheta(\nu)$ denote the element of $\mathscr{E}(\mathfrak{k})$ corresponding to the irreducible representation of $P_{\mathfrak{k}}$-highest weight $\nu$.

Suppose that $A$ is a $\mathfrak{g}$-module belonging to $\mathscr{J}_{\mathfrak{g}}(\mathfrak{k})$. Assume that $\dim(A[\nu]_{\mathfrak{l}}^{\mathfrak{n}^+}) < \infty$ for all integral $\nu \in \mathfrak{h}_1^*$. Then, the modules $A_s$ $(s \in \mathfrak{w}(\mathfrak{k}))$ of the lattice above $A$ have the following property: if $\nu \in \mathfrak{h}_1^*$ is $P_{\mathfrak{k}}$-dominant integral,

$$\dim(A_s[\nu]^{\mathfrak{n}\mathfrak{k}^+}) = \dim(A[(t_0 s^{-1})^{\check{}} \cdot \nu]^{\mathfrak{n}\mathfrak{k}^+}) < \infty.$$

(Proposition 4.7.2). Suppose further that one has an exact resolution

$$0 \to \mathscr{B}^{(r)} \to \mathscr{B}^{(r-1)} \to \cdots \to \mathscr{B}^{(0)} \to \tau(A) \to 0$$

where

$$\mathscr{B}^{(0)} = A_1, \qquad \mathscr{B}^{(i)} = \bigoplus_{l(s)=i} A_s, \qquad (1 \leqslant i \leqslant r).$$

Thus, for any $P_{\mathfrak{k}}$-dominant integral $\nu \in \mathfrak{h}_1^*$ we have (Lemma 2.2.12) the exact sequence of *finite-dimensional* vector spaces

$$0 \to \mathscr{B}^{(r)}[\nu]^{\mathfrak{n}\mathfrak{k}^+} \to \mathscr{B}^{(r-1)}[\nu]^{\mathfrak{n}\mathfrak{k}^+} \to \cdots \to \mathscr{B}^{(0)}[\nu]^{\mathfrak{n}\mathfrak{k}^+} \to \tau(A)[\nu]^{\mathfrak{n}\mathfrak{k}^+} \to 0.$$

Consequently we find, for the multiplicity $[\tau(A): \vartheta(\nu)]$ with which $\vartheta(\nu)$ occurs in $\tau(A)$, the formula

$$\text{(MF)} \qquad [\tau(A): \vartheta(\nu)] = \sum_{s \in \mathfrak{w}(\mathfrak{k})} \varepsilon(s) \dim(A[(t_0 s)' \cdot \nu]^{\mathfrak{n}\mathfrak{k}^+}).$$

This formula reduces our problem to that of calculating the dimensions that appear on the right side when $A$ is a $\mathfrak{g}$-Verma module.

DEFINITION 11.3.1. Let $E = \{\alpha_1, \ldots, \alpha_m\}$ be a sequence of elements of $\mathfrak{h}_1^*$ with the property that for some $H \in \mathfrak{h}_1$, $\alpha_i(H) > 0 \ \forall i = 1, \ldots, m$. Then the partition function $\mathbb{P}_E$ is the function from $\mathfrak{h}_1^*$ to the nonnegative integers given by

$$\mathbb{P}_E(\nu) = \text{number of } m\text{-tuples } (a_1, \ldots, a_m) \text{ of integers } a_i \geqslant 0 \text{ with } \nu = a_1\alpha_1 + \cdots + a_m\alpha_m,$$

for all $\nu \in \mathfrak{h}_1^*$.

It is clear that $\mathbb{P}_E(\nu)$ is finite for all $\nu$.

If $\mathrm{rk}(\mathfrak{g}) = \mathrm{rk}(\mathfrak{k})$, it is usual to define $P_n$ as $P \backslash P_{\mathfrak{k}}$, i.e., as the set of noncompact roots. In the more general case under consideration we define $P_n$ as follows. Since the subspace $\mathfrak{p}$ of the Cartan decomposition $\mathfrak{g} = \mathfrak{k} \oplus \mathfrak{p}$ is stable under $\mathrm{ad}(\mathfrak{h}_1)$, it makes sense to speak, for any $\nu \in \mathfrak{h}_1^*$, the multiplicity with which $\nu$ occurs in the splitting of $\mathfrak{p}$ with respect to $\mathrm{ad}(\mathfrak{h}_1)$. Let $\nu'_1, \ldots, \nu'_p$ be the (distinct) elements of $\mathfrak{h}_1^*$ which are of the form $\alpha|_{\mathfrak{h}_1}$ for some $\alpha \in P$ and which occur in $\mathfrak{p}$. Then we define

$$P_n = \{\nu_1, \ldots, \nu_m\}$$

where each $\nu_i$ is a $\nu'_j$, and $\nu'_j$ occurs in $P_n$ exactly as many times as its multiplicity in $\mathfrak{p}$. It is obvious that $\mathbb{P}_{\pm sP_n}$ is well defined $\forall s \in \mathfrak{w}(\mathfrak{k})$.

The key result for proving the Blattner formulae is the following.

PROPOSITION 11.3.2. *Let* $\lambda \in \mathfrak{h}^*$ *and* $\lambda_1|_{\mathfrak{h}_1}$. *Assume that* (a) $\lambda_1$ *is* $\Delta_{\mathfrak{k}}$*-integral,* (b) $V_{\mathfrak{g},P,\lambda}$ *is irreducible. Then, for any* $\nu \in \mathfrak{h}_1^*$,

$$\dim(V_{\mathfrak{g},P,\lambda}[\nu]^{\mathfrak{n}_{\mathfrak{k}}^+} = \mathbb{P}_{-P_n}(\nu-\lambda_1).$$

The proof is by the method of invariant forms. Let $\{\alpha_1, \ldots, \alpha_d\}$ be a distinct enumeration of the roots in $P_{\mathfrak{k}}$ and let $P_n = \{\nu_1, \ldots, \nu_m\}$ be as above. Let $X_i$ be a root vector in $\mathfrak{k}_{\alpha_i}$ and $Y_i = X_i^\sigma$; here $\sigma$ is an involutive antiautomorphism of $\mathfrak{g}$ that leaves $\mathfrak{k}$ stable and is the identity on $\mathfrak{h}$. Let $Z_j$ $(1 \leqslant j \leqslant m)$ be a set of linearly independent vectors in $\mathfrak{p}$ with $Z_j$ of weight $-\nu_j$ (with respect to $\mathfrak{h}_1$). Then $\{Y_1, \ldots, Y_d, Z_1, \ldots, Z_m\}$ is a basis for $\mathfrak{n}^- = \Sigma_{\alpha \in -P}\mathfrak{g}_\alpha$. The elements

$$Y_1^{a_1} \ldots Y_d^{a_d} Z_1^{b_1} \ldots Z_m^{b_m} \cdot v_{\lambda_1} = v_{\underline{a},\underline{b}}$$

then form a *basis* for $V = V_{\mathfrak{g},P,\lambda}$; $v_{\underline{a},\underline{b}}$ is of weight $\lambda_1 = \underline{a} \cdot \underline{\alpha} - \underline{b} \cdot \underline{\nu}$ where $\underline{a} \cdot \underline{\alpha} = a_1\alpha_1 + \cdots + a_d\alpha_d$ and $\underline{b} \cdot \underline{\nu} = b_1\nu_1 + \cdots + b_m\nu_m$. Put

$$V^- = \sum_{|\underline{b}| \geqslant 0} \mathbb{C}v_{\underline{0},\underline{b}} V^+ = \sum_{|\underline{a}| > 0,\, |\underline{b}| \geqslant 0} \mathbb{C}v_{\underline{a},\underline{b}}$$

where, as is customary, $|\underline{b}| = b_1 + \cdots + b_m$, $|\underline{a}| = a_1 + \cdots + a_d$. It is clear that $V = V^+ \oplus V^-$; moreover, an easy argument shows that

$$V^+ = \sum_{1 \leqslant i \leqslant d} Y_i V.$$

Hence, as $Y_i = X_i^\sigma$ for $1 \leqslant i \leqslant d$, we get easily

$$(V^+)^\perp = V^{\mathfrak{n}_{\mathfrak{k}}^+}$$

using the nondegeneracy of the invariant form on $V$. So, for any $\nu \in \mathfrak{h}_1^*$,

$$\begin{aligned}\dim(V[\nu]^{\mathfrak{n}_{\mathfrak{k}}^+}) &= \dim((V^+)^\perp[\nu]^{\mathfrak{n}_{\mathfrak{k}}^+}) \\ &= \dim(V^-[\nu]^{\mathfrak{n}_{\mathfrak{k}}^+}) \\ &= \mathbb{P}_{-P_n}(\nu-\lambda_1).\end{aligned}$$

From this and the formula (MF) we get Enright's version of the Blattner formulae [1]:

THEOREM 11.3.3. *Let* $\lambda \in \mathfrak{h}^*$, $\lambda_1 = \lambda|_{\mathfrak{h}_1}$. *Assume that* $\lambda_1$ *is* $\Delta_{\mathfrak{k}}$*-integral and that* $V_{\mathfrak{g},P,\lambda}$ *is irreducible. Then, for any* $\nu \in \mathfrak{h}_1^*$ *which is* $P_{\mathfrak{k}}$*-dominant integral, we have the formula*

$$[W_{P,\lambda}: \vartheta(\nu)] = \sum_{s \in \mathfrak{w}(\mathfrak{k})} \varepsilon(s) \mathbb{P}_{-P_n}((t_0 s)(\nu + \delta_{P_{\mathfrak{k}}}) - (\lambda_1 + \delta_{P_{\mathfrak{k}}}))$$

$$= \sum_{s \in \mathfrak{w}(\mathfrak{k})} \varepsilon(s) \mathbb{P}_{-t_0 P_n}(s(\nu + \delta_{P_{\mathfrak{k}}}) - t_0(\lambda_1 + \delta_{P_{\mathfrak{k}}})).$$

COROLLARY 11.3.4. *Let* $\mathrm{rk}(\mathfrak{g}) = \mathrm{rk}(\mathfrak{k})$. *Suppose that* $\lambda$ *is as above and that for some P-simple root* $\beta \in P_{\mathfrak{k}}$ *we have* $\langle \lambda + \delta_P, \beta \rangle = 0$. *Then*

$$W_{P,\lambda} = 0.$$

Note that $\delta_P(\bar{H}_\beta) = \delta_{P_{\mathfrak{k}}}(\bar{H}_\beta) = 1$ so that $\langle \lambda + \delta_P, \beta \rangle = \langle \lambda + \delta_{P_{\mathfrak{k}}}, \beta \rangle = 0$. Write $\check{P} = -t_0 P$, $\check{\beta} = -t_0 \beta$. Then $\check{P}_{\mathfrak{k}} = P_{\mathfrak{k}}$, $\check{\beta} \in P_{\mathfrak{k}}$ and $\check{\beta}$ is $\check{P}$-simple. So the Weyl reflexion $s_{\check{\beta}}$ leaves $\check{P}_n$ stable. As $\lambda + \delta_{P_{\mathfrak{k}}}$ is $s_\beta$-invariant, $t_0(\lambda + \delta_{P_{\mathfrak{k}}})$ is $s_{\check{\beta}}$-invariant. Hence, for any $\nu \in \mathfrak{h}^*$ which is $P_{\mathfrak{k}}$-dominant integral,

$$\mathbb{P}_{\check{P}_n}(s(\nu + \delta_{P_{\mathfrak{k}}}) - t_0(\lambda + \delta_{P_{\mathfrak{k}}})) = \mathbb{P}_{\check{P}_n}(\check{s}_\beta s(\nu + \delta_{P_{\mathfrak{k}}}) - t_0(\lambda + \delta_{P_{\mathfrak{k}}}))$$

while $\varepsilon(s_{\check{\beta}} s) = -\varepsilon(s)$. Hence $[W_{P,\lambda}: \vartheta(\nu)] = 0$. As $\nu$ is arbitrary, $W_{P,\lambda} = 0$.

COROLLARY 11.3.5. *Let* $\mathrm{rk}(\mathfrak{g}) = \mathrm{rk}(\mathfrak{k})$ *and let* $\lambda \in \mathfrak{h}^*$ *be such that* (a) $\lambda$ *is* $\Delta_{\mathfrak{k}}$*-integral,* (b) $\lambda + \delta_P$ *is* $-P$*-dominant. Then* $V_{\mathfrak{g},P,\lambda}$ *is irreducible and we have the dichotomy*:

*either* $W_{P,\lambda} = 0$, *which is equivalent to* $\langle \lambda + \delta_P, \beta \rangle = 0$ *for some P-simple* $\beta \in P_{\mathfrak{k}}$

*or* $W_{P,\lambda}$ *is nonzero and irreducible, which is equivalent to* $t_0' \cdot \lambda$ *being* $P_{\mathfrak{k}}$*-dominant integral.*

Since $(\lambda + \delta_P)(\bar{H}_\alpha)$ is $\leqslant 0$ $\forall\, \alpha \in P$, the irreducibility of $V_{\mathfrak{g},P,\lambda}$ is clear. If $\langle \lambda + \delta_P, \beta \rangle = 0$ for some $P$-simple $\beta \in P_{\mathfrak{k}}$, $W_{P,\lambda} = 0$ by the previous corollary. If $\gamma$ is $P_{\mathfrak{k}}$-simple, $\delta_P(\bar{H}_\gamma) \geqslant 1$ while $\delta_{P_{\mathfrak{k}}}(\bar{H}_\gamma) = 1$ so that $\delta_{P_n}(\bar{H}_\gamma) \geqslant 0$. Thus $(\lambda + \delta_{P_{\mathfrak{k}}})(\bar{H}_\gamma) \leqslant 0$ $\forall\, \gamma \in P_{\mathfrak{k}}$ showing that $\mu + \delta_{P_{\mathfrak{k}}}$ is $P_{\mathfrak{k}}$-dominant integral, $\mu$ being $= t_0' \cdot \lambda$. There are now two cases.

*Case I*: $\mu$ *itself is* $P_{\mathfrak{k}}$*-dominant integral*. By (iii) of Theorem 11.1.2, $W_{P,\lambda}$ is nonzero and irreducible.

*Case II*: $\mu + \delta_{P_{\mathfrak{k}}}$ *is* $\Delta_{\mathfrak{k}}$*-singular*. So $\exists$ a $P_{\mathfrak{k}}$-simple $\gamma$ such that $(\mu + \delta_{P_{\mathfrak{k}}})(\bar{H}_\gamma) = 0$.

Let $\beta = -t_0\gamma$. Then $\beta$ is $P_{\mathfrak{k}}$-simple and $\langle \lambda+\delta_{P_{\mathfrak{k}}}, \beta\rangle = 0$. As $\langle \lambda+\delta_P, \beta\rangle \leqslant 0$, we get $\langle \delta_{P_n}, \beta\rangle \leqslant 0$, hence $\langle \delta_{P_n}, \beta\rangle = 0$, hence $\delta_P(\bar{H}_\beta) = \delta_{P_{\mathfrak{k}}}(\bar{H}_\beta) = 1$. *So $\beta$ is P-simple* and $W_{P,\lambda} = 0$.

## 12. Representations of the Group

**12.1.** In this section I shall discuss very briefly the nature of the representations of the group that are defined by the $\mathfrak{g}$-modules $W_{P,\lambda}$. For reasons of brevity I shall confine myself to the case when $\mathrm{rk}(\mathfrak{g}) = \mathrm{rk}(\mathfrak{k})$ and indicate the connections between the $W_{P,\lambda}$ and the representations of the discrete and mock discrete series. It is possible to treat the fundamental series in a similar manner; see Enright and Wallach [2], we do not take up this case in these notes.

Let $G$ denote a connected s.s. Lie group with finite center and $K \subset G$ a maximal compact subgroup of $G$. We assume that $G$ lies above the real form of the associated simply connected complex group. $\mathfrak{g}_0$ is the Lie algebra of $G$, $\mathfrak{k}_0 \subset \mathfrak{g}_0$ the subalgebra defined by $K$. We also assume $\mathrm{rk}(G) = \mathrm{rk}(K)$ so that $G$ will have a discrete series. $\mathfrak{g}$ (resp. $\mathfrak{k}$) is the complexification of $\mathfrak{g}_0$ (resp. $\mathfrak{k}_0$). Let $\mathfrak{h}_0 \subset \mathfrak{k}_0$ be a CSA; $\mathfrak{h}$, its complexification. Let $T \subset K$ be the CSG corresponding to $\mathfrak{h}_0$. We identify the character group $\hat{T}$ of $T$ with a lattice $\mathbb{L} \subset (-1)^{1/2}\mathfrak{h}_0^*$; for $\nu \in \mathbb{L}$, let $\xi_\nu$ be the corresponding character of $T$, so that $\xi_\nu(\exp R) = e^{\nu(R)}$ $(R \in \mathfrak{h}_0)$. We write $\mathbb{L}'$ for the subset of regular elements of $\mathbb{L}$.

**12.2.** *The discrete series.* For the discrete series we have the following theorem.

THEOREM 12.2.1. *Let $\lambda \in \mathfrak{h}^*$ be $\Delta_{\mathfrak{k}}$-integral, and $P$, a positive system of roots of* $(\mathfrak{g}, \mathfrak{h})$. *Suppose that $\lambda + \delta_P$ is regular, belongs to $\mathbb{L}$, and is $-P$-dominant. Then $W_{P,\lambda}$ is infinitesimally equivalent to the discrete class* $\omega(\lambda+\delta_P)$ (cf. Section 1.5). *All discrete classes of G are obtained this way. Moreover, the class $\omega(\lambda+\delta_P)$ contains the irreducible representation of K having $P_{\mathfrak{k}}$-highest weight $\nu$ with the multiplicity*

$$\sum_{s \in \mathfrak{w}(\mathfrak{k})} \varepsilon(s) \mathbb{P}_{-t_0 P_n}(s(\nu + \delta_{P_{\mathfrak{k}}}) - t_0(\lambda + \delta_{P_{\mathfrak{k}}})).$$

The method of proof is to first establish the identification of $W_{P,\lambda'}$ with $\omega(\lambda'+\delta_P)$ when $\lambda'$ is suitably restricted, and then use tensoring by finite-dimensional representations to go over to the general case.

Let $\lambda' = \lambda - m\delta_P$, $m \gg 0$ being an integer. Then we have, writing $\delta = \delta_P$ for brevity,

$$(F^{m\delta} \otimes V_{\mathfrak{g},P,\lambda'})^{[\lambda+\delta]} \precsim V_{\mathfrak{g},P,\lambda}.$$

This gives

$$(F^{m\delta} \otimes W_{P,\lambda'})^{[\lambda+\delta]} \precsim W_{P,\lambda}.$$

It is a question of proving that $W_{P,\lambda'}$ is infinitesimally equivalent to $\omega(\lambda' + \delta_P)$; once this is done, the argument for $W_{P,\lambda}$ is as in Zuckerman's article [1]. Now $\lambda' + \delta_P$ being regular and $-P$-dominant, $W_{P,\lambda'}$ is irreducible; it is also clear that $W_{P,\lambda'}$ contains the class $\vartheta(\mu')$ where $\mu' = t'_0 \cdot \lambda'$ exactly once; and that if $W_{P,\lambda'}$ contains $\vartheta(\nu)$, then necessarily $\nu = \mu + \zeta$, $\zeta$ is a sum of roots in $-t_0 P$. Results of Wilfried Schmid [2] can now be used to show that $W_{P,\lambda'}$ is infinitesimally equivalent to $\omega(\lambda' + \delta_P)$; or else one can use the results of Enright-Wallach [1] (cf. Theorem 6.2) to reach the same conclusion.

The $K$-multiplicities of the discrete classes are now seen to be given by Theorem 11.3.3.

**12.3.** *The mock-discrete series.* Let $\Lambda \in \mathfrak{h}^*$ be such that (a) $\Lambda$ is $\Delta_{\mathfrak{k}}$-integral, (b) $\Lambda \in \mathbb{L}$, (c) $\Lambda$ is dominant with respect to a positive system $R$. If $\Lambda$ is singular, it can no longer correspond to a discrete class. Nevertheless one can define invariant eigendistributions $\Theta_{R,\Lambda}$, roughly speaking, as follows: one starts with the observation that the coefficients of the exponentials occurring in the character formulae of the discrete classes corresponding to the *regular* $\Lambda'$ that are dominant with respect to $R$, depend only on $R$ and not on $\Lambda'$; therefore, replacing $\Lambda'$ by $\Lambda$ in these formulae will still give an invariant eigendistribution on $G$. We denote it by $\Theta_{R,\Lambda}$ (see Harish-Chandra [4], Varadarajan [1]), and call it the *mock discrete character corresponding to* $\Lambda$ *and* $R$.

The question naturally arises whether the $\Theta_{R,\Lambda}$ are actually characters of representations, as the terminology suggests. Furthermore, it is possible that occasionally $\Theta_{R,\Lambda} = 0$; for instance, when $G$ is compact, this is true for all singular $\Lambda$. So there is also the problem of determining precisely when $\Theta_{R,\Lambda} = 0$. Enright [1] has settled completely both these questions.

THEOREM 12.3.1. *Let* $\lambda \in \mathfrak{h}^*$. *Assume that* (a) $\lambda$ *is* $\Delta_{\mathfrak{k}}$*-integral,* (b) $\lambda + \delta_P \in \mathbb{L}$, (c) $\lambda + \delta_P$ *is dominant with respect to* $-P$. *Then*:

(i) $\Theta_{-P,\lambda+\delta_P} = 0 \Leftrightarrow \langle \lambda + \delta_P, \beta \rangle = 0$ *for some* $P$*-simple root* $\beta \in P_{\mathfrak{k}}$.

(ii) *If* $\Theta_{-P,\lambda+\delta_P} \neq 0$, *it is the character of an irreducible representation which is infinitesimally equivalent to* $W_{P,\lambda}$. *Moreover, this representation contains the irreducible representation of K with highest weight* $\nu$ *(relative to* $P_{\mathfrak{k}}$*), with multiplicity given by*

$$\sum_{s \in \mathfrak{w}(\mathfrak{k})} \varepsilon(s) \mathbb{P}_{-t_0 P_n}(s(\nu + \delta_{P_{\mathfrak{k}}}) - t_0(\lambda + \delta_{P_{\mathfrak{k}}})).$$

By the corollaries to Theorem 11.3.3, $W_{P,\lambda} = 0$ if and only if $\langle \lambda + \delta_P, \beta \rangle = 0$ for some $P$-simple compact root $\beta$. On the other hand, the same argument as in the case of discrete series shows that $W_{P,\lambda}$ has character $\Theta_{-P,\lambda+\delta_P}$.

*Remark* 12.3.2. Under the assumption that $\lambda + \delta_P$ is regular with respect to $P_{\mathfrak{k}}$, Schmid [4] and Zuckerman [1] have shown that $\Theta_{-P,\lambda+\delta_P}$ is the character of an irreducible representation. Theorem 12.3.1 is clearly more general. For instance, let $G = SU(2, 1)$; if we take for $P$ a nonadmissible positive system, there are *no* $P$-simple compact roots, Thus Theorem 12.3.1 is applicable to the case $\lambda = -\delta_P$; however, $\lambda + \delta_P = 0$ is singular with respect to $P_{\mathfrak{k}}$.

## Acknowledgements

I would like to express my gratitude to Professors M. Cahen, M. De Wilde, and J. Wolf, who organized the Nato Advanced Study Institute at Liege and made our stay there so pleasant and rewarding. I would also like to express my deep indebtedness to Thomas Enright of the University of California at San Diego for the many long discussions I had with him concerning various aspects of his work; these conversations were mainly responsible for enabling me to comprehend fully his ideas and to present them in a cogent form at the Institute as well as in the present article.

I would also like to express my deep appreciation to Charlotte Johnson for her beautiful and fast typing and the patience with which she accepted all the demands I made.

*University of California, Los Angeles*

## REFERENCES


Atiyah, M. F., and Schmid, W.

1. 'A geometric construction of the discrete series for semisimple Lie groups', *Inventiones Math.* **42** (1977), 1–62.
2. 'The characters of semisimple Lie groups', Preprint.

Bargmann, W.

1. 'Irreducible unitary representations of the Lorentz group', *Ann. Math.* **48** (1947), 568–640.

Bernstein, I. N., Gel'fand, I. M., and Gel'fand, S. I.

1. 'The structure of representations generated by a vector of highest weight', *Functional Analysis. Appl.* **5** 1971), 1–9.
2. *Differential operators on the base affine space and a study of* $\mathfrak{g}$*-modules*, *Lie groups and their representations*, Summer School of the Bolyai J'anos Math. Soc., edited by I. M. Gel'fand, Halsted Press, Division of John Wiley and Sons, New York, 1975, 21-64.

Borel, A.

1. *Représentations de Groupes localement compacts*, Springer-Verlag Lecture notes, No. 276, Berlin, 1972.

Casselman, W.

1(a) *Matrix coefficients of admissible representations*, to appear.

(b) 'The $\mathfrak{n}$-cohomology of representations with an infinitesimal character', *Comp. Math.* **31** (1975), 219–227. (With Osborne, M. S.).

Dixmier, J.

1. *Algèbres enveloppantes*, Gauthier-Villars, Paris, 1974; English edition: *Enveloping algebras*, North Holland, Amsterdam, 1977.

Enright, T. J.

1. 'On the fundamental series of a real semisimple lie algebra, their irreducibility, resolutions and multiplicity formulae', *Ann. Math.* **110** (1979), 1–82.
2. 'On the construction and classification of irreducible Harish Chandra modules', *Proc. Natl. Acad. Sci. U.S.A.*, **75** (1978).
3. 'Algebraic construction and classification of Harish Chandra modules', Preprint.

Enright, T. J., and Varadarajan, V. S.

1. 'On an infinitesimal characterization of the discrete series', *Ann. Math.*, **102** (1975), 1–15.

Enright, T. J., and Wallach, N.

1. 'The fundamental series of representations of a real semisimple Lie algebra', to appear in *Acta Mathematica*.
2. 'The fundamental series of semisimple Lie algebras and semisimple Lie groups', Preprint. Fomin, A. I., and Shapovalov, N. N.
1. 'A property of the characters of irreducible representations of real semisimple Lie groups', *Funkt. Anal. i Ego. Priloz.* **8** (1974), 87–88; English translation: *Funct. Anal. Appl.*, **8** (1974), 270–271.

Harish-Chandra

1(a) 'Representations of a semisimple Lie group, I, II, III', *Trans. Amer. Math. Soc.* **75** (1953), 185–243; **76** (1954), 26–75; **76** (1954), 234–253.

(b) 'Representations of semisimple Lie groups, I–IV', *Proc. Natl. Acad. Sci. U.S.A.*, **37** (1951), 170–173; **37** (1951), 362–365; **37** (1951), 366–369; **37** (1951), 691–694.

(c) 'Some results on differential equations' (unpublished); 'Differential equations and semisimple Lie groups' (unpublished), Princeton, 1961.
(d) 'Some results on differential equations and their applications', *Proc. Natl. Acad. Sci., U.S.A.*, **45** (1959), 1763–1764.
(e) 'Plancherel formula for the $2 \times 2$ real unimodular group', *Proc. Natl. Acad. Sci., U.S.A.*, **38** (1952), 337–342.
2(a) 'Representations of semisimple Lie groups, IV', *Amer. J. Math.*, **77** (1955), 743–777.
(b) 'Representations of semisimple Lie groups, V', *Amer. J. Math.*, **78** (1956), 1–41.
(c) 'Representations of semisimple Lie groups, VI', *Amer. J. Math.*, **78** (1956), 564–628.
3. 'Invariant eigendistributions on a semisimple Lie group', *Trans. Amer. Math. Soc.*, **119** (1965), 457–508.
4. 'Discrete series for semisimple Lie groups, I, II', *Acta. Math.* **113** (1965), 241–318; **116** (1966), 1–111.
5(a) 'Harmonic analysis on real reductive groups, I. The theory of the constant term', *J. Functional Analysis* **19** (1975), 104–204.
(b) 'Harmonic analysis on real reductive groups, II', *Inventiones Math.* **36** (1975), 1–55.
(c) 'Harmonic analysis on real reductive groups, III', *Ann. Math.* **104** (1976), 117–201.

Hecht, H., and Schmid, W.
1. 'A proof of Blattner's conjecture', *Inventiones Math.* **31** (1975), 129–154.

Jantzen, J. C.
1. 'Kontravariente Formen auf induzierten Darstellungen halbeinfacher Lie-Algebran', *Math. Ann.* **226** (1977), 53–65.

Knapp, A., and Zuckerman, G.
1(a) 'Classification of irreducible tempered representations of semisimple Lie groups', *Proc. Natl. Acad. Sci., U.S.A.*, **73** (1976), 2178–2180.
(b) 'Classification theorems for representations of semisimple Lie groups', Preprint.

Kostant, B.
1. *On the existence and irreducibility of certain series of representations*, *Lie groups and their representations*, Summer School of the Bolyai J'anos Math. Soc., Edited by I. M. Gel'fand, Halsted Press, A division of John Wiley and Sons, 1975, 231–331.

Lang, S.
1. $SL(2, \mathbb{R})$, Addison-Wesley, Reading, Massachusetts, 1975.

Langlands, R. P.
1. 'On the classification of irreducible representations of real reductive groups', Mimeographed notes, Institute for Advanced Study, 1973.

Lepowsky, J.
1. 'Algebraic results on representations of semisimple Lie groups', *Trans. Amer. Math. Soc.*, **176** (1973), 1–44.

Parthasarathy, K. R., Ranga Rao, R., and Varadarajan, V. S.
1. 'Representations of complex semisimple Lie groups and Lie algebras', *Ann. Math.*, **85** (1967), 383–429.

Parthasarathy, R.
1. 'An algebraic construction of a class of representations of a semisimple Lie algebra', *Math. Ann.*, **226** (1977), 1–52.

Rossi, H., and Vergne, M.
1. 'Analytic continuation of the holomorphic discrete series of a semisimple Lie group', *Acta Math.* **136** (1976), 1–59.

Schmid, W.
1(a) 'On the characters of the discrete series (the Hermitian symmetric case)', *Inventiones Math.*, **30** (1975), 47–144.
(b) 'On the realization of the discrete series of a semisimple Lie group', *Rice University Studies*, **56** (1970), 99–108.
2. 'Some properties of square integrable representations of semisimple Lie groups', *Ann. Math.*, **102** (1975), 535–564.
3. 'On a conjecture of Langlands', *Ann. Math.*, **93** (1971), 1–42.
4. *Lectures given at the Institute for Advanced Study*, Princeton, 1976.

Trombi, P. C.
1. 'The tempered spectrum of a real semisimple Lie group', *Amer. J. Math.*, **99** (1977), 57–75.

Varadarajan, V. S.
1. *Harmonic analysis on real reductive groups*, Springer-Verlag Lecture notes, No. 576, Springer-Verlag, New York, 1976.
2. *Lie groups, Lie algebras, and their representations*, Prentice Hall, Englewood Cliffs, New Jersey, 1974.

Verma, Daya-na 'd
1. *Structure of certain induced representations of complex semisimple Lie algebras*, Thesis, Yale University (1966).
1. 'Structure of certain induced representations of complex semisimple Lie algebras', *Bull. Amer. Math. Soc.*, **74** (1968), 160–166, 628.

Warner, G.
1. *Harmonic analysis on semisimple Lie groups, I, II*, Springer-Verlag, New York, 1972.

Wallach, N.
1. 'On the Enright–Varadarajan modules: A construction of the discrete series', *Ann. Ecole Norm. Sup.*, **9** (1976), 171–195.
2. *On the unitarizability of representations with highest weights*, Springer Lecture Notes, No. 466 (1975), 226–231.

Zuckerman, G.
1. 'Tensor products of infinite dimensional and finite-dimensional representations of semisimple Lie groups', *Ann. Math.*, **106** (1977), 295–308.

P. C. TROMBI

# THE ROLE OF DIFFERENTIAL EQUATIONS IN THE PLANCHEREL THEOREM

## CONTENTS



## 1. INTRODUCTION

Let $G$ be a real reductive Lie group with the properties that: (1) $\mathrm{Ad}(G)\subset G_c$, $G_c$ the complex analytic adjoint group of $\mathfrak{g}_c$; (2) Let $G_1$ be the analytic subgroup of $G$ corresponding to $\mathfrak{g}_1=[\mathfrak{g}, \mathfrak{g}]$. Then the center of $G_1$ is finite; (3) $[G: G^0]<\infty$ (for any topological group $H$, $H^0$ will denote its connected component containing the identity). Choose and fix a maximal compact subgroup $K$ of $G$. It follows from condition (3) that we may choose $K$ so that $K$ meets every connected component of $G$. Let $\mathfrak{g}$ denote the Lie algebra of $G$ and $\mathfrak{k}$ the subalgebra of $\mathfrak{g}$ corresponding to $K$. There exists an involution $\theta$ of $\mathfrak{g}$ fixing the elements of $\mathfrak{k}$ which is a Cartan involution on $\mathfrak{g}_1=[\mathfrak{g}, \mathfrak{g}]$. In fact, let $\theta$ be a Cartan involution on $\mathfrak{g}_1$; if $C=\mathrm{center}(G^0)$ then $C=C_1C_2$ where $C_1=C\cap K$, and $C_2$ is a maximal vector subgroup of $C$; if $\mathfrak{l}=\mathfrak{l}_1+\mathfrak{l}_2$, $\mathfrak{l}_i=\mathrm{LA}(C_i)$ $(i=1, 2)$ then we extend $\theta$ to $\mathfrak{g}$ by $\theta(X_1+X_2)=X_1-X_2$ if $X_i\in\mathfrak{l}_i\,(i=1,2)$. We then may write $\mathfrak{g}=\mathfrak{k}+\mathfrak{s}$ (direct sum) where $\mathfrak{s}=\{X\in\mathfrak{g}: \theta(X)=-X\}$. Let $\mathfrak{a}_0\subset\mathfrak{s}$ be a maximal Abelian subspace, $\mathfrak{h}_0$ a $\theta$-stable Cartan subalgebra (CSA) of $\mathfrak{g}$ such that $\mathfrak{h}_0\cap\mathfrak{s}=\mathfrak{a}_0$. Choose compatible orderings on the duals of the real vector spaces $\mathfrak{a}_0$ and $i(\mathfrak{h}_0\cap\mathfrak{k})+\mathfrak{a}_0$. As each root $\alpha\in\Delta(\mathfrak{g}, \mathfrak{h}_0)$ is real valued on $i(\mathfrak{h}_0\cap\mathfrak{k})+\mathfrak{a}_0$ we obtain an ordering of $\Delta(\mathfrak{g}, \mathfrak{h}_0)$. Let $\mathbb{P}_0$ denote the set of positive roots relative to this ordering and set

$$\mathbb{P}_0^-=\{\alpha\in\mathbb{P}_0: \alpha|\mathfrak{a}_0\equiv 0\},\qquad \mathbb{P}_0^+=\mathbb{P}_0\backslash\mathbb{P}_0^-.$$



*J. A. Wolf, M. Cahen, and M. De Wilde (eds.), Harmonic Analysis and Representations of Semi-Simple Lie Groups, 257–315.*

If

$$\mathfrak{n}_0 = \sum_{\alpha \in \mathbb{P}_0^+} (\mathfrak{g}_{c\alpha} \cap \mathfrak{g}) \qquad (\mathfrak{g}_{c\alpha} = \{X \in \mathfrak{g} : [H, X] = \alpha(H)X \text{ for all } H \in \mathfrak{h}_{0c}\},$$

and for any real vector space $V$, $V_c = V \otimes_R \mathbb{C}$), $A_0 = \exp \mathfrak{a}_0$, $N_0 = \exp \mathfrak{n}_0$, $M_0 = \mathrm{Cent}_K(A_0)$ then the mappings $K \times A_0 \times N_0 \to G$ and $K \times \mathfrak{s} \to G$ defined by $(k, a, n) \to kan$ and $(k, X) \to k \exp X$, respectively, are diffeomorphisms and $P_0 = M_0 A_0 N_0$ is a minimal psgp of $G$. Note that

$$\theta(k \exp X) = k \exp(-X)$$

extends $\theta$ to $G$. In particular then if $h = \exp X$ $(X \in \mathfrak{s})$ we shall write $X = \log h$; given $x \in G$ there exists unique $\kappa(x) \in K$, $a(x) \in A_0$, $n(x) \in N_0$ such that $x = \kappa(x) a(x) n(x)$. Let $H_0(x) = H_{P_0}(x) = \log a(x)$.

Put $\rho_0 = \rho_{P_0} = \frac{1}{2}\Sigma_{\alpha \in P_0^+} \alpha | \mathfrak{a}$ and define $\Xi$ and $\sigma$ as follows:

$$\Xi(x) = \int_K \exp[-\rho_0(H_0(xk))]\, dk, \qquad \sigma(x) = \|X\|,$$

where $x = k \exp X$, $\|X\| = -B(X, \theta(X))$ and $B$ is a symmetric bilinear form on $\mathfrak{g}$ such that:

(1) $\quad B([X, Y], Z) = -B(Y, [X, Z]) \qquad (X, Y, Z \in \mathfrak{g})$;

(2) $\quad \|X\| \geqslant 0$, equality holding only if $X = 0$;

(3) $\quad B(\theta(X), \theta(Y)) = B(X, Y)$.

We note then that this class of reductive group behaves very much like semisimple Lie groups. It is in some respects better though than the class of semisimple groups because of the following two facts: Let $X(G)$ denote the group of all (continuous) homomorphisms of $G$ into $\mathbb{R}^\times$ (the nonzero reals). If $\chi \in X(G)$ let $|\chi|$ denote the homomorphism $x \to |\chi(x)|$ $(x \in G)$. Put

$${}^\circ G = \bigcap_{\chi \in X(G)} \ker |\chi|.$$

Then ${}^\circ G$ is a closed normal subgroup of $G$ and $G/{}^\circ G$ is an Abelian Lie group (for future reference we call the $\dim(G/{}^\circ G)$ the parabolic rank of $G$ and denote it $\mathrm{prk}(G)$). Let $Z_G = \mathrm{Center}(G)$, $\mathfrak{z}$ its Lie algebra. Put $\mathfrak{l} = \mathfrak{z} \cap \mathfrak{s}$ and $C = \exp \mathfrak{l}$.

LEMMA 1.1. (1) ${}^\circ G = KG_1$ *and* $C$ *is a split component of* $G$. *Moreover* $C = \mathrm{Center}(G) \cap \exp \mathfrak{s}$.

(2) *$^\circ G$ is $\theta$-stable. Let $^\circ\mathfrak{g}$ denote its Lie algebra and let $^\circ B$ and $^\circ\theta$ denote the restrictions of $B$ and $\theta$ to $^\circ\mathfrak{g}$. Then $^\circ G$ satisfies the same conditions as those imposed on $G$. Moreover if we replace $(G, K, \theta, B)$ by $(^\circ G, K, {}^\circ\theta, {}^\circ B)$, all the above conditions are again fulfilled.*

LEMMA 1.2. *Let $P = M_1N$ be a parabolic subgroup* (psgp) *of $G$. Then $M_1$ satisfies the same conditions as those imposed on $G$. In particular then if $M = {}^\circ M_1$ then so does $M$.*

Hence, for purposes of induction on the dimensions of $G$ we see that this is a good class of groups.

$\mathfrak{G}$ will denote the universal enveloping algebra of $\mathfrak{g}_c$; if $a_1, a_2 \in \mathfrak{G}$ we write for $f \in C^\infty(G)$, $a_1 f a_2(x) = f(a_2 ; x; a_1)$ where if $X_1, \ldots, X_r \in \mathfrak{g}$ then $f(X_1, \ldots, X_r; x) = (\partial^r/(\partial_{t_r} \ldots \partial_{t_1}))_{t_1 = \cdots = t_r = 0} f(\exp t_1X_1 \ldots \exp t_rX_rx)$ and $f(x; X_1 \ldots X_r) = (\partial^r/(\partial_{t_1} \ldots \partial_{t_r}))_{t_1 = \cdots = t_r = 0}\, f(x \exp t_1X_1 \ldots \exp t_rX_r)$. Let $a_1, a_2 \in \mathfrak{G}$, $r \in \mathbb{R}$ and $f \in C^\infty(G)$. Define

$$\nu_{a_1,a_2,r}(f) = \sup_G\, (1+\sigma)^r\, \Xi^{-1} |a_1 f a_2|.$$

Denote by $\mathscr{C}(G)$ the space of all $C^\infty$-functions on $G$ such that $\nu_{a_1,a_2,r}(f) < \infty$ for all $a_1, a_2 \in \mathfrak{G}$ and $r \in \mathbb{R}$. We introduce the structure of a locally convex space on $\mathscr{C}(G)$ by means of the seminorms $\nu_{a_1,a_2,r}$. More generally let $V$ be a complex Frechet space, and let $\mathscr{S}(V)$ denote the set of continuous seminorms on $V$. To shorten the notation let $\mathfrak{G}^{(2)} = \mathfrak{G} \otimes_c \mathfrak{G}$ and if $g_1 \otimes g_2 \in \mathfrak{G}^{(2)}$, $f \in C^\infty(G: V)$ let us write $(g_1 \otimes g_2 f)(x) = f(g_1; x; g_2)$. Then define for $D \in \mathfrak{G}^{(2)}$, $r \in \mathbb{R}$, $s \in \mathscr{S}(V)$, $f \in C^\infty(G: V)$,

$$S_{D,r}(f) = \sup_G\, S(Df)\Xi^{-1}(1+\sigma)^r.$$

If $F \subset \mathfrak{G}^{(2)}$, $|F| < \infty$, then we put

$$S_{F,r}(f) = \sum_{D \in F} S_{D,r}(f).$$

Denote by $\mathscr{C}(G: V)$ the space of all $f \in C^\infty(G: V)$ such that $S_{D,r}(f) < \infty$ for all $D \in \mathfrak{G}^{(2)}$, $r \in \mathbb{R}$, and $s \in \mathscr{S}(V)$. It is not difficult to show that $\mathscr{C}(G: V)$ is a Frechet space in the topology induced by the seminorms $S_{D,r}$ ($D \in \mathfrak{G}^{(2)}$, $r \in \mathbb{R}$). $\mathscr{C}(G: V)$ will henceforth be referred to as the *Schwartz space of V-valued functions on G*. This name is justified in light of the following useful lemma and the remark following it.

LEMMA 1.3. *There exists* $r_0 > 0$ *such that*

$$\int_G \Xi^2(x)(1+\sigma(x))^{-r_0}\,dx < \infty.$$

*Here* dx *is any Haar measure on G.*

*Remark.* It is clear from the lemma that if $f \in \mathscr{C}(G: V)$, $S \in \mathscr{S}(V)$, $D \in \mathfrak{G}^{(2)}$, $r \in \mathbb{R}$, and ${}_S f_{D,r} = S(Df)(1+\sigma)^r$ then ${}_S f_{D,r} \in L^2(G)$.

Let $\mathfrak{h}$ be a $\theta$-stable CSA of $\mathfrak{g}$; $\mathfrak{h} = \mathfrak{h}_I + \mathfrak{h}_R$ where $\mathfrak{h}_I = \mathfrak{h} \cap \mathfrak{k}$, $\mathfrak{h}_R = \mathfrak{h} \cap \mathfrak{s}$, and the sum being direct. We assume that $\mathfrak{a} = \mathfrak{h}_R \subset \mathfrak{a}_0$. Let $A = \exp \mathfrak{a}$ and denote by $\mathscr{P}(\mathfrak{a})$ (resp. $\mathscr{P}(A)$) the set of all parabolic subalgebras (resp. psgps) of $\mathfrak{g}$ (resp. $G$) whose $\theta$-stable split component is $\mathfrak{a}$ (resp. $A$). Let $P \in \mathscr{P}(A)$ and suppose that $P = MAN$ is a Langlands decomposition of $P$. For $\omega \in \mathscr{E}^2(M)$ (the equivalence classes of discrete series representations of $M$) $\nu \in \mathfrak{a}^*$ (for any finite-dimensional vector space $W$ over a field $k$, we shall denote by $W^*$ the dual of $W$ (over $k$)) we have the principal series representation $\pi_{P,\omega,\nu}$. If $Q \in \mathscr{P}(A)$ then $Q = MAN_Q$ and the representation $\pi_{Q,\omega,\nu}$ has the same global character as $\pi_{P,\omega,\nu}$. We denote this character by $\theta_{\omega,\nu}$. If $W(a) = \text{norm}(A)/\text{cent}(A)$, $s \in W(A)$, then $\theta_{s\omega,s\nu} = \theta_{\omega,\nu}$ ($s\omega$ is the class of the representation $s\sigma$, $\sigma \in \omega$, when $s\sigma(m) = \sigma(k_s m k_s^{-1})$), $k_s$ a representative of $s$ in $K$).

For $f \in C_c^\infty(G: V) \subset \mathscr{C}(G: V)$ (actually $C_c^\infty(G: V)$ is dense in $\mathscr{C}(G: V)$) put,

$$\hat{f}_A(\omega: \nu) = \hat{f}(\omega: \nu) = \int_G f(x)\theta_{\omega,-\nu}(x) = \hat{f}(s\omega: s\nu).$$

*We shall refer to* $\hat{f}_A$ *as the A-partial Fourier transform of f.* The properties of $\hat{f}_A$ are made evident by the following lemma. Fix $\omega \in \mathscr{E}^2(M)$ and denote its global character by $\theta_\omega$. Let $M_1 = MA$. By Lemma 1.2 we may define $\Xi_M$. Extend $\Xi_M$ to $M_1$ via the prescription $\Xi_M(ma) = \Xi_M(m)$ ($m \in M$, $a \in A$). We may then define $\mathscr{C}(M_1: V)$. For $P \in \mathscr{P}(A)$ let

$$f^P(m_0) = d_P(m_0) \int_{K \times N} f(km_0 nk^{-1})\,dk\,dn,$$

$$f^P_\omega(m_0) = \int_M \theta_\omega(m) f^P(m_0 m)\,dm \qquad (m_0 \in M_1)$$

where $f \in C_c^\infty(G: V)$, and $d_P(m) = \det(Adm|_{\mathfrak{n}})^{1/2}$, $N = \exp \mathfrak{n}$.

LEMMA 1.4. (1) *Let* $f \in C_c^\infty(G\colon V)$ *then* $f^P \in C_c^\infty(M_1\colon V)$. *Moreover, the map* $f \to f^P$ *lifts to a continuous map of* $\mathscr{C}(G\colon V)$ *into* $\mathscr{C}(M_1\colon V)$.

(2) *Let* $v \in \mathfrak{a}^*$. *Then*

$$\hat{f}_A(\omega\colon v) = \int_A f_\omega^P(a)\, e^{iv(\log a)}\, da$$

*for all* $f \in \mathscr{C}(G\colon V)$.

It follows rather easily from the lemma that $\hat{f}_A(\omega) \in \mathscr{C}(\mathfrak{a}^*\colon V)$ ($\mathscr{C}(\mathfrak{a}^*\colon V)$ denoting the usual Schwartz space of $V$-valued functions on the Euclidean space $\mathfrak{a}^*$) and that $f \to \hat{f}_A(\omega)$ is a continuous map of $\mathscr{C}(G\colon V)$ into $\mathscr{C}(\mathfrak{a}^*\colon V)$. A natural question to ask is whether one can characterize this map. The affirmative answer to this question in the special case to be presently described is the substantive matter for this paper.

Let $\mathscr{E}(K)$ denote the equivalence classes of irreducible representations of $K$. The generic element of $\mathscr{E}(K)$ will be denoted by $d$. Fix $F \subset \mathscr{E}(K)$, $F = F^*$ (here $F^*$ denotes the set of $d \in \mathscr{E}(K)$ which are contragradient to some $d' \in F$), and $|F| < \infty$. In Section 2 we shall define a double unitary $K$-module $(V_F, \tau)$ $(\tau = (\tau_1, \tau_2))$ where $\dim V_F < \infty$. With $V = V_F$, $f \in C^\infty(G\colon V)$ will be said to be $\tau$-*spherical* if

$$f(k_1 x k_2) = \tau_1(k_1) f(x) \tau_2(k_2) = \tau(k_1) f(x) \tau(k_2).$$

Let $\mathscr{C}(G\colon V_F\colon \tau)$ denote the (closed) subspace of all $\tau$-spherical functions in $\mathscr{C}(G\colon V_F)$. Henceforth we shall suppress the $F$ on $V_F$ as it will remain fixed throughout. We shall show that the map $f \to \hat{f}_A$ is a surjection of $\mathscr{C}(G\colon V\colon \tau)$ onto $\mathscr{C}_A(\hat{G}\colon V)$†. Although this result appears to be much more specialized than our original set up we remark that the case of an arbitrary Frechet space $V$ which is a differentiable $K$-module will follow from this result; we do not pretend to prove the more general result in this paper.

In order to prove the surjectivity of the map $f \to \hat{f}_A$ we proceed as follows: if $H$ is a Lie group and $L$ is a subgroup then for any representation $\pi$ of $H$ we denote by $\pi_L$ the restriction of $\pi$ to $L$. Let ${}^\circ\mathscr{C}(M\colon V\colon \tau_M)$ be the subspace of $\mathscr{C}(M\colon V\colon \tau_M)$ of $\tau_M$-spherical cusp forms. In Section 2 we define a product in $V$ which turns $V$ into an associative algebra; if $v_1, v_2 \in V$ we denote their product by $v_1 \cdot v_2$. Let $\psi \in {}^0\mathscr{C}(M\colon V\colon \tau_M)$,

† $\mathscr{C}_A(\hat{G}\colon V) = \{f \in C^\infty(\mathscr{E}^2(M) X \mathfrak{a}^*\colon V)\colon f(\omega) \in \mathscr{C}(\mathfrak{a}^*\colon v)$ for all $\omega \in \mathscr{E}^2(M)$, $f(s\omega\colon sv) =$ and $\int_k \tau(k) f \tau(k^{-1})\, dk = f\}$.

$\beta \in C_c^\infty(\mathfrak{a}^{*\prime}: V)$ ($\mathfrak{a}^{*\prime}$ denotes the set of regular elements of $\mathfrak{a}^*$ which is defined in Section 4. The reason we require supp $\beta \subset \mathfrak{a}^{*\prime}$ will become apparent as we proceed. The removal of this condition is one of the keys to the Plancherel measure which we will presently define), define the Eisenstein integral, $E(P: \psi: \nu)$, as in Section 2 and put,

$$(1.1) \qquad \varphi_\beta(x) = \int_{\mathfrak{a}^*} \beta(\nu) \cdot E(P: \psi: \nu: x)\, d\nu,$$

where $d\nu$ denotes a Lebesgue measure on $\mathfrak{a}^*$. There are two major points to be shown for surjectivity: (A) that $\varphi_\beta \in \mathscr{C}(G: V: \tau)$ and (B) to determine $(\hat{\varphi}_\beta)_A(\omega)$; and in particular how to 'adjust' the definition of $\varphi_\beta$ so that $(\hat{\varphi}_\beta)_A(\omega) = \beta$. Let us now discuss each of these points separately.

For (A) it is clear that we must determine the growth of $E(P: \psi: \nu: x)$ in the variables $\nu$ and $x$. From the integral formula for $E(P: \psi: \nu: x)$ one can obtain certain 'a priori estimates' for $E$; this is done in Section 2. These estimates are in general not sufficient to prove (A). However, they are sufficient to show that by the method of descent (A) can be proved by induction or $\dim(G)$. To describe the method of descent we cannot restrict ourselves to the class of Eisenstein integrals; rather we broaden the class of functions considered to the class of functions of type II($\lambda$). Functions of type II($\lambda$) have the same a priori estimates as the functions $E(P: \psi: \nu)$ and are eigenfunctions for $\mathfrak{Z}$, the center of $\mathfrak{G}$. One should observe that the Eisenstein integrals are eigenfunctions for $\mathfrak{Z}$ when

$$\psi \in {}^\circ\mathscr{C}_\omega(M: \dot{V}: \tau_M) = \left\{ f \in {}^\circ\mathscr{C}(M: V: \tau_M) : \int_M \theta_\omega(m) f(m_0 m)\, dm = d(\omega)^{-1} f(m_0) \right\}$$

($d(\omega)$ = the formal degree of $\omega$). In fact if $\lambda \in i\mathfrak{h}_I^*$ parametrizes $\omega$ (cf. [1]) then for all $\psi \in {}^\circ\mathscr{C}_\omega(M: V: \tau_M)$, $z \in \mathfrak{Z}$,

$$E(P: \psi: \nu: x; z) = \mu_{\mathfrak{h}}(z)(\lambda + i\nu) E(P: \psi: \nu: x)$$

where $\mu_{\mathfrak{h}}: \mathfrak{Z} \to \mathfrak{H}$, $\mathfrak{H}$ the algebra in $\mathfrak{G}$ generated by 1 and $\mathfrak{h}_c$, is the canonical map of $\mathfrak{Z}$ into the Weyl group (of $(\mathfrak{g}_c, \mathfrak{h}_c)$) invariants. Hence, we fix $\lambda \in i\mathfrak{h}_I^*$. The functions of type II($\lambda$) are defined to be all $C^\infty$ functions from $\mathfrak{a}^* \times G$ into $V$ satisfying the estimates already mentioned and which also satisfy the differential equations

$$\varphi(\nu: x; z) = \mu_{\mathfrak{h}}(z)(\lambda + i\nu) \varphi(\nu: x) \qquad (\nu \in \mathfrak{a}^*, x \in G, z \in \mathfrak{Z}).$$

One now shows that if $Q=M_QA_QN_Q$ is an arbitrary psgp of $G$, $Q\supset P_0$ then the differential equations satisfied by any function of type II($\lambda$) can be transcribed to a set of equations on $M_{Q1}=M_QA_Q$ (this is done in Section 3). Let

$$\tilde{\omega}_{\mathfrak{h}}=\prod_{\alpha\in\Delta^+(\mathfrak{g}_c,\mathfrak{h}_c)} H_\alpha$$

where $\Delta(\mathfrak{g}_c,\mathfrak{h}_c)$ denotes the roots of the pair $(\mathfrak{g}_c,\mathfrak{h}_c)$ and $\Delta^+(\mathfrak{g}_c,\mathfrak{h}_c)$ denotes a positive system of roots; $H_\alpha$ has the usual meaning. Then we shall show that if $\varphi$ is a function of type II($\lambda$) on $G$ then $\psi(\nu\colon x)=\tilde{\omega}_{\mathfrak{h}}(\lambda+i\nu)\varphi(\nu\colon x)$ is also, and moreover, from the equations satisfied by $\psi$ on $M_{Q1}$ we show that there exists a unique function

$$\psi_Q\colon \mathfrak{a}^*\times M_{Q1}\to V$$

such that

$$\lim_{\substack{a\to\infty\\ Q}} \{d_Q(ma)\psi(\nu\colon ma)-\psi_Q(\nu\colon ma)\}=0$$

for all $m\in M_{Q1}$, $\nu\in\mathfrak{a}^*$. The important points here are that:

(1) $\quad \psi\in C^\infty(\mathfrak{a}^*\times M_{Q1}: V\colon \tau_Q)(\tau_Q=\tau_{K\cap M_{Q1}})$

and if $W(\mathfrak{a}\colon\mathfrak{a}_Q)$ ($\mathfrak{a}_Q\subset\mathfrak{g}$ and $A_Q=\exp\mathfrak{a}_Q$) denotes the set of injections of $\mathfrak{a}_Q$ into $\mathfrak{a}$ which are of the form $\mathrm{Ad}_x$ for some $x\in G$ then

$$\psi_Q(\nu\colon m)=\sum_{s\in W(\mathfrak{a};\mathfrak{a}_Q)}\psi_{Q,s}(\nu\colon m)\qquad(\nu\in\mathfrak{a}^*,\, m\in M_{Q1})$$

and the functions $\psi^s_{Q,s}$ are functions of type II($\lambda$) on $\mathfrak{a}^*\times M^s_{Q1}$ (cf. Section 1 in [3] for notation); in fact more is true. We can assume without loss of generality that $\mathfrak{a}=\mathfrak{h}_R\subset\mathfrak{h}_{0R}=\mathfrak{a}_0$ and that $Q$ is standard with respect to $P_0$. Let $\Delta_Q=\Delta_Q(\mathfrak{g}_c,\mathfrak{h}_{0c})$ denote the set of roots of $(\mathfrak{g}_c,\mathfrak{h}_{0c})$ which vanish on $\mathfrak{a}_Q$; put $\Delta_Q^+=\Delta_Q\cap\mathbb{P}$

$$\tilde{\omega}_Q=\prod_{\alpha\in\Delta_Q^+} H_\alpha.$$

Further as $\mathfrak{z}_{\mathfrak{a}}=\mathrm{cent}(\mathfrak{a})$ contains both $\mathfrak{h}$ and $\mathfrak{h}_0$ as CSA$^s$ it is clear that there exists $y\in G_c$ such that $\mathrm{Ad}_y$ centralizes $\mathfrak{h}_R$ and $\mathfrak{h}_c^y=\mathfrak{h}_{0c}$. Let $\Lambda_\nu=(\lambda+i\nu)^y$ $(\Lambda_\nu(H)=(\lambda+i\nu)(H^{y^{-1}})$ for all $H\in\mathfrak{h}_{0c})$. Then if we put $\tilde{\omega}=\omega_{\mathfrak{h}}^y$ we have

$$\tilde{\omega}(\Lambda_\nu)=\tilde{\omega}_{\mathfrak{h}}(\lambda+i\nu).$$

One can then show that there is a one-to-one correspondence between $W(\mathfrak{a}: \mathfrak{a}_Q)$ and a subset $W_Q(\mathfrak{g}, \mathfrak{h}_0) \subset W(\mathfrak{g}, \mathfrak{h}_0)$ (the Weyl group of the pair $(\mathfrak{g}, \mathfrak{h}_0)$). We will show that the function

$$(\nu: m) \to \tilde{\omega}_Q(s(t)\Lambda_\nu)^{-1} \psi^t_{Q,t}(\nu: m)$$

is again a function of type II($\lambda$) on $M^t_{Q1}$ (here $s(t)$ is that element of $W_Q(\mathfrak{g}, \mathfrak{h}_0)$ which corresponds to $t \in W(\mathfrak{a}: \mathfrak{a}_Q)$). (2) One can estimate the norm of

$$\mathrm{d}_Q(ma)\psi(\nu: ma) - \psi_Q(\nu: ma)$$

as $\nu$ varies in $\mathfrak{a}^*$, $m$ varies in $M^+_{1Q} = K_Q \mathrm{cl}(A_0^+) K_Q$, $K_Q = K \cap M_Q$, and as $a$ varies in

$$A_Q^+ = \{h \in A_Q : \alpha(\log h) > 0 \qquad \text{for all } \alpha \in \Delta(\mathfrak{q}, \mathfrak{a}_Q)\}$$

(here $Q = \mathrm{Norm}_G(\mathfrak{a})$, $A_Q = \exp \mathfrak{a}_Q$, and $\Delta(\mathfrak{q}, \mathfrak{a}_Q) = \Delta(Q, A_Q)$ denotes the roots of the pair $(\mathfrak{q}, \mathfrak{a}_Q)$ and $A_0^+$ is defined below).

Let us be more specific about this estimate as it relates to point (A). We assume that $\mathrm{Prk}(G) = 0$; the inductive argument mentioned above will have two cases; one for $\mathrm{Prk}(G) > 0$ and another for $\mathrm{Prk}(G) = 0$. We shall only describe the latter case in this section. Let $P_i = M_i A_i N_i$ $(1 \leqslant i \leqslant d)$ be an enumeration of the maximal psgps of $G$ which are standard with respect to $P_0$. Put

$$A_0^+ = \{h \in A_0 : \alpha(\log h) > 0 \qquad \text{for all } \alpha \in \Delta(\mathfrak{p}, \mathfrak{a}_0)\}$$

and if $\mu > 0$, $1 \leqslant i \leqslant d$, set

$$A_i^+(\mu) = \{h \in A_0^+ : \alpha_i(\log h) > \mu \rho_0(\log h)\}.$$

Here $\alpha_i$ is the unique simple root in $\Delta(\mathfrak{p}_i, \mathfrak{a}_i)$, $A_i = \exp \mathfrak{a}_i$, and $P_i = \mathrm{Norm}_G(\mathfrak{p}_i)$. Then if $h \in A_i^+(\mu)$ there exists $h_1 \in M^+_{i1}$, $M_{i1} = M_i A_i$, and $h_2 \in A_i^+$ such that $h = h_1 h_2$. Moreover, it is not difficult to show that for $\mu$ sufficiently small

$$(1.2) \qquad A_0^+ \subseteq \bigcup_{1 \leqslant i \leqslant d} A_i^+(\mu) \qquad \text{and} \qquad G = K\,\mathrm{cl}(A_0^+)K.$$

The form of the inequality then (since $\psi$ is $\tau$-spherical) need only be given on $A_i^+(\mu)$. Roughly this estimate is of the following form: let $\eta_1, \eta_2 \in \mathfrak{M}_{i1}$ ($\mathfrak{M}_{i1}$ is the subalgebra of $\mathfrak{G}$ generated by 1 and $\mathfrak{m}_{i1c}$) then

$$\begin{aligned} &\|\psi(\nu: \eta_1 ; h; \eta_2) - \mathrm{d}_{P_i}^{-1}(h)\psi_{P_i}(\nu: \eta'_1 ; h; \eta'_2)\| \\ &\qquad \leqslant \text{const } \Xi^{1+\varepsilon}(h)(1+\sigma(h))^r \qquad (h \in A_i^+(\mu)). \end{aligned}$$

Here $\|\cdot\|$ denotes the $K$-invariant norm on $V$ and $\eta\to\eta'$ is an automorphism of $\mathfrak{M}_{i1}$. Fix $i$, $1\leqslant i\leqslant d$. Let $\varphi$ be a function of type II($\lambda$) on $G$ and put,

$$\varphi_\beta(x)=\int_{\mathfrak{a}^*}\beta(v)\cdot\varphi(v\colon x)\tilde{\omega}(\Lambda_v)\,\mathrm{d}v \qquad (x\in G)$$

and for $t\in W(\mathfrak{a}\colon\mathfrak{a}_i)$,

$$\varphi_{\beta,t}(m)=\int_{\mathfrak{a}^*}\beta(v)\cdot\varphi_{P_{i},t}(v\colon m)\tilde{\omega}_{P_i}(s(t)\Lambda_v)\,\mathrm{d}v \qquad (m\in M_{i1})$$

where we have written $\varphi_{P_{i},t}$ for $\tilde{\omega}_{P_i}(s(t)\Lambda_v)^{-1}\psi_{P_i,t}(v\colon m)$. As

$$\mathfrak{g}=\mathfrak{k}+\mathfrak{p}_i\Rightarrow\mathfrak{G}=\mathfrak{K}\mathfrak{P}_i$$

(where $\mathfrak{P}_i$ is the subalgebra of $\mathfrak{G}$ generated by 1 and $\mathfrak{p}_{ic}$). We will show that this together with (1.2) implies that in order to show that $\varphi_\beta\in\mathscr{C}(G\colon V\colon\tau)$ it suffices to show that for every $i$, $1\leqslant i\leqslant d$, $D\in\mathfrak{M}_{i1}^{(2)}$, $r\in\mathbb{R}$ then

$$\sup_{A_i^+(\mu)}(1+\sigma)\Xi^{-1}\|D\varphi_\beta\|<\infty.$$

But from the preceding estimate it is easy to see that

$$\sup_{h\in A_i^+(\mu)}(1+\sigma(h))^r\Xi^{-1}(h)\|\varphi_\beta(\eta_1\,;h;\eta_2)-d_{P_i}^{-1}(h)\sum_t\varphi_{\beta,t}(\eta_1'\,;h;\eta_2')\|<\infty.$$

It will follow from standard estimates relating $\mathrm{d}_{P_i}$, $\Xi$, and $\Xi_{M_i}$ that it suffices to show that $\varphi_{\beta,t}\in\mathscr{C}(M_{i1}\colon V\colon\tau_{M_i})$. Recalling that $\varphi_{P_i,t}^t$ is of type II($\lambda$) on $M_{i1}^t$ and observing that $\tilde{\omega}_{P_i}$ is the full product of roots of $M_{i1}$, it is easy to see that we may proceed by induction on dim($G$) to prove point (A); we need only point out that $\tilde{\omega}(\lambda_v)\geqslant c>0$ ($v\in\operatorname{supp}\beta$) for some sufficiently small $c$ to see that (1.1) belongs to $\mathscr{C}(G\colon V\colon\tau)$.

We now come to point (B). Fix $\omega$. We say that a positive Borel measure $\mu$ on $\mathfrak{a}^*$ is the Plancherel measure corresponding to the $A$-partial Fourier transform if for $\beta\in\mathscr{C}(\mathfrak{a}^*\colon V)$, $\beta(sv)=\beta(v)$ (for all $s\in W(\mathfrak{a},\mathfrak{a})=W(\mathfrak{a})=W(A)$), $\psi\in{}^\circ\mathscr{C}_\omega(M\colon V\colon\tau_M)$, and

$$\varphi_\beta^\mu(x)=\int_{\mathfrak{a}^*}\beta(v)E(P\colon\psi\colon v\colon x)\,\mathrm{d}\mu(v) \qquad (x\in G,\ P\in\mathscr{P}(A))$$

then

(1) $\varphi_\beta^\mu \in \mathscr{C}(G\colon V\colon \tau)$

(2) $$\varphi_\beta^\mu(1) = \int_{\mathfrak{a}^*} \hat{\varphi}_\beta^\mu(\omega\colon \nu)\, \mathrm{d}\mu(\nu).$$

We shall see momentarily that (2) is sufficient to show that the map $f \to \hat{f}_A$ of $\mathscr{C}(G\colon V\colon \tau)$ into $\mathscr{C}_A(\hat{G}\colon V)$ is surjective. Recalling Lemma 1.4 it is clear that one should investigate the mapping $\beta \to \varphi_\beta^Q$ for $Q \in \mathscr{P}(A)$ and $\beta \in C_c^\infty(\mathfrak{a}^{*\prime}\colon V)$. For $Q \in \mathscr{P}(A)$, $f \in C_c^\infty(G\colon V\colon \tau)$ put $Q = MAN_Q$ and let,

$$f^{(Q)}(m) = \mathrm{d}_Q(m) \int_{N_Q} f(mn)\, \mathrm{d}n \qquad (m \in M_1).$$

Let $\varphi$ be a function of type II($\lambda$) and let as before $\psi = \tilde{\omega}(\Lambda_\nu)\varphi$. Set

$$\varphi_{\beta,Q}(m) = \int_{\mathfrak{a}^*} \beta(\nu)\psi_Q(\nu\colon m)\, \mathrm{d}\nu \qquad (m \in M_1).$$

Then we shall prove the following surprising equality

(1.3) $$\varphi_\beta^{(\bar{Q})}(m) = \mathrm{d}_Q(m) \int_{\bar{N}_Q} \varphi_\beta(\bar{n}m)\, \mathrm{d}\bar{n} = \int_{\bar{N}_Q} \mathrm{e}^{-\rho_Q(H_Q(\bar{n}))}\varphi_{\beta,Q}(\bar{n}m)\, \mathrm{d}\bar{n}$$

where if $\mathfrak{n}_Q \subseteq \mathfrak{g}$, $N_Q = \exp \mathfrak{n}_Q$, $\rho_Q(H) = \frac{1}{2}\mathrm{tr}(\mathrm{ad}_H|_{\mathfrak{n}_Q})$, $\bar{N}_Q = \theta(N_Q)$, $\bar{Q} = M_Q A_Q \bar{N}_Q$ and $H_Q(x) = \log a_Q(x)$ where $x \in G$ can be written uniquely as

$$x = \kappa_Q(x)\mu_Q(x)a_Q(x)n_Q(x),$$

with $\kappa_Q(x) \in K$, $\mu_Q(x) \in M \cap \exp \mathfrak{s}$, $a_Q(x) \in A$, and $n_Q(x) \in N_Q$. Also

$$\varphi_{\beta,Q}(kmn) = \tau(k)\varphi_{\beta,Q}(m) \qquad (k \in K,\ m \in M_1,\ n \in N).$$

As $\varphi_{\beta,Q}$ is $\tau_M$-spherical this extension is well defined and defines a $C^\infty$-function on $G$.

Let us now apply the above to (1.1). We show in Section 5 that $\psi_Q \in L = {}^\circ\mathscr{C}(M_1\colon V\colon \tau_M)$ (for all $\nu \in \mathfrak{a}^*$). It follows easily from this that with $\varphi(\nu\colon x) = E(P_1\colon \psi\colon \nu\colon x)$ there exists $C_{P_2|P_1}(s\colon \nu) \in \mathrm{End}(L)$ ($s \in W(A)$, $\nu \in \mathfrak{a}^{*\prime}$) such that

$$\begin{aligned}\varphi_{P_2}(\nu\colon ma) &= \tilde{\omega}(\Lambda_\nu)^{-1}\psi_{P_2}(\nu\colon ma)\\ &= \sum_{s \in W(A)} (C_{P_2|P_1}(s\colon \nu)\psi)(m)\mathrm{e}^{is\nu(\log a)}.\end{aligned}$$

Moreover, if we put $\mathfrak{F}=\mathfrak{a}^*$, $\mathfrak{F}_c=\mathfrak{a}_c^*$, and for any $P\in\mathscr{P}(A)$

$$\mathfrak{F}(P)=\{\nu\in\mathfrak{F}_c\colon \langle\alpha, \nu_I\rangle>0 \qquad \text{for all } \alpha\in\Delta(\mathfrak{p}, \mathfrak{a})\}$$

here $\nu\in\mathfrak{F}_c$ can be uniquely written as $\nu=\nu_R+i\nu_I$, $\nu_R$, $\nu_I\in\mathfrak{F}$, and as usual $P=\mathrm{Norm}_G(\mathfrak{p})$, then from the asymptotic analysis one can prove the following result.

LEMMA 1.5. *$C_{\bar{P}|P}(1\colon\nu)$ and $C_{P|P}(1\colon -\nu)$ extend to holomorphic functions of $\nu$ on $\mathfrak{F}(P)$ and they are given by the following integral equations*:

$$(C_{\bar{P}|P}(1\colon \nu)\psi)(m)=\int_{\bar{N}} \tau(\kappa(\bar{n}))\psi(\mu(\bar{n})m)\, \mathrm{e}^{(i\nu-\rho)H(\bar{n})}\, \mathrm{d}\bar{n},$$

$$(C_{P|P}(1\colon -\nu)\psi)(m)=\int_{\bar{N}} \psi(m\mu(\bar{n})^{-1})\tau(\kappa(\bar{n}))^{-1}\, \mathrm{e}^{(i\nu-\rho)H(\bar{n})}\, \mathrm{d}\bar{n},$$

*where $\kappa=\kappa_P$, $\mu=\mu_P$, $H=H_P$, and $\rho=\rho_P$.*

If $T\in\mathrm{End}(L)$ let $T^*$ denote its adjoint relative to the $L^2$-innerproduct on $L$ (i.e. if $f, g\in{}^\circ\mathscr{C}(M\colon V\colon \tau_M)$ put

$$(f, g)=\int_M (f(m), g(m))\, \mathrm{d}m$$

where $(f(m), g(m))$ denotes the invariant inner product in $V$ of $g(m)$ and $f(m)$).

COROLLARY. $C_{P|P}(1\colon\nu)^*=C_{\bar{P}|\bar{P}}(1\colon\nu)$, $C_{\bar{P}|P}(1\colon\nu)^*=C_{P|\bar{P}}(1\colon\nu)$.

Further, for all $P_1, P_2\in\mathscr{P}(A)$, $s, t\in W(A)$, $k=k_s\in K$ such that

$$\mathrm{Ad}_k(H)=s(H) \qquad \text{for all } H\in\mathfrak{a},$$

$\psi\in L$ if we put $\psi^s(m)=\tau(k)\psi(k^{-1}mk)\tau(k)$ and $P_j^k=MAN_j^k$, $N_j^k=kN_jk^{-1}$, ($j=1, 2$ and $P_j=\mathrm{MAN}_j$), then $\psi^s\in L$, $\psi\to\psi^s$ is an endomorphism of $L$, and we have the following easily proved results:

$$sC_{P_2|P_1}(t\colon \nu)=C_{P_2^s|P_1}(st\colon \nu),$$

$$C_{P_2|P_1}(t\colon \nu)s^{-1}=C_{P_2|P_1^s}(ts^{-1}\colon s\nu).$$

The equalities holding on $\mathfrak{F}_c(\delta)=\{\nu\in\mathfrak{F}_c\colon |\nu_I|<\delta\}$. Using the above integral formulas for the $c$-functions together with the identity (1.3) as applied to (1.1) it is not too difficult to prove the following.

$$\varphi_\beta^{(\bar{P})}(ma)=\gamma(P)\int_{\mathfrak{F}} e^{iv(\log a)}\times \tag{1.4}$$
$$\times\sum_{s\in W(A)}\beta(s^{-1}v)(C_{\bar{P}|P}(1:v)C_{P|P}(s:s^{-1}v)\psi)(m)\,dv$$

for $m\in M$, $a\in A$, and $\beta\in C_c^\infty(\mathfrak{F}':V)$. Here $\gamma(P)$ is the constant defined by

$$\gamma(P)=\int_{\bar{N}} e^{-2\rho_P(H_P(\bar{n}))}\,d\bar{n}.$$

We are now very close to identifying the Plancherel measure. We first need some further identities satisfied by the $C$-functions which we shall now state. Let $C_{P_2|P_1}(\omega:s:v)$ denote the restriction of $C_{P_2|P_1}(s:v)$ to ${}^\circ\mathscr{C}_\omega(M:V:\tau_M)=L(\omega)$.

LEMMA 1.6. (1) *Fix $P_1, P_2\in\mathscr{P}(A)$, $s\in W(A)$, $v\in\mathfrak{F}'$ and $\omega\in\mathscr{E}^2(M)$. Then $C_{P_2|P_1}(\omega:s:v)$ defines a bijection of $L(\omega)$ into $L(s\omega)$. Moreover, there exists a $\mu_\omega:\mathfrak{F}\to\mathbb{R}^+\cup\{0\}$ such that $\mu_\omega$ is holomorphic on $\mathfrak{F}_c(\delta)$ for some $\delta>0$, $\mu_\omega(v)>0$ if $v\in\mathfrak{F}'$, $\mu_{s\omega}(sv)=\mu_\omega(v)$, and for all $\psi\in L(\omega)$,*

$$\mu_\omega(v)C_{P_2|P_1}(s:v)^*C_{P_2|P_1}(s:v)\psi=C(A)^2\psi.$$

*Here $C(A)$ is a constant* (Cf. [4], *Section* 11); *if $\sigma\in\omega$, $s\sigma(m)=\sigma(k_s m k_s^{-1})$ ($k_s$ as above) and $s\omega$ denotes the class of $s\sigma$.*

(2) *If $\psi\in L(\omega)$ then*

$$\|\mu_\omega(v)C_{P_2|P_1}(s:v)\psi\|_2^2=C(A)^2\|\psi\|_2^2$$

*where $\|\ \|_2$ denotes the $L^2$-norm on $L^2(M:V)$.*

(3) *Put for $v\in\mathfrak{F}'$,*

$${}^\circ C_{P_3|P_1}(st:v)={}^\circ C_{P_3|P_2}(s:tv){}^\circ C_{P_2|P_1}(t:v).$$

*Then if $\psi\in L(\omega)$, ${}^\circ C_{P_2|P_1}(s:v)\psi\in L(s\omega)$.*

(4) *Let $P_1, P_2, P_3\in\mathscr{P}(A)$, $s, t\in W(A)$, $v\in\mathfrak{F}'$. Then*

$${}^\circ C_{P_3|P_1}(st:v)={}^\circ C_{P_3|P_2}(s:tv){}^\circ C_{P_2|P_1}(t:v).$$

Let $\omega\in\mathscr{E}^2(M)$ and put

$$W_\omega(A)=\{s\in W(A): s\omega=\omega\},$$
$$w_\omega=|W_\omega(A)|,$$

and set

$$(*)\qquad \mu(\omega\colon \nu)=\frac{C(A)^{-2}}{w_\omega}\,\mu_\omega(\nu).$$

THEOREM 1.1. $d(\omega)\mu(\omega\colon \nu)$ *is the Plancherel measure corresponding to the A-partial Fourier transform.*

If $\varphi_\beta=\varphi_\beta^\mu$ (where $\mu$ stands for $d(\omega)\mu(\omega\colon \nu)\,d\nu$) then using point (A) it is not too difficult to show that $\varphi_\beta\in\mathscr{C}(G\colon V\colon \tau)$; this is done in Section 7. We shall now show that (2) in the definition of the Plancherel measure is satisfied. Put for $\nu\in\mathfrak{F}$, $m\in M$

$$[\varphi_\beta^{(P)}]_\nu(m)=\int_A e^{i\nu(\log a)}\varphi_\beta^{(P)}(ma)\,da.$$

Assuming $da$ (the Haar measure on $A$) is fixed, and $d\nu$ has been chosen to be dual to it under the usual Fourier transform on the Abelian group $A$ we have from (1.4) that

$$[\varphi_\beta^{(P)}]_\nu(m)=\sum_{s\in W(A)} d(\omega)\beta(s^{-1}\nu)\mu(\omega\colon \nu)[C_{P|\bar P}(1\colon \nu)C_{\bar P|P}(s\colon s^{-1}\nu)\psi](m).$$

But by definition of ${}^\circ C$ we have

$$C_{\bar P|P}(s\colon s^{-1}\nu)=C_{\bar P|\bar P}(1\colon \nu)^\circ C_{\bar P|P}(s\colon s^{-1}\nu).$$

By (4) of Lemma 1.6 we have

$${}^\circ C_{\bar P|P}(s\colon s^{-1}\nu)={}^\circ C_{\bar P|P}(1\colon \nu)^\circ C_{P|P}(s\colon s^{-1}\nu).$$

Therefore

$$\begin{aligned}C_{\bar P|P}(s\colon s^{-1}\nu)&=C_{\bar P|\bar P}(1\colon \nu)^\circ C_{\bar P|P}(1\colon \nu)^\circ C_{P|P}(s\colon s^{-1}\nu)\\&=C_{\bar P|P}(1\colon \nu)^\circ C_{P|P}(s\colon s^{-1}\nu).\end{aligned}$$

It then follows from the Corollary to Lemma 1.5 and (1) of Lemma 1.6 that

$$\begin{aligned}[\varphi_\beta^{(P)}]_\nu(m)&=\sum_{s\in W(A)}\beta(s^{-1}\nu)d(\omega)\mu(\omega\colon \nu)\times\\&\quad\times\{C_{P|\bar P}(1\colon \nu)C_{\bar P|P}(1\colon \nu)^\circ C_{P|P}(s\colon s^{-1}\nu)\psi\}(m)\\&=\sum_{s\in W(A)}\frac{d(\omega)}{w_\omega}\beta(s^{-1}\nu)\{{}^\circ C_{P|P}(s\colon s^{-1}\nu)\psi\}(m).\end{aligned}$$

Hence applying the character $\theta_\omega$ to both sides of this equality we have on putting $F(v)=\int_K \tau(k)v\tau(k^{-1})\,dk$, for all $v \in V$ and recalling that $\theta_\omega(\psi)=\mathrm{d}(\omega)^{-1}\psi(1)$ if $\psi \in L(\omega)$ and $\theta_\omega(\psi)=0$ if $\psi \notin L(\omega)$ that,

$$\hat{\varphi}_\beta(\omega\colon v)=F(\beta(v)\cdot\psi)(1). \tag{1.5}$$

Hence

$$\mathrm{d}(\omega)\int_{\mathfrak{F}} \hat{\varphi}_\beta(\omega\colon v)\mathrm{d}(\omega)\mu(\omega\colon v)\mathrm{d}v=\mathrm{d}(\omega)\int_{\mathfrak{F}} F(\beta(v)\cdot\psi)(1)\mu(\omega\colon v)\mathrm{d}v.$$

But from the definition of $\varphi_\beta$ we have

$$\varphi_\beta(1)=\mathrm{d}(\omega)\int_{\mathfrak{F}} F(\beta(v)\cdot\psi)(1)\mu(\omega\colon v)\,\mathrm{d}v.$$

The theorem is therefore established.

*Remark.* From (1.5) it is not difficult to show that given $\beta \in \mathscr{C}_A(\hat{G}\colon V)$ then we can find $f \in \mathscr{C}(G\colon v\colon \tau)$ such that $\hat{f}=\beta$.

Let us now recapitulate the essential facts that lead to points (A) and (B).

(1) The identity

$$\hat{f}(\omega\colon v)=\int_A f_\omega^P(a)\,\mathrm{e}^{iv(\log a)}\,\mathrm{d}a.$$

(2) The existence (for any $\varphi$ of type II($\lambda$) and any psgp $Q$) of the limit $\psi_Q$ ($\psi$ the function related to $\varphi$).

(3) The estimate of $\|\psi-\mathrm{d}_Q^{-1}\psi_Q\|$ where $Q$ is a standard maximal psgp of $G$ and the fact that $\psi_Q=\Sigma\psi_{Q,s}$ where $\psi_{Q,s}^s$ was a function of type II($\lambda$) on a reductive group possessing the same properties as $G$. This allowed us to use induction in order to prove point (A).

(4) We then considered $\psi_Q$ for $Q \in \mathscr{P}(\mathfrak{h}_R)$, $\mathfrak{h}_R=\mathfrak{a}$ which introduced the $C$-functions and the integral formulas for $C_{\bar{P}|P}(1\colon v)$ and $C_{P|P}(1\colon -v)$, also we considered the various identities for the $C$-functions.

(5) The relation between the norms of $C_{P_2|P_1}(s\colon v)\psi$, in particular for $\psi \in L(\omega)$ that they are all the same as $P_1, P_2$ vary over $\mathscr{P}(A)$ and $s$ varies in $W(A)$. These are what Harish-Chandra calls the Maass–Selberg relations.

(6) The identification of $\varphi_\beta^{(\bar{P})}$.

We shall now demonstrate the techniques in the theory of differential equations on reductive Lie groups as they pertain to points (2), (3), (4), and (6) listed above.

Although this material is rather technical in nature it should be pointed out that the theory of differential equations has far-reaching effects in both harmonic analysis as well as representation theory of reductive groups. Because of this it is felt that a thorough representation of their use would be of value to those learning the subject.

The material presented in this paper is principally contained in [2], [3], and [4].

## 2. THE EISENSTEIN INTEGRALS

In this section we shall define the double unitary $K$-module $(V, \tau)=(V_F, \tau)$ mentioned in Section 1. Once this module has been defined we shall introduce the Eisenstein integrals, make the connection between them and the matrix coefficients of the cuspidal principal series, and prove they satisfy certain a priori estimates.

Let us fix some notation. As in Section 1, $\mathfrak{a}_0$ is a maximal Abelian subspace of $\mathfrak{s}$; $\mathfrak{h}_0$ is a $\theta$-stable Cartan subalgebra such that $\mathfrak{h}_0 \cap \mathfrak{s} = \mathfrak{a}_0$. We also fix another $\theta$-stable CSA $\mathfrak{h}$ such that $\mathfrak{h}_R = \mathfrak{h} \cap \mathfrak{s} \subset \mathfrak{a}_0$. Put $\mathfrak{m}_1 = \mathrm{cent}_{\mathfrak{g}}(\mathfrak{a})$, $M_1 = \mathrm{cent}_G(\mathfrak{a})$, and as in Section 1, $M = {}^\circ M_1$, $\mathfrak{m} = LA(M)$.

Set $\mathfrak{v} = C^\infty(K \times K)$. Give $\mathfrak{v}$ the topology of uniform convergence of functions and a finite number of their derivatives. Define a double representation $\tau^0$ of $K$ on $\mathfrak{v}$ as follows: for $k_j, k'_j \in K$ $(j=1, 2)$, $v \in \mathfrak{v}$, let

$$\tau^0(k'_1) v \tau^0(k'_2)(k_1 : k_2) = v(k_1 k'_1 : k'_2 k_2).$$

Note that $\tau_0$ is unitary with respect to the usual $L^2$-norm on $K \times K$. Let $d \in \mathscr{E}(K)$, $\xi_d$ denote the character of $d$, and put $\alpha_d = \mathrm{d}(d) \operatorname{conj} \xi_d$ where $\mathrm{d}(d) = \xi_d(1)$. For any finite set $F \subset \mathscr{E}(K)$ we put

$$\alpha_F = \sum_{d \in F} \alpha_d.$$

Fix $F \subset \mathscr{E}(K)$, $F = F^*$, $|F| < \infty$, and put,

$$V = \mathfrak{v}_F = \left\{ v \in \mathfrak{v} : \int_K \alpha_F(k) \tau^0(k) v \, \mathrm{d}k = \int_K \alpha_F(k) v \tau^0(k) \, \mathrm{d}k = v \right\}.$$

Then $V$ is stable under $\tau^0$ and its dimension is finite. We shall denote the restriction of $\tau^0$ to $V$ by $\tau$. If $v_1, v_2 \in \mathfrak{v}$ we put $v = v_1 \cdot v_2$ where

$$v(k_1 : k_2) = \int_K v_1(k_1 : k) v_2(k^{-1} : k_2) \, \mathrm{d}k.$$

Equipped with this product, $V$ becomes an associative algebra. Put

$$\operatorname{tr} v = \int_K v(k: k^{-1})\, dk,$$

$$v^*(k_1: k_2) = \operatorname{conj} v(k_2^{-1}: k_1^{-1}),$$

$v \to v^*$ is an anti-involution (i.e. $(cv)^* = \bar{c}v^*$, $c \in \mathbb{C}$, and $(v_1 \cdot v_2)^* = v_2^* \cdot v_1^*$). Note that if we put

$$(v_1, v_2) = \int_{K \times K} v_1(k_1: k_2)\overline{v_2(k_1: k_2)}\, dk_1\, dk_2$$

then

$$(v_1, v_2) = \operatorname{tr}(v_1 \cdot v_2^*).$$

Let $\omega \in \mathscr{E}^2(M)$, $\sigma \in \omega$, and suppose that $\sigma$ represents $M$ on $U$. Let $\mathscr{H}^\omega$ denote the set of all measurable functions $f: K \to U$ such that

(i) $f(mk) = \sigma(m) f(k)$ for all $m \in K_M = K \cap M$, $k \in K$

and

(ii) $\int_K \|f(k)\|^2\, dk < \infty$, $\|\cdot\|$ denoting the invariant norm on $U$. For $F \subset \mathscr{E}(K)$ as above let $F_M = \{\delta \in \mathscr{E}(K_M): (d{:}\delta) \geqslant 1 \text{ for some } d \in F\}$; define $\alpha_G^M$ analogously to $\alpha_F$ for any $G \subset \mathscr{E}(K_M)$. Put

$$\mathscr{H}_F^\omega = \left\{ f \in \mathscr{H}^\omega: \int_k \alpha_F(k) f(k_1 k)\, dk = f(k_1) \right\},$$

$$U_F = \left\{ v \in U{:} \int_K \alpha_{F_M}^M(m) \sigma(m) v\, dm \right\}.$$

LEMMA 2.1. *For $T \in \operatorname{End}(\mathscr{H}_F^\omega)$ there exists a function*

$$K_T: K \times K \to \operatorname{End}(UF)$$

*such that*

$$T(h)(k_2) = \int_K K_T(k_2: k_1) h(k_1^{-1})\, dk_1$$

*for all $h \in \mathscr{H}_F^\omega$.*

For $T \in \operatorname{End}(\mathscr{H}_F^\omega)$ define $\psi_T \in L(\omega) = {}^\circ\mathscr{C}(M: V: \tau)$ as follows. If $m \in M$,

$\psi_T(m)$ is the element $v$ of $V$ given by

$$v(k_1 : k_2) = \psi_T(k_1 : m : k_2) = \mathrm{tr}\{K_T(k_2 : k_1)\sigma(m)\}.$$

LEMMA 2.2. *The mapping $T \rightarrow \psi_T$ from* $\mathrm{End}(\mathscr{H}_F^\omega)$ *into* $L(\omega)$ *is bijective.*

Let $\psi \in C^\infty(M: V: \tau)$, $P \in \mathscr{P}(A)$, $P = MAN$, $\nu \in \mathfrak{F}_c = \mathfrak{a}_c^*$, and put

$$E(P: \psi: \nu: k_1 : x: k_2) = \int_K \psi(xk)\tau(k^{-1})\mathrm{e}^{(i\nu - \rho_P)H_P(xk)}\,\mathrm{d}k(k_1 : k_2)$$

where $\psi$ is extended to a function on $G$ via the prescription $\psi(kman) = \tau(k)\psi(m)$ ($k \in K$, $m \in M$, $a \in A$, $n \in N$). Since $K \cap P = K_M$ this extension is well defined and is a $C^\infty$-function on $G$. It is easy to check that

$$E(P: \psi) : \mathfrak{F}_c \times G \rightarrow V$$

is a $C^\infty$-function, and for each $x \in G$, $\nu \rightarrow E(P: \psi: \nu: x)$ is holomorphic on $\mathfrak{F}_c$. It is also easy to check that $E(P: \psi: \nu)$ is $\tau$-spherical (cf. Section 1).

Let $C(G)$ denote the space of continuous complex valued functions on $G$. Define,

$${}_FC(G)_F = \left\{ f \in C(G): \int_K \alpha_F(k) f(xk)\,\mathrm{d}k = \int_K \alpha_F(k) f(kx)\,\mathrm{d}k = f(x) \right\}.$$

Let $f \in {}_FC(G)_F$ and put $j(f)(k_1 : x: k_2) = \underline{f}(k_1 : x: k_2) = f(k_1 x k_2)$. One easily checks that $j(f) \in C(G: V: \tau)$. Let $\pi$ be an admissible representation of $G$ on a Hilbert space $\mathscr{H}$. Put

$$\mathscr{H}_F = \left\{ v \in \mathscr{H}: \int_K \alpha_F(k)\pi(k)v\,\mathrm{d}k = v \right\}$$

and let ${}_F\mathscr{A}(\pi)_F$ denote the complex linear span of all functions on $G$ of the form $f(x) = \langle \pi(x)v, v' \rangle$ where $v, v' \in \mathscr{H}_F$. Put $\mathscr{A}(\pi: \nu: \tau) = j({}_F\mathscr{A}(\pi)_F)$.

For $h \in \mathscr{H}^\omega$, $\nu \in \mathfrak{F}_c$, $x \in G$, $k \in K$, define $\pi_{P,\omega,\nu}(x)$ as follows;

$$(\pi_{P,\omega,\nu}(x)h)(k) = \mathrm{e}^{-\{i\nu + \rho_P\}H_P(y)}\sigma(\mu_P(y))^{-1}h(\kappa_P(y)^{-1}k)$$

where $y = kx^{-1}k^{-1}$. Further if $\underline{d} \in \mathscr{E}(K)$ let $E_{\underline{d}} = \pi_{P,\sigma,\nu}(\alpha_{\underline{d}})$, and $E_F = \Sigma_{d \in F} E_d$. Then one easily shows that $\pi_{P,\omega,\nu}(x)h$ again belongs to $\mathscr{H}^\omega$; in fact one can show that $x \rightarrow \pi_{P,\omega,\nu}(x)$ is an admissible representation of $G$ on $\mathscr{H}^\omega$ which has an infinitesimal character. This representation is unitary if $\nu \in \mathfrak{F}$. If $\nu \in \mathfrak{F}_c$, $\omega \in \mathscr{E}^2(M)$, we shall refer to $\pi_{P,\omega,\nu}$ as the *cuspidal principal series* (associated to $P$).

THEOREM 2.1. (1) *Let* $T \in \mathrm{End}(\mathscr{H}_F^{\omega})$. *Then*

$$E(P\colon \psi_T\colon \nu\colon k_1\colon x\colon k_2) = \mathrm{tr}\{TE_F\pi_{P,\omega,\nu}(k_1xk_2)E_F\}.$$

(2) *If* $\pi_{P,\omega,\nu}$ *is irreducible then the mapping* $T \to E(P\colon \psi_T\colon \nu)$ *is a bijection of* $\mathrm{End}(\mathscr{H}_F^{\omega})$ *with* $\mathscr{A}(\pi_{P,\sigma,\nu}\colon V\colon \tau)$.

COROLLARY. *Let* $\psi \in L(\omega)$, $\nu \in \mathfrak{F}_c$, $z \in \mathfrak{Z}$, *and let* $\chi_{\omega,\nu}$ *denote the infinitesimal character of* $\pi_{P,\omega,\nu}$. *Then*

$$E(P\colon \psi\colon \nu\colon x;\, z) = \chi_{\omega,\nu}(z)E(P\colon \psi\colon \nu\colon x), \qquad (x \in G).$$

Our goal now is to estimate the Eisenstein integrals $E(P\colon \psi\colon \nu\colon x)$ for $P \in \mathscr{P}(A)$, $\psi \in L$, $\nu \in \mathfrak{F}_c$, and $x \in G$. We need first some properties of the elementary spherical function $\Xi$. For proofs of these lemmas see [2], Section 30. Let $Q \supset P_0$ be an arbitrary psgp of $G$; we put $Q = M_QA_QN_Q$.

LEMMA 2.3. *Let* $\mathrm{d}_Qk$ *denote the normalized Haar measure on* $K_Q = K \cap M_Q$. *Then*

$$\int_{K_Q} \mathrm{e}^{-\rho_0(H_0(xk))}\,\mathrm{d}_Qk = \mathrm{e}^{-\rho_Q(H_Q(x))}\Xi_{M_Q}(\mu_Q(x)) \qquad (x \in G).$$

COROLLARY. *Let* $\mathrm{d}k$ *denote the normalized Haar measure on* $K$. *Then*

$$\int_K \mathrm{e}^{-\rho_Q(H_Q(xk))}\Xi_{M_Q}(\mu_Q(xk))\,\mathrm{d}k = \Xi(x) \qquad (x \in G).$$

LEMMA 2.4. (1) *There exists constants* $c_0 > 0$, $r_0 \geqslant 0$ *such that if* $M_{Q1}^+ = K_Q\,\mathrm{cl}\,(A_0^+)K_Q$ *then*

$$\mathrm{d}_Q(m)\Xi(m) \leqslant c_0\Xi_{M_Q}(m)(1+\sigma(m))^{r_0} \qquad (m \in M_{Q1}^+).$$

(2) *There exists constants* $D_0 > 0$, $q_0 \geqslant 0$ *such that*

$$\mathrm{e}^{-\rho_0(\log h)} \leqslant \Xi(h) \leqslant D_0\,\mathrm{e}^{-\rho_0(\log h)}(1+\sigma(h))^{q_0} \qquad (h \in A_0^+).$$

as $\mathfrak{g}_c = \mathfrak{k}_c + \mathfrak{m}_{1c} + \mathfrak{n}_c$ and $(\mathfrak{k} + \mathfrak{m}_{1c}) \cap \mathfrak{n}_c = 0$ then

$$\mathfrak{G} = \mathfrak{K}\mathfrak{M}_1 \oplus \mathfrak{G}\mathfrak{n} = \mathfrak{G}\mathfrak{n} \oplus \mathfrak{M}_1\mathfrak{K}$$

Therefore for every $a \in \mathfrak{G}$ there exists unique $X_i \in \mathfrak{n}$, $a_i \in \mathfrak{G}$, $\xi_j \in \mathfrak{K}$, and

$u_j \in \mathfrak{M}_1$ such that

$$a = \sum_i a_i X_i + \sum_j u_j \xi_j.$$

Further we can find (nonuniquely) $u_{ik} \in \mathfrak{M}_1$, $\xi_{ik} \in \mathfrak{K}$, $\eta_{ik} \in \mathfrak{N}$ such that

$$a_i = \sum_k \xi_{ik} u_{ik} \eta_{ik}.$$

For $\delta \in \mathfrak{M}^{(2)} = \mathfrak{M} \otimes \mathfrak{M}$, $f \in \mathscr{C}(M: V: \tau_M)$, $F \subset \mathfrak{M}^{(2)}$, $|F| < \infty$, let

$$v_\delta(f) = \sup_M \|\delta f\| \Xi_M^{-1}, \qquad v_F(f) = \sum_{\delta \in F} v_\delta(f)$$

Here $\|g\|$ means $\|g(m)\|$ where $\|\cdot\|$ denotes the $L^2$-norm on $V$ ($g \in \mathscr{C}(M: V: \tau)$). Put

$$|(v, x)| = (1 + |v|)(1 + \sigma(x)) \qquad (v \in \mathfrak{F}_c,\ x \in G).$$

If $v \in \mathfrak{F}_c$ $v = v_R + i v_I$, it is not too hard to show (cf. [6], Lemma 3.5.2) that we can choose $c_0 \geqslant 0$ so that

$$\operatorname{Re}(iv(H_P(x))| \leqslant c_0 |v_I| \sigma(x) \qquad (v \in \mathfrak{F}_c,\ x \in G).$$

We now give the a priori estimates for the Eisenstein integral.

LEMMA 2.5. *Fix $g_1, g_2 \in \mathfrak{G}$ and $u \in \mathscr{S}(\mathfrak{F}_c)$:*[1] *Then we can choose $r \geqslant 0$ and a finite subset $F \subset \mathfrak{M}^{(2)}$ with the following property.*

$$\|E(P(\psi: v; u: g_1; x; g_2)\| \leqslant v_F(\psi) \Xi(x) |(v, x)|^r \exp\{c_0 |v_I| \sigma(x)\}$$

*for all $v \in \mathfrak{F}_c$, $\psi \in C^\infty(M: v: \tau_M)$, and $x \in G$.*

We shall consider the case $u = 1$; the general result will follow from this case using the holomorphy in $v$ of $E(P: \psi: v)$ and the Cauchy integral formula. Put

$$\psi_v(x) = \psi(x) \exp\{(iv - \rho_P) H_p(x)\} \qquad (x \in G).$$

It is clear that $\psi_v(xn) = \psi_v(x)$ ($x \in G$, $n \in N$), and

$$\|E(P: \psi: v: g_1; x; g_2)\| \leqslant \int_K \|\psi_v(g_1; xk; g_2^k)\|\, dk.$$

---

[1] $\mathscr{S}(\mathfrak{F}_c)$ denotes the symmetric algebra over $\mathfrak{F}_c$. We shall consider elements of $\mathscr{S}(\mathfrak{F}_c)$ as derivatives acting on $C^\infty$ functions defined on $\mathfrak{F}_c$.

Also note that

$$\psi_\nu(g_1 ; kman; g_2) = \tau(k)\psi_\nu(g_1^k ; man; g_2).$$

Let $g \in \mathfrak{G}$. From the Poincaré–Birkhoff–Witt theorem there exists $g_i \in \mathfrak{G}$, and $f_i \in C^\infty(K)$ $(1 \leqslant i \leqslant s)$ such that $g^k = \Sigma_{1 \leqslant i \leqslant s} f_i(k) g_i$ for all $k \in K$. Combining this with the decomposition $\mathfrak{G} = \mathfrak{K}\mathfrak{M}_1 \oplus \mathfrak{G}\mathfrak{n}$ and the fact that $\tau$ is unitary with respect to the norm $\|\cdot\|$ we see that there exists $u_i, v_i \in \mathfrak{M}$ $(1 \leqslant i \leqslant P)$ such that

$$\|\psi_\nu(g_1 ; xk; g_2^k)\| \leqslant \sum_{1 \leqslant i \leqslant P} \|\psi(u_i ; \mu_P(xk); v_i)\| (1 + |\nu|)^r \, \mathrm{e}^{-(\nu_I + \rho_P) H_P(xk)}.$$

Setting $F = \{u_i \otimes v_i : 1 \leqslant i \leqslant P\}$ and recalling the definition of $\nu_F$ we have

$$\|E(P : \psi ; \nu : g_1 ; x; g_2)\| \leqslant \nu_F(\psi) \int_K \mathrm{e}^{-(\nu_I + \rho_P) H_P(xk)} \Xi_M(\mu_P(xk))\, \mathrm{d}k (1 + |\nu|)^r.$$

The result now follows from the corollary to Lemma 2.3.

## 3. Functions of type II($\lambda$) and their limits

In this section we introduce the functions of type II($\lambda$). There are eigenfunctions of $\mathfrak{Z}$, the center of $\mathfrak{G}$, which satisfy estimates analogous to those satisfied by the Eisenstein integrals (cf. Lemma 2.5). We shall show that if $Q$ is an arbitrary psgp of $G$ containing $P_0$ then the eigenfunction equations can be transcribed into a first-order vector equation on the reductive part of $Q$. Recall that ${}^\circ G = K \text{ cl } (A_0^+) K$, and ${}^\circ G^+ = K A_0^+ K$ is an open dense subset of ${}^\circ G$. If then, $\varphi$ is $\tau$-spherical and is bounded on $C$ (cf. Section 1) then its behavior at infinity on $G$ is completely determined by its restriction to $A_0^+$. $A_0$ is a vector group (i.e. isomorphic to $\mathbb{R}^d$ for some integer $d$) and hence $A_0^+$ is an $d'$-dimensional manifold $(d' \leqslant d)$. There are then many ways to approach infinity in $\mathrm{Cl}(A_0^+)$. In fact if $H \in \mathrm{Cl}(A_0^+)$, $a_t = \exp tH$, and $H \neq 0$, then letting $t$ tend to $+\infty$ gives us such a direction where possibly some of the coordinates on $A_0^+$ remain bounded and some tend to $+\infty$. Let

$$\mathfrak{m}_{1H} = \mathrm{Cent}_{\mathfrak{g}}(H),$$

$$\Delta_H = \Delta_H(\mathfrak{g}, \mathfrak{a}_0) = \{\alpha \in \Delta(\mathfrak{g}, \mathfrak{a}_0) : \alpha(H) = 0\},$$

and

$$\Delta_H^+ = \Delta_H(\mathfrak{g}, \mathfrak{a}_0) \cap \Delta(\mathfrak{p}_0, \mathfrak{a}_0).$$

Put

$$\mathfrak{n}_H = \sum_{\alpha \in \Delta(\mathfrak{p}_0, \mathfrak{a}_0)/\Delta_H^+} \mathfrak{g}_\alpha, \qquad \text{and} \qquad N_H = \exp \mathfrak{n}_H.$$

Set

$$\mathfrak{a}_H = \{H \in \mathfrak{a}_0 : \alpha(H) = 0 \quad \forall \alpha \in \Delta_H\}, \qquad \text{and} \qquad A_H = \exp \mathfrak{a}_H.$$

If we put $M_{1H} = \mathrm{Cent}_G(\mathfrak{a}_H)$, $M_H = {}^\circ M_{1H}$ (cf. Section 1 for notation) then $M_{1H} = M_H A_H$ and $P_H = M_H A_H N_H$ is a psgp of $G$, $P_0 \subset P_H$, and $A_0 \supset A_H$. Hence the various directions to infinity on $A_0^+$ are indexed by parabolic pairs $(Q, A)$ ($A$ is the unique maximal vector subgroup lying in the center of $Q \cap \theta(Q)$) such that $P_0 \subset Q$, $A_0 \supset A$; henceforth we shall write $(Q, A) \succ \succ (P_0, A_0)$. Fix a $P$-pair $(Q, A) \succ (P_0, A_0)$, $Q = MAN$ (note the use of $A$ as the split component of $Q$ here should not be confused with the usage of $A$ in Sections 1 and 2; it is *not* intended that $A$ equal $\exp \mathfrak{h}_R$. We shall fix notation below). We show using the vector differential equation already mentioned that if $\varphi$ is a function of type II($\lambda$) and $\psi$ is related to $\lambda$ (cf. Section 1) then there exists a unique function $\psi_Q \in C^\infty(\mathfrak{F} \times MA : V : \tau_M)$ such that

$$\lim_{a \underset{Q}{\to} \infty} \{d_Q(ma)\psi(v : ma) - \psi_Q(v : ma)\} = 0 \qquad (v \in \mathfrak{F}, m \in MA)$$

where the symbol $a \underset{Q}{\to} \infty$ means that $a$ is a variable element of

$$A^+ = \{h \in A : \alpha(\log h) > 0 \qquad \text{for all } \alpha \in \Delta(Q, A)\}$$

and there exists $\varepsilon > 0$ so that $\alpha(\log h) > \varepsilon\sigma(h)$ and $\sigma(h) \to +\infty$. Moreover, if $H \in \mathfrak{a}^+$, $A^+ = \exp \mathfrak{a}^+$, and $a_t = \exp tH$, $m \in M_1^+ = K_M A_0^+ K_M$ ($K_M = K \cap M$) we shall estimate

$$\|d_Q(ma_t)\psi(v : ma_t) - \psi_Q(v : ma_t)\|.$$

The final important point of this section is that $\psi_Q$ can be written as a finite sum of functions which suitably translated are of type II($\lambda$).

Let us begin by setting notation and then introduce some invariant theory which leads to the vector differential equation on $M_1 = MA$. Recall we have fixed $\theta$-stable CSA$^s$ $\mathfrak{h}$ and $\mathfrak{h}_0$ such that $\mathfrak{h}_R \subset \mathfrak{h}_{0R} = \mathfrak{a}_0$ and a pair $(Q, A) \succ (P_0, A_0)$. Let $\mathfrak{q} = \mathfrak{m} + \mathfrak{a} + \mathfrak{n}$, $\mathfrak{m}_1 = \mathfrak{m} + \mathfrak{a}$, where $LA(M) = \mathfrak{m}$,

$A = \exp \mathfrak{a}$, $N = \exp \mathfrak{n}$. Set $W_1$ equal to the subgroup of $W_0 = W(\mathfrak{g}_c, \mathfrak{h}_{0c})$ (the Weyl group of the pair $(\mathfrak{g}_c, \mathfrak{h}_{0c})$) which is generated by the reflections $s_\alpha$ where $\alpha|_{\mathfrak{a}} \equiv 0$. Note that we may identify $W_1$ with $W(\mathfrak{m}_{1c}, \mathfrak{h}_{0c})$. Let $\mathfrak{H}_0$ be the subalgebra of $\mathfrak{G}$ generated by 1 and $\mathfrak{h}_{0c}$. Denote by $\mathfrak{H}_0^{W_0}$(resp $\mathfrak{H}_0^{W_1}$) the set of $W_0$ (resp $W_1$) invariants in $\mathfrak{H}_0$. Let $\mathfrak{M}$, $\mathfrak{M}_1$, $\mathfrak{A}$, $\mathfrak{N}$ be the subalgebras of $\mathfrak{G}$ generated by 1 and $\mathfrak{m}_c$, $\mathfrak{m}_{1c}$, $\mathfrak{a}_c$, and $\mathfrak{n}_c$, respectively. Let $\mathfrak{Z}_1$ denote the center of $\mathfrak{M}_1$. There exists a canonical isomorphism

$$\mu_0 = \mu_{\mathfrak{g}/\mathfrak{h}_0} (\text{resp } \mu_1 = \mu_{\mathfrak{m}_1/\mathfrak{h}_0})$$

of $\mathfrak{Z}$ (resp $\mathfrak{Z}_1$) onto $\mathfrak{H}_0^{W_0}$ (resp $\mathfrak{H}_0^{W_1}$). If $z \in \mathfrak{Z}$ then there exists $z_1 \in \mathfrak{Z}_1$ such that $z - z_1 \in \mathfrak{G}\mathfrak{n}$. It is known that $z - z_1 \in \theta(\mathfrak{n})\mathfrak{G}\mathfrak{n}$ and that, if we write $\mu_{01}(z) = d_Q \circ z_1 \circ d_Q^{-1}$,[2] then $\mu_{01}$ is an algebra injection of $\mathfrak{Z}$ into $\mathfrak{Z}_1$, and $\mu_0(z) = \mu_1(\mu_{01}(z))$ for all $z \in \mathfrak{Z}$. It follows from this that $\mathfrak{Z}_1$ is a finite free module over $\mu_{01}(\mathfrak{Z})$ of rank $r = [W_0 : W_1]$. Let $v_1 = 1, v_2, \ldots, v_r$ be a free basis for this module. Then if $v \in \mathfrak{Z}_1$ there exists unique $z_{r;ij} \in \mathfrak{Z}$ such that

$$vv_i = \sum_{1 \leq j \leq r} \mu_{01}(z_{v:ij})v_j \qquad (1 \leq i \leq r).$$

Fix a complex Hilbert space $T$ of dimension $r$ and an orthonormal basis $\{e_1, \ldots, e_r\}$. We shall identify endomorphisms of $T$ with their matrices in this fixed basis. Put $\underline{V} = V \otimes_c T$, $\underline{\tau}_i(k) = \tau_i(k) \otimes 1$, and $\underline{\tau} = (\underline{\tau}_1, \underline{\tau}_2)$. $\underline{V}$ is a Hilbert space in the usual way when we consider $V$ as a Hilbert space with the $L^2$-norm; we denote the norm on $V$ and $\underline{V}$ by $\|\cdot\|$.

For $\Lambda \in \mathfrak{h}_{0c}^*$, $v \in \mathfrak{Z}_1$, let $\Gamma(\Lambda : v)$ be the endomorphism of $T$ with matrix $(\mu_0(z_{v;ji} : \Lambda))_{1 \leq i,j \leq r}$;[3] then $\Gamma(s\Lambda : v) = \Gamma(\Lambda : v)$ $(s \in W)$ and $v \to \Gamma(\Lambda : v)$ is a representation of $\mathfrak{Z}_1$ in $T$. It is known that $\Gamma(\Lambda : v)$ has the numbers $\mu_1(v : s\Lambda)$ $(s \in W_0)$ as its eigenvalues, and that it is semisimple if $\Lambda$ is regular (i.e. $\Lambda(H_\alpha) \neq 0$ $\forall \alpha \in \Delta(\mathfrak{g}_c, \mathfrak{h}_{0c})$). Let $\mathfrak{h}_{0c}^{*\prime}$ be the set of regular elements of $\mathfrak{h}_{0c}^*$. Since $\mathfrak{Z}_1 \supset \mathfrak{A}$ it is clear that if $\Lambda \in \mathfrak{h}_{0c}^{*\prime}$ and $H \in \mathfrak{A}$ then $\Gamma(\Lambda : H)$ is semisimple with eigenvalues $s\Lambda(H)$ $(s \in W_0)$. In fact the following lemma is valid.

---

[2]Identify $\mathfrak{m}_{1c}$ with its image in $\mathfrak{M}_1$. If $X \in \mathfrak{m}_{1c}$, $X = X_1 + H$ $(X_1 \in \mathfrak{m}_c, H \in \mathfrak{a}_c)$. The map $X \to X - \rho_Q(H)$ (resp $X \to X + \rho_Q(H)$) is an isomorphism of $\mathfrak{m}_{1c}$. Let $\eta \to {}'\eta = d_Q \circ \eta \circ d_Q^{-1}$ (resp $\eta \to \eta' = d_Q^{-1} \circ \eta \circ d_Q$) denote the lifted map to $\mathfrak{M}_1$.

[3]Elements of $\mathfrak{H}_0$ can be considered as polynomial functions on $\mathfrak{h}_{0c}^*$. $\mu_0(z : \Lambda) = \mu_0(z)(\Lambda)$ denotes evaluation at $\Lambda$; similarly for $\mu_1(v : \Lambda)$.

Fix a positive system of roots for $\Delta(\mathfrak{g}_c, \mathfrak{h}_{0c})$ call it $P_0$; put

$$P_1 = \{\alpha \in P_0 : \alpha|_{\mathfrak{a}} \equiv 0\},$$

$$\tilde{\omega}_0 = \prod_{\alpha \in P_0} H_\alpha, \qquad \tilde{\omega}_1 = \prod_{\alpha \in P_1} H_\alpha, \quad \tilde{\omega}_{01} = \tilde{\omega}_0 / \tilde{\omega}_1.$$

Fix a complete set of representatives of the cosets $W_0/W_1$; call them $s_1 = 1, \ldots, s_r$.

LEMMA 3.1. *Let* $u_j = \mu_1(v_j)$ $(1 \leqslant j \leqslant r)$ *and* $e_k(\Lambda) = \Sigma_{1 \leqslant j \leqslant r} u_j(s_k\Lambda) e_j \in T$. *Then if* $\Lambda \in \mathfrak{h}_{0c}^{*\prime}$, *the* $e_j(\Lambda)$ *form a basis for* $T$, *and* $\Gamma(\Lambda : v) e_j(\Lambda) = \mu_1(v : s_j\Lambda) e_j(\Lambda)$ $(v \in \mathfrak{Z}_1,\ 1 \leqslant j \leqslant r)$. *Moreover, there is an* $r \times r$ *matrix* $E$ *with entries in the quotient field of* $\mathfrak{H}_0^{W_1}$ *having the following properties*: (i) $\tilde{\omega}_{01} E$ *has entries in* $\mathfrak{H}_0^{W_1}$, (ii) *for* $\Lambda \in \mathfrak{h}_{0c}^{*\prime}$, $E(s_k\Lambda)$ *are the projections* $T \to \mathbb{C} e_k(\Lambda)$ *corresponding to the direct sum* $T = \Sigma_{1 \leqslant k \leqslant r} \mathbb{C} e_k(\Lambda)$.

An element $\lambda \in (\mathfrak{h}_I)_c^*$ is called singular if $\lambda(H_\alpha) = 0$ for some imaginary root $\alpha$ of $(\mathfrak{g}_c, \mathfrak{h}_c)$. Otherwise we call it regular. Put $\mathfrak{F} = \mathfrak{h}_R^*$ and $\tilde{\omega} = \tilde{\omega}_{\mathfrak{g}/\mathfrak{h}} = \Pi_{\alpha > 0} H_\alpha$ (here $\Pi_{\alpha > 0}$ means product over some positive system for $\Delta(\mathfrak{g}_c/\mathfrak{h}_c)$). Fix a regular element $\lambda \in i\mathfrak{h}_I^*$, extend it to $\mathfrak{h}$ by requiring $\lambda(H) = 0$ for $H \in \mathfrak{h}_R$, and let $\mathfrak{F}_c'(\lambda)$ denote the set of $v \in \mathfrak{F}_c$ such that $\tilde{\omega}(\lambda + iv) \neq 0$. Put $\mathfrak{F}'(\lambda) = \mathfrak{F} \cap \mathfrak{F}_c'(\lambda)$. Then $\mathfrak{F}'(\lambda)$ is an open dense subset of $\mathfrak{F}$. If $v \in \mathfrak{F}_c$ we put $v(H) = 0$ if $H \in \mathfrak{h}_I$.

Let $\mathfrak{z} = \mathrm{cent}_{\mathfrak{g}}(\mathfrak{h}_R)$. Then $\mathfrak{h}$ and $\mathfrak{h}_0$ are two Cartan subalgebras of $\mathfrak{z}$. Hence there exists $y \in G_c$ such that $y$ centralizes $\mathfrak{h}_R$ and $\mathfrak{h}_c^y = \mathfrak{h}_{0c}$. Put $\Lambda_v = (\lambda + iv)^y$ for $v \in \mathfrak{F}_c$. Then if $v \in \mathfrak{F}_c'(\lambda)$, it is clear that $\tilde{\omega}_0(\Lambda_v) \neq 0$.

Let $\mathfrak{D} = \mathscr{S}(\mathfrak{a}_c) \otimes \mathscr{S}(\mathfrak{F}_c)$,[4] $g_1 \otimes g_2 \in \mathfrak{G}^{(2)}$, and $\varphi \in C^\infty(\mathfrak{F} \times G : V)$. We define for $P \otimes u \in \mathfrak{D}$

$$\varphi(v; P \otimes u : x) = P(v) \varphi(v; u : x)$$

$$(P \otimes u \otimes g_1 \otimes g_2 \varphi)(v : x) = \varphi(v; P \otimes u : g_1 ; x; g_2).$$

Further we put $\tilde{\mathfrak{G}} = \mathfrak{D} \otimes \mathfrak{G}^{(2)}$, and for $D \in \tilde{\mathfrak{G}}$, $r \in \mathbb{R}$, we put

$$S_{D,r}(\varphi) = \sup_{\mathfrak{F} \times G} \|D\varphi\| \Xi^{-1} |(v, x)|^{-r}.$$

If $F \subset \tilde{\mathfrak{G}}$, $|F| < \infty$, then we put

$$S_{F,R}(\varphi) = \sum_{D \in F} S_{D,r}(\varphi).$$

[4] $\mathscr{S}(\mathfrak{a}_c)$ denotes the symmetric algebra over $\mathfrak{a}_c^*$, we consider elements of $\mathscr{S}(\mathfrak{a}_c)$ to be polynomial functions on $\mathfrak{F}_c$. Then $\mathfrak{D}$ is the set of polynomial differential operators.

A function $\varphi : \mathfrak{F}\times G\to V$ will be called a function of type II($\lambda$) if the following conditions hold:

(1) $\varphi$ is of class $C^\infty$.

(2) For any $\nu\in\mathfrak{F}$, the function $\varphi_\nu=\varphi(\nu)$ is a $\tau$-spherical function on $G$.

(3) $z\varphi_\nu=\mu_0(z:\Lambda_\nu)\varphi_\nu$ $(z\in\mathfrak{Z})$.

(4) For any $D\in\tilde{\mathfrak{G}}$, we can choose a number $r\geqslant 0$ such that $S_{D,r}(\varphi)<\infty$.

*Remark.* Let $\lambda\in i\mathfrak{h}_I^*$, $\omega=\omega(\lambda)$ (cf. [1]), $\psi\in L(\omega)$, $\mathfrak{p}\in\mathscr{P}(\mathfrak{h}_R)$, $P=\mathrm{Norm}_G(\mathfrak{p})$. Then for all $\nu\in\mathfrak{F}_c$, $z\in\mathfrak{Z}$, $E(P:\psi:\nu:x;z)=\mu_0(z:\Lambda_\nu)E(P:\psi:\nu:x)$. Hence, from Lemma 2.5 we see that $E(P:\psi)$ is a function of type II($\lambda$).

Fix $f$, a function of type II($\lambda$), and put

$$\Phi(f:\nu:m)=\sum_{1\leqslant i\leqslant r} f(\nu:m;v_i\circ d_Q)\otimes e_i \qquad (m\in M_i).$$

For $v\in\mathfrak{Z}$, choose $z_{v;ij}\in\mathfrak{Z}$ so that

$$vv_i=\sum_{1\leqslant j\leqslant r}\mu_{01}(z_{v;ij})v_j$$
$$\Rightarrow vv_i\circ d_Q=\sum_{1\leqslant j\leqslant r} d_Q\circ\mu_{01}(z_{v;ij})'v_j'=\sum_{1\leqslant j\leqslant r} d_Q\circ v_j'\mu_{01}(z_{v;ij})_0'^2.$$

Put

$$u_i(v:\nu)=\sum_{1\leqslant j\leqslant r}\mu_{01}(z_{v;ij}-\mu_0(z_{v;ij}:\Lambda_\nu))v_j,$$

and

$$\Psi_v(f:\nu:m)=\sum_{1\leqslant i\leqslant r} f(\nu:m;u_i(v:\nu)\circ d_Q)\otimes e_i.$$

Let 1 denote the identity transformation on $V$ and put

$$\underline{\Gamma}(\Lambda:v)=1\otimes\Gamma(\Lambda:v) \qquad (\Lambda\in\mathfrak{h}_{0c}^*,\ v\in\mathfrak{Z}_1).$$

Then $v\to\underline{\Gamma}(\Lambda:v)$ is a representation of $\mathfrak{Z}_1$ on $\underline{V}$.

LEMMA 3.2. *Let $\eta\in\mathfrak{M}_1$, $v\in\mathfrak{Z}_1$. Then*

$$\Phi(f:\nu:m;\eta v)=\underline{\Gamma}(\Lambda_\nu:v)\Phi(f:\nu:m;\eta)+\Psi_v(f:\nu:m;\eta)$$

*for all $m\in M_1$, and $\nu\in\mathfrak{F}$.*

By definition,

$$\Phi(f:\nu:m;\eta v)=\sum_{1\leqslant i\leqslant r} f(\nu:m;\eta vv_i\circ d_Q)\otimes e_i$$

$$= \sum_{1\leqslant i\leqslant r} \sum_{1\leqslant j\leqslant r} f(v\colon m; \eta \circ d_Q \circ \mu_{01}(z_{v:ij})' v_j') \otimes e_i$$

$$= \sum_{1\leqslant i\leqslant r} \sum_{1\leqslant j\leqslant r} f(v\colon m; \eta \circ d_Q \circ v_j'[z_{v:ij} - (z_{v;ij} - \mu_{01}(z_{v:ij})')]) \otimes e_i$$

$$= \sum_{1\leqslant i\leqslant r} \sum_{1\leqslant j\leqslant r} f(v\colon m; \eta \circ d_Q \circ v_j' \circ z_{v:ij}) \otimes e_i -$$

$$- \sum_{1\leqslant i\leqslant r} \sum_{1\leqslant j\leqslant r} f(v\colon m; \eta \circ d_Q \circ (z_{v:ij} - \mu_{01}(z_{v:ij})') \circ v_j') \otimes e_i.$$

But

$$z_{v:ij} - \mu_{01}(z_{v:ij})' = z_{v:ij} - \mu_0(z_{v:ij}\colon \Lambda_v) - \mu_{01}(z_{v:ij} - \mu_0(z_{v:ij}\colon \Lambda_v))'.$$

As $z_{v;ij} - \mu_0(z_{v;ij}\colon \Lambda_v)$ kills every function of type II($\lambda$) we have

$$\Phi(f : v\colon m; \eta v) = \sum_{1\leqslant j\leqslant r} f(v\colon m; \eta v_j\, d_Q) \otimes \sum_{1\leqslant i\leqslant r} \mu_0(z_{v:ij}\colon \Lambda_v) e_i +$$

$$+ \sum_{1\leqslant i\leqslant r} \sum_{1\leqslant j\leqslant r} f(v\colon m; \eta \mu_{01}(z_{v:ij} - \mu_0(z_{v:ij}\colon \Lambda_v)) v_j \circ d_Q) \otimes e_i.$$

The lemma is now obvious.

COROLLARY. *Let* $\eta_1, \eta_2 \in \mathfrak{M}_1$, $v \in \mathfrak{Z}_1$. *Then*

$$\Phi(f : v\colon \eta_1 ; m; \eta_2 v) = \underline{\Gamma}(\Lambda_v\colon v) \Phi(f : v\colon \eta_1 ; m; \eta_2) + \\ + \Psi_v(f : v\colon \eta_1 ; m; \eta_2).$$

COROLLARY. *Let* $\eta_1, \eta_2 \in \mathfrak{M}_1$, $H \in \mathfrak{a}$. *Then for all* $T \in \mathbb{R}$, $m \in M_1$, *we have*

$$\Phi(f : v\colon \eta_1 ; m \exp TH; \eta_2)$$
$$= \exp\{T \underline{\Gamma}(\Lambda_v\colon H)\}\, \Phi(f : v\colon \eta_1 ; m; \eta_2)$$
$$+ \int_0^T \exp\{(T-t)\underline{\Gamma}(\Lambda_v\colon H)\}\, \Psi_H(f : v\colon \eta_1 ; m \exp tH; \eta_2)\, dt.$$

To prove the second corollary, we put

$$\Phi(f:\nu:m:T)=\exp\{-T\underline{\Gamma}(\Lambda_\nu:H)\}\Phi(f:\nu:m\exp TH).$$

Then it is easy to show that as $H \in \mathfrak{z}_1$,

$$\begin{aligned}\frac{\mathrm{d}}{\mathrm{d}T}\Phi(f:\nu:\eta_1;m;\eta_2:T) &= -\underline{\Gamma}(\Lambda_\nu:H)\Phi(f:\nu:\eta_1;m;\eta_2:T)+\\ &\quad+\exp\{-T\underline{\Gamma}(\Lambda_\nu:H)\}\Phi(f:\nu:\eta_1;m\exp TH;\eta_2H)\\ &=\exp\{-T\underline{\Gamma}(\Lambda_\nu:H)\}\Psi_H(f:\nu:\eta_1;m\exp TH;\eta_2).\end{aligned}$$

Hence,

$$\Phi(f:\nu:\eta_1;m;\eta_2:T)=\Phi(f:\nu:\eta_1;m;\eta_2)+\\ +\int_0^T\exp\{-t\underline{\Gamma}(\Lambda_\nu:H)\}\Psi_H(f:\nu:\eta_1;m\exp tH;\eta_2)\,\mathrm{d}t.$$

Multiplying both sides of this last equation by $\exp\{T\underline{\Gamma}(\Lambda_\nu:H)\}$ gives the result.

Let us recall the notation of Lemma 2.4 and put

$$|(\nu,x,y)|=(1+|\nu|)(1+\sigma(x))(1+\sigma(y))\quad (x,y\in G,\ \nu\in\mathfrak{F}).$$

LEMMA 3.3. *Fix $D\in\mathfrak{D}$, $g_1$, $g_2\in\mathfrak{G}$, $X\in\mathfrak{n}$, and $f$ as above. Then we can choose a finite subset $F\subset\mathfrak{G}$ such that,*

$$d_Q(ma)\{\|f(\nu;D:g_1X;ma;g_2)\|+\|f(\nu;D:g_1;ma;\theta(X)g_2)\|\}\\ \leqslant S_{F,r}(f)\Xi_M(m)|(\nu,m,a)|^{r+r_0}\,\mathrm{e}^{-\beta_Q(\log a)}$$

*and*

$$d_Q(ma)\|f(\nu;D:g_1;m;g_2)\|\leqslant S_{F,r}(f)\Xi_M(m)|(\nu,m)|^{r+r_0}$$

*for all $m\in M_1^+=K_M\,\mathrm{Cl}(A_0^+)K_M$, $a\in A^+$, and $r\in\mathbb{R}^+$. Here*

$$\beta_Q(H)=\inf\{\alpha(H):\alpha\in\Delta(Q,A)\}$$

The lemma follows easily from the following observations. If $m\in M_1^+$, $a_0\in\mathrm{Cl}(A_0^+)$ then $ma_0=k_1^{-1}hk_2$ where $h=a_0a\in\mathrm{Cl}(A_0^+)$; $k_1$, $k_2\in K_M$. Therefore if $\varphi\in C^\infty(G)$ then with $Y=\theta(X)$ we have

$$\varphi(g_1X;ma;g_2)=\tau(k_1^{-1})\varphi(g_1^{k_1};h;X^{h^{-1}k_1}g_2^{k_2})\tau(k_2)$$
$$\varphi(g_1;ma;Yg_2)=\tau(k_1^{-1})\varphi(g_1^{k_1}Y^{hk_2};h;g_2^{k_2})\tau(k_2).$$

As in the proof of Lemma 2.5 there exists $f_{ij} \in C^\infty(K)$, $g_{ij} \in \mathfrak{G}$ $(1 \leqslant i \leqslant \alpha$, $j=1, 2)$ such that $g_j^{k_j} = \Sigma_i f_{ij}(k_j) g_{ij}$; if $X = X_\alpha$, $\alpha \in \Delta(Q, A)$ then $X^{h^{-1}k_1} = e^{-\alpha(\log h)} X^{k_1}$, $Y^{hk_2} = e^{-\alpha(\log h)} Y^{k_2}$. Again writing $X^{k_1}$, $Y^{k_2}$ in terms of a basis for $\mathfrak{g}$ (with $C^\infty$-functions on $K$ as coefficients) we see the lemma follows.

Put

$$|(v, x, Y)| = (1+|v|)(1+\sigma(x))(1+|Y|) \qquad (v \in \mathfrak{F}, x \in G, Y \in \mathfrak{g}).$$

COROLLARY. *Fix $D \in \mathfrak{D}$, $v \in \mathfrak{Z}$, $\eta_1, \eta_2 \in \mathfrak{M}_1$. Then we can choose a finite subset $F \subset \tilde{\mathfrak{G}}$ such that*

$$\|\Psi_v(f: v; D: \eta_1; m \exp H; \eta_2)\| \leqslant S_{F,r}(f) \Xi_M(m) |(v, m, H)|^{r+r_0} e^{-\beta_Q(H)}$$

$$\|\Phi(f: v; D: \eta_1; m \exp H; \eta_2)\| \leqslant S_{F,r}(f) \Xi_M(m) |(v, m, H)|^{r+r_0}$$

*The two inequalities hold for $m \in M_1^+$, $H \in \mathrm{Cl}(\mathfrak{a}^+)$, $r \in \mathbb{R}^+$, and $v \in \mathfrak{F}$*

Let $E$ be the operator defined in Lemma 3.1 and put

$$\underline{E}_1 = 1 \otimes \tilde{\omega}_0 E.$$

Further if $1 \leqslant i \leqslant r$ and $v \in \mathfrak{Z}_1$, put,

$$\Phi_i(f: v: m) = \underline{E}_1(s_i \Lambda_v) \Phi(f: v: m),$$
$$\Psi_{v,i}(f: v: m) = \underline{E}_1(s_i \Lambda_v) \Psi_v(f: v: m).$$

LEMMA 3.4. *Let $\eta_1, \eta_2 \in \mathfrak{M}_1$, $m \in M_1$, $H \in \mathfrak{a}$. Then for all $T \in \mathbb{R}$,*

$$\Phi_i(f: v: \eta_1; m \exp TH; \eta_2) = e^{Ts_i\Lambda_v(H)} \Phi_i(f: v: \eta_1; m; \eta_2) + \int_0^T e^{(T-t)s_i\Lambda_v(H)} \Psi_{H,i}(f: v: \eta_1; m \exp tH; \eta_2)\, dt.$$

The lemma follows immediately from the second corollary to Lemma 3.2.

In view of Lemma 3.1 it is clear that there exists $E \geqslant 0$, $e \geqslant 0$ such that,

$$\text{(3.1)} \qquad \|\underline{E}_1(s_i \Lambda_v)\| \leqslant E[(1+|\lambda|)(1+|v|)]^e \qquad (1 \leqslant i \leqslant r, v \in \mathfrak{F}).$$

Let $1 \leqslant i \leqslant r$ and let $\lambda_i$ denote the restriction of $s_i \lambda^y$ on $\mathfrak{a}$. Set $I = \{1, \ldots, r\}$

and define $I^0, I^+, I^-$ as follows:

$$I^0 = \{i \in I : \lambda_i \equiv 0\}$$
$$I^+ = \{i \in I : \lambda_i(H) > 0 \quad \text{for some } H \in \mathfrak{a}^+\}$$
$$I^- = \{i \in I : \lambda_i(H) < 0 \quad \text{for some } H \in \mathfrak{a}^+\}.$$

If $\mathfrak{l}$ is a $\theta$-stable Cartan subalgebra of $\mathfrak{g}$ we put $\mathfrak{l}^R = i\mathfrak{l}_I + \mathfrak{l}_R$; $\mathfrak{l}_I = \mathfrak{l} \cap \mathfrak{k}$, $\mathfrak{l}_R = \mathfrak{l} \cap \mathfrak{s}$. Then $\mathfrak{l}^R$ is $W(\mathfrak{g}, \mathfrak{l})$-stable. Hence as $\mathfrak{a} \subset \mathfrak{h}_0^R$ it is clear that $s_i \lambda^\nu$ is real valued on $\mathfrak{h}_0^R$ and hence $I = I^+ \cup I^0 \cup I^-$. It also follows from this that $s_j \Lambda_\nu(H) \in \mathbb{R}$ for all $j \in I^0$.

LEMMA 3.5. *Let notation be as above. Then if* $D \in \mathfrak{D}$, $\eta_1, \eta_2 \in \mathfrak{M}_1$, *and* $i \in I^0$ *the integral*

$$\int_0^\infty \|\Psi_{H,i}(f : \nu; D \circ \mathrm{e}^{-ts_i\Lambda_\nu(H)} : \eta_1 ; m \exp tH ; \eta_2)\| \,dt$$

*converges uniformly for* $\nu$ *and* $m$ *varying in compact subsets of* $\mathfrak{F}$ *and* $M_1$ *respectively.*

The lemma follows easily from the corollary to Lemma 3.3, (3.1), together with the following result.

LEMMA 3.6. *Let $C$ be a compact subset of $M_1$, $\Omega$ a compact subset of $\mathfrak{a}^+$. Then we can choose $T_0 \geq 0$ so that $m \exp TH \in M_1^+$ for $m \in C$ and $T \geq T_0$.*

See [1], Lemma 54, for a proof of this result.

LEMMA 3.7. *Let $i \in I^0$ and the remaining notation be as in Lemma 3.5. Then*

$$\Phi_{i\infty}(f : \nu : m : H) = \lim_{T \to \infty} \Phi_i(f : \nu : m \exp TH)\, \mathrm{e}^{-Ts_i\Lambda_\nu(H)}$$

*exists. Moreover, $\Phi_{i\infty}$ is of class $C^\infty$ on $\mathfrak{F} \times M_1$ and*

$$(3.2) \qquad \Phi_{i\infty}(f : \nu; D : \eta_1 ; m; \eta_2 : H) = \Phi_i(f : \nu; D : \eta_1 ; m; \eta_2) \times$$
$$\times \int_0^\infty \Psi_{H,i}(f : \nu; D \circ \mathrm{e}^{-ts_i\Lambda_\nu(H)} : \eta_1 ; m \exp tH ; \eta_2)\, \mathrm{d}t.$$

The lemma follows easily on combining Lemmas 3.4 and 3.5.

Let $H_1, H_2 \in \mathfrak{a}^+$. The following argument shows that $\Phi_{i\infty}(f: v: m: H_1) = \Phi_{i\infty}(f: v: m: H_2)$ for all $i \in I^0$. Put $m_1 = m \exp T_1 H$ $(T_1 \geqslant 0)$ for a fixed $m \in M_1$. Then

$$\begin{aligned}\Phi_i(f: v: m \exp(T_1 H_1 + T_2 H_2))\, e^{-T_1 s_i \Lambda_v(H_1) - T_2 s_i \Lambda_v(H_2)} &= \Phi_i(f: v: m \exp T_1 H_1)\, e^{-T_1 s_i \Lambda_v(H_1)} + \\ &+ \int_0^{T_2} \Psi_{H_2, i}(f: v: m \exp(T_1 H_1 + t_2 H_2)) \times \\ &\times e^{-(T_1 s_i \Lambda_v(H_1) + t_2 s_i \Lambda_v(H_2))}\, dt_2.\end{aligned}$$

But from the corollary to Lemma 3.3

$$\lim_{T_1 \to \infty} \int_0^{\infty} \|\Psi_{H_2, i}(f: v: m \exp(T_1 H_1 + t_2 H_2))\|\, dt_2 = 0.$$

Hence,

$$\lim_{T_1, T_2 \to \infty} \Phi_i(f: v: m \exp(T_1 H_1 + T_2 H_2))\, e^{-(T_1 s_i \Lambda_v(H_1) + T_2 s_i \Lambda_v(H_2))} = \Phi_{i\infty}(f: v: m: H_1).$$

As the argument is symmetric in $H_1$ and $H_2$ the claim follows. *As $\Phi_{i\infty}(f: v: m: H)$ does not depend on $H \in \mathfrak{a}^+$ we shall henceforth denote it by $\Phi_{i\infty}(f: v: m)$.*

For $i \in I^+ \cup I^-$ put

$$\Phi_{i\infty}(f) \equiv 0.$$

Choose $\delta$ so that $\frac{1}{2} > \delta > 0$ and $\lambda_i(H) < -\delta \beta_Q(H)$ $(h \in \mathfrak{a}^+, i \in I^-)$.

LEMMA 3.8. (1) $\Phi_{i\infty}(f: v: m; v) = \mu_1(v: s_i \Lambda_v) \Phi_{i\infty}(f: v: m)$, $(v \in \mathfrak{Z}_1)$.

(2) *Given $\eta_1, \eta_2 \in \mathfrak{M}_1$, $D \in \mathfrak{D}$, we can choose a finite subset $F \subset \tilde{\mathfrak{G}}$ such that for every $r \in \mathbb{R}^+$ there exists $C$ so that,*

$$\|\Phi_{i\infty}(f: v; D: \eta_1; m; \eta_2)\| \leqslant C S_{F,r}(f) \Xi_M(m) |(v, m)|^{r + r_0 + e}$$

$(m \in M_1)$.

*Finally,*

(3) $$\|\Phi_i(f: v: \eta_1; m \exp TH; \eta_2) - \Phi_{i\infty}(f: v: \eta_1; m \exp TH; \eta_2)\|$$

$$\leqslant e^{-T\delta\beta\varrho(H)}\{\|\Phi_i(f:v:\eta_1;m;\eta_2)\| + \int_0^\infty \|\Phi_{H,i}(f:v:\eta_1;m\exp tH;\eta_2)\|\, e^{t\beta\varrho(H)/2}\,dt\}.$$

*This last inequality holding for all $\eta_1, \eta_2 \in \mathfrak{M}_1$, $m \in M_1$, $H \in \mathfrak{a}^+$, $v \in \mathfrak{F}$, $i \in I$, and $T \geqslant 0$.*

From Lemma 3.1 and Lemma 3.2 we have

$$\Phi_i(f:v:m\exp TH;v) = \mu_1(v:s_i\Lambda_v)\Phi_i(f:v:m\exp TH) + \Psi_{v,i}(f:v:m\exp TH).$$

But by the corollary to Lemma 3.3 we have,

$$\lim_{T\to\infty} e^{-Ts_i\Lambda_v(H)}\Psi_{v,i}(f:v:m\exp TH) = 0 \qquad (i \in I^0).$$

The first result follows then from Lemma 3.7.

(2) follows on combining (3.2) with the corollary to Lemma 3.3 (note as $\Phi_{i\infty}(f) \equiv 0$ if $i \in I^+ \cup I^-$, we need only consider the case when $i \in I^0$).

It remains then to show (3). But it follows from Lemma 3.7 that for all $i \in I^0$,

(3.3) $$\Phi_i(f:v:\eta_1;m\exp TH;\eta_2) = \Phi_{i\infty}(f:v:\eta_1;m;\eta_2) - \int_T^\infty \Psi_{H,i}(f:v:\eta_1;m\exp tH;\eta_2)\, e^{(T-t)s_i\Lambda_v(H)}\,dt.$$

Hence

$$\|\Phi_i(f:v:\eta_1;m\exp TH;\eta_2) - \Phi_{i\infty}(f:v:\eta_1;m;\eta_2)\| \leqslant \int_0^\infty \|\Psi_{H,i}(f:v:\eta_1;m\exp(tH);\eta_2)\|\,dt.$$

It is easy to show that the right-hand side of the above inequality is

$$\leqslant e^{-T\beta\varrho(H)/2} \int_0^\infty \|\Psi_{H,i}(f:v:\eta_1;m\exp tH;\eta_2)\|\, e^{t\beta\varrho(H)/2}\,dt.$$

Let now $i \in I^-$. Then

$$\|\Phi_i(f:v:\eta_1;m\exp TH;\eta_2)\| \leqslant e^{\delta\beta_Q(H)T}\|\Phi_i(f:v:\eta_1;m;\eta_2)\| + \\ + \int_0^T e^{(T-t)\lambda_i(H)}\|\Psi_{H,i}(f:v:\eta_1;m\exp tH;\eta_2)\|\,dt \\ \leqslant e^{-\delta\beta_Q(H)T}\{\|\Phi_i(f:v:\eta_1;m;\eta_2)\| + \\ + \int_0^T e^{\delta\beta_Q(H)T+(T-t)\lambda_i(H)}\|\Psi_{H,i}(f:v:\eta_1;m\exp G;\eta_2)\|\,dt\Big\}.$$

as $(T-t)\lambda_i(H) \leqslant -\delta(T-t)\beta_Q(H)$ and $\delta < \frac{1}{2}$ the result follows in this case.

Fix $m \in M_1$, $i \in I^+$, $H \in \mathfrak{a}^+$. Choose $H_0 \in \mathfrak{a}^+$ so that $\operatorname{Re} s_i\Lambda_v(H_0) > 0$. From the corollary to Lemma 3.3 and Lemma 3.6 we have

$$\lim_{T\to\infty} e^{-Ts_i\Lambda_v(H_0)}\Phi_i(f:v:\eta_1;m\exp TH_0;\eta_2) = 0.$$

Hence, replacing $m$ by $m\exp sH$ we have

$$\Phi_i(f:v:\eta_1;m\exp(sH+TH_0);\eta_2) \\ = -\int_T^\infty \Psi_{H_0,i}(f:v:\eta_1;m\exp(sH+tH_0);\eta_2)\,e^{-(t-T)s_i\Lambda_v(H_0)}\,dt.$$

Setting $T=0$ and recalling the corollary to Lemma 3.3 shows that

$$\lim_{s\to\infty}\Phi_i(f:v:\eta_1;m\exp(sH);\eta_2) = 0 \\ \Rightarrow \lim_{s\to\infty} e^{-ss_i\Lambda_v(H)}\Phi_i(f:v:\eta_1;m\exp sH;\eta_2) = 0 \\ \Rightarrow \Phi_i(f:v:\eta_1;m\exp sH;\eta_2) \\ = -\int_s^\infty \Psi_{H,i}(f:v:\eta_1;m\exp tH;\eta_2)\,e^{-(t-s)s_i\Lambda_v(H)}\,dt \\ \Rightarrow \|\Phi_i(f:v:\eta_1;m\exp sH;\eta_2)\| \\ \leqslant \int_s^\infty \|\Psi_{H,i}(f:v:\eta_1;m\exp tH;\eta_2)\|\,e^{t\beta_Q(H)/2}\,dt.$$

LEMMA 3.9. *Let $H \in \mathfrak{a}$, $m \in M_1$. Then*

$$\Phi_{i\infty}(f: v: m \exp H) = e^{s_i \Lambda_v(H)} \Phi_{i\infty}(f: v: m).$$

The lemma follows on observing that for $v \in \mathfrak{F}(\lambda)$.

$$\frac{d}{dT} \Phi_{i\infty}(f: v: m \exp TH) = \Phi_{i\infty}(f: v: m \exp TH; H)$$

$$= s_i \Lambda_v(H) \Phi_{i\infty}(f: v: m \exp TH).$$

The last equality by (1) of Lemma 3.8. But this implies that

$$\frac{d}{dT} \{e^{-Ts_i\Lambda_v(H)} \Phi_{i\infty}(f: v: m \exp TH)\} = 0$$

$$\Rightarrow \Phi_{i\infty}(f: v: m \exp TH) = e^{Ts_i\Lambda_v(H)} \Phi_{i\infty}(f: v: m).$$

Setting $T=1$ gives the result for $v \in \mathfrak{F}(\lambda)$. By continuity the equality persists on $\mathfrak{F}$.

LEMMA 3.10. *Fix $i \in I$ and suppose that $v \in \mathfrak{F}(\lambda)$. Then $\Phi_{i\wp}(f) \equiv 0$ unless $s_i^{-1} \mathfrak{a} \subset \mathfrak{h}_R$.*

The proof of this lemma is rather long and would require us to consider the differential equations of functions other than those of type II($\lambda$). By necessity we do not include a proof of this result; rather we refer the reader to [3], Lemma 6.3.

Let $W(\mathfrak{h}_R; \mathfrak{a})$ denote the set of all linear mappings of $\mathfrak{a}$ into $\mathfrak{h}_R$ such that there exists $x \in G$ for which $s(H) = \mathrm{Ad}_x(H)$ for all $H \in \mathfrak{a}$. Note that if $k \in K$ and we define $s(H) = \mathrm{Ad}_k(H)$ (for all $H \in \mathfrak{a}$) then $s$ completely determines the coset $kK_M$. If $H$ is any subgroup of $G$ normalized by $K_M$ we shall define $H^s = kHk^{-1} = H^k$. In particular, we shall write $Q^s = M^s A^s N^s$; $M_1^s = (MA)^s$. If $\varphi$ is a $\tau_M$-spherical function on $M_1$ we shall define $\varphi^s$ on $M_1^s$ as follows:

$$\varphi^s(m^k) = \tau(k)\varphi(m)\tau(k^{-1}).$$

Note that $\varphi^s$ is $\tau_{M^s}$-spherical ($\tau_{M^s} = \tau|_{K \cap M^s}$).

LEMMA 3.11. *Given $s \in W(\mathfrak{h}_R: \mathfrak{a})$ there exists unique index $i \in I$ such that $s(H) = s_i(H)$ for all $H \in \mathfrak{a}$.*

The lemma follows from Corollary 2 to Lemma 5.1 of [2]. We shall denote for $s \in W(\mathfrak{h}_R\colon \mathfrak{a})$ the index $i \in I$ corresponding to $s$ by $i(s)$; hence $s(H) = s_{i(s)}(H)$ for all $H \in \mathfrak{a}$.

Let $\psi_f(v\colon x) = \tilde{\omega}_0(\Lambda_v) f(v\colon x)$. There exists for $i \in I$ unique functions $\varphi_{ij}(f)\colon \mathfrak{F} \times M_1 \to V$ such that

$$\Phi_{i\infty}(f : v : m) = \sum_{1 \leqslant j \leqslant r} \varphi_{ij}(f : v : m) \otimes e_j.$$

Put

$$\psi_{f,i}(v : m) = \varphi_{i1}(f : v : m)$$

and

$$(\psi_f)_{Q,s} = \psi_{f,s} = \psi_{f,i(s)}.$$

Recalling that $v_1 = 1$ and $\tilde{\omega}_0(s\Lambda) = \varepsilon(s)\tilde{\omega}_0(\Lambda)$ ($\varepsilon(s) = \pm 1$) we obtain the following result. Let $\Omega$ be a compact subset of $\mathfrak{a}^+$. Choose $\varepsilon_0 > 0$ so that $\beta_Q(H) \geqslant 2\varepsilon_0$ for all $H \in \Omega$. Put $\varepsilon = \delta\varepsilon_0$ where $\delta$ is as in Lemma 3.8.

THEOREM 3.1. *Given $\eta_1, \eta_2 \in \mathfrak{M}_1$, we can choose a finite subset $F \subset \tilde{\mathfrak{G}}$ and for every $r \in \mathbb{R}^+$ a constant $c > 0$ so that*

$$\begin{aligned} &\|\mathrm{d}_Q(m \exp TH)\psi_f(v : \eta_1' ; m \exp TH ; \eta_2') - \\ &\qquad - \sum_{s \in W(\mathfrak{h}_R;\mathfrak{a})} \psi_{f,s}(v : \eta_1 ; m \exp TH ; \eta_2)\| \\ &\leqslant c S_{F,r}(f)\, \mathrm{e}^{-\varepsilon T} \Xi_M(m) |(v, m)|^{r + r_0 + e}. \end{aligned}$$

*The inequality holding for all $m \in M_1^+$, $H \in \Omega$, and $T \geqslant 0$.*

It is clear that

$$\begin{aligned} &\|d_Q(m \exp TH)\psi_f(v : \eta_1' ; m \exp TH ; \eta_2') - \\ &\qquad - \sum_{s \in W(\mathfrak{h}_R;\mathfrak{a})} \psi_{f,s}(v : \eta_1 ; m \exp TH ; \eta_2)\| \\ &\leqslant \sum_{1 \leqslant i \leqslant r} \|\Phi_i(f : v : \eta_1 ; m \exp TH ; \eta_2) - \\ &\qquad - \Phi_{i\infty}(f : v : \eta_1 ; m \exp TH ; \eta_2)\| \\ &\leqslant \mathrm{e}^{-T\delta\beta_Q(H)} \sum_{1 \leqslant i \leqslant r} \|\Phi_i(f : v : \eta_1 ; m ; \eta_2)\| + \end{aligned}$$

$$+\int_0^\infty \|\Psi_{H,i}(f:v:\eta_1;m\exp tH;\eta_2)\|\,e^{(t/2)\beta\varrho(H)}\,dt.$$

The last inequality following from (3) of Lemma 3.8. From (3.1) and the corollary to Lemma 3.3 we see that there exists $F\subset\tilde{\mathfrak{G}}$ such that for all $r'\in\mathbb{R}$,

$$\leqslant e^{-T\varepsilon}\cdot r\cdot S_{F,r'}(f)\Xi_M(m)|(v,h,H)|^{r'+r_0+}\times$$

$$\times\left[\int_0^\infty e^{-t\varepsilon_0}(1+t)^{r'+r_0+e}\,dt+1\right].$$

The theorem follows easily from this.

*Remark.* Let $\Lambda\in\mathfrak{h}_{0c}^*$ and let $W_0(\Lambda)$ be the subgroup of all $s\in W_0$ which leave $\Lambda$ fixed. Let $P_0$ be a set of positive roots of $\Delta(\mathfrak{g}_c,\mathfrak{h}_{0c})$ and $P_0(\Lambda)$ the set of those $\alpha\in P_0$ such that $\Lambda(H_\alpha)\neq 0$. Put

$$\tilde{\omega}_{0,\Lambda}=\prod_{\alpha\in P_0(\Lambda)} H_\alpha.$$

Let $W_0(s,\lambda)=W_0(s\lambda^y)$. Then if $s\in W_0(s_i,\lambda)$, $i\in I^0$ and if $s_j\equiv ss_i(\text{mod } W_1)$ then $j\in I^0$. Let us denote by $j(s,i)=j$ the index in $I^0$ such that $s_j\equiv ss_i(\text{mod } W_1)$ $(s\in W_0(s_i,\lambda))$. Put

$${}_i\psi_f(v:m)=|W_1\cap W_0(s_i,\lambda)|^{-1}\sum_{t\in W_0(s_i,\lambda)}\tilde{\omega}_0(s_i\Lambda_v)^{-1}\psi_{f,j(t,i)}(v:m)$$

$$(v\in\mathfrak{F}(\lambda),\ m\in M_1)$$

then it is easy to see that ${}_i\psi_f={}_j\psi_f$ if $s_i\lambda^y=s_j\lambda^y$; in fact suppose that $s_i\lambda^y=s_j\lambda^y$. Then as $W_0s_i\ni s_j$ we see upon putting $W_0^\lambda=W_0(s_i,\lambda)=W_0(s_j,\lambda)$, and choosing $t_k\in W_0$ so that $W_0=\bigcup_k t_kW_0^\lambda$ (distinct representatives) that

$$t_kts_i=s_j$$

for some $k$ and $t\in W_0^\lambda$. But this implies that

$$ts_i\lambda=t_k^{-1}s_j\lambda\Rightarrow s_i\lambda=t_k^{-1}s_i\lambda.$$

Hence

$$t_k^{-1}\in W_0^\lambda\Rightarrow t_k=1.$$

This shows that $W_0^\lambda s_i \ni s_j$ and hence that ${}_i\psi_f = {}_j\psi_f$. Let $^\circ I$ be a maximal set of indices such that if $i, j \in {}^\circ I$ then $s_i\lambda^y \neq s_j\lambda^y$. Then we can show that

$$\sum_{i \in {}^\circ I^\circ} {}_i\psi_f(v: m) = \sum_{i \in I} \tilde{\omega}(s_i\Lambda_v)^{-1}\psi_{f,i}(v: m), \qquad (v \in \mathfrak{F}(\lambda), m \in M_1).$$

Here $^\circ I^\circ = {}^\circ I \cap I^\circ$. Moreover, we can show that if

$$\tilde{\omega}_{si,\lambda} = \prod_{\alpha \in P_0(s_i\lambda y)} H_\alpha.$$

Then

$$\tilde{\omega}_{si,\lambda}(s_i\Lambda_v)^{-1}{}_i\psi_f \in C(\mathfrak{F} \times G: V).$$

But

$$|\tilde{\omega}_{si,\lambda}(s_i\Lambda_v)| \geqslant |\tilde{\omega}_{si,\lambda}(s_i\lambda^y)| > 0.$$

Hence we see that if

$$f_Q(v: m) = \sum_{i \in {}^\circ I^\circ} {}_i\psi_f(v: m)$$

then $f_Q$ is a continuous function and

$$\lim_{\substack{a \to \infty \\ Q}} \{\mathrm{d}_Q(ma) f(v: ma) - f_Q(v: ma)\} = 0, \qquad (v \in \mathfrak{F}, m \in M_1).$$

Let $s \in W(\mathfrak{h}_R: \mathfrak{a})$ and put

(3.4) $$\psi^1_{f,s}(v: m) = \tilde{\omega}_1(s_{i(s)}\Lambda_v)^{-1}\psi_{f,s}(v: m), \qquad (v \in \mathfrak{F}(\lambda), m \in M_1).$$

THEOREM 3.2. (1) *The function* $\psi^1_{f,s}$ $(s \in W(\mathfrak{h}_R: \mathfrak{a}))$ *has a continuous extension to* $\mathfrak{F} \times M_1$.

(2) *If* $\eta_1, \eta_2 \in \mathfrak{M}_1$, $D \in \mathfrak{D}$, *there exists* $F \subset \tilde{\mathfrak{G}}$, $|F| < \infty$ *such that for all* $r \in \mathbb{R}^+$ *there exists* $c > 0$ *so that*

$$\|\psi^1_{f,s}(v; D: \eta_1; m; \eta_2)\| \leqslant c S_{F,r}(f) \Xi_M(m) |(v, m)|^{r + r_0 + e}.$$

(3) *Let* $v \in \mathfrak{Z}_1$. *Then*

$$\psi^1_{f,s}(v: m; v) = \mu_{\mathfrak{m}_1^s/\mathfrak{h}}(v^s: \lambda + iv)\psi^1_{f,s}(v: m).$$

*Here* $\mu_{\mathfrak{m}_1^s/\mathfrak{h}}: \mathfrak{Z}_1^s \to \geqslant \mathfrak{H}^{W_{1,s}}$, $W_{1,s} = W(\mathfrak{m}_1^s, \mathfrak{h})$, *is the canonical map.*

For (1) note that if $v \in \mathfrak{F}(\lambda)$

$$\tilde{\omega}_1(s_{i(s)}\Lambda_v)^{-1}\underline{E}_1(s_i\Lambda_v) = 1 \otimes \tilde{\omega}_{01}(s_{i(s)}\Lambda_v^{-1})E(s_{i(s)}\Lambda_v).$$

Recalling Lemma 3.1 we see easily by (3.2) that $\tilde{\omega}_1(s_{i(s)}\Lambda_\nu)^{-1}\Phi_{i\infty}(f: \nu: m)$ has a continuous extension to $\mathfrak{F}\times M_1$. The result (1) now is immediate.

For (2) we note that

$$\|\psi^1_{f,s}(\nu; D: \eta_1; m; \eta_2)\| \leqslant \|\tilde{\omega}_1(s_{i(s)}\Lambda_\nu)^{-1}\Phi_{i\infty}(f: \nu; D: \eta_1; m; \eta_2)\|$$

Using the above observation the estimate follows in a similar manner as (2) of Lemma 3.8.

From (1) of Lemma 3.8 it is clear that if $v \in \mathfrak{Z}_1$ then

$$\psi^1_{f,s}(\nu: m; v) = \mu_1(v: s_{i(s)}\Lambda_\nu)\psi^1_{f,s}(\nu: m)$$

(3) is immediate from the following result.

LEMMA 3.12. ([3], Lemma 7.2).

$$\mu_1(v: s_{i(s)}\Lambda_\nu) = \mu_{\mathfrak{m}_i^s/\mathfrak{h}}(v^s: \lambda + i\nu), \qquad (v \in \mathfrak{Z}_1).$$

We now have the following corollaries to Theorem 3.2.

COROLLARY. *Let $\nu \in \mathfrak{F}'(\lambda)$, $s \in W(\mathfrak{h}_R: \mathfrak{a})$. Then*

$$\psi^1_{f,s}(\nu: ma) = e^{i\nu(\log a^S)}\psi^1_{f,s}(\nu: m)$$

*for all $m \in M$, $a \in A$.*

COROLLARY. *$(\psi^1_{f,s})^s$ is a function of type $\mathrm{II}(\lambda)$ on $M_1^s$.*

## 4. Schwartz space wave packets

Retain the notation of Section 3. If

$$\alpha \in \mathscr{C}(\mathfrak{F}: V) = \{f \in C^\infty(\mathfrak{F}: V) : sup_{\mathfrak{F}}(1+|\nu|)^r \|f(\nu; u)\| < \infty \;\; r \in \mathbb{R}, u \in \mathscr{S}(\mathfrak{F}_c)\},$$

and $\varphi$ is a function of type $\mathrm{II}(\lambda)$ we put

$$\varphi_\alpha(x) = \int_{\mathfrak{F}} \alpha(\nu)\cdot\varphi(\nu: x)\tilde{\omega}_0(\Lambda_\nu)\, d\nu.$$

We shall in this section prove the following.

THEOREM 4.1. *$\varphi_\alpha \in \mathscr{C}(G: V: \tau)$ and the map $\alpha \to \varphi_\alpha$ is a continuous map of $\mathscr{C}(F: V)$ into $\mathscr{C}(G: V: \tau)$.*

Because we wish to prove this theorem by induction it will be necessary to first introduce some further notation and definitions. We will then state Theorem 4.1 in a different form, one which lends itself to the method of descent as described in the introduction.

For $D \in \tilde{\mathfrak{G}}$, $r \in \mathbb{R}$, $r \geqslant 0$ define

$$^{\circ}S_{D,r}(f) = \sup_{\mathfrak{F} \times G} \|Df\| \Xi^{-1}(1+\sigma)^{-r}, \qquad (f \in C^{\infty}(\mathfrak{F} \times G : V)).$$

As before if $F \subset \tilde{\mathfrak{G}}$, $|F| < \infty$, we write

$$^{\circ}S_{F,r}(f) = \sum_{D \in F} {}^{\circ}S_{D,r}(f).$$

$\varphi : \mathfrak{F} \times G \to V$ will be said to be of type I($\lambda$) if:

(1) $\varphi$ is of type II($\lambda$).

(2) For any $D \in \tilde{\mathfrak{G}}$ we can choose $r \geqslant 0$ such that $^{\circ}S_{D,r}(\varphi) < \infty$. Let $\mathscr{I}_G(\lambda)$ denote the space of functions of type I($\lambda$) on $G$. Pick $(Q, A) \succ (P_0, A_0)$.

LEMMA 4.1. *Let $\varphi$ be a function of type* II($\lambda$), $\alpha \in \mathscr{C}(\mathfrak{F} : V)$, $\psi_{\varphi}(v : x) = \tilde{\omega}_0(\Lambda_v)\varphi(v : x)$, *and for $s \in W(\mathfrak{h}_R : \mathfrak{a})$ (in the notation of* (3.4)) *put,*

$$\psi_{\varphi,\alpha}(v : x) = \alpha(v) \cdot \psi_{\varphi}(v : x), \qquad \psi^1_{\varphi,\alpha,s}(v : x) = \alpha(v) \cdot \psi^1_{\varphi,s}(v : x).$$

*Then $\psi_{\varphi,\alpha}\ \mathscr{I}_G(\lambda)$, and $(\psi^1_{\varphi,\alpha,s})^s \in \mathscr{I}_{M_1^s}(\lambda)$.*

The fact that $\psi_{\varphi,\alpha} \in \mathscr{I}_G(\lambda)$ follows easily from the definitions of functions of type II($\lambda$) and $\mathscr{C}(\mathfrak{F} : V)$. The fact that $(\psi^1_{\varphi,\alpha,s})^s \in \mathscr{I}_{M_1^s}(\lambda)$ follows from the second corollary of Lemma 3.12.

THEOREM 4.2. *Let $\varphi$ be a function of type* II($\lambda$) *and put*

$$\iota(\varphi : \alpha : x) = \int_{\mathfrak{F}} \alpha(v) \cdot \psi_{\varphi}(v : x)\, dv = \int_{\mathfrak{F}} \psi_{\varphi,\alpha}(v : x)\, dv.$$

*Then $\iota_{\varphi} \in \mathscr{C}(G : V : \tau)$. Fix $g_1, g_2 \in \tilde{\mathfrak{G}}$ and $r' \geqslant 0$. Then we can choose a finite subset $F \subset \tilde{\mathfrak{G}}$ with the following property. Given $r \geqslant 0$, there exists a number $c > 0$ such that*

$$\|\iota(\varphi : \alpha : g_1 ; x ; g_2)\| \leqslant c\, {}^{\circ}S_{F,r}(\psi_{\varphi,\alpha}) \Xi(x)(1+\sigma(x))^{-r'} \qquad (x \in G)$$

*for all functions of type* II($\lambda$) and $\alpha \in \mathscr{C}(\mathfrak{F} : V)$.

Let us assume Theorem 4.2 and show that Theorem 4.1 follows. But this is obvious as we may choose $P'_i \otimes u'_i$, $P_i \otimes u_i \in \mathfrak{D}$, $g_i$, $g'_i \in \mathfrak{G}$, such that

$$^\circ S_{F,r}(\psi_{\varphi,\alpha}) \leqslant \sup_{\mathfrak{F} \times G} \Xi(x)^{-1}(1+\sigma(x))^{-r} \times$$
$$\times \sum_i \|\alpha(v; P_i \otimes u_i) \cdot \varphi(v; P'_i \otimes u'_i; g_i; x; g'_i)\|.$$

But for $F = \{P'_i \otimes u'_i \otimes g_i \otimes g'_i\}$ we can choose $r'' \in \mathbb{R}^+$ so that

$$S_{F',r''}(\varphi) < \infty.$$

Therefore

$$^\circ S_{F,r}(\psi_{\varphi,\alpha}) \leqslant S_{F',r''}(\varphi) \sup_{\mathfrak{F}} (1+|v|)^{r''} \sum_i \|\alpha(v; P_i \otimes u_i)\|.$$

Obviously if $C''$ is the constant corresponding to $r''$ then

$$\nu(\alpha) = S_{F',r''}(\varphi) C'' \sup_{\mathfrak{F}} (1+|v|)^{r''} \sum_i \|\alpha(v; P_i \otimes u_i)\|$$

is a continuous seminorm on $\mathscr{C}(\mathfrak{F}: V)$. Theorem 4.1 is then clear.

We shall now proceed to prove Theorem 4.2 by induction on $\dim(G)$. If $\dim(G)=0$ the theorem is trivial. Hence let us assume the result for $\dim(G) \leqslant N-1$, $N \geqslant 1$ and suppose $\dim(G)=N$. We shall consider two cases; corresponding to $\mathrm{Prk}(G)=0$ or not. First let us assume $\mathrm{Prk}(G)>0$. Let $M={}^\circ G$, $A$ the split component of $G$, $\mathfrak{m}=LA(M)$, $\mathfrak{a}=LA(A)$. Then $\mathfrak{h}_R = \mathfrak{m} \cap \mathfrak{h}_R + \mathfrak{a}$ where the sum is direct. Let $\mathfrak{F}_1$ and $\mathfrak{F}_2$ be the subspaces consisting of all $v \in \mathfrak{F}$ which vanish identically on $\mathfrak{m} \cap \mathfrak{h}_R$ and $\mathfrak{a}$, respectively. Then $\mathfrak{F} = \mathfrak{F}_1 + \mathfrak{F}_2$ where the sum is direct. We note that $\mathfrak{D}_i = \mathfrak{D}(\mathfrak{F}_{ic}) \subset \mathfrak{D} = \mathfrak{D}(\mathfrak{F}_c)$. Let $dv_i$ denote the Euclidean measure on $\mathfrak{F}_i$ so normalized that $dv = dv_1 dv_2$ ($v = v_1 + v_2$, $v_i \in \mathfrak{F}_i$, $i=1, 2$). Since $\mathfrak{a} \subset \mathfrak{Z}$, it follows from our assumptions that if $\varphi$ is a function of type II($\lambda$) then

$$\varphi(v_1+v_2: ma) = \varphi(v_1+v_2: ma)\, e^{iv_1(\log a)} \qquad (m \in M, a \in A).$$

Fix $\eta_1, \eta_2 \in \mathfrak{M}$ and $u \in \mathfrak{A}$ (as usual $\mathfrak{M}$ and $\mathfrak{A}$ denote the subalgebras of $\mathfrak{G}$ generated by 1 and $\mathfrak{m}_c$ and $\mathfrak{a}_c$ respectively). Then

$$\iota(\varphi: \alpha: \eta_1; ma; \eta_2 u) = \int \psi_{\varphi,\alpha}(v_1+v_2: \eta_2; m; \eta_2) u(iv_1)\, e^{iv_1(\log a)}\, dv_1\, dv_2$$

(we regard $u$ as a polynomial function on $\mathfrak{F}_{1c}$ in the right-hand side of this expression). Fix $r' \geqslant 0$ and choose $P, P_1 \in \mathscr{S}(\mathfrak{F}_{1c})$ such that

$$P(iH) \geqslant (1+\|H\|)^{r'} \quad (H \in \mathfrak{a}), \quad P_1 \geqslant 1, \quad \text{and} \quad \int_{\mathfrak{F}_1} P_1^{-1}\, dv_1 < \infty.$$

From this it is clear that there exists an element $D_1 \in \mathfrak{D}$ such that

$$\|\iota(\varphi:\alpha:\eta_1;ma;\eta_2)\|(1+\sigma(a))^{r'} \leqslant$$
$$\leqslant \sup_{\nu_1\in\mathfrak{F}_1} \Big\| \int_{\mathfrak{F}_2} \psi_{\varphi,\alpha}(\nu_1+\nu_2;D_1:\eta_1;m;\eta_2)\,\mathrm{d}\nu_2\Big\|.$$

On the other hand, $\dim(M)<\dim(G)$ and so the induction hypothesis is applicable to $M$. This implies that corresponding to $\eta_1$, $\eta_2$ and $r'$ we can choose $F_2 \subset \tilde{\mathfrak{M}} = \mathfrak{D}_2 \otimes \mathfrak{M}^{(2)}$ so that if $r \in \mathbb{R}^+$ there exists $c>0$ for which

$$\|\iota(\mu:\beta:\eta_1;m;\eta_2)\| \leqslant c^\circ S_{F_2,r}(\psi_{\mu,\beta})\Xi(m)(1+\sigma(m))^{-r'}$$

for all functions $\mu$ of type II($\lambda$) on $M$ and $\beta \in \mathscr{C}(\mathfrak{F}: V)$. We must remember that $\tilde{\omega}_{0,G}=\tilde{\omega}_{0,M}$. Fix $\varphi$ of type II($\lambda$), $\nu_1 \in \mathfrak{F}_1$, and put for $D_1', D_1'' \in \mathfrak{D}_1$

$$\mu(\nu_2:m)=\varphi(\nu_1+\nu_2;D_1':m) \qquad (\nu_2\in\mathfrak{F}_2,\, m\in M)$$
$$\beta(\nu_2)=\alpha(\nu_1+\nu_2;D_1'').$$

Applying the above to $\mu$ and $\beta$ and observing that $\psi_{\varphi,\alpha}(\nu_1+\nu_2;\,D)$ can be written as a sum of functions of the type $\mu$, $\beta$ defined above it is clear that for every $r\geqslant 0$ there exists $c$ such that

$$\|\iota(\varphi:\alpha:\eta_1;ma;\eta_2 u)\| \leqslant cS_{F,r}(\psi_{\varphi,\alpha})\Xi(m)(1+\sigma(m))^{-r'}(1+\sigma(a))^{-r'}$$

for each $m\in M$, $a \in A$, and $\varphi$ a function of type II($\lambda$), $\alpha \in \mathscr{C}(\mathfrak{F}: V)$. Hence Theorem 4.2 is proved in this case.

Now consider the case when $\mathrm{Prk}(G)=0$. Again we assume that $\dim(G)=N$. Let us now specialize the results of Section 3 to the case when $Q$ is a standard (with respect to $P_0=M_0A_0N_0$) maximal psgp of $G$. Let $\Sigma_0=\{\alpha_1,\ldots,\alpha_d\}\subset\Delta(P_0,A_0)$ denote a simple system of roots for $\Delta(\mathfrak{g},\mathfrak{a}_0)$. For $1\leqslant i\leqslant d$ let $\mathfrak{a}_i=\{H\in\mathfrak{a}_0: \alpha_j(H)=0 \text{ for all } j\neq i\}$, $A_i=\exp\mathfrak{a}_i$, $\mathfrak{m}_{i1}=\mathrm{Cent}_{\mathfrak{g}}(\mathfrak{a}_i)$, $M_{i1}=\mathrm{Cent}_G(\mathfrak{a}_i)$. Put $\Delta_i^+=\Delta(P_0,A_0)\backslash(\Delta(P_0,A_0)\cap\Delta(\mathfrak{M}_{i1},\mathfrak{a}_0))$, and $\mathfrak{n}_i=\Sigma_{\alpha\in\Delta_i^+}\mathfrak{g}_\alpha$, $N_i=\exp\mathfrak{n}_f$. Then $\mathfrak{p}_i=\mathfrak{m}_{i1}+\mathfrak{n}_i$ is a parabolic subalgebra, $P_i=M_{i1}N_i$ a psgp. Further, if $M_i={}^\circ M_{i1}$, $\mathfrak{m}_i=LA(M_i)$, then $M_{i1}=M_iA_i$, $\mathfrak{m}_{i1}=\mathfrak{m}_i\oplus\mathfrak{a}_i$ and $P_i=M_iA_iN_i$, $\mathfrak{p}_i=\mathfrak{m}_i+\mathfrak{a}_i+\mathfrak{n}_i$ are Langlands decompositions. As usual we let $\mathfrak{M}_{i1}$, $\mathfrak{M}_i$ and $\mathfrak{A}_i$ denote the subalgebras of $\mathfrak{G}$ generated by 1 and $\mathfrak{m}_{i1c}$, $\mathfrak{m}_{ic}$, and $\mathfrak{a}_{ic}$, respectively. $\mathfrak{Z}_{i1}$ will denote the center of $\mathfrak{M}_{i1}$. For $H\in\mathfrak{a}_0$ set

$$\rho^i(H)=\tfrac{1}{2}\,\mathrm{tr}(\mathrm{ad}_H)_{\mathfrak{n}_i}, \qquad \rho_i(H)=\tfrac{1}{2}\,\mathrm{tr}(\mathrm{ad}_H)_{\mathfrak{m}_i\cap\mathfrak{n}_0}.$$

Then $\rho_0=\rho_i+\rho^i$, $\rho_i|\mathfrak{a}_i\equiv 0$, and $\rho^i|_{\mathfrak{a}_0\cap\mathfrak{m}_i}\equiv 0$. $\beta_i=\beta_{P_i}=\alpha_i$ where as before $\beta_i(H)=\min_{\alpha\in\Delta(P_i,A_i)}\{\alpha(H)\}$. Put

$$\mathfrak{a}_i^+=\{H\in\mathfrak{a}_i\colon \alpha_i(H)>0\},\qquad A_i^+=\exp\mathfrak{a}_i^+.$$

If $\mu>0$ let

$$A_i^+(\mu)=\{h\colon h\in A_0^+,\ \alpha_i(\log h)>\mu\rho_0(\log h)\}.$$

Then $A_i^+(\mu)\subseteq A_i^+(\mu')$ if $0<\mu'\leqslant\mu$, and $A_0^+\subseteq\bigcup_{1\leqslant i\leqslant d}A_1^+(\mu)$ for $\mu$ sufficiently small. Fix for each i, $1\leqslant i\leqslant d$, $H_i\in\mathfrak{a}_i$ such that $\alpha_i(H_i)=1$, and put $\Omega_i=\{H_i\}$. Put $L_i(\lambda)=\{\lambda_j(H_i)\colon j\in I_i^-\}$. (Recall that $\lambda_j=s_j\lambda^y|\mathfrak{a}_i$, and $I_i=\{1,\ldots,r_i\}$, where $r_i=W_0/W_{i1}$, and $W_{i1}=W(\mathfrak{m}_{i1c},\mathfrak{h}_{0c})$.) Choose $\delta>0$ so that

$$\alpha<-\delta\ \left(\text{for all }\alpha\in\bigcup_{1\leqslant i\leqslant d}L_i(\lambda)\right).$$

Fix $\mu>0$ so that $A_0^+\subseteq\bigcup_{1\leqslant i\leqslant d}A_i^+(\mu)$, and put $\varepsilon_0=1/4$, $\varepsilon=\varepsilon_0\delta$, and $\bar{\varepsilon}=\varepsilon\mu$. For any function $\varphi$ of type II($\lambda$) we put as in Section 3

$$\psi_\varphi(v\colon x)=\tilde{\omega}_0(\Lambda_v)\varphi(v\colon x).$$

LEMMA 4.2. *Fix j, $1\leqslant j\leqslant d$, and let $\eta_1,\eta_2\in\mathfrak{M}_{j1}$. Then there exists a finite subset $F\subset\tilde{\mathfrak{G}}$ with the following property: for every $r\in\mathbb{R}^+$ there exists $c>0$ such that*

$$\left\|\psi_\varphi(v\colon\eta'_1;h;\eta'_2)-d_{P_j}^{-1}(h)\sum_{s\in W(\mathfrak{h}_R;\mathfrak{a}_j)}\psi_{\varphi,s}(v\colon\eta_1;h;\eta_2)\right\|$$
$$\leqslant CS_{F,r}(\varphi)\Xi^{1+\bar{\varepsilon}}(h)|(v,h)|^{r+a_j},\qquad (h\in A_j^+(\mu)).$$

*Here $a_j=r_{0j}+q_{0j}+e_j$ where $q_{0j}$, $r_{0j}$ are the constants in* (1) *of Lemma* 2.4 *and $e_j$ is as in* (3.1) *with $Q=P_j$. The inequality holds for all functions $\varphi$ of type* II($\lambda$).

Let $h\in A_j^+(\mu)$; then $h=\exp{}_jH\ \exp\alpha_j(\log h)H_j$ where ${}_jH\in{}_j\mathfrak{a}$ and ${}_j\mathfrak{a}=\{H\in\mathfrak{a}_0\colon\alpha_j(H)=0\}$. As $h\in A_0^+$ then ${}_jH\in\mathrm{Cl}(\mathfrak{a}_0^+)$, which implies that $\exp{}_jH\in M_{j1}^+=K_j\mathrm{Cl}(A_0^+)K_j$, $K_j=M_j\cap K$. Applying Theorem 3.1 with the above definitions of $\varepsilon_0$ and $\delta$ we see that there exists $F\subset\tilde{\mathfrak{G}}$, $|F|<\infty$, so that for all $r\in\mathbb{R}^+$ there exists $C>0$ for which

$$\left\|\psi_\varphi(v\colon\eta'_1;h;\eta'_2)-d_j^{-1}(h)\sum_{s\in W(\mathfrak{h}_R;\mathfrak{a}_j)}\psi_{\varphi,s}(v\colon\eta_1;h;\eta_2)\right\|$$
$$\leqslant CS_{F,r}(\varphi)\,\mathrm{e}^{-\varepsilon\alpha_j(\log h)}d_j^{-1}(h)\Xi_j(\exp{}_jH)|(v,h)|^{r+r_{0j}+e_j}$$

where $\Xi_j = \Xi_{M_j}$, $d_j = d_{P_j}$, and the inequality holds for all $h \in A_j^+(\mu)$, and functions $\varphi$ of type II($\lambda$). By Lemma 2.4 there exists constants $D_{0j} > 0$, $q_{0j} \geqslant 0$ so that

$$e^{-\rho j(\log a)} \leqslant \Xi_j(a) \leqslant D_{0j}\, e^{-\rho j(\log a)}(1+\sigma(a))^{q_{0j}}$$

the inequalities holding for all $a \in \exp(\mathrm{Cl}(\mathfrak{a}_0^+) \cap \mathfrak{m}_j)$. Hence, using the fact that if $h \in A_j^+(\mu)$ then $\varepsilon\alpha_j(\log h) \geqslant \varepsilon\mu\rho_0(\log h)$ the above inequality becomes

$$\leqslant CD_{0j} S_{F,r}(\varphi)\, e^{-\bar{\varepsilon}\rho_0(\log h)}\, e^{-(\rho_j(_jH)+\rho j(\log h))}|(v, h)|^{r+a_j}.$$

As $\rho_j(_jH) + \rho^j(\log h) = \rho_0(\log h)$ we have on using Lemma 2.4

$$\leqslant CD_{0j} S_{F,r}(\varphi)\Xi^{1+\bar{\varepsilon}}(h)|(v, h)|^{r+a_j}.$$

As a corollary to the proof of Lemma 4.1 we have the following; let $g_1 \in \mathfrak{G}$, $(1 \leqslant i \leqslant 4)$ and assume that $g_1 \in \mathfrak{G}\mathfrak{n}_j$, $g_4 \in \theta(\mathfrak{n}_j)\mathfrak{G}$. Let $D \in \mathfrak{D}$. Then there exists $F \subset \tilde{\mathfrak{G}}$, $|F| < \infty$, so that for all $r \in \mathbb{R}^+$,

$$\text{(4.1)} \qquad \sum_{i \in \{1,3\}} \|\psi_\varphi(v; D: g_i; h; g_{i+1})\| \leqslant S_{F,r}(\varphi)\Xi^{1+\bar{\varepsilon}}(h)|(v, h)|^{r+a_j}$$

the inequality holding on $A_j^+(\mu)$. (4.1) is, in fact, an immediate consequence of the above proof and Lemma 3.3. It follows then that we can choose a finite subset $F \subset \tilde{\mathfrak{G}}$ such that for all $r \in \mathbb{R}^+$ there exists $C$ so that

$$\text{(4.2)} \qquad \sum_{i \in \{1,3\}} \| j(\varphi: \alpha: g_i; h; g_{i+1})\| \leqslant C^\circ S_{F,r}(\Psi_{\phi,\alpha})\Xi^{1+\bar{\varepsilon}}(h)(1+\sigma(h))^{r+a_j},$$

holds for $h \in A_j^+(\mu)$.

In the introduction it was pointed out that if $g_1, g_2 \in \mathfrak{G}$ then there exists $g_{\alpha\beta} \in \mathfrak{G}$, $f_{\alpha\beta} \in C^\infty(K)$ $(1 \leqslant \alpha \leqslant S,\ \beta = 1, 2)$ so that if $k_1, k_2 \in K$, $a \in A_0$, and $F \in C^\infty(G: V: \tau)$ then

$$\|F(g_1; k_1 a k_2; g_2)\| \leqslant \sum_{\gamma,\delta} |f_{\gamma 1}(k_1) f_{\delta 2}(k_2)|\ \|F(g_{\gamma 1}; a; g_{\delta 2})\|.$$

Hence there exists $C > 0$ so that if $\Xi_r = \Xi(1+\sigma)^{-r}$ then

$$\sup_G \Xi_r^{-1}(x)\|\iota(\varphi: \alpha: g_1; x; g_2)\|$$

$$\leqslant C \sup_{\mathrm{Cl}(A_0^+)} \Xi_r^{-1}(a) \sum_{\gamma,\delta} \|\iota(\varphi: \alpha: g_{\gamma 1}; a; g_{\delta 2})\|$$

$$\leqslant C \max_{1 \leqslant j \leqslant d}\ \sup_{\mathrm{Cl}(A_j^+(\mu))} \Xi_r^{-1}(h) \sum_{\gamma,\delta} \|\iota(\varphi: \alpha: g_{\gamma 1}; h; g_{\delta 2})\|$$

But $\mathfrak{G}=\mathfrak{K}\mathfrak{M}_{j1}\mathfrak{N}_j=\theta(\mathfrak{N}_j)\mathfrak{M}_{j1}\mathfrak{K}$. Hence for each $\gamma$, $\delta$ there exists $\eta^j_{\gamma 1}\in\mathfrak{K}\mathfrak{M}_{j1}$, $\eta^j_{\delta 2}\in\mathfrak{M}_{j1}\mathfrak{K}$ so that $g_{\gamma 1}-\eta^j_{\gamma 1}\in\mathfrak{G}\mathfrak{n}_j$, and $g_{\delta 2}-\eta^j_{\delta 2}\in\theta(\mathfrak{n}_j)\mathfrak{G}$. It follows from (4.2) together with the unitarity of $\tau$ that Theorem 4.2 will follow in this particular case if we can show that given $r'\in\mathbb{R}^+$, $\eta_1$, $\eta_2\in\mathfrak{M}_{j1}$ we can choose $F\subset\tilde{\mathfrak{G}}$, $|F|<\infty$ with the property that for every $r\in\mathbb{R}^+$ there exists $C_j>0$ so that (cf. (*) below)

$$\|\iota(\varphi:\alpha:\eta_1;h;\eta_2)\|\leqslant C_j^\circ S_{F,r}(\psi_{\varphi,\alpha})\Xi(h)(1+\sigma(h))^{-r'},\quad (h\in A_j^+(\mu)).$$

We put then (with $Q=P_j$ we define $\tilde{\omega}_1$, $\tilde{\omega}_{01}$ as in Section 3)

$$\iota(\psi^1_{\varphi,s}:\alpha:m)=\int_{\mathfrak{F}}\alpha(v)\cdot\psi^1_{\varphi,s}(v:m)\tilde{\omega}_1(s_{i(s)}\Lambda_v)\,dv,\quad (m\in M_{j1}),\tag{4.3}$$

$$\iota(\psi^1_{\varphi,j}:\alpha:m)=\sum_{s\in W(\mathfrak{h}_R;\mathfrak{a}_j)}\iota(\psi^1_{\varphi,s}:\alpha:m).\tag{4.4}$$

From Theorem 3.2, and the fact that $\varphi$ is a function of type II($\lambda$) it is easily seen that the integrals defining $\iota(\varphi:\alpha)$, $\iota(\psi^1_{\varphi,s}:\alpha)$, converge absolutely and uniformly as long as $x$ and $m$ vary in compact subsets of $G$ and $M_{j1}$, respectively. One also sees that $j(\varphi:\alpha)$ and $j(\psi_{\varphi,s}:\alpha)$ are $C^\infty$-functions whose derivatives can be computed by differentiation under the integral sign. Further from Lemma 4.2 we see that given $\eta_1$, $\eta_2\in\mathfrak{M}_{j1}$ there exists $F\subset\tilde{\mathfrak{G}}$, $|F|<\infty$, and for every $r\in\mathbb{R}$, $r\geqslant 0$, $C>0$ so that

$$\|\iota(\varphi:\alpha:\eta'_1;h;\eta'_2)-d_j^{-1}(h)\iota(\psi^1_{\varphi,j}:\alpha:\eta_1;h;\eta_2)\|$$

$$\leqslant CS_{F,r}(\varphi)\Xi^{1+\bar{\varepsilon}}(h)(1+\sigma(h))^{r+a_j}\int_{\mathfrak{F}}(1+|v|)^{r+a_j}\|\alpha(v)\|\,dv$$

for all $h\in A_j^+(\mu)$. But we can choose $k_0>0$ so that

$$\int_{\mathfrak{F}}(1+|v|)^{-k_0}\,dv<\infty.$$

Moreover, for every $l\geqslant 0$ there exists $L>0$ such that

$$\sup_{h\in A_j^+}(1+\sigma(h))^l\Xi^{\bar{\varepsilon}}(h)\leqslant L.\tag{*}$$

Hence it is clear that given $\eta_1$, $\eta_2\in\mathfrak{M}_{j1}$, $r'\in\mathbb{R}^+$ there exists $F\subset\tilde{\mathfrak{G}}$ with the following property. Given $r\in\mathbb{R}^+$ there exists a constant $C>0$ such that

$$\|\iota(\varphi:\alpha:\eta'_1;h;\eta'_2)-\iota(\psi^1_{\varphi,j}:\alpha:\eta_1;h;\eta_2)d_j^{-1}(h)\|$$

$$\geqslant C^{\circ} S_{F,r}(\psi_{\varphi,\alpha})\Xi(h)(1+\sigma(h))^{-r'}, \qquad (h \in A_j^+(\mu)).$$

But we note that $\tilde{\omega}_1^s$ is the product of positive roots of $(\mathfrak{m}_{j1}^s, \mathfrak{h}_0)$ and that

$$\tilde{\omega}_1^s(\Lambda_v) = \tilde{\omega}_1(s_{i(s)}\Lambda_v).$$

Hence as $(\psi_{\varphi,s}^1)^s$ is a function of type II($\lambda$) on $M_{j1}^s$ it follows easily from the induction hypothesis and (2) of Theorem 3.2 that we can choose $F_1 \subset \tilde{\mathfrak{G}}$ such that for every $r$ there exists $C$ so that

$$\|\iota(\psi_{\varphi,j}^1 \colon \alpha \colon \eta_1 ; m; \eta_2)\| \leqslant C^0 S_{F_1,r}(\psi_{\varphi,\alpha})\Xi_j(m)(1+\sigma(m))^{-r'}(m \in M_{j1}^+).$$

Combining this with Lemma 2.4 we see that the theorem follows in this case and hence is completely proved.

*Remark.* Let $\beta \in C^\infty(\mathfrak{F})$ and suppose that $\beta_0(v) = \tilde{\omega}_0(\Lambda_v)^{-1}\beta(v)$ also is of class $C^\infty$. Further, let us assume that for every $u \in \mathscr{S}(\mathfrak{F}_c)$ there exists $C>0$ and $l \geqslant 0$ so that

$$|\beta_0(v; u)| < C(1+|v|)^l.$$

Then defining,

$$\iota_\beta(\varphi \colon \alpha \colon x) = \int_{\mathfrak{F}} \alpha(v) \cdot \varphi(v \colon x)\beta(v)\, dv = \int_{\mathfrak{F}} \alpha(v)\beta_0(v) \cdot \varphi(v \colon x)\tilde{\omega}_0(\Lambda_v) dv$$

and using the fact that $\alpha \to \beta_0\alpha$ is a continuous map of $\mathscr{C}(\mathfrak{F} \colon V)$ into itself it will follow from the proof of Theorem 4.2 that $\iota_\beta(\varphi \colon \alpha) \in \mathscr{C}(G \colon V \colon \tau)$ for all $\varphi$ of type II($\lambda$) on $G$ and $\alpha \in \mathscr{C}(\mathfrak{F} \colon V)$. Moreover, for each $\varphi$ of type II($\lambda$), the continuity of the map $\alpha \to j_\beta(\varphi \colon \alpha)$ can also be established.

## 5. The $C$-functions

Let the notation and assumptions be as in Sections 3 and 4. Hence $\lambda \in i\mathfrak{h}_I^*$, $\lambda$ regular. For $(Q, A) > (P_0, A_0)$ we have shown in Section 3 that if $\varphi$ is a function of type II($\lambda$) then there exists $\varphi_Q \colon \mathfrak{F} \times M_1 \to V$ such that

$$\lim_{\substack{a \to \infty \\ Q}} \{d_Q(ma)\varphi(v \colon ma) - \varphi_Q(v \colon ma)\} = 0.$$

Further if $\psi_\varphi(v \colon x) = \tilde{\omega}_0(\Lambda_v)\varphi(v \colon x)$ then we showed that there exists $\psi_{\varphi,s} \colon \mathfrak{F} \times M_1 \to V$ ($s \in W(\mathfrak{h}_R \colon \mathfrak{a})$) such that

$$(\psi_\varphi)_Q = \sum_{s \in W(\mathfrak{h}_R;\mathfrak{a})} \psi_{\varphi,s}.$$

Moreover if $\psi^1_{\varphi,s}(v: m)=\omega_1(s_{i(s)}\Lambda_v)^{-1}\psi_{\varphi,s}(v: m)$ then $(\psi^1_{\varphi,s})^s$ is again of type $\mathrm{II}(\lambda)$ on $M^s_1$.

Let $(Q, S)>(Q', A')>(P_0, A_0)$. Put $Q=MAN$, $Q'=M'A'N'$, and $^*Q=M_1\cap Q'$. Then $(^*Q, A')$ is a $p$-pair of $M_1 \Rightarrow$ if $s\in W(\mathfrak{h}_R: \mathfrak{a})$ then $(^*Q^s, (A')^s)$ is a $p$-pair of $M^s_1$. For any $s\in W(\mathfrak{h}_R: \mathfrak{a})$ let $W_s(\mathfrak{h}_R: \mathfrak{a}')$ be the set of all $t\in W(\mathfrak{h}_R: \mathfrak{a}')$ such that $t=s$ on $\mathfrak{a}\subset\mathfrak{a}'$. Put $(\psi_\varphi)_{Q'}=\Sigma_{t\in W(\mathfrak{h}_R:\mathfrak{a}')}\psi_{\varphi,t}$.

LEMMA 5.1. *Fix $s\in W(\mathfrak{h}_R: \mathfrak{a})$ and choose a representative $k\in K$ for $s$. Put*

$$f(v: m^k)=\tau(k)\psi^1_{\varphi,s}(v: m)\tau(k^{-1}),\qquad \psi_f(v: m^k)=\tilde{\omega}^s_1(\Lambda)f(v: m^k),$$
$$(m\in M_1).$$

*Then*:

$$(1)\qquad (\psi_f)_{*Q^k}(v: m)=\sum_{t\in W_s(\mathfrak{h}_R:\mathfrak{a}')}\psi_{f,t}(v: {}^*m),\qquad ({}^*m\in({}^*MA')^k).$$

(2) *For any $t\in W_s(\mathfrak{h}_R: \mathfrak{a}')$ if we put $\tilde{\omega}^*_1=\Pi H_\alpha$ where the product is taken over $\alpha\in P_0$ such that $\alpha|_{\mathfrak{a}'}\equiv 0$ (cf. Section 3) and put*

$$\psi^1_{f,t}(v: m)=\tilde{\omega}^*_1(s_{i(t)}\Lambda_v)^{-1}\psi_{f,t}(v: m)$$

*then $(\psi^1_{f,t})^t$ is of type $\mathrm{II}(\lambda)$ on $(^*MA')^t$.*

(3) *For any $t\in W_s(\mathfrak{h}_R: \mathfrak{a}')$, $k'$ is a representative of $t$, $v\in\mathfrak{F}$,*

$$\tau(k')\psi_{\varphi,t}(v: m)\tau(k'^{-1})=\tau(k'k)\psi_{f,t}(v:m)\tau(k^{-1}k'^{-1}),\qquad (m\in {}^*MA').$$

The lemma follows easily from Section 3 with $(f, M^s_1)$ replacing $(f, G)$ together with the fact that if $W_1(\mathfrak{h}_R: \mathfrak{a}'^k)$ denotes the set of all injective maps $t$ of $\mathfrak{a}'^k$ into $\mathfrak{h}_R$ such that $t=\mathrm{Ad}_m k$ for some $m\in M$ then $t\to t\circ \mathrm{Ad}_k^{-1}$ is a bijection of $W_s(\mathfrak{h}_R: \mathfrak{a}')$ with $W_1(\mathfrak{h}_R: \mathfrak{a}'^k)$

LEMMA 5.2. *If* $\mathrm{Prk}(Q)=\dim(\mathfrak{h}_R)$ *then* $\psi_{\varphi,t}(v)\in{}^\circ\mathscr{C}(M: V: \tau_M)$ *for all* $t\in W(\mathfrak{h}_R: \mathfrak{a})$ *and* $v\in\mathfrak{F}$.

Recall that $\psi_{\varphi,t}$ is an eigenfunction for $\mathfrak{Z}_1$ ((3) of Theorem 3.2). One can then define the constant term $(\psi_{\varphi,t})_{*Q}$ of $\psi_{\varphi,t}$ along an arbitrary psgp $^*Q$ of $M$ (cf. the remark following Theorem 3.1). To show that

$$\psi_{\varphi,t}(v)\in{}^\circ\mathscr{C}(M: V: \tau_M)$$

it suffices to show that $(\psi_{\varphi,t})_{*Q}\equiv 0$ for all psgps $^*Q$ of $M$ such that $\mathrm{Prk}(^*Q)\geqslant 1$. But if $^*Q$ is a psgp of $M$ there exists $(Q', A')<(Q, A)$ such that

$^*Q = Q' \cap M$. Moreover,

$$\mathrm{Prk}(Q') = \mathrm{Prk}(^*Q) + \mathrm{Prk}(Q) > \mathrm{Prk}(Q).$$

Hence $W(\mathfrak{h}_R\colon \mathfrak{a}') = \phi$. Fix a representative for $t$ in $K$, say $k$, and put $f(v\colon m) = \psi^t_{\varphi,t}(v\colon m)$ $(m \in (AM))$. Let $P = (Q' \cap M_1)^k$. It follows from Lemma 5.1 that $f_P \equiv 0$. It is easy to show that

$$(\psi_{\varphi,t})_{^*Q}(v\colon m) = (f_P)^{k^{-1}}(v\colon m) \qquad (m \in M).$$

Hence $(\psi_{\varphi,t})_{^*Q} \equiv 0$.

Now let $L = {}^\circ\mathscr{C}(M\colon V\colon \tau_M)$. Then $\dim(L) < \infty$. For $\psi \in L$ let

$$\|\psi\|_2^2 = \int_M \|\psi(m)\|^2 \, dm$$

where as usual $\|v\|$ denotes the $\tau$-invariant norm of $v \in V$. Then $L$ is a finite dimensional Hilbert space with this norm. Let $\omega \in \mathscr{E}^2(M)$ and as before let $L(\omega) = L \cap L^2_\omega(M) \otimes V$, where $L^2_\omega(M)$ denotes the smallest closed subspace of $L^2(M)$ containing the matrix coefficients of $\omega$. Then

$$L = \sum_{\omega \in \mathscr{E}^2(M)} L(\omega).$$

(The sum being an orthogonal direct sum.)

For the remainder of this section we let

$$\mathfrak{a} = \mathfrak{h}_R, \qquad A = \exp \mathfrak{a}, \qquad W(A) = W(\mathfrak{a}, \mathfrak{a}).$$

Let $P \in \mathscr{P}(A)$, $\alpha_1, \ldots, \alpha_r$ denote the set of distinct roots of $\Delta(\mathfrak{p}, \mathfrak{a})$, $m_i$ the multiplicity of $\alpha_i$, and put $\pi(v) = \Pi_{1 \leqslant i \leqslant r} \langle v, \alpha \rangle^{m_i}$ for all $v \in \mathfrak{F}_c$. Then we define $\mathfrak{F}_c' = \{v \in \mathfrak{F}_c\colon \pi(v) \neq 0\}$, and $\mathfrak{F}' = \mathfrak{F}_c' \cap \mathfrak{F}$.

THEOREM 5.1. *Fix $v \in \mathfrak{F}'$ and $P_1, P_2 \in \mathscr{P}(A)$. Then there exists*

$$C_{P_2|P_1}(s\colon v) \in \mathrm{End}(L) \qquad (s \in W(A))$$

*such that*

$$E_{P_2}(P_1\colon \psi\colon v\colon ma) = \sum_{s \in W(A)} (C_{P_2|P_1}(s\colon v)\psi)(m)\, e^{isv(\log a)}$$

*for $\psi \in L$, $m \in M$ and $a \in A$. Moreover, we can choose $\delta > 0$ such that for every $s \in W(A)$, $\pi(v) C_{P_2|P_1}(s\colon v)$ extends to a holomorphic function of $v$ on $\mathfrak{F}_c(\delta) = \{v \in \mathfrak{F}_c\colon |v_I| < \delta\}$ for some sufficiently small $\delta > 0$.*

*Remark.* Although $E(P_1\colon \psi\colon \nu)$ is not in general an eigenfunction for $\psi \in L$ we note that

$$\psi = \sum_{\omega \in \mathcal{E}^2(M)} \psi_\omega \ldots\ldots (\psi_\omega \in L(\omega))$$

and hence

$$E(P_1\colon \psi\colon \nu) = \sum_{\omega \in \mathcal{E}^2(M)} E(P_1\colon \psi_\omega\colon \nu).$$

(As dim $L < \infty$ this sum is finite.) We have already mentioned that $E(P_1\colon \psi_\omega\colon \nu)$ is a function of type II($\lambda$) for a suitably chosen $\lambda$. Hence $E_{P_2}(P_1\colon \psi_\omega\colon \nu)$ make sense by the remark following Theorem 3.1. We define then $E_{P_2}(P_1\colon \psi\colon \nu) = \Sigma_\omega E_{P_2}(P_1\colon \psi_\omega\colon \nu)$.

It follows from Lemma 22.3 of [3] that if $\nu \in \mathfrak{F}'$ and $t, s \in W(A)$ then $s\nu \neq t\nu$ hence the uniqueness follows easily. By the above remark in order to define the functions $C_{P_2|P_1}(s\colon \nu)$ it suffices to take $\psi \in L(\omega)$, for which $L(\omega) \neq 0$. But there exists $\lambda \in i\mathfrak{h}_I^*$, $\lambda$ regular, such that $v\psi = \mu_{\mathfrak{m}_1/\mathfrak{h}}(v\colon \lambda)\psi$ for all $v \in \mathfrak{Z}_M = \text{center}(\mathfrak{M})$. It follows then that $E(P_1\colon \psi\colon \nu)$ is a function of type II($\lambda$). It is not hard to show that $\mathfrak{F}' \subset \mathfrak{F}'(\lambda)$. Hence by Theorem 3.1 and the remark following it we have for

$$\varphi(\nu\colon x) = E(P_1\colon \psi\colon \nu\colon x), \qquad \psi_\varphi = \tilde{\omega}_0(\Lambda_\nu)\varphi,$$

$$E_{P_2}(P_1\colon \psi\colon \nu\colon ma) = \sum_{s \in W(A)} \tilde{\omega}_0^{-1}(\Lambda_\nu)\psi_{\varphi,s}(\nu\colon ma).$$

By Lemma 5.2 $\tilde{\omega}_0^{-1}(\Lambda_\nu)\psi_{\varphi,s}(\nu) \in {}^\circ\mathscr{C}(M\colon V\colon \tau_M)$. Hence define

$$(C_{P_2|P_1}(s^{-1}\colon \nu)\psi)(m) = \tilde{\omega}_0^{-1}(\Lambda_\nu)\psi_{\varphi,s}(\nu\colon m).$$

Then the first part of the theorem follows from the first corollary to Theorem 3.2.

For the second part of the theorem we observe that (in the notation of Section 3) $\Phi_i(\psi_\varphi\colon \nu\colon x)$ can be written in terms of integrals (3.2) where all functions (for $\psi_\varphi$ as above) are meromorphic on $\mathfrak{F}_c$. Taking a little more care in our estimates it follows that the defining integrals converge if $\nu \in \mathfrak{F}_c(\delta)$, $\delta$ sufficiently small. From this one can deduce the holomorphy of $C_{P_2|P_1}(s\colon \nu)$.

*Remark.* From our asymptotic analysis it is not difficult to prove the integral formulae for $C_{\bar{P}|P}(1\colon \nu)$ and $C_{P|P}(1\colon \nu)$ given in Lemma 1.5.

## 6. MAASS–SELBERG RELATIONS. A SKETCH

Let $A$ be a special vector subgroup of $G$ ([2], p. 114), $A = \exp \mathfrak{a}$. A root $\alpha \in \Delta(\mathfrak{g}, \mathfrak{a})$ is called reduced if $r\alpha \notin \Delta(\mathfrak{g}, \mathfrak{a})$ if $0 < r < 1$ ($r \in \mathbb{R}$). If $P \in \mathscr{P}(A)$ we denote by $\Sigma(P)$ the reduced roots of $(P, A)$. If $P_1, P_2 \in \mathscr{P}(A)$, put $\Sigma(P_2|P_1) = \Sigma(\bar{P}_2) \cap \Sigma(P_1)$ ($\bar{P}_2 = \theta(P_2)$) and $d(P_1, P_2) = |\Sigma(P_2|P_1)|$. Then $d$ is a metric on the finite set $\mathscr{P}(A)$.

LEMMA 6.1. ([4], Lemma 2.1). *Fix $P_1$, $P_2$ and $P$ in $\mathscr{P}(A)$. Then the following conditions are equivalent.*

(1) $d(P_1, P_2) = d(P_1, P) + d(P, P_2)$

(2) $\Sigma(P_2|P_1) \supset \Sigma(P|P_1)$

(3) $\Sigma(P_2|P_1)$ *is the disjoint union of* $\Sigma(P_2|P)$ *and* $\Sigma(P|P_1)$.

We say that $P$ *lies between* $P_1$ *and* $P_2$ if (1) holds. Moreover, $P_1$, $P_2$ are said to be *adjacent* if $d(P_1|P_2) = 1$.

Let $P \in \mathscr{P}(A)$, $P = MAN$, $\omega \in \mathscr{E}^2(M)$, $\nu \in \mathfrak{a}_c^*$, and denote by $\mathscr{H}_{P,\omega,\nu}$ the Hilbert space of all measurable functions $f: G \to U$ ($\sigma \in \omega$, $\sigma$ represents $M$ on $U$) such that:

(1) $f(\text{man } x) = e^{(\rho_P + i\nu)(\log a)} \sigma(m) f(x) \qquad (m \in M, a \in A, n \in N, x \in G)$

(2) $$\|f\|^2 = \int_K |f(k)|^2 \, dk < \infty$$

where $|f(k)|$ denotes the invariant norm on $U$.

Define $J_{P_2|P_1}(\nu): \mathscr{H}_{P_1,\omega,\nu} \to \mathscr{H}_{P_2,\omega,\nu}$ ($P_1, P_2 \in \mathscr{P}(A)$) as follows. If $h \in \mathscr{H}_{P_1,\omega,\nu}$ put,

$$(J_{P_2|P_1}(\nu)h)(x) = \gamma_{P_2|P_1}^{-1} \int_{N_2 \cap \bar{N}_1} h(\bar{n}x) \, d\bar{n}.$$

Here $\gamma_{P_a|P_1}$ is a constant whose definition can be found in [4], p. 125. Whenever the integral defining $J_{P_2|P_1}(\nu)$ converges we have the following. Let $\pi_{P,\omega,\nu}$ denote the right regular representation of $G$ on $\mathscr{H}_{P,\omega,\nu}$.

PROPOSITION 6.1. $J_{P_2|P_1}(\nu)\pi_{P_1,\omega,\nu}(x) = \pi_{P_2,\omega,\nu}(x) J_{P_2|P_1}(\nu)$ ($x \in G$).

Put $\mathfrak{F}_c = \mathfrak{a}_c^*$, $\mathfrak{F} = \mathfrak{a}^*$.

LEMMA 6.2. *Let $\mathfrak{F}_c(P_2|P_1) = \{\nu \in \mathfrak{F}_c : \langle \nu_I + \rho_{P_i}, \alpha \rangle < 0$ for all $\alpha \in \Sigma(P_2|P_1)\}$. Then for a fixed $h \in \mathscr{H}_{P,\omega,\nu}$ the mapping $\nu \to J_{P_2|P_1}(\nu)h$ from $\mathfrak{F}(P_2|P_1)$ into $\mathscr{H}_{P,\omega,\nu}$ is holomorphic.*

Let $C(A) \in \mathbb{R}$ be defined as in Section 11 of [4]. For $F \subset \mathscr{E}(K)$, $|F| < \infty$, $F = F^*$ set $j_{P_2|P_1}(v) = E_F J_{P_2|P_1}(v) E_F$, $E_F = \pi_{P,\omega,v}(\alpha_F)$ (cf. Section 2 for notation).

LEMMA 6.3. *Fix $P \in \mathscr{P}(A)$ and $v \in \mathfrak{F}(P|\bar{P})$. Then*

$$C_{P|P}(1: v)\psi_T = C(A)\psi_{j_{\bar{P}|P}(v)T}$$

*for all $T \in \mathrm{End}(E_F \mathscr{H}_{P,\omega,v})$.*

Using the fact that if $v \in \mathfrak{F}$, $\pi_{P,\omega,v}$ is unitary it is easy to prove the following. Let $\mathfrak{F}''$ be the set of all $v \in \mathfrak{F}'$ (defined as in Section 5) where $j_{\bar{P}|P}(v)$ is bijective.

LEMMA 6.4. *Fix $v \in \mathfrak{F}''$. Then there exists a scalar $C > 0$ such that*

$$j_{\bar{P}|P}(v)^* j_{\bar{P}|P}(v) = C$$

*where for $T \in \mathrm{End}(E_F \mathscr{H}_{P,\omega,v}) T^*$ denotes its adjoint.*

LEMMA 6.5. *Let $v \in \mathfrak{F}'$. Then*

$$j_{P|\bar{P}}(v) = (j_{\bar{P}|P}(v))^*$$

*and*

$$C_{\bar{P}|P}(1: v)\psi_T = C(A)\psi_{T j_{P|\bar{P}}(v)}.$$

Assume now that $\dim A = \mathrm{Prk}(G) + 1$. Then $\mathscr{P}(A) = \{P, \bar{P}\}$.

LEMMA 6.6. *Suppose that $E_F \mathscr{H}_{P,\omega,v} \neq 0$ and fix $v \in \mathfrak{F}$. Then there exists a number $\mu_\omega(v) > 0$ such that*

$$\mu_\omega(v) j_{P|\bar{P}}(v) j_{\bar{P}|P}(v) = \mu_\omega(v) j_{\bar{P}|P}(v) j_{P|\bar{P}}(v) = 1.$$

*Moreover, $\mu_\omega(v)$ is independent of $F$.*

Fix $P \in \mathscr{P}(A)$ and $\alpha \in \Sigma(P)$. Let $\sigma_\alpha$ denote the hyperplane $\alpha = 0$ in $\mathfrak{a}$ and $Z_\alpha$ be the centralizer of $\sigma_\alpha$ in $G$. Put $M_\alpha = {}^\circ Z_\alpha$ (cf. Section 1 for notation) and ${}^*P_\alpha = M_\alpha \cap P$. For $\lambda \in \mathfrak{a}^*$ let $\mathfrak{g}_\lambda = \{X \in \mathfrak{g}: [H, X] = \lambda(H)X\}$.

LEMMA 6.7. *$M_\alpha$ is a reductive subgroup of $G$ satisfying* (1), (2), (3) *of the introduction, and ${}^*P_\alpha$ is a psgp of $M_\alpha$. Moreover, the Langlands decompo-*

*sition of* $^*P_\alpha$ *is given by* $^*P_\alpha = MA_\alpha N_A$ *where* $\mathfrak{a}_\alpha = \mathbb{R}H_\alpha$, $A_\alpha = \exp\,\mathfrak{a}_\alpha$, *and* $\mathfrak{n}_\alpha = \Sigma_{k\geq 1}\mathfrak{g}_{k\alpha}$.

Let $\alpha$ be a reduced root of $(\mathfrak{g}, \mathfrak{a})$. Define $M_\alpha$, $\mathfrak{a}_\alpha$ as in the preceding lemma. For $\nu \in \mathfrak{F}$, $^*\nu$ denotes the restriction of $\nu$ on $\mathfrak{a}_\alpha$. If we replace $(G, A)$ by $(M_\alpha, A_\alpha)$ then $\dim(A_\alpha) = \mathrm{Prk}(M_\alpha)+1$. Hence put $\mu_{\omega,\alpha}(\nu) = \mu_\omega(^*\nu)$ where $\mu_\omega$ is defined as in Lemma 6.6 for the pair $(M_\alpha, M)$ in place of $(G, M)$.

Suppose $P_1$, $P_2$ are two adjacent elements in $\mathscr{P}(A)$ and $\alpha$ is the unique root in $\Sigma(P_2|P_1)$. Then with the notation of Lemma 6.7 we put $^*P = P_1 \cap M_\alpha$.

LEMMA 6.8. $j_{P_2|P_1}(\nu)$ *and* $j_{P_1|P_2}(\nu)$ *commute and*

$$\mu_{\omega,\alpha}(\nu) j_{P_1|P_2}(\nu) j_{P_2|P_1}(\nu) = 1.$$

*Moreover,* $j_{P_2|P_1}(\nu) = (j_{P_1|P_2}(\nu))^*$ *for* $\nu \in \mathfrak{F}$.

Put

$$\mu_{\omega,P_2|P_1}(\nu) = \prod_{\alpha\in\Sigma(P_2|P_1)} \mu_{\omega,\alpha}(\nu)$$

and

$$\mu_\omega(\nu) = \mu_{\omega,\bar{P}|P}(\nu).$$

Then $\mu_\omega(\nu)$ is independent of $P \in \mathscr{P}(A)$. Combining Lemmas 6.4, 6.5, and 6.6 it is easy to see that $\mu_\omega(\nu) > 0$ on $\mathfrak{F}$.

LEMMA 6.9. *If* $\omega \in \mathscr{E}^2(M)$, $\nu \in \mathfrak{F}$, $s \in W(A)$ *then*

$$\mu_\omega(\nu)\| C_{P_2|P_1}(s\colon \nu)\psi \|_2^2 = C(A)^2 \|\psi\|_2^2.$$

THEOREM 6.1. $\mu_\omega$ *extends to a meromorphic function on* $\mathfrak{F}_c$. *Moreover we can choose* $\delta > 0$ *such that the following two conditions hold*:

(1) $\mu_\omega$ *is holomorphic on* $\mathfrak{F}_c(\delta) = \{\nu \in \mathfrak{F}_c\colon |\nu_I| < \delta\}$

(2) *there exists* $C, r \geq 0$ *such that*

$$|\mu_\omega(\nu)| \leq C(1 + |\nu_R|)^r, \qquad (\nu \in \mathfrak{F}_c(\delta)).$$

## 7. Proof of the Plancherel Theorem

Let $\mu(\omega\colon \nu)$ be defined as in (*) of Section 1. We wish to show that if

$\alpha \in \mathscr{C}(\mathfrak{F}: V)$ and

$$\varphi_\alpha(x) = \varphi_a^\mu(x) = \int_{\mathfrak{F}} \alpha(v) \cdot E(P: \psi: v: x)\mu(\omega: v)d(\omega)\, dv$$

then $\varphi_\alpha \in \mathscr{C}(G: V: \tau)$. We have indicated in Section 1 how this leads to the fact (assuming the formula for $\varphi_\alpha^{(\bar{P})}$ which we will prove in the next section) that $d(\omega)\varphi(\omega: v)$ is the Plancherel measure corresponding to the $A$-partial Fourier transform. Put

$$\beta_0(v) = \mu(\omega: v)/\tilde{\omega}_0(\Lambda_v).$$

By the remarks at the end of Section 4 it suffices to show that $\beta_0$ is slowly increasing; that is for all $u \in \mathscr{S}(\mathfrak{F}_c)$ there exists $C>0$, and $l \geqslant 0$ so that

$$|\beta_0(v; u)| \leqslant C(1+|v|)^l.$$

By the product formula for $\mu_\omega$ we are reduced to showing this when dim $A=1$, $\text{Prk}(G)=0$. But in this case $\mu_\omega(v)$ is known explicitly (cf [4]) and the result follows from the formula.

## 8. The Determination of $\varphi_\alpha^{(\bar{P})}$

Fix a $\theta$-stable Cartan subalgebra $\mathfrak{h}$; put $\mathfrak{a} = \mathfrak{h}_R$, $\mathfrak{m}_1 = \text{Cent}_{\mathfrak{g}}(\mathfrak{a})$, $M_1 = \text{Cent}_G(\mathfrak{a})$, $M = {}^\circ M$, $\mathfrak{m} = LA(M)$, and $A = \exp \mathfrak{a}$. Let $\lambda \in i\mathfrak{h}_I^*$ be regular and assume that $\varphi$ is a function of type II($\lambda$) on $G$. With $\mathfrak{F}_c = \mathfrak{a}_c^*$, $\mathfrak{F} = \mathfrak{a}^*$, and $\alpha \in \mathscr{C}(\mathfrak{F}: V)$ put,

$$\varphi_\alpha(x) = \int_{\mathfrak{F}} \alpha(v) \cdot \varphi(v: x)\tilde{\omega}_0(\Lambda_v)\, dv$$

where we use the notation of Section 1. We have already seen in Section 4 that $\varphi_\alpha \in \mathscr{C}(G: V: \tau)$; it follows from Lemma 1.4 that $\varphi_\alpha^{(P)} \in \mathscr{C}(M_1: V: \tau_M)$ for $P \in \mathscr{P}(A)$, $P = MAN$. Moreover if we put in the notation of Section 3 (see the remark following Theorem 3.1) ($k \in K$, $m \in M_1$, $n$)

$$\varphi_{\alpha,P}(kmn) = \int \alpha(v) \cdot \tau(k)\varphi_P(v: m)\tilde{\omega}_0(\Lambda_v)\, dv$$

then $\varphi_{\alpha,P} \in \mathscr{C}(M_1: V: \tau_M)$. We shall now prove the following remarkable identity:

$$\varphi_\alpha^{(\bar{P})}(m) = d_P(m) \int_{\bar{N}} \varphi_\alpha(\bar{n}m)\, d\bar{n} = \int_{\bar{N}} e^{-\rho P(H_P(\bar{n}))}\varphi_{\alpha,P}(\bar{n}m)\, d\bar{n}. \tag{8.1}$$

This identity is proven using differential equations. We consider the function $\psi_\varphi\colon \mathfrak{F} \times \bar{N} \times G \to V$ defined by

$$\psi(v\colon \bar{n}\colon x) = \psi_\varphi(v\colon \bar{n}\colon x) = \varphi(v\colon \bar{n}x)\tilde{\omega}_0(\Lambda_v).$$

Let the notation be as in Section 3 and put for $m \in M_1$, $\bar{n} \in \bar{N}$, $v \in \mathfrak{F}$,

$$\Phi(\psi\colon v\colon \bar{n}\colon m) = \sum_{1 \leqslant i \leqslant r} \psi(v\colon \bar{n}\colon m;\, v_i \circ d_P) \otimes e_i.$$

It is easy to see that if $v \in \mathfrak{Z}_1$ then

$$\Phi(\psi\colon v\colon \bar{n}\colon m;\, v) = \underline{\Gamma}(\Lambda_v\colon v)\Phi(\psi\colon \bar{n}\colon m) + \Psi_v(\psi\colon v\colon \bar{n}\colon m) \tag{8.2}$$

where

$$\Psi_v(\psi\colon v\colon \bar{n}\colon m) = \sum_{1 \leqslant i \leqslant r} \psi(v\colon \bar{n}\colon m;\, u_i(v) \circ d_P) \otimes e_i$$

here $u_i(v)$ is defined in Section 3. Further by the left invariance of the Haar measure on $\bar{N}$ it is easy to show that if $\alpha \in \mathscr{C}(\mathfrak{F}\colon V)$, $X \in \bar{\mathfrak{n}}$, $g \in \mathfrak{G}$ then,

$$\int_{\bar{N}} \mathrm{d}\bar{n} \int_{\mathfrak{F}} \alpha(v) \cdot \varphi(v\colon \bar{n}m;\, Xg)\, \mathrm{d}v \equiv 0 \qquad (m \in M_1). \tag{8.3}$$

define $I^0$, $I^+$, and $I^-$ as in Section 3 and put

$$\underline{E}^0(\Lambda_v) = \sum_{i \in I^0} \underline{E}(s_i\Lambda_v).$$

Fix $H_0 \in \mathfrak{a}^+$ and put for $v \in \mathfrak{F}$, $\bar{n} \in \bar{N}$, $m \in M_1$, and $t \in \mathbb{R}$,

$$\Phi^0(\psi\colon v\colon \bar{n}\colon m\colon t) = \exp\{-t\underline{\Gamma}(\Lambda_v\colon H_0)\}\underline{E}^0(\Lambda_v)\Phi(\psi\colon v\colon \bar{n}\colon m \exp tH_0),$$

$$\Psi^0(\psi\colon v\colon \bar{n}\colon m\colon t) = \exp\{-t\underline{\Gamma}(\Lambda_v\colon H_0)\}\underline{E}^0(\Lambda_v)\Psi_{H_0}(\psi\colon v\colon \bar{n}\colon m \exp tH_0),$$

$$\Phi^0(\psi\colon v\colon \bar{n}\colon m) = \underline{E}^0(\Lambda_v)\Phi(\psi\colon v\colon \bar{n}\colon m).$$

LEMMA 8.1.

$$\frac{\mathrm{d}}{\mathrm{d}t}\Phi^0(\psi\colon v\colon \bar{n}\colon m\colon t) = \Psi^0(\psi\colon v\colon \bar{n}\colon m\colon t),$$

*and therefore*

$$\Phi^0(\psi\colon v\colon \bar{n}\colon m\colon t) = \Phi^0(\psi\colon v\colon \bar{n}\colon m) + \int_0^t \Psi^0(\psi\colon v\colon \bar{n}\colon m\colon s)\, \mathrm{d}s.$$

The lemma follows easily from (8.2).

We treat $\underline{V}=V\otimes T$ (cf. Section 3 for notation) as a $V$-module by defining for $v,\ v'\in V,\ e\in T,\ v\cdot v'\otimes e=(v\cdot v')\otimes e$. Further if $s\in\mathrm{End}(\underline{V})$ we put $v\otimes s\cdot v'\otimes e=v\cdot s(v'\otimes e)$. Let $\alpha\in\mathscr{C}(\mathfrak{F}:V)$, $s\in\mathrm{End}(\underline{V})$ and put,

$$\Phi(\alpha\otimes s:\bar{n}:m)=\int_{\mathfrak{F}}\alpha(v)\otimes s\cdot\Phi(\psi:v:\bar{n}:m)\,\mathrm{d}v,$$

$$\Phi^0(\alpha\otimes s:\bar{n}:m)=\int_{\mathfrak{F}}\alpha(v)\otimes s\cdot\Phi^0(\psi:v:\bar{n}:m)\,\mathrm{d}v,\quad(\bar{n}\in\bar{N},\ m\in M_1)$$

$$\Phi^0(\alpha\otimes s:\bar{n}:m:t)=\int_{\mathfrak{F}}\alpha(v)\otimes s\cdot\Phi^0(\psi:v:\bar{n}:m:t)\,\mathrm{d}v.$$

If $U:\bar{N}\times M\to\underline{V}$ we put

$$F(U:m)=\int_{\bar{N}}U(\bar{n}:m)\,\mathrm{d}\bar{n}.$$

We shall now list the steps in the proof of the identity (8.1). They are;

(1) $F(\Phi(\alpha\otimes I):m)=F(\Phi^0(\alpha\otimes I):m)$ for all $m\in M_1$. Here $I$ denotes the identity transformation on $\underline{V}$.

(2) Let $\beta\in C_c^\infty(\bar{N})$ and put for $\bar{n}_0\in\bar{N}$, $m\in M_1$, $t\in\mathbb{R}$,

$$\Phi^0(\alpha:\beta:\bar{n}_0:m:t)=\int_{\bar{N}}\beta(\bar{n})\,\Phi^0(\alpha\otimes I:\bar{n}_0\bar{n}:m:t)\,\mathrm{d}\bar{n}$$

and

$$\Phi^0(\alpha:\beta:\bar{n}:m)=\Phi^0(\alpha:\beta:\bar{n}:m:0).$$

Then

$$\Phi^0_\infty(\alpha:\beta:\bar{n}:m)=\lim_{t\to\infty}\Phi^0(\alpha:\beta:\bar{n}:m:t)$$

exists.

(3) Fix $\beta\in C_c^\infty(\bar{N})$ such that $\int_{\bar{N}}\beta(\bar{n})\,\mathrm{d}\bar{n}=1$. Then

$$F(\Phi(\alpha\otimes I):m)=\int_{\bar{N}}\Phi^0_\infty(\alpha:\beta:\bar{n}:m)\,\mathrm{d}\bar{n}.$$

If we write both sides of this last equation in the form $\Sigma_{1 \leqslant i \leqslant r} f_i \otimes e_i$ and equate the $f^{s/}$ corresponding to $i=1$, then on recalling that $v_1 = 1$ we see that (3) is very close to being the desired identity. The $\beta$ appearing on the right hand side will disappear if we show we can interchange the orders of integration. All that remains is to show that $\Phi^0_\infty(\alpha\colon \beta\colon \bar{n}\colon m)$ can be connected to the limit defined in Section 3.

(4) $\Phi^0_\infty(\psi\colon v\colon \bar{n}\colon m) = \lim_{t\to\infty} \Phi^0(\psi\colon v\colon \bar{n}\colon m\colon t)$ exists. Moreover,

$$\Phi^0_\infty(\alpha\colon \bar{n}\colon m) = \lim_{t\to\infty} \Phi^0(\alpha \otimes \exp\{-t\underline{\Gamma}(H_0)\}\colon \bar{n}\colon m \exp tH_0)$$

and

$$\Phi^0_\infty(\alpha\colon \bar{n}\colon m) = \int_{\mathfrak{F}} \alpha(v) \cdot \Phi^0_\infty(\psi\colon v\colon \bar{n}\colon m)\, \mathrm{d}v.$$

(5) Put

$$\Phi^0_\infty(\psi\colon v\colon m) = \Phi^0_\infty(\psi\colon v\colon 1\colon m).$$

Extend $\Phi^0_\infty(\psi\colon v\colon m)$ to a function on $G$ via the prescription $\Phi^0_\infty(\psi\colon v\colon kmn) = \underline{\tau}(k)\Phi^0_\infty(\psi\colon v\colon m)$ $(k \in K,\ m \in M_1,\ n \in N)$. Then for $\bar{n} \in \bar{N}$, $m \in M_1$,

$$\Phi^0_\infty(\psi\colon v\colon \bar{n}\colon m) = \mathrm{e}^{-\rho_P(H_0(n))}\Phi^0_\infty(\psi\colon v\colon \bar{n}m).$$

With the above described procedure of taking $f^{s/}$ with $i=1$, and using (3) and (4) the identity follows.

We shall now proceed with the proofs of steps 1 through 5.

(1) Let $i \in I \backslash I^0$ and put for $v \in \mathfrak{F}$, $\bar{n} \in \bar{N}$, $m \in M_1$, and $s \in \mathrm{End}(\underline{V})$

$$\Phi_i(\psi\colon v\colon \bar{n}\colon m) = \underline{E}(s_i\Lambda_v)\Phi(\psi\colon v\colon \bar{n}\colon m),$$

$$\Phi_i(\alpha \otimes s\colon \bar{n}\colon m) = \int_{\mathfrak{F}} \alpha(v) \otimes s \cdot \Phi_i(\psi\colon v\colon \bar{n}\colon m)\, \mathrm{d}v$$

and as before

$$F(\Phi_i(\alpha \otimes s)\colon m) = \int_{\bar{N}} \Phi_i(\alpha \otimes s\colon \bar{n}\colon m)\, \mathrm{d}\bar{n}.$$

We shall show that $F(\Phi_i(\alpha \otimes I)) \equiv 0$ which will prove (1). Let $\mu \in \mathfrak{F}$ and put

$$F(\Phi_i(\alpha \otimes s)\colon m\colon \mu) = \int_{\mathfrak{F}} F(\Phi_i(\alpha \otimes s)\colon m \exp H)\, \mathrm{e}^{-i\mu(H)}\, \mathrm{d}H$$

$dH$ denoting the Euclidean measure on $\mathfrak{a}$. It follows from the results of Section 3 and 4 that

$$\zeta_s : \alpha \to \| F(\Phi_i(\alpha \otimes s): m: \mu)\|$$

is a continuous distribution on $\mathscr{C}(\mathfrak{F}: V)$.

Let $v \in \mathfrak{Z}_1$ *and let* $\Gamma(v)$: $\mathfrak{F} \to \mathrm{End}(V)$ be defined by $\Gamma(v)(v) = \Gamma(\Lambda_v: v)$, $\alpha \otimes \Gamma(v)$ has the obvious definition as a function of $\mathfrak{F}$ into $\mathrm{End}(\underline{V})$. From (8.2) and (8.3) we have,

$$F(\Phi_I(\alpha \otimes \zeta): m; v) = F(\Phi_i(\alpha \otimes \Gamma(v)): m), \qquad (m \in M_1).$$

Hence,

$$F(\Phi_i(\alpha \otimes \Gamma(H)): m: \mu) = i\mu(H) F(\Phi_i(\alpha \otimes I): m: \mu). \tag{8.4}$$

Recall that if $\Lambda \in \mathfrak{h}_c^*$ we put $W_0(\Lambda) = \{s \in W_0: s\Lambda = \Lambda\}$, and if $1 \leqslant i \leqslant r$, $W_0(s_i, \lambda) = W_0(\Lambda)$ where $\Lambda = s_i \lambda^\nu$. Fix $H \in \mathfrak{a}$ and put

$$\eta(v) = \prod_{t \in W_0(s_i, \lambda)} \{i\mu(H) - t s_i \Lambda_v(H)\}$$

then

LEMMA 8.2.

$$\zeta_I(\eta\alpha) = 0$$

*for all* $\alpha \in \mathscr{C}(\mathfrak{F}: V)$.

The lemma is immediate from (8.4) on noting that

$$\Phi_i(\alpha \otimes \underline{\Gamma}(H)) = \Phi_i(P_i \alpha \otimes I)$$

where $P_i(v) = s_i \Lambda_v(H)$. Hence if $\zeta_I \not\equiv 0$ then there exists $v_0 \in \mathrm{supp}(\zeta_I)$ $(v_0 \in \mathfrak{F}) \Rightarrow \eta(v_0) \equiv 0$. Since $\mathrm{Re}(t s_i \Lambda_v) = s_i \lambda^\nu(H)$ must equal 0 then this implies that $i \in I^0$. Hence if $i \notin I^0 \Rightarrow F(\Phi_i(\alpha \otimes I)) \equiv 0$.

(2) Define for $\beta \in C_c^\infty(\bar{N})$, $\Phi^0(\alpha: \beta: \bar{n}_0: m: t)$ as above and put for $\bar{n}_0 \in \bar{N}$, $m \in M_1$, $t \in \mathbb{R}$,

$$\Psi^0(\alpha: \beta: \bar{n}_0: m: t) = \int_{\bar{N}} \beta(\bar{n}) \Psi^0(\alpha \otimes I: \bar{n}_0 \bar{n}: m: t)\, d\bar{n}.$$

Then from Lemma 8.1 we have

$$\Phi^0(\alpha: \beta: \bar{n}: m: t) = \Phi^0(\alpha: \beta: \bar{n}: m) + \int_0^t \Psi^0(\alpha: \beta: \bar{n}: m: s)\, ds.$$

Note that if we put

$$\varphi_\alpha(\beta:\bar n_0:x)=\int_{\bar N}\beta(\bar n)\varphi_\alpha(\bar n_0\bar n x)\,\mathrm{d}\bar n,$$

and let $X\in\bar{\mathfrak{n}}$, $g\in\mathfrak{G}$ then

$$\varphi_\alpha(\beta:\bar n:m;Xg)=-\varphi_\alpha(X^m\beta:\bar n:m;g).$$

From this observation and the fact that $\varphi$ is of type II($\lambda$) we can show that fixing $m\in M_1$, we can choose $\varepsilon>0$ and $C>0$, $r_0\geqslant 0$ so that

$$\|\Psi^0(\alpha:\beta:\bar n:m:s)\|\leqslant C\,\mathrm{e}^{-\varepsilon s}d_P(m\exp sH_0)\int_{\operatorname{supp}\beta}\Xi_{r_0}(\bar n\bar n_0 m\exp tH_0)\,\mathrm{d}\bar n_0$$

where $\Xi_r(x)=\Xi(x)(1+\sigma(x))^r$. Hence,

$$\int_0^\infty\|\Psi^0(\alpha:\beta:\bar n:m:s)\|\,\mathrm{d}s<\infty.$$

Put

$$\Phi^0_\infty(\alpha:\beta:\bar n:m)=\Phi^0(\alpha:\beta:\bar n:m)+\int_0^\infty\Psi^0(\alpha:\beta:\bar n:m:s)\,\mathrm{d}s.$$

Then

$$\Phi^0_\infty(\alpha:\beta:\bar n:m)=\lim_{t\to\infty}\Phi^0(\alpha:\beta:n:m:t).$$

(3) It follows from the estimates for $\Xi$ restricted to $\bar N$ (cf. [2], Section 10) that

$$\int_{\bar N}\mathrm{d}\bar n\int_0^\infty\|\Psi^0(\alpha:\beta:\bar n:m:s)\|\,\mathrm{d}s<\infty\Rightarrow\int_{\bar N}\|\Phi^0_\infty(\alpha:\beta:\bar n:m)\|\,\mathrm{d}\bar n<\infty.$$

On the other hand by (8.3) we have

$$\int_{\bar N}\Psi^0(\alpha:\beta:\bar n:m:s)\,\mathrm{d}\bar n=0.$$

Therefore by Fubini's Theorem

$$\int_{\bar N}\Phi^0_\infty(\alpha:\beta:\bar n:m)\,\mathrm{d}\bar n=\int_{\bar N}\Phi^0(\alpha:\beta:\bar n:m)\,\mathrm{d}\bar n$$

$$= \int_{\bar{N}} d\bar{n} \int_{\bar{N}} \beta(\bar{n}_0)\Phi^0(\alpha \otimes I: \bar{n}_0\bar{n}: m)\, d\bar{n}_0$$

$$= \int_{\bar{N}} \beta(\bar{n}_0)\, d\bar{n}_0 \int_{\bar{N}} \Phi^0(\alpha \otimes I: \bar{n}_0\bar{n}: m)\, d\bar{n}$$

$$= F(\Phi^0(\alpha \otimes I): m) = F(\Phi(\alpha \otimes I): m).$$

The last equality by (1).

(4) From Lemma 8.1 we have

$$\Phi^0(\psi: \nu: \bar{n}: m: t) = \Phi^0(\psi: \nu: \bar{n}: m) + \int_0^t \Psi^0(\psi: \nu: \bar{n}: m: s)\, ds.$$

Fix $x \in G$, $X \in \bar{\mathfrak{n}}$, $g \in \mathfrak{G}$. Then we can choose $C, r \geq 0$ so that

$$\|\psi(\nu: x \exp tH_0;\, Xg)\| \leq C(1+|\nu|)^r\, e^{-2\varepsilon t}\Xi(x \exp tH_0)(1+t)^r$$

where $\varepsilon = 2\beta_P(H_0)$. This follows since if $x_t = x \exp tH_0$ then

$$\psi(\nu: x_t;\, Xg) = \psi(\nu: X^{x_t};\, x_t;\, g).$$

It follows from this and the fact that $\exp\{-t\underline{\Gamma}(\Lambda_\nu: H_0)\}\underline{E}^0(\Lambda_\nu)$ has only purely imaginary eigenvalues that for fixed $\bar{n}$, $m \in M_1$, there exists $C>0$, $r \geq 0$ such that

$$\|\Psi^0(\psi: \nu: \bar{n}: m: t)\| \leq C\, e^{-\varepsilon t}(1+|\nu|)^r$$

$$\Rightarrow \int_0^\infty \|\Psi^0(\psi: \nu: \bar{n}: m: s)\|\, ds < \infty.$$

Hence, set

$$\Phi^0_\infty(\psi: \nu: \bar{n}: m) = \Phi^0(\psi: \nu: \bar{n}: m) + \int_0^\infty \Psi^0(\psi: \nu: \bar{n}: m: s)\, ds.$$

Then

$$\Phi^0_\infty(\psi: \nu: \bar{n}: m) = \lim_{t \to \infty} \Phi^0(\psi: \nu: \bar{n}: m: t).$$

Moreover, from Lemma 8.1 we have

$$\Phi^0(\psi: \nu: \bar{n}: m: t) = \Phi^0_\infty(\psi: \nu: \bar{n}: m) - \int_t^\infty \Psi^0(\psi: \nu: \bar{n}: m: s)\, ds$$

$$\Rightarrow \|\Phi^0(\psi: \nu: \bar{n}: m \exp tH_0) - \exp\{t\underline{\Gamma}(\Lambda_\nu: H_0)\}\Phi^0_\infty(\psi: \nu: \bar{n}: m)\|$$

$$\leqslant \left\| \exp\{t\underline{\Gamma}(\Lambda_\nu: H_0)\} \int_t^\infty \Psi^0(\psi: \nu: \bar{n}: m: s)\, ds \right\|.$$

It follows from this that there exists $\varepsilon' > 0$, $r' \geqslant 0$, so that

$$\|\Phi^0(\psi: \nu: \bar{n}: m \exp tH_0) - \exp\{t\underline{\Gamma}(\Lambda_\nu: H_0)\Phi^0_\infty(\psi: \nu: \bar{n}: m)\| \\ \leqslant C\, e^{-\varepsilon' t}(1 + |\nu|)^{r'}.$$

Put

$$\Phi^0_\infty(\alpha: \bar{n}: m) = \lim_{t \to \infty} \Phi^0(\alpha \otimes \exp\{-t\underline{\Gamma}(H_0)\}: \bar{n}: m \exp tH_0).$$

Then it is clear from the above that the limit exists and in fact that

$$\Phi^0_\infty(\alpha: \bar{n}: m) = \int_{\mathfrak{F}} \alpha(\nu) \cdot \Phi^0_\infty(\psi: \nu: \bar{n}: m)\, d\nu.$$

(5) Put

$$\Phi^0_\infty(\psi: \nu: m) = \Phi^0_\infty(\psi: \nu: 1: m).$$

Extend $\Phi^0_\infty(\psi: \nu: m)$ to a function on $G$ by setting

$$\Phi^0_\infty(\psi: \nu: kmn) = \underline{\tau}(k)\Phi^0_\infty(\psi: \nu: m) \qquad (k \in K, m \in M_1, n \in N).$$

Then claim

$$\Phi^0_\infty(\psi: \nu: \bar{n}: m) = e^{-\rho_P(H_P(n))}\Phi^0_\infty(\psi: \nu: \bar{n}m).$$

Since

$$\bar{n}m = \kappa(\bar{n}m)\mu(\bar{n}m) \exp(H(\bar{n}m))n(\bar{n}m)$$

where $\kappa(x) \in K$, $\mu(x) \in M \cap \exp \mathfrak{s}$, $H(x) = \log a(x)$, $a(x) \in A$, and $n(x) \in N$. Whereas,

$$\bar{n}m = \kappa(\bar{n})\mu(\bar{n}) \exp(H(\bar{n}))n(\bar{n})\, m \\ = \kappa(\bar{n})\mu(\bar{n})m \exp(H(\bar{n}))n' \qquad (n' \in N).$$

Hence $H(\bar{n}m) = H(\bar{n}) + H(m)$. Recalling then the definitions of $\Phi(\psi: \nu: \bar{n}: m)$ and $\Phi(\psi: \nu: m)$ we see that $(\rho = \rho_P)$

$$\Phi(\psi: \nu: \bar{n}: m) = e^{-\rho(H(\bar{n}))}\Phi(\psi: \nu: \bar{n}m)$$

from which (5) follows.

To finally obtain the identity let

$$\Phi^0_\infty(\alpha: m) = \int_{\mathfrak{F}} \alpha(\nu) \cdot \Phi^0_\infty(\psi: \nu: m)\, d\nu.$$

Then extend $\Phi_\infty(\alpha)$ to a function on $G$ by the prescription $\Phi^0_\infty(\alpha: kmn) = \underline{\tau}(k)\Phi^0_\infty(\alpha: m)$ $(k \in K,\ m \in M_1,\ n \in N)$. We have then that

$$F(\Phi(\alpha \otimes I): m) = F(\Phi^0(\alpha \otimes I): m) = \int_{\bar{N}} \Phi^0_\infty(\alpha: \beta: \bar{n}: m)\, d\bar{n}.$$

Hence taking a Dirac sequence $\beta_n \in C_c^\infty(\bar{N})$ $(1 \leqslant n < \infty)$ filtering to the identity and using the fact that $\Phi^0_\infty(\alpha: \bar{n}: m)$ exists and equals

$\int_{\mathfrak{F}} \alpha(\nu) \cdot \Phi^0_\infty(\psi: \nu: \bar{n}: m)\, d\nu$ then we have

$$F(\Phi(\alpha \otimes I): m) = \int_{\bar{N}} \Phi^0_\infty(\alpha \otimes I: \bar{n}: m)\, d\bar{n}$$

$$= \int_{\bar{N}} e^{-\rho_P(H_P(\bar{n}))}\Phi^0_\infty(\alpha: \bar{n}m)\, d\bar{n}.$$

## References


[1] Harish-Chandra, 'Discrete series for Semisimple Lie Groups', II, *Acta Math.* **116** (1966), 1–111.

[2] Harish-Chandra, 'Harmonic Analysis on Real Reductive groups, I', *J. Funct. Anal.* **19** (1975), 104–204.

[3] Harish-Chandra, 'Harmonic Analysis on Real Reductive groups, II', *Inv. Math.* **36** (1976), 1–55.

[4] Harish-Chandra, 'Harmonic Analysis on Real Reductive groups, III', *Ann. of Math.* **104** (1976), 117–201.

[5] Trombi, P. C., 'On Harish-Chandra's Theory of the Eisenstein Integral for Real Semisimple Lie groups', Lecture notes, University of California, Los Angeles.

[6] Trombi, P. C., and Varadarajan, V. S., 'Spherical Transforms on Semisimple Lie groups', *Ann. of Math.* **94** (1971), 246–303.

[7] Trombi, P. C., and Varadarajan, V. S., 'Asymptotic Behavior of Eigenfunctions on a Semisimple Lie group: The Discrete Spectrum', *Acta Math.* **129** (1972), 237–280.

[8] Varadarajan, V. S. *Harmonic Analysis on Real Reductive Groups*, Springer Verlag, Lecture Notes 576, Berlin, New York (1977).

# A Geometric Construction of the Discrete Series for Semisimple Lie Groups

Michael Atiyah[1] and Wilfried Schmid[2]

[1] Mathematical Institute, 24–29 St Giles, Oxford OX1 3LB, England
[2] Department of Mathematics, Columbia University, New York, NY 10027, USA

Dedicated to Friedrich Hirzebruch

## § 1. Introduction

In the representation theory of a compact group $K$, a major role is played by the Peter-Weyl theorem, which asserts that the regular representation $L^2(K)$ decomposes as a countable direct sum of irreducibles with finite multiplicity. For compact connected Lie groups this becomes much more concrete: the irreducibles are explicitly known, their characters are given by the famous Hermann Weyl formula, and there is a uniform geometrical construction for them due to Borel and Weil.

For a general locally compact group $G$, the general Plancherel theorem, which goes back essentially to von Neumann and is a very sophisticated generalization of the Peter-Weyl theorem, gives a direct integral decomposition of $L^2(G)$. For real semisimple Lie groups (connected and with finite center) the work of Harish-Chandra makes the Plancherel theorem quite explicit. The first and basic step is the identification of the discrete part of $L^2(G)$: the irreducible representations which enter here are (by definition) those of the discrete series. For these Harish-Chandra has given very precise results in terms of their (generalized) characters [14, 15]. Moreover, a geometrical realization of the discrete series, analogous to the Borel-Weil theorem, was conjectured by Langlands and subsequently established in various forms (cf. [21, 23, 22, 25]).

The purpose of this paper is to give a new and, to a large extent, self-contained account of the principal results concerning the discrete series. The main novelty in our presentation is that we use (a weak form of) the geometric realization to construct the discrete series representations and to obtain information about their characters. Previously things were done in the reverse order, the existence of the discrete series, proved by Harish-Chandra, being used to get the geometric realization. Analytically our proof of the existence of the discrete series rests on the $L^2$-index theorem [1], which gives a suitable generalization to non-compact manifolds of the Atiyah-Singer index theorem. The way in which the $L^2$-index theorem is used is quite analogous to the role of the Riemann-Roch theorem in the Borel-Weil-Bott approach to the representations of compact Lie groups.



*J. A. Wolf, M. Cahen, and M. De Wilde (eds.), Harmonic Analysis and Representations of Semi-Simple Lie Groups*, 317–378.

In broad outline the main results concerning the discrete series may be listed as follows:

(1) *existence* of discrete series representations $\mathbf{H}_\lambda$, indexed by a suitable lattice parameter $\lambda$,

(2) *exhaustion*—proof that the representations in (1) give all the discrete series,

(3) *Geometric realization*—identification of $\mathbf{H}_\lambda$ with the space of $L^2$-solutions of a certain elliptic differential equation (of Dirac type) on the symmetric space $G/K$,

(4) *Character behavior*—discrete series characters decay at $\infty$ (in an appropriate sense) and are determined by their restriction to (regular points of) a compact Cartan subgroup.

(5) *Character formula*—explicit formula for the character of $\mathbf{H}_\lambda$ on a compact Cartan subgroup.

One of the features of our approach is that (1), (2) and (3) are all treated together in a rather natural manner. The character properties (4) and (5) are needed for the proof of (2). The results given in this paper are complete for (1), (2), (4) and (5), and almost complete for (3). We refer to Section 9 for a more detailed description of the situation concerning (3).

To a great extent this paper is a synthesis of existing results and methods. Its aim is to demonstrate how the theory of the discrete series may be established on geometric-analytic foundations. Because the literature is diverse, extensive and highly technical, we have thought it worthwhile to present here a reasonably complete account, hopefully written so as not to make extensive demands on the reader's knowledge of Lie group theory. As a result many relevant results are reproved (or sketched) in the version which we need—frequently this is much simpler than in the published form.

Naturally some basic results are going to be needed. In order to clarify the situation it is perhaps best to list the sort of results which we shall assume. These belong to several different categories beginning with some generalities:

(a) algebraic structure of semisimple Lie groups,

(b) basic generalities about unitary representations, including the existence of the Harish-Chandra distributional characters,

(c) representation theory of compact Lie groups,

(d) abstract Plancherel theorem.

Next come the results which are crucial to our approach:

(e) existence of uniform discrete subgroups $\Gamma$ of real semisimple Lie groups,

(f) $L^2$-index theorem,

(g) index theorem for compact manifolds,

(h) differential-geometric computations involving curvature.

The existence of $\Gamma$ is essential for the application of (f), which via (g) leads to the curvature computations in (h), exactly as in the work of Hirzebruch [17]. Finally there are three more specific results of Harish-Chandra:

(i) $\mathbf{H} \otimes V$ has a finite composition series when $\mathbf{H}$ is an irreducible unitary representation of $G$ and $V$ is a finite-dimensional representation,

(j) structure of the algebra of bi-invariant differential operators on $G$,

(k) local integrability of the Harish-Chandra characters.

Of these (i) follows from the fact that **H** appears as a subrepresentation of an induced representation (see the simple proof by Casselman [9]). The algebraic result (j) is required as the first step in the proof of the much deeper theorem (k). An alternative proof of (k), incorporating also a simple proof of (j), will be given in [3], which should therefore be viewed as foundational for the present paper.

A number of purely algebraic arguments which we reproduce in suitable form are relegated to an Appendix. These include Parasarathy's computation of the spinor Laplacian, an estimate for the action of the Casimir operator on a unitary representation, and an algebraic characterization of certain representations in terms of their $K$-decomposition.

We turn now to a description of the contents of the various sections, highlighting the main features. We begin in Section 2 with a general review of the abstract Plancherel theorem. This is essentially standard material included for the benefit of the reader, but it also sets the scene for Section 3, where we apply the $L^2$-index theorem of [1] to Dirac operators on the symmetric space $G/K$, with $G$ and $K$ having the same rank. The final result of Section 3 is Theorem (3.16), which gives an explicit formula for the difference of the Plancherel measures of two spaces $\mathscr{H}_\mu^+$ and $\mathscr{H}_\mu^-$ of square-integrable, harmonic spinors on $G/K$, with values in a vector bundle $\mathscr{V}_\mu$. The space $\mathscr{H}_\mu^+$ will eventually turn out to be the representation space of a discrete series representation, but the important fact at this stage is that $\mathscr{H}_\mu^+ \neq 0$, for suitable parameters $\mu$.

In Section 4 we show that the formal difference $\mathscr{H}_\mu^+ - \mathscr{H}_\mu^-$ is a finite linear combination of irreducibles, which therefore belong to the discrete series. This is made precise by the difference formula (4.24), which in particular contains an existence theorem for the discrete series. From the identity (4.24) we derive a formula, valid on the elliptic set, for the sum of the discrete series characters which correspond to a given infinitesimal character, each multiplied by its formal degree (Theorem (4.41)). The arguments in § 4 depend on an analysis of the $K$-characters of representations of $G$, and the one non-trivial fact we use is that only finitely many irreducible representations can have a particular infinitesimal character (a consequence of the local integrability of characters).

In Section 5 we assume the parameter $\mu$ is sufficiently positive. Using an algebraic version of Parasarathy's formula (explained in the Appendix) we deduce that $\mathscr{H}_\mu^- = 0$ and that any irreducible constituent $\mathbf{H}_j$ of $\mathscr{H}_\mu^+$ has a $K$-decomposition with a lowest highest weight. Combined with a purely algebraic result concerning such representations (Proposition (5.14), proof sketched in Appendix), this leads rapidly to Theorem (5.20), which asserts that $\mathscr{H}_\mu^+$ is irreducible and gives its formal degree, as well as the character formula on the elliptic set.

Sections 6 and 7 are devoted to a study of growth conditions at infinity for characters, or more generally, for invariant eigendistributions. Unlike Harish-Chandra, who uses a certain Schwartz space for this purpose, we carry out our analysis in the framework of Sobolev spaces. The Sobolev spaces are technically simple to deal with, and they are already used in [1], on which our construction of the discrete series rests. We first show (Lemma (6.3)) that a discrete series character extends from $C_0^\infty(G)$ to some Sobolev space. We then enunciate the main results (Proposition (6.10) and (6.11) and Lemma (6.15)), which give the precise form of (4) above, and in particular show that the discrete series occurs if and only if rank $G$ = rank $K$. As a further consequence of these results, in combination with Theorem (4.41), we also obtain an explicit description of the infinitesimal characters of the discrete series (Corollary (6.13)). Here we use (4.41), which came from the $L^2$-index theorem, for singular as well as nonsingular weights. Thus the $L^2$-index theorem, as applied to $G/K$, embodies both an existence theorem (leading to the existence of the discrete series) and a nonexistence theorem, which enters our exhaustion proof for the discrete series.

The proofs of (6.10), (6.11) and (6.15) are given in Section 7. In the first place we verify (Corollary (7.12)) that the Sobolev spaces of $G$ and of a Cartan subgroup $B$ are compatible in the way one would expect. By splitting $B$ into its toroidal part and its vector part, (6.10) is eventually reduced to an easy property of Sobolev spaces in Euclidean space. The proofs of (6.11) and (6.15) are then straightforward applications of the Harish-Chandra "matching conditions", which describe the behavior of an invariant eigendistribution as one moves between two adjacent Cartan subgroups. These matching conditions are an essential supplement to the local integrability, and they will also be established in [3].

In Section 8 we use the properties of characters just described, together with Zuckerman's tensor product technique (which we explain to the extent that it is used here), to deal with those parameters $\mu$ not covered by Theorem (5.20). Zuckerman's method enables us to "shift" $\mu$, making it sufficiently nonsingular to apply Theorem (5.20). This makes it possible to refine Theorem (4.41), which gave a formula for certain linear combinations of discrete series characters on a compact Cartan subgroup, into the corresponding formula for individual discrete series characters. The resulting character formula constitutes one part of our main theorem (8.1), the others being the computation of the formal degree and an exhaustion statement. These last two parts are easy and purely formal consequences of the same arguments which produced the character formula. Finally, at the end of Section 8, we prove Theorem (8.5), which extends to all discrete series representations the results on the "lowest highest weight" of their $K$-decompositions, previously established for sufficiently nonsingular parameters. Again the proof uses the Zuckerman technique to shift the parameter.

Section 9 deals with the problem of realizing the discrete series geometrically. The enumeration of the discrete series representations and the results about their $K$-decompositions, as described in Section 8, make it a simple matter to identify the discrete part of the spaces of square-integrable, harmonic spinors $\mathcal{H}_\mu^+$, $\mathcal{H}_\mu^-$ (Lemma (9.4)). In order to eliminate the continuous part from the Plancherel decomposition of $\mathcal{H}_\mu^+$, $\mathcal{H}_\mu^-$, we must appeal to Harish-Chandra's results about the explicit form of the Plancherel formula. The crucial statement appears as Lemma (9.8); although the details of its proof go well beyond the framework of this paper, we indicate at least the main ideas. The final conclusion is Theorem (9.3), which describes the spaces $\mathcal{H}_\mu^+$, $\mathcal{H}_\mu^-$, and which in particular provides a geometric realization for every discrete series representation. We end the section with some comments about ways to avoid the use of delicate properties of Plancherel measure.

The Appendix, finally, contains proofs of four technical statements, which have appeared elsewhere, and which are collected here for the convenience of the reader.

An approach to the discrete series similar to ours has recently been developed by de George and Wallach. They start in the same way by choosing a uniform discrete subgroup $\Gamma$ and then computing the index of the Dirac operator on $\Gamma \backslash G/K$. However, instead of using the $L^2$-index theorem to relate this to $G/K$, they use a descending sequence $\Gamma \supset \cdots \supset \Gamma_n \supset \cdots$ of normal subgroups of finite index, intersecting in the identity. In a sense their argument is more elementary, since they only work with compact manifolds. On the other hand it appears to yield somewhat weaker results, and it does not tie in so directly with the geometric realization. A connecting link between the de George-Wallach approach and ours is to be found in a paper by Kazdan [18], which investigates $L^2(G/K)$ via the limit of $L^2(\Gamma_n \backslash G/K)$, as $n \to \infty$.

## § 2. Review of the Plancherel Theorem

Since we shall be making essential use of the Plancherel theorem, we review here the basic facts, with particular reference to those properties which we shall be using. We restrict ourselves to connected real semisimple Lie groups $G$ with finite center and recall that these are unimodular (left and right Haar measure coincide).

The first basic fact is that any irreducible unitary representation $\mathbf{H}$ of $G$, when restricted to a maximal compact subgroup $K$, has a direct sum decomposition

$$(2.1) \quad \mathbf{H} = \bigoplus_{i \in \hat{K}} n_i V_i,$$

where the $V_i$ are the irreducible representations of $K$, and the multiplicities $n_i$ satisfy the bound

$$n_i \leqq \dim V_i.$$

This in turn implies that for any $f \in C_0^\infty(G)$, the corresponding operator

$$\pi(f) = \int_G f(g)\, \pi(g)\, dg$$

($g \mapsto \pi(g)$ denotes the action of $G$ on $\mathbf{H}$) is of trace class, and that $f \mapsto \operatorname{trace} \pi(f)$ is continuous, hence defines a distribution $\Theta$, called the (Harish-Chandra) character of the representation. Moreover the fact that $\pi(f)$ is in particular compact implies that $G$ is of type $I$, i.e. that every factor representation involves only a type $I$ factor. This means that every unitary representation of $G$ can be decomposed in an appropriate sense into irreducibles. We let $\hat{G}$ denote the set of (equivalence classes of) irreducible unitary representations of $G$.

The Plancherel theorem is concerned with decomposing the left (right) regular representation, i.e. the action on $L^2(G)$ induced by left (right) translation. In the first instance it asserts that we have a direct integral decomposition

$$(2.2)\quad L^2(G) \cong \int_{\hat{G}} \mathbf{H}_j \hat{\otimes} \mathbf{H}_j^* \, dj,$$

where $dj$ is a positive measure on $\hat{G}$, $\mathbf{H}_j$ is the irreducible representation indexed by $j \in \hat{G}$, and $\mathbf{H}_j \hat{\otimes} \mathbf{H}_j^*$ is the Hilbert space tensor product of $\mathbf{H}_j$ and its dual. The isomorphism (2.2) is compatible with both the left and right actions of $G$. Moreover, if $\mathscr{A}$ denotes the von Neumann algebra generated by the left translations on $L^2(G)$, we have an isomorphism

$$(2.3)\quad \mathscr{A} \cong \int_{\hat{G}} \mathscr{L}(\mathbf{H}_j) \, dj,$$

where $\mathscr{L}(\mathbf{H}_j)$ denotes the algebra of all bounded operators on $\mathbf{H}_j$. There is a similar statement for right translations which generate the commutant $\mathscr{A}'$ of $\mathscr{A}$.

A further aspect of the Plancherel theorem (which determines the Plancherel measure $dj$ uniquely) involves the consideration of traces. On the $C^\infty$ group algebra ($C_0^\infty(G)$ under convolution), evaluation at the identity $e$ of $G$ defines a natural "trace". It turns out that this can be extended to the von Neumann algebra $\mathscr{A}$. More precisely, it extends to a map

$$\mathrm{trace}_G \colon \mathscr{A}^+ \to \mathbb{R}^+ \cup \{\infty\},$$

where $\mathscr{A}^+$ denotes the cone of positive operators in $\mathscr{A}$. The elements $A \in \mathscr{A}^+$ with $\mathrm{trace}_G(A)$ finite are the positive part $\mathfrak{m}^+$ of an ideal $\mathfrak{m}$ of $\mathscr{A}$, on which $\mathrm{trace}_G$ extends as a linear functional. The elements of $\mathfrak{m}$ will be said to be of $G$-trace class. If $A \in \mathfrak{m}$ then its "Fourier components" $A_j$ in (2.3) are of trace class almost everywhere, and

$$(2.4)\quad \mathrm{trace}_G A = \int_{\hat{G}} \mathrm{trace}\, A_j \, dj.$$

In particular, taking $A = l(f)$ to be the operator representing $f \in C_0^\infty(G)$ in the left regular representation, we obtain

$$(2.5)\quad f(e) = \int_{\hat{G}} \Theta_j(f) \, dj$$

($\Theta_j =$ Harish-Chandra character of $\mathbf{H}_j$).

If we apply (2.5) to $h = f * \tilde{f}$, where $\tilde{f}(g) = \overline{f(g^{-1})}$, then since

$$h(e) = \int_G |f(g)|^2 \, dg = \|f\|^2$$

and $\pi_j(\tilde{f}) = \pi_j(f)^*$, the result can be written in the form

$$(2.6)\quad \|f\|^2 = \int_{\hat{G}} \mathrm{trace}(\pi_j(f) \cdot \pi_j(f)^*) \, dj \qquad \text{(Plancherel formula).}$$

The restrictions on $f$ can be relaxed somewhat. If $f \in L^2(G)$ and $l(f) \in \mathscr{A}$ (i.e. $l(f)$ is bounded), then $\pi(f) \cdot \pi(f)^*$ is of $G$-trace class, and so we can apply (2.4). Moreover

$$\mathrm{trace}_G(l(f) \cdot l(f)^*) = \|f\|^2$$

follows by continuity from the corresponding statement with $f \in C_0^\infty(G)$. The easiest way to insure that $l(f)$ is bounded is to take $f \in L^2(G) \cap L^1(G)$. Thus (2.6) holds for all $f \in L^2(G) \cap L^1(G)$.

The preceeding results are all standard, but in addition we shall need:

(2.7) $f \in C^\infty(G)$ and $l(f) \in \mathscr{A}^+ \Rightarrow l(f)$ is of $G$-trace class and $\text{trace}_G(f) = f(e)$.

As we shall explain in the next section, a proof of (2.7) is essentially given in [1].

We shall use (2.7) in the particular case that $l(f)$ is an orthogonal projection onto a subspace $W$ of $L^2(G)$. We shall write $\dim_G W$ for $\text{trace}_G\, \pi(f)$ which, by (2.6), is then equal to $f(e)$. Now $W$ has a decomposition

(2.8) $$W = \int_{\hat{G}} W_j \hat{\otimes} \mathbf{H}_j^*\, dj,$$

where $W_j \subset \mathbf{H}_j$ is the image of the projection operator $\pi_j(f)$. Since, by (2.4) and (2.7), $\pi_j(f)$ is of trace class (for almost all $j$), this means that $\dim W_j$ is finite (for almost all $j$), and (2.4) becomes

(2.9) $$f(e) = \dim_G W = \int_{\hat{G}} \dim W_j\, dj.$$

In our applications the space $L^2(G)$ will be replaced by $L^2(G/K, \mathscr{F})$ – the $L^2$-sections of a homogeneous vector bundle $\mathscr{F}$ over the symmetric space $G/K$. It is a fairly simple matter to modify the above formulae involving the Plancherel measure to cover this case, as we shall now explain.

If $F$ is a finite-dimensional unitary $K$-module, then $\mathscr{F} = G \times_K F$ is a homogeneous vector bundle over $G/K$, and $L^2(G/K, \mathscr{F})$ may be identified with the space of right $K$-invariants in $L^2(G) \otimes F$. Because of (2.2) this gives

(2.10) $$L^2(G/K, \mathscr{F}) \cong \int_{\hat{G}} \mathbf{H}_j \otimes W_j\, dj,$$

where $W_j$ is the $K$-invariant part of $\mathbf{H}_j^* \otimes F$, which is finite-dimensional by (2.1). From (2.3) we see that the algebra $\mathscr{B}$ of $G$-invariant bounded operators on $L^2(G/K, \mathscr{F})$ corresponds under (2.10) to the direct integral

(2.11) $$\int_{\hat{G}} \mathscr{L}(W_j)\, dj.$$

In the algebra $\mathscr{A}' \otimes \mathscr{L}(F)$ there is a natural trace given by the tensor product of $\text{trace}_G$ in $\mathscr{A}'$ (the algebra generated by right translation in $L^2(G)$) and the ordinary trace in $\mathscr{L}(F)$: for brevity we still denote this by $\text{trace}_G$ and the corresponding dimension function by $\dim_G$. Restricting to the subalgebra $\mathscr{B}$ which acts on the subspace (2.10), we see that

$$\text{trace}_G A = \int_{\hat{G}} \text{trace}\, A_j\, dj;$$

here $A_j \in \mathscr{L}(W_j)$ are the components (given by (2.11)) of an operator $A$ of $G$-trace class in $\mathscr{B}$. In particular, if $U \subset L^2(G/K, \mathscr{F})$ is any closed $G$-invariant subspace, then

$$U=\int_{\hat{G}} \mathbf{H}_j \otimes U_j \, dj, \quad \text{with } U_j \subset W_j,$$

and

$$\text{(2.12)} \quad \dim_G U = \int_{\hat{G}} \dim U_j \, dj.$$

Suppose moreover that orthogonal projection $B$ onto $U$ is given by a smooth kernel $b$ – a section of $\operatorname{Hom}(\mathscr{F}, \mathscr{F})$ over $G/K \times G/K$. The corresponding $A$ on $G$ given by $A = p^* B p_*$ ($p$: $G \to G/K$ being the projection) then also has a smooth kernal $a$, with $a(x, y) = b(p(x), p(y))$. Hence, by (2.7), $A$ is of $G$-trace class and

$$\text{(2.13)} \quad \dim_G U = \operatorname{trace}_G A = \operatorname{tr} a(e, e) = \operatorname{tr} b(0, 0),$$

where $p(e) = 0$ is the base point of $G/K$ and tr denotes the usual trace in $\mathscr{L}(F)$: note that the fibre $\mathscr{F}_0$ may be identified with $F$.

*Remark.* In the above we have implicitly assumed that Haar measure on $K$ is normalized to have total volume equal to 1, and that $G/K$ is then given the quotient of the two Haar measures. On $G$ Haar measure is supposed fixed once and for all – the Plancherel measure on $\hat{G}$ depends on this choice.

Returning to the general Plancherel theorem, we recall that $\mathbf{H}_{j_0}$ is said to belong to the discrete series if $j_0 \in \hat{G}$ has positive measure $d_{j_0}$. Then $\mathbf{H}_{j_0}$ occurs as a direct summand of the left (or right) regular representation, the corresponding projection $P_0$ is of $G$-trace class in $\mathscr{A}'$ (or $\mathscr{A}$), and (2.9) reduces to $\dim_G \mathbf{H}_{j_0} = d_{j_0}$. For this reason, $d_{j_0}$ is called the $G$-dimension, or formal degree, of $\mathbf{H}_{j_0}$.

Finally we consider the universal enveloping algebra $\mathfrak{U}(\mathfrak{g}^{\mathbb{C}})$ of the complexified Lie algebra $\mathfrak{g}^{\mathbb{C}}$ of $G$. For any unitary representation $\mathbf{H}$ of $G$, we get an action of $\mathfrak{U}(\mathfrak{g}^{\mathbb{C}})$ on the space of $C^\infty$ vectors in $\mathbf{H}$. If $\mathbf{H}$ is irreducible, this space of $C^\infty$ vectors is dense in $\mathbf{H}$, the elements of $\mathfrak{U}(\mathfrak{g}^{\mathbb{C}})$ are represented by unbounded operators, and the center $\mathfrak{Z}$ of $\mathfrak{U}(\mathfrak{g}^{\mathbb{C}})$ acts by scalars. Thus every element $Z \in \mathfrak{Z}$ defines a scalar function on $\hat{G}$. On $L^2(G)$, the operator $l(Z)$ and its formal adjoint $l(Z^*)$ are both defined on the dense domain $C_0^\infty(G)$ (here $*$ denotes the standard anti-automorphism of $\mathfrak{U}(\mathfrak{g}^{\mathbb{C}})$, given by $X^* = -\bar{X}$ for $X \in \mathfrak{g}^{\mathbb{C}}$). Hence we may take the closed operator $T$ = closure of $l(Z)$, and form the bounded operator $A = T(1 + T^* T)^{-1/2}$, which commutes with both right and left translation and hence belongs to the center of the von Neumann algebra $\mathscr{A}$. Thus, in the decomposition (2.2), $A$ is represented by a diagonalized operator, i.e. the Fourier components $A_j$ are scalars (almost everywhere), and the function $j \to A_j$ is measurable on $\hat{G}$. From this it follows easily that the function which $Z$ defines on $\hat{G}$ is also measurable (but unbounded) on $\hat{G}$. In particular these remarks apply to the Casimir operator of $\mathfrak{g}^{\mathbb{C}}$.

## § 3. The $L^2$-Index Theorem

In this section we shall apply the $L^2$-index theorem of [1] to the symmetric space $G/K$ of a semisimple Lie group $G$. Formally this is quite analogous to the way in which the index theorem for compact manifolds may be used to derive the dimension formulae for irreducible representations of compact Lie groups. Computationally it is also closely related to the manner in which Hirzebruch [17] computed the dimensions of spaces of automorphic forms.

For the convenience of the reader we shall first recall the statement of the $L^2$-index theorem. We suppose given a discrete group $\Gamma$ acting smoothly and freely on a manifold $\tilde{X}$ with $X=\Gamma\backslash\tilde{X}$ compact, and an elliptic differential operator $\tilde{D}$ on $\tilde{X}$ which is $\Gamma$-invariant (and so is the lift of an elliptic differential operator $D$ on $X$). Hilbert spaces are defined by using $\Gamma$-invariant hermitian metrics, and we put

$$\mathscr{H}^+ = \text{space of } L^2\text{-solution of } \tilde{D}u=0,$$

$$\mathscr{H}^- = \text{space of } L^2\text{-solutions of } \tilde{D}^* v=0,$$

where $\tilde{D}^*$ is the adjoint differential operator. The orthogonal projections onto $\mathscr{H}^\pm$ have $C^\infty$ kernels $K^\pm(\tilde{x},\tilde{y})$, which are $\Gamma$-invariant, i.e.

$$K^\pm(\gamma\tilde{x},\gamma\tilde{y}) = K^\pm(\tilde{x},\tilde{y}) \quad \text{for } \gamma\in\Gamma.$$

Hence we can define a (real-valued) $\Gamma$-dimension of $\mathscr{H}^+$ by

$$\dim_\Gamma \mathscr{H}^+ = \operatorname{trace}_\Gamma K^+ = \int_{\Gamma\backslash\tilde{X}} \operatorname{tr} K^+(\tilde{x},\tilde{x})\, d\tilde{x},$$

and similarly for $\mathscr{H}^-$. Here tr denotes the pointwise trace of the matrix $K^\pm(\tilde{x},\tilde{x})$ – acting on the fibers of the vector bundle involved – and the integral is taken over a fundamental domain in $\tilde{X}$ for $\Gamma$. Finally we put

$$\operatorname{index}_\Gamma \tilde{D} = \dim_\Gamma \mathscr{H}^+ - \dim_\Gamma \mathscr{H}^-.$$

The $L^2$-index theorem then asserts

$$(3.2) \quad \operatorname{index}_\Gamma \tilde{D} = \operatorname{index} D.$$

Note that index $D$ is an integer, being the difference of ordinary dimensions. By the index theorem for compact manifolds [4], there is an explicit formula for index $D$ in cohomological terms. If index $D>0$ then (3.2) implies in particular that the space $\mathscr{H}^+$ of $L^2$-solutions of $\tilde{D}u=0$ is non-zero. Moreover it says that $\mathscr{H}^+$ is in a certain precise sense "bigger" than $\mathscr{H}^-$. In our application to $G/K$ these consequences will be spelled out in greater detail.

Fundamental to our applications is the following result of Borel [5] (see also Borel and Harish-Chandra [6]): every real semisimple Lie group has a discrete torsion-free subgroup $\Gamma$ with $\Gamma\backslash G$ compact. If $K$ is a maximal compact subgroup of $G$ then $\Gamma$ meets no conjugate of $K$, and so $\Gamma$ acts freely on the symmetric space $\tilde{X}=G/K$. The quotient $X=\Gamma/\tilde{X}=\Gamma\backslash G/K$ is then a smooth manifold. We shall pick such a $\Gamma$ once and for all. In our final results the group $\Gamma$ will disappear; it enters only as a conveninent tool. This is to be contrasted with the superficially similar situations arising in the study of $\Gamma$-automorphic forms where $\Gamma$ is the main object of interest.

The differential operators $\tilde{D}$ on $G/K$ to which we shall apply the $L^2$-index theorem are the Dirac operators with coefficients in a (homogeneous) vector bundle. We begin therefore by recalling the basic facts concerning these operators; for details we refer to [8, 22].

Recall first that an oriented Riemannian manifold $M$ is said to be a spin manifold if the structure group of its principal tangent bundle can be lifted from $SO(n)$ to $\mathrm{Spin}(n)$. A choice of such a lifting defines a spin structure, and the basic Spin representation $S$ of $\mathrm{Spin}(n)$ then defines an associated vector bundle $\mathcal{S}$ on $M$, whose sections are "spinor fields". The Dirac operator is a first order formally selfadjoint, elliptic differential operator acting on spinor fields. If $\dim M = n$ is even, then the spin representation $S$ breaks up into two half-spin representations $S^+, S^-$ and correspondingly $\mathcal{S} = \mathcal{S}^+ \oplus \mathcal{S}^-$. The Dirac operator switches the two factors, so that it consists of an operator

$$D^+ : C^\infty(M, \mathcal{S}^+) \to C^\infty(M, \mathcal{S}^-);$$

together with $D^- = (D^+)^*$. If $\mathcal{V}$ is any complex vector bundle on $M$ with a Hermitian connection, then one can define a Dirac-type operator $D_{\mathcal{V}}$ on $\mathcal{S} \otimes \mathcal{V}$ which again, when $n$ is even, decomposes into $D_{\mathcal{V}}^+$ and its adjoint.

We now take $M = X = G/K$. The standard Riemannian metric of $G/K$ is certainly $G$-invariant, but in order for the Spin structure to be $G$-invariant it is necessary that the representation $K \mapsto \mathrm{Aut}(\mathfrak{g}/\mathfrak{k})$, induced by the adjoint representation of $G$, should lift to the Spin covering of $\mathrm{Aut}(\mathfrak{g}/\mathfrak{k}) \cong SO(n)$, so that $S$ becomes a $K$-module. We shall assume for the moment that this is the case. Since the Dirac operator $D$ is canonically associated to the metric and Spin structure, it follows that it will be $G$-invariant. Similarly, if $\mathcal{V}$ is any homogeneous vector bundle with connection on $G/K$, the operator $D_{\mathcal{V}}$ will be $G$-invariant. Now a homogeneous vector bundle on $G/K$ is associated to a representation $V$ of $K$, and it inherits a homogeneous connection from that of the principal bundle $G \to G/K$ (defined by the orthogonal complement $\mathfrak{p}$ of $\mathfrak{k}$ in $\mathfrak{g}$). In this manner every finite-dimensional representation $V$ of $K$ defines a $G$-invariant operator $D_{\mathcal{V}}$.

We now assume that $\mathrm{rank}\, K = \mathrm{rank}\, G$, i.e. that $G$ has a compact Cartan subgroup. This implies in particular that $\dim G/K$ is even, and so the spin representation $S$ decomposes into a direct sum $S = S^+ \oplus S^-$. Thus for every $V$, we have a $G$-invariant operator

$$D_{\mathcal{V}}^+ : C^\infty(G/K, \mathcal{V} \otimes \mathcal{S}^+) \to C^\infty(G/K, \mathcal{V} \otimes \mathcal{S}^-),$$

and its adjoint $D_{\mathcal{V}}^-$ in the opposite direction. We now apply the $L^2$-index theorem to this operator: to conform with our previous notation we shall relabel it $\tilde{D}_{\mathcal{V}}^+$. Since $\tilde{D}_{\mathcal{V}}^+$ is $G$-invariant, it is certainly $\Gamma$-invariant, for any $\Gamma \subset G$, and $D_{\mathcal{V}}^+$ will denote the corresponding operator on $\Gamma \backslash G/K$: it is the Dirac operator on this double coset space with coefficients in the bundle $\Gamma \backslash \mathcal{V}$.

The fact that $D_{\mathcal{V}}^+$ is actually $G$-invariant enables us to simplify both sides of (3.2). On the one hand the $\Gamma$-dimensions can be replaced essentially by $G$-dimensions and related to the Plancherel measure. On the other hand the index $D_{\mathcal{V}}^+$ on $\Gamma \backslash G/K$ can be computed in terms of invariant differential forms. We proceed to describe these two aspects in detail.

For simplicity we consider first $L^2(G)$, with the two von Neumann algebras $\mathcal{A}, \mathcal{A}'$ generated by left and right translations. The operators that commute with left translation by all elements of $\Gamma$ form a von Neumann algebra $\mathscr{B}$ which contains $\mathscr{A}'$. On $\mathscr{B}$ we have the $\Gamma$-trace defined in [1]. For operators $T$ with smooth kernel $t(x, y)$, which is compactly supported on $\Gamma\backslash(G\times G)$ we have

$$\operatorname{trace}_\Gamma T = \int_{\Gamma\backslash G} t(x,x)\,dx.$$

In particular, if $T=r(f)$ is right convolution by $f\in C_0^\infty(G)$,

$$(3.3)\quad \begin{aligned}\operatorname{trace}_\Gamma r(f) &= \int_{\Gamma\backslash G} f(e)\,dg = \operatorname{vol}(\Gamma\backslash G)\,f(e)\\ &= \operatorname{vol}(\Gamma\backslash G)\operatorname{trace}_G r(f).\end{aligned}$$

Since the $r(f)$ are dense in $\mathscr{A}'$, this shows that, up to a volume factor, $\operatorname{trace}_G$ on $\mathscr{A}'$ is the restriction of $\operatorname{trace}_\Gamma$ on $\mathscr{B}$. In particular, for any operator $T\in\mathscr{A}'$

$$T \text{ is of } G\text{-trace class} \Leftrightarrow T \text{ is of } \Gamma\text{-trace class.}$$

Thus (2.7) follows from [1, (4.8)].

An entirely similar situation holds on replacing $L^2(G)$ by $L^2(G/K, \mathscr{F})$, where $\mathscr{F}$ is the vector bundle associated to a $K$-module $F$. We now take $F=V\otimes S^\pm$ and $T=T^\pm$ = projection onto the space $\mathscr{H}^\pm$ of $L^2$-solutions of the appropriate Dirac equation on $G/K$. We get

$$\begin{aligned}(\operatorname{vol}(\Gamma\backslash G))^{-1}\dim_\Gamma\mathscr{H}^\pm &= \operatorname{tr} t^\pm(0,0) &&(t^\pm = \text{kernel of } T^\pm)\\ &= \dim_G\mathscr{H}^\pm &&(\text{by } (2.13))\\ &= \int_{\hat G}\dim U_j^\pm\,dj &&(\text{by } (2.12)),\end{aligned}$$

where

$$(3.4)\quad \mathscr{H}^\pm = \int_{\hat G}\mathbf{H}_j\otimes U_j^\pm\,dj.$$

Hence

$$(3.5)\quad (\operatorname{vol}(\Gamma\backslash G))^{-1}\operatorname{index}_\Gamma\tilde D_{\mathscr{V}}^+ = \int_{\hat G}(\dim U_j^+ - \dim U_j^-)\,dj.$$

We turn now to the other side of the problem, namely the computation of the index of the generalized Dirac operators on the compact manifold $\Gamma\backslash G/K$. Since the work of Hirzebruch [17] this has become a standard type of computation, and so we shall review it only briefly. For further details see for example [17].

The index formula of [4] gives an explicit expression in terms of the Pontrjagin classes of $\Gamma\backslash G/K$ and the Chern classes of the bundle $\mathscr{V}$. Using the differential forms which represent these characteristic classes, the index formula takes the form

$$\operatorname{index} D_{\mathscr{V}}^+ = \int_{\Gamma\backslash G/K} f(\Theta,\Phi),$$

where $\Theta$ is the curvature of $\Gamma\backslash G/K$, $\Phi$ the curvature of $\mathscr{V}$, and $f$ an explicitly known polynomial. Since we are dealing with homogeneous spaces and homogeneous bundles, the integrand is a constant multiple of the volume form $dx$:

(3.6) $f(\Theta, \Phi) = C(V)\,dx,$

and hence

(3.7) $\operatorname{index} D_{\mathscr{V}}^{+} = C(V)\operatorname{vol}(\Gamma\backslash G/K).$

It remains to calculate the constant $C(V)$, which depends only on the $K$-module $V$. This is a purely algebraic computation in the Lie algebra of $G$, and the details are notationally complicated but not difficult. There is however a proportionality principle due to Hirzebruch which enables us to by-pass the computations, or rather to reduce them to well-known results for compact groups. The basic idea is to compare the index formula (3.7) with the corresponding formula for $M/K$ where $M$ is a "compact dual" of $G$.

We assume for the moment that $G$ is a real form of the simply-connected complex semisimple group $G^{\mathbb{C}}$; we shall show later how to drop this assumption as well as the earlier spin assumption on $K$. The compact dual $M$ of $G$ is the maximal compact subgroup of $G^{\mathbb{C}}$ containing $K$, whose Lie algebra $\mathfrak{m}$ is related to that of $G$ by the orthogonal decompositions

$$\mathfrak{g} = \mathfrak{k} \oplus \mathfrak{p}, \qquad \mathfrak{m} = \mathfrak{k} \oplus i\,\mathfrak{p}.$$

The Killing form is positive definite on $\mathfrak{p}$, negative definite on $i\,\mathfrak{p}$, and so (if one chooses the appropriate sign) it induces natural invariant metrics on $G/K$ and $M/K$. The corresponding curvature tensors essentially coincide up to sign: in making the comparison we work only at the identity coset (since they are $G$- or $M$-invariant), and then use the correspondence $\mathfrak{p} \leftrightarrow i\,\mathfrak{p}$. Similar remarks apply to the curvature tensors of the two bundles associated to the $K$-module $V$. Hence the index of the corresponding Dirac operator on $Y = M/K$, which we denote by $D_{\mathscr{V}}^{+}(Y)$, is given by

(3.8) $\operatorname{index} D_{\mathscr{V}}^{+}(Y) = (-1)^q\, C(V)\operatorname{vol}(Y).$

Here $C(V)$ is the same constant as in (3.7), and $2q = \dim G/K = \dim Y$.

Comparing (3.7) with (3.8), we get

(3.9) $\operatorname{index} D_{\mathscr{V}}^{+} = \sigma_{\Gamma} \cdot \operatorname{index} D_{\mathscr{V}}^{+}(Y),$

where $\sigma_{\Gamma}$ is a constant independent of $V$, but depending on $\Gamma$ and given by

(3.10) $$\sigma_{\Gamma} = (-1)^q \frac{\operatorname{vol}(\Gamma\backslash G/K)}{\operatorname{vol}(M/K)}.$$

In this identity both volumes are determined by the Riemannian metrics defined by the Killing form. Since the volume of $K$ was normalized to be 1, the invariant measures on $G$ and $G/K$ are related in a definite manner. It will now be convenient to renormalize Haar measure on $G$, and along with it the metrics on

$G/K$ and $M/K$, by requiring that the total volume of $M$ should also be equal to 1. With this choice of Haar measure, (3.10) becomes

$$(3.11) \quad \sigma_\Gamma = (-1)^q \operatorname{vol}(\Gamma \backslash G).$$

If we now apply the $L^2$-index theorem to $D^+_{\mathscr{V}}$ and use (3.5), (3.9) and (3.11), we get

$$(3.12) \quad \int_{\hat{G}} (\dim U_j^+ - \dim U_j^-)\, dj = (-1)^q \operatorname{index} D^+_{\mathscr{V}}(Y),$$

with $U_j^\pm$ as in (3.4).

Thus we have reduced the problem to that of computing the index of a homogeneous elliptic operator on the compact homogeneous space $Y$. This question has been extensively studied from many different points of view, but a good account for our purposes is that given in [8]. Clearly it is sufficient to take $\mathscr{V} = \mathscr{V}_\mu$ associated to an irreducible $K$-module $V_\mu$ with highest weight $\mu$. We then have the following simple result:

$$(3.13) \quad \operatorname{index} D^+_{\mathscr{V}_\mu}(Y) = (-1)^q \dim W_{\mu - \rho_n},$$

where $W_\lambda$ is the irreducible $M$-module with highest weight $\lambda$ and $\rho_n$ the half-sum of the positive noncompact roots. The ordering of the roots must be chosen so as to make $\mu + \rho_c$ dominant ($\rho_c$ = half-sum of the positive compact roots), and the labelling of $\mathscr{S}^+, \mathscr{S}^-$ is pinned down by requiring that $\rho_n$ should occur as a weight for the $K$-module $S^+$. If $\mu - \rho_n$ is not a possible highest weight (i.e. if $\mu + \rho_c$ is singular), $W_{\mu - \rho_n}$ is to be interpreted as zero.

We shall explain in outline how (3.13) arises, referring to [8] for further details. First we note that the index of the $M$-invariant operator $D^+_{\mathscr{V}_\mu}(Y)$ can be refined to give a (virtual) character of $M$: the ordinary integer index is then obtained by evaluating this character index at the identity of $M$. Next we note that the spaces of sections $L^2(Y, \mathscr{V}_\mu \otimes \mathscr{S}^\pm)$ have well-defined formal $M$-characters (i.e. formal series $\sum_{i \in \hat{M}} m_i \chi_i$, with $m_i \in \mathbb{Z}$ and $\chi_i$ = character of $i \in \hat{M}$). Hence the character index of $D^+_{\mathscr{V}_\mu}(Y)$ can be computed as the difference of these two formal characters. The computations give precisely the Hermann Weyl character formula for $W_{\mu - \rho_n}$, multiplied by the sign factor $(-1)^q$. The dimension formula (3.13) then follows. In fact (3.13) is closely related to the Borel-Weil-Bott construction for the irreducible representations of $M$, except that the flag manifold of $M$ is here replaced by $Y$.

From (3.12) and (3.13) we deduce

$$(3.14) \quad \int_{\hat{G}} (\dim U_j^+ - \dim U_j^-)\, dj = \dim W_{\mu - \rho_n}.$$

This is then the explicit form for the $L^2$-index theorem on $G/K$. It remains now to remove the restrictions imposed on $G$ and $K$.

If the representation $K \to \operatorname{Aut}(\mathfrak{g}/\mathfrak{k})$ does not lift to Spin, then the bundles $\mathscr{S}^\pm$ on $G/K$ are not $G$-homogeneous, though they are $\tilde{G}$-homogeneous for a suitable double cover of $G$. The bundles $\mathscr{S}^\pm \otimes \mathscr{V}_\mu$ will be $G$-homogeneous provided the

$\tilde{K}$-modules $S^{\pm} \otimes V_\mu$ descend to $K$. This is easily seen to be guaranteed by assuming that $\mu - \rho_n$ is a weight of $K$ (cf. §4 below). Thus the operators $\tilde{D}^+_{\mathscr{V}_\mu}$ are defined, as $G$-invariant operators, under this assumption on $\mu$.

Next we drop the requirement that the real form $G_1$ of the simply connected complex group $G^{\mathbb{C}}$ coincides with $G$. The maximal compact subgroup $K_1$ is then only locally isomorphic to $K$. If $\mu - \rho_n$ is also a weight of $K_1$, our previous argument still goes through and we get (3.13).

To extend all this when $\mu - \rho_n$ is a weight of $K$ but not a weight of $K_1$, we observe first that the constant $C(V_\mu)$ in (3.6) is a polynomial in $\mu$. This follows easily from the Weyl character formula and the relations between characters and characteristic classes. Hence (3.14) continues to hold provided we replace $\dim W_{\mu-\rho_n}$ by $d(\mu - \rho_n)$, the explicit polynomial in $\mu$ which gives the dimension formula. Note that Haar measure in $G$ is now normalized by requiring

(3.15) $\operatorname{vol} K = \operatorname{vol} M/K_1 = 1.$

Collecting all our results together, we see that we have proved the following:

(3.16) **Theorem.** Let $\mu - \rho_n$ be a weight for $K$, such that $(\mu + \rho_c, \alpha) \geqq 0$ for all positive roots $\alpha$. Let $\mathscr{H}_\mu^+, \mathscr{H}_\mu^-$ be the $L^2$ null spaces of the Dirac operators $D^+_{\mathscr{V}_\mu}, D^-_{\mathscr{V}_\mu}$ on $G/K$, and

$$\mathscr{H}_\mu^{\pm} = \int_{\hat{G}} \mathbf{H}_j \otimes U_j^{\pm} \, dj$$

their Plancherel decompositions. Then

$$\int_{\hat{G}} (\dim U_j^+ - \dim U_j^-) \, dj = d(\mu - \rho_n),$$

where $d(\lambda)$ is the polynomial in $\lambda$ giving the dimension of the irreducible finite dimensional representation with highest weight $\lambda$. In this identity, Haar measure is normalized so that

$$\operatorname{vol}(K) = \operatorname{vol}(M/K_1) = 1,$$

where $M/K_1$ denotes the simply connected compact dual of the symmetric space $G/K$.

The spaces $\mathscr{H}_\mu^{\pm}$ in (3.16) are also the $L^2$ kernels of the spinor Laplacians $D^- D^+$ and $D^+ D^-$ (we omit here the various subscripts). This is so because the minimal and maximal domains of $D^+$, and similarly of $D^-$, coincide, as is proved in [1]. This re-interpretation of $\mathscr{H}_\mu^{\pm}$ has the advantage that the spinor Laplacians take a particularly simple form, namely

(3.17) $D^2 = -\Omega + (\mu - \rho_n, \mu - \rho_n + 2\rho),$

where $\Omega$ represents the Casimir operator, acting on $L^2(G/K, \mathscr{V}_\mu \otimes \mathscr{S}^{\pm})$. The formula (3.17) is due to Parthasarathy [22] and will be proved in the Appendix. Thus $\mathscr{H}_\mu^{\pm}$ is just the eigenspace of the Casimir operator on $L^2(G/K, \mathscr{V}_\mu \otimes \mathscr{S}^{\pm})$, corresponding to the eigenvalue

(3.18) $c_\mu = (\mu - \rho_n, \mu - \rho_n + 2\rho)$.

Hence, if we consider the Plancherel decomposition

$$L^2(G/K, \mathscr{V}_\mu \otimes \mathscr{S}^\pm) = \int_{\hat{G}} \mathbf{H}_j \otimes V_j^\pm \, dj,$$

we see that the subspaces $U_j^\pm$ describing the decomposition of $\mathscr{H}_\mu^\pm$ are given by

$$\begin{aligned} U_j^\pm &= V_j^\pm \quad \text{if } \Omega \text{ acts on } \mathbf{H}_j \text{ by } c_\mu, \\ &= 0 \quad \text{otherwise.} \end{aligned}$$

The formula of Theorem (3.16) can therefore be re-written as

(3.19) $$\int_{\hat{G}_\mu} (\dim V_j^+ - \dim V_j^-) \, dj = d(\mu - \rho_n),$$

where $\hat{G}_\mu \subset \hat{G}$ is the subspace consisting of all $j \in \hat{G}$ at which the Casimir operator takes the value $c_\mu$.

## § 4. Existence of the Discrete Series

We now use the results of the preceding section, in particular the identity (3.19), to prove the existence of discrete series representations. The crucial step consists of showing that the integrand in (3.19) vanishes for all but finitely many classes $j \in \hat{G}$. Any finite subset of $\hat{G}$, outside of the discrete series, has zero Plancherel measure. Hence only the discrete series contributes to the integral (3.19). The integrand, for any $j \in \hat{G}$, is related to the global character of $j$, restricted to the maximal compact subgroup $K$. These observations lead to an explicit formula, on $K$, for certain linear combinations of discrete series characters; the formula is stated as Theorem (4.41), at the end of this section. We shall deduce concrete information about individual discrete series representations in subsequent sections.

Recall the Plancherel decomposition of the spaces of $\mathscr{V}_\mu$-valued $L^2$-spinors:

$$L^2(G/K, \mathscr{S}^\pm \otimes \mathscr{V}_\mu) = \int_{\hat{G}} \mathbf{H}_j \otimes V_j^\pm \, dj,$$

with $V_j^\pm = K$-invariant part of $\mathbf{H}_j^* \otimes S^\pm \otimes V_\mu$; cf. (2.10). Here $V_\mu$ stands for the irreducible $K$-module of highest weight $\mu$, on which the homogeneous vector bundle $\mathscr{V}_\mu$ is modelled. The half spin modules $S^+, S^-$ are self-dual if $q = \frac{1}{2} \dim G/K$ is even, and dual to each other if $q$ is odd. Thus we can identify the integrand in (3.19) as

(4.1) $$\begin{aligned} &\dim V_j^+ - \dim V_j^- \\ &= \dim \operatorname{Hom}_K(V_\mu^*, \mathbf{H}_j^* \otimes S^+) - \dim \operatorname{Hom}_K(V_\mu^*, \mathbf{H}_j^* \otimes S^-) \\ &= (-1)^q \{\dim \operatorname{Hom}_K(V_\mu, \mathbf{H}_j \otimes S^+) - \dim \operatorname{Hom}_K(V_\mu, \mathbf{H}_j \otimes S^-)\}. \end{aligned}$$

Restricted to $K$, the $G$-module $\mathbf{H}_j$ decomposes into a direct sum of $K$-irreducibles,

$$\mathbf{H}_j = \bigoplus_{i \in \hat{K}} n_i V_i .$$

We denote the character of $V_i$ by $\chi_i$. Then, as follows from the bound $n_i \leqq \dim V_i$, the formal series

$$(4.2) \quad \tau_j = \sum_{i \in \hat{K}} n_i \chi_i$$

converges to a distribution on $K$, the so-called $K$-character of $\mathbf{H}_j$. It should be noticed that

(4.3) $\dim V_j^+ - \dim V_j^- = (-1)^q \times$ multiplicity with which $\chi_\mu$ occurs in the formal series $(\sigma^+ - \sigma^-)\,\tau_j$,

with $\chi_\mu$ = character of the irreducible $K$-module $V_\mu$, and $\sigma^\pm$ = character of the $K$-module $S^\pm$.

According to a fundamental theorem of Harish-Chandra [13, 3] the global character $\Theta_j$ of $\mathbf{H}_j$ is (integration against) a locally $L^1$ function on $G$, which moreover is real-analytic on $G'$, the set of regular, semisimple elements of $G$. For more elementary reasons [12, 3], the $K$-character $\tau_j$ restricts to a real-analytic function on $G' \cap K$ (not necessarily an $L^1$ function on $K$, however), and

(4.4) $\Theta_j$ and $\tau_j$ agree as functions on $G' \cap K$.

We now appeal to Lemma 4.10 of [24], whose proof we shall sketch in the Appendix:

(4.5) $(\sigma^+ - \sigma^-)\,\tau_j$ is a finite integral linear combination of irreducible characters of $K$.

From (4.3–4.5), one deduces:

(4.6) $(\sigma^+ - \sigma^-)\,\Theta_j|_{G' \cap K}$ is a finite linear combination of irreducible characters of $K$, in which $\chi_\mu$ occurs with coefficient $(-1)^q(\dim V_j^+ - \dim V_j^-)$.

As we shall argue next, the coefficient of $\chi_\mu$ can be non-zero for only finitely many classes $j \in \hat{G}$.

We denote the complexified Lie algebra of $G$ by $\mathfrak{g}^{\mathbb{C}}$, and the center of its universal enveloping algebra by $\mathfrak{Z}$. Then $\mathfrak{Z}$ can be naturally identified with the algebra of all bi-invariant linear differential operators on $G$. By its very definition, the character $\Theta_j$ is an invariant eigendistribution, i.e. a conjugation-invariant distribution, on which the algebra of bi-invariant differential operators $\mathfrak{Z}$ acts by scalars. Hence there exists a character $\varphi_j \colon \mathfrak{Z} \to \mathbb{C}$, such that

$$(4.7) \quad Z\Theta_j = \varphi_j(Z)\,\Theta_j, \qquad \text{for all } Z \in \mathfrak{Z}.$$

In Harish-Chandra's terminology, $\varphi_j$ is the infinitesimal character of the $G$-module $\mathbf{H}_j$.

Corresponding to any Cartan subalgebra $\mathfrak{b}^{\mathbb{C}}$ of $\mathfrak{g}^{\mathbb{C}}$, Harish-Chandra has constructed a canonical isomorphism

$$(4.8)\quad \gamma\colon \mathfrak{Z} \xrightarrow{\sim} I(\mathfrak{b}^{\mathbb{C}}),$$

between $\mathfrak{Z}$ and the algebra $I(\mathfrak{b}^{\mathbb{C}})$ of all Weyl group invariants in the symmetric algebra $S(\mathfrak{b}^{\mathbb{C}})$ [12, 26, 3]. If $\mathfrak{b}^{\mathbb{C}}$ arises as the complexified Lie algebra of a Cartan subgroup $B \subset G$, every $X \in S(\mathfrak{b}^{\mathbb{C}})$, and therefore every $\gamma(Z)$, may be viewed as a translation invariant, linear differential operator on the group $B$. Going to a two-fold covering of $B$, if necessary, one can select a square-root $\Delta_B \in C^\infty(B)$ of

$$(4.9)\quad \Delta_B^2 = (-1)^d \det\{1 - \mathrm{Ad}|_B\colon \mathfrak{g}^{\mathbb{C}}/\mathfrak{b}^{\mathbb{C}} \to \mathfrak{g}^{\mathbb{C}}/\mathfrak{b}^{\mathbb{C}}\},$$

with $d = \frac{1}{2}(\dim G - \mathrm{rk}\, G)$. Since $\Delta_B^2$ vanishes precisely on $B \cap G'$, $\Delta_B$ makes sense also as a smooth function on each connected component of $B \cap G'$. The isomorphism (4.8) has the following crucial property:

$$(4.10)\quad (ZF)|_{B\cap G'} = (\Delta_B^{-1} \cdot \gamma(Z) \cdot \Delta_B)\, F|_{B\cap G'},$$

for any conjugation invariant function $F$ [12, 3].

Since the maximal compact subgroup $K$ was assumed to have the same rank as $G$, it contains a Cartan subgroup $H$ of $G$. Via exponentiation, the dual group $\hat{H}$ of the torus $H$ becomes isomorphic to a lattice $\Lambda$,

$$(4.11)\quad \hat{H} \cong \Lambda \subset i\mathfrak{h}^*,$$

contained in $i\mathfrak{h}^*$, the real vector space of all those linear functions on $\mathfrak{h}^{\mathbb{C}}$, which assume purely imaginary values on the Lie algebra $\mathfrak{h}$. In particular, the root system $\Phi = \Phi(\mathfrak{g}^{\mathbb{C}}, \mathfrak{h}^{\mathbb{C}})$ lies inside the lattice $\Lambda$. A root $\alpha \in \Phi$ is said to be compact or noncompact, depending on whether or not it is a root of the pair $(\mathfrak{k}^{\mathbb{C}}, \mathfrak{h}^{\mathbb{C}})$. Thus $\Phi$ decomposes into a disjoint union

$$(4.12)\quad \Phi = \Phi^c \cup \Phi^n$$

of the sets $\Phi^c$, $\Phi^n$ of all compact and noncompact roots, respectively.

As $\beta$ ranges over $\Phi^n$, $e^\beta$ exhausts the set of characters by which $H$ acts on $\mathfrak{g}^{\mathbb{C}}/\mathfrak{k}^{\mathbb{C}}$; each of these has multiplicity one. Thus $\Phi^n$ is the set of weights of the standard representation of $\mathrm{SO}(\mathfrak{g}/\mathfrak{k})$, pulled back to $K$ via $K \to \mathrm{SO}(\mathfrak{g}/\mathfrak{k})$. Expressed in terms of the weights $\pm\mu_1, \pm\mu_2, \ldots, \pm\mu_n$ of the standard representation of $\mathrm{SO}(2n)$, the weights of the two half spin modules are

$$\tfrac{1}{2}(\pm\mu_1 \pm \mu_2 \pm \cdots \pm \mu_n),$$

with an even number of minus signs in one case, and an odd number in the other. Hence, if one appropriately chooses a positive root system $\Psi \subset \Phi$,

$$(4.13)\quad (\sigma^+ - \sigma^-)|_H = \prod_{\beta \in \Phi^n \cap \Psi} (e^{\beta/2} - e^{-\beta/2});$$

a "wrong" choice of $\Psi$ would introduce a minus sign. It should be observed that the labelling of the half spin modules $S^+$, $S^-$ is determined by (3.13), and that a positive root system $\Psi$ gives the correct sign if it makes $\mu+\rho_c$ dominant ($\rho_c$ = half-sum of the positive, compact roots). Once and for all, we fix[1] such a $\Psi$.

The function $\Delta_H$ of (4.9), corresponding to the Cartan subgroup $H$, can be expressed as

$$(4.14) \quad \Delta_H = \prod_{\alpha\in\Psi} (e^{\alpha/2} - e^{-\alpha/2}),$$

at least by passing to a covering of $H$, if necessary. Thus $\Delta_H$ equals the product of $(\sigma^+ - \sigma^-)$ with the denominator of Weyl's character formula for $K$. We define

$$(4.15) \quad W = N_G(H)/H = \text{Weyl group of } \mathfrak{h}^{\mathbb{C}} \text{ in } \mathfrak{k}^{\mathbb{C}}$$

(the normalizer of $H$ in $G$ is actually contained in $K$!). Because of (4.6), Weyl's character formula leads to an identity

$$(4.16\text{a}) \quad \Delta_H\, \Theta_j|_{H\cap G'} = \sum_\nu n_\nu\, e^\nu,$$

where $\nu$ ranges over a finite subset of $\Lambda$, and

$$(4.16\text{b}) \quad n_{w\nu} = \varepsilon(w)\, n_\nu, \qquad \text{for } w\in W$$

($\varepsilon(w)$ = sign of $w$). Moreover, (4.6) allows us to identify the coefficient of $e^{\mu+\rho_c}$ as

$$(4.16\text{c}) \quad n_{\mu+\rho_c} = (-1)^q (\dim V_j^+ - \dim V_j^-).$$

For the usual reasons, (4.16) may hold, strictly speaking, only on a finite covering of $H$.

The group $W$ is contained in $W_{\mathbb{C}}$, the Weyl group of $\mathfrak{h}^{\mathbb{C}}$ in $\mathfrak{g}^{\mathbb{C}}$. Via the mapping $\gamma$ of (4.8), $\mathfrak{Z}$ becomes isomorphic to $I(\mathfrak{h}^{\mathbb{C}})$, which may be thought of as the algebra of all $W_{\mathbb{C}}$-invariant polynomial functions on $\mathfrak{h}^{\mathbb{C}*}$. Thus every $\nu\in\mathfrak{h}^{\mathbb{C}*}$, and especially every $\nu\in\Lambda$, defines a character

$$(4.17) \quad \varphi_\nu : \mathfrak{Z} \to \mathbb{C},$$

with $\varphi_\nu(Z) = \gamma(Z)(\nu)$. The $W_{\mathbb{C}}$-invariant polynomials separate any two $W_{\mathbb{C}}$-orbits in $\mathfrak{h}^{\mathbb{C}*}$; hence

$$(4.18) \quad \varphi_\nu = \varphi_\mu \Leftrightarrow \mu = w\,\nu, \qquad \text{for some } w\in W_{\mathbb{C}}.$$

If $\Theta_j|_{H\cap G'} \neq 0$, the identity (4.10) makes it possible to relate the infinitesimal character $\varphi_j$ to the character formula (4.16):

$$(4.19) \quad n_\nu \neq 0 \Rightarrow \varphi_j = \varphi_\nu.$$

Because of (4.16c), this implies

$$(4.20) \quad \dim V_j^+ - \dim V_j^- = 0, \quad \text{unless} \quad \varphi_j = \varphi_{\mu+\rho_c}.$$

---

[1] This is possible: one first orders $\Phi^c$, so that $\mu$ becomes dominant for the resulting system of positive, compact roots; $\rho_c$ is then determined, and $\mu+\rho_c$ is not only $\Phi^c$-dominant, but also $\Phi^c$-nonsingular. Thus, if a system of positive roots $\Psi$ makes $\mu+\rho_c$ dominant, it necessarily induces the original ordering on $\Phi^c$

In particular, the classes $j \in \hat{G}$ which contribute a non-zero integrand in (3.19) all have the same infinitesimal character.

A result of Harish-Chandra [12] asserts that only finitely many classes $j \in \hat{G}$ can have a given infinitesimal character. One may see this, for example, as follows. Since the global characters of non-isomorphic representations are linearly independent, it is enough to prove that the space of invariant eigendistributions, on which $\mathfrak{Z}$ acts according to a given character, has finite dimension. As a consequence of Harish-Chandra's regularity theorem [13, 3], an invariant eigendistribution $\Theta$ is completely determined by its restriction to $G'$ – or equivalently, by the restrictions $\Theta|_{B \cap G'}$, with $B$ running over a set of representatives for the finitely many conjugacy classes of Cartan subgroups. The intersections $B \cap G'$ have only finitely many connected components. One can now deduce the finite-dimensionality from (4.10), provided one knows: for any character $\varphi: I(\mathfrak{b}^{\mathbb{C}}) \to \mathbb{C}$, the system of differential equations

$$Xf = \varphi(X)f, \qquad X \in I(\mathfrak{b}^{\mathbb{C}}),$$

on any connected open subset of $B$, has a finite-dimensional solution space. This is indeed the case, since $S(\mathfrak{b}^{\mathbb{C}})$ is finitely generated as a module over $I(\mathfrak{b}^{\mathbb{C}})$ [12, 26, 3].

As we have just finished arguing, only finitely many classes $j \in \hat{G}$ make a non-zero contribution to the integral (3.19). We denote the discrete series of $G$ by $\hat{G}_d$; in other words, $\hat{G}_d \subset \hat{G}$ is the subset consisting of all square-integrable classes. A single point $j \in \hat{G}$ has positive Plancherel mass precisely when $j$ belongs to $\hat{G}_d$. The integral (3.19) therefore remains unchanged if we integrate only over the set $\hat{G}_\mu \cap \hat{G}_d$, on which the measure is discrete:

$$\text{(4.21)} \qquad \sum_{j \in \hat{G}_\mu \cap \hat{G}_d} (\dim V_j^+ - \dim V_j^-)\, d(j) = d(\mu - \rho_n);$$

here $d(j)$ denotes the Plancherel measure of $\{j\}$, which is also called the formal degree of the class $j \in \hat{G}_d$.

Via the isomorphism $\gamma$, the Casimir operator $\Omega$ corresponds to

$$\text{(4.22)} \quad \gamma(\Omega) = \sum_i X_i^2 - (\rho, \rho) \cdot 1,$$

where $\{X_i\}$ is a basis for $\mathfrak{h}^{\mathbb{C}}$, orthonormal with respect to the Killing form. Hence

$$\varphi_{\mu+\rho_c}(\Omega) = \gamma(\Omega)(\mu+\rho_c) = (\mu+\rho_c, \mu+\rho_c) - (\rho, \rho) = c_\mu$$

(cf. (3.18); $\rho = \rho_c + \rho_n$), which means that $\hat{G}_\mu$ contains the set

$$\text{(4.23)} \quad \{j \in \hat{G} \mid \varphi_j = \varphi_{\mu+\rho_c}\}.$$

According to (4.20), the summands in (4.21) vanish outside of this set. Thus, instead of summing over $\hat{G}_\mu \cap \hat{G}_d$, we may sum over the set (4.23), intersected with $\hat{G}_d$. Weyl's dimension formula gives

$$d(\mu - \rho_n) = \prod_{\alpha \in \Psi} \frac{(\mu + \rho_c, \alpha)}{(\rho, \alpha)}.$$

We conclude:

$$(4.24)\quad \sum_j (\dim V_j^+ - \dim V_j^-)\, d(j) = \prod_{\alpha\in\Psi} \frac{(\mu+\rho_c,\alpha)}{(\rho,\alpha)},$$

with $j$ ranging over the finite set $\{j\in\hat{G}_d \mid \varphi_j = \varphi_{\mu+\rho_c}\}$.

For the moment, we let $\rho$ denote the half sum of the positive roots, relative to an arbitrary ordering of the roots. Then

$$(4.25)\quad \Lambda_\rho = \Lambda + \rho$$

does not depend on the particular ordering: any two possible choices for $\rho$ differ by a sum of roots, and hence by an element of $\Lambda$. Roughly speaking, the discrete series can be parametrized in a natural manner by the $W$-orbits in $\Lambda_\rho$; this is the reason for introducing $\Lambda_\rho$. The Weyl group $W$ does indeed operate on $\Lambda_\rho$, since $W$ preserves the lattice $\Lambda$, and since any $W$-translate of $\rho$ differs from $\rho$ be a sum of roots. The complex Weyl group $W_{\mathbb{C}}$, on the other hand, need not act on $\Lambda_\rho$, unless $G$ is linear.

For the remainder of this section, we keep fixed a particular $\lambda\in\Lambda_\rho$, and we define

$$(4.26)\quad \tilde{\Theta}_\lambda = \sum_{j\in\hat{G}_d,\ \varphi_j=\varphi_\lambda} d(j)\,\Theta_j.$$

Thus $\tilde{\Theta}_\lambda$ is a finite linear combination of discrete series characters. We shall use the identity (4.24) corresponding to various parameters $\mu$, to compute the restriction of $\tilde{\Theta}_\lambda$ to $H\cap G'$; (4.16c) provides the link between (4.24) and the formula for $\tilde{\Theta}_\lambda$ on $H$.

We let $\Psi$ denote a positive root system, which makes $\lambda$ dominant. If $\lambda$ is singular, there will of course be more than one possible choice. The character formulas (4.16), for the summands $\Theta_j$, lead to an identity

$$(4.27\text{a})\quad \tilde{\Theta}_\lambda|_{H\cap G'} = \frac{\sum_\nu a_\nu e^\nu}{\prod_{\alpha\in\Psi}(e^{\alpha/2}-e^{-\alpha/2})};$$

here $\nu$ runs over a finite set, and the $a_\nu$ are real constants. Moreover

$$(4.27\text{b})\quad a_{w\nu} = \varepsilon(w)\,a_\nu, \qquad \text{for } w\in W.$$

Because of this skew-symmetry with respect to $W$,

$$(4.28)\quad a_\nu = 0, \qquad \text{whenever } \nu \text{ is } \Phi^c\text{-singular}.$$

All the summands which make up $\tilde{\Theta}_\lambda$ have infinitesimal character $\varphi_\lambda$. Hence (4.18) and (4.19) imply

$$(4.29)\quad a_\nu = 0, \quad \text{unless } \nu \text{ is } W_{\mathbb{C}}\text{-conjugate to } \lambda.$$

Although neither the numerator nor the denominator in (4.27) may be well-defined on $H$, the quotient necessarily is well-defined. The denominator, multiplied by $e^{-\rho}$, equals

$$\prod_{\alpha\in\Psi}(1-e^{-\alpha}),$$

which makes sense on $H$. It follows that the product of the numerator with $e^{-\rho}$ must also be well-defined on $H$, and hence equal to a finite linear combination of characters of $H$. In other words,

(4.30) $a_\nu \neq 0 \Rightarrow \nu \in \Lambda_\rho$.

We remark that the preceeding observations apply equally to any linear combination of irreducible characters, provided all of the summands have the same infinitesimal character.

We ennumerate the set of all those $W_{\mathbb{C}}$-conjugates of $\lambda$ in $\Lambda_\rho$, which are both $\Phi^c$-nonsingular and dominant with respect to $\Phi^c \cap \Psi$, as

(4.31) $\lambda_1, \lambda_2, \ldots, \lambda_N$.

Every $\nu \in \Lambda_\rho$, if it is $\Phi^c$-nonsingular and $W_{\mathbb{C}}$-conjugate to $\lambda$, is then $W$-conjugate to precisely one of the $\lambda_i$. Hence there exist constants $a_i$, $1 \leqq i \leqq N$, such that

(4.32) $$\tilde{\Theta}_\lambda|_{H\cap G'} = \sum_{i=1}^{N} a_i \frac{\sum_{w\in W} \varepsilon(w)\, e^{w\lambda_i}}{\prod_{\alpha\in\Psi}(e^{\alpha/2}-e^{-\alpha/2})}.$$

The set (4.31) may be empty, in which case $\tilde{\Theta}_\lambda$ vanishes on $H$. Otherwise, replacing $\lambda$ by one of its $W_{\mathbb{C}}$-conjugates, if necessary, we can arrange

(4.33) $\lambda = \lambda_1$.

We shall assume that this has been done.

The elements of the weight lattice $\Lambda$ are integral with respect to the root system $\Phi^c$, as is $\rho$, which is integral even with respect to all of $\Phi$. Thus $\lambda \in \Lambda_\rho$ must also be $\Phi^c$-integral. Like every $\lambda_i$, $\lambda$ lies in the interior of the positive Weyl chamber for $\Phi^c \cap \Psi$. Hence

(4.34) $\mu = \lambda - \rho_c$

is at least integral and dominant with respect to $\Phi^c \cap \Psi$. In particular, $\mu$ arises as the highest weight of an irreducible $\mathfrak{k}^{\mathbb{C}}$-module $V_\mu$. The $\mathfrak{k}^{\mathbb{C}}$-action on $V_\mu$ lifts to a representation of $K$ if and only if $\mu$ belongs to $\Lambda$, which need not be the case. However, $\Lambda$ contains $\lambda+\rho=\mu+\rho_n$ ($\rho_n$ = half sum of the positive, noncompact roots). Since every weight of the half spin modules $S^+$, $S^-$ differs from $\rho_n$ by a sum of roots, $K$ acts on the tensor products $V_\mu \otimes S^+$, $V_\mu \otimes S^-$. This makes it possible to define the vector bundles $\mathscr{V}_\mu \otimes \mathscr{S}^+$, $\mathscr{V}_\mu \otimes \mathscr{S}^-$ on $G/K$. We conclude that the identity (4.24) applies in our present context.

The summations in (4.24) and (4.26) range over the same set. Hence, using (4.16c), we can identify the coefficient $a_1$ in (4.32) as

(4.35) $$a_1 = (-1)^q \prod_{\alpha\in\Psi} \frac{(\alpha, \lambda)}{(\alpha, \rho)}.$$

In particular, if $\lambda$ is singular with respect to $\Phi$, the coefficient $a_1$ vanishes. Of course, $\lambda=\lambda_1$ is not really distinguished among the $\lambda_i$. We can therefore let each $\lambda_i$ play the role of $\lambda$. If $\lambda$ is singular, then so are all of the $\lambda_i$; hence

(4.36) $\tilde{\Theta}_\lambda|_{H\cap G'}=0,\quad$ if $\lambda$ is singular.

We now suppose that $\lambda$, and therefore its conjugates, are nonsingular. Every $\lambda_i$ is made dominant by a unique positive root system $\Psi_i$, namely

(4.37) $\Psi_i=\{\alpha\in\Phi\,|\,(\lambda_i, a)>0\}$.

In analogy to (4.3), we find

(4.38) $a_i=\varepsilon_i(-1)^q \prod_{\alpha\in\Psi_i} \dfrac{(\alpha,\lambda_i)}{(\alpha,\rho_i)}$

($\rho_i=$half-sum of the roots in $\Psi_i$). The sign factor $\varepsilon_i=\pm 1$ is determined by

(4.39) $\prod_{\alpha\in\Psi_i}(e^{\alpha/2}-e^{-\alpha/2})=\varepsilon_i \prod_{\alpha\in\Psi}(e^{\alpha/2}-e^{-\alpha/2})$;

its presence in (4.38) is due to the fact that the denominator in (4.32) was defined in terms of $\Psi$, rather than $\Psi_i$. Each $\lambda_i$ is conjugate to $\lambda$ by the unique $w\in W_{\mathbb{C}}$ which maps $\Psi_i$ onto $\Psi$; hence

(4.40) $\prod_{\alpha\in\Psi_i}\dfrac{(\alpha,\lambda_i)}{(\alpha,\rho_i)}=\prod_{\alpha\in\Psi}\dfrac{(\alpha,\lambda)}{(\alpha,\rho)}$.

Combining (4.36–4.40), we may conclude:

(4.41) **Theorem.** If $\lambda\in\Lambda_\rho$ is nonsingular, the restriction of $\tilde{\Theta}_\lambda$ to $H\cap G'$ equals

$$(-1)^q\left(\prod_{\alpha\in\Psi}\frac{(\alpha,\lambda)}{(\alpha,\rho)}\right)\sum_{i=1}^{N}\frac{\sum_{w\in W}\varepsilon(w)\,e^{w\lambda_i}}{\prod_{\alpha\in\Psi_i}(e^{\alpha/2}-e^{-\alpha/2})}.$$

Whenever $\lambda$ is singular, $\tilde{\Theta}_\lambda$ vanishes on $H$.

## § 5.1. The "Sufficiently Nonsingular" Case

In the last two sections, we used the $L^2$-index theorem to study the formal difference $\mathscr{H}_\mu^+-\mathscr{H}_\mu^-$, of the two spaces of harmonic, $\mathscr{V}_\mu$-valued $L^2$-spinors $\mathscr{H}_\mu^\pm$. This difference is in particular a finite integral linear combination of discrete series representations. To obtain information about $\mathscr{H}_\mu^+$ and $\mathscr{H}_\mu^-$ individually, we shall now combine the index theorem with a suitable "vanishing theorem". Vanishing theorems in various contexts, or rather the proofs of the vanishing theorems, tend to work only in the "generic" situation. This is also the case here: we shall have to assume that the parameter $\mu$ lies far away from all of the root hyperplanes. For such values of $\mu$, certain algebraic arguments will show that $\mathscr{H}_\mu^-$ vanishes, whereas $\mathscr{H}_\mu^+$ is irreducible. Because of what is already known about the formal difference $\mathscr{H}_\mu^+ - \mathscr{H}_\mu^-$, $\mathscr{H}_\mu^+$ must then belong to the discrete series. The algebraic arguments also lead to a formula for the character of $\mathscr{H}_\mu^+$, restricted to the maximal compact subgroup $K$. As a result, we obtain explicit realizations and character formulas for "most" of the discrete series. The full description of the discrete series will have to wait until § 8, following a discussion of the growth properties of discrete series characters in § 6 and § 7.

Throughout this section, we freely use the notation of § 4. A particular system of positive roots $\Psi$ in $\Phi$ will be kept fixed. As in the past, $\rho_c$ and $\rho_n$ stand for the half-sums of all positive compact and noncompact roots, respectively; $\rho = \rho_c + \rho_n$ is the half-sum of all positive roots. When we talk of the highest weight of an irreducible $K$-module, it will always be with respect to the positive root system $\Phi^c \cap \Psi$ in $\Phi^c$. We recall that the vector bundles $\mathscr{V}_\mu \otimes \mathscr{S}^+$, $\mathscr{V}_\mu \otimes \mathscr{S}^-$ on $G/K$ can be defined whenever $\mu + \rho_n$ lies in the weight lattice $\Lambda$, or equivalently, whenever

(5.1) $\mu + \rho_c \in \Lambda_\rho$;

cf. (4.25). Once and for all, we require that

(5.2) $(\mu + \rho_c - B, \alpha) > 0, \quad$ for every $\alpha \in \Psi$,

if $B$ is any sum of distinct positive, noncompact roots. Since there exist only finitely many possibilities for $B$, (5.2) would certainly be implied by a condition of the form

(5.3) $(\mu, \alpha) > c, \quad$ for $\alpha \in \Psi$,

with a suitably chosen constant $c$.

For the moment, $\pi$ shall denote an arbitrary irreducible unitary representation of $G$, and $V_\nu$ an irreducible $K$-module, of highest weight $\nu$, such that

(5.4a) $V_\nu$ occurs in the restriction of $\pi$ to $K$,

and

(5.4b) $(\nu - \rho_n, \alpha) \geq 0, \quad$ if $\alpha \in \Phi^c \cap \Psi$.

Since $\pi$ is irreducible, the Casimir operator $\Omega$ acts under $\pi$ by some constant $\pi(\Omega)$. According to Lemma 4.11 of [24], the hypotheses (5.4) imply

(5.5) $\pi(\Omega) \leq (\nu - \rho_n + \rho_c, \nu - \rho_n + \rho_c) - (\rho, \rho)$.

We shall outline a proof of this estimate, which is based on an algebraic version of Parthasarathy's formula (3.17), in the Appendix.

As was shown in § 3, the spaces of square-integrable, $\mathscr{V}_\mu$-valued, harmonic spinors have Plancherel decompositions

(5.6) $\mathscr{H}_\mu^\pm = \int_{\hat{G}_\mu} \mathbf{H}_j \otimes V_j^\pm \, dj$,

with

(5.7) $V_j^\pm = K$-invariant part of $\mathbf{H}_j^* \otimes S^\pm \otimes V_\mu$

$\cong \mathrm{Hom}_K(\mathbf{H}_j, S^\pm \otimes V_\mu)$,

and $\hat{G}_\mu$ = set of all classes $j \in \hat{G}$ on which $\Omega$ acts as multiplication by

(5.8) $c_\mu = (\mu - \rho_n, \mu - \rho_n + 2\rho)$.

In the tensor product $V_\nu \otimes W$ of an irreducible $K$-module $V_\nu$, of highest weight $\nu$, with an arbitrary finite-dimensional $K$-module $W$, every irreducible constituent has a highest weight which is the sum of $\nu$ and some weight $\tau$ of $W$. Moreover, $\nu + \tau$ occurs as highest weight in $V_\nu \otimes W$ at most as often as the multiplicity of the weight $\tau$ in $W$. According to the discussion which preceeds (4.13), every weight of $S^+ \oplus S^-$ can be expressed as $\rho_n - B$, where $B$ stands for a sum of distinct positive, noncompact roots; the weight $\rho_n$ has multiplicity one and occurs in $S^+$. Thus $V_j^+$ and $V_j^-$ can be non-zero only if $\mathbf{H}_j$ contains an irreducible $K$-module with a highest weight of the form $\mu + \rho_n - B$, and, in the case of $V_j^-$, $B \neq 0$.

Because of the assumption (5.2), any such highest weight $\nu = \mu + \rho_n - B$ satisfies the hypothesis (5.4b): if a weight $\tau$, e.g. $\tau = \mu + \rho_c - B$, is dominant and nonsingular with respect to $\Phi^c \cap \Psi$, then $\tau - \rho_c$ must at least be dominant. Thus we may apply the estimate (5.5) to any class $j \in \hat{G}_\mu$, for which $V_j^\pm$ does not vanish, to conclude

(5.9) $(\mu - \rho_n, \mu - \rho_n + 2\rho) = c_\mu \leqq (\mu + \rho_c - B, \mu + \rho_c - B) - (\rho, \rho)$,

or equivalently,

(5.10) $2(\mu + \rho_c, B) \leqq (B, B)$.

On the other hand, $B$ is a sum of positive roots, so (5.2) insures that both $\mu + \rho_c - B$ and $\mu + \rho_c$ ($B = 0$ is not excluded in (5.2)!) have a strictly positive inner product with $B$, unless $B = 0$. But then

$$0 < (\mu + \rho_c - B, B) + (\mu + \rho_c, B) = 2(\mu + \rho_c, B) - (B, B)$$

which contradicts (5.10).

We have shown: among the irreducible constituents of $V_\mu \otimes S^\pm$, only the one having highest weight $\mu + \rho_n$ can appear in $\mathbf{H}_j$, for any class $j \in \hat{G}_\mu$. This irreducible constituent has multiplicity one in $V_\mu \otimes S^+$, and multiplicity zero in $V_\mu \otimes S^-$. In particular, for $j \in \hat{G}_\mu$,

(5.11) $\dim V_j^- = 0$, $\dim V_j^+ =$ multiplicity, in $\mathbf{H}_j$, of the irreducible $K$-module of highest weight $\mu + \rho_n$.

The estimate (5.9) also proves:

(5.12) no irreducible $K$-module which occurs in $\mathbf{H}_j$ can have a highest weight of the form $\mu + \rho_n - \beta$, with $\beta \in \Phi^n \cap \Psi$,

again for any $j \in \hat{G}_\mu$. As one consequence of (5.11), we obtain the vanishing theorem

(5.13) $\mathscr{H}_\mu^- = 0$

(Parthasarathy [22]).

We should remark that the arguments leading up to (5.13) are really curvature estimates, in algebraic disguise. The decomposition of the $K$-modules $V_\mu \otimes S^\pm$ into irreducibles determines an analogous decomposition of the bundles $\mathscr{V}_\mu \otimes \mathscr{S}^\pm$. Under the assumption (5.2), the curvature properties of the bundles and of the manifold $G/K$ force all square-integrable, harmonic spinors to take values in a certain sub-bundle of $\mathscr{V}_\mu \otimes \mathscr{S}^+$, namely the one that corresponds to the $K$-submodule of highest weight $\mu + \rho_n$ in $V_\mu \otimes S^+$. As happens often with differential-geometric arguments of this nature, the resulting vanishing theorem fails to be precise: the hypothesis (5.2) is unnecessarily stringent.

We shall now appeal to some algebraic results about representations of $G$, which can be found, for example, in [24]. The arguments of [24] work with much weaker hypotheses than (5.2), and can be simplified considerably in our more special situation. For this reason, we shall present a proof of the relevant result in the Appendix.

(5.14) **Proposition.** Suppose that $\mu$ satisfies the inequalities $(\mu + \rho - B, \alpha) \geqq 0$, for every $\alpha \in \Phi^c \cap \Psi$, and every sum $B$ of distinct positive, noncompact roots. Up to isomorphism, there exists at most one irreducible unitary representation $\pi$ of $G$, such that

a) $\pi|_K$ has an irreducible constituent of highest weight $\mu + \rho_n$, and

b) no irreducible constituent of $\pi|_K$ has a highest weight of the form $\mu + \rho_n - \beta$, $\beta \in \Phi^n \cap \Psi$.

In such a representation $\pi$, the irreducible $K$-module of highest weight $\mu + \rho_n$ occurs exactly once. The highest weight of any irreducible constituent of $\pi|_K$ can be expressed as $\mu + \rho_n + \sum n_i \beta_i$, with $\beta_i \in \Phi^n \cap \Psi$, $n_i \geqq 0$.

We first observe that (5.2) implies the hypothesis of the proposition. Indeed, as the highest weight of an irreducible $K$-module (which occurs in $S^+$), $\rho_n$ has a non-negative inner product with every $\alpha \in \Phi^c \cap \Psi$; hence

$$(\mu + \rho - B, \alpha) \geqq (\mu + \rho_c - B, \alpha),$$

which is positive because of (5.2). According to (5.11–5.12), any class $j \in \hat{G}_\mu$ which contributes to the Plancherel decomposition of $\mathscr{H}_\mu^\pm$ has the two properties a) and b). The Plancherel measure of the totality of these classes $j \in \hat{G}_\mu$ is non-zero, as follows from (3.19). The proposition now guarantees that there can be only one such $j \in \hat{G}_\mu$, which necessarily belongs to the discrete series; also

$$(5.15) \quad V_j^- = 0, \quad \dim V_j^+ = 1.$$

Since $j$ alone enters the Plancherel decomposition of $\mathscr{H}_\mu^+$, one can identify $\mathscr{H}_\mu^+$ with $\mathbf{H}_j \otimes V_j^+ \cong \mathbf{H}_j$. We recall that the Plancherel measure of a class in the discrete series is also called the formal degree. The difference formula (3.19) gives the formal degree of $j$ as

$$d(\mu - \rho_n) = \prod_{\alpha \in \Psi} \frac{(\alpha, \mu + \rho_c)}{(\alpha, \rho)}.$$

We summarize: in the situation (5.2), $\mathscr{H}_\mu^-$ vanishes, and $\mathscr{H}_\mu^+$ is a non-zero, irreducible, unitary $G$-module, belonging to the discrete series, whose formal degree is $d(\mu-\rho_n)$; moreover, $\mathscr{H}_\mu^+$ has the two properties a) and b) in the statement of Proposition (5.14).

We denote the global character of $\mathscr{H}_\mu^+ = \mathbf{H}_j$ by $\Theta$. According to (4.16), the restriction of $\Theta$ to $H \cap G'$ can be expressed as

(5.16a) $$\Delta_H \Theta|_{H\cap G'} = \sum_{\nu\in\Lambda} n_\nu e^\nu$$

(finite sum), with

(5.16b) $$n_{w(\mu+\rho_c)} = (-1)^q \varepsilon(w), \qquad \text{for } w \in W$$

(cf. (5.15)). The significance of the coefficients $n_\nu$ is that

(5.17) $$\tau(\sigma^+ - \sigma^-)|_{H\cap G'} = \frac{\sum_\nu n_\nu e^\nu}{\prod_{\alpha\in\Phi^c\cap\Psi} (e^{\alpha/2} - e^{-\alpha/2})};$$

here $\tau$ stands for the $K$-character of $\mathbf{H}_j$. In a formal sense, $\tau(\sigma^+ - \sigma^-)$ is the character of the virtual $K$-module $\mathbf{H}_j \otimes S^+ - \mathbf{H}_j \otimes S^-$. We set $C_+$ = closed cone spanned by the positive roots. As follows from (5.14), the highest weight of any $K$-invariant, $K$-irreducible component of $\mathbf{H}_j$ lies in $\mu+\rho_n+C_+$. Every weight of $S^+ \oplus S^-$ can be expressed as $-\rho_n+\beta_1+\cdots+\beta_k$, with $\beta_i \in \Phi^n \cap \Psi$. If the character of an irreducible $K$-module appears in $\tau(\sigma^+ - \sigma^-)$, its highest weight must be the sum of a highest weight occuring in $\mathbf{H}_j$ and some weight of $S^+ \oplus S^-$, and hence lies in $\mu + C_+$. Combining this information with Weyl's character formula, we find

(5.18) $$n_\nu \neq 0 \Rightarrow w\nu \in \mu+\rho_c+C_+, \qquad \text{for some } w\in W.$$

Any two weights $\nu$ which occur in (5.16) with non-zero coefficient $n_\nu$ are $W_{\mathbb{C}}$-conjugate; this follows from (4.18–4.19). In particular,

(5.19) $$n_\nu \neq 0 \Rightarrow (\nu,\nu) = (\mu+\rho_c, \mu+\rho_c).$$

Since $\mu+\rho_c$ was assumed dominant with respect to $\Psi$, it has a strictly smaller length than all other elements of the cone $\mu+\rho_c+C_+$. Comparing (5.18) and (5.19), we may conclude that $n_\nu = 0$, unless $\nu$ is $W$-conjugate to $\mu+\rho_c$. This proves:

(5.20) **Theorem.** Subject to the condition (5.2), $\mathscr{H}_\mu^-$ vanishes, whereas $\mathscr{H}_\mu^+$ is a non-zero Hilbert space, on which $G$ acts unitarily and irreducibly. The resulting representation belongs to the discrete series and has formal degree $d(\mu-\rho_n)$. Its character $\Theta$ satisfies

$$\Theta|_{H\cap G'} = (-1)^q \frac{\sum_{w\in W} \varepsilon(w)\, e^{w(\mu+\rho_c)}}{\prod_{\alpha\in\Psi} (e^{\alpha/2} - e^{-\alpha/2})}.$$

Every irreducible $K$-module which occurs in $\mathscr{H}_\mu^+$ has a highest weight of the form $\mu+\rho_n+\sum n_i\beta_i$, with $\beta_i \in \Phi^n \cap \Psi$, $n_i \geq 0$. The irreducible $K$-module with highest weight $\mu+\rho_c$ occurs exactly once in $\mathscr{H}_\mu^+$.

## § 6. Characters and Sobelev Spaces

The characters of discrete series representations, unlike those of a general irreducible, unitary representation, extend continuously from $C_0^\infty(G)$ to much larger function spaces. Harish-Chandra has shown that they are in particular tempered, i.e. their domain of definition includes a suitably defined Schwartz space of rapidly decreasing functions. He used this fact to describe the discrete series characters: within the class of tempered invariant eigendistributions, a discrete series character is completely determined by its restriction to a compact Cartan subgroup.

In this section, we present Harish-Chandra's results in somewhat modified form. The vehicle for our arguments will be certain global Sobolev spaces, rather than the Schwartz space. This is not only consistent with our emphasis on $L^2$-methods elsewhere in our construction, but allows us also to prove the completeness of the parametrization of the discrete series within the framework of the existence proof. Sobolev spaces usually serve as a tool for studying local regularity properties of functions and distributions. Not so in our context, where they are used to measure global growth properties.

By infinitesimal right translation, the complexified Lie algebra $\mathfrak{g}^{\mathbb{C}}$ of $G$ acts on $C^\infty(G)$ as the Lie algebra of left-invariant complex vector fields. When this action is extended to the universal enveloping algebra $\mathfrak{U}(\mathfrak{g}^{\mathbb{C}})$, one obtains an isomorphism

(6.1) $r\colon \mathfrak{U}(\mathfrak{g}^{\mathbb{C}}) \xrightarrow{\sim} \mathscr{D}_l$,

between $\mathfrak{U}(\mathfrak{g}^{\mathbb{C}})$ and the algebra $\mathscr{D}_l$ of all left-invariant linear differential operators. As a quotient of the tensor algebra of $\mathfrak{g}^{\mathbb{C}}$, $\mathfrak{U}(\mathfrak{g}^{\mathbb{C}})$ has a natural filtration. We shall say that $X \in \mathfrak{U}(\mathfrak{g}^{\mathbb{C}})$ has degree at most $n$ if it lies in the image of $\bigoplus_{k=0}^{n} (\bigotimes^{k} \mathfrak{g}^{\mathbb{C}})$. For each positive integer $n$, we define the $n$-th (left) Sobolev space $H_n(G)$ as

(6.2) $H_n(G) = \{f \in L^2(G) \mid r(X) f \in L^2(G)$, for every $X \in \mathfrak{U}(\mathfrak{g}^{\mathbb{C}})$ of degree at most $n\}$;

here $r(X) f$ is to be interpreted in the sense of distributions.

One can turn $H_n(G)$ into a Banach space, in an essentially natural manner: although the Banach norm is not intrinsic, the resulting topology is. The group $G$ acts continuously on $H_n(G)$, by left translation, but $H_n(G)$ is not right invariant. When one topologizes $C_0^\infty(G)$ in the usual fashion, the inclusion of $C_0^\infty(G)$ into $H_n(G)$ becomes continuous. We remark that $C_0^\infty(G)$ lies densely in $H_n(G)$; this fact is proven, in effect, in [1], but it will not be needed here.

The next result is implicit in the proof of Harish-Chandra's lemma 76 [15].

(6.3) **Lemma.** Let $\Theta_\pi$ be the character of an irreducible, unitary representation $\pi$, which belongs to the discrete series. Then $\Theta_\pi$ extends continuously from $C_0^\infty(G)$ to $H_n(G)$, for every sufficiently large integer $n$.

At first glance, the statement of the lemma appears to be asymmetric, since it prefers the left Sobolev spaces over their right counterparts. However, the two possible versions of the lemma, corresponding to choices of either left or right Sobolev spaces, are quite immediately equivalent: the distribution $\Theta_\pi$ remains invariant under conjugation; hence, if one lets it act on $r(X)f$, $f \in C_0^\infty(G)$, the differentiation $r(X)$ can be shifted to the left, without affecting the result.

*Proof* of (6.3). The assertion of the lemma amounts to saying that the linear functional

$$f \mapsto \Theta_\pi(f) = \operatorname{tr} \pi(f), \quad f \in C_0^\infty(G),$$

is continuous, relative to the topology which $H_n(G)$ induces on its subspace $C_0^\infty(G)$. It therefore suffices to prove the following two statements:

(6.4) $\pi(f)$ is a Hilbert-Schmidt operator, and $\|\pi(f)\|_{\text{H.S.}} \leqq c\|f\|_2$, with $c = c(\pi)$,

and

(6.5) there exists a Hilbert-Schmidt operator $T$, and some $Z \in \mathfrak{U}(\mathfrak{g}^{\mathbb{C}})$, such that $\pi(f) = \pi(r(Z)f) \cdot T$

for every $f \in C_0^\infty(G)$. Indeed, (6.4) and (6.5) give the estimate

$$\begin{aligned} |\Theta_\pi(f)| &= |\operatorname{trace} \pi(f)| = |\operatorname{trace}(\pi(r(Z)f) \cdot T)| \\ &\leqq \|\pi(r(Z)f)\|_{\text{H.S.}} \|T\|_{\text{H.S.}} \leqq c\|T\|_{\text{H.S.}} \|r(Z)f\|_2; \end{aligned}$$

this bounds $\Theta_\pi$ in terms of the seminorm $f \mapsto \|r(Z)f\|_2$, which is continuous with respect to the topology induced by $H_n(G)$ on $C_0^\infty(G)$, provided $n \geqq \deg Z$.

The Plancherel theorem asserts that for any given $f \in L^2(G) \cap L^1(G)$, the operators $\pi_j(f)$, $j \in \hat{G}$, are Hilbert-Schmidt operators, except possibly on a set of Plancherel measure zero, and

$$\|f\|_2^2 = \int_{\hat{G}} \|\pi_j(f)\|_{\text{H.S.}}^2 \, dj.$$

Since $\pi$ belongs to the discrete series, its class in $\hat{G}$ has positive Plancherel measure $d(\pi)$. Hence

$$\|f\|_2^2 \geqq d(\pi) \|\pi(f)\|_{\text{H.S.}}^2,$$

which implies (6.4), with $c = d(\pi)^{-1/2}$.

We now turn to (6.5) – which, incidentally, holds for any irreducible unitary representation. The various $K$-invariant, $K$-irreducible subspaces of the representation space $\mathbf{H}$ span a dense linear subspace $\mathbf{H}_\infty \subset \mathbf{H}$, which consists entirely of analytic vectors. In particular, the complexified Lie algebra $\mathfrak{g}^{\mathbb{C}}$ of $G$ acts on $\mathbf{H}_\infty$. This infinitesimal representation turns $\mathbf{H}_\infty$ into a module over the universal enveloping algebra $\mathfrak{U}(\mathfrak{g}^{\mathbb{C}})$; we refer to the action of $\mathfrak{U}(\mathfrak{g}^{\mathbb{C}})$ on $\mathbf{H}_\infty$ also by the symbol $\pi$. For $f \in C_0^\infty(G)$, $v \in \mathbf{H}_\infty$, and $Z \in \mathfrak{U}(\mathfrak{g}^{\mathbb{C}})$, the identity

$$\pi(r(Z)f)\, v = \pi(f)\, \pi(Z)\, v$$

amounts to a tautology. Since $\Omega_K$, the Casimir operator of $K$, is positive semi-definite, $\pi(1 + \Omega_K)$ has a bounded inverse. Hence, setting $Z = (1 + \Omega_K)^n$, one finds

(6.6) $\pi(f)=\pi(r(Z)f)\cdot\pi(1+\Omega_K)^{-n}$,

for any $f\in C_0^\infty(G)$, $n\in\mathbb{N}$. The set $\hat{K}$ of isomorphism classes of irreducible $K$-modules has a natural parametrization in terms of highest weights, which range over a lattice, intersected with a cone. On the irreducible $K$-module of highest weight $\mu$, $\Omega_K$ acts by a constant approximately equal to $\|\mu\|^2$. Each class $i\in\hat{K}$ occurs in $\mathbf{H}$ at most as often as its degree, and this degree can be bounded by a polynomial in the length of the highest weight. Conclusion: for every sufficiently large $n\in\mathbb{N}$, $\pi(1+\Omega_K)^{-n}$ is a Hilbert-Schmidt operator. Because of (6.6), the assertion (6.5), and hence the lemma, follow.

According to Harish-Chandra's fundamental regularity theorem [13, 3], every invariant eigendistribution – in particular, every character – can be represented as integration against a locally $L^1$ function; this locally $L^1$ function is actually real-analytic on $G'$, the set of regular, semisimple elements. We shall not distinguish between the invariant eigendistribution and the real-analytic function on $G'$ that represents it.

We recall the definition of the rank of $G$: it is the minimal possible multiplicity of the eigenvalue one for the automorphisms $\mathrm{Ad}\,g$ of $\mathfrak{g}^{\mathbb{C}}$, as $g$ ranges over $G$. Any particular $g\in G$ realizes this minimal multiplicity precisely when $g$ is both regular and semisimple. Thus, writing

(6.7) $$\det(\lambda+1-\mathrm{Ad}\,g)=\sum_{k\geqq 0} D_k(g)\,\lambda^{r+k}$$

($r$ = rank of $G$), one finds

(6.8) $$G'=\{g\in G\,|\,D_0(g)\neq 0\}.$$

Incidentally, $D_0$ assumes only real values, since $\mathrm{Ad}\,G$ preserves the real form $\mathfrak{g}\subset\mathfrak{g}^{\mathbb{C}}$. After passage to some finite covering of $G$, if necessary, the function $D_0$, restricted to any Cartan subgroup, admits a smooth square-root; $D_0^{1/2}$ appears as a universal denominator in character formulas. We shall amplify on these statements later. To motivate the definitions which follow, we merely remark that the singularities of a general invariant eigendistribution near the complement of $G'$ are comparable to those of $D_0^{-1/2}$. In particular, multiplication by $|D_0|^{1/2}$ renders any invariant eigendistribution $\Theta$ locally bounded on $G$. For this reason, the growth properties of such a distribution $\Theta$ tend to be reflected by the behavior at infinity, along the various Cartan subgroups[1], of the function $|D_0|^{1/2}\Theta$, rather than by the behavior of the function $\Theta$ itself.

For lack of better terminology, we shall call an invariant eigendistribution $\Theta$ "bounded at infinity" if, for any Cartan subgroup $B$,

(6.9) $$\sup_{b\in B\cap G'} |D_0(b)|^{1/2}\,|\Theta(b)|<\infty.$$

Similarly, $\Theta$ will be said to "decay at infinity" if the restriction of $|D_0|^{1/2}\Theta$ to any Cartan subgroup $B$ tends to zero outside of compact subsets. According to a

[1] Since $|D_0|^{1/2}\Theta$ is invariant under conjugation, its behavior at infinity must be measured transversely to the conjugacy classes, i.e. along the various Cartan subgroups

criterion of Harish-Chandra, the property of being bounded at infinity is essentially equivalent to his notion of temperedness. When $\Theta$ arises as the character of a representation, the equivalence becomes precise, as follows from a result of Fomin-Shapovalov [11]. However, we shall not use the notion of a tempered distribution.

Until the present section, $G$ was assumed to contain a compact Cartan subgroup. This hypothesis did not play a role in Lemma (6.3), nor will it enter the next two propositions. The first of these amounts to a modified version of one direction of Harish-Chandra's temperedness criterion; the second is the analogue, in our context, of the uniqueness statement in [14, Theorem 3].

(6.10) **Proposition.** An invariant eigendistribution, which extends continuously from $C_0^\infty(G)$ to $H_n(G)$, for some $n \in \mathbb{N}$, decays at infinity.

(6.11) **Proposition.** A non-zero invariant eigendistribution which decays at infinity has a non-trivial restriction on some compact Cartan subgroup. In particular, no such invariant eigendistributions exist on $G$, unless $G$ contains a compact Cartan subgroup.

The proofs of the two propositions will be given in § 7. We conclude this section with some fairly immediate corollaries.

According to the results of § 4, if $G$ has a compact Cartan subgroup, then its discrete series is not empty. The preceeding two propositions, in conjunction with Lemma (6.3), also imply the converse:

(6.12) **Corollary** (Harish-Chandra). For the existence of a non-empty discrete series it is necessary, as well as sufficient, that $G$ contain a compact Cartan subgroup.

Let us assume then that $G$ does contain a compact Cartan subgroup, which we may choose to lie in $K$. We denote this group by $H$. As in § 4, $\Lambda$ shall refer to the weight lattice of the torus $H$, and $\Lambda_\rho$ to the weight lattice translated by the half-sum of the positive roots; cf. (4.25). We also recall the definition (4.17) of the characters $\varphi_\nu$ of $\mathfrak{Z}$ (=center of the universal enveloping algebra $\mathfrak{U}(\mathfrak{g}^{\mathbb{C}})$).

(6.13) **Corollary.** The infinitesimal character of any given discrete series representation is equal to $\varphi_\lambda$, for some nonsingular $\lambda \in \Lambda_\rho$. Conversely, every such $\varphi_\lambda$ arises as the infinitesimal character of some discrete series representation.

*Proof.* We consider a particular discrete series character $\Theta_j$, and we express its restriction to $H$ as in (4.16a). The argument which proves (4.30) also applies in the present situation; thus

(6.14) $n_\nu \neq 0 \Rightarrow \nu \in \Lambda_\rho$.

Because of (6.3) and (6.10–6.11), not all of the coefficients $n_\nu$ can vanish (any two compact Cartan subgroups are conjugate!). Combining this knowledge with (4.19) and (6.14), we find that $\Theta_j$ has infinitesimal character $\varphi_\lambda$, for some $\lambda \in \Lambda_\rho$. As a linear combination of discrete series characters, the invariant eigendistribution $\tilde{\Theta}_\lambda$ of (4.26) must decay at infinity, and hence

$$\tilde{\Theta}_\lambda = 0 \Leftrightarrow \tilde{\Theta}_\lambda|_H = 0.$$

We now appeal to Theorem (4.41):

$\tilde{\Theta}_\lambda = 0 \Leftrightarrow \lambda$ is singular.

On the other hand, $\tilde{\Theta}_\lambda$ is a linear combination, with non-zero coefficients, of all discrete series characters which correspond to the infinitesimal character $\varphi_\lambda$. Characters of non-isomorphic irreducible unitary representations are linearly independent. Consequently $\tilde{\Theta}_\lambda$ vanishes if and only if no discrete series representation has infinitesimal character $\varphi_\lambda$. This concludes the proof of the corollary.

The two Propositions (6.10–6.11) make it possible to describe the discrete series characters, uniquely within the class of invariant eigendistributions which decay at infinity, in terms of their restriction to a compact Cartan subgroup. More generally, one can give such a description within the larger class of invariant eigendistributions which are merely bounded at infinity. For this purpose, we state a lemma, due to Harish-Chandra [14], whose proof is similar to that of (6.11). It will be proved in the next section, along with the two propositions.

(6.15) **Lemma.** For every nonsingular $\lambda \in \Lambda_\rho$, there exists at most one invariant eigendistribution $\Theta_\lambda$, such that $\Theta_\lambda$ is bounded at infinity, and

$$\Theta_\lambda|_{H \cap G'} = (-1)^q \frac{\sum_{w \in W} \varepsilon(w)\, e^{w\lambda}}{\prod_{\substack{\alpha \in \Phi,\\ (\alpha,\lambda) > 0}} (e^{\alpha/2} - e^{-\alpha/2})}\,.$$

## § 7. Proofs of the Preceding Statements

We begin with the proof of Proposition (6.10). The crucial step will be to relate growth properties of invariant eigendistributions on $G$ to their behavior on the various Cartan subgroups.

Until further notice, $\Theta$ shall denote an arbitrary invariant eigendistribution, and $B$ a Cartan subgroup of $G$. We did not require $G$ to have a faithful finite-dimensional representation, and hence $B$ need not be abelian. Nevertheless, the identity component $B^0$ lies in the center of $B$, so that one can define a mapping

(7.1) $\xi\colon G/B^0 \times (B \cap G') \to (B \cap G')^G$, with $\xi(gB^0, b) = g\,b\,g^{-1}$;

as usual, $(B \cap G')^G$ stands for the open subset

$\{g\,b\,g^{-1} \mid g \in G,\ b \in B \cap G'\}$

of $G$. Then $\xi$ is a covering mapping, with fibre $N_G(B)/B^0$, which is finite. At any coset $gB^0$, the complexified tangent space of $G/B^0$ may be identified, via left translation by $g$, with

$\mathfrak{b}^{\mathbb{C}\perp} = \{X \in \mathfrak{g}^{\mathbb{C}} \mid B(X, \mathfrak{b}^{\mathbb{C}}) = 0\}$

($B(\ ,\ )$ = Killing form of $\mathfrak{g}^{\mathbb{C}}$). Similarly, for $b \in B$, left translation by $b$ identifies the complexified tangent space of $B$ at $b$ with $\mathfrak{b}^{\mathbb{C}}$. In terms of these conventions, the differential $\xi_*$ of the mapping $\xi$ at a point $(gB^0, b)$ of $G/B^0 \times (B\cap G')$ is given by the formula

$$(7.2)\quad \xi_*(l_{g*}X, l_{b*}Y) = l_{h*}\operatorname{Ad} g\{(\operatorname{Ad} b^{-1}-1)(X)+Y\},$$

for $X\in \mathfrak{b}^{\mathbb{C}\perp}$, $Y\in \mathfrak{b}^{\mathbb{C}}$, $h=gbg^{-1}=\xi(gB^0,b)$. To verify the identity, one may as well suppose that the tangent vectors $X$ and $Y$ are real, in which case $\xi_*(l_{g*}X, l_{b*}Y)$ becomes the tangent vector of the curve

$$\begin{aligned} t &\mapsto \xi(g\exp(tX)B^0, b\exp(tY)) \\ &= g\exp(tX)\,b\exp(tY)\exp(-tX)\,g^{-1} \\ &= gbg^{-1}\exp(t\operatorname{Ad} g\cdot \operatorname{Ad} b^{-1}X)\exp(t\operatorname{Ad} g\,Y)\exp(-t\operatorname{Ad} g\,X) \\ &= h\exp(t\operatorname{Ad} g\{(\operatorname{Ad} b^{-1}-1)(X)+Y\}+O(t^2)), \end{aligned}$$

at $t=0$.

The identity (7.2) implies, in particular, the following standard integration formula: if the invariant measures $dg$ on $G$, $dg^*$ on $G/B^0$, and $db$ on $B$ are suitably normalized,

$$(7.3)\quad \int_{(B\cap G')^G} f\,dg = \int_B |D_0(b)| \int_{G/B^0} f(gbg^{-1})\,dg^*\,db,$$

for every continuous function $f$ with compact support in $(B\cap G')^G$; cf. (6.7). Indeed, $\operatorname{Ad} g$ operates as the identity on the top exterior power of $\mathfrak{g}^{\mathbb{C}}$, whereas

$$\operatorname{Ad} b^{-1}-1\colon \mathfrak{b}^{\mathbb{C}\perp}\to \mathfrak{b}^{\mathbb{C}\perp}$$

has determinant $\pm D_0(b)$. The top exterior power of $\xi_*$ is therefore represented by the function $\pm D_0$, relative to translation-invariant sections of the top exterior powers of the various tangent bundles; this proves the integration formula. As one consequence of the formula,

$$(7.4)\quad \Theta(f) = \int_{B\cap G'} |D_0(b)|\,\Theta(b) \int_{G/B^0} f(gbg^{-1})\,dg^*\,db,$$

whenever $f\in C_0^\infty(G)$ has support in $(B\cap G')^G$,

for every invariant eigendistribution $\Theta$.

Corresponding to any $\varphi\in C_0^\infty(G/B^0)$, we define a mapping

$$(7.5)\quad T_\varphi\colon C_0^\infty(B\cap G')\to C_0^\infty((B\cap G')^G)$$

as follows: for $f\in C_0^\infty(B\cap G')$ and $g\in (B\cap G')^G$, $T_\varphi f(g)$ is to be the average, extended over the fibre $\xi^{-1}(g)$, of the values of the function

$$(gB^0, b)\mapsto \varphi(gB^0)\,f(b)\,|D_0(b)|^{-1/2}.$$

An invariant eigendistribution $\Theta$, when restricted to $B\cap G'$, remains invariant under the conjugation action of $N_G(B)$. Hence

$$\begin{aligned} (7.6)\quad \Theta(T_\varphi f) &= \int_{B\cap G'} |D_0(b)|\,\Theta(b) \int_{G/B^0} \varphi(gB^0)\,f(b)\,|D_0(b)|^{-1/2}\,dg^*\,db \\ &= \int_{G/B^0} \varphi\,dg^* \int_{B\cap G'} f\,\Theta\,|D_0|^{1/2}\,db. \end{aligned}$$

Every root $\alpha$ in $\Phi_B$, the root system of $(\mathfrak{g}^{\mathbb{C}}, \mathfrak{b}^{\mathbb{C}})$, lifts to a character $e^{\alpha}$ of $B$. Since the rank of $G$ coincides with the dimension of its Cartan subgroups, one finds

(7.7) $$D_0(b) = \prod_{\alpha \in \Phi_B} (1 - e^{\alpha}(b)) \qquad (b \in B).$$

In particular, $B \cap G' = \{b \in B \mid e^{\alpha}(b) \neq 1, \text{ for } \alpha \in \Phi_B\}$. For $\varepsilon > 0$, we define

(7.8) $$B_{\varepsilon} = \{b \in B \mid |e^{\alpha}(b) - 1| > \varepsilon, \text{ for } \alpha \in \Phi_B\}.$$

The sets $B_{\varepsilon}$ are open in $B$, and they exhaust $B \cap G'$. As a Cartan subgroup, $B$ centralizes its own Lie algebra. Consequently every right-invariant vector field on $B$ is automatically left-invariant, and vice versa. More generally, the two notions of invariance agree for any linear differential operator. Just as for $G$, we introduce global Sobolev spaces $H_n(B)$, $n \in \mathbb{N}$:

(7.9) $f \in H_n(B) \Leftrightarrow Xf \in L^2(B)$, whenever $X$ is a translation-invariant differential operator, of order at most $n$.

(7.10) **Lemma.** For any fixed $\varepsilon > 0$ and $n \in \mathbb{N}$,

$$T_{\varphi}\colon C_0^{\infty}(B_{\varepsilon}) \to C_0^{\infty}((B \cap G')^G)$$

is continuous with respect to the topologies induced by $H_n(B)$ and $H_n(G)$.

*Proof.* We consider a vector field $r(Z)$ on $G$, with $Z \in \mathfrak{g}^{\mathbb{C}}$; cf. (6.1). The mapping $\xi$ pulls back $r(Z)$ to a vector field $\xi^* Z$ on $G/B^0 \times (B \cap G')$. We denote the projections of $\mathfrak{g}^{\mathbb{C}}$ onto $\mathfrak{b}^{\mathbb{C}\perp}$ and $\mathfrak{b}^{\mathbb{C}}$ by $p$ and $q$, respectively. According to (7.2), at points $(gB^0, b)$ of $G/B^0 \times (B \cap G')$, $\xi^* Z$ takes the value $(l_{g*}X, l_{b*}Y)$, with

$$X = (\operatorname{ad} b^{-1} - 1)^{-1} \cdot p \cdot \operatorname{Ad} g^{-1}(Z), \qquad Y = q \cdot \operatorname{Ad} g^{-1}(Z).$$

The automorphism $(\operatorname{ad} b^{-1} - 1)^{-1}$ of $\mathfrak{b}^{\mathbb{C}\perp}$ is semisimple with eigenvalues $m_{\alpha}(b) = (e^{-\alpha}(b) - 1)^{-1}$, indexed by $\alpha \in \Phi_B$. As functions on $B_{\varepsilon}$, the $m_{\alpha}$ and all their derivatives with respect to translation-invariant differential operators are uniformly bounded. For the purposes of this proof, the space of all such functions will be referred to as $U^{\infty}(B_{\varepsilon})$. As follows from our observations, there exist smooth vector fields $X_i$ on $G/B^0$, translation-invariant vector fields $Y_j$ on $B$, and functions $u_i \in U^{\infty}(B_{\varepsilon})$, $v_j \in C^{\infty}(G/B^0)$, such that

$$\xi^* Z = \sum_i u_i X_i + \sum_j v_j Y_j.$$

To understand the effect of $r(Z)$ on $T_{\varphi} f$, for $f \in C^{\infty}(B_{\varepsilon})$, we average the product

$$\varphi f \in C_0^{\infty}(G/B^0 \times B_{\varepsilon})$$

with respect to the finite group $N_G(B)/B^0$, which operates on $G/B^0$ by right translation, and on $B$ by conjugation; it should be observed that the action preserves $B_{\varepsilon} \subset B$. The averaged product can be expressed as $\sum_l \tilde{\varphi}_l \tilde{f}_l$, with $\tilde{\varphi}_l \in C_0^{\infty}(G/B^0)$, $\tilde{f}_l \in C_0^{\infty}(B_{\varepsilon})$. Moreover,

$$(T_\varphi f)(g b g^{-1}) = \sum_l \tilde{\varphi}_l(g B^0) |D_0(b)|^{-1/2} \tilde{f}_l(b),$$

since $D_0$ already is $N_G(B)/B^0$-symmetric. In particular,

$$r(Z)(T_\varphi f)(g b g^{-1}) = \sum_{i,l} (X_i \tilde{\varphi}_l)(g B^0)\, u_i(b) |D_0(b)|^{-1/2} \tilde{f}_l(b) + \sum_{j,l} v_j(g B^0)\, \tilde{\varphi}_l(g B^0)\, Y_j(|D_0|^{-1/2} \tilde{f}_l)(b).$$

Because of (7.7), the functions $|D_0|^{1/2}(Y_j |D_0|^{-1/2})$ lie in $U^\infty(B_\varepsilon)$. Hence one obtains an identity

$$r(Z)(T_\varphi f)(g b g^{-1}) = \sum_{j,l} \varphi_{jl}(g B^0) |D_0(b)|^{-1/2} (Y_j \tilde{f}_l)(b) + \sum_{k,l} \psi_{kl}(g B^0)\, h_k(b) |D_0(b)|^{-1/2} \tilde{f}_l(b),$$

with suitably chosen $\varphi_{j,l}, \psi_{k,l} \in C_0^\infty(G/B^0)$, $h_k \in U^\infty(B_\varepsilon)$, which depend on $Z$ and $\varphi$. Inductively this procedure leads to a formula of the following type: if $Z \in \mathfrak{U}(\mathfrak{g}^{\mathbb{C}})$ has degree $n$,

$$r(Z)(T_\varphi f)(g b g^{-1}) = \sum_{i,j,l} \varphi_{ijl}(g B^0)\, h_i(b) |D_0(b)|^{-1/2} (Y_j \tilde{f})(b).$$

Here $Y_j$ runs over a basis of translation-invariant differential operations on $B$, of order up to $n$, $\{\varphi_{ijl}\}$ is a collection of functions in $C_0^\infty(G/B^0)$, and the $h_i \in C^\infty(B \cap G')$ are uniformly bounded on $B_\varepsilon$. Except for a constant factor, the mapping $f \to \tilde{f}_l$ is translation by some element of $N_G(B)/B^0$, hence continuous in the topology of $H_n(B)$. The lemma now follows from an application of the integration formula (7.3).

We select an ordering $>$ on $\Phi_B$, and we let $\rho_B$ denote one-half of the sum of the positive roots. Then

$$D_0(b) = \prod_{\substack{\alpha \in \Phi_B, \\ \alpha > 0}} \{(1 - e^\alpha(b))(1 - e^{-\alpha}(b))\} = (-1)^d\, e^{2\rho_B}(b) \prod_{\substack{\alpha \in \Phi_B, \\ \alpha > 0}} (1 - e^{-\alpha}(b))^2,$$

with $d$ equal to half of the cardinality of $\Phi_B$. Passing to a finite covering of $G$, if necessary, one can arrange that $\rho_B$ lifts to a character $e^{\rho_B}$ of $B$. In this situation,

$$\text{(7.11)} \quad \Delta_B = e^{\rho_B} \prod_{\substack{\alpha \in \Phi_B, \\ \alpha > 0}} (1 - e^{-\alpha}) = \prod_{\substack{\alpha \in \Phi_B, \\ \alpha > 0}} (e^{\alpha/2} - e^{-\alpha/2})$$

becomes a well-defined function on $B$, such that

$$D_0(b) = (-1)^d\, \Delta_B(b)^2 \qquad (b \in B).$$

On any connected component of $B \cap G'$, $|D_0|^{1/2}$ coincides with a constant multiple of $\Delta_B$ ($D_0$ is real-valued!). Hence (7.6) and (7.10) imply:

(7.12) **Corollary.** If the invariant eigendistribution $\Theta$ extends continuously from $C_0^\infty(G)$ to $H_n(G)$, then the linear functional

$$f \mapsto \int_B f \Delta_B \Theta \, db$$

on $C_0^\infty(B_\varepsilon)$ is continuous with respect to the topology induced by $H_n(B)$, for any $\varepsilon > 0$.

At this point, we have to recall certain facts about the local structure of an invariant eigendistribution. As before, $B$ will denote a Cartan subgroup, with Lie algebra $\mathfrak{b}$, and $\Phi_B$ the root system of $(\mathfrak{g}^{\mathbb{C}}, \mathfrak{b}^{\mathbb{C}})$. Every root $\alpha$ of the sub-root system

(7.13) $\Phi_{B,\mathbb{R}} = \{\alpha \in \Phi_B \mid \alpha \text{ is real-valued on } \mathfrak{b}\}$

lifts to a character $e^\alpha$ of $B$, which assumes only real values[1]. If $\Theta$ is an invariant eigendistribution, the function $\Delta_B \Theta$ on $B \cap G'$ has a real-analytic extension to the larger subset

(7.14) $B'' = \{b \in B \mid e^\alpha(b) \neq 1, \text{ for } \alpha \in \Phi_{B,\mathbb{R}}\}$

of $B$; this is part of Harish-Chandra's "matching condition" [13, 3]. The group $B$ can be expressed as a direct product

(7.15a) $B = B_+ B_-$,

such that $B_+$ is a compact group, and $B_-$ a vector group. Via the exponential map, $B_-$ becomes isomorphic to its own Lie algebra $\mathfrak{b}_-$, i.e.

(7.15b) $\exp\colon \mathfrak{b}_- \xrightarrow{\sim} B_-$.

The identity component $B_+^0$ of $B_+$ is a torus, with Lie algebra $\mathfrak{b}_+$, so that

(7.15c) $B_+^0 = \exp(\mathfrak{b}_+)$.

We observe that

(7.16) $e^\alpha(B_+) \subset \{\pm 1\}$, if $\alpha \in \Phi_{B,\mathbb{R}}$,

since $e^\alpha$ takes only real values.

We enumerate the connected components of $B''$ as $B''_1, \ldots, B''_N$. For each $j$,

(7.17) $\Phi_{B,\mathbb{R},j} = \{\alpha \in \Phi_{B,\mathbb{R}} \mid e^\alpha > 0 \text{ on } B''_j\}$

is then a sub-root system of $\Phi_{B,\mathbb{R}}$, and

(7.18) $\Phi^+_{B,\mathbb{R},j} = \{\alpha \in \Phi_{B,\mathbb{R},j} \mid e^\alpha > 1 \text{ on } B''_j\}$

---

[1] This is not totally obvious, since $B$ may have several connected components. One should observe that the statement really concerns the adjoint group, which is an algebraic group over $\mathbb{R}$, and which contains the image of $B$ as an algebraic subgroup. The character $e^\alpha$ is also defined over $\mathbb{R}$, and hence must assume real values on all of $B$

a system of positive roots in $\Phi_{B,\mathbb{R},j}$, which corresponds to the Weyl chamber

(7.19) $C_j = \{X \in \mathfrak{b}_- \mid \langle \alpha, X \rangle > 0, \text{ for } \alpha \in \Phi^+_{B,\mathbb{R},j}\}$

in $\mathfrak{b}_-$. We claim that $B''_j$ can be decomposed into a product

(7.20) $B''_j = b_j B^0_+ \exp C_j, \quad$ for some fixed $b_j \in B_+$.

Indeed, since $B_+$ meets every connected component of $B$, one can choose a $b_j \in B_+$, such that $B''_j$ lies in the connected component $b_j B^0_+ \exp \mathfrak{b}_-$. But then $B''_j$ must be a connected component of the open subset

(7.21) $\{b \in b_j B^0_+ \exp \mathfrak{b}_- \mid e^\alpha(b) \neq 1, \text{ for } a \in \Phi_{B,\mathbb{R}}\}$

of $B$. According to (7.16), for any $\alpha \in \Phi_{B,\mathbb{R}}$, $e^\alpha$ is identically equal to $+1$ or $-1$ on $b_j B^0_+$, depending on whether or not $\alpha$ belongs to $\Phi_{B,\mathbb{R},j}$. The set (7.21) therefore coincides with

$$\{b_j b_0 \exp X \mid b_0 \in B^0_+, X \in \mathfrak{b}_-, \langle \alpha, X \rangle \neq 0 \text{ for } \alpha \in \Phi_{B,\mathbb{R},j}\},$$

i.e. the disjoint union of the connected, open subsets $b_j B^0_+ \exp C$, where $C$ runs over the collection of Weyl chambers, in $\mathfrak{b}_-$, of the root system $\Phi_{B,\mathbb{R},j}$. One of these is $B''_j$; the description (7.18) of $\Phi^+_{B,\mathbb{R},j}$ shows that it can only be $b_j B^0_+ \exp C_j$.

We now focus our attention on a particular invariant eigendistribution $\Theta$, and we keep fixed a connected component $B''_j$ of $B''$, as in (7.20). The Weyl group $W_{B,\mathbb{C}}$ of $(\mathfrak{g}^{\mathbb{C}}, \mathfrak{b}^{\mathbb{C}})$ operates on $\mathfrak{b}^{\mathbb{C}}$ in the usual manner, and by duality also on the dual space $\mathfrak{b}^{\mathbb{C}*}$. As a preliminary step in the proof of Harish-Chandra's regularity theorem [13, 3], one obtains the following result: there exist polynomial functions $p_{j,w}$ on $\mathfrak{b}_-$, indexed by $w \in W_{B,\mathbb{C}}$, and a linear function $\mu$ on $\mathfrak{b}^{\mathbb{C}}$, such that

(7.22) $(\Delta_B \Theta)(b_j \exp(X+Y)) = \sum_{w \in W_{B,\mathbb{C}}} p_{j,w}(X)\, e^{\langle w\mu, X+Y \rangle},$

whenever $X \in C_j$, $Y \in \mathfrak{b}_+$.

If $\mu$ is nonsingular, i.e. $w\mu \neq \mu$ for $w \neq 1$, the $p_{j,w}$ are actually constants, and they are uniquely determined. In general, to make the $p_{j,w}$ unique, one should sum not over $W_{B,\mathbb{C}}$, but over the quotient of $W_{B,\mathbb{C}}$ by the stabilizer of $\mu$.

According to (7.12), if $\Theta$ extends continuously to $H_n(G)$, the linear functional

(7.23) $f \mapsto \int_B f \Delta_B \Theta \, db, \quad f \in C^\infty_0(B''_j \cap B_\varepsilon),$

becomes bounded in the topology which $H_n(B)$ induces on $C^\infty_0(B''_j \cap B_\varepsilon)$; $\varepsilon > 0$ is arbitrary. To complete the proof of Proposition (6.10), we must deduce:

(7.24) $p_{j,w} \neq 0 \Rightarrow \operatorname{Re}\langle w_\mu, X \rangle < 0, \quad$ for any non-zero $X$ in the closure of $C_j$.

By separating out the toroidal variable, we shall reduce the problem to one about functions on Euclidean spaces. For this purpose, we re-interpret the identity (7.22): there exist distinct characters $\eta_1, \ldots, \eta_N$ of the torus $B_+^0$, and functions $h_1, \ldots, h_N \in C^\infty(B_-)$, such that

$$(7.25) \quad (\Delta_B \Theta)(b_j\, b \exp X) = \sum_i \eta_i(b)\, h_i(\exp X),$$

if $b \in B_+^0$, $X \in C_j$. Moreover,

$$(7.26) \quad h_i(\exp X) = \sum_{j=1}^{M_i} p_{ij}(X)\, e^{\langle v_{ij}, X\rangle}, \qquad X \in \mathfrak{b}_-,$$

where the $p_{ij}$ are polynomial functions on $\mathfrak{b}_-$, and $v_{ij} \in \mathfrak{b}_-^{\mathbb{C}*}$. We let $C_j'$ denote the translate of $C_j$ by some fixed $X_0 \in C_j$; thus $C_j$ contains the closure of $C_j'$. As will be argued shortly, one can select a non-empty open subset $U \subset B_+^0$, and some $\varepsilon > 0$, which have the property that

$$(7.27) \quad b_j\, U \exp C_j' \subset B_j'' \cap B_\varepsilon.$$

No non-trivial linear combination of the $\eta_i$ is perpendicular to $C_0^\infty(U)$ in $L^2(B_+^0)$. Hence, for suitably chosen functions $f_i \in C_0^\infty(U)$,

$$\int_{B_+^0} f_i\, \eta_j\, db = \delta_{ij}.$$

Testing the linear functional (7.23) against products $f_i f$, with $f \in C_0^\infty(\exp C_j')$, one finds that

$$(7.28) \quad f \mapsto \int_{B_-} f\, h_i\, db, \quad f \in C_0^\infty(\exp C_j'),$$

is bounded, relative to the topology of $H_n(B_-)$,

for $1 \leq i \leq N$; the Sobolev space $H_n(B_-)$ for the vector group $B_- \cong \mathfrak{b}_-$ is defined in the usual fashion.

We still must produce $U$ and $\varepsilon$, as in (7.27). For any real root $\alpha$, $e^\alpha$ is uniformly bounded away from one, on the entire set $b_j\, B_+^0 \exp C_j'$. Every $\alpha \in \Phi_B$, whether or not it lies in $\Phi_{B,\mathbb{R}}$, assumes only real values on $\mathfrak{b}_-$, so that $e^\alpha > 0$ on $B_-$. On the other hand, $|e^\alpha| \equiv 1$ on $B_+$. Hence, if $\delta$ is sufficiently small,

$$|e^\alpha(b_j\, b) - 1| > \delta \Rightarrow |e^\alpha(b_j\, b \exp X) - 1| > \tfrac{1}{2}\delta,$$

whenever $b \in B_+^0$, $X \in \mathfrak{b}_-$. Thus (7.27) will be satisfied for any open, relatively compact subset $U$ of

$$(7.29) \quad \{b \in B_+^0 \mid e^\alpha(b_j\, b) \neq 1 \text{ for } \alpha \in \Phi_B,\ \alpha \notin \Phi_{B,\mathbb{R}}\},$$

if only $\varepsilon > 0$ is small enough. No character $e^\alpha$, with $\alpha \in \Phi_B$, $\alpha \notin \Phi_{B,\mathbb{R}}$, remains constant on $B_+^0$. The set (7.29) is therefore non-empty, and we can indeed choose $U$ and $\varepsilon$.

Because of (7.28), the verification of the assertion (7.24) amounts to a problem about functions on Euclidean space. In order to state the relevant result, we let $Q$ denote the positive quadrant in $\mathbb{R}^d$,

$$Q=\{(x_1,\dots,x_d)\in\mathbb{R}^d \mid x_i>0,\ 1\leqq i\leqq d\}.$$

Via the inclusion $\mathbb{R}^d\hookrightarrow\mathbb{C}^d$ and the natural pairing $\mathbb{C}^d\times\mathbb{C}^d\mapsto\mathbb{C}$, every $\xi\in\mathbb{C}^d$ defines a complex-valued linear function $x\mapsto\langle\xi,x\rangle$ on $\mathbb{R}^d$.

(7.30) **Lemma.** Let $\xi_1,\dots,\xi_N$ be distinct elements of $\mathbb{C}^d$, and $p_1,\dots,p_N$ non-zero polynomial functions on $\mathbb{R}^d$. The distribution

$$f\mapsto\int_{\mathbb{R}^d} f(x)\sum_i p_i(x)\,e^{\langle\xi_i,x\rangle}\,dx,\qquad f\in C_0^\infty(Q),$$

does not extend continuously to the closure of $C_0^\infty(Q)$ in any Sobolev space $H_n(\mathbb{R}^d)$, unless the real part of each $\xi_i$ lies in $-Q$.

In terms of suitable linear coordinates $x_1,\dots,x_d$ on $\mathfrak{b}_-$, the Weyl chamber $C_j$ can be described as

$$\{(x_1,\dots,x_d)\in\mathbb{R}^d \mid x_i>0,\ 1\leqq i\leqq k\},$$

for some $k\leqq d$. Except for a finite number of hyperplanes, this set decomposes into $2^{d-k}$ copies of $Q$. We also note that translation by some $x_0\in\mathbb{R}^d$ transforms the distribution described in the lemma into another one, of the same type, with the same exponents $\xi_i$. In particular, the conclusion remains unchanged if the domain $C_0^\infty(Q)$ of the distribution in question is replaced by $C_0^\infty(Q+x_0)$. For these reasons, the lemma implies the statement (7.24).

*Proof* of (7.30). If a distribution is continuous in the topology of the $n$-th Sobolev space, then its image under a constant coefficient differential operator, of order $k$, is continuous in the topology of the $(n+k)$-th Sobolev space. For any polynomial $q$ of $d$ variables,

$$q\left(\frac{\partial}{\partial x_i}\right)\Big(\sum_j p_j(x)\,e^{\langle\xi_j,x\rangle}\Big)=\sum_j\sum_I\frac{1}{I!}\left(\frac{\partial}{\partial x_I}p_j\right)(x)\left(\frac{\partial}{\partial x_I}q\right)(\xi_j)\,e^{\langle\xi_j,x\rangle},$$

with $I$ running over all $d$-tuples of indices $I=(i_1,\dots,i_d)$ and $I!=\prod_l i_l!$. Hence, by an appropriate choice of the constant coefficient operator $q\left(\frac{\partial}{\partial x_i}\right)$, we can arrange that

$$q\left(\frac{\partial}{\partial x_i}\right)\Big(\sum_j p_j(x)\,e^{\langle\xi_j,x\rangle}\Big)=e^{\langle\xi_1,x\rangle}.$$

In this fashion, we reduce the problem to the case of a single exponential term $e^{\langle\xi,x\rangle}$, without a polynomial coefficient. A further reduction is possible: by testing the distribution $e^{\langle\xi,x\rangle}$ against products

$$f(x_1,\dots,x_d)=f_1(x_1)\cdot f_2(x_2)\cdot\dots\cdot f_d(x_d),\qquad f_i\in C_0^\infty(\mathbb{R}^+),$$

one can deal with one variable at a time. In other words, we may and shall assume that $d=1$. We now evaluate the distribution $e^{\xi x}$ on a sequence of functions $f_k(x)=f(x-k)$, for a fixed $f\in C_0^\infty(\mathbb{R}^+)$. The boundedness of the distribution in some Sobolev topology gives, at least, the estimate $\operatorname{Re}\xi\leqq 0$; we must exclude the possibility that $\operatorname{Re}\xi=0$. If $\xi$ were purely imaginary, translating the distribution would change it only by a multiplicative constant, of modulus one. Every $f\in C_0^\infty(\mathbb{R})$ has a translate whose support lies in $\mathbb{R}^+$. Consequently, the distribution $e^{\xi x}$ would be bounded, in the topology of some Sobolev space, not only on $C_0^\infty(\mathbb{R}^+)$, but actually on all of $C_0^\infty(\mathbb{R})$. That is absurd: evaluation of the Fourier transform $\hat{f}$ at some point $y\in\mathbb{R}$, for $f\in C_0^\infty(\mathbb{R})$, fails to be continuous with respect to any Sobolev topology.

The proofs of Proposition (6.11) and of Lemma (6.15) are straightforward applications of Harish-Chandra's "matching conditions". We shall briefly recall what is involved. Until further notice, we keep fixed the following data: an invariant eigendistribution $\Theta$, a Cartan subgroup $B$ of $G$, a connected component $B_j''$ of $B''$, as in (7.20), and a root $\gamma\in\Phi_{B,\mathbb{R},j}$ which is simple with respect to the positive root system $\Phi_{B,\mathbb{R},j}^+$. To any real root, and in particular to $\gamma$, one can associate a so-called Cayley transform, an inner automorphism of $\mathfrak{g}^{\mathbb{C}}$, which maps $\mathfrak{b}^{\mathbb{C}}$ onto the complexification of a Cartan subalgebra $\mathfrak{b}_\gamma\subset\mathfrak{g}$; the corresponding Cartan subgroup $B_\gamma$ meets $B$ along

$$B\cap B_\gamma=\{b\in B\mid e^\gamma(b)=1\},$$

and the dimension of its compact part $B_{\gamma,+}$ exceeds that of $B_+$ by one. The open subsets $(B\cap G')^G$ and $(B_\gamma\cap G')^G$ of $G$ have a hypersurface $S_\gamma$ as common boundary, which contains $B\cap B_\gamma$. As an invariant eigendistribution, $\Theta$ satisfies certain differential equations, which are used, in particular, to obtain expressions of the type (7.22). An investigation of the same differential equations, near the hypersurface $S_\gamma$, leads to relations between the restrictions of $\Theta$ to $B$ and $B_\gamma$, respectively. These are the matching conditions [13, 3].

Since $\gamma$ is simple with respect to $\Phi_{B,\mathbb{R},j}^+$, the intersection $B\cap B_\gamma$ – equivalently, the kernel of the character $e^\gamma$ on $B$ – contains a whole "wall" of $B_j''$, namely

(7.31) $\{b_j b\exp X\mid b\in B_+^0,\ X\in\text{closure of } C_j,\ \langle\gamma,X\rangle=0\}.$

According to (7.22), the restriction of $\Delta_B\Theta$ to $B_j''$ extends as an analytic function across the wall (7.31); for simplicity, the extension shall be referred to as $\varphi$. We now choose a non-zero, translation-invariant vector field $X_\gamma$ on $B$, which is normal to the codimension one subgroup $B\cap B_\gamma$. The matching conditions assert that, for each odd integer $n$, $X_\gamma^n\varphi$ coincides, on the subset (7.31) of $B\cap B_\gamma$, with a similar expression derived from the restriction of $\Theta$ to $B_\gamma$. In particular, if $\Theta\equiv 0$ on $B_\gamma$, all of the odd derivatives $X_\gamma^n\varphi$ vanish on the wall (7.31). Equivalently, $\Delta_B\Theta|_{B_j''}$ must then be symmetric with respect to the reflection about the root $\gamma$. The reflections about all simple roots in $\Phi_{B,\mathbb{R},j}^+$ generate the full Weyl group of the root system $\Phi_{B,\mathbb{R},j}$. Hence, if $\Theta$ vanishes on $B_\gamma$, for every simple root $\gamma\in\Phi_{B,\mathbb{R},j}^+$, then $\Delta_B\Theta|_{B_j''}$ will be symmetric with respect to this entire Weyl group. The compact factor of every such $B_\gamma$ has dimension one greater than the compact factor of $B$. We conclude:

(7.32) $\Delta_B\Theta|_{B_j''}$ is symmetric with respect to the Weyl group of $\Phi_{B,\mathbb{R},j}$, provided $\dim B_+$ is maximal, among all Cartan subgroups on which $\Theta$ does not vanish identically.

Our arguments depend only on this one consequence of the matching conditions.

Let us suppose now that $\Theta$ decays at infinity. The exponents $w\mu$ which appear with non-zero coefficient $p_{j,w}$ in (7.22) then satisfy $\langle w\mu, X\rangle < 0$, for every $X \neq 0$ in the closure of $C_j$. If $\Delta_B \Theta|_{B_j''}$ were symmetric with respect to the Weyl group of $\Phi_{B,\mathbb{R},j}$, each $w\mu$ would have to assume strictly negative values also on the various translates of the Weyl chamber $C_j$, i.e. on all of $\mathfrak{b}_-$. This is impossible unless $\Theta|_{B_j''}=0$, or $\mathfrak{b}_-=0$, in which case $B$ is compact. The assertion of Proposition (6.11) follows.

We argue similarly to prove Lemma (6.15). Let us suppose that $\Theta_1$, $\Theta_2$ are two distinct invariant eigendistributions, which both have the properties mentioned in the lemma. As was described in § 4, the explicit formula for $\Theta_i|_H$ makes it possible to identify the infinitesimal character of $\Theta_i$, namely

(7.33) $Z \mapsto \gamma_H(Z)(\lambda), \quad Z \in \mathfrak{Z};$

$\gamma_H: \mathfrak{Z} \xrightarrow{\sim} I(\mathfrak{h}^{\mathbb{C}})$ is the isomorphism (4.8) corresponding to the Cartan subalgebra $\mathfrak{h}^{\mathbb{C}}$, and $\gamma_H(Z)$ is viewed as a polynomial function on $\mathfrak{h}^{*\mathbb{C}}$. Since $\Theta_1$ and $\Theta_2$ have the same infinitesimal character, their difference $\Theta = \Theta_1 - \Theta_2$ is again an invariant eigendistribution. We now choose a Cartan subgroup $B$ and a connected component $B_j''$ of $B''$, such that $\Theta$ does not vanish identically on $B_j''$, but does vanish on any Cartan subgroup whose compact factor has a larger dimension than $B_+$. Then $B$ cannot be compact: any two compact Cartan subgroups are conjugate, and $\Theta|_H \equiv 0$. Because of our assumptions, $\Theta$ remains bounded at infinity. Hence, when the restriction of $\Delta_B \Theta$ to $B_j''$ is expressed as in (7.22), every exponent $w\mu$ which occurs with a non-trivial coefficient $p_{j,w}$ satisfies $\operatorname{Re} w\mu \leqq 0$, on $C_j$. The assertion (7.32) allows us to conclude

(7.34) $p_{j,w} \neq 0 \Rightarrow \operatorname{Re} w\mu \equiv 0 \quad \text{on } \mathfrak{b}_-.$

To complete the proof of (6.15), we must derive a contradiction.

The reasoning which led to the description (7.33) of the infinitesimal character of $\Theta$ also gives the alternate description

$$Z \mapsto \gamma_B(Z)(\mu), \quad Z \in \mathfrak{Z}$$

in terms of the isomorphism $\gamma_B: \mathfrak{Z} \xrightarrow{\sim} I(\mathfrak{b}^{\mathbb{C}})$. Because of the canonical nature of the Harish-Chandra isomorphism (4.8), $\gamma_B$ and $\gamma_H$ are related: $\gamma_H = c \cdot \gamma_B$, whenever $c: \mathfrak{b}^{\mathbb{C}} \xrightarrow{\sim} \mathfrak{h}^{\mathbb{C}}$ is induced by an inner automorphism of $\mathfrak{g}^{\mathbb{C}}$. Hence $c^* \lambda$ and $\mu$ lie in the same Weyl group orbit, i.e.

(7.35) $\mu = u\, c^* \lambda, \quad \text{for some } u \in W_{B,\mathbb{C}}.$

The elements of the weight lattice $\Lambda$ of $H$ can be expressed as $\mathbb{Q}$-linear combinations of roots $\alpha \in \Phi$. Since every root $\alpha \in \Phi_B$ assumes real values on the split part $\mathfrak{b}_-$ of $\mathfrak{b}$, and since $c^* \Phi = \Phi_B$, the Weyl group translates of $\mu$ must be real on $\mathfrak{b}_-$. Thus (7.34) can be sharpened:

(7.36) $p_{j,w} \neq 0 \Rightarrow w\mu \equiv 0 \quad \text{on } \mathfrak{b}_-.$

The real roots $\alpha \in \Phi_{B,\mathbb{R}}$ vanish on $\mathfrak{b}_+$, and the decomposition $\mathfrak{b} = \mathfrak{b}_+ \oplus \mathfrak{b}_-$ is orthogonal with respect to the Killing form. Thus (7.36) implies

$$(w\mu, \alpha) = 0, \qquad \text{for } \alpha \in \Phi_{B,\mathbb{R}},$$

provided $p_{j,w} \neq 0$. This is the case with at least one $w \in W_{B,\mathbb{C}}$, because $\Theta|_{B_j'} \neq 0$. A Cartan subgroup has an empty system of real roots precisely when it is fundamental, i.e. when its compact part is of maximal possible dimension. In our situation, $G$ contains a compact Cartan subgroup, so that $B$ cannot be fundamental. We conclude that $\mu$ is singular. In view of (7.35), this contradicts our assumption on $\lambda$.

## § 8. Complete Description of the Discrete Series

As was shown in § 6, $G$ has a non-empty discrete series if and only if it contains a compact Cartan subgroup. We assume that this is the case, and we select a particular compact Cartan subgroup $H \subset G$. We shall follow the notation of § 4. In particular, $\Lambda_\rho \subset i\mathfrak{h}^*$ is the lattice formed by the differentials of characters of $H$, translated by the half-sum of the positive roots; cf. (4.25). We recall the notion of an invariant eigendistribution which is bounded at infinity, as defined in § 6. With these ingredients, we can now state and prove Harish-Chandra's fundamental result on the discrete series [15]:

(8.1) **Theorem.** Corresponding to any nonsingular $\lambda \in \Lambda_\rho$, there exists exactly one invariant eigendistribution $\Theta_\lambda$, which is bounded at infinity, and such that

$$\Theta_\lambda|_{H \cap G'} = (-1)^q \frac{\sum_{w \in W} \varepsilon(w)\, e^{w\lambda}}{\prod_{\substack{\alpha \in \Phi, \\ (\alpha, \lambda) > 0}} (e^{\alpha/2} - e^{-\alpha/2})}$$

($q = \frac{1}{2} \dim G/K$). Every such $\Theta_\lambda$ arises as the character of a discrete series representation, whose formal degree equals

$$d(\lambda - \rho) = \prod_{\substack{\alpha \in \Phi, \\ (\alpha, \lambda) > 0}} \frac{(\alpha, \lambda)}{(\alpha, \rho)}, \qquad \text{with} \quad \rho = \tfrac{1}{2} \sum_{(\alpha, \lambda) > 0} \alpha.$$

Conversely every discrete series character occurs among the $\Theta_\lambda$.

Because of the uniqueness which is asserted in the theorem, two of the $\Theta_\lambda$ coincide precisely when they have the same restriction to $H$, and this is the case whenever their parameters are related by the action of $W$, the Weyl group of $H$ in $G$. In particular, then, the theorem provides a one-to-one parametrization of the discrete series, in terms of the quotient

(8.2) $\{\lambda \in \Lambda_\rho \mid \lambda \text{ is nonsingular}\}/W$.

Our proof of the theorem also leads to information about the decomposition of discrete series representations under the maximal compact subgroup $K$. We consider a particular discrete series representation $\pi_\lambda$, with character $\Theta_\lambda$. The parameter $\lambda$ determines a system of positive roots $\Psi$ in $\Phi$, namely

(8.3) $\Psi=\{\alpha\in\Phi \mid (\lambda,\alpha)>0\}$.

As before, we set

(8.4) $\rho_c=\frac{1}{2}\sum_{\Phi^c\cap\Psi}\alpha, \quad \rho_n=\frac{1}{2}\sum_{\Phi^n\cap\Psi}\alpha.$

The next statement is a weaker, but useful version of Blattner's conjecture.

(8.5) **Theorem.** In the restriction of $\pi_\lambda$ to $K$, the irreducible $K$-module of highest weight $\lambda+\rho_n-\rho_c$ occurs exactly once. Any irreducible constituent of $\pi_\lambda|_K$ has a highest weight of the form $\lambda+\rho_n-\rho_c+A$, where $A$ stands for a sum of roots in $\Psi$.

Turning to the proof of the two theorems, we note that the uniqueness of the $\Theta_\lambda$ was established in § 6. We now fix a nonsingular $\lambda\in\Lambda_\rho$, and we define $\Psi, \rho_c, \rho_n$ in terms of $\lambda$, as above. As was argued below (4.34),

(8.6) $\mu=\lambda-\rho_c$

is the highest weight of an irreducible $\mathfrak{k}^{\mathbb{C}}$-module $V_\mu$, and the action of $\mathfrak{k}^{\mathbb{C}}$ on the tensor products $V_\mu\otimes S^\pm$ lifts to $K$. Thus we can apply Theorem (5.20): if

(8.7) $\min_{\alpha\in\Phi}|(\alpha,\lambda)|>c,$

for some appropriately chosen constant $c$, there does exist a discrete series representation $\pi_\lambda$, which has the properties described in (8.5), and whose character $\Theta_\lambda$ satisfies the conditions of (8.1). In the case of a general $\lambda$, Theorem (4.41) provides at least a partial answer. We shall use a method of Zuckerman [27] to deduce the assertions of (8.1) and (8.5). Since our arguments depend only on a specialized version of Zuckerman's technique, we shall develop his ideas to the extent that they are needed here.

Although we are concerned with unitary representations, it is necessary at this point to work in the wider context of representations on Banach spaces. We briefly recall the important properties of such representations; details can be found in [26], for example. A representation $\pi$ of $G$ on a Banach space is said to be admissible if each class $i\in\hat{K}$ occurs with finite multiplicity $n_i$ in $\pi|_K$. Every irreducible unitary representation has this property; it is unknown whether irreducibility implies admissibility in general. Each admissible representation $\pi$ gives rise to an infinitesimal representation of $\mathfrak{U}(\mathfrak{g}^{\mathbb{C}})$, on the space of $K$-finite vectors. The sub-representations of $\pi$ correspond precisely to the invariant subspaces for the infinitesimal representations. If the infinitesimal representations attached to global representations are isomorphic, one calls the global representations infinitesimally equivalent. Informally, one may think of infinitesimally equivalent representations as being identical, except for a modification of the topology on the underlying vector spaces. Among unitary representations, the notions of infinitesimal equivalence and unitary equivalence coincide. Whenever $\pi$ is both admissible and irreducible, the $K$-multiplicities $n_i$ satisfy the bound $n_i \leqq$ degree of $i$, just as in the unitary case; moreover, the center $\mathfrak{Z}$ of $\mathfrak{U}(\mathfrak{g}^{\mathbb{C}})$ then acts by scalars. Conversely, if $\pi$ is admissible, and if $\mathfrak{Z}$ operates on the space of $K$-finite vectors according to a character, then $\pi$ has at least a finite composition series.

Admissible, irreducible representations have global characters, for essentially the same reasons as in the unitary case. More generally, the definition of the character of an admissible representation $\pi$ makes sense, provided only $\pi$ has a finite composition series. One can describe the character explicitly as the sum, in the sense of distributions, of the diagonal matrix coefficients, relative to a "basis" consisting of vectors in the various $K$-invariant, $K$-irreducible subspaces. Characters cannot distinguish between infinitesimally equivalent representations: two admissible, irreducible representations have the same character precisely when they are infinitesimally equivalent. In fact, the characters corresponding to any finite set of infinitesimally distinct, irreducible representations are linearly independent.

We now consider a particular admissible, irreducible representation $\pi$, and a finite-dimensional representation $\tau$, whose characters we denote by $\Theta_\pi$ and $\chi_\tau$. As is not hard to check, the tensor product will then be admissible, too. Furthermore, it is known that

(8.8) $\pi \otimes \tau$ has a finite composition series.

One can see this, for example, as follows. Like any admissible, irreducible representation, $\pi$ is infinitesimally equivalent to a sub-representation of some principal series representation[1], which need not be unitary, of course. Hence, in verifying (8.8), we may as well assume that $\pi$ itself is a principal series representation, instead of being irreducible. Thus $\pi$ is induced, from a minimal parabolic subgroup $P \subset G$, by a finite-dimensional, irreducible representation $\sigma$ of $P$, i.e. $\pi = \mathrm{ind}_P^G\, \sigma$. For essentially formal reasons,

$$\pi \otimes \tau \cong \mathrm{ind}_P^G(\sigma \otimes \tau).$$

Any composition series of the finite-dimensional representation $\sigma \otimes \tau$ of $P$ determines a filtration of $\mathrm{ind}_P^G(\sigma \otimes \tau)$, of finite length, whose quotients are again principal series representations. A principal series representation, finally, does have a finite composition series, as follows from the fact that it has an infinitesimal character, and that it is admissible.

As the character of the tensor product $\pi \otimes \tau$, $\chi_\tau \Theta_\pi$ equals the sum of the characters of the composition factors. For each character $\varphi$ of the algebra $\mathfrak{Z}$, we let $(\chi_\tau \Theta_\pi)_\varphi$ denote the sum of the characters of those composition factors on which $\mathfrak{Z}$ acts according to $\varphi$. Then

(8.9) $$\chi_\tau \Theta_\pi = \sum_\varphi (\chi_\tau \Theta_\pi)_\varphi,$$

with $\varphi$ ranging over a finite set of characters of $\mathfrak{Z}$.

---

1 Casselman [9] has given a simple proof of this fact

For the purpose of stating the next lemma, we fix a system of positive roots $\Psi \subset \Phi$, and some $\lambda \in \Lambda_\rho$, such that $\lambda$ is dominant with respect to $\Psi$, but not necessarily nonsingular. Although $G$ itself need not be linear, it is at least a finite covering of a linear group. Hence arbitrarily large positive multiples of $\lambda$ occur as highest weights of irreducible, finite-dimensional representations of $G$. We suppose that $\tau$ is such an irreducible, finite-dimensional representation, of highest weight $(m-1)\lambda$, $m \geqq 1$; $\chi_\tau$ denotes the character of $\tau$, and $\chi_\tau^*$ the character of the representation dual to $\tau$. We shall use the symbols $\mathscr{C}_\lambda$ and $\mathscr{C}_{m\lambda}$ to refer to the sets of characters of irreducible, admissible representations, with infinitesimal character $\varphi_\lambda$ and $\varphi_{m\lambda}$, respectively; cf. (4.17).

(8.10) **Lemma** (Zuckerman [27]): The mapping

$$S\colon \Theta \mapsto (\chi_\tau \Theta)_{\varphi_{m\lambda}}$$

establishes a bijection between the sets $\mathscr{C}_\lambda$ and $\mathscr{C}_{m\lambda}$, whose inverse is given by

$$T\colon \Theta \mapsto (\chi_\tau^* \Theta)_{\varphi_\lambda}.$$

If $\Theta \in \mathscr{C}_\lambda$ decays at infinity, then so does $S\Theta$, and conversely.

*Proof.* In the obvious manner, $S$ and $T$ extend to mappings between the linear spans of $\mathscr{C}_\lambda$ and $\mathscr{C}_{m\lambda}$, in the appropriate spaces of invariant eigendistributions. By virtue of their definition, both $S$ and $T$ associate to each irreducible character an integral linear combination of irreducible characters, with non-negative coefficients. We shall show that

$$(8.11)\quad T\Theta \neq 0, \quad \text{if } \Theta \in \mathscr{C}_{m\lambda}$$

and

$$(8.12)\quad T\cdot S = \text{identity}, \quad \text{on } \mathscr{C}_\lambda,$$

as well as the corresponding statements with reversed roles for $S$ and $T$. Thus, if $S\Theta$ were a sum of more than one irreducible character, $\Theta = T\cdot S\Theta$ would also have to be a sum of several irreducible characters, contradicting the irreducibility of $\Theta$. This will prove the first half of the lemma. The second assertion will follow from an explicit description of $S$.

In order to understand $S$ and $T$, we consider a particular $\Theta \in \mathscr{C}_\lambda$, restricted to a Cartan subgroup $B$. On any connected component $B_j''$ of $B''$, as in (7.20), $\Theta$ can be expressed by a formula like (7.22):

$$(8.13)\quad (\Delta_B \Theta)(b_j \exp(X+Y)) = \sum_{w \in W_{B,\mathbb{C}}} p_{j,w}(X)\, e^{\langle w\mu, X+Y\rangle}, \quad \text{for } X \in C_j,\ Y \in \mathfrak{b}_+.$$

Since $\Theta$ is a character, the coefficients $p_{j,w}$ are known to be constants, not polynomials [11], but this fact turns out to be irrelevant for our purposes. The Cartan subalgebras $\mathfrak{b}^{\mathbb{C}}$ and $\mathfrak{h}^{\mathbb{C}}$ are related by an inner automorphism of $\mathfrak{g}^{\mathbb{C}}$, say

$$(8.14)\quad c\colon \mathfrak{b}^{\mathbb{C}} \xrightarrow{\sim} \mathfrak{h}^{\mathbb{C}}.$$

The arguments which precede (7.35) also apply in the present situation. Thus, modifying $c$ by an element of $W_{\mathbb{C}}$, if necessary, one can arrange

$$(8.15) \quad \mu = c^* \lambda.$$

Every weight $\nu$ of $\tau$, relative to the Cartan subalgebra $\mathfrak{b}^{\mathbb{C}}$, lifts to a character $e^{\nu}$ of $B$. We set $n_{\nu}$ = multiplicity of the weight $\nu$; then

$$\chi_{\tau}|_B = \sum_{\nu} n_{\nu} e^{\nu}.$$

The Weyl group $W_{B,\mathbb{C}}$ of $(\mathfrak{g}^{\mathbb{C}}, \mathfrak{b}^{\mathbb{C}})$ leaves $\chi_{\tau}$ invariant, so that

$$(\Delta_B \chi_{\tau} \Theta)(b_j \exp(X+Y)) = \sum_{\nu} \sum_{w \in W_{B,\mathbb{C}}} n_{\nu} e^{w\nu}(b_j) \, p_{j,w}(X) \, e^{\langle w(\mu+\nu), X+Y \rangle},$$

if $X \in C_j$, $Y \in \mathfrak{b}_+$. In this formula, the contribution of $S\Theta$ consists of those terms which belong to the infinitesimal character $\varphi_{m\lambda}$, i.e. the terms for which $\mu+\nu$ is $W_{B,\mathbb{C}}$-conjugate to $c^*(m\lambda) = m\mu$.

We claim: $\mu+\nu$ lies in the $W_{B,\mathbb{C}}$-orbit of $m\mu$ only if $\nu = (m-1)\mu$, in which case $n_{\mu} = 1$. Indeed, $(m-1)\mu$ occurs as an extreme weight of the finite-dimensional, irreducible representation $\tau$, and hence has multiplicity one. The weights of $\tau$ all lie in the convex body spanned by the extreme weights, i.e. by the $W_{B,\mathbb{C}}$-translates of $(m-1)\mu$. When this convex body is shifted by $\mu$, among the new vertices, $m\mu$ lies farthest from the origin. The claim follows, and we deduce:

$$(8.16) \quad (\Delta_B(S\Theta))(b_j \exp(X+Y)) = \sum_{w \in W_{B,\mathbb{C}}} e^{(m-1)w\mu}(b_j) \, p_{j,w}(X) \, e^{m\langle w\mu, X+Y \rangle},$$

whenever $X \in C_j$, $Y \in \mathfrak{b}_+$. Similarly, if the restriction of some $\Theta \in \mathscr{C}_{m\lambda}$ to $B_j''$ is given by

$$(\Delta_B \Theta)(b_j \exp(X+Y)) = \sum_{w \in W_{B,\mathbb{C}}} q_{j,w}(X) \, e^{m\langle w\mu, X+Y \rangle},$$

then

$$(8.17) \quad (\Delta_B(T\Theta))(b_j \exp(X+Y)) = \sum_{w \in W_{B,\mathbb{C}}} e^{-(m-1)w\mu}(b_j) \, q_{j,w}(X) \, e^{\langle w\mu, X+Y \rangle},$$

again for all $X \in C_j$, $Y \in \mathfrak{b}_+$; the verification is entirely analogous to that of (8.16).

Every invariant eigendistribution is completely determined by its restrictions to the various Cartan subgroups. Hence the explicit formulas (8.16–8.17) imply the two assertions (8.11–8.12). We recall that an invariant eigendistribution $\Theta$ decays at infinity if, and only if, in terms of the notation of (8.13),

$$(8.18) \quad p_{j,w} \neq 0 \Rightarrow w\mu \text{ assumes strictly negative values on the closure of } C_j,$$

except of course at the origin, for all choices of $B$ and $B_j''$. The condition (8.18) remains unchanged by an application of $S$ or $T$, which has the effect of multiplying the exponents by a positive constant. This completes the proof of the lemma.

When the arguments leading up to (8.17) are carried out for the compact Cartan subgroup $H$, one finds:

(8.19) **Corollary.** If $\Theta\in\mathscr{C}_{m\lambda}$ satisfies

$$\Delta_H\Theta|_H=\sum_{w\in W_{\mathbb{C}}} a_w\, e^{mw\lambda},$$

then

$$\Delta_H(T\Theta)|_H=\sum_{w\in W_{\mathbb{C}}} a_w\, e^{w\lambda}.$$

We return to the proof of the main theorem. Thus a nonsingular $\lambda\in\Lambda_\rho$ is given, and $\Psi$ is the system of positive roots which makes $\lambda$ dominant. We enumerate as

(8.20) $\lambda_1=\lambda, \lambda_2, \ldots, \lambda_N$

those $W_{\mathbb{C}}$-conjugates of $\lambda$ which lie in $\Lambda_\rho$ and are dominant with respect to $\Phi^c\cap\Psi$; this is consistent with the notation of Theorem (4.41). The correspondences $S$ and $T$ of (8.10) depend on the choice of the integer $m$. We make $m$ so large that the multiples $m\lambda_i$ of the various $\lambda_i$ become sufficiently nonsingular, in the sense of (8.7). By assumption, $(m-1)\lambda$ occurs as weight of a finite-dimensional representation of $G$, hence lies in $\Lambda$, as do its $W_{\mathbb{C}}$-conjugates. Thus $\Lambda_\rho$ contains not only the various $\lambda_i$, but also their multiples $m\lambda_i$. Theorem (8.1) has already been established for every nonsingular parameters. In particular, there exist discrete series characters $\Theta_{m\lambda_i}$, such that

$$\Theta_{m\lambda_i}|_{H\cap G'}=(-1)^q\frac{\sum_{w\in W}\varepsilon(w)\, e^{mw\lambda_i}}{\prod_{\substack{\alpha\in\Phi,\\(\alpha,\lambda_i)>0}}(e^{\alpha/2}-e^{-\alpha/2})}.$$

We define

(8.21) $\Theta_{\lambda_i}=T\Theta_{m\lambda_i}$.

Then, because of (8.10) and (8.19), $\Theta_{\lambda_i}$ is the character of an irreducible representation,

$$(8.22)\quad \Theta_{\lambda_i}|_{H\cap G'}=(-1)^q\frac{\sum_{w\in W}\varepsilon(w)\, e^{w\lambda_i}}{\prod_{\substack{\alpha\in\Phi,\\(\alpha,\lambda_i)>0}}(e^{\alpha/2}-e^{-\alpha/2})},$$

and $\Theta_{\lambda_i}$ decays at infinity.

The invariant eigendistribuition $\tilde{\Theta}_\lambda$ of (4.26) was defined as the sum of the discrete series characters which correspond to the infinitesimal character $\varphi_\lambda$, each multiplied by its formal degree. As a linear combination of discrete series characters, $\tilde{\Theta}_\lambda$ decays at infinity. If two invariant eigendistribution agree on a compact Cartan subgroup, and if they both decay at infinity, than they coincide; this follows from (6.11) (any two compact Cartan subgroups are conjugate!). Thus, comparing (4.41) and (8.22), we find

$$\text{(8.23)}\quad \tilde{\Theta}_\lambda = \left(\prod_{\alpha\in\Psi} \frac{(\alpha,\lambda)}{(\alpha,\rho)}\right) \sum_{i=1}^{N} \Theta_{\lambda_i}.$$

An invariant eigendistribution can be expressed as a linear combination of irreducible characters in only one way, if at all. Also, the coefficient appearing in (8.23) is non-zero. We conclude: the $\Theta_{\lambda_i}$, $1 \leqq i \leqq N$, are precisely the discrete series characters corresponding to the infinitesimal character $\varphi_\lambda$; each has formal degree

$$d(\lambda-\rho) = \prod_{\alpha\in\Psi} \frac{(\alpha,\lambda)}{(\alpha,\rho)}.$$

The infinitesimal character of any discrete series representation equals $\varphi_\lambda$, for some nonsingular $\varphi \in \Lambda_\rho$, as was shown in § 6. The proof of Theorem (8.1) is therefore complete.

For very nonsingular parameters $\lambda \in \Lambda_\rho$, the assertion of Theorem (8.5) is part of Theorem (5.20). If $\lambda \in \Lambda_\rho$ is nonsingular, but otherwise arbitrary, we choose the integer $m$ as in the previous argument, so that

$$\Theta_\lambda = T\Theta_{m\lambda} = (\chi_\tau^* \Theta_{m\lambda})_{\varphi_\lambda}.$$

Thus, if $\pi_\lambda$ and $\pi_{m\lambda}$ are discrete series representations with global character $\Theta_\lambda$ and $\Theta_{m\lambda}$, respectively, then $\pi_\lambda$ can be realized as a composition factor of $\pi_{m\lambda} \otimes \tau^*$, up to infinitesimal equivalence. Here $\tau^*$ stands for the representation dual to $\tau$, i.e. the finite-dimensional, irreducible representation which has $-(m-1)\lambda$ as lowest weight. In particular, all irreducible constituents of $\pi_\lambda|_K$ are among those of $(\pi_{m\lambda} \otimes \tau^*)|_K$. The highest weight of any irreducible constituent of the tensor product can be expressed as the sum of the highest weight of a constituent of $\pi_{m\lambda}|_K$ with some weight of $\tau^*$. We already know that every highest weight in $\pi_{m\lambda}|_K$ is of the form $m\lambda + \rho_n - \rho_c + A$, with $A$ equal to a sum of positive roots. Any weight of $\tau^*$ differs from the lowest weight by a sum of positive roots. All this implies the second assertion of (8.5).

Since $\pi_{m\lambda}|_K$ contains the irreducible $K$-module of highest weight $m\lambda + \rho_n - \rho_c$ only once, and since the lowest weight $-(m-1)\lambda$ of $\tau^*$ has multiplicity one, the irreducible $K$-module of highest weight $\lambda + \rho_n - \rho_c$ cannot occur more than once in $\pi_\lambda|_K$. We must check that it does occur. The results in the beginning of § 4, especially (4.16c), coupled with the explicit formula for $\Theta_\lambda|_{H\cap G'}$, show that

(8.24) $\lambda - \rho_c$ is the highest weight of an irreducible constituent of $(\pi_\lambda|_K) \otimes (s^+ \oplus s^-)$

($s^+, s^-$ refer to the action of $K$ on the half spin modules $S^+, S^-$). Every weight of $S^+ \oplus S^-$ can be written as $-\rho_n + \beta_1 + \cdots + \beta_s$, with $\beta_i \in \Phi^n \cap \Psi$, and every highest weight in $\pi_\lambda|_K$ as $\lambda + \rho_n - \rho_c + A$, $A$ being a sum of positive roots. Hence (8.24) is possible only if $\pi_\lambda|_K$ contains the irreducible $K$-module of highest weight $\lambda+\rho_n-\rho_c$. This we have proved Theorem (8.6).

## § 9. Realization of the Discrete Series

Although the spaces $\mathscr{H}_\mu^+, \mathscr{H}_\mu^-$ of square-integrable, harmonic spinors play a crucial role in our construction, we have not yet described them completely, except for very nonsingular parameters $\mu$. It is known that each $\mathscr{H}_\mu^\pm$ either vanishes, or is an irreducible unitary $G$-module, which belongs to the discrete series; moreover, every discrete series representation can be realized in this manner [22, 24]. For the sake of completeness, we now recall the precise statement and discuss its proof.

As in the past, $\mu$ shall denote the highest weight of an irreducible $\mathfrak{k}^{\mathbb{C}}$-module $V_\mu$, and $\Psi$ a system of positive roots, such that

(9.1) $(\mu+\rho_c, \alpha) \geqq 0, \quad \text{for } \alpha\in\Psi.$

We also assume

(9.2) $\mu+\rho_n \in \Lambda,$

which insures that the twisted spin bundles $\mathscr{V}_\mu \otimes \mathscr{S}^\pm$ can be defined on $G/K$.

(9.3) **Theorem.** Both $\mathscr{H}_\mu^+$ and $\mathscr{H}_\mu^-$ vanish whenever $\mu+\rho_c$ is singular. Otherwise, for nonsingular $\mu+\rho_c$, only $\mathscr{H}_\mu^-$ vanishes, whereas $G$ acts irreducibly on $\mathscr{H}_\mu^+$, according to the discrete series representation with character $\Theta_{\mu+\rho_c}$.

If $\lambda\in\Lambda_\rho$ is nonsingular, and if $\Psi$ is the positive root system which makes $\lambda$ dominant, then $\mu=\lambda-\rho_c$ has the required properties. In particular, the theorem provides a concrete realization for every discrete series representation.

Turning to the proof of the theorem, we recall the Plancherel decomposition (5.6) of $\mathscr{H}_\mu^\pm$. The discrete series contributes to it discretely; any $j\in\hat{G}_d$ occurs as often as the dimension of $V_j^\pm$, if $j$ lies in $\hat{G}_\mu$, and does not occur at all for $j\notin\hat{G}_\mu$. The explicit enumeration of the discrete series representations, combined with the information about their $K$-decompositions in Theorem (8.5), makes it possible to determine these multiplicities:

(9.4) **Lemma.** Suppose $j$ is a class in the discrete series, which belongs to $\hat{G}_\mu$. Then $V_j^+ + V_j^- = 0$, except in the following situation: $\mu+\rho_c$ is nonsingular, and $j$ has character $\Theta_{\mu+\rho_c}$, in which case $\dim V_j^+ = 1$, $V_j^- = 0$.

**Proof.** We consider a particular class $j\in\hat{G}_d\cap\hat{G}_\mu$, with character $\Theta_\lambda$. For the time being, $\Psi$ shall denote the system of positive roots

$$\{\alpha\in\Phi \mid (\alpha,\lambda)>0\},$$

which may be inconsistent with (9.1). However, replacing $\lambda$ by one of its $W$-translates, if necessary, we can arrange that

(9.5) $(\mu+\rho_c, \alpha) > 0, \quad \text{for } \alpha\in\Phi^c\cap\Psi.$

As was remarked in § 5, $V_j^{\pm}$ can be non-zero only if $\mathbf{H}_j$ contains a $K$-invariant, $K$-irreducible subspace with a highest weight of the form $\mu+\rho_n-B$; here $B$ stands for a sum of distinct positive, noncompact roots and, in the case of $V_j^-$, $B \neq 0$. This statement, incidentally, depends only on the property (9.5) of $\Psi$, except for the labelling of $V_j^{\pm}$, which is determined by the stronger condition (9.1). According to (8.5), every such highest weight can be expressed as $\lambda+\rho_n-\rho_c+A$, with $A$ equal to a sum of positive roots. Thus $V_j^{\pm}=0$, unless

$$\mu+\rho_c=\lambda+A+B.$$

The preceding equality implies

$$(9.6)\quad (\mu+\rho_c, \mu+\rho_c) \geqq (\lambda, \lambda)+2(\lambda, A+B);$$

because of our choice of $\Psi$, $(\lambda, A+B)$ is strictly positive, except for $A=B=0$.

Since $j$ lies in $\hat{G}_\mu$, and since $\varphi_\lambda$ is its infinitesimal character, we find

$$(\mu+\rho_c, \mu+\rho_c)-(\rho, \rho)=c_\mu=\varphi_\lambda(\Omega)=(\lambda, \lambda)-(\rho, \rho);$$

cf. (4.22). Equivalently,

$$(\mu+\rho_c, \mu+\rho_c)=(\lambda, \lambda),$$

which is compatible with (9.6) only if $A=B=0$, $\lambda=\mu+\rho_c$. Thus we may as well assume that $\mu+\rho_c$ is nonsingular and equal to $\lambda$. In this situation, the positive root system $\Psi$ does have the property (9.1), and hence gives the correct labelling of $V_j^+$, $V_j^-$. If $V_j^-$ were non-zero, the inequality (9.6) would have to hold with $B \neq 0$, which is impossible. Finally,

$$\dim V_j^+ = \dim V_j^+ - \dim V_j^- = 1,$$

as follows from (4.16c).

In order to complete the proof of Theorem (9.3), we must show that the complement of the discrete series does not contribute to the Plancherel decomposition (5.6) – or equivalently, that the set

$$(9.7)\quad \{j \in \hat{G}_\mu \mid j \notin \hat{G}_d,\ V_j^+ \oplus V_j^- \neq 0\}$$

has zero Plancherel measure. At present, only one argument is known which proves this last statement in full generality. It depends on Harish-Chandra's work on the explicit form of Plancherel measure, as we shall now explain.

To each conjugacy class of Cartan subgroups, Harish-Chandra attaches a series of unitary representations, which are induced from a parabolic subgroup; the inducing representations, restricted to the semisimple part of a Levi component, belong to the discrete series. At one extreme, the discrete series corresponds to the conjugacy class of compact Cartan subgroups (here the inducing process becomes trivial: $G$ is viewed as a parabolic subgroup of itself), at the other extreme lies the unitary principal series. The union of these so-called "nondegenerate series" supports the Plancherel measure. Roughly speaking, the series attached to a Cartan subgroup $B$ is parametrized by the dual group $\hat{B}$, modulo the action of the Weyl group. In terms of this parametrization, the

Plancherel measure is completely continuous with respect to Haar measure on $\hat{B}$. The action of the Casimir operator can be described, in terms of the same parametrization, as the Killing form plus a constant, transferred to a function on $\hat{B}$ via the exponential map. The set of points where this function assumes a given value has zero Haar measure, except if $\hat{B}$ is discrete, in which case $\hat{B}$ parametrizes the discrete series. The preceding results, which go well beyond the bounds of this paper, directly imply the following statement:

(9.8) **Lemma** (Harish-Chandra [16]). The set of classes $j \in \hat{G}$, outside of the discrete series, at which the Casimir operator takes any given value, has measure zero.

In particular, the set (9.7) has zero Plancherel measure, and this completes the proof of Theorem (9.3). The proof is not quite satisfactory, of course, since it uses almost the full strength of Harish-Chandra's explicit Plancherel formula. We shall now mention two alternate approaches to the problem of realizing the discrete series geometrically, which are independent of Lemma (9.8).

Instead of working with the Dirac operator, we could have equally well carried out our construction in the framework of $L^2$-cohomology, as originally suggested by Langlands [20]. This has the disadvantage of making some of the arguments technically more difficult. On the other hand, a comparatively elementary result of Casselman and Osborne [10] then takes the place of Lemma (9.8). Because of it, all classes $j \in \hat{G}$ which enter the Plancherel decomposition of one of the $L^2$-cohomology groups have the same infinitesimal character, and consequently only the discrete series can contribute. For details the reader is referred to [25].

The second alternative applies only in the case of a linear group $G$, and it can deal only with nonsingular values of $\mu+\rho_c$ (which however is enough to realize all discrete series representations). In this situation, because $\mu+\rho_c$ is a dominant and nonsingular weight, $\mu-\rho_n=(\mu+\rho_c)-\rho$ is at least dominant. Thus $\mu-\rho_n$ occurs as the highest weight of an irreducible $\mathfrak{k}^{\mathbb{C}}$-module $V_{\mu-\rho_n}$. Since $\rho_n$ is the highest weight of an irreducible constituent of $S^+$, there exists an injective $K$-homomorphism

$$(9.9) \quad V_\mu \hookrightarrow V_{\mu-\rho_n} \otimes (S^+ \oplus S^-).$$

The direct sum of the two half-spin modules may be viewed as a square-root of the exterior algebra of the standard representation. In our context, this means that

$$(S^+ \oplus S^-) \otimes (S^+ \oplus S^-) \cong \wedge \mathfrak{p}^{\mathbb{C}}$$

($\mathfrak{p}$ = orthogonal complement of $\mathfrak{k}$ in $\mathfrak{g}$). Tensoring both sides of (9.9) with $(S^+ \oplus S^-)$, one finds

$$(9.10) \quad V_\mu \otimes (S^+ \oplus S^-) \hookrightarrow V_{\mu-\rho_n} \otimes \wedge \mathfrak{p}^{\mathbb{C}}.$$

We recall the isomorphism

$$(9.11) \quad V_j^\pm \cong \mathrm{Hom}_K(\mathbf{H}_j, V_\mu \otimes S^\pm).$$

Because of (9.10–9.11), the set (9.7) becomes a subset of

(9.12) $\{j \in \hat{G}_\mu \mid j \notin \hat{G}_d,\ \mathrm{Hom}_K(\mathbf{H}_j, V_{\mu-\rho_n} \otimes \wedge \mathfrak{p}^{\mathbb{C}}) \neq 0\}$.

It will be enough to prove that this set has zero Plancherel measure, or more specifically, that it is finite.

If $M$ is a $(\mathfrak{g}^{\mathbb{C}}, K)$-module (i.e. a $\mathfrak{g}^{\mathbb{C}}$-module such that the action of $\mathfrak{k}^{\mathbb{C}}$ lifts to the group $K$), the relative Lie algebra cohomology groups $H^*(\mathfrak{g}^{\mathbb{C}}, \mathfrak{k}^{\mathbb{C}}; M)$ can be computed in terms of the standard complex $C^*(\mathfrak{g}^{\mathbb{C}}, \mathfrak{k}^{\mathbb{C}}; M)$, with

(9.13) $C^p(\mathfrak{g}^{\mathbb{C}}, \mathfrak{k}^{\mathbb{C}}; M) = \mathrm{Hom}_K(\wedge^p \mathfrak{p}^{\mathbb{C}}, M)$.

We apply this remark to the tensor product $M = \mathbf{H}_\infty \otimes F$, where $\mathbf{H}_\infty$ is the space of $K$-finite vectors in an irreducible, unitary $G$-module $\mathbf{H}$, and $F$ an irreducible, finite-dimensional $G$-module. The cochain groups (9.13) are then finite-dimensional and carry a natural inner product. With respect to this inner product, the formal Laplacian turns out to be the difference of the constants by which the Casimir operator $\Omega$ acts on $\mathbf{H}_\infty$ and $F$ [7]. In particular,

(9.14) $H^p(\mathfrak{g}^{\mathbb{C}}, \mathfrak{k}^{\mathbb{C}}; \mathbf{H}_\infty \otimes F) \cong \mathrm{Hom}_K(\wedge^p \mathfrak{p}^{\mathbb{C}}, \mathbf{H} \otimes F)$, provided $\Omega$ acts on $\mathbf{H}_\infty$ and $F$ by the same constant.

We let $F_{\mu-\rho_n}$ denote the irreducible, finite-dimensional $G$-module of highest weight $\mu - \rho_n$, and $F^*_{\mu-\rho_n}$ its contragredient. Then $F_{\mu-\rho_n}$ contains $V_{\mu-\rho_n}$ as a $K$-submodule, and $\Omega$ operates on both $F_{\mu-\rho_n}$ and $F^*_{\mu-\rho_n}$ as multiplication by the constant $c_\mu$ of (3.18). Hence, for any $j \in \hat{G}_\mu$,

(9.15) $\mathrm{Hom}_K(\mathbf{H}_j, V_{\mu-\rho_n} \otimes \wedge^p \mathfrak{p}^{\mathbb{C}}) \neq 0 \Rightarrow H^p(\mathfrak{g}^{\mathbb{C}}, \mathfrak{k}^{\mathbb{C}}; \mathbf{H}_{j,\infty} \otimes F^*_{\mu-\rho_n}) \neq 0$,

as follows from (9.14).

The category of $(\mathfrak{g}^{\mathbb{C}}, K)$-modules contains enough projectives, so that one can define the derived functors $\mathrm{Ext}^*_{\mathfrak{g}^{\mathbb{C}}}$ of the functor

$M \mapsto \mathrm{Hom}_{\mathfrak{g}^{\mathbb{C}}}(M, N)$.

Arguing purely formally, one obtains isomorphisms

(9.16) $\mathrm{Ext}^p_{\mathfrak{g}^{\mathbb{C}}}(M, N) \cong H^p(\mathfrak{g}^{\mathbb{C}}, \mathfrak{k}^{\mathbb{C}}; \mathrm{Hom}_{\mathbb{C}}(M, N))$,

since both sides agree for $p = 0$. The Ext groups classify exact sequences of $(\mathfrak{g}^{\mathbb{C}}, K)$-modules, beginning with $N$ and ending with $M$, modulo a certain equivalence relation (Yoneda equivalence). As was pointed out by D. Wigner, this implies in particular: if $M, N$ are $(\mathfrak{g}^{\mathbb{C}}, K)$-modules with infinitesimal characters $\chi_M$ and $\chi_N$, then

(9.17) $\mathrm{Ext}^p_{\mathfrak{g}^{\mathbb{C}}}(M, N) = 0 \quad$ whenever $\chi_M \neq \chi_N$.

Combining (9.15–9.17), we see that the elements of the set (9.12) all have the same infinitesimal character as the $G$-module $F_{\mu-\rho_n}$. Consequently the set must be finite.

To our knowledge, the preceding argument was originally put together by Zuckerman, although others may have been aware of it independently. The details which we left out can be found in the first two sections of Borel-Wallach [7].

## Appendix

In order to keep this paper reasonably self-contained, we shall sketch the proofs of some auxiliary results which were used in our construction, and which have already appeared elsewhere, namely:

a) Parthasarathy's formula (3.17) for the square of the spinor Laplacian [22],
b) the characterization (4.5) of the singularities of the $K$-character [24],
c) the bound (5.5) for the action of the Casimir operator [24], and
d) Proposition (5.14).

The first three of these are closely related and have fairly simple proofs. Proposition (5.14) is a special case of the main result of [24]; in the situation which we consider, its verification can be simplified quite a bit.

We begin with the proof of (3.17). The Lie algebra of the maximal compact subgroup $K$ has a unique $\mathrm{Ad}K$-invariant complement $\mathfrak{p}$ in $\mathfrak{g}$. Since the $K$-modules $S^+, S^-$ were introduced as the half-spin representations of the orthogonal group of $\mathfrak{p}$, there exist distinguished $K$-homomorphisms

$$\text{(A.1)} \quad \mathfrak{p}^{\mathbb{C}} \otimes S^+ \to S^-, \quad \mathfrak{p}^{\mathbb{C}} \otimes S^- \to S^+$$

which we write as

$$X \otimes s \mapsto c(X)\, s;$$

$c(X) \in \mathrm{Hom}(S^{\pm}, S^{\mp})$ is "Clifford multiplication" by $X$. The mappings (2.1) are induced by an action of the Clifford algebra on $S^+ \oplus S^-$, which implies

$$\text{(A.2)} \quad c(X)^2 = -B(X, X), \quad \text{for } X \in \mathfrak{p}^{\mathbb{C}}$$

($B$ = Killing form of $\mathfrak{g}^{\mathbb{C}}$). As irreducible modules for the orthogonal group of $\mathfrak{p}$ – or more precisely, for the corresponding spin group – $S^+$ and $S^-$ carry essentially unique Hermitian metrics. When these metrics are suitably normalized,

$$\text{(A.3)} \quad -c(\bar{X}) \text{ is the adjoint of } c(X);$$

here $X$ denotes the complex conjugate of $X \in \mathfrak{p}^{\mathbb{C}}$, relative to the real form $\mathfrak{p}$. Since $B$ is positive definite on $\mathfrak{p}$, we can choose an orthonormal basis $\{X_i\}$. The action $s^{\pm}(Z)$ of any $Z \in \mathfrak{k}^{\mathbb{C}}$ on $S^{\pm}$ is then given by

$$\text{(A.4)} \quad s^{\pm}(X) = -\tfrac{1}{4} \sum_i c([Z, X_i])\, c(X_i).$$

In fact, the analogous identity holds for every $Z$ in the Lie algebra of $SO(\mathfrak{p})$ [2].

The twisted spin bundles $\mathscr{V}_\mu \otimes \mathscr{S}^{\pm}$ are associated to the principal bundle

$$K \to G \to G/K$$

by the action of $K$ on $V_\mu \otimes S^{\pm}$. Hence there exist natural $G$-isomorphisms

$$\text{(A.5)} \quad C^\infty(G/K, \mathscr{V}_\mu \otimes \mathscr{S}^{\pm}) \cong (C^\infty(G) \otimes V_\mu \otimes S^{\pm})_K;$$

$(\ldots)_K$ refers to the subspace of $K$-invariants, with $K$ acting on $C^\infty(G)$ by right translation, and on $V_\mu \otimes S^\pm$ in the obvious manner.

We again let $\{X_i\}$ be an orthonormal basis of $\mathfrak{p}$. Each $X_i$ determines a left-invariant vector field $r(X_i)$, by infinitesimal right translation. In terms of the isomorphisms (A.5), the Dirac operators on $\mathscr{V}_\mu \otimes \mathscr{S}^\pm$ can be expressed as

$$\text{(A.6)}\quad \sum_i r(X_i) \otimes 1 \otimes c(X_i)\colon (C^\infty(G) \otimes V_\mu \otimes S^\pm)_K \to (C^\infty(G) \otimes V_\mu \otimes S^\mp)_K.$$

Indeed, one can check that these operators preserve the $K$-invariants in $C^\infty(G) \otimes V_\mu \otimes S^\pm$, and that they commute with the action of $G$. Consequently they define $G$-invariant first order operators between $\mathscr{V}_\mu \otimes \mathscr{S}^+$ and $\mathscr{V}_\mu \otimes \mathscr{S}^-$. When the fibres of $\mathscr{V}_\mu \otimes \mathscr{S}^\pm$ at the identity coset are identified with $V_\mu \otimes S^\pm$, and the cotangent space with $\mathfrak{p}^{\mathbb{C}*} \cong \mathfrak{p}^{\mathbb{C}}$ (this latter isomorphism comes from the Killing form), the symbols of the operators (A.6) are given by the linear maps

$$\mathfrak{p}^{\mathbb{C}} \otimes V_\mu \otimes S^\pm \to V_\mu \otimes S^\mp, \qquad X \otimes v \otimes s \mapsto v \otimes c(X)\, s,$$

which are also the symbols of the twisted Dirac operators. A translation-invariant first order operator between $\mathscr{V}_\mu \otimes \mathscr{S}^+$ and $\mathscr{V}_\mu \otimes \mathscr{S}^-$ is completely determined by its symbol: since $S^+$ and $S^-$ have no weights in common [1],

$$\text{(A.7)}\quad \operatorname{Hom}_K(V_\mu \otimes S^+, V_\mu \otimes S^-) = 0,$$

which implies that there exist no non-trivial, translation-invariant bundle maps between $\mathscr{V}_\mu \otimes \mathscr{S}^+$ and $\mathscr{V}_\mu \otimes \mathscr{S}^-$.

The spinor Laplacian on $\mathscr{V}_\mu \otimes \mathscr{S}^\pm$ is the composition of the two Dirac operators on $\mathscr{V}_\mu \otimes \mathscr{S}^\pm$, which are adjoint to each other; it therefore equals

$$\begin{aligned}
&\Big(\sum_i r(X_i) \otimes 1 \otimes c(X_i)\Big)^2 \\
&\quad = \sum_{i,j} r(X_j)\, r(X_i) \otimes 1 \otimes c(X_j)\, c(X_i) \\
&\quad = \tfrac{1}{2} \sum_{i,j} \{ r(X_j)\, r(X_i) \otimes 1 \otimes c(X_j)\, c(X_i) + r(X_i)\, r(X_j) \otimes 1 \otimes c(X_i)\, c(X_j) \} \\
&\quad = \tfrac{1}{2} \sum_{i,j} r(X_j)\, r(X_i) \otimes 1 \otimes (c(X_j)\, c(X_i) + c(X_i)\, c(X_j)) \\
&\qquad + \tfrac{1}{2} \sum_{i,j} r([X_i, X_j]) \otimes 1 \otimes c(X_i)\, c(X_j) \\
&\quad = -\sum_i r(X_i)^2 \otimes 1 \otimes 1 + \tfrac{1}{2} \sum_{i,j} r([X_i, X_j]) \otimes 1 \otimes c(X_i)\, c(X_j).
\end{aligned}$$

In the last step, we have used the identity

$$c(X_i)\, c(X_j) + c(X_j)\, c(X_i) = -2\delta_{ij},$$

[1] According to the discussion above (4.13), if this were not the case, there would exist an odd number of noncompact roots which add up to zero. When a noncompact root, positive or negative, is expressed as an integral linear combination of simple roots, the sum of the coefficients of the noncompact simple roots is an odd integer. An odd number of noncompact roots can therefore never add up to zero

which follows from (A.2). We now choose an orthonormal basis $\{Z_l\}$ for $\mathfrak{k}^{\mathbb{C}}$. The Casimir operators $\Omega$ of $\mathfrak{g}^{\mathbb{C}}$ and $\Omega_K$ of $\mathfrak{k}^{\mathbb{C}}$ can then be expressed as

$$\Omega=\sum_i X_i^2+\sum_l Z_l^2, \quad \Omega_K=\sum_l Z_l^2.$$

Hence

(A.8) $$\Big(\sum_i r(X_i)\otimes 1\otimes c(X_i)\Big)^2$$
$$=-r(\Omega)\otimes 1\otimes 1+r(\Omega_K)\otimes 1\otimes 1+\tfrac{1}{2}\sum_{i,j} r([X_i,X_j])\otimes 1\otimes c(X_i)\,c(X_j);$$

as in the past, $r$ refers to the right infinitesimal action of $\mathfrak{U}(\mathfrak{g}^{\mathbb{C}})$ on $C^\infty(G)$.

Since $\mathfrak{g}=\mathfrak{k}\oplus\mathfrak{p}$ is a Cartan decomposition, $[\mathfrak{p},\mathfrak{p}]$ lies in $\mathfrak{k}$, and $[\mathfrak{k},\mathfrak{p}]$ in $\mathfrak{p}$. Thus we find:

$$\tfrac{1}{2}\sum_{i,j} r([X_i,X_j])\otimes 1\otimes c(X_i)\,c(X_j)$$
$$=\tfrac{1}{2}\sum_{i,j,l} B([X_i,X_j],Z_l)\,r(Z_l)\otimes 1\otimes c(X_i)\,c(X_j)$$
($\{Z_l\}$ is an orthonormal basis of $\mathfrak{k}^{\mathbb{C}}$)
$$=-\tfrac{1}{2}\sum_{i,j,l} B([Z_l,X_j],X_i)\,r(Z_l)\otimes 1\otimes c(X_i)\,c(X_j)$$
($B$ is $\operatorname{Ad}G$-invariant)
$$=-\tfrac{1}{2}\sum_{j,l} r(Z_l)\otimes 1\otimes c([Z_l,X_j])\,c(X_j)$$
($\{X_i\}$ is an orthonormal basis of $\mathfrak{p}^{\mathbb{C}}$)
$$=2\sum_l r(Z_l)\otimes 1\otimes s^{\pm}(Z_l)$$
(because of (2.4)).

The various operators act on the $K$-invariants in $C^\infty(G)\otimes V_\mu\otimes S^{\pm}$. Hence, if $\tau_\mu$ denotes the representation of $\mathfrak{k}^{\mathbb{C}}$ on $V_\mu$,

$$\sum_l r(Z_l)\otimes 1\otimes s^{\pm}(Z_l)=-\sum_l 1\otimes\tau_\mu(Z_l)\otimes s^{\pm}(Z_l)-1\otimes 1\otimes s^{\pm}(\Omega_K).$$

Similarly,

$$r(\Omega_K)\otimes 1\otimes 1=1\otimes(\tau_\mu\otimes s^{\pm})(\Omega_K)$$
$$=1\otimes\tau_\mu(\Omega_K)\otimes 1+1\otimes 1\otimes s^{\pm}(\Omega_K)+2\sum_l 1\otimes\tau_\mu(Z_l)\otimes s^{\pm}(Z_l).$$

All this allows us to rewrite the identity (A.8) as follows:

(A.9) $$\Big(\sum_i r(X_i)\otimes 1\otimes c(X_i)\Big)^2$$
$$=-r(\Omega)\otimes 1\otimes 1+1\otimes\tau_\mu(\Omega_K)\otimes 1-1\otimes 1\otimes s^{\pm}(\Omega_K).$$

On the irreducible $\mathfrak{k}^{\mathbb{C}}$-module with highest weight $\mu$, $\Omega_K$ acts as multiplication by

(A.10) $\tau_\mu(\Omega_K)=(\mu+\rho_c,\mu+\rho_c)-(\rho_c,\rho_c)$.

In order to complete the proof of (3.17), we must identify $s^\pm(\Omega_K)$.

We recall the character formula (4.13). Applying Weyl's denominator formula for $G$, one finds

(A.11)
$$(\operatorname{trace} s^+ - \operatorname{trace} s^-)|_H = \prod_{\beta\in\Phi^n\cap\Psi}(e^{\beta/2}-e^{-\beta/2})$$
$$=\Big(\prod_{\alpha\in\Phi^c\cap\Psi}(e^{\alpha/2}-e^{-\alpha/2})\Big)^{-1}\prod_{\alpha\in\Psi}(e^{\alpha/2}-e^{-\alpha/2})$$
$$=\Big(\prod_{\alpha\in\Phi^c\cap\Psi}(e^{\alpha/2}-e^{-\alpha/2})\Big)^{-1}\sum_{w\in W}\varepsilon(w)\,e^{w\rho}.$$

The $K$-modules $S^+,S^-$ have no irreducible constituent in common; this is a special case of (A.7), with $\mu=0$. Since there can be no cancellation in (A.11), every irreducible summand of $S^+\oplus S^-$ must have a highest weight of the form $w\rho-\rho_c$, for some $w\in W_{\mathbb{C}}$. The Casimir operator $\Omega_K$ therefore acts as multiplication by

$$s^\pm(\Omega_K)=(w\rho,w\rho)-(\rho_c,\rho_c)=(\rho,\rho)-(\rho_c,\rho_c).$$

Combining this with (A.9) and (A.10), we obtain Parthasarathy's formula (3.17).

For the proof of (4.5) and (5.5), we consider an irreducible unitary representation $\pi$ of $G$, on a Hilbert space $\mathbf{H}$. The various $K$-invariant, $K$-irreducible subspaces span a dense linear subspace $\mathbf{H}_\infty\subset\mathbf{H}$, which consists entirely of analytic vectors. Thus $\mathfrak{g}^{\mathbb{C}}$, and hence $\mathfrak{U}(\mathfrak{g}^{\mathbb{C}})$, act on $\mathbf{H}_\infty$ by differentiation. We shall let $\pi$ denote also this infinitesimal action. It is irreducible, and determines $\pi$ up to unitary equivalence. Again we choose an orthonormal basis $\{X_i\}$ of $\mathfrak{p}^{\mathbb{C}}$. The two $K$-invariant linear mappings

(A.12)
$$d_\pm:\ \mathbf{H}_\infty\otimes S^\pm\to\mathbf{H}_\infty\otimes S^\mp,$$
$$d_\pm:\ v\otimes s\mapsto\sum_i\pi(X_i)\,v\otimes c(X_i)\,s,$$

become adjoint to each other, when $\mathbf{H}_\infty$ is viewed as a pre-Hilbert space. This follows from (A.3) and is formally analogous to the self-adjointness of the Dirac operators. Parthasarathy's formula has a direct analogue in the present context:

(A.13)
$$d_-d_+=d_+d_-=(\pi\otimes s^\pm)(\Omega_K)-\pi(\Omega)\otimes 1-(\rho,\rho)\,1+(\rho_c,\rho_c)\,1;$$

its proof is virtually identical to that of (3.17).

As a function on the dual $\hat{K}$ of $K$, $\Omega_K$ tends to $+\infty$ outside of finite subsets. Since each irreducible $K$-module occurs only finitely often in $\mathbf{H}_\infty$, and hence in $\mathbf{H}_\infty\otimes S^\pm$, the identity (A.13) forces $d_+$ to have finite-dimensional kernel and cokernel. For purely formal reasons

$$\mathbf{H}_\infty\otimes S^+-\mathbf{H}_\infty\otimes S^-=\ker d_+-\operatorname{coker} d_+,$$

as virtual $K$-modules. Thus, if $\tau$ denotes the $K$-character of $\pi$, and $\sigma^\pm$ the character of $S^\pm$,

$$\tau(\sigma^+ - \sigma^-) = \text{char}(\mathbf{H}_\infty \otimes S^+ - \mathbf{H}_\infty \otimes S^-) = \text{char}(\ker d_+ - \text{coker}\, d_+).$$

In particular, $\tau(\sigma^+ - \sigma^-)$ is a finite integral linear combination of characters of irreducible $K$-modules, as asserted by (4.5). We should remark that the preceding argument does not really use the unitary structure. Thus (4.5) holds more generally for irreducible, admissible representations on Banach spaces.

If $V$ and $W$ are irreducible $K$-modules, with $V$ having highest weight $\nu$ and $W$ lowest weight $\eta$, then $\nu + \eta$ occurs as the highest weight of an irreducible constituent of $V \otimes W$, provided

$$(\nu + \eta, \alpha) \geqq 0, \quad \text{for } \alpha \in \Phi^c \cap \Psi;$$

this follows, for example, from Weyl's character formula. With respect to any ordering compatible with $\Psi$, $-\rho_n$ is lowest among weights of $S^+ \oplus S^-$, and hence is the lowest weight of an irreducible summand of $S^+ \oplus S^-$. The conditions (5.4) therefore guarantee that $\mathbf{H}_\infty \otimes (S^+ \oplus S^-)$ contains the irreducible $K$-module of highest weight $\nu - \rho_n$ at least once. On this $K$-module, $\Omega_K$ operates as multiplication by

$$(\nu - \rho_n + \rho_c, \nu - \rho_n + \rho_c) - (\rho_c, \rho_c).$$

If one applies the positive semi-definite operator (A.13) to the submodule in question, one is led to the estimate (5.5).

Let us turn to the proof of Proposition (5.14)! We shall show that the two conditions a) and b) determine the infinitesimal representation of $\mathfrak{U}(\mathfrak{g}^{\mathbb{C}})$ on $\mathbf{H}_\infty$. The unitary structure will again be irrelevant, and hence (5.14) applies more generally to irreducible, admissible representations on Banach spaces: up to infinitesimal equivalence, there exists at most one such representation which has the two properties a) and b).

The irreducible $K$-module of highest weight $\mu + \rho_n$, $V_{\mu+\rho_n}$, is a left $\mathfrak{U}(\mathfrak{k}^{\mathbb{C}})$-module, and $\mathfrak{U}(\mathfrak{g}^{\mathbb{C}})$ is a right $\mathfrak{U}(\mathfrak{k}^{\mathbb{C}})$-module, via right multiplication. Thus one can form the tensor product

$$\text{(A.14)} \quad M = \mathfrak{U}(\mathfrak{g}^{\mathbb{C}}) \otimes_{\mathfrak{U}(\mathfrak{k}^{\mathbb{C}})} V_{\mu+\rho_n},$$

on which $\mathfrak{U}(\mathfrak{g}^{\mathbb{C}})$ operates by left multiplication. The $\mathfrak{U}(\mathfrak{g}^{\mathbb{C}})$-module $M$ has the universal mapping property

$$\text{(A.15)} \quad \text{Hom}_{\mathfrak{U}(\mathfrak{g}^{\mathbb{C}})}(M, L) \cong \text{Hom}_{\mathfrak{U}(\mathfrak{k}^{\mathbb{C}})}(V_{\mu+\rho_n}, L),$$

for any $\mathfrak{U}(\mathfrak{g}^{\mathbb{C}})$-module $L$. From $\mathfrak{U}(\mathfrak{g}^{\mathbb{C}})$, $M$ inherits a filtration

$$0 \subset M_0 \subset M_1 \subset \cdots \subset M_n \subset M_{n+1} \subset \cdots \subset M,$$

with

$$M_0 = 1 \otimes V_{\mu+\rho_n}, \qquad M_{n+1} = M_n + \mathfrak{g}^{\mathbb{C}} M_n;$$

it is $\mathfrak{k}^{\mathbb{C}}$-stable, because the adjoint action of $\mathfrak{k}^{\mathbb{C}}$ on $\mathfrak{U}(\mathfrak{g}^{\mathbb{C}})$ preserves the filtration of the latter. As a consequence of the Birkhoff-Witt theorem, there exist natural $\mathfrak{k}^{\mathbb{C}}$-isomorphisms

(A.16) $M_n/M_{n-1} \cong \mathfrak{p}^{\mathbb{C}(n)} \otimes V_{\mu+\rho_n}$

($\mathfrak{p}^{\mathbb{C}(n)} = n$-th symmetric power of $\mathfrak{p}^{\mathbb{C}}$). In particular,

(A.17) $M_1 \cong (\mathfrak{p}^{\mathbb{C}} \otimes V_{\mu+\rho_n}) \oplus V_{\mu+\rho_n}$;

this splitting is canonical: the irreducible $\mathfrak{k}^{\mathbb{C}}$-module $V_{\mu+\rho_n}$ does not embed into $\mathfrak{p}^{\mathbb{C}} \otimes V_{\mu+\rho_n}$.

Every irreducible component of $\mathfrak{p}^{\mathbb{C}} \otimes V_{\mu+\rho_n}$ has a highest weight of the form $\mu+\rho_n+\beta$, $\beta \in \Phi^n$. Lumping together these components for which $\beta$ is, respectively, positive or negative, one obtains submodules $U_+$ and $U_-$, which decompose $\mathfrak{p}^{\mathbb{C}} \otimes V_{\mu+\rho_n}$:

(A.18) $\mathfrak{p}^{\mathbb{C}} \otimes V_{\mu+\rho_n} = U_+ \oplus U_-$.

The isomorphism (A.17) provides an inclusion $U_- \hookrightarrow M_1$. We set

$N = \mathfrak{U}(\mathfrak{g}^{\mathbb{C}})$-submodule of $M$ generated by $U_-$,

$Q = M/N$.

The $\mathfrak{k}^{\mathbb{C}}$-invariant filtration of $M$ induces a filtration $\{Q_n\}$ of the quotient $Q$.

We now suppose that the representation $\pi$ has the two properties a) and b) mentioned in Proposition (5.14). The former, in conjunction with the mapping property (A.15), guarantees the existence of a non-zero $\mathfrak{U}(\mathfrak{g}^{\mathbb{C}})$-homomorphism $M \to \mathbf{H}_\infty$. Because of the latter, any such homomorphism must annihilate $N$. Thus one can produce a non-trivial $\mathfrak{U}(\mathfrak{g}^{\mathbb{C}})$-map $Q \to \mathbf{H}_\infty$, which is necessarily surjective, since $\mathbf{H}_\infty$ is known to be irreducible. Consequently,

(A.19) $\mathbf{H}_\infty$ is isomorphic, as $\mathfrak{U}(\mathfrak{g}^{\mathbb{C}})$-module, to an irreducible quotient of $Q$,

and this reduces the proposition to:

(A.20) **Lemma.** The $\mathfrak{U}(\mathfrak{g}^{\mathbb{C}})$-module $Q$ has a unique irreducible quotient. Under the action of $\mathfrak{k}^{\mathbb{C}}$, $Q$ breaks up into a direct sum of irreducible $\mathfrak{k}^{\mathbb{C}}$-modules, each occuring with finite multiplicity. Every irreducible constituent has a highest weight which can be expressed as $\mu+\rho_n+\beta_1+\cdots+\beta_m$, with $\beta_1,\ldots,\beta_m \in \Phi^n \cap \Psi$ and $m \geqq 0$; the highest weight $\mu+\rho_n$ appears at most once.

To begin with, we shall argue that the last two assertions imply the first. Indeed, since $M_0$ generates the $\mathfrak{U}(\mathfrak{g}^{\mathbb{C}})$-module $M$, $Q_0$ must generate $Q$. As a quotient of $M_0 \cong V_{\mu+\rho_n}$, $Q_0$ either vanishes, in which case $Q=0$, or must itself be isomorphic to $V_{\mu+\rho_n}$, and hence irreducible under the action of $\mathfrak{k}^{\mathbb{C}}$. In particular, no proper submodule of $Q$ can meet $Q_0$. Since the irreducible $\mathfrak{k}^{\mathbb{C}}$-module of highest weight $\mu+\rho_n$ is known to occur only once in $Q$, it cannot lie in any proper submodule, nor in the linear span of any number of proper submodules. Hence $Q$ has a unique maximal proper submodule, or equivalently, a unique irreducible quotient.

The adjoint action of $\mathfrak{k}^{\mathbb{C}}$ on $\mathfrak{U}(\mathfrak{g}^{\mathbb{C}})$ lifts to the group $K$, which operates by conjugation. Since $K$ operates on $V_{\mu+\rho_n}$ as well, the action of $\mathfrak{k}^{\mathbb{C}}$ on $M$, and hence on $Q$, also lifts to $K$. The finite-dimensional $\mathfrak{k}^{\mathbb{C}}$-submodules $Q_n$ exhaust $Q$. It follows that $Q$ breaks up, under $\mathfrak{k}^{\mathbb{C}}$, into a direct sum of irreducibles, although conceivably with infinite multiplicities. In particular,

(A.21) $Q \cong \bigoplus_{n=0}^{\infty} Q_n/Q_{n-1}$, as $\mathfrak{k}^{\mathbb{C}}$-module.

The positive roots lie in a closed cone, which is properly contained in a half-space. Each weight can therefore be expressed as a sum of positive roots in at most finitely many ways, and a non-empty sum of positive roots can never equal the zero weight. Hence the lemma becomes a consequence of the following statement,

(A.22) each irreducible constituent of the $\mathfrak{k}^{\mathbb{C}}$-module $Q_n/Q_{n-1}$ has a highest weight of the form

$$\mu+\rho_n+\beta_1+\cdots+\beta_n, \quad \text{with} \quad \beta_1, \ldots, \beta_n \in \Phi^n \cap \Psi,$$

which will be verified next.

The inclusion of $U_-$ in $\mathfrak{p}^{\mathbb{C}} \otimes V_{\mu+\rho_n}$, tensored with the identity on $\mathfrak{p}^{\mathbb{C}(n-1)}$, and followed by multiplication, determines a $\mathfrak{k}^{\mathbb{C}}$-homomorphism

(A.23) $h$: $\mathfrak{p}^{\mathbb{C}(n-1)} \otimes U_- \to \mathfrak{p}^{\mathbb{C}(n)} \otimes V_{\mu+\rho_n}$.

Under the isomorphism (A.16), the image of $h$ will certainly go into $N \cap M_n/N \cap M_{n-1}$, i.e., into the kernel of the projection $M_n/M_{n-1} \to Q_n/Q_{n-1}$. Thus:

(A.24) $Q_n/Q_{n-1}$ is isomorphic, as $\mathfrak{k}^{\mathbb{C}}$-module, to a quotient of the cokernel of $h$.

The homomorphism $h$ is induced, in a certain sense, from a homomorphism between modules of a Borel subalgebra of $\mathfrak{k}^{\mathbb{C}}$. To describe the induction process, we use a sequence of functors, which were introduced in [24]. We shall briefly summarize their definition and main properties.

The root spaces in $\mathfrak{g}^{\mathbb{C}}$ corresponding to all negative, compact roots span a maximal nilpotent subalgebra $\mathfrak{n} \subset \mathfrak{k}^{\mathbb{C}}$, which is normalized by $\mathfrak{h}^{\mathbb{C}}$. Hence

$$\mathfrak{b} = \mathfrak{h}^{\mathbb{C}} \oplus \mathfrak{n}$$

becomes a Borel subalgebra of $\mathfrak{k}^{\mathbb{C}}$. For any $\mathfrak{b}$-module $E$, the cohomology groups $H^p(\mathfrak{n}, E)$ have natural $\mathfrak{h}^{\mathbb{C}}$-module structures, since $\mathfrak{h}^{\mathbb{C}}$ acts on both $E$ and $\mathfrak{n}$. The subspace of $\mathfrak{h}^{\mathbb{C}}$-invariants will be denoted by $H^p(\mathfrak{n}, E)_{\mathfrak{h}^{\mathbb{C}}}$. In the following, we only consider finite-dimensional $\mathfrak{b}$-modules $E$, such that the action of $\mathfrak{h}^{\mathbb{C}}$ on $E$ lifts to the torus $H$. In this situation, $E$ and its $\mathfrak{n}$-cohomology groups become completely reducible, as $\mathfrak{h}^{\mathbb{C}}$-modules. In particular, any short exact sequence of such $\mathfrak{b}$-modules

$$0 \to E' \to E \to E'' \to 0$$

gives rise to a long exact sequence

$$\to H^p(\mathfrak{n}, E')_{\mathfrak{h}^{\mathbb{C}}} \to H^p(\mathfrak{n}, E)_{\mathfrak{h}^{\mathbb{C}}} \to H^p(\mathfrak{n}, E'')_{\mathfrak{h}^{\mathbb{C}}} \to H^{p+1}(\mathfrak{n}, E')_{\mathfrak{h}^{\mathbb{C}}} \to .$$

For each $i \in \hat{K}$, we select a $\mathfrak{k}^{\mathbb{C}}$-module $W_i$ which represents the isomorphism class $i$, and we define

$$\text{(A.25)} \quad I^p(E) = \bigoplus_{i \in \hat{K}} W_i \otimes H^p(\mathfrak{n}, W_i^* \otimes E)_{\mathfrak{h}^{\mathbb{C}}};$$

here $W_i^*$, the $\mathfrak{k}^{\mathbb{C}}$-module dual to $W_i$, is regarded as $\mathfrak{b}$-module by restriction. With $\mathfrak{k}^{\mathbb{C}}$ acting trivially on the right factors $H^p(\mathfrak{n}, W_i^* \otimes E)_{\mathfrak{h}^{\mathbb{C}}}$, $I^p(E)$ becomes a completely reducible $\mathfrak{k}^{\mathbb{C}}$-module. The definition (A.25) is functorial in $E$. Hence:

(A.26) $I^p$, $0 \leqq p \leqq \dim \mathfrak{n}$, is a sequence of functors from the category of finite-dimensional $\mathfrak{b}$-modules, for which the action of $\mathfrak{h}^{\mathbb{C}}$ lifts to $H$, to the category of completely reducible $\mathfrak{k}^{\mathbb{C}}$-modules.

Moreover,

(A.27) every short exact sequence $0 \to E' \to E \to E'' \to 0$ determines a long exact sequence

$$0 \to I^0(E') \to I^0(E) \to E^0(E'') \to I^1(E') \to \cdots;$$

this follows from the analogous property of the functors $E \to H^p(\mathfrak{n}, E)_{\mathfrak{h}^{\mathbb{C}}}$. Loosely speaking, $I^0(E)$ is obtained by inducing $E$ holomorphically from the complex Lie group with Lie algebra $\mathfrak{b}$ to the complexification of $K$. This process is left exact, and the functors $I^1, I^2, \ldots$ measure the obstruction to its exactness on the right.

Three properties of the functors $I^p$ will be crucial. The first of these is directly implied by Kostant's Lie algebra version of the Borel-Weil-Bott theorem [19]. For any weight $\mu \in \Lambda$, $L_\mu$ shall denote the one-dimensional $\mathfrak{b}$-module on which $\mathfrak{h}^{\mathbb{C}}$ acts according to the linear functional $\mu$. Then

(A.28) $I^0(L_\mu)$ is irreducible, with highest weight $\mu$, provided $\mu$ is dominant with respect to $\Phi^c \cap \Psi$; in all remaining cases, $I^0(L_\mu) = 0$; if $(\mu + \rho_c, \alpha) \geqq 0$ for every $\alpha \in \Phi^c \cap \Psi$, $I^p(L_\mu)$ vanishes for $p > 0$.

Since $\mathfrak{b}$ is solvable, every finite-dimensional $\mathfrak{b}$-module $E$ has a composition series

$$\text{(A.29a)} \quad 0 \subset E_0 \subset E_1 \subset \cdots \subset E_m = E,$$

with one-dimensional quotients. In this situation, there exists an injective $\mathfrak{k}^{\mathbb{C}}$-homomorphism

$$\text{(A.29b)} \quad I^p(E) \hookrightarrow \bigoplus_{l=0}^{m} I^p(E_l / E_{l-1}),$$

which need not be functorial, however. Indeed, for each $l$,

$$I^p(E_{l-1}) \to I^p(E_l) \to I^p(E_l / E_{l-1})$$

is an exact sequence of completely reducible $\mathfrak{k}^{\mathbb{C}}$-modules, so that the assertion can be verified inductively. Finally,

(A.30) for each finite-dimensional $K$-module $V$, there exist $\mathfrak{k}^{\mathbb{C}}$-isomorphisms $I^p(V \otimes E) \cong V \otimes I^p(E)$, which are functorial in both $V$ and $E$.

To see this, one should observe that

$$W_i^* \otimes V \otimes E \cong \bigoplus_{j \in \hat{K}} (W_i^* \otimes V \otimes W_j)_K \otimes (W_j^* \otimes E),$$

where $(\ldots)_K$ denotes the subspace of $K$-invariants, equipped with the trivial $\mathfrak{b}$-action. Hence

$$\begin{aligned} I^p(V \otimes E) &\cong \bigoplus_{i,\, j \in \hat{K}} W_i \otimes (W_i^* \otimes V \otimes W_j)_K \otimes H^p(\mathfrak{n}, W_j^* \otimes E)_{\mathfrak{h}^{\mathbb{C}}} \\ &\cong \bigoplus_{i \in \hat{K}} W_i \otimes (W_i^* \otimes V \otimes I^p(E))_K \cong V \otimes I^p(E), \end{aligned}$$

which proves (A.30).

As a particular consequence of (A.28) and (A.30),

$$\text{(A.31)} \quad I^0(L_{\mu+\rho_n}) \cong V_{\mu+\rho_n}, \qquad I^0(\mathfrak{p}^{\mathbb{C}} \otimes L_{\mu+\rho_n}) \cong \mathfrak{p}^{\mathbb{C}} \otimes V_{\mu+\rho_n};$$

these isomorphisms will be regarded as identities. The root spaces indexed by the various negative, noncompact roots span a $\mathfrak{b}$-submodule $\mathfrak{p}_-$ of $\mathfrak{p}^{\mathbb{C}}$. In any composition series of $\mathfrak{p}_-$, precisely the $\mathfrak{b}$-modules $L_{-\beta}$, $\beta \in \Phi^n \cap \Psi$, occur as the one-dimensional quotients; similarly, the $\mathfrak{b}$-modules $L_\beta$, $\beta \in \Phi^n \cap \Psi$, decompose $\mathfrak{p}^{\mathbb{C}}/\mathfrak{p}_-$. Appealing to (A.28) and (A.29), one finds that every irreducible summand of $I^0(\mathfrak{p}_- \otimes L_{\mu+\rho_n})$ has a highest weight of the form $\mu+\rho_n-\beta$, with $\beta$ positive and noncompact. None of these highest weights occur in the $\mathfrak{k}^{\mathbb{C}}$-module $U_+$ of (A.18), so that

$$\operatorname{Hom}_{\mathfrak{k}^{\mathbb{C}}}(I^0(\mathfrak{p}_- \otimes L_{\mu+\rho_n}), U_+) = 0.$$

For completely analogous reasons

$$\operatorname{Hom}_{\mathfrak{k}^{\mathbb{C}}}(U_-, I^0(\mathfrak{p}^{\mathbb{C}}/\mathfrak{p}_- \otimes L_{\mu+\rho_n})) = 0.$$

Since

$$0 \to I^0(\mathfrak{p}_- \otimes L_{\mu+\rho_n}) \to I^0(\mathfrak{p}^{\mathbb{C}} \otimes L_{\mu+\rho_n}) \to I^0(\mathfrak{p}^{\mathbb{C}}/\mathfrak{p}_- \otimes L_{\mu+\rho_n})$$

is exact, the subspace $U_- \subset \mathfrak{p}^{\mathbb{C}} \otimes V_{\mu+\rho_n} = I^0(\mathfrak{p}^{\mathbb{C}} \otimes L_{\mu+\rho_n})$ must coincide with the image of $I^0(\mathfrak{p}_- \otimes L_{\mu+\rho_n})$. We conclude: under the identifications

$$\begin{aligned} &\mathfrak{p}^{\mathbb{C}(n-1)} \otimes U_- \cong \mathfrak{p}^{\mathbb{C}(n-1)} \otimes I^0(\mathfrak{p}_- \otimes L_{\mu+\rho_n}) \cong I^0(\mathfrak{p}^{\mathbb{C}(n-1)} \otimes \mathfrak{p}_- \otimes L_{\mu+\rho_n}), \\ &\mathfrak{p}^{\mathbb{C}(n)} \otimes V_{\mu+\rho_n} \cong I^0(\mathfrak{p}^{\mathbb{C}(n)} \otimes L_{\mu+\rho_n}), \end{aligned}$$

the $\mathfrak{k}^{\mathbb{C}}$-homomorphism $h$ of (A.23) corresponds to the mapping

$$\text{(A.32)} \quad I^0(\mathfrak{p}^{\mathbb{C}(n-1)} \otimes \mathfrak{p}_- \otimes L_{\mu+\rho_n}) \to I^0(\mathfrak{p}^{\mathbb{C}(n)} \otimes L_{\mu+\rho_n}),$$

which arises from the inclusion $\mathfrak{p}_- \hookrightarrow \mathfrak{p}^{\mathbb{C}}$, followed by multiplication $\mathfrak{p}^{\mathbb{C}(n-1)} \otimes \mathfrak{p}^{\mathbb{C}} \to \mathfrak{p}^{\mathbb{C}(n)}$. We must identify the cokernel of this homomorphism.

The decomposition $\mathfrak{p}^{\mathbb{C}} = \mathfrak{p}_- \oplus \bar{\mathfrak{p}}_-$ ($\bar{\mathfrak{p}}_-$ = complex conjugate of $\mathfrak{p}_-$) determines a complex structure on the real vector space $\mathfrak{p}$. When the polynomial version of the Dolbeaut lemma is dualized, one obtains an exact sequence

$$0 \to \mathfrak{p}^{\mathbb{C}(n-q)} \otimes \Lambda^q \mathfrak{p}_- \to \mathfrak{p}^{\mathbb{C}(n-q+1)} \otimes \Lambda^{q-1} \mathfrak{p}_- \to \cdots$$
$$\cdots \to \mathfrak{p}^{\mathbb{C}(n-1)} \otimes \mathfrak{p}_- \to \mathfrak{p}^{\mathbb{C}(n)} \to (\mathfrak{p}^{\mathbb{C}}/\mathfrak{p}_-)^{(n)} \to 0,$$

in which all arrows are $\mathfrak{b}$-homomorphisms. It remains exact when it is tensored with $L_{\mu+\rho_n}$:

$$\text{(A.33)} \quad 0 \to \mathfrak{p}^{\mathbb{C}(n-q)} \otimes \Lambda^q \mathfrak{p}_- \otimes L_{\mu+\rho_n} \to \mathfrak{p}^{\mathbb{C}(n-q+1)} \otimes \Lambda^{q-1} \mathfrak{p}_- \otimes L_{\mu+\rho_n} \to \cdots$$
$$\cdots \to \mathfrak{p}^{\mathbb{C}(n-1)} \otimes \mathfrak{p} \otimes L_{\mu+\rho_n} \to \mathfrak{p}^{\mathbb{C}(n)} \otimes L_{\mu+\rho_n} \to (\mathfrak{p}^{\mathbb{C}}/\mathfrak{p}_-)^{(n)} \otimes L_{\mu+\rho_n} \to 0.$$

The one-dimensional quotients in a composition series of the $\mathfrak{b}$-module $\Lambda^s \mathfrak{p}_- \otimes L_{\mu+\rho_n}$ all belong to weights $\mu+\rho_n-B$, where $B$ stands for a sum of $s$ distinct positive, noncompact roots. Hence the statements (A.28–A.30), coupled with the hypothesis of Proposition (5.14), guarantee that

$$\text{(A.34)} \quad I^p(\mathfrak{p}^{\mathbb{C}(n-s)} \otimes \Lambda^s \mathfrak{p}_- \otimes L_{\mu+\rho_n}) \cong \mathfrak{p}^{\mathbb{C}(n-s)} \otimes I^p(\Lambda^s \mathfrak{p}_- \otimes L_{\mu+\rho_n}) = 0,$$
$$\text{for } 0 \leqq s \leqq q, \ p \geqq 1.$$

Any exact $\mathfrak{b}$-module sequence $0 \to E_n \to E_{n-1} \to \cdots \to E_0 \to E \to 0$, which has the property that $I^p(E_s)=0$ if $0 \leqq s \leqq n$ and $p \geqq 1$, is transformed into an exact sequence by the functor $I^0$; this is purely formal, and can be checked by induction. In particular,

$$I^0(\mathfrak{p}^{\mathbb{C}(n-1)} \otimes \mathfrak{p}_- \otimes L_{\mu+\rho_n}) \to I^0(\mathfrak{p}^{\mathbb{C}(n)} \otimes L_{\mu+\rho_n}) \to I^0((\mathfrak{p}^{\mathbb{C}}/\mathfrak{p}_-)^{(n)} \otimes L_{\mu+\rho_n}) \to 0$$

is exact.

To complete the proof of Proposition (5.14), we must verify the statement (A.22). According to (A.24), $Q_n/Q_{n-1}$ is isomorphic to a quotient of the cokernel of $h$, hence to a quotient of the cokernel of the homomorphism (A.32), hence finally to a quotient of $I^0((\mathfrak{p}^{\mathbb{C}}/\mathfrak{p}_-)^{(n)} \otimes L_{\mu+\rho_n})$. The $\mathfrak{b}$-module $(\mathfrak{p}^{\mathbb{C}}/\mathfrak{p}_-)^{(n)}$ has a composition series with quotients $L_{\beta_1+\cdots+\beta_n}$, $\beta_1, \ldots, \beta_n \in \Phi^n \cap \Psi$. Thus (A.28) and (A.29) allow us to identify the highest weights of the potential irreducible constituents of $I^0((\mathfrak{p}^{\mathbb{C}}/\mathfrak{p}_-)^{(n)} \otimes L_{\mu+\rho_n})$: they all can be expressed as $\mu+\rho_n+\beta_1+\cdots+\beta_n$, with $\beta_i \in \Phi^n \cap \Psi$. This proves (A.22), and along with it Proposition (5.14).

## References


1. Atiyah, M.F.: Elliptic operators, discrete groups and von Neumann algebras. Asterisque **32/33**, 43–72 (1976)
2. Atiyah, M.F., Bott, R., Shapiro, A.: Clifford modules. Topology **3**, 3–38 (1964)
3. Atiyah, M.F., Schmid, W.: A new proof of the regularity theorem for invariant eigendistributions on semisimple Lie groups. To appear
4. Atiyah, M.F., Singer, I.M.: The index of elliptic operators on compact manifolds. Bull. Amer. math. Soc. **69**, 422–433 (1963)
5. Borel, A.: Compact Clifford-Klein forms of symmetric spaces. Topology **2**, 111–122 (1963)
6. Borel, A., Harish-Chandra: Arithmetic subgroups of algebraic groups. Ann. of Math. **75**, 485–535 (1962)

7. Borel, A., Wallach, N.: Seminar notes on the cohomology of discrete subgroups of semi-simple groups. To appear in Springer Lecture Notes in Mathematics
8. Bott, R.: The index theorem for homogeneous differential operators. In: Differential and Combinatorial Topology (A Symposium in Honor of Marston Morse) pp. 167–186. Princeton: Princeton University Press 1965
9. Casselman, W.: Matrix coefficients of representations of real reductive matrix groups. To appear
10. Casselman, W., Osborne, M.S.: The n-cohomology of representations with an infinitesimal character. Compositio Math. **31**, 219–227 (1975)
11. Fomin, A.I., Shapovalov, N.N.: A property of the characters of real semisimple Lie groups. Functional Analysis and its Applications **8**, 270–271 (1974)
12. Harish-Chandra: The characters of semisimple Lie groups. Trans. Amer. math. Soc. **83**, 98–163 (1956)
13. Harish-Chandra: Invariant eigendistributions on a semisimple Lie group. Trans. Amer. math. Soc. **119**, 457–508 (1965)
14. Harish-Chandra: Discrete series for semisimple Lie groups I. Acta Math. **113**, 241–318 (1965)
15. Harish-Chandra: Discrete series for semisimple Lie groups II. Acta Math. **116**, 1–111 (1966)
16. Harish-Chandra: Harmonic Analysis on semisimple groups. Bull. Amer. Math. Soc. **76**, 529–551 (1970)
17. Hirzebruch, F.: Automorphe Formen und der Satz von Riemann-Roch. In: Symp. Intern. Top. Alg. 1956, 129–144, Universidad de Mexico 1958
18. Kazdan, D.A.: On arithmetic varieties. In: Proc. Summer School on Group Representations, Budapest 1971, 151–217, New York: Halsted Press 1975
19. Kostant, B.: Lie algebra cohomology and the generalized Borel-Weil theorem. Ann. of Math. **74**, 329–387 (1961)
20. Langlands, R.P.: The dimension of spaces of automorphic forms. In: Algebraic Groups and Discontinuous Subgroups, Proc. of Symposia in Pure Mathematics, vol. IX, 253–257, Amer. Math. Soc., Providence 1966
21. Narasimhan, M.S., Okamoto, K.: An analogue of the Borel-Weil-Bott theorem for Hermitian symmetric pairs of noncompact type. Ann. of Math. **91**, 486–511 (1970)
22. Parthasarathy, R.: Dirac operators and the discrete series. Ann. of Math. **96**, 1–30 (1972)
23. Schmid, W.: On a conjecture of Langlands. Ann. of Math. **93**, 1–42 (1971)
24. Schmid, W.: Some properties of square-integrable representations of semisimple Lie groups. Ann. of Math. **102**, 535–564 (1975)
25. Schmid, W.: $L^2$-cohomology and the discrete series. Ann. of Math. **103**, 375–394 (1976)
26. Warner, G.: Harmonic Analysis on Semi-Simple Lie groups, vols. I and II. Berlin-Heidelberg-New York: Springer 1972
27. Zuckerman, G.: Tensor products of finite and infinite dimensional representations of semisimple Lie groups. To appear in Ann. of Math.

*Received June 10, 1977*

Inventiones math. 42, 1 – 62 (1977)

MICHAEL ATIYAH AND WILFRIED SCHMID

# ERRATUM TO THE PAPER: A GEOMETRIC CONSTRUCTION OF THE DISCRETE SERIES FOR SEMISIMPLE LIE GROUPS

In the above paper [2] a key role is played by a result of Borel [3], concerning discrete subgroups $\Gamma$ of semisimple Lie groups $G$. They prove that if $G$ is linear, one can find a torsion-free $\Gamma$ with $\Gamma \backslash G$ compact. Unfortunately we applied this result in [2] even for non-linear $G$, in which case the existence of such $\Gamma$ is seriously in doubt, as pointed out to us by P. Deligne and J. P. Serre. The difficulty is that a torsion-free subgroup of the adjoint group lifts to a cocompact subgroup $\Gamma \subset G$ which contains the (finite) center $Z$ of $G$, and there may be an obstruction to removing this torsion subgroup. As it stands, [2] is correct only for linear $G$, and we shall now indicate how to extend the proof to cover all $G$.

As mentioned above, the result of [3] gives us a discrete subgroup $\Gamma$ acting on the symmetric space $G/K$ with constant isotropy group $Z$. An irreducible representation $\rho$ of $K$ determines a homogeneous vector bundle $\tilde{\mathscr{V}}_\rho$ on $G/K$. Similarly the spin bundles $\tilde{\mathscr{S}}^+$, $\tilde{\mathscr{S}}^-$ correspond to representations $s^+$, $s^-$ of $K$. Whenever $\rho \otimes s^+$, $\rho \otimes s^-$ are trivial on $Z$, the bundles $\tilde{\mathscr{V}}_\rho \otimes \mathscr{S}^\pm$ descend to vector bundles $\mathscr{V}_\rho \otimes \mathscr{S}^\pm$ on $\Gamma \backslash G/K$, and the twisted Dirac operator

$$\tilde{D}_\rho^+ : C^\infty(\tilde{\mathscr{V}}_\rho \otimes \tilde{\mathscr{S}}^+) \to C^\infty(\tilde{\mathscr{V}}_\rho \otimes \tilde{\mathscr{S}}^-)$$

descends to an elliptic differential operator $D_\rho^+$ on this compact manifold. This is the starting point of [2]. If $\rho \otimes s^+$, $\rho \otimes s^-$ are non-trivial on $Z$, they act according to a one-dimensional character of $Z$ (the same character in both cases) and define projective representations of $K/Z$. Thus on $\Gamma \backslash G/K$ we only have projective bundles, and in general there is a cohomological obstruction to lifting these to actual vector bundles. More concretely, there exist local vector bundles $\tilde{\mathscr{V}}_{\rho,i} \otimes \mathscr{S}_i^\pm$ over small open sets $U_i \subset \Gamma \backslash G/K$, and transition functions $g_{ij}$ which are not quite consistent: on $U_i \cap U_j \cap U_k$

$$(1) \qquad c_{ijk} = g_{ij} g_{jk} g_{ki}$$

is a constant scalar multiple of the identity. Since scalar constants commute with differential operators, the inconsistency (1) does not prevent the



*J. A. Wolf, M. Cahen, and M. De Wilde (eds.), Harmonic Analysis and Representations of Semi-Simple Lie Groups, 379–383.*

consistency of the local differential operators $D_i$ defined by $\tilde{D}_\rho^+$. Thus $D_i$ acts on sections of $\mathscr{V}_{\rho,i} \otimes \mathscr{S}_i^+$ and, restricted to $U_i \cap U_j$, $D_i$ and $D_j$ coincide via the identifications $g_{ij}$. The inconsistency does prevent us however from constructing a global space of sections on which a differential operator $D_\rho$ would act, and hence we do not appear to have an index to work with.

The way around this difficulty is to introduce a consistent set of parametrices $P_i$ for the $D_i$ and to *define* a real-valued index in terms of these parametrices, along the lines explained in [1, §1]. More precisely, we choose open, relatively compact subsets $U'_i \subset U_i$ which still cover $\Gamma \backslash G/K$, and parametrices

$$P_i: C_0^\infty(U'_i, \mathscr{V}_{\rho,i} \otimes \mathscr{S}_i^-) \to C_0^\infty(U_i, \mathscr{V}_{\rho,i} \otimes \mathscr{S}_i^+)$$

for the local operators $D_i$ on $U'_i$, which on overlapping sets $U'_i \cap U'_j$ satisfy the same consistency condition as the $D_i$. We may also assume that the Schwartz kernels of the $P_i$ are supported very close to the diagonal. By definition of a parametrix, the operators

$$S'_i = 1 - P_i D_i, \qquad S''_i = 1 - D_i P_i,$$

have $C^\infty$ kernels $S'_i(x, y)$, $S''_i(x, y)$, again with support near the diagonal, and consistent in the same sense as the $D_i$. In particular the local kernels can be pieced together to give globally defined kernels $S'(x, y)$, $S''(x, y)$. If $D_\rho^+$ exists as a global differential operator on $\Gamma \backslash G/K$, its index is given by

$$\text{(2)} \qquad \operatorname{index} D_\rho^+ = \int_{\Gamma \backslash G/K} \{\operatorname{trace} S'(x, x) - \operatorname{trace} S''(x, x)\}\, dx$$

In our present situation we take (2) as the definition of the index. The main result of [1], relating the index on $\Gamma \backslash G/K$ to a $\Gamma$-index on $G/K$, then goes over with only trivial modifications to the proof.

We must still compute the index (2) as a function of $\rho$. If the representations $\rho \otimes s^+$, $\rho \otimes s^-$ are trivial on $Z$, the index (2) is the index of the global operator $D_\rho^+$ in the usual sense. As explained in [2, §3], the index of $D_\rho^+$ can then be expressed as a polynomial function of the highest weight of $\rho$. We must show that the same polynomial gives the index (2) for every $\rho$. Since the weight lattice of $K/Z$ has finite index in the weight lattice of $K$, it is enough to know that the dependence of the index on the highest weight is given by *some* polynomial. Specifically we shall prove the existence of an element $Y$ in the center of $U(\mathfrak{k})$ (=universal enveloping algebra of the complexified Lie algebra $\mathfrak{k}$ of $K$), such that

(3) $\qquad$ index $D_\rho = \text{trace}\ \rho(Y)$,

uniformly in $\rho$. Via the Harish-Chandra isomorphism, $Y$ corresponds to a polynomial function on the weight lattice, whose value at the highest weight of $\rho$ equals the constant by which $Y$ acts in the irreducible representation $\rho$. The Weyl dimension formula expresses the degree of $\rho$ as a polynomial in the highest weight, and thus also the quantity (3) depends on the highest weight in a polynomial fashion.

As a first step in the verification of (3) we remark that it is not necessary to work with local parametrices $P_i$ of infinite order – in other words, the operators $S_i'$, $S_i''$ need not have $C^\infty$ kernels. If $S_i'$, $S_i''$ are of order $-N$, as pseudodifferential operators, they have integral kernels of class $C^k$, with $k = N - 2 - \frac{1}{2}\dim G/K$; this follows from the Sobolev lemma. Hence, for $N$ sufficiently large in relation to the dimension of $G/K$, the operators $S_i'$, $S_i''$ are still of trace class, the traces can be calculated by integrating their kernels over the diagonal, and (2) remains unchanged.

To emphasize the dependence of the local differential operators and parametrices on $\rho$, we now write $D_{\rho,i}$, $P_{\rho,i}$, etc., and $D_{0,i}$, $P_{0,i}$, etc., in the particular case of the trivial one-dimensional representation $\rho$. We shall think of the $P_{0,i}$ as fixed, and we shall construct concrete $N$-th order parametrices $P_{\rho,i}$ in terms of the $P_{0,i}$. By choosing the original open cover $\{U_i\}$ of $\Gamma\backslash G/K$ fine enough, we can make the local bundles $\mathscr{V}_{\rho,i}$ trivial, with constant fiber

$$V_\rho = \text{representation space of } \rho,$$

simultaneously for all $\rho$. If one tensors the $D_{0,i}$ with the identity on $V_\rho$, one obtains local operators which act on $\mathscr{V}_{\rho,i}$-valued spinors, and which have the same principal symbols as the $D_{\rho,i}$. The differences

$$A_{\rho,i} = D_{0,i} \otimes 1 - D_{\rho,i}$$

thus are 0-th order operators. Putting

(4) $$P_{\rho,i} = \sum_{l=0}^{N-1} \{(P_{0,i} \otimes 1) A_{\rho,i}\}^l (P_{0,1} \otimes 1),$$

one finds

$$(5)\qquad S'_{\rho,i} = 1 - P_{\rho,i} D_{\rho,i}$$

$$= \{(P_{0,i} \otimes 1) A_{\rho,i}\}^N + \sum_{l=0}^{N-1} \{(P_{0,i} \otimes 1) A_{\rho,i}\}^l (S'_{0,i} \otimes 1)$$

and

$$(6)\qquad S''_{\rho,i} = \{A_{\rho,i}(P_{0,i} \otimes 1)\}^N + (S''_{0,i} \otimes 1) \sum_{l=1}^{N-1} \{A_{\rho,i}(P_{0,i} \otimes 1)\}^l.$$

An appropriate choice of cutoff functions in the construction of the local parametrices $P_{0,i}$ insures that the $N$-fold compositions (4–6) can be applied to compactly supported sections on $U'_i$. The pseudodifferential operators $S'_{\rho,i}$, $S''_{\rho,i}$ have order $-N$, and so the $P_{\rho,i}$ are indeed $N$th order parametrices which may be used in the definition of the index (2).

The first order operators $D_{0,i} \otimes 1$ and $D_{\rho,i}$ correspond to the same symbol and the same choice of connection on the spin bundle $\mathscr{S}_i^+$, but to different connections on the bundle $\mathscr{V}_{\rho,i}$: in one case the trivial flat connection, in the other case the canonical connection induced by the $G$-invariant connection of $\mathscr{V}_\rho$. The difference $A_{\rho,i}$ is the difference of the connection forms, contracted with the symbol of $D_{0,i}$. To see how this depends on $\rho$, we regard the $\mathscr{V}_{\rho,i}$ as vector bundles associated to local principal bundles $\mathscr{K}_i$ on $U_i$, which are derived from the principal bundle

$$(7)\qquad K \to G \to G/K$$

The simultaneous trivialization of the various $\mathscr{V}_{\rho,i}$ amounts to a trivialisation of these local principal bundles. Each $\mathscr{K}_i$ inherits a canonical connection from the bundle (7), which in terms of the chosen trivialisation corresponds to a $\mathfrak{k}$-valued 1-form $\omega_i$. This is the universal connection form for the canonical connections on the bundles $\mathscr{V}_{\rho,i}$, in the sense that the connection form of any $\mathscr{V}_{\rho,i}$ equals the composition of $\omega_i$ with $\rho$. The final conclusion about the operators $A_{\rho,i}$ is that they can be expressed as

$$A_{\rho,i} = \sum_\alpha A_{i,\alpha} \otimes \rho(X_\alpha)$$

with $A_{i,\alpha}$ independent of $\rho$, and $X_\alpha \in \mathfrak{k}$. The dependence on $\rho$ of the operators (5–6) and of their integral kernels can be isolated in the same manner, but with $X_\alpha$ ranging over a finite subset of $U(\mathfrak{k})$. Taking traces and performing the integration (3), one obtains a $Y \in U(\mathfrak{k})$ which satisfies the identity (3). To put $Y$ into the center of $U(\mathfrak{k})$, we replace it by

$$\int_K \mathrm{Ad}\, k(Y)\, \mathrm{d}k$$

without affecting the identity (3).

## References


[1] Atiyah, M. F., 'Elliptic operators, discrete groups and von Neumann algebras', *Soc. Math. France, Astérisque* 32–33 (1976), 43–72.

[2] Atiyah, M. F., and Schmid, W., 'A geometric construction of the discrete series for semisimple Lie groups', *Inventiones Math.* 42 (1977), 1–62.

[3] Borel, A., 'Compact Clifford-Klein forms of symmetric spaces', *Topology* 2 (1963), 111–122.

M. FLATO AND D. STERNHEIMER

# DEFORMATIONS OF POISSON BRACKETS, SEPARATE AND JOINT ANALYTICITY IN GROUP REPRESENTATIONS, NONLINEAR GROUP REPRESENTATIONS AND PHYSICAL APPLICATIONS

## CONTENTS





*J. A. Wolf, M. Cahen, and M. De Wilde (eds.), Harmonic Analysis and Representations of Semi-Simple Lie Groups, 385–448.*

## CHAPTER 1

# 1-DIFFERENTIABLE DEFORMATIONS OF THE POISSON BRACKET LIE ALGEBRA

## 0. INTRODUCTION

This chapter and the following will deal with deformations of the Lie algebra structure defined by the Poisson brackets on the space of differentiable functions over a symplectic manifold (general phase space) i.e. with Lie algebra structures on the same space, the bracket of which is 'close' (in a sense to be made precise later) to the Poisson bracket. The Poisson bracket itself being a bidifferential operator of order 1 (an operator which acts as a differential operator of order 1 on each function) when acting upon a couple of functions, we shall discuss only those deformations which are given by a series of bidifferential operators. In this first chapter we shall restrict ourselves further, and suppose that these bidifferential operators are of order $\leqslant 1$, in which case a complete theory can be developed. We shall begin with a short outline of the Gerstenhaber theory of Lie algebra deformations and its connection with the Chevalley–Eilenberg cohomology of Lie algebras.

## 1. LIE ALGEBRA DEFORMATIONS AND COHOMOLOGY

(a) Let $N$ be a Lie algebra, with bracket denoted by $\{\ ,\ \}$. $N$ acts on itself through the adjoint representation:

$$\mathrm{ad}\colon N \ni u \mapsto \mathrm{ad}\, u \in \mathscr{L}(N, N), \qquad \mathrm{ad}\, u\colon N \ni v \mapsto \{u, v\} \in N.$$

The $p$-cochains $C$ of $N$ (with values in $N$) are the $p$-linear alternate maps $N^p \to N$, the 0-cochains being identified with elements of $N$ (constant maps). The coboundary of the $p$-cochain $C$ (in the Chevalley–Eilenberg cohomology with values in the adjoint representation) is the $(p+1)$ cochain $\partial C$ defined by [1]:

$$(1.1) \qquad \partial C(u_0, \ldots, u_p) = \sum_{\alpha=0}^{p} (-1)^{\alpha}\{u_\alpha, C(u_0, \ldots, \hat{u}_\alpha, \ldots, u_p)\} + \\ + \sum_{\alpha<\beta} (-1)^{\alpha+\beta} C(\{u_\alpha, u_\beta\}, u_0, \ldots, \hat{u}_\alpha, \ldots, \hat{u}_\beta, \ldots, u_p)$$

where $\hat{u}_\alpha$ means that $u_\alpha \in N$ has to be omitted. Similar formulas define the coboundary with values in any other representation, with the obvious changes in the first sum of the right-hand side.

A $p$-cochain $C$ is said to be a *p-cocycle* if $\partial C=0$. The space of the $p$-cocycles will be denoted by $Z^p(N)$. One easily checks that $\partial^2=0$, which means that $Z^p(N)$ contains special $p$-cocycles $C$, those for which there exists a $(p-1)$ cochain $\gamma$ (defined up to a $(p-1)$-cocycle) such that $C=\partial\gamma$: these are the *p-coboundaries* (of $(p-1)$-cochains), the space of which will be denoted by $B^p(N)$. The quotient space $H^p(N)=Z^p(N)/B^p(N)$ is called the $p$th-cohomology space of $N$ (with values in the adjoint representation).

(b) Let $E(N, \lambda)$ denote the space of formal series $\Sigma_{p=0}^{\infty} \lambda^p u_p$ in a parameter $\lambda$ with coefficients in $N$; we shall, in general, not deal with the question of how (through convergence or otherwise) such a series can represent an element in $N$ but consider only the formal series aspect. A *formal deformation* [2] of the Lie algebra $N$ is a new bracket law:

(1.2) $$N \times N \ni (u, v) \mapsto [u, v]_\lambda = \sum_{r=0}^{\infty} \lambda^r C_r(u, v) \in E(N; \lambda), \qquad C_0(u, v)=\{u, v\},$$

(which can be naturally extended from $N^2$ to $E(N; \lambda)^2$) satisfying formally the requirements of a Lie algebra, namely skew-symmetry which means that the $C_r$ are 2-cochains, and Jacobi identity:

(1.3) $$S[[u, v]_\lambda, w]_\lambda = 0$$

where $S$ denotes the sum over cyclic permutations of $u, v, w \in N$. Developing (1.3) into powers of $\lambda$ and expressing that the coefficient of $\lambda^t$ vanishes identically we obtain:

(1.4) $$\partial C_t(u, v, w) = E_t(u, v, w), \qquad (t=1, 2, \ldots),$$

where the 3-cochain $E_t$ is defined by

(1.5) $$E_t(u, v, w) = S \sum_{r+s=t, rs\neq 0} C_r(C_s(u, v), w)$$

and has thus to be a coboundary. For $t=1$, (1.4) reduces to $\partial C_1=0$, namely it has to be a 2-cocycle.

(c) An *infinitesimal deformation* is a law

(1.6) $$(u, v) \mapsto [u, v]_\lambda = \{u, v\} + \lambda C_1(u, v)$$

such that (1.3) is satisfied modulo $\lambda^2$, namely where $C_1$ is a 2-cocycle. Such a deformation will be a rigorous deformation of $N$ iff (if and only if) the

$\lambda^2$ term also vanishes, i.e.

(1.7) $\frac{1}{2}E_2(u, v, w) = SC_1((C_1(u, v), w) = 0.$

Moreover one can show (cf. [2]) that if (1.4) is satisfied for $t = 1, \ldots, q-1$, then $E_q$ is in fact a 3-cocycle of $N$, i.e. $\partial E_q = 0$. Therefore if $C_1, \ldots, C_{q-1}$ are given such that (1.3) is satisfied modulo $\lambda^q$, one can continue one step further the deformation (find a $C_q$) iff the class of $E_q$ in $H^3(N)$ (the obstruction at order $q$) vanishes. In particular, if $H^3(N) = 0$, any infinitesimal deformation can be extended to a formal deformation.

(d) Finally we shall say that two deformations $[u, v]_\lambda^{(0)}$ and $[u, v]_\lambda^{(1)}$ of the type (1.2) are *equivalent* (or cohomologically equivalent) [3] if there exists a formal series of endomorphisms of $N$ (1-cochains), with $T_0 = \text{identity}$:

(1.8) $T_\lambda = \sum_{r=0}^{\infty} \lambda^r T_r$

such that the following identity of formal series holds:

(1.9) $T_\lambda [u, v]_\lambda^{(1)} = [T_\lambda u, T_\lambda v]_\lambda^{(0)}.$

One can show that if there exist $T_1, \ldots, T_{q-1}$ such that (1.9) holds modulo $\lambda^q$, a certain 2-cochain (expressed in terms of the $C$'s and preceding $T$'s) will be a 2-cocycle, and has to be a 2-coboundary $\partial T_q$ in order to be able to continue the equivalence one step further. $H^2(N)$ thus classifies the *obstructions to equivalence* of deformations. In particular if one takes $[u, v]_\lambda^{(0)} = \{u, v\}$, then the deformation $[u, v]_\lambda$ will be said *trivial* if (1.9) holds with $[u, v]_\lambda^{(1)} = [u, v]_\lambda$. Accordingly, with the same notations, an infinitesimal deformation (1.6) will be said *infinitesimally trivial* if (1.9) is satisfied modulo $\lambda^2$; computing the $\lambda$-term, this means that there exists a 1-cochain $T_1$ such that

(1.10) $\partial T_1 = C_1$

Therefore $H^2(N)$ represents exactly the infinitesimal deformations of $N$ modulo the trivial ones.

(e) All that has been said here applies to general deformations of general Lie algebras. One should perhaps say here that a contraction of Lie algebras is the inverse procedure, namely going to the limit

$$[u, v]_\lambda \xrightarrow[\lambda \to 0]{} \{u, v\}.$$

The differential character of the Poisson bracket leads to further specializations. But first we have to recall some basic facts on symplectic structures.

## 2. Symplectic manifolds. Poisson brackets and 1-differentiable cohomology

(a) A symplectic manifold is a (connected and paracompact) differentiable manifold $W$ of even dimension $2n$, on which is given a *closed* 2-form $F$ such that $F^n$ be everywhere $\neq 0$ (it is the symplectic volume $2n$-form). All cotangent bundles $W = T^*M$ to an $n$-dimensional manifold $M$ have a natural symplectic structure. We shall denote by $\{x^i\}$ $(i=1,\ldots,2n)$ a local chart defined on some domain $U \subset W$. It is known that $W$ has atlases of local charts $\{x^\alpha, x^{\bar\alpha}\}$ $(\alpha=1,\ldots,n;\ \bar\alpha=\alpha+n)$, called canonical, such that $F$ can be written $\Sigma_{\alpha=1}^n \mathrm{d}x^\alpha \wedge \mathrm{d}x^{\bar\alpha}$ on the domain of such a chart. In particular, for the phase-space $W=\mathbb{R}^{2n}$ with coordinates $(p^\alpha, q^\alpha)$ one usually takes $F$ to be $\Sigma_\alpha \mathrm{d}p^\alpha \wedge \mathrm{d}q^\alpha$.

A vector bundle isomorphism $\mu\colon TW \to T^*W$ between the tangent and cotangent bundles can be defined by the interior product with $F$, namely $\mu(X) = -i(X)F \in T^*W$ for $X \in TW$ (tangent vector field), where $i$ denotes the interior product. This isomorphism extends naturally to tensor bundles, and in particular we can introduce the (contravariant skew-symmetric) 2-tensor $G = \mu^{-1}(F)$.

The Poisson bracket of two functions $u, v \in N = C^\infty(W)$, the space of (real-valued) differentiable functions on $W$, can now be defined by

$$\{u, v\} = i(G)(\mathrm{d}u \wedge \mathrm{d}v). \tag{2.1}$$

This is a Lie algebra structure on $N$. On the domain $U$ of a canonical chart, one can write (2.1) in a more familiar form, where $\partial_\alpha = \partial/\partial x^\alpha$, $\partial_i = \partial/\partial x^i$:

$$\{u, v\}|_U = \sum_{\alpha=1}^{n} (\partial_\alpha u \partial_{\bar\alpha} v - \partial_\alpha v \partial_{\bar\alpha} u) = \sum_{i,j=1}^{2n} G^{ij} \partial_i u \partial_j v. \tag{2.2}$$

The Poisson bracket appears thus as a bidifferential operator of order 1 (in each variable), what we call a 1-differentiable 2-cochain. Therefore, if $C$ is a *differentiable* $p$-cochain (the action of which on each function is that of a differential operator), $\partial C$ will be a differentiable $(p+1)$-cochain. This makes natural the introduction of the differentiable cohomology $H^p_{\mathrm{diff}}(N)$, the space of the differentiable $p$-cocycles modulo the coboundaries of differentiable $(p-1)$-cochains.

(b) We shall say that a $p$-cochain $C$ is 1-*differentiable* if it acts as a differential operator of order $\leqslant 1$ on each function of $N$. A direct computation shows [4] that *if* $C$ *is* 1-*differentiable, so is* $\partial C$. It therefore makes sense to introduce the 1-differentiable cohomology $H^p_{1\text{-diff}}(N)$, defined in the same way. In contradistinction with the preceding cohomologies of the Lie algebra $N$ of $C^\infty$ functions on a symplectic manifold, for which only partial results have been obtained, the latter cohomology can be completely calculated in terms of the (de Rham) cohomology of the manifold.

The result is the following [4]. We denote by $H^p(W)$ the $p$th space of (real) cohomology classes of the manifold $W$, namely the quotient of the space of the closed differential $p$-forms $\omega$ ($d\omega=0$) by the space of the exact $p$-forms ($\omega=d\omega'$). We recall that by Poincaré lemma, for an open ball in a Euclidean space, this space is trivial (for $p\geqslant 1$). The exterior product by the closed 2-form $F$ induces a map $H^p(W)\to H^{p+2}(W)$, the kernel and image of which we shall denote respectively by $P^p(W; F)$ and $Q^{p+2}(W; F)$. Then:

$$H^p_{1\text{-diff}}(N)=P^{p-1}(W; F)\oplus H^p(W)/Q^p(W; F) \tag{2.3}$$

which, when $F$ is exact, reduces to

$$H^p_{1\text{-diff}}(N)=H^{p-1}(W)\oplus H^p(W). \tag{2.4}$$

To prove the result, one uses a decomposition of a $p$-cochain $C$ into $C=A+B$, where $A$ is a $p$-tensor (a skew-symmetric contravariant tensor of order $p$) and $B$ a $(p-1)$-tensor, so that on the domain $U$ of a local chart $\{x^k\}$ one has (with summation over repeated indices):

$$\begin{cases} A(u_1,\ldots,u_p)=A^{k_1\ldots k_p}\partial_{k_1}u_1\ldots\partial_{k_p}u_p \\ B(u_1,\ldots,u_p)=\dfrac{1}{(p-1)!}\,\varepsilon^{\alpha_1\ldots\alpha_p}_{1\ldots p}B^{k_2\ldots k_p}u_{\alpha_1}\partial_{k_2}u_{\alpha_2}\ldots\partial_{k_p}u_{\alpha_p} \end{cases} \tag{2.5}$$

where $\varepsilon$ is the skew-symmetric Kronecker indicator. When $B=0$, the cochain $C=A$ is said *pure*. In the decompositions (2.3) and (2.4), the first summand corresponds to the non-pure part of the cocycles, the second one to the pure part. A useful formula, which stresses the 1-differentiable character of $\partial C$, is:

$$\mu(\partial C)=(F\wedge\mu(B)-d\mu(A))+d\mu(B). \tag{2.6}$$

(c) In a way similar to 1-differentiability, one can consider *m-differentiable* $p$-cochains (defined by differentiable operators of order $\leqslant m$). Such cochains $C$ are *local*, i.e. if one of the $u_k$'s vanishes on some domain $U$ in $W$, then $C(u_1, \ldots, u_p)$ also vanishes on $U$. An important (and useful) result is the following

THEOREM 1.1. *Let $T$ be a 1-cochain of $N$ such that $\partial T$ is an m-differentiable ($m \geqslant 1$) 2-cochain of $N$. Then if the symplectic manifold $(W, F)$ is non-compact, or if $T$ is local, the cochain $T$ is m-differentiable (i.e. is a differential operator of order $\leqslant m$).*

To prove the result (cf. [5]) one first shows that $T$ coincides with a differential operator of order $m$ on any local chart, and then glues the charts together under the above hypotheses.

In particular since a derivation of the Lie algebra $N$ is a 1-cocycle, it follows immediately that, under the above hypotheses, they are all 1-differentiable (in the compact case, there are however non-local derivations [6]).

### 3. FORMAL 1-DIFFERENTIABLE DEFORMATIONS

(a) We still consider a symplectic manifold $(W, F)$ and the Lie algebra $N = C^\infty(W)$ with the Poisson bracket. Obviously, all the theory developed in Section 1 makes still sense if one limits oneself to differentiable cochains to define the deformation, the relevant cohomology being then the differentiable cohomology. It is less obvious that a further restriction to 1-differentiable cochains $C_t$ in (1.2) is consistent. But this is true:

LEMMA 1.1. *If $C_1$ and $C_2$ are two 1-differentiable 2-cochains, then*

$$SC_1(C_2(u, v), w) + SC_2(C_1(u, v), w)$$

*is a 1-differentiable 3-cochain.*

Therefore, if the 2-cochains $C_t$ defining the deformation (1.2) are 1-differentiable, so are the cochains $E_t$ defined by (1.5) and their class in $H^3_{1\text{-diff}}(N)$ is the obstruction at order $t$ to the construction of a formal 1-differentiable deformation of $N$.

We are therefore in position to apply the general deformation theory to the 1-differentiable case and study the 1-*differentiable deformations of* $N$. Theorem 1.1 above will be useful for the triviality of these deformations. Indeed, we shall say that a 1-differentiable deformation (resp. an infinitesimal 1-differentiable deformation) is *trivial* if there exist differential operators $T_r$ of order $r$ (resp. a 1-differentiable cochain $T_1$) such that, with $t=\Sigma_{r=0}^{\infty}\lambda^r T_r$ as in (1.8) one has:

$$(3.1) \qquad T_\lambda[u, v]_\lambda = \{T_\lambda u, T_\lambda v\}$$

(resp. $\partial T_1 = C_1$). (3.1) can be expressed as:

$$(3.2) \qquad \partial T_t(u, v) = C_t(u, v) - \sum_{\substack{r+s=t,\\ rs\neq 0}} (\{T_r u, T_s v\} - T_s C_r(u, v)).$$

In view of Theorem 1.1, if $W$ is non-compact, necessarily $T_r$ is $r$-differentiable and in particular $T_1$ is a 1-differentiable cochain: this justifies the definition of triviality we just gave, for 1-differentiable deformations, and (2.3) gives:

PROPOSITION 1.1. *The space of infinitesimal 1-differentiable deformations of $N$, modulo the trivial ones, is isomorphic to $P^1(W; F)\oplus H^2(W)/Q^2(W; F)$.*

(b) The symplectic structure of $W$ can equally well be defined in terms of the 2-tensor $G=\mu^{-1}(F)$. The relation $\mathrm{d}F=0$ can then be expressed $[G, G]=0$ in terms of the Schouten–Nijenhuis bracket [7], and is equivalent to the Jacobi identity for the Poisson bracket (2.1). We are therefore led to introduce a special kind of 1-differentiable deformations, those defined by the bracket

$$(3.3) \qquad \{u, v\}_{G_\lambda} = i(G_\lambda)(\mathrm{d}u \wedge \mathrm{d}v), \qquad u, v \in N,$$

where

$$(3.4) \qquad G_\lambda = \sum_{r=0}^{\infty} \lambda^r G_r, \qquad G_0 = G = \mu^{-1}(F),$$

satisfies formally (in terms of Schouten brackets)

$$(3.5) \qquad [G_\lambda, G_\lambda] = 0.$$

We shall thus say that a formal 1-differentiable deformation $[u, v]_\lambda$ is *inessential* (i.e. up to a trivial one, given by a deformation of the symplectic structure) if there exist $G_\lambda$ and $T_\lambda$ (given by (1.8) with differential operators $T_r$) such that:

$$(3.6) \qquad T_\lambda[u, v]_\lambda = \{T_\lambda u, T_\lambda v\}_{G_\lambda}.$$

Now (2.6) can be expressed, in terms of the Schouten bracket:

$$\partial C = (G \wedge B - [G, A]) + [G, B]$$

while (3.6) and (3.5) give, at the first order in $\lambda$:

$$(3.7) \qquad C_1 = G_1 + \partial T_1, \qquad [G, G_1] = 0.$$

Therefore $G_1$ is necessarily a *pure* 1-differentiable cocycle. By analogy with the trivial case, we shall say that an *infinitesimal* 1-differentiable deformation

$$(3.8) \qquad [u, v]_\lambda = \{u, v\} + \lambda C_1(u, v)$$

is *inessential* if (3.6) is satisfied modulo $\lambda^2$, i.e. if (3.7) holds. We say that it is *essential* if it is not inessential. Conversely an infinitesimal deformation defined by a cocycle homologous in $H^2_{1\text{-diff}}(N)$ to some pure 1-differentiable cocycle $G_1$ is inessential. Now the space of classes of pure 1-differentiable 2-cocycles is $P^1(W, F)$ $(=H^1(W)$ when $F$ is exact).

Therefore:

PROPOSITION 1.2. *The space of infinitesimal 1-differentiable deformations, modulo the inessential deformations, is isomorphic to* $P^1(W; F)$.

Let us denote by $b_p = b_p(W)$ the dimension of $H^p(W)$ (the $p$th Betti number). It follows that if $b_1 = 0$, all infinitesimal deformations of $N$ are inessential, and if in addition $b_2 = 0$ they are all trivial. On the other hand, if $F$ is exact, $b_1 \neq 0, b_2 = 0 = b_3$, it follows from (2.4) and the theory of deformations that there exist *formal* 1-*differentiable essential* deformations of $N$. The conditions $b_2 = 0 = b_3$ ensure that there will be no obstructions to the continuation of an infinitesimal essential deformation. We shall now show that, in important cases (such as cotangent bundles) they are not required, namely that there exist rigorous essential deformations of the type (3.8) whenever $b_1 \neq 0$.

## 4. Examples of Rigorous Deformations of $N$

(a) Let $W$ be a symplectic manifold with *exact* 2-form $F = \mathrm{d}\omega$ ($W$ being therefore non-compact). We define a vector field $Z$ by $\mu(Z) = -\omega$, or $i(Z)\,\mathrm{d}\omega = \omega$, whence $i(Z)\omega = 0$. If $\mathscr{L}$ denotes the Lie derivative we thus have $\mathscr{L}(Z)\omega = \omega$, $\mathscr{L}(Z)\,F = F$.

Let $C_1 = C$ be a 1-differentiable 2-cocycle. It defines a rigorous deformation of $N$ iff the 3-cocycle $E_2 = E$ defined in (1.7) vanishes.

On a local chart $U$ one can write, if $C = A + B$ (where $B$ is a vector-field):

$$(4.1) \qquad E(u, v, w) = [A, A]^{jkl} \partial_j u \partial_k v \partial_l w - 2[B, A]^{kl} S(w \partial_k u \partial_l v).$$

If we write $\mu(A) = \alpha$, $\mu(B) = \beta$, $C$ will be a cocycle, in view of (2.6), iff

$$(4.2) \qquad d\beta = 0, \qquad d\alpha = F \wedge \beta.$$

This will hold if $\beta$ is any closed 1-form and $\alpha = \omega \wedge \beta$, or $A = -Z \wedge \beta$. Now

$$[B, A] = \mathscr{L}(B)A = -\mathscr{L}(B)(Z \wedge B) = \mathscr{L}(Z)B \wedge B$$

and

$$\begin{aligned} \mathscr{L}(Z)B = \mathscr{L}(Z)\mu^{-1}(\beta) &= (\mathscr{L}(Z)\mu^{-1})(\beta) + \mu^{-1}(\mathscr{L}(Z)\beta) \\ &= -B + \mu^{-1}(d i(Z)\beta). \end{aligned}$$

If we choose the closed form $\beta$ such that $i(Z)\beta = \text{constant}$, we obtain $\mathscr{L}(Z)B = -B$, hence $[B, A] = 0$. Moreover in this case a direct computation shows that we have also $[A, A] = 0$. Indeed, on a local chart $U$ where $Z$ has constant components (which can always be found around any point $x$ where $Z(x) \neq 0$):

$$\begin{aligned} [A, A]^{abc} &= \varepsilon^{abc}_{jkl}(Z^r B^j - Z^j B^r)\partial_r (Z^k B^l - Z^l B^k) \\ &= \varepsilon^{abc}_{jkl}(B^j Z^k \mathscr{L}(Z)B^l - B^j Z^l \mathscr{L}(Z)B^k) \\ &= -\varepsilon^{abc}_{jkl}(B^j B^l Z^k - B^j B^k Z^l) = 0. \end{aligned}$$

Therefore, if there exists a closed non-exact 1-form $\beta$ such that $i(Z)\beta = 0$, (3.8) with on a local chart $U$:

$$(4.3) \qquad C_1(u, v)|_U = (Z^j B^i - Z^i B^j)\partial_i u \partial_j v + B^i(u \partial_i v - v \partial_i u)$$

defines a rigorous essential deformation of $N$.

(b) Let $M$ be any differentiable manifold of dimension $n$, $W = T^*M$ its cotangent bundle, with projection $\pi$: $W \to M$. On $W$ we have the Liouville form $\omega$, which writes in canonical local coordinates:

$$\omega|_U = \sum_{\alpha=1}^{n} p_\alpha \, dq^\alpha$$

and therefore $Z$ writes locally on $(W, d\omega)$: $Z|_U = \Sigma_\alpha p_\alpha (\partial / \partial p_\alpha)$.

Now if $\gamma$ is a closed non-exact 1-form of $M$, the closed 1-form $\beta = \pi^* \gamma$ of $W$ is non-exact and satisfies $i(Z)\beta = 0$. Moreover, the family of such $\beta$'s is exactly parametrized by a basis of classes of $\gamma$'s in $H^1(M)$. Therefore

PROPOSITION 1.3. *Let M be a differentiable manifold with* $b_1(M) \neq 0$. *There exists a* $b_1$*-dimensional space of inequivalent essential rigorous 1-differentiable deformations of the Lie algebra* $N(T^*M)$.

Examples of such deformations will be given in Chapter 3, in connection with 'non-conservative' systems.

## CHAPTER 2

# DIFFERENTIABLE DEFORMATIONS OF THE POISSON BRACKET LIE ALGEBRA

## 0. INTRODUCTION

In the first chapter, we dealt only with Lie algebra deformations. Here, for differentiable deformations, it will be appropriate to treat also deformations of associative algebras – which of course will generate Lie algebra deformations. The theory of deformations for associative algebras follows exactly the same pattern as for Lie algebras, the connection with the second and third (associative algebra) Hochschild cohomology groups, the algebra acting on itself by left and right multiplication, being the same; so we shall not repeat it. (Actually the original paper of Gerstenhaber [2] was formulated in the associative algebra context, with a remark that the Lie algebra case was similar.) The main difference is that these groups are so large in the associative case that it is then even more difficult to obtain concrete results.

We shall start with a simple constatation. The vector-space $N$ of $C^\infty$ functions over a symplectic manifold $W$ can be endowed with an associative algebra structure with the ordinary (commutative) product of functions; we shall denote by $N_a$ this associative algebra. Now it is easily checked that the product law

$$(0.1)\qquad (u \cdot v)_\nu = uv + \nu\{u, v\}, \qquad u, v \in \mathbb{N},$$

where $\{u, v\} = P(u, v)$ denotes the Poisson bracket of $u$ and $v$, defines an infinitesimal (differentiable) deformation of $N_a$ (i.e. associativity is satisfied modulo $\nu^2$). It turns out that this deformation is non trivial, and we shall look for formal deformations of $N_a$ which extend this infinitesimal deformation; this will give us formal deformations of the Lie algebra $N$ by taking $\nu^{-1}$ times the commutator. Though this study is quite difficult in the general case, partial (physically interesting) results can be achieved. The direct study of Lie algebra differentiable deformations of $N$ is somewhat easier, and more complete (less explicit, however) results can be obtained. We shall just mention here that in view of Theorem 1, Chapter 1, under the same hypotheses, the cochains $T_r$ defining the (cohomological) equivalence of deformations (see Section 1(d), Chapter 1) are necessarily differential operators when the $C_r$ are differentiable cochains – whence the relevance of the differentiable cohomology for this study.

## 1. Differentiable Lie Algebra Deformations of $N$

(a) We shall start by the 'analogue of the $SL(2, \mathbb{R})$' case in our theory, namely the case of the symplectic manifold $W = \mathbb{R}^2$ or $\mathbb{R}^{2n}$ (both cases can be treated exactly along the same lines except for one question, that we shall indicate later). Let $(p, q)$ be the coordinates in $W$ and consider the bidifferential operator (Poisson bracket operator relative to the form $F = \mathrm{d}p \wedge \mathrm{d}q$) $P = {}^1\partial_p{}^2\partial_q - {}^1\partial_q{}^2\partial_p$, where the superscript 1 (resp. 2) indicates that when taking $P(u, v)$, the differentiation with respect to $p$ (noted $\partial_p$) or $q$ (noted $\partial_q$) acts on the first function $u$ (resp. the second function $v$).

This can also be written in a more intrinsic way $P = G^{ij\,1}\partial_i{}^2\partial_j$ $(i, j = 1, 2,$ $\partial_1 = \partial_p$, $\partial_{\bar{1}} = \partial_2 = \partial_q$ on $\mathbb{R}^2$ in the usual coordinates) as in Equation (2.2) of Chapter 1. We can take the powers of this bidifferential operator:

$$(1.1) \qquad P^r = ({}^1\partial_p{}^2\partial_q - {}^1\partial_q{}^2\partial_p)^r = G^{i_1 j_1} \ldots G^{i_r j_r\,1}\partial_{i_1, \ldots, i_r}{}^2\partial_{j_1, \ldots, j_r}$$

which are $r$-differentiable 2-cochains of the associative algebra $N_a$ of $C^\infty$ functions on $W$, the odd powers $P^{2r+1}$ being $(2r+1)$-differentiable 2-cochains of the Lie algebra $N$ (with Poisson bracket $P$). Since we are on the formal level, we may deal here with the formal power series in $p$ and $q$, $\mathscr{P}' = \mathbb{R}[[p, q]]$, or restrict ourselves to the polynomials in $p$ and $q$, $\mathscr{P} = \mathbb{R}[p, q]$ on which everything will be well-defined. Now it is easy to check (cf. [3, 9]; we shall come back to this point later) that

$$(1.2) \qquad (u, v) \mapsto M(u, v) \equiv \nu^{-1} \sin(\nu P)(u, v),$$
$$M(u, v) = \sum_{r=0}^{\infty} \frac{\lambda^r}{(2r+1)!} P^{2r+1}(u, v) = [u, v]_\lambda,$$

with $\lambda = -\nu^2$ is a formal deformation of the Lie algebra $\mathscr{P}'$ with bracket $P$, in the sense of Equation (1.2) of Chapter 1. The cochains of the deformation are of the type $C_t = P^{2t+1}$, i.e. this is a *differentiable deformation* of $\mathscr{P}'$. This bracket is that introduced in 1949 by Moyal [10]. The corresponding infinitesimal deformation, defined by $P + (\lambda/6)P^3$, is *non trivial* because (cf. [11]) if $P^3$ were a coboundary, it would be that of a differential operator $T_3$ of order $\leqslant 3$ (in view of Theorem 1.1 of Chapter 1), which is impossible since $\partial T_3$ contains no term of bidifferential type (3, 3) and $P^3$ is exactly of type (3, 3) in the chosen coordinates – the type being defined as the couple (order on $u$, order on $v$).

(b) For a more complete study, we have to know the differentiable cohomology $H^*_{\text{diff}}(\mathscr{P}')$, at least in degrees 2 and 3. This has been done by J. Vey [9] who showed:

(i) $H^*_{\text{diff}}(\mathscr{P}')$ is isomorphic to the tensor product of the cohomology $H^*(S, \mathbb{R})$ of the Lie algebra $S$ of those formal symplectic (also said Hamiltonian) vector fields (i.e. vector fields $X$ satisfying $\mathscr{L}(X)F = \mathrm{d}i(X)F = 0$, $\mathscr{L}$ being the Lie derivative) vanishing at the origin, with values in the trivial representation space $\mathbb{R}$, by the outer (1-differentiable) derivation defined with the homotheties (dilation). This result makes use of the Gelfand–Fuks theory.

(ii) The dimension of $H^p(S, \mathbb{R})$ is 0 for $p=1$, 1 for $p=2$; if $W=\mathbb{R}^{2n}$ with $n>1$, it is 1 for $p=3$ and $\geqslant 2$ for $p=4$. (The obstructions vanish, already for the associative algebra, when $W$ is any 2-dimensional symplectic manifold).

(iii) For the differentiable cohomology of the Lie algebra $N$ of $C^\infty$ functions on a symplectic manifold $W$, there is in addition a factorization with the cohomology of the manifold (as we noticed in the 1-differentiable case). A precise formulation can be found in [9]. Cf. also [12].

(c) It follows from this study that, for $W=\mathbb{R}^{2n}$, there are only 2 classes of infinitesimal differentiable Lie algebra deformations of $N$: the trivial one, and the one obtained above $(P+(\lambda/6)P^3)$. The same of course holds locally on any symplectic manifold $W$. The question therefore arises whether one can 'lift' the locally defined deformation (1.2) to a formal differentiable deformation of $N$, globally defined on $W$. A natural idea would be to replace in (1.1) the ordinary differentiation by the covariant derivative $\nabla$ with respect to some well-chosen connection $\Gamma$. We shall come back to this idea in the next section, and just mention here that it does not work in general. The basic reason is that, in order to 'glue together' the cochains defined on intersecting symplectic charts, one has to add lower-order terms at every order of $\lambda$, and each such addition influences the higher orders of $\lambda$ due to the deformation conditions. However, by tracing inductively the possible obstructions in the (finite-dimensional) space $H^3_{\text{diff}}(N)$, J. Vey was able to prove the following existence theorem:

THEOREM 2.1. *Let $(W, G)$ be a symplectic manifold such that $H^3(W,\mathbb{R})=0$. There exists a formal deformation of the Lie algebra $N$:*

$$(1.3) \qquad Q(u, v; \lambda) = \sum_{r=0}^{\infty} \frac{\lambda^r}{(2r+1)!} Q^{2r+1}(u, v),$$

*where the 2-cochain $Q^{2r+1}$ is defined by a bidifferential operator of order $(2r+1)$ on each argument, null on the constant functions and having the same principal symbol as $P^{2r+1}$ on the domain of any canonical chart.*

We say that a *connection* $\Gamma$ on $W$ is *symplectic* if it is without torsion and satisfies $\nabla G = 0$, i.e. iff on any canonical chart $\Gamma_{ijk} = F_{il}\Gamma^{l}_{jk}$ is completely symmetric. There are many such connections locally, and they can be glued together (by partition of unity) on any symplectic manifold. Then [3] one can take for $Q^3$ any 2-cocycle in the cohomology class of the cocycle $S^3_\Gamma$ defined on any canonical chart $U$ by:

$$(1.4) \qquad S^3_\Gamma(u, v)\big|_U = G^{i_1 j_1} G^{i_2 j_2} G^{i_3 j_3} (\mathcal{L}(X_u)\Gamma)_{i_1 i_2 i_3} (\mathcal{L}(X_v)\Gamma)_{j_1 j_2 j_3}$$

where $X_u = \mu^{-1}(du)$ is the Hamiltonian vector field defined by the 'Hamiltonian' $u \in N$.

The cohomology class of $S^3_\Gamma$ is ([3, 11]) independent of the choice of the symplectic connection $\Gamma$, and for any $Q^3 \in \beta$ there exists a unique $\Gamma$ such that $Q^3 = S^3_\Gamma + \partial K$, $K$ being a differential operator of order $\leqslant 2$ (instead of 3).

Unfortunately, there is no explicit formula for the higher order terms in general. The technical condition $H^3(W; \mathbb{R}) = 0$ can certainly be weakened, if not dropped altogether, but was needed in the proof.

We shall now deal with the associative case, which is in general even more complicated.

## 2. Flat Poisson Manifolds and Related Associative and Lie Algebra Deformations of $N$

(a) Many of the results valid for symplectic manifolds extend without change to 'degenerate' symplectic manifolds, what we call *Poisson manifolds*.

These are $m$-dimensional (differentiable) manifolds $W$ endowed with a 2-tensor $G$ of rank $2n$ ($2n \leqslant m$, but here the rank may be smaller than the dimension) satisfying $[G, G] = 0$ in terms of the Schouten bracket [7], which is the necessary and sufficient condition for

$$(2.1) \qquad P(u, v) = \{u, v\} = i(G)(du \wedge dv), \qquad u, v \in N,$$

to define a Lie algebra structure (which we shall call the *Poisson Lie algebra*) on the space $N$ of differentiable functions over $W$. The Poisson manifold will be said *regular* if $G$ has constant rank $2n$ (i.e. $G^n \neq 0$, $G^{n+1}=0$, everywhere). A regular Poisson manifold admits [8] a foliation by $2n$-dimensional symplectic manifolds.

As example of Poisson manifold it is worthwhile to mention the dual space $\mathscr{A}^*$ of a real Lie algebra $\mathscr{A}$, the Poisson structure being defined by the *coadjoint* (= dual of the adjoint) *representation*; that is, the 2-tensor $G$ is defined by $G_\xi(a, b) = \langle \xi, [a, b] \rangle$ where $a, b \in \mathscr{A}$ with bracket $[a, b]$ and $\xi \in \mathscr{A}^*$, the duality between $\mathscr{A}$ and $\mathscr{A}^*$ being expressed by $\langle\ ,\ \rangle$. The orbits in $\mathscr{A}^*$ under the coadjoint representation of a connected Lie group with Lie algebra $\mathscr{A}$ are symplectic manifolds, with the structure defined by $G$.

A *Poisson connection* on a Poisson manifold $(W, G)$ is a linear connection without torsion such that $\nabla G = 0$, $\nabla$ being the covariant differentiation defined by $\Gamma$ (cf. e.g. [13]). There exist infinitely many Poisson connections on a Poisson manifold; they can be characterized by the fact that, on any natural chart (a chart for which the components of $G$ are constants), if $\Gamma^i_{jk}$ are the usual coefficients of $\Gamma$, the quantities $\Gamma^{ijk} = G^{jl}G^{km}\Gamma^i_{lm}$ are completely symmetric. *The Poisson manifold* $(W, G)$ is said *flat* if it admits a *Poisson connection without curvature.*

(b) As was indicated by Vey, the $p$th Hochschild cohomology group of the associative algebra $\mathscr{P}_a$ of polynomials in $p_\alpha$ and $q_\alpha$ ($\alpha = 1, \ldots, n$) is isomorphic to $\mathscr{P}_a \otimes_R \wedge^p \mathbb{R}^{2n}$, and accordingly the differentiable cohomology of $N_a(W)$ is isomorphic to the global sections of the fibre bundle $\Lambda^p TW$. These are huge spaces and it is very difficult to trace obstructions into the zero class; except of course when dim $W = 2$ because then $H^3(N_a(W)) = 0$ and there is no obstruction to prolong the infinitesimal deformation (0.1) of $N_a$. Indeed, in the general case, Vey could do this only up to the fourth order. A case of special interest which includes $\mathbb{R}^{2n}$ with its usual symplectic structure is that of the above defined flat Poisson manifolds.

Indeed, let us consider on a natural chart U the following operator:

$$P^r(u, v)|_U = G^{i_1 j_1} \ldots G^{i_r j_r} \nabla_{i_1, \ldots, i_r} u \nabla_{j_1, \ldots, j_r} v. \tag{2.2}$$

This defines a bidifferential operator $P^r$ on $W$, and we set:

$$u *_\nu v = \exp(\nu P)(u, v) = \sum_{r=0}^{\infty} \frac{\nu^r}{r!} P^r(u, v) \tag{2.3}$$

where $P^0(u, v)$ is defined as $uv$.

We thus have a *formal associative deformation* of $N_a$, which is *unique* in the following sense (cf. [3]): $f(\nu P)=\Sigma_{r=0}^{\infty} a_r(\nu^r/r!)P^r$ defines a formal deformation of $N_a$ iff $f$ is (up to a constant factor) the exponential function.

The corresponding Lie algebra deformation of $N$ is given by (1.2), with $P^r$ defined by (2.2), and it is unique in the sense that the only formal function of $P$ which defines a Lie algebra deformation of $N$ is the sine function (cf. [3] and [14] for another proof when $W=\mathbb{R}^2$). Since dim $H^2_{\text{diff}}(N)=1$ for $W=\mathbb{R}^{2n}$, this deformation (the Moyal bracket) is also *infinitesimally locally the only non-trivial* differentiable deformation of $N$, up to equivalence. In view of the results mentioned in Section 1(b) for terms of higher order in $\lambda$ we have at each step at most 2 choices (for $W=\mathbb{R}^{2n}$, and a finite number for a general symplectic manifold) of nonequivalent differentiable deformations.

Finally, it is worthwhile mentioning that, by a direct computation, one sees that (2.2)–(2.3) defines an associative algebra deformation of $N_a$ up to second order in $\lambda$ iff the torsion of $\Gamma$ vanishes, and up to the third order in $\lambda$ (and then to all orders) iff in addition the curvature of the connection $\Gamma$ vanishes.

(It can be shown that when $b_2(W)=0$, all Moyal-Vey deformations (1.3) of $N$ are equivalent, and that when $F$ is exact and dim $H^2_{\text{diff}}(N)=1$, all non trivial differentiable deformations of $N$ can be reduced to the Moyal type by equivalence and a monomial change of the deformation parameter $\lambda$. If $b_3(W)=0$, there always exist differentiable associative deformations of $N_a$ starting as in (0.1). Cf. [15] for details.

## 3. Derivations of the Lie algebra $N$ on a symplectic manifold

The derivations of a Lie algebra are its infinitesimal automorphisms. In the case of $N$, a Lie algebra of functions on $W$, one may also study which infinitesimal transformations of the manifold $W$ preserve the Lie algebra structure.

When $N$ is endowed with the *Poisson bracket structure*, the answer to the latter question is well-known: the Poisson bracket structure is preserved by those transformations which leave the symplectic form $F$ invariant, i.e. the Lie algebra $L$ of symplectic (or Hamiltonian) vector fields characterized by $\mathscr{L}(Z)F=0$; they are also called infinitesimal canonical transformations (time-independent by construction). Note that $L$ is the quotient of $N$ by the constant functions. The answer to the former question has been given by Avez and Lichnerowicz [6]. If $W$ is non-compact, the derivations algebra of $(N, P)$ is the Lie algebra $L^c$ of conformal symplectic vector fields, i.e. vector-fields $Z$ such that $(\mathscr{L}(Z)+k(Z))F=0$ for some constant $k(Z)$; if $W$ is compact, in addition to $L$ we have the (outer, non-local) derivation $u \mapsto \int u\eta$, where $\eta = F^n$ is the symplectic volume element.

Concerning the deformed algebras $(N, Q)$ where $Q$ is given by (1.3), one has to distinguish between derivations (or vector-fields) not depending on the deformation parameter $\lambda$, or depending on it. This study has been performed in [3], and here we shall present only the results.

THEOREM 2.2. *Let $W$ be a symplectic manifold and $(N, Q)$ a deformation of $(N, P)$ of the Vey type, given by* (1.3).

(i) *The Lie algebras $L(Q)$ of derivations of $(N, Q)$ independent of $\lambda$, and of vector fields independent of $\lambda$ preserving the brackets $Q$, are subalgebras of the finite-dimensional Lie algebra $L_{SA}$ of the symplectic vector fields, affine for the unique symplectic connection $\Gamma$ associated with the cocycle $Q^3$. In the flat case with bracket $M$, given by* (1.2), *$L(M) = L_{SA}$; when $W = \mathbb{R}^{2n}$ with usual Moyal bracket, $L(M)$ is the inhomogeneous symplectic Lie algebra $\mathfrak{sp}(n, \mathbb{R}) \cdot \mathbb{R}^{2n}$; and the formal series of vector fields $Z = \Sigma_{r=0}^{\infty} \lambda^r Z_r$ preserves $M$ iff all $Z_r \in L(M)$.*

(ii) *The Lie algebra of derivations depending on $\lambda$ of a Vey deformation $(N, Q)$ is isomorphic to the Lie algebra $E(L; \lambda)$ of formal series in $\lambda$ with coefficients in $L$, while the inner derivations are isomorphic to $E(L^*; \lambda)$ where $L^* = [L, L]$ ($\mu(Z)$ is a closed 1-form iff $Z \in L$ and is exact iff $Z \in L^*$; thus all derivations are inner if $H^1(W, \mathbb{R}) = 0$).*

To prove the first part of the theorem one studies the Lie algebra $L(Q_1)$ of derivations (modulo $\lambda^2$) of the infinitesimal deformation $Q_1 = P + (\lambda/6)Q^1$. For the second part one shows inductively that the conformal part of the derivations appear only at the highest order in $\lambda$, but that all of $L$ does appear at any order.

## 4. Associative deformations (a star product is born)

As we mentioned in Section 2, the construction of an associative algebra deformation of $N_a$ (a $*$-product) is quite difficult in the non-flat case. However a producedure can be given, which allows the construction of such $*$-products using the flat case (cf. [3]).

Indeed, let $\hat{W}$ be a flat symplectic manifold, with 2-tensor $\hat{G}$ and connection $\hat{\Gamma}$ (e.g. an open set in some $\mathbb{R}^{2n+2}$ with the usual structure and connection). Formulas (2.2) and (2.3) define, on the $C^\infty$ functions $\hat{N}(\hat{W})$, cochains $\hat{P}^r$ and a deformation $\exp(\nu\hat{P})$. Let a symplectic manifold $W$ be the quotient of $\hat{W}$ by some group $S$ of diffeomorphisms of $\hat{W}$ with Lie algebra contained in $L_{SA}$, namely preserving both $G$ and $\hat{\Gamma}$, the symplectic structure $G$ on $W$ being defined by quotient from $\hat{G}$. Then $N(W)$ can be identified with the space $\hat{N}_S$ of $S$-invariant $C^\infty$ functions on $\hat{W}$. By construction, if $\hat{u}, \hat{v} \in \hat{N}_S$, then $\exp(\nu\hat{P})(\hat{u}, \hat{v}) \in \hat{N}_S$ also (since $S$ preserves also $\hat{\Gamma}$). Therefore, if we denote by $\phi$ the isomorphism $N(W) \to \hat{N}_S$, we can define an associative deformation of $N_a$ by

$$u * v = \phi^{-1}(\exp(\nu\hat{P})(\phi u, \phi v)), \qquad u, v \in N(W). \tag{4.1}$$

For instance, take for $W$ the cotangent bundle $T^*S^n$ to an $n$-dimensional sphere, imbedded in $\hat{W} = T^*(\mathbb{R}^{n+1} - \{0\})$ by two (second-class Dirac) constraints:

$$p^2 \equiv \sum_{\alpha=1}^{n+1} p_\alpha^2 = 1, \qquad pq \equiv \sum_{\alpha=1}^{n+1} p_\alpha q_\alpha = 0.$$

$W$ is the quotient of $\hat{W}$ under the 2-dimensional group

$$(p_\alpha, q_\alpha) \mapsto (\rho p_\alpha, \rho^{-1} q_\alpha + \sigma p_\alpha), \qquad \rho > 0, \sigma \in \mathbb{R}.$$

In this way we define a $*$-product on $W$, which will be very useful for physical applications (hydrogen atom problem).

Analogous procedures (more involved) can be used (cf. [16]) to construct explicit $*$-products on the cotangent bundles of most real classical groups and Stiefel manifolds (such as $SU(m, n)/SU(m', n')$, *etc.*).

Such constructions, and the finite-dimensionality of the invariance group of a Vey deformation, lead us to develop a more abstract theory of $*$-products, to which we are now coming.

## 5. Invariant $*$-Products

Let $(W, G)$ be a Poisson manifold. Any 'Hamiltonian' $u \in N(W)$ defines a vector field $X_u \in L$ by $X_u v = \{u, v\}$ which is an inner derivation (infinitesimal automorphism) of the Poisson bracket $\{\cdot, \cdot\} = P$. Inspired by our study of Section 3, we shall consider an associative algebra structure on $N$, the product of which will be denoted $*$ and say that it is *invariant under* $u$ if $X_u$ is a derivation of $(N, *)$, which can be written:

$$\{u, f * g\} = \{u, f\} * g + f * \{u, g\}, \qquad \forall f, g \in N. \tag{5.1}$$

Obviously, the set of $u \in N$ under which $*$ is invariant is a Lie subalgebra of $(N, P)$. From Section 3, we know that in the flat case with the 'usual Moyal' $*$-product given by (2.3), this is a finite-dimensional Lie algebra (the semi-direct product of $\mathfrak{sp}(n, \mathbb{R})$ with the nilpotent Heisenberg algebra $H_n$ generated by $p_\alpha$, $q_\alpha$ and $I$, $\alpha = 1, \ldots, n$, in the case of $\mathbb{R}^{2n}$ with the usual structure).

We are thus led to start with a finite-dimensional Lie algebra $\mathscr{A}$, look at the Poisson structure in its dual $\mathscr{A}^*$, restrict ourselves to some submanifold $W$ (defined by some polynomial equation e.g.) and look at $*$-products on $N(W)$ or on $N(\mathscr{A}^*)$ invariant under $\mathscr{A}$. More precisely:

DEFINITION. Let $\mathscr{P}$ be the polynomials on $\mathscr{A}^*$ (in other words, the symmetric algebra of $\mathscr{A}$), and $\mathscr{I}$ some ideal in $\mathscr{P}$, the zeros of which define an algebraic variety $M \subset \mathscr{A}^*$. Then an *invariant $*$-product* on $M$ is an associative algebra structure on $\mathscr{P}/\mathscr{I}$ defined by a product law $*$ such that ($\hbar$ being 'some' parameter)

(5.2) $$k * \bar{f} = f * k = kf, \qquad a * b - b * a = i\hbar\{a, b\},$$

(5.3) $$\{a, f * g\} = \{a, f\} * g + f * \{a, g\}$$

for all $k \in \mathscr{P}$, $a, b \in \mathscr{A}$; $f, g \in \mathscr{P}/\mathscr{I}$.

The simplest example is perhaps $\mathscr{A} = H_n$ with basis $p_\alpha$, $q_\alpha$ and $I$, $M$ being defined by the equation $I = 1$, which is a simple $2n$-dimensional (symplectic) orbit in $\mathscr{A}^*$, and the $*$-product defined by (2.3) on polynomials in $p_\alpha$'s and $q_\alpha$'s (this corresponds to the so-called symmetric ordering in Weyl quantization).

More generally, if $\phi$ is any vector-space isomorphism between the symmetric algebra $\mathscr{P}$ of a Lie algebra $\mathscr{A}$ and its enveloping algebra $\mathscr{U}(\mathscr{A})$, then

(5.4) $$f * g = \phi^{-1}(\phi f \cdot \phi g)$$

is an invariant $*$-product on $\mathscr{A}$ (provided $\phi$ is normalized so that (5.2) is satisfied), $\phi f \cdot \phi g$ being the usual product in $\mathscr{U}(\mathscr{A})$. (5.4) defines also an invariant $*$-product on the manifold in $\mathscr{A}^*$ defined by fixing the values of the elements of the center $\mathscr{Z}(\mathscr{A})$ of $\mathscr{U}(\mathscr{A})$, $\phi$ being a vector-space isomorphism between the corresponding quotients of $\mathscr{P}$ and of $\mathscr{U}(\mathscr{A})$.

For $\mathscr{A} = H_n$, defining $\phi$ by the choice of a basis in $\mathscr{U}(\mathscr{A})$, one gets in this way the various so-called orderings (e.g. the standard ordering is obtained by choosing the basis $q^k p^l I^m$; cf. [17]).

Now we can extend what was done here on $\mathscr{P}$ to the algebra $\mathscr{P}'$ of formal power series on $\mathscr{A}^*$. This allows us to define (cf. [3]) the so-called $*$-exponential function Exp: $\mathscr{A} \to \mathscr{P}'$ by

$$\text{(5.6)} \qquad \operatorname{Exp}(a) = \sum_{n=0}^{\infty} ((i\hbar)^n n!)^{-1} (a*)^n, \qquad (a*)^n = (a* \cdots * a)$$

(*n* factors $a \in \mathscr{A}$),

which is a kind of formal group element. Then Exp can be characterized by some properties, the main one being that Exp is a solution of some differential equations (related to the elements of $\mathscr{L}(\mathscr{A})$), and any function satisfying these properties is the Exp for some invariant $*$-product on $\mathscr{A}^*$. This gives a procedure for computing explicitly such products, which has been carried out in several important examples (including $\mathfrak{sl}(2, \mathbb{R})$ of course). In general, these products will be given by pseudodifferential operators.

And one is not surprised to find a very strong relation between the various $*$-products and the various irreducible representations. In fact, the $*$-products theory can be viewed as a 'representation theory without operators', in the same way as we shall use $*$-products in Chapter 4 to perform an autonomous 'quantum mechanics without operators'. This seems to pave the way for a new 'geometric representation theory', which looks promising.

## Notes and references to chapters 1 and 2

[1] Chevalley, C., and Eilenberg, S., *Trans. Amer. Math. Soc.* **63** (1948), 85–124.
[2] Gerstenhaber, M., *Ann. Math.* **79** (1964), 59–103.
[3] Bayen, F., Flato, M., Fronsdal, C., Lichnerowicz, A., and Sternheimer, D., 'Deformation theory and quantization: I. Deformations of symplectic structures', *Ann. Phys.* (*N.Y.*) 111 (1978), 61–110; *cf.* also *Lett. Math. Phys.* 1 (1977), 521–530.
[4] Lichnerowicz, A., *J. Math. Pures et Appl.* **53** (1974), 459–484.
[5] Flato, M., Lichnerowicz, A., and Sternheimer, D., *Compositio Mathematica* **31** (1975), 47–82.
[6] Avez, A., and Lichnerowicz, A., *C.R. Acad. Sc. Paris* **275** (1972), A. 113–118.
[7] Nijenhuis, A., *Indag. Math.* **17** (1955), 390–403.
The bracket can be defined in the following way: if $A$ (resp. $B$) is a $p$-tensor (resp. $q$-tensor), $[A, B]$ is the $(p+q-1)$ tensor defined by the relation:

$$i([A, B])\beta = (-1)^{pq+q} i(A)\, \mathrm{d}i(B)\beta + (-1)^p i(B)\, \mathrm{d}i(A)\beta$$

for every closed $(p+q-1)$ form $\beta$; in particular, for $p=1$, $[A, B] = \mathscr{L}(A)B$, where $\mathscr{L}$ is the Lie derivative.
[8] Lichnerowicz, A., *J. Diff. Geom.* 12 (1977), 253–300.
[9] Vey, J., *Commentarii Math. Helvet.* 50 (1975), 421–454.

[10] Moyal, J., *Proc. Cambridge Phil. Soc.* **45** (1949), 99.
[11] Flato, M., Lichnerowicz, A., and Sternheimer, D., *C.R. Acad. Sc. Paris* **283** (1976), A 19–24.
[12] Gutt, S., '2e and 3e espace de cohomologie differentiable de l'algèbre de Lie de Poisson d'une variété symplectique', Brussels University preprint (1979).
[13] Kobayashi, S., and Nomizu, K., *Foundations of Differential Geometry*, Interscience N.Y., 1963.
[14] Mehta, C. L., *J. Math. Phys.* **5** (1964) 677.
[15] Gutt, S., *Lett. Math. Phys.* 3 (1979), 297.
Lichnerowicz, A., *C. R. Acad. Sci Paris* 289 (1979).
Neroslavsky, O. M and Vlasov, A. T. (to be published).
[16] Lichnerowicz, A., *Lett. Math. Phys.* 2 (1977), 133.
[17] Agarwal, G. S., and Wolf, E. *Phys. Rev.* D2 (1970), 2161–2225.

## CHAPTER 3

# PHYSICAL APPLICATIONS RELATED TO 1-DIFFERENTIABLE DEFORMATIONS OF POISSON BRACKETS

## 0. INTRODUCTION

When do we utilize the structure of a deformed Lie algebra in Physics? As a matter of fact and a posteriori one can say that almost always when we passed from one level of description of phenomena to another one the deformation structure was present. This is the case when one passes from non-relativistic Galilean mechanics to relativistic (Lorentz invariant) mechanics, as the Poincaré Lie algebra is a deformation of the Galilean Lie algebra (when the light velocity $c$ becomes finite). This is also the case, as will be shown in the next chapter, when one passes from classical mechanics, which is described by the infinite-dimensional Lie algebra of $C^\infty$ functions on phase-space endowed with the Poisson bracket structure, to a phase-space formulation of quantum mechanics (in this case it is $\hbar$ – the Planck constant – which becomes a finite positive number).

In this chapter, we shall show some applications to physics of the 1-differentiable deformations only. It is quite evident that if one is concerned with questions of the kind of whether there exists in a certain sense a 'close' and non-equivalent formalism to the Hamiltonian formalism of dynamics (which might describe a different type of dynamics from the usual one), or the question of stability of a given Hamiltonian system under *natural instabilities*, the deformation structure is a very natural one to use. It is also clear that one might be interested to know under which conditions a perturbed Hamilton system will be equivalent to a non-perturbed (or free) deformed system.

Another important application will be the Dirac constraints formalism: in order to treat and quantize systems for which position and momentum are restricted by some constraints, Dirac had to introduce a new bracket (the so-called Dirac bracket) which, as it turns out, has a natural geometric interpretation with symplectic manifolds, and can formally be expressed as a 1-differentiable deformation of the Poisson bracket and be considered as the Poisson bracket for the natural Poisson structure on the constraints manifolds. But first let us briefly describe what is the original Dirac approach.

## 1. Dirac Constraints Formalism

(a) In classical mechanics, one has coordinates $q_\alpha$ ($\alpha=1,\ldots,n$), velocities $\dot{q}_\alpha=\mathrm{d}q_\alpha/\mathrm{d}t$, and a Lagrangian $L$, a function of $q$ and $\dot{q}$. From the variation of the action integral $\int L\,\mathrm{d}t$ one gets the Lagrange equations of motion

$$\frac{\mathrm{d}}{\mathrm{d}t}\left(\frac{\partial L}{\partial \dot{q}_\alpha}\right)=\frac{\partial L}{\partial q_\alpha}.$$

One then defines momentum as $p_\alpha=\partial L/\partial \dot{q}_\alpha$, whence the phase-space $\mathbb{R}^{2n}$ with coordinates $(p_\alpha, q_\alpha)$. The Hamiltonian is defined as $H=p_\alpha\dot{q}_\alpha-$ $-L(q,\dot{q})$ where we sum over repeated indices; since its variation is:

$$\delta H=\dot{q}_\alpha\cdot\delta p_\alpha-(\partial L/\partial q_\alpha)\cdot\delta q_\alpha, \tag{1.1}$$

$H$ is a function of the $p_\alpha$ and $q_\alpha$. The evolution of any dynamical variable $f(p,q)$ is now given by $\dot{f}=\{H,f\}$, the Poisson bracket (Equation (2.2), Chapter 1) of $H$ and $f$ ($x_\alpha=p_\alpha$, $x_{\bar{\alpha}}=q_\alpha$): for $f=p_\alpha$ and $q_\alpha$, we get the usual Hamilton equations of motion, whence one deduces the evolution of any $f$.

Usually, one assumes that the momenta $p$ are independent functions of the velocities $\dot{q}$. But this need not be always true, and we may have a priori some (independent) relations $\varphi_r(p,q)=0$ ($r=1,\ldots,k_1$), which restrict the possible values of the $p$'s and $q$'s. Bergmann [2] called these relations *primary constraints* to distinguish them from the consistency relations derived (as we shall see later) from expressing $\dot{\varphi}_r=0$ (for a given system), called *secondary constraints*. Because of the restrictions on the variation of $p$'s and $q$'s imposed by the constraints, we deduce from (1.1) and Lagrange equations that

$$\dot{q}_\alpha=\frac{\partial H}{\partial p_\alpha}+u_r\frac{\partial \varphi_r}{\partial p_\alpha},\qquad \dot{p}_\alpha=\frac{\partial L}{\partial q_\alpha}=-\frac{\partial H}{\partial q_\alpha}-u_r\frac{\partial \varphi_r}{\partial q_\alpha} \tag{1.2}$$

where the $u_r$'s are some (unknown) 'coefficients (which are not functions of $p$ and $q$ but may depend on $t$)'. If we write $\approx$ for an equality which is satisfied on the surface defined by the constraints (a 'weak equality' in the terminology of Dirac) one deduces from (1.2) that $\dot{f}\approx\{H_T, f\}$, where $H_T=H+u_r\cdot\varphi_r$ is called the total Hamiltonian. In fact, in Dirac's formalism, $\approx 0$ means equal to some 'linear' combination of the constraints, with arbitrary functions as coefficients (i.e. belongs to the ideal generated by the constraints). A more intrinsic formulation will be given later.

(b) If we want our theory to be consistent, the time trajectories in phase-space must remain on the surface defined by the constraints, that is we must have $\dot{\varphi}_s \approx 0$ or

(1.3) $$\{H, \varphi_s\} + u_r\{\varphi_r, \varphi_s\} \approx 0.$$

These relations lead to one of the following four alternatives:

(i) An inconsistency. We exclude this possibility, that is we assume (as in the non-constrained theory) that the Lagrangian (which is given by a concrete physical situation) does not give rise to inconsistencies in the development of the theory.

(ii) An identity, in which case there is nothing more to say.

(iii) A new equation involving only the $p$'s and $q$'s (but not the $u$'s) independent of the $\varphi$'s (otherwise we are in case (ii)), namely a new constraint $\chi(p, q) = 0$. Such constraints are called *secondary*. Expressing $\dot{\chi} \approx 0$ we may get a further secondary constraint, etc. In this way we get finally a family $\varphi_s (s = 1, \ldots, k \geqslant k_1)$ of constraints, primary for $s = 1, \ldots, k_1$ and secondary for $k_1 + 1 \leqslant s \leqslant k$.

(iv) A condition on the $u$'s, which may be obtained also from secondary constraints. That is, in (1.3), we take $1 \leqslant r \leqslant k_1$ and $1 \leqslant s \leqslant k$. The general solution of this system (linear in the $u$'s) which exists (if not, we are in case (i) above which was excluded) is of the form $u_r = U_r + v_a V_{ar}$ where the $U$'s and $V$'s are functions of $p$ and $q$, the $V_{ar}$ being independent solutions of the homogeneous equations $V_r\{\varphi_r, \varphi_s\} \approx 0$ and the coefficients $v_a$ $(a = 1, \ldots, h)$ arbitrary functions of time. The total Hamiltonian can then be written $H_T = H' + v_a \varphi_a$, where we define

(1.4) $$H' = H + U_r \varphi_r, \qquad \varphi_a = V_{ar} \varphi_r.$$

(c) Using (1.3) one sees that these quantities $H'$ and $\varphi_a$ have Poisson brackets with the constraints $\varphi_s$ that vanish weakly. Dirac [1] called *first-class* all functions $f$ (of $p$ and $q$) such that $\{f, \varphi_s\} \approx 0$ for all $s = 1, \ldots, k$; that is $\{f, \varphi_s\} = g_{ss'} \varphi_{s'}$. All others are called *second-class*. Thus $H'$ is first-class and the $\varphi_a$ $(a = 1, \ldots, h)$ are *first-class constraints*. Jacobi identity shows that the Poisson bracket of two first-class functions is still first-class; these functions form therefore a Lie subalgebra of the Poisson bracket algebra of functions.

(d) The arbitrariness of the $v$'s, which means that the physical content of the theory is independent of their choice, shows [1] that the symplectic vector fields (infinitesimal contact transformations in Dirac's terminology), $\mu(\mathrm{d}\varphi_a)$ in the notations of Chapter 1, generated by the primary first-class constraints $\varphi_a$ and their Poisson brackets between themselves or with $H'$ lead to changes in $p$'s and $q$'s which 'do not affect the physical state'. We shall come back later to the geometric meaning of this fact.

Let us just mention here that the distinction between first-class and second-class constraints is more natural and quite disjoint from the distinction between primary and secondary constraints. The latter follows in fact from our possible clumsiness in the choice of the original Lagrangian and more generally in the expression, in a given situation, of all the limitations which have to be taken into account. These limitations manifest themselves through the fact that everything occurs on some 'surface' in phase-space, and this will be the framework of the geometric formulation.

(e) Let us now say a few words on the quantization of a constrained system, which will lead to the so-called Dirac brackets. If we have only first-class constraints, we quantize as usual the $p$'s and $q$'s by mapping them into a linear representation of the Heisenberg algebra $H_n$ (integrable to a unitary group representation, in the sense of Chapter 5), and write the Schrödinger equation $i\hbar(\mathrm{d}\psi/\mathrm{d}t)=H'\psi$, the wave-function $\psi$ being a (unit) vector in the representation space. Then the (first-class) constraints give us additional conditions on the wave function, namely $\Phi_j\psi=0$. These relations are consistent if $[\Phi_i, \Phi_j]=C_{ijl}\Phi_l$ where the $C$'s may be operators. For first-class constraints one does have on the classical level $\{\varphi_i, \varphi_j\}=$ $=c_{ijl}\varphi_l$; however there is both an ordering problem for the right-hand side when one quantizes, and the fact that it is not the Poisson bracket which goes into the commutator of two operators in the Weyl quantiza-ation procedure but the Moyal bracket (as we shall see in the next chapter); so that one needs 'a bit of luck' (in Dirac's terminology) to get the consistency conditions in the form written above, with the $C$'s on the left. For the additional conditions to be consistent with the Schrödinger equation one gets similarly $[\Phi_j, H']\psi=0$, and thus the fact that the classical $H'$ is first-class has to be translated by $[\Phi_j, H']=B_{jl}\Phi_l$ with the $B$'s on the left; this also requires a bit of luck.

(f) Now if we have second-class constraints, the straightforward procedure described above fails, no matter how lucky we are. Indeed, for the simplest example of only two canonically conjugate constraints $p_1 \approx 0$ and $q_1 \approx 0$, which are second-class since their Poisson bracket is equal to 1 (and not 0 even weakly), the additional conditions given by the constraints are inconsistent because (due to the commutation relations) the operators $P_1$ and $Q_1$ cannot have common eigenvectors. However this example shows us the way: here one has to modify the Poisson bracket so as to get the Poisson bracket induced on the symplectic submanifold defined by the constraints, and then quantize as usual.

We thus first put (by linear combinations, with functions as coefficients) 'as many constraints as possible into the first class', and call $\chi_s$ $(s=1,\ldots,k-h)$ the remaining second-class constraints. That is, in the ideal (of the algebra $N$ of functions) generated by the constraints, we take generators of a (non-uniquely determined in general) supplementary to the subideal of first-class constraints. Then Dirac proves (and we shall see below why this works) that $\det(\{\chi_s, \chi_{s'}\})\neq 0$ which shows that (because it is skew-symmetric) the number $k-h$ of second-class constraints is even. This allows Dirac to take the inverse matrix $(C_{ss'})$ defined by $C_{ss'}\{\chi_{s'}, \chi_{s''}\}= =\delta_{ss''}$, and to introduce his famous brackets:

$$\{f, g\}^* = \{f, g\} - \{f, \chi_s\} C_{ss'} \{\chi_{s'}, g\}. \tag{1.5}$$

By a cumbersome 'straightforward' computation, Dirac shows that (1.5) is a Lie algebra law. The profound reason for this is, as we shall see, that the new bracket is nothing but the Poisson bracket on the (even dimensional) symplectic submanifold defined by the second-class constraints.

One then notices that, since the total Hamiltonian $H_T$ is first-class, one has $\dot{f}=\{H_T,f\}\approx\{H_T,f\}^*$ and also $\{f, \chi_s\}^*=0$ for all functions $f$. This shows that we can describe the evolution by means of the new bracket and put $\chi_s=0$ before working out the new brackets. Therefore in the quantization procedure, if we use the new brackets, we can translate the equations $\chi_s=0$ into relations between operators (and not into conditions on the wave functions). The remaining conditions are all of the first-class type dealt above.

(g) We are now in position to give a geometric description of Dirac's formalism on the classical level ([3, 4], see also [5] for a partial treatment). This description also paves the way towards quantization (in a phase-space formulation) at least on the bracket level for which we have the Vey deformations; the associative product is a more difficult question, but it can be solved in many instances as we mentioned in Chapter 2 – and this justifies a posteriori the 'bit of luck' of Dirac.

Let $(W, F)$ be a symplectic manifold of dimension $2n$, $N$ its dynamical algebra ($C^\infty$ functions with Poisson bracket). Let $M$ be a submanifold of $W$ of dimension $2n-k$, $U$ any chart domain on $W$ intersecting $M$. Denote by $C_U$ the space of all functions (in $N(U)$) which are constant on $M \cap U$: thus, $M$ is given by the set of constraints $\varphi(x)=\varphi(x_0)$, $\varphi \in C_U$, $x_0$ being any point in $M \cap U$. We now say that a function $f$ is first-class if $\{f, \varphi\}$ is zero on $M \cap U$ for all $\varphi \in C_U$ and denote by $B_U$ the Lie subalgebra of $N$ of all such functions. The first-class constraints $A_U = B_U \cap C_U$ form an ideal in $B_U$.

The 2-tensor $\mu^{-1}(F)=G$ defines by restriction to $M$ a structure of Poisson manifold. We may, for our purpose, restrict ourselves to the case where $M$ is a regular Poisson manifold, i.e. suppose that the 2-form $F$ has fixed rank $2n-k-h$ when restricted to $M$. Then for all $x \in M$ the $h$-planes (in the tangent space $T_x(M)$ to $M$ at $x$) defined by

$$N_x = \{v \in T_x(M);\ i(v)F|_M(x)=0\}$$

are an integrable distribution (cf. e.g. [6]) and define a foliation in $M$, thus a quotient space $\hat{M}$. We suppose that $\hat{M}$ is a manifold (of dimension $2n-k-h$), the projection $M \to \hat{M}$ being a submersion. Then $G$ defines in a natural manner a symplectic structure $\hat{G}$ on $\hat{M}$, the manifold of physical states in the sense of Dirac.

We shall say that $M$ is first-class if $h=k$, second-class if $h=0$. Since $A_U$ is an ideal in $C_U$, it makes sense to "put the maximum number of constraints into the first-class", but the choice of a supplementary is not unique. Nevertheless we have a splitting, that is (cf. [4, 5]) there are second-class symplectic submanifolds $(\tilde{W}, \tilde{G})$ of $(W, G)$ of dimension $2n-k+h$ such that $M$ is a first-class submanifold of $(\tilde{W}, \tilde{G})$. The Dirac bracket on $N(W)$ is then nothing else but the Poisson bracket of the Poisson manifold $W$ with 2-tensor $\tilde{G}$ (easily expressed in terms of $G$ and the $\chi$'s), and coincides on $\tilde{W}$ with the Poisson bracket of the symplectic manifold $(\tilde{W}, \tilde{G})$.

(h) As an application of this procedure, we shall show that Nambu's proposal [7] of a generalized mechanics is equivalent [8] to a degenerate Hamiltonian system with Dirac constraints. Let $\vec{r}=(x_1, x_2, x_3) \in \mathbb{R}^3$ be a vector in a 'generalized Nambu phase-space', and $H(\vec{r})$, $G(\vec{r})$ be two 'Hamiltonians' ($C^\infty$ functions on this space). Nambu's equations of motion (for which $H$ and $G$ are first integrals of motion) are then written as $\mathrm{d}\vec{r}/\mathrm{d}t = \vec{\nabla} H \times \vec{\nabla} G$. We propose to imbed these equations in singular (Dirac) Hamiltonian formalism in a six-dimensional phase-space, $x$ being identified with the position $q$. Let $\Omega \subset \mathbb{R}^3$ be an open set in $\mathbb{R}^3$ such that $(\vec{\nabla} H \times \vec{\nabla} G)_3 \neq 0$ in $\Omega$. Define the Lagrangian

$$L_N(\vec{q}, \dot{\vec{q}}) = H(\vec{q}) + \sum_{i=1}^{3} \dot{q}_i(\partial G(\vec{q})/\partial q_i).$$

It is easy to see that under our hypotheses we have *for this Lagrangian* both $H$ and $G$ as first integrals of motion. With the aid of Dirac's formalism we get 3 constraints (two second-class and one first-class), and compatibility equations translating the fact that the constraints should not change with time, written as $\vec{q}=v(t, \vec{q})(\vec{\nabla} H \times \vec{\nabla} G)$ with an arbitrary function $v$ (a 'Dirac gauge'). After rescaling the time in these equations in a position-dependent way, we get rid of the $v$ and obtain the above mentioned Nambu equations. Actually, since $v$ is arbitrary, we have infinitely many copies of these equations (when we vary $v$ in a suitable function space), but it is quite easy to interpret this construction both mathematically and physically in such a way that this infinity of copies of Nambu's equations contains the *same dynamical informations* as the original equations.

Note that the use of a time rescaling suggests (at least in presence of first-class constraints) that it would be worthwhile to write Dirac's formalism in the framework of canonical manifolds [9], which are regular Poisson manifolds $W$ of dimension $2n+1$ with 2-tensor $G$ of rank $2n$ everywhere satisfying $[G, G]=0$ in terms of Schouten brackets, endowed with the additional structure defined by a regular function $t(x)$ ($dt \neq 0$ everywhere) satisfying $[G, t]=0$. The automorphisms of such a structure are (when $W$ is the product of a symplectic manifold of dimension $2n$ by $\mathbb{R}$) the (time-dependent) canonical transformations in the sense of [10].

## 2. Applications of 1-differentiable deformations

(a) *Dirac brackets as a deformation.* Let $W$ be a symplectic manifold of dimension $2n$, $N$ its dynamical Lie algebra (functions with Poisson brackets), and define Dirac bracket by (1.5) for a given family of independent second-class constraints. We may write for $u, v \in N$:

$$\{u, v\}^* = \{u, v\} + C(u, v) \tag{2.1}$$

the cochain $C$ being given by (1.5). One remarks first that the Jacobi identity for the bracket

$$\{u, v\} + \lambda C(u, v) \tag{2.2}$$

gives the condition ($S$ being summation over circular permutations of $u, v, w$ and with summation over $s, s' = 1, \ldots, h$):

$$\lambda(\lambda-1) S\{u, \chi_s\}\{v, \chi_{s'}\}\{C_{ss'}, w\} = 0 \tag{2.3}$$

which shows that in general the bracket (2.2) will be a Lie algebra law iff

$\lambda=0$ (original Poisson bracket) or $\lambda=1$ (Dirac bracket). However, if the constraints are canonically conjugate (that is, can be taken as some of the $p$'s and the corresponding $q$'s on any symplectic chart for suitably chosen coordinates), all $C_{ss'}$ are constant and (2.2) defines a Lie algebra for any value of $\lambda$: in this case, (2.2) is a rigorous 1-differentiable deformation. (For a given second-class submanifold $\tilde{W}$ of codimension $h=2$ in $\tilde{W}$, one may choose two defining constraints $\chi$ so that this will be true; in the general case one has to reduce the dimension by steps).

Nevertheless it is possible to consider any Lie algebra law (2.1) where $C$ is a pure 1-differentiable 2-cochain – e.g. the Dirac bracket – as an *instant of a formal* 1-*differentiable deformation* when $u, v \in \mathscr{P}'$, the space of formal series on $\mathbb{R}^{2n}$. More precisely one proves [3] that *there exists a formal deformation of* $\mathscr{P}'$ (i.e. there exist cochains $C_r$ defining a deformation):

$$(2.4) \qquad [u, v]_\lambda = \{u, v\} + \sum_{r=1}^{\infty} \lambda^r C_r(u, v)$$

*such that* (2.1) *is a specialization of* (2.4) *for, say,* $\lambda=1$

$$\text{(i.e. } C(u, v) = \sum C_r(u, v) \qquad \text{for } u, v \in \mathscr{P}'\text{).}$$

(b) *Deformations and perturbations.* We denote by $E(N; \lambda)$ the Lie algebra of formal series $u_\lambda = u + \Sigma_{r=1}^{\infty} \lambda^r u_r$ with coefficients $u, u_r \in N$.

Let $H \in N$ be a Hamiltonian and

$$(2.6) \qquad H_\lambda = H + \sum_{r=1}^{\infty} \lambda^r V_r = H + V \in E(N; \lambda)$$

a formal perturbation of $H$. We ask the question of the equivalence between usual mechanics with perturbed Hamiltonian $H_\lambda$, for which the evolution of $u_\lambda \in E(N; \lambda)$ is given by Hamilton equations

$$(2.7) \qquad \dot{u}_\lambda = \{H_\lambda, u_\lambda\}$$

and a deformed mechanics with free Hamiltonian $H$ but deformed bracket (2.4), for which, in local symplectic coordinates, the deformed Hamilton equations give the evolution of $u_\lambda$:

$$(2.8) \qquad \dot{u}_\lambda = [H, p_\alpha]_\lambda \partial_\alpha u_\lambda + [H, q_\alpha]_\lambda \partial_{\bar{\alpha}} u_\lambda.$$

One may remark that (2.8) is equivalent to

$$(2.9) \qquad \dot{u}_\lambda = [H, u_\lambda]_\lambda$$

for all $H \in N$ iff the cochains $C_G$ are pure. This can be shown by direct computation or by noticing that the product law

(2.10) $\quad [u, vw]=[u, v]w+v[u, w]$

holds for all $u, v, w \in N$ with the bracket (2.4) iff all $C_r$ are pure.

Comparing both values of $u_\lambda$, one sees that (2.7) and (2.8) are equivalent iff, for all $r \geqslant 1$:

(2.11) $\quad C_r(H, q_\alpha)\, \mathrm{d}p_\alpha - C_r(H, p_\alpha)\, \mathrm{d}q_\alpha = \mathrm{d}V_r$

or, if the $C_r$ are *pure*, $\{V_r, u\} = C_r(H, u)$. For $u=H$, one sees that *the allowed perturbations must be constants of motion for the free Hamiltonian.*

For example, when $H=\frac{1}{2}\Sigma_{\alpha=1}^3 p_\alpha^2$ (squared linear momentum) which generates the center of the enveloping algebra $\mathscr{U}$ of the Euclidean group $SO(3)\cdot\mathbb{R}^3$, any element of $\mathscr{U}$ (such as the angular momentum given by $L^2$ or one of its components $L_\alpha$) can be a perturbation.

If we denote $\alpha_r = \mu(C_r)$ and $Z=\mu^{-1}(\mathrm{d}H)$, the $C_r$ being pure, (2.11) writes $i(Z)\, \mathrm{d}\alpha_r = -\mathrm{d}V_r$, which can be verified iff the integrability conditions

(2.12) $\quad \mathscr{L}(Z)\alpha_r = i(Z)\, \mathrm{d}\alpha_r$

are satisfied. Conversely, locally, given $V_r$'s commuting with $H$, one can build [3] by induction the $C_r$'s:

PROPOSITION 3.1. *Any sequence of pure* 1-*differentiable* 2-*cochains* $C_r$ *such that* (2.4) *is a deformation and* (2.12) *is satisfied can be associated with a perturbation* (2.6) *of the free Hamiltonian* $H$ *by* $V_r$*'s commuting with* $H$ *in such a way that* (2.7) *and* (2.9) *are equivalent.*

*Conversely, on any open chart, given a perturbation* (2.6) *of* $H$ *by* $V_r$*'s commuting with* $H$*, one can find pure cochains* $C_r$ *defining a deformation* (2.4) (*for which* (2.12) *will hold*) *such that* (2.7) *and* (2.9) *are equivalent.*

(c) *Applications of rigorous essential* 1-*differentiable deformations.* We have seen that for a cotangent bundle $W=T^*M$ to a manifold with $H^1(M) \neq \{0\}$, formula (4.3) of Chapter 1 defines a rigorous essential deformation. In canonical coordinates, if $\beta = B^\alpha\, \mathrm{d}q_\alpha$ on $M$ (summation on $\alpha=1,\ldots,n$), this writes:

(2.13) $\quad [u, v]_\lambda = \{u, v\} - \lambda(p^\alpha B^{\alpha'} - p^{\alpha'} B^\alpha)\partial_\alpha u \partial_{\alpha'} v + \lambda B^\alpha(u\partial_\alpha v - v\partial_\alpha u).$

We shall apply this to the simplest cases, $M=T^1$ (circle) and $M=T^2$ (torus). In the first case, (2.13) writes for the choice $\beta=\mathrm{d}q$ (writing $\partial_1=\partial/\partial_p$

as usual):

(2.14) $$[u, v]_\lambda = \{u, v\} + \lambda(u\partial_1 v - v\partial_1 u).$$

In the second case (2.13) writes (summation on $\alpha = 1, 2$)

(2.15) $$[u, v]_\lambda = \{u, v\} + \lambda^\alpha C_\alpha(u, v)$$

where $\lambda^\alpha = \lambda B^\alpha$, a combination of two deformations where (with $\alpha + 1 = 1$ for $\alpha = 2$):

(2.16) $$C_\alpha(u, v) = p_{\alpha+1}(\partial_\alpha u \partial_{\alpha+1} v - \partial_{\alpha+1} u \partial_\alpha v) + (u\partial_\alpha v - v\partial_\alpha u).$$

The free Hamiltonian $H = \frac{1}{2} p_\alpha p^\alpha$ ($\alpha = 1$ for $T^1$; $\alpha = 1, 2$ for $T^2$) gives for the deformed Hamilton equation in both cases:

(2.17) $$\dot{p}_\alpha = -\lambda_\alpha H, \qquad \dot{q}_\alpha = p_\alpha + q_\alpha \dot{H}/H.$$

For the circle, $p \to 0$ and $q \to \lambda^{-1}$ when $t \to \infty$: we have a motion qualitatively similar to the asynchrone pendulum; the deformation adds a kind of friction which ultimately stops the motion. For the torus we have a first integral of motion $k = \lambda_1 p_2 - \lambda_2 p_1$ (and we suppose $\lambda_1^2 + \lambda_2^2 \neq 0$); if $k \neq 0$, the integration of (2.17) gives that for $t \to t_0$ (some fixed value), at least one (both if $\lambda_1 \lambda_2 \neq 0$) $p_\alpha \to \infty$ and $q_\alpha \to \infty$ but $q_1/q_2 \to \lambda_1 K_1/\lambda_2 K_2$ ($K_1$, $K_2$ being some integration constants) and $p_1/p_2 \to \lambda_1/\lambda_2$ (if $\lambda_2 \neq 0$); the motion is increasingly accelerated towards a straight line (in the $\mathbb{R}^2$-picture of the torus) but for $t \geqslant t_0$ we have $k = 0$, in which case we have separation of variables, and for each projection the same kind of motion as for the circle: we have a kind of classical picture of particle creation (at $t_0$), then a decay.

For the circle, the physical pendulum $H = p^2/2I + R(1 - \cos q)$ with deformed bracket (2.14) gives the equation (for $\lambda = 0$, we have the usual one)

(2.18) $$\ddot{q} + \tfrac{3}{2}\lambda \dot{q}^2 (1 - \lambda q)^{-1} + RI^{-1}(1 - \lambda q)(\sin q - \lambda(1 - \cos q)) = 0.$$

We have here (for appropriate values of $\lambda$) a kind of 'viscosity' or 'friction' term in $\dot{q}^2$, which is not so surprising since in the case of essential deformations (2.8) and (2.9) are not equivalent in general, hence $\dot{H} \neq [H, H]_\lambda = 0$ and energy is not conserved.

These examples suggest that the usual Hamilton equations with brackets deformed by essential deformations may be appropriate to describe open systems (where energy is not conserved). Such systems could thus be treated in a way parallel to the usual treatment of closed systems.

## References

[1] Dirac, P. A. M., *Lectures on Quantum Mechanics*, Belfer Graduate School of Sciences Monograph Series No 2 Yeshiva University, New-York, 1964. See also *Canadian J. Math.* 3 (1950), 129–148 and 3 (1951); 1–33. *Proc. Roy. Soc.* **A246** (1958), 326.

[2] Bergmann, P. G., *Phys. Rev.* **75** (1949), 680–685. Bergmann, P. G., and Brunings, J. H., *Rev. Mod. Phys.* **21** (1949), 480–487. Anderson, J. L., and Bergmann, P. G., *Phys. Rev.* **83** (1951), 1018–1025 and other references in Bergmann, P. G., and Goldberg, I., *Phys. Rev.* **98** (1955), 531–538.

[3] Flato, M., Lichnerowicz, A., Sternheimer, D., *J. Math. Phys.* **17** (1976), 1754–1762.

[4] Lichnerowicz, A., *C.R. Acad. Sc. Paris*, **A280** (1974), 523–527.

[5] Sniatycki, J., *Ann. Inst. H. Poincaré*, **A20** (1974), 365–372.

[6] Chevalley, C., *Theory of Lie Groups*, Princeton University Press, 1946.

[7] Nambu, Y., *Phys. Rev.* **D7** (1973), 2405.

[8] Bayen, F., and Flato, M., *Phys. Rev.* **D11** (1975), 3049–3053.

[9] Flato, M., Lichnerowicz, A., and Sternheimer, D., *J. Math. Pures Appl.* 54 (1975), 445–480; Lichnerowicz, A., *C.R. Acad. Sc. Paris* A280 (1975), 37–40, and *J. Diff. Geom.* 12 (1977), 253–300.

[10] Abraham, R., *Foundations of Mechanics*, Benjamin N.Y., 1967.

## Chapter 4

# PHYSICAL APPLICATIONS RELATED TO DIFFERENTIABLE DEFORMATIONS OF POISSON BRACKETS: QUANTUM MECHANICS

## 0. Introduction

Quantum mechanics is usually formulated in terms of linear operators on a Hilbert space of physical states, while classical mechanics relies on the algebra $N$ of functions over phase-space. In one of the most naive presentations, it is said that the passage from classical to quantum mechanics is realized through the correspondence principle of Bohr, according to which one replaces the position variable $q$ by the operator of multiplication by $q$, momentum $p$ by $i\hbar\partial/\partial q$ where $\hbar$ is the Planck constant, 'and the Poisson bracket by the commutator'; the classical limit of a quantum theory being obtained by letting $\hbar \to 0$. This smells of deformation theory – but how can one 'deform' a function into an operator? Due to the fundamental difference in the nature of the observables in classical and quantum theories, it may seem at first sight hopeless to interpret quantum mechanics as a deformation. We shall nevertheless show in this chapter that one can give an *autonomous* phase-space formulation of quantum mechanics, in the framework of which computations can be made, and for which quantum mechanics will appear naturally as a differentiable deformation of classical mechanics. Quantization will manifest itself in a deformation of the algebra of observables rather than in a radical change of their nature. The link between this formulation of quantum mechanics and the usual one is provided by the Weyl application, that we shall explain in Section 1. This application, which maps functions to operators, and its inverse, have been used in the past in attempts to interpret quantum mechanics as a statistical theory over phase space, interpretation which was suggested by its probabilistic nature. This was the starting point of Moyal, and led him to the discovery of the so-called Moyal bracket. Our point of view is, however, different: we do not translate quantum mechanics from the operator formulation to a phase-space formulation, we consider the latter as an autonomous theory, which translates to the former in special cases (such as the flat case of $\mathbb{R}^{2n}$ through Weyl application), but could have a much wider range.

## 1. Weyl Application

Let $a$ be a function on the phase-space $\mathbb{R}^{2l}$. Then Weyl suggested [1] in 1927 the following (one-to-one) correspondence $\Omega$ between the function $a$ of $p_\alpha$ and $q_\alpha$ ($\alpha=1,\ldots,l$) and an operator $A$:

$$(1.1)\qquad a\to A=\int \tilde{a}(\xi^\alpha,\eta^\alpha)\exp\left(i\hbar^{-1}(\xi^\alpha P_\alpha+\eta^\alpha Q_\alpha)\right)\mathrm{d}\xi^\alpha\,\mathrm{d}\eta^\alpha=\Omega(a)$$

where $\tilde{a}$ is the inverse Fourier transform of $a$ and $P_\alpha$, $Q_\alpha$ are the self-adjoint operators representing the canonical commutation relations in the Heisenberg representation. This formula is exactly parallel to that of inverse Fourier transform, with the difference that usual exponential is replaced by a unitary operator representing a group element in the 3-dimensional non-Abelian nilpotent Lie group. The representation space is usually taken as $L^2(\mathbb{R}^l)$, and the operator $A$ may be given by an integral kernel (the coefficients of $A$ in configuration space), in terms of which the inverse mapping $A\to A_W=a$ can be written (this mapping has been first considered by Wigner [2] in 1932). To a vector $\varphi$ in Hilbert space one associates the function corresponding to the orthogonal projection on the one-dimensional vector space generated by $\varphi$.

The Weyl application and its inverse have been given several alternative forms, and their domain of validity has been studied by several authors; see for example the references listed in [3–5] and references quoted therein. Let us just mention that, on the domain of differentiable vectors for the representation of the Heisenberg group, the operator defined by (1.1) makes sense for functions $a\in\mathcal{O}_M$ (the space of slowly increasing functions in the notation of L. Schwartz); if $a\in L^2$, $\Omega(a)$ is a Hilbert–Schmidt operator and conversely (in this case, the inverse mapping is given by a trace formula).

Looking for the inverse image of the quantum commutator $(i\hbar)^{-1}[A,B]$ of two operators $A=\Omega(a)$ and $B=\Omega(b)$, Moyal [6] found that it is not the Poisson bracket $P(a,b)=\{a,b\}$ but rather what is now called the Moyal bracket (Equation (1.2) of Chapter 2) and has been rediscovered later by J. Vey, namely:

$$(1.2)\qquad M(a,b)=(a*b-b*a)/i\hbar$$

where

$$(1.3)\qquad a*b=\exp\left(\tfrac{1}{2}i\hbar P\right)(a,b)$$

corresponds to the product $AB$ of the two operators (a similar formula for the product, in a different context, had been obtained earlier by Groenewold [7]). The relation between quantum mechanics and the deformations considered in Chapter 2 becomes now obvious.

## 2. Some remarks on quantum mechanics on phase space

(a) One has to be a little careful in the interpretation of quantum mechanics in phase-space. It is well-known that position and momentum cannot be simultaneously measured, i.e. (due to the commutation relations) they do not have common eigenvectors (one has the famous uncertainly relations). In our language, we shall say that when $\hbar \neq 0$ there does not exist a function $\rho \neq 0$ such that, for some $(p', q') \in \mathbb{R}^{2l}$, one has $(p-p') * \rho = 0 = (q-q') * \rho$. Thus while it is possible to speak (somewhat incorrectly perhaps) of trajectories in phase-space for classical mechanics, this is impossible on the quantum level. However trajectories in $N$ do make sense in both cases, and we shall thus write the evolution equation for a dynamical variable $f \in N$ in a system with Hamiltonian $H \in N$:

(2.1) $\quad (\mathrm{d}/\mathrm{d}t) f_t = P(H, f_t)$

in classical mechanics (consequence of Hamilton equations), and

(2.2) $\quad (\mathrm{d}/\mathrm{d}t) f_t = M(H, f_t)$

in quantum mechanics (image of the Schrödinger equation of motion).

A more detailed discussion will be found in [8] and references quoted there. We would also like to mention here that it is the non-locality of the equations of motion which provided the basis for the introduction of 'fuzzy phase-spaces' [9]. It is also interesting to note that if $H$ is a polynomial of order $\leqslant 2$ (e.g. the harmonic oscillator Hamiltonian), Equations (2.1) and (2.2) are the same, and therefore the classical and quantum trajectories in $N$ coincide for these preferred Hamiltonians, which generate the invariance algebra $\mathfrak{sp}(l, \mathbb{R}) \cdot H_l$ mentioned in Chapter 2, Section 5.

(b) One of the advantages of working out quantum mechanics on the classical phase-space is that, using Moyal-Vey brackets and deformed products, we have a natural framework for 'global' quantization when the phase-space is not $\mathbb{R}^{2l}$. This happens in several concrete situations (see below the hydrogen atom problem) and is generally the case when we have constraints.

A natural question is then that of the uniqueness of quantum mechanics, considered as a deformation. And a related question is that of the physical equivalence of mathematically equivalent deformations. From what we have seen in Chapter 2 follows that mathematically the quantum mechanical bracket is locally infinitesimally unique and (on a flat Poisson manifold) unique as a function of Poisson bracket – the latter result being also true for the (quantum mechanical) star-product. A more refined study of the higher cochains in the deformation would be necessary in order to have a more complete answer to the mathematical uniqueness of quantum mechanics in the framework of deformation theory. However, for e.g. $\mathbb{R}^{2l}$, we already know it is unique (up to equivalence) among differentiable deformations of the Vey type.

We shall come back to the question of physical equivalence after we have dealt with the spectral problem in star-calculus. Let us just mention here one aspect of this problem. Weyl quantization corresponds to the so-called Weyl ordering. Other orderings are also used, e.g. the standard ordering where one writes all $q$'s on the left and all $p$'s on the right. If we denote by $H_l$ the Heisenberg Lie algebra mentioned in Section 5 of Chapter 2, the choice of an ordering amounts to the choice of a vector-space isomorphism between its symmetric algebra and its enveloping algebra. One may also get the orderings by introducing a weight function $\Omega$ in the integrand of formula (1.1), $\Omega \equiv 1$ for the Weyl ordering [3]; $\Omega$ can be expressed for the orderings used, as a sine or exponential function of a second-order polynomial times $\hbar$: it is an entire series in the parameter $\hbar$ of the deformation, with polynomial coefficients in phase-space. The Fourier image of the multiplication by $\Omega$ will therefore be a power series $T_h = \Sigma_{r=0}^{\infty} \hbar^r T_r$ where the $T_r$ are differential operators; this series $T_h$ will realize an equivalence in the sense of Section 1(d) of Chapter 1 of the two Lie algebra deformations corresponding to the Weyl ordering and the $\Omega$-ordering; it will also realize an equivalence (in a similar sense) of the two associative algebra deformations. This equivalence (a formal isomorphism between two product structures on the algebra of functions) will enable us to translate any treatment in one ordering to a similar treatment in another ordering: in this sense, the two formalisms are equivalent.

## 3. Methods for Calculating Spectra

(a) Let $\hat{H} = \Omega(H)$ be the self-adjoint Hamiltonian operator of a physical system, $\Omega$ being (e.g.) the Weyl application. The spectrum $I$ of $\hat{H}$ gives the energy levels of the system; supposing for simplicity that they are discrete, we have the spectral decomposition $\hat{H}=\Sigma_{\lambda\in I}\lambda\hat{\pi}_\lambda$ of $\hat{H}$ in terms of the orthogonal projectors $\hat{\pi}_\lambda=\hat{\pi}_\lambda^*$ on the eigenspace with eigenvalue $\lambda$, and thus $\hat{\pi}_\lambda\hat{H}=\hat{H}\hat{\pi}_\lambda=\lambda\hat{\pi}_\lambda$, $\hat{\pi}_\lambda\hat{\pi}_\mu=\delta_{\lambda\mu}\hat{\pi}_\lambda$; the multiplicity of the level $\lambda$ is then given by $\mathrm{Tr}(\hat{\pi}_\lambda)$.

In our formalism, we shall be looking directly for functions $\pi_\lambda=\bar{\pi}_\lambda$ over phase-space $\mathbb{R}^{2l}$ such that $\pi_\lambda * H=H * \pi_\lambda=\lambda\pi_\lambda$, $\pi_\lambda * \pi_\mu=\delta_{\lambda\mu}\pi_\lambda$, where $*$ is the deformed product law in our quantization procedure. It turns out that this is not easy to do (except in the most simple case of the 1-dimensional harmonic oscillator). However, the evolution of our quantum system is given by the unitary operator $\exp(i\hat{H}t)=\Sigma\, e^{it\lambda}\hat{\pi}_\lambda$, and this gives us the hint how to proceed.

(b) We build the $n$th $*$-powers of $H$, $(H\,*)^n=H*\cdots*H$ ($n$ times) and define the $*$-exponential (Sections 5 and 6, Chapter 2)

$$\text{(3.1)}\qquad \operatorname{Exp}(Ht)=\sum_{n=0}^{\infty}(i\hbar)^{-n}(n!)^{-1}(Ht\,*)^n.$$

We then look for a Fourier-Dirichlet expansion of this quantity

$$\text{(3.2)}\qquad \operatorname{Ext}(Ht)=\sum_{\lambda\in I}\pi_\lambda\exp(\lambda t/i\hbar).$$

If (3.1) converges as a series in $t$ for $|t|<\rho$ and fixed $(p,q)\in\mathbb{R}^{2l}$ and as a distribution in $(p,q)$ for fixed $t$, $|t|<\rho$, and if (3.2) is the Fourier-Dirichlet expansion of this distribution, then this treatment makes sense and the frequencies $\lambda\in I$ can be interpreted in our autonomous formalism as the spectrum of $H$. The multiplicity will then be given by $(2\pi\hbar)^{-l}\int\pi_\lambda\,dp\,dq$. More generally, we may look for an expansion that we shall write (in the distribution sense)

$$\text{(3.3)}\qquad \operatorname{Exp}(Ht)=\int\exp(\lambda t/i\hbar)\,d\mu(\lambda)$$

i.e. we look for $d\mu(\lambda)$ (which depends also on $(p,q)\in\mathbb{R}^{2l}$), the Fourier transform (in $t$) of the distribution $\operatorname{Exp}(Ht)$. Then the spectrum of $H$ will be defined as the spectrum of this distribution in the sense of L. Schwartz [10], that is the support of its Fourier transform $d\mu(\lambda)$. In the discrete case,

$$d\mu(\lambda)=\sum_{\lambda'\in I}\pi_{\lambda'}\delta(\lambda-\lambda').$$

Of course, the functions $\pi_\lambda$ that we obtain in this way will be solutions of our original eigenvalue problem. Fourier analysis of the 'evolution function' is here just a device to get all the spectrum in a single stroke.

In some instances, it will also be easier to work, not with the original $H$, but with some $*$-function of it. The spectrum of $H$ will then be given by the inverse (ordinary) function of the obtained spectrum. Also, one has to be careful in the choice of the $*$-product one uses, in order to have the simplest calculations possible. For this purpose, a careful study of the symmetry of our problem will be useful, as we shall see below.

(c) We are now in position to come back to the question of the physical equivalence of two orderings, or more generally of two equivalent deformations. Suppose we have two products $*$ and $*'$ on our algebra $N$ of functions (or on the formal series $E(N;\hbar)$), and an isomorphism $\tau:(N,*)\rightarrow(N,*')$ which realizes the equivalence between the two products: that is, we have $\tau(u*v)=\tau u *' \tau v$. Then it is obvious that the spectrum of $H$ in the $*$-formalism will be the same as the spectrum of $\tau H$ in the $*'$-formalism.

Some people look at the question from another point of view. They start with a classical function $H$, get two operators $\hat{H}_1$ and $\hat{H}_2$ by using two different orderings; of course, except in most simple cases, one finds that the two operators are not the same, and then one sometimes concludes that the two orderings are physically inequivalent. This is an 'active' way to look at things, to distinguish from the 'passive' way which says that (provided we are in the domains of both Weyl applications) everything which is done in one formalism can be translated 'isomorphically' in the other – and that both have therefore the same physical content.

In our autonomous formulation, the map $\tau$ expresses the passive equivalence.

## 4. Example 1: The Harmonic Oscillator

This is the basic tool for other examples also (in the discrete spectrum case): the 'trick', in these examples, is somehow to reduce the problem to that of the harmonic oscillator, for which the procedure described in Section 3(b) is relatively easy to work out.

(a) *One-dimensional case.* The Hamiltonian is $H(p,q)=\frac{1}{2}(p^2+q^2)$. Now if $f$ is a $C^\infty$ function of one variable with derivatives $f', f'', \ldots$, one checks that:

$$H*f(H)=Hf(H)-(\hbar/2)^2 f'(H)-(\hbar/2)^2 Hf''(H) \tag{4.1}$$

where $*$ is the usual $*$ of Moyal, that is $\exp(\frac{1}{2}i\hbar P)$ where $P$ is the Poisson bracket. This shows that $(H*)^n$ is a polynomial $K_n(H)$ of degree $n$ in $H$ and same parity as $n$.

One then proves [8] that (with $l=1$ here), the power series $\operatorname{Exp}(Ht)$, given by (3.1), converges for $|t|<\pi$ for fixed $(p,q)\in\mathbb{R}^{2l}$ to

$$\left(\cos\frac{t}{2}\right)^{-l}\exp\left(\frac{p^2+q^2}{i\hbar}\operatorname{tg}\frac{t}{2}\right)=F_l(p,q;t) \tag{4.2}$$

and for fixed $|t|<\pi$ to the same limit in the weak topology of $\mathscr{D}'(\mathbb{R}^{2l})$. Now the map $t\to F_l(p,q;t)$ is weakly analytic from the disk $|t|<\pi$ into $\mathscr{D}'(\mathbb{R}^{2l})$ and has continuation to the complementary $U'$ of $\{(2k+1)\pi, k\in\mathbb{Z}\}$ in $\mathbb{C}$. This will define $\operatorname{Exp}(Ht)$ for fixed $t\in U'$, and it turns out that $\operatorname{Exp}(Ht)\in\mathscr{S}'(\mathbb{R}^{2l})$ if $\operatorname{Im}t\leqslant 0$. Moreover, for fixed $(p,q)\neq\{0\}$, $F_1(p,q;t)$ is a periodic distribution in $\mathscr{D}'(\mathbb{R})$ with Fourier expansion given by (for $l=1$).

$$F_l(p,q;t)=\sum_{n=0}^{\infty}\pi_n^{(l)}(p,q)\,\mathrm{e}^{-i(n+(l/2))t}, \tag{4.3}$$

$$\pi_n^{(l)}(p,q)=2^l\exp\left(-(2/\hbar)H(p,q)\right)(-1)^n L_n^{(l-1)}((4/\hbar)H(p,q)),$$

where $L_n^0$ (resp. $L_n^{(l-1)}$) is the usual (resp. generalized) Laguerre polynomial of degree $n$. In addition, for fixed $t\in U'$ with $\operatorname{Im}t\geqslant 0$, the series in the right-hand side of (4.3) converges weakly in $\mathscr{S}'(\mathbb{R}^{2l})$ to the left-hand side (and to $(\mp i\pi\hbar)^l$ for $t=\pm\pi$).

We have thus found the energy levels $E_n=(n+\frac{1}{2})\hbar$ here, and the multiplicity is found by $(2\pi\hbar)^{-l}\int\pi_n\,\mathrm{d}p\,\mathrm{d}q=1$ in this case.

(b) *l-dimensional case.* This case is instructive by the appearance of the $\mathfrak{sl}(2,\mathbb{R})$ Lie algebra. Indeed, we have $H=\frac{1}{2}(p^2+q^2)$ with $p^2=\Sigma_{\alpha=1}^{l}p_\alpha^2$, $q^2=\Sigma_{\alpha=1}^{l}q_\alpha^2$. Let us denote by $pq=\Sigma_{\alpha=1}^{l}p_\alpha q_\alpha$. Then we have a $\mathfrak{sl}(2,\mathbb{R})$ Lie algebra (for both Poisson and Moyal brackets) generated by

$$\{X=ap^2+2bpq+cq^2;\ a,b,c\in\mathbb{R}\},$$

and for any $C^\infty$ function $f$ of a real variable we have, with $d=ac-b^2$:

$$X*f(X)=Xf(X)-ld\hbar^2f'(X)-d\hbar^2f''(X)$$

The analysis is therefore exactly the same as before (where now $l$ is allowed to take any integer value), the development (4.2) being valid for $\operatorname{Exp}(Ht)$, and we find the levels $E_n=(n+\frac{1}{2}l)\hbar$ and the correct multiplicity (e.g. $\frac{1}{2}(n+1)(n+2)$ for $l=3$).

Moreover, for an hyperbolic element such as $X=pq$, for which $d=-\frac{1}{4}$, one finds easily an expression similar to (4.2) with trigonometric replaced by hyperbolic functions, and the continuous spectrum expressed in terms of some generalized 'projectors' $\pi(\lambda, X)$ (cf. [8] for their expressions) by:

$$(4.4) \qquad \operatorname{Exp}(Xt) = \int_{-\infty}^{\infty} \exp(\lambda t/i\hbar)\pi(\lambda, X)\,d\lambda.$$

Finally one can deduce from (4.3) that the functions $\operatorname{Exp}(Xt)$, $X = ap^2 + 2bpq + cq^2$, generate a $*$-representation of the metaplectic group, the twofold covering of $SL(2, \mathbb{R})$. The spectrum $\{n + \frac{1}{2}l\}$ of $H$ (twice the 'usual' elliptic generator) splits into $\{2m + \frac{1}{2}l\} \cup \{2m + 1 + \frac{1}{2}l\}$ with integer $m$. This $*$-representation corresponds, for $l = 1$, to the sum of two irreducible faithful representations of this covering group – and for this sum the character is regular at the origin. It is interesting to see that the metaplectic group is here 'built in' our formalism, without any need to introduce e.g. half-densities.

## 5. Example 2: Angular Momentum

Let

$$M_{jk} = q_j p_k - q_k p_j \qquad (1 \leqslant j < k \leqslant l,\ l > 2)$$

be generators of a $\mathfrak{so}(l)$ Lie algebra (for Poisson and Moyal brackets). The Casimir is here

$$(5.1) \qquad C = \sum (M_{jk}*)^2 = g^2(p, q) - l(l-1)(\hbar/2)^2$$

where $g^2(p, q) = p^2 q^2 - (pq)^2$ in our notations, $g \geqslant 0$. In particular for $l = 3$ one has $C = \Sigma M_{jk}^2 - \frac{3}{2}\hbar^2$. One then gets a formula of the type of (4.1) for $g^2 * f(g^2)$ but more involved (with fourth-order derivatives). Then one finds [8] that $\gamma = g - (l-2)\hbar^2/4g$, defined where $g \neq 0$, satisfies there $(\gamma *)^2 = g^2 - (3l-4)(\hbar/2)^2$ and that more generally for $n \geqslant 1$:

$$(5.2) \qquad (\gamma *)^n = g^{-1} G_{n+1}(g)$$

where the polynomial $G_n$ of degree $n$ may be defined by

$$(5.3) \qquad F_{l-2}(H, t) = \left(\cos\frac{t}{2}\right)^{2-l} \exp\left(\frac{2H}{i\hbar}\operatorname{tg}\frac{t}{2}\right) = \sum_{n=0}^{\infty} (n!)^{-1}(t/i\hbar)^n G_n(H).$$

From there one deduces, using the preceding section, that for fixed $s \in \mathbb{C}$, $\operatorname{Im} s<0$, one has (with weak convergence in $\mathscr{S}'(\mathbb{R}^{2l})$) functions $\pi_n$ (of $g$) such that

$$\text{(5.4)} \qquad \operatorname{Exp}(\gamma s)=\sum_{n=0}^{\infty} \pi_n(g) \exp(-is(n+\tfrac{1}{2}(l-2))).$$

The spectrum of $\gamma$ is thus $\hbar(n+\frac{1}{2}(l-2))$ and that of $C=(\gamma *)^2-(l-2)(\hbar/2)^2$ is $\{n(n+l-2)\hbar^2, n=0, 1, \ldots\}$.

## 6. Example 3: Hydrogen Atom and Kepler Problem

Here we take $(p, q) \in \mathbb{R}^{2l}$ ($l=3$ in the physical case) with $H=\frac{1}{2}p^2-(1/r)$ $r=(q_1^2+\cdots+q_l^2)^{1/2}$. A treatment with the usual Moyal product $*$ seems hopeless. However the physical phase-space is in fact $W=T^*S^l$ (with momentum in $S^l$), with the $SO(l+1)$ symmetry noticed by Fock in the physical case. We shall therefore use the star-product adapted to this situation, which was built in Section 4 of Chapter 2 and is $\mathfrak{so}(l+1)$ invariant in the sense of Section 5, Chapter 2, and denote by $*'$ its expression (via a stereographic projection) in $\mathbb{R}^{2l}$.

We consider the generators of the $\mathfrak{so}(l+1)$ symmetry:

$$M_{jk}=q_jp_k-q_kp_j, \qquad M_{j,l+1}=-\tfrac{1}{2}(p^2-1)q_j+(pq)p_j, \qquad (1 \leqslant j, k \leqslant l).$$

Using the preceding section, the spectrum of the Casimir (expressed with $*'$) $C=\Gamma^2-l(l+1)(\hbar/2)^2$, where $\Gamma=\frac{1}{2}r(p^2+1)$, is found to be $n(n+l-1)\hbar^2$, and the $*'$-square root of $C+(l-1)^2(\hbar/2)^2$ is

$$\gamma=\Gamma-(l-1)(\hbar/2)^2\Gamma^{-1},$$

the $*'$-spectrum of which is given by the following development of the $*'$-exponential:

$$\operatorname{Exp}(\gamma s)=\sum \pi_n'(\Gamma) \exp(-is(n+\tfrac{1}{2}(l-1)))$$

with multiplicity

$$(2\pi h)^{-l} \int \pi_n' \, dp \, dq=(2n+l-1)\frac{(n+l-2)!}{n!(l-1)!},$$

the number of spherical harmonics of degree $n$ on $S^l$.

Using a standard procedure (energy twist) the finding of energy eigenvalues, i.e. of solutions of $(H-E)*\Phi=0$ with $\Phi=\bar{\Phi}$ and $E<0$, is transformed by Exp $(-Ts)*(H-E)*$ Exp $(Ts)$ where $e^s=(-2E)^{1/2}$, $T=pq$, to the finding of real solutions of

$$(1-e^{-s}(\gamma_0 *)^{-1})*\tilde{\phi}=0,$$

where

$$\gamma_0=\tfrac{1}{2}(p^2+1)^{1/2}*r*(p^2+1)^{1/2}.$$

Now the transformation $\tau$, which transforms the $*$-product algebra to the $*'$-product algebra, transforms $(\gamma_0 *)^2$ into $(\gamma *')^2$. These two functions have thus the same spectrum, wherefrom one deduces that the $*$-spectrum of $\gamma_0$ is identical with the $*'$-spectrum of $\gamma$. It follows that

$$(-2E)^{-1/2}=e^{-s}=(n+\tfrac{1}{2}(l-1))\hbar, \qquad (n=0,1,\dots).$$

In particular, for $l=3$ one finds the usual levels of the hydrogen atom $E=-\tfrac{1}{2}n^{-2}\hbar^{-2}$ with multiplicity $n^2$ $(n=1,2,\dots)$.

The continuous spectrum is reduced by the same 'twist', with here $e^s=(2E)^{1/2}$ to the $*$-spectrum of $\tfrac{1}{2}r^{1/2}*(p^2-1)*r^{1/2}$; the image of this function by $\tau$ is then shown to have the same $*'$-spectrum as the function $T$, namely the real line. Therefore one has $e^{-s}=\lambda\in\mathbb{R}$, whence $E=(2\lambda)^{-1}>0$.

## References

[1] Weyl, H., *The Theory of Groups and Quantum Mechanics*, Dover, 1931.
[2] Wigner, E. P., *Phys. Rev.* **40** (1932), 749.
[3] Agarwal, G. S., and Wolf, E., *Phys. Rev.* **D2** (1970), 2161.
[4] Liu, K. C., *J. Math. Phys.* **17** (1976), 859.
[5] Voros, A., *Développements semi-classiques*, Thèse Université de Paris Sud, No. 1843, Mai 1977.
[6] Moyal, J. E., *Proc. Cambridge Phil. Soc.* **45** (1949), 99.
[7] Groenewold, H. J., *Physica* **12** (1946), 405.
[8] Bayen, F., Flato, M., Fronsdal, C., Lichnerowicz, A., and Sternheimer, D., *Ann. Phys. (N.Y.)* **111** (1978), 111–151.
[9] Ali, S. T., and Prugovecki, E., *J. Math. Phys.* **18** (1977), 219; and references quoted.
[10] Schwartz, L., *Theorie des Distributions. II*, Hermann, Paris, 1951.

## CHAPTER 5

# SEPARATE AND JOINT ANALYTICITY OF VECTORS IN GROUP REPRESENTATIONS AND AN APPLICATION TO FIELD THEORY

## 0. INTRODUCTION

From the beginning of quantum mechanics, it was assumed that an observable quantity should be represented by a self-adjoint operator in Hilbert space, the various values which could be obtained by measuring this quantity (when the system is in an eigenstate) being the (real) eigenvalues of that operator and the corresponding eigenvectors being the particular states of the physical system for which these values are obtained. In the absence of the so-called superselection rules (a nontrivial operator commuting with all observables), it was also assumed that every self-adjoint operator on the Hilbert space of states corresponds to an observable.

Now it almost always occurs (of course in quantum field theory, but also in quantum mechanics – cf. the hydrogen atom, treated with a different formalism in Chapter 4) that one has to deal with an infinite-dimensional Hilbert space. In this case, we may have a continuous spectrum – with no corresponding eigenstates in the Hilbert space. Also, the operators will in general be unbounded, and the sum of two observables will then in general not be self-adjoint, and can very well be not essentially self-adjoint (not have a self-adjoint closure) and even not defined. Moreover, even if two observables $A$ and $B$ (self-adjoint operators in Hilbert spaces) commute on a dense domain on which $\alpha A+\beta B$ is essentially self adjoint for all $\alpha, \beta \in \mathbb{R}$, an example of Nelson [1] shows that one may have $\exp(i\alpha A) \exp(i\beta B) \neq \exp(i\beta B) \exp(i\alpha A)$, which is not a very desirable feature for the composition of these two evolution operators; this of course cannot occur if $A$ and $B$ have commuting spectral resolutions, and in particular have a total set of common eigenvectors – neither, as we shall see, if they have a total set of common analytic vectors. Moreover, this last condition will also ensure that if the observables $A$ and $B$ generate a finite-dimensional Lie algebra under commutation and linear combination, this Lie algebra of operators can be exponentiated to the corresponding (simply connected) Lie group, and all linear combinations of $A$ and $B$ will be essentially self adjoint on their common domain; in particular, if $A$ is some 'free' Hamiltonian $H_0$ and $B$ some 'interaction' Hamiltonian $H_{\text{int}}$, then in this case the perturbed Hamiltonian $H_0+\lambda H_{\text{int}}$ will be essentially self adjoint (for any $\lambda \in \mathbb{R}$ here).

The notion of analytic vector appears thus as a generalization of that of eigenvector, and can replace it for a number of purposes. For the sake of completeness, we shall now give a rapid summary of the 'classical' theory of analytic vectors.

## 1. Analytic and differentiable vectors in Lie group representations

(a) If $A$ is an operator (with dense domain) in a Banach space $H$, we say that a vector $\varphi \in H$ is an *analytic vector* for $A$ if for some $t>0$,

$$(1.1) \qquad \sum_{n=0}^{\infty} t^n (n!)^{-1} \|A^n \varphi\| < \infty.$$

If $A$ is a self-adjoint operator in a Hilbert space $H$, then one shows easily that $\varphi \in H$ is analytic for $A$ iff it belongs to the domains of both $e^{tA}$ and $e^{-tA}$ for some $t>0$. More generally, if $A$ generates a one-parameter (Lie) group $\exp(tA)$ in the Banach space $H$, then one sees easily that $\varphi$ is analytic for $A$ iff the map $\mathbb{R} \ni t \to \exp(tA)\varphi \in H$ is an analytic vector-valued function.

If $\varphi \in H$ is an eigenvector for the operator $A$, or a finite combination of eigenvectors, then obviously $\varphi$ is an analytic vector for $A$. The same holds if $\varphi$ is a suitable infinite combination of eigenvectors (with sufficiently rapidly decreasing coefficients). In general, an analytic vector will be a kind of 'nice' combination of (possibly generalized) eigenvectors.

The importance of analytic vectors for Lie group representations can be seen, already in the one-dimensional case, from the following result of Nelson: a closed symmetric operator $A$ in Hilbert space is self-adjoint iff it has a dense set of analytic vectors. The connection with the group follows from the Stone theorem, which says that a linear operator $A$ in Hilbert space is self-adjoint iff $iA$ is the infinitesimal generator of a strongly continuous one-parameter unitary group $\exp(itA)$.

(b) In the case of a finite-dimensional Lie group or Lie algebra, one may either start with a Lie group representation and study analyticity properties of vectors related to this representation or to its differential on the Lie algebra, or one can start with a Lie algebra representation and, using analyticity properties, study its integrability (i.e. if it is the differential of a

Lie group representation). In this section we shall deal with the first aspect, in the next section with the second one, and finally come to the question of the relation between separate and joint analyticity.

Given a continuous representation $U$ of a Lie group $G$ in a Banach (or locally convex quasi-complete) space $H$, we say that a vector $\varphi \in H$ is *analytic* (resp. *differentiable*) for the representation if the map $G \ni g \to \to U(g)\varphi \in H$ is analytic (resp. differentiable).

It has been shown by Gårding [2] in Banach spaces, and by the same method by Bruhat for more general spaces [3] that the space of *differentiable* vectors for a continuous representation $U$ of a Lie group $G$ in a quasi-complete locally convex space is a *dense subspace* $H^\infty$. Indeed, $H^\infty$ contains (and when $H$ is a Fréchet space, according to a recent result of Dixmier and Malliavin, is in fact identical with) the 'Gårding domain' of finite sums of vectors of the type

$$(1.2) \qquad \int_G f(g)U(g)\varphi \, \mathrm{d}g$$

with $f \in \mathscr{D}(G)$ (the space of $C^\infty$ functions with compact support), $\mathrm{d}g$ being the Haar measure on $G$ and $\varphi \in H$ being arbitrary; taking for $f$ an approximation of $\delta_e$ one sees that the Gårding domain is dense in $H$. Moreover, $H^\infty$ has a natural (quasi-complete) topology induced by the space $\mathscr{E}(G; H)$ of $C^\infty$ functions on $G$ with values in $H$, which when $H$ is a Banach or Fréchet space is a Fréchet topology that can be explicitly defined [4] with a countable set of semi-norms.

(c) When $U$ is a unitary representation of $G$ in a Hilbert space $H$, if $X_1, \ldots, X_r$ denote the skew-adjoint operators representing a basis of the Lie algebra $\mathfrak{g}$ of $G$, then the Laplacian (Nelson operator) $\Delta = X_1^2 + \cdots + X_r^2$ is essentially self-adjoint on $H^\infty$ (=the Gårding domain) and more generally on any dense domain invariant under $U(G)$. The Fréchet space $H^\infty$ can be viewed as the intersection of the domains of $\bar{\Delta}^n$ ($n = 1, 2, \ldots$, the bar denoting the closure) with topology defined by the semi-norms $\|\bar{\Delta}^n\varphi\|$. Moreover, the topology of $H^\infty$ is then nuclear for all unitary irreducible representations when $G$ is semi-simple or nilpotent (and in many other cases, e.g. when $G$ is the Poincaré group and $U$ is unitary irreducible with non-zero mass; see e.g. [5] and references quoted).

(d) It is not true that the space of analytic vectors is always dense. For instance, the translation one-parameter group $\mathbb{R} \ni t \to \varphi_t \in \mathscr{D}(\mathbb{R})$,

$\varphi_t(x)=\varphi(x+t)$, has obviously no non zero analytic vector (which would necessarily be a real analytic function here) – but the contragredient representation in the dual space $\mathscr{D}'(\mathbb{R})$ has a dense subspace of analytic vectors.

However, a continuous representation $U$ of a real Lie group $G$ in a *Banach space* $H$ does have a *dense subspace* $H^\omega$ *of analytic vectors* (with a natural (LF)-topology [4]).

This was proved with increasing generality and simplicity by Harish-Chandra, Cartier and Dixmier, Nelson [1] and finally Gårding [6]. Indeed, $H^\omega$ contains the dense 'Gårding domain of analytic vectors' of the type

$$\int_G f_t(g)U(g)\varphi \, \mathrm{d}g \tag{1.3}$$

where $\varphi \in H$ and $f_t(g)=(\mathrm{e}^{t\Delta'}f)(g)$ is any solution of the heat equation $(\partial f_t/\partial t)=\Delta' f_t$ on the group with initial value $f_0=f \in \mathscr{D}(G)$, $\Delta'$ being the above mentioned Laplacian for the regular representation of $G$ in $L^2(G)$ – the density of these vectors is straightforward; the main difficulty is to show [6] that the integral converges and defines an analytic vector.

(3) The proof given by Nelson [1] is somewhat different. In short, he shows that a vector $\varphi$ is analytic iff for some $t>0$ and the representatives $\{X_1,\ldots,X_r\}$ of some linear basis of the Lie algebra $\mathfrak{g}$ of $G$, the series

$$\sum_{n=0}^{\infty} t^n(n!)^{-1} \sum_{1\leqslant i_1,\ldots,i_n\leqslant n} X_{i_1}\ldots X_{i_n}\varphi \tag{1.4}$$

is absolutely convergent in the Banach space $H$. Then (using among other things the Brownian motion) he shows that this is true if $\varphi$ is an analytic vector for the Laplacian $\Delta$ and that the latter has indeed a dense set of analytic vectors.

When $U$ is a unitary representation in a Hilbert space $H$, in which case $\bar{\Delta}$ is self-adjoint and negative, the analytic vectors for the representation are exactly [4] those for the self-adjoint operator $(I-\bar{\Delta})^{1/2}$ where $I$ denotes the identity operator. Moreover in this case, when $H^\infty$ is a nuclear Fréchet space, the semi-norms defining its topology show that $(I-\bar{\Delta})^{-k}$ is a Hilbert–Schmidt operator for some integer $k>0$ and that therefore $\bar{\Delta}$ has a pure discrete spectrum.

## 2. Integrability criteria

In this section, we shall suppose given a representation $T$ of a Lie algebra $\mathfrak{g}$ and give sufficient conditions for its integrability to a Lie group representation.

(a) NELSON CRITERION. *Let $T$ be a representation of a real finite-dimensional Lie algebra $\mathfrak{g}$ on a Hilbert space $H$ by skew-symmetric operators on a dense invariant domain $D$. If the Laplacian $\Delta$ relative to some basis is essentially self-adjoint on $D$, $T$ is integrable to a unique unitary representation $U$ of the connected and simply connected group $G$ with Lie algebra $\mathfrak{g}$, the differential* d$U$ *of which coincides with $T$ on $D$.*

Indeed, from the hypothesis and what we saw in Section 1(a) follows that $\bar{\Delta}$ has a dense set of analytic vectors, hence [1] that there is a dense set of analytic vectors for the Lie algebra (i.e. the series (1.4) is absolutely convergent), whence (by an argument made more precise by Goodman [4]), using the Campbell–Hausdorff–Dynkin formula, the integrability of the representation.

This criterion, which can be useful in cases for which $\Delta$ has a simple expression (e.g. in terms of some Casimir operators) is however not practical in many cases (even for semi-simple Lie algebras), when the group dimension is large. It would be more practical to have a criterion which would 'separate' the coordinates on the group, namely use analyticity in the sense (1.1) separately for each generator of the Lie algebra instead of the joint analyticity in the sense of the convergence of the series (1.4) (which follows from the hypothesis on $\Delta$ in the Nelson criterion). We may indeed distinguish four types of analyticity for a vector $\varphi$ in a representation $T$ of a Lie algebra $\mathfrak{g}$.

(A1) Joint analyticity: convergence of (1.4)

(A2) Analyticity (in the sense of (1.1)) for all $A \in T(\mathfrak{g})$.

(A3) Separate analyticity: analyticity for all $X_i (i=1,\dots,r)$ in a basis of $T(\mathfrak{g})$.

(A4) Weak separate analyticity: analyticity for all $X_i (i=1,\dots,s \leqslant r)$ in a set of Lie generators of $T(\mathfrak{g})$, i.e. which generate all $T(\mathfrak{g})$ by linear combinations and commutators.

In the following criterion (named $\mathrm{FS}^3$ in [7]), only separate analyticity will be needed.

(b) $FS^3$ CRITERION. *Let $T$ be a representation of a Lie algebra $\mathfrak{g}$ by skew-symmetric operators defined on an invariant domain $D$ in a Hilbert space $H$, $(x_1, \ldots, x_s)$ be a set of Lie generators of $\mathfrak{g}$, and $D_1$ be a set of analytic vectors for each of the $T(x_1), \ldots, T(x_s)$ separately. Then there exists a unique unitary $U$ representation of the connected and simply connected Lie group $G$, the Lie algebra of which is $\mathfrak{g}$, defined on the closure $H_1$ (in $H$) of the smallest subspace $D'_1$ containing $D_1$ and invariant under $T(\mathfrak{g})$, such that* $\mathrm{d}U$ *coincides with $T$ on $D'_1$.*

This criterion was proved for a vector-space basis of $\mathfrak{g}$ and assuming the invariance of $D_1$ in the original $FS^3$-paper [8]. The extension to Lie generators (in a reflexive Banach space) was done in [9] and the invariance requirement lifted in [10] for the skew-symmetric representations. To prove it one first shows that $D'_1$ consists of common analytic vectors, which allows us to define the one-parameter groups corresponding to the $T(x_i)$ on $H_1$. Then one deduces the existence of a linear basis $(y_1, \ldots, y_r)$ of $\mathfrak{g}$ represented by essentially skew-adjoint operators on $D'_1$ and that $\exp(ty_i)y_j \exp(-ty_i) \in \mathfrak{g}$ has the same expression as a product in the representation on $D'_1$. Using the uniqueness of solutions of differential equations one then shows (without the Campbell–Hausdorff formula, the application of which requires joint analyticity) that, if one defines the representative of a group element in a neighbourhood of the identity in $G$ as the product of the representatives of the one-parameter coordinate groups (using coordinates of the second kind on this neighbourhood), the group law is valid in the representation of this neighbourhood and therefore on all of $G$.

An extension of this criterion to *semi-reflexive locally convex spaces* can be found in [8] (formulated with a basis, and with Lie generators on a reflexive Banach space in [9]). The main modifications are that one has to assume the existence of one-parameter groups for the generators in the *contragredient* representation, and that the invariant domain of analytic vectors is required for this contragredient representation (if one wants to integrate the original representation). A similar extension to general Banach spaces (not necessarily reflexive) was given by J. Kisynski.

(c) We shall now see a simple application of this criterion. On $L^2(\mathbb{R})$, let us consider the representation of a nilpotent four-dimensional Lie algebra by $Y_1 = \partial/\partial x$, $Y_2 = ix$, $Y_3 = i\lambda x^2$ $(\lambda \in \mathbb{R})$, $Y_4 = iI$, these operators being defined (e.g.) on the space $\mathscr{S}(\mathbb{R})$ of rapidly decreasing functions. They obviously have a common dense set of analytic vectors (take e.g. functions which are analytic in a small strip around the real line [4] and decrease at infinity like $\exp(-x^2)$, or the space $S_{1/2}^{1/2}$ of Gelfand and Shilov). Therefore this representation is integrable to a unitary (evidently irreducible) representation of the corresponding group, the nuclear space of differentiable vectors being $\mathscr{S}(\mathbb{R})$. Therefore the Nelson operator $\Delta$, and also

$$-\Delta - I = -(\partial^2/\partial x^2) + x^2 + \lambda^2 x^4$$

which is the Hamiltonian of the anharmonic oscillator, are essentially self-adjoint and have a pure point spectrum.

Other applications have been given by J. M. Maillard to show the integrability of representations of $\mathfrak{so}(p, q)$ built by Nikolov in terms of the Gelfand–Zeitlin patterns, by J. Niederle and A. Kotecky to the cases of $\mathfrak{so}(p, q)$ and $\mathfrak{su}(p, q)$, etc. The integrability of the compact subalgebra being rather straightforward, one has essentially to show that one (non-compact) Lie generator has a dense set of analytic vectors, which is easier to handle.

Another physical application (which we mentioned in the introduction) is the following: if two 'observables' $X$ and $Y$ are symmetric on a common invariant dense domain $D$ in a Hilbert space $H$, have a common dense set of analytic vectors in $D$ and generate by commutators and linear combinations a finite-dimensional Lie algebra, then the operators $\alpha X + \beta Y$ are essentially self-adjoint on $D$ for all $\alpha, \beta \in \mathbb{R}$ (and the corresponding unitary exponentials have the correct composition law).

## 3. Separate versus joint analyticity of vectors

(a) From what we have seen in the preceding section follows that (under the preceding hypotheses) the existence of a dense domain of analytic vectors for each generator separately implies the existence of a dense domain of analytic vectors for the whole Lie algebra (convergence of (1.4), in short (A.1) analytic), which follows from the integrability. It is however not clear that the original separately analytic vectors were (A.1) analytic. This is a kind of Hartogs theorem, which – in contradistinction with the case of complex variables – requires some conditions in the case of real-analytic functions (see e.g. [12]). It seems, however, that the existence of a representation provides a structure for which separate analyticity is somewhat more than just real analyticity in separate coordinate directions, and that this will imply joint analyticity (for example, the analytic vectors for $\partial/\partial x$ in $L^2(\mathbb{R})$ are functions which have analytic continuation in a strip of constant width around the real line [4]).

In the case of differentiable vectors, one can prove [4], using techniques of partial differential equations, that a vector which is differentiable for the representatives $X_1, \ldots, X_r$ of any basis of a Lie algebra $\mathfrak{g}$ (i.e. belongs to the domains of all $X_i^n$, $i=1, \ldots, r$; $n=1, 2, \ldots$) in the case of a group representation in Banach space, is a differentiable vector for that group representation. The same result for analytic vectors is still a conjecture. However, in the case of Hilbert space representations significant results have been obtained ([10, 13], and [4] in the solvable case). More precisely one can show (by means of precise estimates involving the Nelson operators relative to various bases, in the reductive case):

THEOREM 5.1. (i) *Let $U$ be a (continuous) unitary representation of a reductive Lie group $G$ in a Hilbert space $H$ and $(x_1, \ldots, x_r)$ any basis of its Lie algebra $\mathfrak{g}$. Denote by $H_i^\omega$ the space of analytic vectors for $\mathrm{d}U(x_i)$ and by $H^\omega$ the space of analytic vectors for $U(G)$. Then $H^\omega = \bigcap_{i=1}^r H_i^\omega$.*

(ii) *Let $U$ be a (continuous) representation of a Lie group $G$ in a Hilbert space $H$. There exists a basis $(x_1, \ldots, x_r)$ of the Lie algebra $\mathfrak{g}$ of $G$ such that, in the above notations, $H^\omega = \bigcap_{i=1}^r H_i^\omega$.*

The second part of the theorem follows from the first part (applied to a compact Lie algebra) and from the following lemma (see [10] and [4] for ii)), which is basically a consequence of structural considerations:

LEMMA 5.1. (i) *Let $G$ be a Lie group with Lie algebra $\mathfrak{g}$, $\mathfrak{g}_1$ and $\mathfrak{g}_2$ two subalgebras such that $\mathfrak{g}$ is a unification of $\mathfrak{g}_1$ and $\mathfrak{g}_2$ (i.e. $\mathfrak{g}=\mathfrak{g}_1+\mathfrak{g}_2$, not necessarily direct sum of vector spaces), $G_1$ and $G_2$ the corresponding subgroups of $G$, $T_0$ a (continuous) representation of $G$ in a Banach space $H$, $T_1$ and $T_2$ the restrictions of $T_0$ to $G_1$ and $G_2$ (respectively), and $H^\omega(T_i)$ $(i=0, 1, 2)$ the corresponding spaces of analytic vectors. Then $H^\omega(T_0)=H^\omega(T_1)\cap H^\omega(T_2)$.*

(ii) *In the above notations, if $G$ is a solvable group, $(x_1, \ldots, x_r)$ a Jordan basis, and $U$ a (continuous) representation of $G$ in a Banach space $H$, then $H^\omega(U)=\bigcap_{i=1}^r H_i^\omega$.*

It would be of interest to prove a result similar to part (i) of Theorem 5.1 in the general case, or at least for unitary representations of solvable groups.

(b) To indicate possible *physical applications* of this result, we shall just mention that it was used by H. Snellman [14] to prove that in Wightman's general theory of quantized fields with $\mathscr{S}$ as space of test functions, the domain $D_0$ obtained by the action of the polynomial ring in the field operators on the vacuum (the distinguished 1-dimensional subspace in the Fock space) contains a dense set of analytic vectors for the representation of the Poincaré group which enters in the theory: these vectors are proved to be separately analytic for every (infinitesimally unitary) representative of a suitably chosen basis of the Poincaré Lie algebra, which ensures their joint analyticity.

## 4. Essential skew-adjointness and integrability

To end this chapter we shall show that essential skew adjointness of the generators is not sufficient to ensure integrability, even for a compact semi-simple Lie algebra. Such examples stress the importance of analytic vectors for the integrability problem.

A first example was provided by E. Nelson [1] who showed that, by an appropriate 'surgery' on the realization of the 2-dimensional torus as a rectangle in $\mathbb{R}^2$ with the usual identifications, one could obtain two operators $A$ and $B$ (the derivation with respect to each coordinate in this realization) such that the operators $\alpha A+\beta B$ ($\alpha, \beta \in \mathbb{R}$) are essentially skew-adjoint and commuting on a common invariant dense domain but the unitary operators $\exp(\alpha A)$ and $\exp(\beta B)$ do not commute (they generate a kind of infinite-dimensional group): their spectral resolutions do not commute, and they do not have nontrivial common analytic vectors.

Similar examples can be found in the noncommutative case [15]. Let us, for instance, take the regular representation of $G=SU(2)\approx S^3$ on $L^2(S^3)$, and think of $S^3$ as two unit balls in $\mathbb{R}^3$ glued together on their surface $S^2$, or as a unit ball for which each inner point counts twice. Take out one diameter $V$ of the ball in the latter realization: the manifold $G_V$ thus obtained is no more simply connected, but still one-parameter trajectories of the deleted submanifold have zero Haar measure. We may now take a covering $\tilde{G}_V$, twofold for instance, of $G_V$ and define $L^2(\tilde{G}_V)=H$ by lifting the Haar measure. On $H$ we have a representation of the Lie algebra $\mathfrak{su}(2)$ by taking the same expression for the generators as in the regular representation; all these operators are essentially skew-adjoint on the common dense domain $\mathscr{D}(\tilde{G}_V)$ because of the preceding remark on the trajectories and of an argument of Nelson [1]. However this representation is not integrable since the periodicity of at least some of the 1-parameter groups has been doubled, and we would thus need a covering group of $SU(2)$ which does not exist. Of course, there are not enough common analytic vectors here.

The same procedure can be done for all compact semi-simple Lie groups. It has to be mentioned that all these representations are highly reducible. A somewhat less reducible representation of $\mathfrak{su}(2)$ can be found [16] by taking the usual (quasi-regular) representation of $SO(3)$ on $L^2(S^2)$, deleting two poles on $S^2$, taking a fourfold covering of the manifold thus obtained, and applying the same construction.

Examples of Schur-irreducible Lie algebra representations (every operator commuting with the spectral resolutions of the skew-adjoint generators is a multiple of identity) which are not integrable to the group are known [17], but in such examples the linear combinations of the generators are not all essentially skew-adjoint on a common invariant domain. It is not known whether a Lie algebra representation can be Schur-irreducible, all generators being essentially skew-adjoint on a common invariant dense domain, and the representation be not integrable.

## References


[1] Nelson, E., *Ann. Math.* **81** (1959), 547–560.
[2] Gårding, L., *Proc. Nat. Acad. Sci. U.S.A.*, **33** (1947), 331–332.
[3] Bruhat, F., *Bull. Soc. Math. Fr.* **84** (1956), 97–205.
[4] Goodman, R., *Trans. Amer. Math. Soc.* **143** (1969), 55–76.
[5] Nagel, B., in A. O. Barut (Ed.), Proceedings NATO ASI on Math. Phys., Istanbul 1970, D. Reidel, 1971.
See also Arnal, D., *J. Math. Phys.* **19** (1978), 1881.
[6] Gårding, L., *Bull. Soc. Math. Fr.* **88** (1960), 77–93.
[7] Barut, A. O., and Raczka, R., *Theory of Group Representations and Applications*, Polish Scientific Publishers, Warsaw, 1977.
[8] Flato, M., Simon, J., Snellman, H., and Sternheimer, D., *Ann. Scient. Ec. Norm. Sup. 4e serie*, **5** (1972), 423–434.
[9] Simon, J., *Commun. Math. Phys.* **2** (1972), 39–46.
[10] Flato, M., Simon, J., *J. Funct. Anal.* **13** (1973), 268–276.
[11] Kisynski, J., Preprint IC/74/130, ICTP, Trieste, 1974.
[12] Browder, F., *Canad. J. Math.* **13** (1961), 650–656.
[13] Simon, J., *C.R. Acad. Sc. Paris*, **285** (1977), A 199–202.
[14] Snellman, H., *J. Math. Phys.* **15** (1974), 1054–1059.
[15] Nagel, B., and Snellman, H., *J. Math. Phys.* **15** (1974), 245–246.
Flato, M., Simon, J., and Sternheimer, D., *C. R. Acad. Sc. Paris* **277** (1973), A939–942.
[16] Maillard, J. M., and Sternheimer, D., *C.R. Acad. Sc. Paris* **280** (1975), A 73–75.
[17] Flato, M., and Sternheimer, D., *Commun. Math. Phys.* **12** (1969), 296–303; *Phys. Rev. Letters*, **16** (1966), 1185–1186.

## Chapter 6

# NONLINEAR REPRESENTATIONS OF LIE GROUPS AND APPLICATIONS

## 0. Introduction

Many of the developments in modern mathematics have to do with linear structures: linear spaces, linear operators and representations, linear algebra, etc. But many of the problems posed by Nature make use of nonlinearities in their expression and require adequate treatment of these nonlinearities. Still, what more specific motivations do we have to study nonlinear representations of Lie groups in linear spaces? We may of course reverse the argument and ask why in the past did we study mainly linear representations of a nonlinear object?!

More seriously, we know that for one analytic vector field, the solution to the problem of linearizability goes back to Henri Poincaré [1]. Then Palais and Smale posed the problem of the linearizability of the action of a semi-simple Lie group around a fixed point on a (finite-dimensional) manifold. Robert Hermann showed that a semi-simple Lie algebra is formally linearizable around a singular point; the argument is in fact the observation that the non-linear terms give rise to 1-cocycles (for some representations of the Lie algebra) and can be formally successively annihilated if these 1-cocycles are 1-coboundaries, which is of course true here (Whitehead's lemma; cf. e.g. [2]). This paper of Hermann [3] was followed by a 'Remark' by Guillemin and Sternberg [4] who attacked by another method the convergence problem and showed using among other things the result of Poincaré and 'Weyl unitary trick', that there is an analytic diffeomorphism which linearizes a representation of a (finite-dimensional) semi-simple Lie algebra by analytic vector fields around a common fixed point on an analytic manifold of finite dimension. The connection with 1-cohomology gives a hint on how to tackle the infinite-dimensional case; but this requires to develop first in this case the machinery (which is trivial in the finite-dimensional case) of passage from the Lie group representation to the Lie algebra and vice-versa. This was the main technical difficulty met in [5], the results of which we shall present in this chapter.

Physically, we know that, in the presence of interactions, we have to deal with non-linear field (or wave) equations. Such equations have been dealt with extensively in the past decade (and even before), e.g. for scalar fields with polynomial self-interaction $P(\varphi)$, namely equations of the type $(\Box+m^2)\varphi=P(\varphi)$; significant results could be obtained in 2 (and 3) space-time dimensions, but the physical case (4 dimensions) seems still far away. It would of course be important to know if they can be transformed (by a possibly non-trivial transformation) to the corresponding free equations ($P=0$). The fact that these equations are covariant under Poincaré group and that something is known on the 1-cohomology of its representations does provide a starting point for such a study, and this was done in [6], on which we shall say a few words at the end. It is worthwhile to mention that infinitesimal methods prove themselves useful in the computation of 1-cohomology of representations [7], and to be able to use them for nonlinear representations one has to develop the machinery of differentiation and exponentiation, which is indeed a basic tool in all this theory.

## 1. Formal representations

(a) *Formal series.* Let $E$ be a Banach space. In most applications $E$ will be a Hilbert space (a one-particle space for instance). Since we want to differentiate to the Lie algebra we shall also have to consider the case when $E$ is a Fréchet space (the differentiable vectors for the linear part of the representation). $L_n(E)$ will denote the space of $n$-linear symmetric continuous mappings from $E^n$ to $E$, which we shall identify with elements of $L(\hat{\otimes}^n E, E)$, the space of linear continuous mappings from the projective tensor product [8] $\hat{\otimes}^n E$ of $n$ copies of $E$ into $E$ (this identification requires the completion of the algebraic tensor product with the projective topology, which is smaller than the Hilbert space tensor product when $E$ is an infinite-dimensional Hilbert space).

If $f^n \in L_n(E)$, one associates with it an homogeneous polynomial $\hat{f}^n$ of degree $n$ on $E$ by $\hat{f}^n(\varphi)=f^n(\varphi,\ldots,\varphi)$ for all $\varphi \in E$ (and conversely). An analytic function on $E$ can then be defined [9] as a convergent sum $\Sigma \hat{f}^n$ of homogeneous polynomials of various degrees. Non-linearities of this type (at least in a region of $E$) can thus be 'spread out' in the space $F(E)$ of formal 'power' series of the type

$$f=\sum_{n\geqslant 1} f^n, \qquad f^n \in L_n(E).$$

Let us write for simplicity

$$\hat{h}^i \otimes \hat{h}^j(\varphi) \equiv \hat{h}^i(\varphi) \otimes \hat{h}^j(\varphi).$$

Then the composition law [10] of formal series in $F(E)$ may be defined by

$$\sum_{p\geqslant 1} f^p \circ \sum_{q\geqslant 1} h^q = \sum_{n\geqslant 1} (fh)^n,$$

where

$$(\widehat{fh})^n = \sum_{1\leqslant p\leqslant n} f^p \sum_{i_1+\cdots+i_p=n} \hat{h}^{i_1} \otimes \ldots \otimes \hat{h}^{i_p};$$

(b) *Formal group representations.* We can now define a (non linear) *formal representation* $(S, E)$ of a real Lie group $G$ in a *Fréchet* space $E$ as a homomorphism $g \mapsto S_g$ from $G$ into the group of invertible elements of $F(E)$ such that the maps

$$G \ni g \mapsto S_g^n(\varphi_1, \ldots, \varphi_n) \in E$$

are measurable for all $\varphi_1, \ldots, \varphi_n \in E$ and all $n \geqslant 1$. In particular $S^1$ is a measurable linear representation of $G$ in $E$, called the *free part* (or linear part) of $S$. From the definition, it is obvious that the group action on $E$ leaves the origin invariant.

Consider the algebraic direct sum $\tilde{E} = \bigcup_{n\geqslant 1} E_n$, where $E_n = \bigoplus_{p=1}^n (\hat{\otimes}^p E)$ in a Fréchet (resp. Banach) space together with $E$. A one-to-one algebra homomorphism $\Lambda$: $f \mapsto \Lambda(f)$ between $F(E)$ and the algebra $L(\tilde{E})$ of linear endomorphisms of $\tilde{E}$, the restriction of which to each $E_n$ belongs to $L(E_n)$, can be defined as follows:

$$\Lambda(f)(\varphi_1 \otimes \ldots \otimes \varphi_n) = \sum_{1\leqslant p\leqslant n} \sum_{i_1+\cdots+i_p=n} f^{i_1} \otimes \ldots \otimes f^{i_p}(\sigma_n(\varphi_1 \otimes \ldots \otimes \varphi_n))$$

where $\sigma_n$ is the symmetrization operator on $\hat{\otimes}^n E$ defined by

$$\sigma_n(\varphi_1 \otimes \ldots \otimes \varphi_n) = \frac{1}{n!} \sum_{\sigma\in\mathfrak{S}_n} \varphi_{\sigma(1)} \otimes \ldots \otimes \varphi_{\sigma(n)}$$

($\varphi_1, \ldots, \varphi_n \in E$; $\mathfrak{S}_n$ is the group of their permutations). The map $g \mapsto \tilde{S}_g = \Lambda(S_g)$ will be called the *linear representation* of $G$ (in the 'Fock space' $\tilde{E}$) *associated with* $(S, E)$.

When $E$ is a Banach space, $(S^1, E)$ is a continuous representation of $G$

[11] and one sees by induction using $\tilde{S}$ that the maps

$$(g; \varphi_1, \ldots, \varphi_n) \mapsto S_g^n(\varphi_1, \ldots, \varphi_n)$$

are also continuous (analytic if $E$ is finite-dimensional) from $G \times E^n$ to $E$.

(c) *Formal Lie algebra representations.* Given $A = \Sigma_{n \geqslant 1} A^n$ and $B = \Sigma_{n \geqslant 1} B^n$ in $F(E)$, we define $A \times B \in F(E)$ by

$$(A \times B)^n = \sum_{1 \leqslant p \leqslant n} A^p \left( \sum_{0 \leqslant q \leqslant p-1} I_q \otimes B^{n-p+1} \otimes I_{p-q-1} \right) \sigma_n$$

and a one-to-one map $\mathrm{d}\Lambda$: $F(E) \to L(\tilde{E})$ by

$$\mathrm{d}\Lambda(A) | \hat{\otimes}^n E = \sum_{1 \leqslant p \leqslant n} \left( \sum_{0 \leqslant q \leqslant p-1} I_q \otimes A^{n-p+1} \otimes I_{p-q-1} \right) \sigma_n.$$

We now define $[A, B] = A \times B - B \times A$ for $A, B \in F(E)$. Since

$$\mathrm{d}\Lambda([A, B]) = [\mathrm{d}\Lambda(A), \mathrm{d}\Lambda(B)]$$

where on the right-hand side we have the commutator of two linear maps, $[A, B]$ defines a Lie algebra structure on $F(E)$. *A formal representation* $(\mathrm{d}S, E)$ of a real *Lie algebra* $\mathfrak{g}$ in a Fréchet space $E$ will therefore be defined as a Lie algebra homomorphism $\mathrm{d}S$ from $\mathfrak{g}$ to $F(E)$. In particular, the *free part* $(\mathrm{d}S^1, E)$ of $(\mathrm{d}S, E)$ is a linear representation of $\mathfrak{g}$ in $E$. $\mathrm{d}\tilde{S} = \mathrm{d}\Lambda \circ \mathrm{d}S$ will be the associated linear representation of $\mathfrak{g}$ in $\tilde{\mathrm{E}}$.

Since we need convergence properties for physical applications, we shall now introduce analytic structures.

## 2. Analytic representations

(a) *Definition.* A formal representation $(S, E)$ of a real Lie group $G$ in a Banach space $E$ is said to be *analytic* if there exists a neighbourhood $V$ of the identity in $G$ such that $S_g$ is analytic in a neighbourhood $U$ of the origin in $E$ for all $g \in V$.

One shows that $U$ can be taken independent of $g$ in $V$. Using Gårding's proof of the exponential growth of $\|S_g^1\|$ on $G$ [12], one can show that $V$ can contain any given bounded subset of $G$ (hence all $G$ if $G$ is compact) if $U$ is chosen small enough.

Let us denote by $H_r(E)$ the Banach space [10] of 'analytic functions'

$f=\Sigma_{n\geq 1} f^n \in F(E)$ with the norm $\|f\|_r=\Sigma r^n\|f^n\|<\infty$ $(r>0)$;

$$(f, \varphi)\mapsto f(\varphi)=\Sigma \hat{f}^n(\varphi)$$

is obviously a $C^\infty$ map: $H_r(E)\times B_r \to E$, where $B_r$ is the ball $\{\varphi \in E; \|\varphi\|<r\}$.

(b) *Equivalence.* A map $A=\Sigma_{n\geq 1} A^n$, $A^n \in L_n(E)$, analytic around the origin of the Banach space $E$, is an analytic isomorphism in a neighbourhood of the origin if and only if $A^1$ is an automorphism of $E$. We then say that two analytic representations $(S, E)$ and $(S', E')$ of $G$ in Banach spaces $E$ and $E'$ are *equivalent* if there exists an analytic map $A=\Sigma_{n\geq 1} A^n$, $A^1$ being an isomorphism of $E$ onto $E'$, such that the power series equality $S'_g=AS_gA^{-1}$ holds for every $g \in G$.

(c) *Smoothness.* If $E$ is infinite-dimensional, the space $E_\infty$ of differentiable vectors of a non-trivial $S^1$ is a Fréchet space strictly included in $E$. It is important to know whether or not the 'non linear terms' of $S$ introduce additional differentiability conditions.

We shall therefore say that an *analytic representation* $(S, E)$ of $G$ in a Banach space $E$ is *smooth* if there exist $r>0$ and an open neighbourhood $V$ of the identity in $G$ such that $g\mapsto S^1_{g^{-1}}S_g\equiv R_g$ is $C^\infty$ from $V$ to $H_r(E)$. One shows that $V$ and $r$ can be chosen small enough so that $(g, \varphi)\mapsto R_g(\varphi)$ and $(g, \varphi)\mapsto L_g(\varphi)\equiv S_g(S^1_{g^{-1}}\varphi)$ are differentiable maps $V\times B_r \to E$, and that $g\mapsto R^n_g=S^1_{g^{-1}}S^n_g$ is also a differentiable map $V\to L_n(E)$.

(d) *Differentiable vectors.* In a way similar to linear representations, a *differentiable vector* of an analytic representation $(S, E)$ of $G$ in a Banach space $E$ is a vector $\varphi$ such that $g\mapsto S_g(\varphi)$ is $C^\infty$ from a neighbourhood of the identity of $G$ into $E$.

When $(S, E)$ is smooth, one shows that there exists a neighbourhood $V$ of the identity in $G$ and $r>0$ such that $E_\infty \cap B_r$ is the set of differentiable vectors of $(S, E)$ contained in $B_r$, and that for $\varphi \in E_\infty \cap B_r$ the map $g\mapsto S_g(\varphi)$ is $C^\infty$ from $V$ to the Fréchet space $E_\infty$. This justifies the introduction of smooth representations. We shall now see that it is possible to restrict ourselves to smooth representations.

(e) *Smoothing of analytic representations.*

PROPOSITION 6.1. *Given an analytic representation $(S, E)$ of a real Lie group $G$ in a Banach space $E$ and a compact subgroup $K$ of $G$, there exists a* *smooth representation* $(S', E)$ *equivalent to* $(S, E)$ *such that in addition the restriction of* $S'$ *to* $K$ *is linear.*

The idea behind this result is that the nonlinear representation $S$ may be built by successive extensions of the linear representation $S^1$ by its successive tensor powers, the extension 1-cocycles being defined by the $R^n_g$. Then, up to a coboundary (which gives the equivalence operator step by step) each of these 1-cocycles *can* be chosen differentiable.

For a proof, see [5]. As a consequence, there exist $r>0$ and a neighbourhood $V$ of the identity in $G$ such that $(\mathfrak{g}, \varphi)\mapsto S_g(\varphi)$ is continuous from $V\times B_r$ to $E$. The (known) linearizability of analytic representations $(S, E)$ when $G$ is compact is also a consequence of this proposition.

(f) *Banal representations.* An analytic representation $(S, E)$ of a real Lie group in a Banach space $E$ is said *banal* if it is linearizable, i.e. equivalent to a linear representation.

## 3. Passage from the Lie group to the Lie algebra and vice-versa

(a) *Differentiation.* Let $(S, E)$ be a Banach space *formal* representation of $G$, the map $G\ni g\mapsto R^n_g\equiv S^1_{g^{-1}}S^n_g\in L_n(E)$ being $C^\infty$ in some open neighbourhood of the identity of $G$.

Define $\mathrm{d}S^n_x=((\mathrm{d}/\mathrm{d}t)R^n_{\exp tx})_{t=0}$ for $n\geqslant 2$, $x$ in the Lie algebra $\mathfrak{g}$ of $G$, and $\mathrm{d}S^1$ as the differential of the free part $S^1$ of $S$ on the Fréchet space $E_\infty$. Then $x\mapsto \mathrm{d}S_x=\mathrm{d}S^1_x+\Sigma_{n\geqslant 2}\,\mathrm{d}S^n_x$ defines a formal representation of $\mathfrak{g}$ in $E_\infty$, called the differential of $(S, E)$.

We shall say that a formal representation $(T, E_\infty)$ of a Lie algebra $\mathfrak{g}$ is an *analytic representation compatible with a given linear continuous representation* $(U, E)$ of $G$ in a Banach space $E$ if $E_\infty$ is the Fréchet space of differentiable vectors of $(U, E)$, the free part $T^1$ of $T$ coincides with $\mathrm{d}U$, and $T^n_x\in L_n(E)$ for $n\geqslant 2$ in such a way that $\Sigma_{n\geqslant 2}\, T^n_x$ is analytic around the origin of $E$ for every $x\in\mathfrak{g}$. This definition is justified by the following:

PROPOSITION 6.2. *In the above notations, if* $(S, E)$ *is smooth,* $(\mathrm{d}S, E_\infty)$ *is an analytic representation of* $\mathfrak{g}$ *compatible with* $(S^1, E)$; *moreover, in this case,* $\mathrm{d}S_x$ *maps* $E_\infty\cap B_r$ *into* $E_\infty$ *for some* $r>0$ *and all* $x\in\mathfrak{g}$.

One of the advantages of this differentiation procedure is that the Lie

algebra representation extends trivially to the complexified $\mathfrak{g}_c$ of $\mathfrak{g}$, which will enable us to use the Weyl unitary trick in the semi-simple case to pass to a compact form. However we need also the inverse procedure (integration to the group representation).
(b) *Integration.* The passage to the group is performed first at the formal level, then for analytic representations.

PROPOSITION 6.3. *Let $(U, E)$ be a continuous linear representation of a connected and simply connected Lie group $G$ in a Banach space $E$, and let $E_\infty$ be the Fréchet space of its differentiable vectors.*

(a) *Given a formal representation $(T, E_\infty)$ of the Lie algebra $\mathfrak{g}$ of $G$ such that $T_x^1 = dU_x$ and $T_x^n \in L_n(E_\infty)$ [resp. $T_x^n \in L_n(E_\infty) \cap L_n(E)$] for $n \geqslant 2$, where $T_x = \Sigma_{n\geqslant 1} T_x^n$ for $x \in \mathfrak{g}$, then there exists a unique formal representation $(S, E_\infty)$ [resp. $(S, E)$] of $G$ in $E_\infty$ [resp. $E$] such that for $\varphi_1, \ldots, \varphi_n \in E_\infty$, the map*

$$G \ni g \mapsto S_g^n(\varphi_1, \ldots, \varphi_n) \in E$$

*where $S_g = \Sigma_{n\geqslant 1} S_g^n$, has a derivative and*

$$T_x^n(\varphi_1, \ldots, \varphi_n) = ((d/dt) S^n_{\exp tx}(\varphi_1, \ldots, \varphi_n))_{t=0}$$

(b) *If $(T, E_\infty)$ is an analytic representation of $\mathfrak{g}$ compatible with $(U, E)$, the above defined formal representation $(S, E)$ is analytic. Moreover, for $\varphi \in E_\infty$ and $x \in \mathfrak{g}$, the map $t \mapsto S_{\exp tx}(\varphi)$ has a derivative around 0, the value of which at $t = 0$ is exactly $T_x(\varphi)$.*

The proof of this crucial result is rather involved (cf [5]). It makes use (for the first part) of the cohomological theory of extensions of representations (cf. [7]) and (for the second part) of the implicit functions theorem [10]. It makes possible the use of infinitesimal methods to linearize representations. The problem is to find an intertwining (nonlinear) operator $A$ that 'kills' the non-linear parts $S^n$ $(n \geqslant 2)$ of $S$, and this is basically a question of 1-cohomology, for which such methods are powerful.

## 4. Some applications, including covariant evolution equations

THEOREM 6.1. *Every analytic representation $S$ of a connected real semi-simple [resp. nilpotent] Lie group $G$ in a finite-dimensional vector space $E$ [resp. in a Hilbert space $E$, the free part $S^1$ of $S$ being unitary, irreducible and non-trivial] is banal.*

As an example, we shall sketch the proof of the semi-simple case. In view of Proposition 6.1, one may assume that $S$ is smooth and that therefore (Proposition 6.2) $(\mathrm{d}S, E)$ is analytic and compatible with $(S^1, E)$. Now $(\mathrm{d}S, E)$ extends to the complexified $\mathfrak{g}_c$ of $\mathfrak{g}$, and (Proposition 6.3) can be exponentiated to a representation of the corresponding real Lie group $G'$, equivalent (Proposition 6.1) to a smooth representation linear on the maximal compact subgroup of $G'$ (which is the compact form of $G$). This equivalent representation gives a linear representation of $\mathfrak{g}$, and therefore of $G$, equivalent to $(S, E)$.

In the nilpotent case one first shows that $S$ and $S^1$ have the same kernel when $S^1$ is a representation by non-constant homotheties, and then one builds directly an intertwining operator that kills the non-linear part.

More generally one can prove [6], building by induction the linearization formal series $A$, the following result (which however gives only a formal linearization):

PROPOSITION 6.4. *Let $(S, E)$ be a formal representation of a Lie group $G$ in a Fréchet space $E$ such that $G \ni g \mapsto \hat{S}^n_g \varphi \in E$ is $C^\infty$ for every $\varphi \in E$. Suppose that the differentiable* 1*-cohomology (cf.* [6, 7]) $H^1_\infty(G, L(\hat{\otimes}^n E, E))$ *of $G$ with values in $L(\hat{\otimes}^n E, E)$, $G$ acting on $\hat{\otimes}^n E$ through the nth tensor power of the* free *part $(S^1, E)$ of $(S, E)$, vanishes for every $n \geqslant 2$. Then there exists $A \in F(E)$, invertible, such that $S_g = AS^1_g A^{-1}$.*

The physical interest of this result follows from the fact that, when $G$ is Poincaré group and $E$ is the space of differentiable vectors for a unitary irreducible representation $(S^1, H)$ of the Poincaré group with mass $m^2 > 0$, the hypotheses – and therefore the conclusion – of Proposition 6.4 are satisfied. Indeed [7] already the Lie algebra 1-cohomology vanishes in this case.

Let now $\mathfrak{g}$ be the Lie algebra of a Lie group $G$, represented in a Fréchet space $H_\infty$ through an analytic representation $(T, H_\infty)$ compatible with a continuous linear representation $(S^1, H)$ of $G$ in a Banach space $H$. Then we say that the evolution equation

$$(6.1) \qquad (\mathrm{d}/\mathrm{d}t)\varphi_t = B\varphi_t \qquad (\varphi_t \in H_\infty)$$

is *covariant* under the action of $\mathfrak{g}$ if $B = T_X$ for some $X \in \mathfrak{g}$.

From Propositions 6.3 and 6.4 we now deduce the following corollary, which shows the formal linearizability of some covariant nonlinear equations:

COROLLARY 6.1. *Let* (6.1) *be a nonlinear differential equation, covariant under the Poincaré Lie algebra* $\mathfrak{g}$ *where (in the above notations)* $H_\infty$ *is the space of differentiable vectors for a unitary irreducible representation* $(S^1, H)$ *of* $G$ *with mass* $m^2>0$ *and* $B$ *is the representative* $T_{P_0}$ *of the energy generator* $P_0 \in \mathfrak{g}$. *Then there exists a unique formal representation* $(S, H_\infty)$ *of* $G$ *and* $A \in F(H_\infty)$ *invertible such that, if* $S_t$ *denotes* $S_{\exp tP_0}$,

$$t \mapsto \varphi_t = S_t\varphi \quad (\varphi \in H_\infty)$$

*has a derivative (as formal series), is a solution of* (1), *and* $S_t = A^{-1} S^1_{\exp tP_0} A$ on $H_\infty$.

Finally we may indicate a very trivial example of application of Theorem 6.1. Take $G=\mathbb{R}$, $\varphi \in E=\mathbb{C}$, $S^1_t = e^{it}$ $(t \in \mathbb{R})$ and look at the corresponding Equation (6.1) of the form $\mathrm{d}\varphi/\mathrm{d}t = i\varphi + P(\varphi)$ where $P$ is an entire function with lower term $\varphi^n$ $(n \geqslant 2)$. Then Theorem 6.1 tells us that for small enough initial data $\varphi_0$, these equations are linearizable and admit solutions which are global in time.

**Note added in proof.** Since this chapter was written, important developments have occurred in the theory. We shall mention here only results which have already been published. Pinczon and Simon studied finite-dimensional representations of inhomogeneous classical groups [13] and in particular showed that, in the cases studied, nonlinear terms can be combined only with a quantized 'coupling constant'. Corollary 6.1 has been extended [6b] to the case when both energy signs are present, formal linearization being obtained on the subspaces of positive or negative energy initial data. When $S^1$ is a unitary representation of Poincaré group with zero mass and any finite number of irreducible components (discrete helicities, any energy sign), then $T_{P_0}$, and therefore the corresponding Equation (6.1), can be formally linearized [14]; for instance, Yang–Mills equations supplemented with a relativistic gauge condition (in *Minkowski* space) are formally linearizable.

*Université de Dijon, France*

## REFERENCES

[1] Poincaré, H., Oeuvres, Gauthier-Villars, Paris, 1951.
[2] Bourbaki, N., *Algebres de Lie*, Hermann, Paris, 1971.
Jacobson, N., *Lie Algebras*, Interscience NY, 1962.

[3] Hermann, R., *Trans. Amer. Math. Soc.* **130** (1968), 105–109.
[4] Guillemin, V., and Sternberg, S., *Trans. Amer. Math. Soc.* **130**, 110–
[5] Flato, M., Pinczon, G., and Simon, J., *Ann. Ec. Norm. Sup.* **10** (1977), 405–418.
[6] Flato, M., and Simon, J., (a) *Lett. Math. Phys.* **2** (1977), 155–160; (b) *J. Math. Phys.* **21** (1980).
[7] Pinczon, G., and Simon, J., *Lett. Math. Phys.* **1** (1975), 83; *C.R. Acad Sci. Paris* **277** (1974), 455; *Rep. Math. Phys.* (1979) (in press).
Simon, J., 'Introduction to 1-cohomology of Lie groups', this volume.
[8] Schaeffer, H. H., *Topological Vector Spaces*, Macmillan, New York, 1966.
[9] Nachbin, L., *Topology on Spaces of Holomorphic Mappings*, Springer, Berlin, 1969.
[10] Bourbaki, N., *Variétés différentielles et analytiques*, Hermann, Paris, 1967.
[11] Moore, R. T., *Memoirs Amer. Math. Soc.*, no. 78, (1968).
[12] Gårding, L., *Bull. Soc. Math. Fr.* **88** (1960), 77–93.
[13] Pinczon, G., and Simon, J., *Lett. Math. Phys.* **2** (1978), 499–504.
[14] Flato, M., and Simon, J., *Lett. Math. Phys.* **3** (1979), 279–283.

JACQUES SIMON

# INTRODUCTION TO THE 1-COHOMOLOGY OF LIE GROUPS

CONTENTS



## 1. INTRODUCTION

Let $G$ be a real Lie group, $\mathscr{H}$ a locally convex barreled complete topological vector space (C.T.V.S.) on $\mathbb{C}$. Let $(U, \mathscr{H})$ be a continuous representation of $G$ in $\mathscr{H}$. A 1-cocycle on $G$ with values in the $G$-module $\mathscr{H}$ is a continuous mapping $a: G \to \mathscr{H}$ such that

$$(1.1) \qquad a_{gg'} - a_g = U_g a_{g'}$$

for every $g, g' \in G$. We denote by $Z_c^1(G, \mathscr{H})$ the space of the 1-cocycles on $G$ with values in $\mathscr{H}$ and by $B^1(G, \mathscr{H})$ the space of the coboundaries (i.e. the space of the 1-cocycles of the form $a_g = U_g\varphi - \varphi$ for a given $\varphi \in \mathscr{H}$).

Set $H_c^1(G, \mathscr{H}) = Z_c^1(G, \mathscr{H})/B^1(G, \mathscr{H})$.

We shall give three examples where these cocycles play a role.

### (a) *Affine Representations*

A continuous affine representation $V$ of $G$ in $\mathscr{H}$ is necessarily of the form $V_g\varphi = U_g\varphi + a_g$ where $(U, \mathscr{H})$ is a continuous linear representation and $a \in Z_c^1(G, \mathscr{H})$.

Moreover, $H_c^1(G, \mathscr{H})$ can be considered as the space of the (affine) equivalence classes of affine representations (the linear part of which is $(U, \mathscr{H})$).



*J. A. Wolf, M. Cahen, and M. De Wilde (eds.), Harmonic Analysis and Representations of Semi-Simple Lie Groups, 449–465.*

(b) *Extensions of Representations*

Given two representations $(U^1, \mathscr{H}^1)$ and $(U^2, \mathscr{H}^2)$ of $G$ in C.T.V.S., an *extension* $(V, E)$ of $(U^1, \mathscr{H}^1)$ by $(U^2, \mathscr{H}^2)$ is a continuous representation of $G$ in a C.T.V.S. $E$ such that we have the following exact sequence of $G$-morphisms.

$$0 \to \mathscr{H}^1 \overset{i}{\to} E \overset{\pi}{\to} \mathscr{H}^2 \to 0$$

and where $i(\mathscr{H}^1)$ has a topological supplementary in $\mathscr{H}$. Let $\mathscr{L}_f(\mathscr{H}^2, \mathscr{H}^1)$ be the space of the linear continuous operators from $\mathscr{H}^2$ to $\mathscr{H}^1$ endowed with the topology of the convergence on finite sets. One defines a continuous representation $R$ of $G$ in $\mathscr{L}_f(\mathscr{H}^2, \mathscr{H}^1)$ by $R_g A = U_g^1 A U_{g-1}^2$ and one can define $Z_c^1(G, \mathscr{L}_f(\mathscr{H}^2, \mathscr{H}^1))$, $B^1(G, \mathscr{L}_f(\mathscr{H}^2, \mathscr{H}^1))$ and

$$H_c^1(G, \mathscr{L}_f(\mathscr{H}^2, \mathscr{H}^1)).$$

Given a continuous section $\sigma$ of $\pi$, $\sigma(\mathscr{H}^2)$ is a topological supplementary of $i(\mathscr{H}^1)$. Every vector $\varphi \in \mathscr{H}$ writes $\varphi = i(a) + \sigma(b)$ $(a \in \mathscr{H}^1,\ b \in \mathscr{H}^2)$. Then, there exists a unique continuous function $f: G \to \mathscr{L}_f(\mathscr{H}^2, \mathscr{H}^1)$ such that

$$V_g(i(a) + \sigma(b)) = i(U_g^1(a) + f_g(b)) + \sigma(U_g^2(b)),$$

and $f_{gg'} = U_g^1 f_{g'} + f_g U_{g'}^2$. Define $F_g = f_g U_{g-1}^2$. Obviously,

$$F \in Z_c^1(G, \mathscr{L}_f(\mathscr{H}^2, \mathscr{H}^1)).$$

*$F$ is the cocycle associated to the extension by the section $\sigma$.*

Conversely, if $F \in Z_c^1(G, \mathscr{L}_f(\mathscr{H}^2, \mathscr{H}^1))$, this is the cocycle of an extension $(V, E)$ where $E = \mathscr{H}^1 \times \mathscr{H}^2$ and $V_g(a+b) = U_g^1(a) + F_g U_g^2(b) + U_g^2(b)$. Two extensions $(V, E)$ and $(V', E')$ of $(U^1, \mathscr{H}^1)$ by $(U^2, \mathscr{H}^2)$ are strongly equivalent if there exists a linear continuous $G$-morphism $T: E \to E'$ such that the following diagram commutes

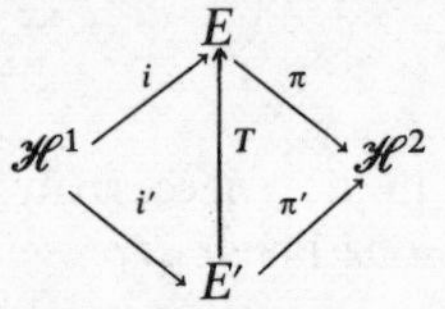


Consequently, $T$ is an isomorphism of $E$ into $E'$ and therefore we have an equivalence relation.

We denote by $\mathrm{Ext}_s(\mathscr{H}^1, \mathscr{H}^2)$ the set of the strong equivalence class of extensions of $(U^1, \mathscr{H}^1)$ by $(U^2, \mathscr{H}^2)$. One easily sees that

$$\mathrm{Ext}_s(\mathscr{H}^1, \mathscr{H}^2) \simeq H_c^1(G, \mathscr{L}_f(\mathscr{H}^2, \mathscr{H}^1)).$$

Equivalence of extensions will be studied later.

(c) *Non-linear Representations.*

Given a Banach space $\mathscr{H}$ and a continuous mapping $(g, \varphi) \to T_g(\varphi)$ from $G \times \mathscr{H}$ to $\mathscr{H}$ such that $T_g(0) = 0$ which is moreover analytic in $\varphi \in \mathscr{H}$, write

$$T_g(\varphi) = T_g^1(\varphi) + T_g^2(\varphi, \varphi) + o(\|\varphi\|^3),$$

where $T_g^1 \in \mathscr{L}(\mathscr{H}, \mathscr{H})$, $T_g^2 \in \mathscr{L}_2(\mathscr{H}, \mathscr{H})$ (space of the bilinear continuous symmetric functions from $\mathscr{H} \times \mathscr{H}$ to $\mathscr{H}$. We say that $T$ is a non-linear representation of $G$ is $T_{gg'} = T_g T_{g'}$. This implies that $T^1$ is a linear representation of $G$ and that

$$T_{gg'}^2 = T_g^1 \circ T_g^2 + T_g^2 \circ (T_{g'}^1 \times T_{g'}^1). \tag{1.2}$$

Identifying $\mathscr{L}_2(\mathscr{H}, \mathscr{H})$ with a subspace of $\mathscr{L}(\mathscr{H} \hat{\otimes}_\pi \mathscr{H}, \mathscr{H})$, and defining on $\mathscr{L}_f(\mathscr{H} \hat{\otimes}_\pi \mathscr{H}, \mathscr{H})$ a continuous linear representation by

$$V_g^2 A = T_g^1 \circ A \circ (T_{g-1}^1 \otimes T_{g-1}^1), \qquad A \in \mathscr{L}_f(\mathscr{H} \hat{\otimes}_\pi \mathscr{H}, \mathscr{H})$$

the equation (1.2) means that, if

$$t_g^2 = T_g^2 \circ (T_{g-1}^1 \otimes T_{g-1}^1), \qquad t^2 \in Z_c^1(G, \mathscr{L}_f(\mathscr{H} \hat{\otimes}_\pi \mathscr{H}, \mathscr{H})).$$

This example will be exposed with more details in the paper by Flato and Sternheimer in this Volume.

## 2. Differentiation of the Cocycles

We keep the hypotheses of the last section. Denote by $\mathscr{H}_\infty$ (resp. $\mathscr{H}_\omega$) the space of the differentiable (resp. analytic) vectors of $(U, \mathscr{H})$. We denote by $Z_\infty^1(G, \mathscr{H}_\infty)$ (resp. $Z_\omega^1(G, \mathscr{H}_\omega)$ the space of the elements in $Z_c^1(G, \mathscr{H})$ which are $C^\infty$ functions (resp. analytic functions) from $G$ to $\mathscr{H}$. It follows from relation (1.1) that the range of such a cocycle is contained in $\mathscr{H}_\infty$ (resp. $\mathscr{H}_\omega$). We denote by $B^1(G, \mathscr{H}_\infty)$ (resp. $B^1(G, \mathscr{H}_\omega)$) the space

$$Z_\infty^1(G, \mathscr{H}_\infty) \cap B^1(G, \mathscr{H})$$

(resp. $Z^1_\omega(G, \mathscr{H}_\omega) \cap B^1(G, \mathscr{H})$. We denote by $H^1_\infty(G, \mathscr{H}_\infty)$ (resp. $H^1_\omega(G, \mathscr{H}_\omega)$) the quotient space $Z^1_\infty(G, \mathscr{H}_\infty)/B^1(G, \mathscr{H}_\infty)$ (resp. $Z^1_\omega(G, \mathscr{H}_\omega)/B^1(G, \mathscr{H}_\omega)$).

PROPOSITION 1.1. *$H^1_c(G, \mathscr{H}) \simeq H^1_\infty(G, \mathscr{H}_\infty)$. Moreover, when $\mathscr{H}$ is a Banach space, $H^1_c(G, \mathscr{H}) \simeq H^1_\omega(G, \mathscr{H}_\omega)$*

*Proof.* Suppose that $\mathscr{H}$ is a Banach space. Define a linear representation $(V, \mathscr{H} \oplus \mathbb{C})$ of $G$ by $U'_g(\varphi, t) = (U_g\varphi + ta_g, t)$. Since the function $g \to \|U'_g(0, t)\|$ is at most of exponential increase on $G$, it follows that the function $g \to \|a_g\|$ is at most of exponential increase on $G$. Let $(t, g) \to u(t, g)$, from $\mathbb{R}_+ \times G$ to $\mathbb{C}$, be a solution of the heat equation on $G$, with arbitrary fast decrease [Sections 6 and 7 of Ref. 2] and such that $\int_G u(t_0, g^{-1})\, dg = 1$ for a given $t_0 > 0$ ($dg$ being a left invariant measure on $G$).

$g \to \int_G u(t_0, g'^{-1}g)\, dg'$ exists and is analytic from $G$ to $\mathscr{H}$ [Lemma 1, Section 8 of Ref. 2]. Define

$$a'_g = \int_G u(t_0, g'^{-1}g) a_{g'}\, dg' - \int_G u(t_0, g^{-1}) a_{g'}\, dg'.$$

We have

$$a'_g = a_g + (U_g - I) \int_G u(t_0, g'^{-1}) a_{g'}\, dg'.$$

Therefore $a \equiv a' \bmod B^1(G, \mathscr{H})$. Since $a' \in Z^1(G, \mathscr{H}_\omega)$, we get the second equality. For the first equality, in the general case where $\mathscr{H}$ is a C.T.V.S., the proof is easier (replace $g \to u(t_0, g)$ by a $C^\infty$ function on $G$ with compact support).

Denote by $\mathfrak{g}$ the Lie algebra of $G$. When $\varphi \in \mathscr{H}_\infty$, we can define

$$dU_X(\varphi) = (d/dt)[U_{\exp tX}\varphi]_{t=0}, \text{ for every } X \in \mathfrak{g}.$$

$(dU, \mathscr{H}_\infty)$ is a $\mathfrak{g}$-module leaving $\mathscr{H}$-invariant. Given $a \in Z'_\infty(G, \mathscr{H}_\infty)$ and $X \in \mathfrak{g}$ define $\Delta a_X = (d/dt)[a_{\exp tX}]_{t=0}$.

Given a representation $(\pi, V)$ of $\mathfrak{g}$ in a complex vector space $V$, a 1-cocycle on $\mathfrak{g}$ with values in the $\mathfrak{g}$-module $V$ is a linear mapping $\xi : \mathfrak{g} \to V$ such that

$$(2.1) \qquad \xi_{[X,Y]} = \pi_X \xi_Y - \pi_Y \xi_X$$

for every $X, Y \in \mathfrak{g}$. We denote by $Z^1(\mathfrak{g}, V)$ the space of the 1-cocycles on $\mathfrak{g}$ with values in $V$ and by $B^1(G, V)$ the space of the 1-coboundaries (i.e.

the space of the 1-cocycles of the form $\xi_X = \pi_X \varphi$, $\varphi \in V$). Define $H^1(\mathfrak{g}, V) = Z^1(\mathfrak{g}, V)/B^1(\mathfrak{g}, V)$. Since

$$dU'_X(\varphi, t) = (dU_X \varphi + t\Delta a_X, 0), \qquad (\varphi \in \mathscr{H}_\infty, t \in \mathbb{C}),$$

defines a $\mathfrak{g}$-module on $\mathscr{H}_\infty \times \mathbb{C}$, one sees that $\Delta a \in Z^1(\mathfrak{g}, \mathscr{H}_\infty)$ when $a \in Z'_\infty(G, \mathscr{H}_\infty)$ and that $\Delta a \in Z^1_\omega(\mathfrak{g}, \mathscr{H}_\omega)$ when $a \in Z^1_\omega(G, \mathscr{H}_\omega)$.

The following result is then obvious.

PROPOSITION 2.2. *Suppose that G is connected. The map*

$$\Delta\colon Z^1_r(G, \mathscr{H}_r) \to Z^1(\mathfrak{g}, \mathscr{H}_r), \qquad (r = \infty, \omega),$$

*is a canonical imbedding.*

## 3. Extensions of representations

### (a) *Regularisation of the* 1-*cocycles*

Given two continuous representations $(U^1, \mathscr{H}^1)$ and $(U^2, \mathscr{H}^2)$ of a Lie group $G$ in Banach spaces, we define $R_g(A) = U^1_g A U^2_{g-1}$, $A \in \mathscr{L}_f(\mathscr{H}^2, \mathscr{H}^1)$. We denote by $\mathscr{L}_b(\mathscr{H}^2, \mathscr{H}^1)$ the space $\mathscr{L}(\mathscr{H}^2, \mathscr{H}^1)$ endowed with the topology of the uniform convergence on bounded sets. In general, $R$ is not continuous on $\mathscr{L}_b(\mathscr{H}^2, \mathscr{H}^1)$. We denote by $\mathscr{D}(\mathscr{H}^2, \mathscr{H}^1)$ the closed subspace of $\mathscr{L}_b(\mathscr{H}^2, \mathscr{H}^2)$ where $R$ is continuous. We suppose that $\mathscr{D}(\mathscr{H}^2, \mathscr{H}^1)$ is endowed with the Banach space topology induced by that of $\mathscr{L}_b(\mathscr{H}^2, \mathscr{H}^1)$.

PROPOSITION 3.1. $H^1_c(G, \mathscr{D}(\mathscr{H}^2, \mathscr{H}^1)) \simeq H^1_c(G, \mathscr{L}_f(\mathscr{H}^2, \mathscr{H}^1))$.

*Proof.* Given $F \in Z^1_c(G, \mathscr{L}_f(\mathscr{H}^2, \mathscr{H}^1))$, the function $g \to \|F_g\|$ is lower semi-continuous. It then results from the cocycle equation that $g \to \|F_g\|$ is bounded above on every compact set of $G$. Given a continuous function $u$ on $G$, with compact support and such that $\int_G u(g^{-1})\,dg = 1$, the mapping $g \to \int_G u(g'^{-1}g) F_{g'}\,dg'$ is continuous from $G$ to $\mathscr{L}_b(\mathscr{H}^2, \mathscr{H}^1)$.

Define

$$F'_g = \int_G u(g'^{-1}g) F_{g'}\,dg' - \int_G u(g'^{-1}) F_{g'}\,dg'$$

We have

$$F'_g = F_g + (R_g - I) \int_G u(g^{-1}) F_{g'}\,dg'$$

Consequently, $F'_g \equiv F_g \bmod B^1(G, \mathscr{L}_f(\mathscr{H}^2, \mathscr{H}^1))$ and $F \in Z_c^1(G, \mathscr{D}(\mathscr{H}^2, \mathscr{H}^1))$.

(b) *Differentiable and Analytic Vectors of an Extension*

Suppose given an extension $(V, E)$ of $(U^1, \mathscr{H}^1)$ by $(U^2, \mathscr{H}^2)$ ($\mathscr{H}^1$ and $\mathscr{H}^2$ being Banach spaces), a section $\sigma$ and a cocycle $F$ associated to the extension by the section $\sigma$. We put $f_g = F_g U_g^2$.

LEMMA 3.2. *If the mapping $g \to f_g(b)$ is $C^\infty$ for every $b \in \mathscr{H}^2_\infty$, we have $E_\infty = i(\mathscr{H}^1_\infty) + \sigma(\mathscr{H}^2_\infty)$ (topological direct sum). If moreover $g \to f_g(b)$ is analytic for every $b \in \mathscr{H}^2_\omega$, we have $E_\omega = i(\mathscr{H}^1_\omega) + \sigma(\mathscr{H}^2_\omega)$.*

*Proof.* We have $i(\mathscr{H}^1_\infty) \subset E_\infty$ ($i$ continuous $G$ morphism). If $b \in \mathscr{H}^2_\infty$, the mapping

$$g \to V_g(\sigma(b)) = i(f_g(b)) + \sigma(U_g^2(b))$$

is $C^\infty$ from $G$ to $E$. Therefore, $\sigma(\mathscr{H}^2_\infty) \subset E_\infty$ and therefore $i(\mathscr{H}^1_\infty) + \sigma(\mathscr{H}^2_\infty) \subset E_\infty$. Suppose that $\varphi = i(a) + \sigma(b) \in E_\infty$. The mapping

$$g \to \pi(V_g(\varphi)) = U_g^2(b)$$

is $C^\infty$ from $G$ to $\mathscr{H}^2$. Therefore $b \in \mathscr{H}^2_\infty$. The mapping $g \to i(f_g(b))$ is then $C^\infty$ from $G$ to $E$. Consequently

$$g \to i(U_g^1(a)) = V_g(\varphi) - (i(f_g(b)) + \sigma(U_g^2(b)))$$

is $C^\infty$. Therefore $a \in \mathscr{H}^1_\infty$ and $E_\infty = i(\mathscr{H}^1_\infty) + \sigma(\mathscr{H}^2_\infty)$.

The proof is similar for analytic vectors. It remains to prove that the direct sum is topological for the spaces of differentiable vectors.

We have $\pi(E_\infty) = \mathscr{H}^2_\infty$ and $\sigma(\mathscr{H}^2_\infty) \subset E_\infty$. Therefore $\pi|_{E_\infty} \in \mathscr{L}(E_\infty, \mathscr{H}^2_\infty)$ and $\sigma|_{\mathscr{H}^2_\infty} \subset \mathscr{L}(\mathscr{H}^2_\infty, E_\infty)$. Therefore, the projection $\rho = \sigma|_{\mathscr{H}^2_\infty} \circ \pi|_{E_\infty}$ on $\sigma(\mathscr{H}^2_\infty)$ according to $i(\mathscr{H}^1_\infty)$ belongs to $\mathscr{L}(E_\infty, E_\infty)$.

PROPOSITION 3.3. *There exists a section $\sigma$ such that*

$$E_\infty = i(\mathscr{H}^1_\infty) + \sigma(\mathscr{H}^2_\infty),$$

*(topological direct sum).*

$$E_\omega = i(\mathscr{H}^1_\omega) + \sigma(\mathscr{H}^2_\omega),$$

*Proof.* This results from Proposition 3.1., Proposition 2.1. and Lemma 3.2.

(c) *Weakly Equivalent Extensions*

DEFINITION 3.4. An extension $(V, E)$ is *weakly equivalent* to an extension $(V', E')$ if there exists a closed linear operator $T: E \to E'$ with dense domain $D_T$ invariant under $V$ such that

(a) $T$ is a $G$-morphism.
(b) $D_T \supset i(\mathscr{H}^1_\infty)$.
(c) $\pi(D_T) \supset \mathscr{H}^2_\infty$.
(d) The following diagram is commutative

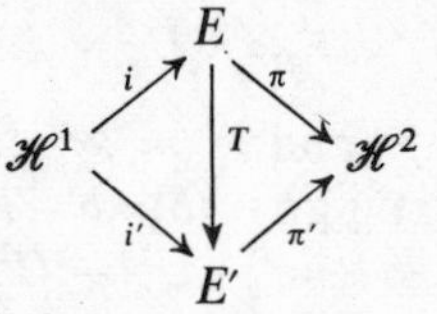


Obviously strong equivalence implies weak equivalence.

It can be proved, by standard arguments that weak equivalence is an equivalence relation. In fact, two extensions $(V, E)$ and $(V', E')$ are equivalent if and only if there exists a continuous $G$-linear morphism $T: E_\infty \to E'_\infty$ such that the following diagram is commutative

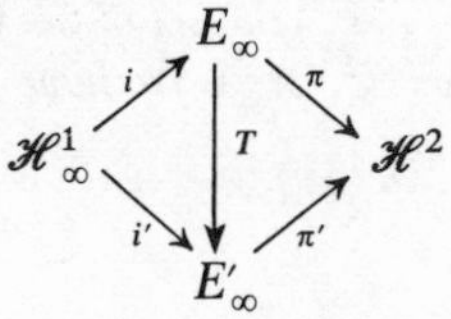


(see [6]).

We shall now introduce a cohomology adapted to the classification of the weak equivalence class of extensions.

For that purpose we shall introduce a new class of coboundaries.

DEFINITION 3.5. A cocycle $F \in Z^1_\infty(G, \mathscr{D}(\mathscr{H}^2, \mathscr{H}^1))$ is a *weak coboundary* if there exists $l \in \mathscr{L}(\mathscr{H}^2_\infty; \mathscr{H}^1_\infty)$ such that $F_g = U_g^{-1} l U^2_{g^{-1}} - l$ on $\mathscr{H}^2_\infty$. We denote by $B'_W(G, \mathscr{D}(\mathscr{H}^2, \mathscr{H}^1)_\infty)$ the space of the weak coboundaries in $Z^1_\infty(G, \mathscr{D}(\mathscr{H}^2, \mathscr{H}^1)_\infty)$ and

$$H^1_W(G, \mathscr{D}(\mathscr{H}^2, \mathscr{H}^1)_\infty) = Z^1_\infty(G, \mathscr{D}(\mathscr{H}^2, \mathscr{H}^1)_\infty)/B^1_W(G, \mathscr{D}(\mathscr{H}^2, \mathscr{H}^1)_\infty).$$

PROPOSITION 3.6. *Given two extensions* $(E, V)$ *and* $(E', V')$, $\sigma$ *and σ′ two continuous sections of these extensions such that the associated cocycles F and F′, respectively, belong to* $Z^1_\infty(G, \mathscr{D}(\mathscr{H}^2, \mathscr{H}^1)_\infty)$ *then (E, V) and (E′, V′) are weakly equivalent if and only if*

$$F - F' \in B^1_W(G, \mathscr{D}(\mathscr{H}^2, \mathscr{H}^1)_\infty).$$

*Proof.* Suppose that $(E, V)$ and $(E', V')$ are weakly equivalent. We can suppose that $E = E' = \mathscr{H}^1 \times \mathscr{H}^2$ and that

$$V_g(a+b) = U^1_g(a) + f_g(b) + U^2_g(b), \qquad f_g = F_g U^2_g,$$

$$V'_g(a+b) = U'_g(a) + f'_g(b) + U^2_g(b), \qquad f'_g = F'_g U^2_g.$$

But, $E_\infty = E'_\infty = \mathscr{H}^1_\infty \times \mathscr{H}^2_\infty$ (Proposition 3.3.) and $E'_\infty = \mathscr{H}^1_\infty \times T(\mathscr{H}^2_\infty)$. If $b \in \mathscr{H}^2_\infty$, we have $b - T(b) \in \mathscr{H}^1_\infty$ and the mapping $b \to l(b) = b - T(b)$ belongs to $\mathscr{L}(\mathscr{H}^2_\infty, \mathscr{H}^1_\infty)$. Moreover on $\mathscr{H}^2_\infty$, we have $f'_g = f_g + U^1_g l - l U^2_g$. Therefore $F_g - F'_g$ is a weak coboundary. Conversely, suppose that $f'_g = f_g + U^1_g l - l U^2_g$ with $l \in \mathscr{L}(\mathscr{H}^2_\infty, \mathscr{H}^1_\infty)$. One then defines $T\colon \mathscr{H}^1_\infty \times \mathscr{H}^2_\infty \to \mathscr{H}^1_\infty \times \mathscr{H}^2_\infty$ by $T(a+b) = (a - l(b)) + b$. □

At this level the following proposition is obvious.

PROPOSITION 3.7. *Denote by* $\mathrm{Ext}_W(\mathscr{H}^1, \mathscr{H}^2)$ *the set of the weak equivalence classes of extensions of* $(U^1, \mathscr{H}^1)$ *by* $(U^2, \mathscr{H}^2)$. *We have*

$$\mathrm{Ext}_W(\mathscr{H}^1, \mathscr{H}^2) = H^1_W(G, \mathscr{D}(\mathscr{H}^2, \mathscr{H}^1)_\infty).$$

## 4. Exponentiations of the Cocycles

Denote by

$$\mathscr{P}(\mathscr{H}^2, \mathscr{H}^1) = \mathscr{L}(\mathscr{H}^2, \mathscr{H}^1) \cap \mathscr{L}(\mathscr{H}^2_\infty, \mathscr{H}^1_\infty)$$

and by

$$Z^1(\mathfrak{g}, \mathscr{P}(\mathscr{H}^2, \mathscr{H}^1))$$

the space of the linear mappings

$$\xi\colon \mathfrak{g} \to \mathscr{P}(\mathscr{H}^2, \mathscr{H}^1)$$

satisfying

$$\xi_{[X,Y]} = (dU^1_X \xi_Y - \xi_Y\, dU^2_X) - (dU^1_Y \xi_X - \xi_X\, dU^2_Y)$$

on $\mathscr{H}^2_\infty$. Denote by $B^1_W(\mathfrak{g}, \mathscr{P}(\mathscr{H}^2, \mathscr{H}^1))$ the subspace of the cocycles of the form $\xi_X = dU^1_X T - T\, dU^2_X$ with $T \in \mathscr{L}(\mathscr{H}^2_\infty, \mathscr{H}^1_\infty)$.

Finally we denote by

$$H^1_W(\mathfrak{g}, \mathscr{P}(\mathscr{H}^2, \mathscr{H}^1)) = Z^1(\mathfrak{g}, \mathscr{P}(\mathscr{H}^2, \mathscr{H}^1)/B^1_W(\mathfrak{g}, \mathscr{P}(\mathscr{H}^2, \mathscr{H}^1)).$$

PROPOSITION 4.1. *We have* $H^1_W(G, \mathscr{D}(\mathscr{H}^2, \mathscr{H}^1)_\infty) \subset H^1_W(\mathfrak{g}, \mathscr{P}(\mathscr{H}^2, \mathscr{H}^1))$. *The inclusion being an equality when G is simply connected.*

*Proof.* The proof of the inclusion is straightforward. We shall prove the equality. Denote by $\mathscr{L}'$ the set of the linear forms on $\mathscr{L}(\mathscr{H}^2, \mathscr{H}^1)$ of the type

$$T \to u(T) = c(T(b)), \qquad c \in (\mathscr{H}^1_\infty)', \qquad b \in \mathscr{H}^2_\infty.$$

This set separates points on $\mathscr{L}(\mathscr{H}^2_\infty, \mathscr{H}^1_\infty)$.

Given $\xi \in Z^1(\mathfrak{g}, \mathscr{P}(\mathscr{H}^2, \mathscr{H}^1))$ and $u \in \mathscr{L}'$, one defines $\omega^u_X(g) = u(R_g \xi_X)$ with $R_g \xi_X = U'_g \xi_X U^2_{g-1}$. The mapping $g \to \omega^u_X(g)$ belongs to $C^\infty(G)$. Define $X(l)(g) = (d/dt)(l(g \exp tX))_{t=0}$ with $X \in \mathfrak{g}$ and $l \in C^\infty(G)$. Then

$$\omega^u_{[X,Y]} = X(\omega^u_Y) - Y(\omega^u_X)$$

(i.e. $\omega^u$ is a closed form).

Therefore, there exists $f^u \in C^\infty(G)$ such that $\omega^u_X = X(f^u)$ and we can suppose that $f^u(e) = 0$ since it is defined up to an additive constant. We have

$$f^u(\exp X) = \int_0^1 \omega^u_X(\exp tX)\, dt.$$

Now, $G$ acts on $\mathscr{L}'$ by $g \cdot u(T) = u(R_{g^{-1}}(T))$ and

$$f^u(gh) = f^u(g) + f^{g^{-1}u}(h) \tag{4.1}$$

Suppose that $u(T) = c(T(b))$. Then

$$f^u(\exp X) = c(F_{\exp X}(b))$$

with

$$F_{\exp X} = \int_0^1 u^1_{\exp tX} \xi_X U^2_{\exp(-tX)}\, dt.$$

We have $F_{\exp X} \in \mathscr{P}(\mathscr{H}^2, \mathscr{H}^1)$ and relation (4.1) can be written

$$u(F_{gh}) = u(F_g + R_g F_h).$$

And, since $\mathscr{L}'$ separates points in $\mathscr{L}(\mathscr{H}^2_\infty, \mathscr{H}^1_\infty)$, we have $F_{gh} = F_g + R_g F_h$. The differentiability properties on $G$ are then easily verified. □

Given a representation $(U, \mathscr{H})$ of $G$ in a Banach space $\mathscr{H}$, we have $H^1_c(G, \mathscr{H}) \simeq H^1_c(G, \mathscr{L}(\mathbb{C}, \mathscr{H}))$ (we identify $\mathscr{H} \simeq \mathscr{L}(\mathbb{C}, \mathscr{H})$ with the trivial representation acting on $\mathbb{C}$). The previous result rewrites

PROPOSITION 4.2. *When $G$ is connected and simply connected, we have*

$$H^1_c(G, \mathscr{H}) \simeq H^1(\mathfrak{g}, \mathscr{H}_\infty).$$

## 5. Applications to Semisimple Lie Groups

Let $G$ be a connected semisimple Lie group. Fix a Cartan decomposition $\mathfrak{g} = \mathfrak{k} + \mathfrak{p}$ of its Lie algebra and let $\mathfrak{g}_c = \mathfrak{k}_c + \mathfrak{p}_c$ be its complexification. (We denote by $K$ the analytic subgroup of $G$ corresponding to $\mathfrak{k}$). We say that a differentiable vector of a representation $(U, \mathscr{H})$ is a $K$-finite vector of it is contained in a finite-dimensional invariant space of $U_K$. Let $\mathscr{H}_K$ be the space of all $K$-finite vectors that we call the Harish-Chandra module of $U$ relative to $K$. $\mathscr{H}_K$ is a $\mathfrak{g}$-module dense in $\mathscr{H}$. Let $Z^1_K(\mathfrak{g}, \mathscr{H}_K)$ (resp. $B^1_K(\mathfrak{g}, \mathscr{H}_K)$, resp. $H^1_K(\mathfrak{g}, \mathscr{H}_K)$) be the space of the cocycles vanishing on $\mathfrak{k}$ (resp. the space of the coboundaries vanishing on $\mathfrak{k}$; resp. the space $Z^1_K(\mathfrak{g}, \mathscr{H}_K)/B^1_K(\mathfrak{g}, \mathscr{H}_K)$). Let $Z^1_K(G, \mathscr{H}_\infty)$ (resp. $B^1_K(G, \mathscr{H}_\infty)$, resp. $H^1_K(G, \mathscr{H}_\infty)$) be the space of the differentiable cocycles vanishing on $K$ (resp. of the differentiable coboundaries vanishing on $K$, resp. the space $Z^1_K(G, \mathscr{H}_\infty)/B^1_K(G, \mathscr{H}_\infty)$). Suppose that $K$ is compact

PROPOSITION 5.1. *We have the following inclusions*:

$$H^1_\infty(G, \mathscr{H}_\infty) = H^1_K(G, \mathscr{H}_\infty) \subset H^1_K(\mathfrak{g}, \mathscr{H}_K) = H^1(\mathfrak{g}, \mathscr{H}_K) \subset H^1(\mathfrak{g}, \mathscr{H}_\infty).$$

*If $G$ is simply connected, these inclusions are equalities.*

*Proof.* To prove that $H^1(\mathfrak{g}, \mathscr{H}_K) \subset H^1(\mathfrak{g}, \mathscr{H}_\infty)$, it is sufficient to prove that $H^1(\mathfrak{g}, \mathscr{H}_K) \cap B^1(\mathfrak{g}, \mathscr{H}_\infty) = B^1(\mathfrak{g}, \mathscr{H}_K)$. If $\xi$ belongs to this intersection, it satisfies $\xi_X = dU_X \psi$, $\psi \in \mathscr{H}_\infty$ and $\xi_X \in \mathscr{H}_K$, $\forall X \in \mathfrak{g}$.

Given a basis $\{X_1, \dots X_p\}$ of $\mathfrak{k}$ we have then

$$dU(\mathscr{U}(\mathfrak{k}))\psi \subset \mathbb{C}\psi + \sum_{k=1}^{p} dU(\mathscr{U}(k)) \cdot \xi(X_i)$$

($\mathscr{U}(\mathfrak{k})$ the enveloping algebra of $\mathfrak{k}$).

It follows that $\dim(dU(\mathscr{U}(\mathfrak{k}))\psi) < +\infty$, and therefore $\psi \in \mathscr{H}_K$. Consequently $\psi \in B^1(\mathfrak{g}, \mathscr{H}_K)$.

If $a \in Z^1_\infty(G, \mathscr{H}_\infty)$, define $\varphi = \int_K a(x)\,dx$. We have $a(x) = (I - U_x)\varphi$, $\forall x \in K$. Since the mapping $x \to U_x\varphi = \int_K a(xk)\,dk - a_x$ is $C^\infty$ on $G$, $\varphi \in \mathscr{H}_\infty$. Therefore, the cocycle $a'_x = a_x + (U_x - I)\varphi$ belongs to $Z'_K(G, \mathscr{H}_\infty)$. The first equality follows. Take $a \in Z^1_K(G, \mathscr{H}_\infty)$. $\Delta a$ vanishes on $\mathfrak{k}$ and $\Delta a_{[X,Y]} = dU_X \Delta a_Y$, $\forall X \in \mathfrak{k}$, $\forall Y \in \mathfrak{p}$. Therefore $\Delta a$ is a $\mathfrak{k}$-morphism from $\mathfrak{p}$, with the adjoint representation, into $\mathscr{H}_\infty$. It results from this that $\Delta a(\mathfrak{p})$ is a finite dimensional $\mathfrak{k}$-invariant subspace of $\mathscr{H}_\infty$. Therefore $\Delta a \in Z^1_K(\mathfrak{g}, \mathscr{H}_K)$; the mapping $a \to \Delta a$ from $Z^1_K(G, \mathscr{H}_\infty)$ to $Z^1_K(\mathfrak{g}, \mathscr{H}_K)$ induces then an imbedding from $H^1_K(G, \mathscr{H}_\infty)$ to $H^1_K(\mathfrak{g}, \mathscr{H}_K)$. The last assertion results from Proposition 4.2.

PROPOSITION 5.2. *Let $U$ be a quasi-simple representation of $G$ on a Banach space $\mathscr{H}$. If the infinitesimal character of $U$ is not trivial, then* $H^1(\mathfrak{g}, \mathscr{H}_\infty) = \{0\}$.

*Proof.* Let $\xi \in Z^1(\mathfrak{g}, \mathscr{H}_\infty)$. Extending the representation

$$dU'_X(\varphi, t) = (dU_X\varphi + t\xi_X, 0)$$

of $\mathfrak{g}$ on $\mathscr{H}_\infty \times \mathbb{C}$ to $\mathscr{U}(\mathfrak{g})$, we see that $\xi$ extends to $\mathscr{U}(\mathfrak{g})$ such that $\xi_{xy} = dU_x\xi_y + \mathscr{E}_y\xi_x$, where $\mathscr{E}$ is the unitary morphism which extends the trivial representation of $\mathfrak{g}$ to $\mathscr{U}(\mathfrak{g})$. Denote by $\theta$ the infinitesimal character of $U$. If $\mathscr{Z}(\mathfrak{g})$ is the center of $\mathscr{U}(\mathfrak{g})$, fix $Q \in \mathscr{Z}(\mathfrak{g})$ such that $\theta(Q) \neq \mathscr{E}(Q)$. Writing $\xi_{xQ} = \xi_{Qx}$ we get

$$\xi_x = dU_x \frac{1}{\theta(Q - \mathscr{E}_Q \cdot 1)} \xi_{(Q - \mathscr{E}_Q \cdot 1)}.$$

Therefore, $\xi \in B^1(\mathfrak{g}, \mathscr{H}_\infty)$.

Let $\mathscr{C}$ be the center of $G$.

PROPOSITION 5.3. *If $U$ is a topologically completely irreducible (T.C.I.) representation of $G$ on a Banach space $\mathscr{H}$, we have the following alternative*:

(1) *$\mathscr{C}$ is not represented trivially and then $H_c^1(G, \mathscr{H})=\{0\}$,*
(2) *$\mathscr{C}$ is represented trivially, and then*

$$H_c^1(G, \mathscr{H})=H_c^1(G/\mathscr{C}, \mathscr{H}).$$

*Proof.* (1) There exists $c \in \mathscr{C}$ such that $U_c=\lambda I_{\mathscr{H}}(\lambda \neq 1)$. Let $a \in Z_c^1(G, \mathscr{H})$. Writing $a_{gc}=a_{cg}$, we get $a_g=(U_g-I)a_c/(\lambda-1)$. Hence $a \in B^1(G, \mathscr{H})$.

(2) If $U$ is the trivial representation this is obvious. If not, fix $a \in Z_c^1(G, \mathscr{H})$ and $c \in \mathscr{C}$. We have $(U_g-I)a_c=0$ $\forall g \in G$. Hence $a_c=0$. □

COROLLARY 5.4. *If $(U, \mathscr{H})$ is a T.C.I. representation of G we have*

$$H_c^1(G, \mathscr{H})=H^1(\mathfrak{g}, \mathscr{H}_\infty).$$

*Proof.* Let $\tilde{G}$ be the universal covering group of $G$. By Proposition 5.3 we can assume that $G=\tilde{G}/\mathscr{C}$. This result comes then from Proposition 4.2.

PROPOSITION 5.5. *There exists at most a finite number of T.C.I. representations (up to Naimark equivalence) of a connected semisimple Lie group G, on a Banach space $\mathscr{H}$ such that $H^1(G, \mathscr{H})\neq\{0\}$.*

*Proof.* We can suppose from Proposition 5.3 that $G$ has a finite center. We can moreover suppose the cocycles vanish on $K$. But, such a cocycle is a $\mathfrak{k}$ morphism from the $\mathfrak{k}$-module $\mathfrak{p}$ to the $\mathfrak{k}$-module $\mathscr{H}_K$. Therefore, if

$$\xi \in Z_K^1(\mathfrak{g}, \mathscr{H}_K)-B_K^1(\mathfrak{g}, \mathscr{H}_K),$$

$\mathscr{H}_K$ must contain an isotypic component of the $\mathfrak{k}$-module $\mathfrak{p}$ and have a trivial infinitesimal character trivial (Proposition 5.2). There are a finite number of such representations.

*Example of Representation with Cohomology*

Take $\mathfrak{g}$ semisimple with rank $\mathfrak{g}=$rank $\mathfrak{k}$ (Ex $\mathfrak{g}=\mathfrak{so}(p, 2)$) Let $\mathfrak{h}$ be a compack Cartan subalgebra and the corresponding root system $(\beta_1, \ldots, \beta_r)$. There exists one and only one non-compact root say $\beta_r$. Let $(\pi_{-\beta_r}, V)$ be the irreducible representation of dominant weight $-\beta_r$. It comes from a group representation in a Banach space and it can be proved that $H^1(\mathfrak{g}, V)\neq\{0\}$ (this representation is not unitary).

PROPOSITION 5.6. *If $(U, \mathscr{H})$ is a spherical unitary representation of $G$, we have $H_c^1(G, \mathscr{H})=\{0\}$.*

*Proof.* Take a basis $\{X_1, \ldots X_n\}$ of $\mathfrak{g}$ such that $\{X_1, \ldots, X_k\}$ is a basis of $\mathfrak{k}$, $\{X_{k+1}, \ldots, X_n\}$ is a basis of $\mathfrak{p}$ and

$$Q=-\sum_{i=1}^{k} X_i^2+\sum_{i=k+1}^{n} X_i^2=-Q_k+\Omega \in \mathscr{Z}(\mathfrak{g})$$

$Q_k=-\Sigma_{i=1}^k X_i^2 \in \mathscr{Z}(\mathfrak{k})$. There exists $\varphi \neq 0$ such that $\mathrm{d}U(\mathscr{U}(\mathfrak{k}))\varphi=0$. Then $\Sigma_{i=k+1}^n \mathrm{d}U(X_i)^2\varphi=0$. So, $\Sigma_{i=k+1}^n \|\mathrm{d}U(X_i)^2\varphi\|^2=0$; then $\mathrm{d}U$ vanishes on $\varphi$, and since $U$ is irreducible, $U$ is the trivial representation. □

PROPOSITION 5.7. *Let $G$ be a connected and simply connected semi-simple Lie group. Assume that $G$ is not simple. Given a unitary faithful representation $(U, \mathscr{H})$ of $G$ on Hilbert space $\mathscr{H}$, we have $H_c^1(G, \mathscr{H})=\{0\}$.*

*Proof.* From Proposition 5.3, we can assume that $G$ has a finite center. The Lie algebra $\mathfrak{g}$ is the direct product of two semisimple ideals $\mathfrak{g}_1$ and $\mathfrak{g}_2$. We write the Cartan decompositions: $\mathfrak{g}_i=k_i+\mathfrak{p}_i$, $(i=1, 2)$, $\mathfrak{g}=\mathfrak{k}+\mathfrak{p}$ with $\mathfrak{k}=\mathfrak{k}_1+\mathfrak{k}_2$ and $\mathfrak{p}=\mathfrak{p}_1+\mathfrak{p}_2$.

The Harish-Chandra module of $U$ writes $V_1 \otimes V_2$ where $V_i$ is a $K$-finite $\mathfrak{g}_i$-module. If $\mathfrak{p}_1=\Sigma_{i=1}^n \mathfrak{q}_i$ (resp. $\mathfrak{p}_2=\Sigma_{j=1}^n \mathfrak{q}'_j$) is the reduction of the adjoint representation of $\mathfrak{k}_i$ on $\mathfrak{p}_i$ $(i=1, 2)$, the reduction of the adjoint representation of $\mathfrak{k}$ on $\mathfrak{p}$ writes

$$\mathfrak{p}=\sum_{i=1}^{n} \mathfrak{q}_i \otimes \mathbb{C}+\sum_{j=1}^{n} \mathbb{C} \otimes \mathfrak{q}'_i.$$

If the cohomology of $U$ is not trivial $V_1 \otimes V_2$ must contain at least an isotypic component of one of these types. But the isotypic components of $V_1 \otimes V_2$ can be written $V_{1\delta} \otimes V_{2\delta}$ where $V_{i\delta}$ is an isotypic component of $V_i (i=1, 2)$. Therefore one of the two modules $V_1$ or $V_2$ is spherical. Since we can assume that the infinitesimal character is trivial, this spherical representation is trivial, in contradiction to the fact that $U$ is faithful. □

Let $\mathfrak{g}=\mathfrak{k}+\mathfrak{a}+\mathfrak{n}$ be an Iwasawa decomposition of $\mathfrak{g}$. The following result due to Delorme will be given without proof [1].

PROPOSITION 5.8. *Suppose that $G$ is simple, that* $\dim \mathfrak{a} \geqslant 2$ *and that $(U, \mathscr{H})$ is an unitary irreducible representation of $G$. Then $H_c^1(G, \mathscr{H})=\{0\}$.*

## 6. Cohomology of Induced Representations

Let $G$ be a connected Lie group, $P$ a closed subgroup of $G$, with a finite number of connected components and $(L, \mathscr{H})$ a unitary representation of $P$. We denote by $(U^L, \mathscr{H}^L)$ the representation of $G$ induced by $(L, \mathscr{H})$. We denote by $\delta$ the module of $P$ and by $\Delta$ the module of $G$. If $\rho = (\delta/\Delta)^{1/2}$, we denote by $L'$ the representation defined by $L' = \rho L$, and $\mathscr{H}'$ the corresponding $P$-module.

PROPOSITION 6.1. *If $G/P$ is compact, we have $H_c^1(G, \mathscr{H}') = H_c^1(P, \mathscr{H}')$.*

For the proof see [5]. For a generalisation of this result when $G$ is a locally compact group see [3].

These hypotheses apply for the unitary $P$-principal series of a semi-simple Lie group $G$.

## 7. Extensions and Cohomology of the Lie Algebra

Let $(U^1, \mathscr{H}^1)$ and $(U^2, \mathscr{H}^2)$ be two continuous representations of $G$ in Banach spaces. We keep the notations of Section 4.

Given $\xi \in Z^1(\mathfrak{g}, \mathscr{P}(\mathscr{H}^2, \mathscr{H}^1))$ we consider the extension of the $\mathfrak{g}$-module $(dU^1, \mathscr{H}^1_\infty)$ by the $\mathfrak{g}$-module $(dU^2, \mathscr{H}^2_\infty)$ associated to $\xi$[7, exposé 4, §4] defined on $\mathscr{H}^1_\infty \times \mathscr{H}^2_\infty$ by

$$W_x(a+b) = dU^1_X(a) + \xi_X(b) + dU^2_X(b), \qquad X \in \mathfrak{g}.$$

It results from Proposition 4.2 that $W_X = dV_X$ for a certain extension of $(U^1, \mathscr{H}^1)$ by $(U^2, \mathscr{H}^2)$.

PROPOSITION 7.1. *Any $\xi \in Z^1(\mathfrak{g}, \mathscr{P}(\mathscr{H}^2, \mathscr{H}^1)$ has an extension to the universal envelopping algebra $\mathscr{U}(\mathfrak{g})$ of $\mathfrak{g}$, such that $\xi_{uv} = dU^1_u \xi_v + \xi_u\, dU^2_v$ on $\mathscr{H}^2_\infty$, $u, v \in \mathscr{U}(\mathfrak{g})$, and $\xi_u \in \mathscr{L}(\mathscr{H}^2_\infty, \mathscr{H}^1_\infty)$. Moreover, if $\xi_X = dU^1_X l - l\, dU^2_X$, $X \in \mathfrak{g}$ and $l \in \mathscr{L}(\mathscr{H}^2_\infty, \mathscr{H}^1_\infty)$, we have $\xi_u = dU^1_u l - l\, dU^2_u$.*

*Proof.* Write that $\xi$ defines an extension $(W, \mathscr{H}^1_\infty \times \mathscr{H}^2_\infty)$ of $(dU^1, \mathscr{H}^1_\infty)$ by $(dU^2, \mathscr{H}^2_\infty)$ and extend $W$ to $\mathscr{U}(\mathfrak{g})$; one gets the result writing

$$W_u(a+b) = dU^1_u(a) + \xi_u(b) + dU^2_u(b).$$

PROPOSITION 7.2. *Suppose that $(U^1, \mathscr{H}^1)$ and $(U^2, \mathscr{H}^2)$ are two quasi-*

*simple representations of $G$. If the infinitesimal character of these representations are different one has $H^1(\mathfrak{g}, \mathscr{P}(\mathscr{H}^2, \mathscr{H}^1))=\{0\}$ (and therefore every extension of $(U^1, \mathscr{H}^1)$ by $(U^2, \mathscr{H}^2)$ is a direct sum).*

*Proof.* Take $q \in Z(\mathfrak{g})$ such that $uU^1(q)=q_1 I_{\mathscr{H}^1}$ and $\mathrm{d}U^2(q)=q_2 I_{\mathscr{H}^2}$ with $q_1 \neq q_2$. If $\xi \in Z^1(\mathfrak{g}, \mathscr{P}(\mathscr{H}^2, \mathscr{H}^1))$, write that $\xi_{QX}=\xi_{XQ}$, $X \in \mathfrak{g}$. Then,

$$\xi_x=\frac{1}{q_1-q_2}\,(\mathrm{d}U^1_X\xi_Q-\xi_Q\,\mathrm{d}U^2_X). \qquad \square$$

## 8. Extensions of Representations of Semisimple Lie Groups

Suppose from now on that $G$ is a connected semisimple Lie group; we keep the notations of Section 5.

DEFINITION 8.1. Given two $K$-finite representations $(U^1, \mathscr{H}^1)$ and $(U^2, \mathscr{H}^2)$ of $G$, $(V, E)$ and $(V', E')$ two extensions of $(U^1, \mathscr{H}^1)$ by $(U^2, \mathscr{H}^2)$, one says that $(V, E)$ and $(V', E')$ are Naimark equivalent if there exists a closed linear operator which is a $G$-morphism $T: E \to E'$ such that $D_T$ is dense, $D_T \supset i(\mathscr{H}^1_K)$ and the following diagram is commutative

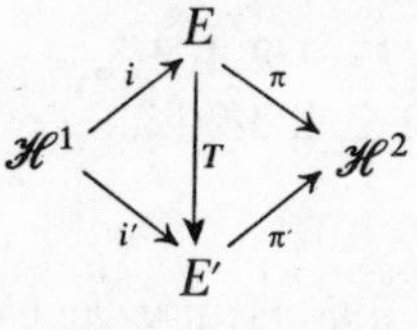


We denote by $\mathrm{Ext}_N(\mathscr{H}^1, \mathscr{H}^2)$ the set of the Naimark equivalence class of extensions of $(U^1, \mathscr{H}^1)$ by $(U^2, \mathscr{H}^2)$.

As before we can suppose that the cocycles of the group or of the Lie algebra vanish on $K$ and $\mathfrak{k}$, respectively, and we shall denote the corresponding spaces by adding an exponent $K$ and $\mathfrak{k}$, respectively, $(Z^1(G,.)^K$, $Z^1(\mathfrak{g},.)^{\mathfrak{k}}, \ldots)$.

Given an extension $(V, E)$ of two $K$-finite representations $(U^1, \mathscr{H}^1)$ and $(U^2, \mathscr{H}^2)$ of $G$, there corresponds an extension $(\mathrm{d}V, E_K)$ of the $\mathfrak{g}$-module $(\mathrm{d}U^1, \mathscr{H}^1_K)$ by the $\mathfrak{g}$-module $(\mathrm{d}U^2, \mathscr{H}^2_K)$.

One can prove the following propositions [6].

PROPOSITION 8.2. *Given two extensions $(V, E)$ and $(V', E')$ of two $K$-finite representations $(U^1, \mathscr{H}^1)$ and $(U^2, \mathscr{H}^2)$ of $G$, we denote by $(\mathrm{d}V, E_K)$ and $(\mathrm{d}V', E'_K)$ the corresponding $\mathfrak{g}$-module extensions. The following two properties are equivalent:*

(a) *$(V, E)$ and $(V', E')$ are Naimark equivalent extensions.*

(b) *$(\mathrm{d}V, E_K)$ and $(\mathrm{d}V', E'_K)$ are algebraically equivalent extensions* [7, exposé 4 §4].

PROPOSITION 8.3. *Suppose that $(U^1, A)$ and $(U^2, A)$ are $K$-finite representations of $G$ and that $\Delta = \{\delta \in \hat{K} | (\mathscr{H}^1_K)_\delta \neq \{0\}$ and $(\mathscr{H}^2_K)_\delta \neq \{0\}\}$ where $(\mathscr{H}^1_K)_\delta$ (resp. $(\mathscr{H}^2_K)_\delta$) is the isotypic component of type $\delta$ in $\mathscr{H}^1_K$ (resp. $\mathscr{H}^2_K$). Suppose that $\Delta$ is finite. Then, two extensions of $(U^1, \mathscr{H}^1)$ by $(U^2, \mathscr{H}^2)$ are strongly equivalent if and only if they are Naimark equivalent.*

Suppose now that $G = SL(2, \mathbb{R})$. Let $Y$, $F$, $G$ be a Weyl basis of $\mathfrak{sl}(2, \mathbb{R})$ associated to a maximal compact subgroup ($iY \in \mathfrak{k}$).

$$[Y, F] = F, \quad [Y, G] = -G, \quad [F, G] = 2Y; \quad Q = GF + Y + Y^2$$

is the generator of $\mathscr{Z}(\mathfrak{g}_c) \simeq \mathbb{C}[Q]$. Given an unitary irreducible representation $(U, \mathscr{H})$, there exists an orthonormal basis $\varphi_n$ of $\mathscr{H}_K$ such that $\mathrm{d}U_Y\varphi_n = \lambda_n\varphi_n$, $\mathrm{d}U_F\varphi_n = \alpha_{n+1}\varphi_{n+1}$, $\mathrm{d}U_G\varphi_n = -\alpha_n\varphi_{n-1}$, $\mathrm{d}U_Q\varphi_n = q\varphi_n$, where $\alpha_{n+1} = (\lambda_n\lambda_{n+1} - q)^{1/2}$. The V. Bargmann classification of $U$ is the following.

(1) $C^0_p$, $q < 0$, $\lambda_n = 0, \pm 1 \pm 2, \ldots,$

(2) $C^{1/2}_q$, $q < -\frac{1}{4}$, $\lambda_n = \pm 1/2, \pm 3/2, \ldots,$

(3) $D^-_k$, $q = k(k+1)$, $\lambda_n = -k, -k-1, \ldots, k = 1/2, 1, 3/2, \ldots,$

(4) $D^+_k$, $q = k(k+1)$, $\lambda_n = k, k+1, \ldots, k = 1/2, 1, 3/2 \ldots.$

We have the following result.

PROPOSITION 8.4. *All extensions of unitary irreducible or finite-dimensional representations of $SL(2, \mathbb{R})$ are Naimark equivalent to the trivial extension (direct sum), except in the following cases where*

$$\dim \operatorname{Ext}_N(\mathscr{H}^2, \mathscr{H}^1) = \dim \operatorname{Ext}_W(\mathscr{H}^2, \mathscr{H}^1) = 1$$

(a) *$C^0_q$ or $C^{1/2}_q$ by itself,*

(b) *$D^+_{1/2}$ by $D^-_{1/2}$ or $D^-_{1/2}$ by $D^+_{1/2}$.*

(c) *$D^\pm_k$ by $D(k-1)$ or $D(k-1)$ by $D^\pm_k$.*

Let $(U^1, \mathscr{H}^1)$ and $(U^2, \mathscr{H}^2)$ be representations of type $C^0_q$. It can be proved (for more details see [6]) that there exists an extension of $(U^1, \mathscr{H}^1)$ by

$(U^2, \mathscr{H}^2)$ which is weakly equivalent (hence Naimark equivalent) to the trivial extension but is not strongly equivalent to it.

*Université de Dijon, France*

## References


[1] P. Delorme, *Bull. Soc. Math. Fr.* **105**, (1977) 281–336.

[2] L. Gårding, 'Vecteurs analytiques dans les représentations de groupes de lie', *Bull Soc. Math. France*, **88** (1960), 73.

[3] J. Pichaud, '1-cohomologie de représentations induites', These de 3e cycle, Paris.

[4] G. Pinczon, and J. Simon, 'On the 1-cohomology of Lie groups', *Letters Math. Phys.*, **1** (1975), 83–91.

[5] G. Pinczon, and J. Simon, 'Sur la 1-cohomologie des groupes de Lie semi-simples' *C.R. Acad. Sc.* (*Paris*), A279 (1974), 455.

[6] G. Pinczon, and J. Simon, 'Extensions of group representations and cohomology', to be published in *Reports on Mathematical Physics* (1979).

[7] Séminaire Sophus Lie, Ecole Normale Supérieure, 1955.

HARRY FURSTENBERG

# RANDOM WALKS ON LIE GROUPS

## CONTENTS



We present in this paper a limited selection of topics intended to introduce the reader to the subject of random walks on Lie groups. The selection has been highly individual and makes no pretense of completeness. For a more comprehensive view of the area the reader should consult [1, 8–10].

The setup we work with is the following. Let $G$ be a locally compact, $\sigma$-compact topological group and let $\mu$ be a probability measure on Borel subsets of $G$. We form the product space $\Omega = G \times G \times G \times \ldots$ and the measure $P = \mu \times \mu \times \mu \times \ldots$ and let $X_i$: $\Omega \to G$ denote the $i$th coordinate function, $i = 1, 2, 3, \ldots$. $\Omega$ together with $P$ forms a probability space and the $X_n$ form a sequence of 'independent, identically distributed $G$-valued random variables.' The partial products $W_n = X_n X_{n-1} \ldots X_1$ constitutes a 'random walk' on $G$. We are interested in studying the behavior of this random walk, mostly with regard to what happens as $n \to \infty$.

Traditionally, $G$ was the group of reals and $\mu$ the distribution of an ordinary numerical random variable. The behavior of the series $S_n = X_1 + X_2 + \cdots + X_n$ is the principal object of study of classical probability theory. It is useful to distinguish between three kinds of results. To begin with there is the weak law of large numbers and the central limit theorem both of which relate to the asymptotic distribution of $S_n$ for $n \to \infty$. Namely, under suitable hypotheses, $S_n/n$ converges to a constant $c$ in distribution and $(S_n - nc)\sqrt{n}$ tends in distribution to a normal, or Gaussian variable. These 'large number' phenomena relate to the behavior of a large, but finite, number of random variables $X_1, X_2, \ldots X_n$. The strong law of large numbers, on the other hand, requires that the entire infinite series $\{X_n\}$ be simultaneously defined on a probability space in



*J. A. Wolf, M. Cahen, and M. De Wilde (eds.), Harmonic Analysis and Representations of Semi-Simple Lie Groups, 467–489.*

which case (under the hypothesis of finite first movement) it asserts that with probability 1, $S_n/n \to c = E(X_1)$. All of these are results of a quantitative nature and, moreover, refer to a normalized version of $S_n$. A third category of results considers the variables $S_n$ themselves and studies their qualitative as well as quantitative behavior. A striking result in the classical theory is that if $E(X_1)=0$ then, with probability 1, the random walk $S_n$ comes arbitrarily close to 0. By contrast, if $G=R^3$ and $\mu$ is not concentrated on a two-dimensional subspace, then with probability 1, the random walk $S_n$ wanders to X.

In our discussion of random walks on more general groups we shall ignore the first category of results pertaining to distributions of large products of random variables. These are treated in [8], [10], and [14]. We will, however, discuss the quantitative almost everywhere behavior of $W_n = X_n X_{n-1} \ldots X_1$ in the form of laws of large numbers as well as the qualitative behavior of $W_n$, a topic which entails development of the so-called boundary theory of random walks.

## 1. Laws of Large Numbers

Let us begin by assuming that $G$ is a group of complex matrices, $G \subset GL(m, C)$, and let $\|g\|$ be some Banach algebra norm on the space of $m \times m$ matrices. One then has the following generalization of the strong law of large numbers:

THEOREM 1.1. *If $\mu$ is a measure on G satisfying*

$$(1.1 \qquad \int \log^+ \|g\| \, d\mu(g) < \infty$$

*and $\{X_n\}$ is a G-valued sequence of independent identically distributed random variables (i.i.d.r.v.'s) with distributions $\mu$, then with probability* 1,

$$\|X_n \ldots X_1\|^{1/n}$$

*converges to a constant.*

The easiest way to prove Theorem 1.1 is to invoke Kingman's subadditive ergodic theorem [11]:

THEOREM (Kingman). *Let $W_{mn}$, $m, n=1, 2, 3, \ldots, m<n$ be a family of real-valued random variables satisfying*

(i) $W_{mn}$ *and* $W_{m'n'}$ *have the same distribution whenever* $n-m=n'-m'$.
(ii) *For* $m<n<p$, $W_{mp}\leqslant W_{mn}+W_{np}$.
(iii) $E(W_{12}^{+})<\infty$.
*Then with probability* 1, $W_{1n}/n$ *converges to a constant* $\geqslant -\infty$.

To obtain Theorem 1.1 simply set $W_{mn}=\log \|X_nX_{n-1}\dots X_{m+1}\|$. We shall however discuss another proof which will be more instructive for our purposes. First let us notice that Theorem 1.1 give us not one but a variety of laws of large numbers. For let $G$ be a group, $\mu$ a distribution on $G$, and let $p$ be a representation of $G$ into a matrix group $GL(m, C)$, satisfying

$$\int \log^{+} \|p(g)\| \, \mathrm{d}(g)<\infty. \tag{1.2}$$

Then by Theorem 1.1,

$$\lim \|p(X_n)p(X_{n-1})\dots p(X_1)\|^{1/n}=\gamma_p$$

exists. These laws are not independent. For example, if $p=p'+p''$, clearly $\gamma_p=\max\{\gamma_{p'}, \gamma_{p''}\}$. In fact one can see that to obtain all such laws it suffices to consider only irreducible representation $p$, and to decompose an arbitrary representation into a composition series.

Now the foregoing examples of laws of large numbers represent a special case of a general procedure for obtaining such laws. This procedure involves a notion of 'cocycles' for a group action.

DEFINITION 1.1. Let $M$ be a compact $G$-space. We denote by $Z(G, M)$ the set of continuous real-valued functions on $G\times M$ satisfying

$$r(g_1g_2, u)=r(g_1, g_2u)+r(g_2, u). \tag{1.3}$$

The functions in $Z(G, M)$ will be called *cocycles*.

For example if $p\colon G\to GL(m, C)$ we can consider the complex projective space $CP^{m-1}$ as a $G$-space with $gu=\overline{p(g)\tilde{u}}$ where $\tilde{u}$ is a vector of $C^m$ in the direction $u$ and $v\to\bar{v}$ is the map of $C^m\to CP^{m-1}$. We then obtain an example of a cocycle by setting

$$r(g, u)=\log \|p(g)\tilde{u}\|/\|\tilde{u}\|. \tag{1.4}$$

Another example of a cocycle occurs when $M$ is a differentiable manifold on which $G$ acts by diffeomorphisms. Letting $m$ denote a smooth measure on $M$, for each $g\in G$, $g^{-1}m$ will again be a smooth measure and the Radon–Nikodym derivative $(dg^{-1}m/dm)(u)=r(g, u)$ will be a continuous function on $G\times M$. It is not hard to verify that $r(g, u)$ satisfies the cocycle equation (1.3).

Now let $f(u)$ be a continuous function on $M$ and form

$$r(g, u)=f(gu)-f(u).$$

this is again a cocycle which will be referred to as a coboundary. The group of these is denoted $B(G, M)$ and the quotient $Z(G, M)/B(G, M)$ is denoted $H(G, M)$, the 'cohomology of $G$ on $M$'. We will not need cohomology in any other dimensions.

We need one more notion before formulating a general law of large numbers. If $M$ is a $G$-space we have a map $G\times M\to M$. Now let $\nu$ be a measure on $M$, $\mu$ a measure on $G$, then the above map sends the measure $\mu\times\nu$ on $G\times M$ to a measure on $M$. This measure is the convolution $\mu*\nu$ of $\mu$ and $\nu$.

DEFINITION 1.2. A measure $\nu$ on $M$ is called a $(\mu)$-*stationary* measure if $\mu*\nu=\nu$.

When $\nu$ is stationary then for every continuous function $f(u)$ we have

$$\iint f(gu)\,d\mu(g)\,d\nu(u)=\int f(u)\,d\nu(u).$$

We remark that if $M$ is compact, then for any probability measure $\mu$ there exists a $\mu$-stationary measure $\nu$ on $M$.

We now formulate a general law of large numbers.

THEOREM 1.2. *Let $M$ be a $G$-space, $\mu$ a measure on $G$, $\nu$ a $\mu$-stationary measure on $M$ and $r\in Z(G, M)$. Assume*

$$\iint |r(g, u)|\,d\mu(g)\,d\nu(u)<\infty, \tag{1.5}$$

*and set*

$$\alpha(r)=\iint r(g, u)\,d\mu(g)\,d\nu(u).$$

*Then if $\{X_n\}$ is a sequence of i.i.d.r.v.'s with distribution $\mu$ we will have, for almost all (with respect to $\nu$) $u\in M$, and with probability* 1,

$$r(X_nX_{n-1}\ldots X_1, u)/n\to\alpha(r). \tag{1.6}$$

Theorem 1.2 is a straightforward consequence of the ordinary ergodic theorem. If we let $Z_0$ denote an $M$-valued r.v. with distribution $\nu$ independent of the variables $X_1, X_2, \ldots$, then $Z_1 = X_1 Z_0$ has distribution $\mu * \nu = \nu$ and similarly if $Z_n = X_n X_{n-1} \cdots X_1 Z_0$ then the sequence $\{Z_n, X_n\}$ is 'stationary' and the ergodic theorem gives

$$(1.7) \qquad (r(X_n, Z_{n-1}) + \cdots + r(X_1, Z_0))/n \to \alpha(r).$$

But

$$\sum_1^n r(X_k, Z_{k-1}) = \sum_1^n r(X_k, X_{k-1} \ldots X_1 Z_0)$$
$$= r(X_n X_{n-1} \ldots X_1, Z_0)$$

so that (1.7) is equivalent to (1.6).

We mention a slightly more subtle result which is proved in [4].

THEOREM 1.3. *Assume in addition to the hypotheses of Theorem 1.2 that $M$ is compact and that the integral $\alpha(r)$ is independent of the $\mu$-stationary measure $\nu$ (there may be more than one). Then for any $u \in M$,*

$$r(X_n X_{n-1} \ldots X_1, u)/n \to \alpha(r).$$

We now combine Theorems 1.2 and 1.3 to obtain a general law of large numbers for an irreducible representation of $G$.

THEOREM 1.4. *Let $\mu$ be a probability measure on $G$, $p$ an irreducible representation of $G$ into $GL(m, C)$ which remains irreducible when restricted to the smallest subgroup of $G$ containing the support of $\mu$. Assume that*

$$\int \log (\|p(g)\| \, \|p(g^{-1})\| \, d\mu(g) < \infty.$$

*Then there exists a constant $\beta_p(\mu)$ with*

$$\frac{1}{n} \log \|p(X_n)p(X_{n-1}) \ldots p(X_1)v\| \to \beta_p(\mu)$$

*for any vector $v \in C^m$.*

*Proof.* Take $M = CP^{m-1}$ and let $r(g, u) = \log \|p(g)\tilde{u}\| / \|\tilde{u}\|$. Let $\nu$ be a stationary measure in $CP^{m-1}$. Suppose $\nu$ is supported in a linear subvariety of the projective space. Since $\nu$ is stationary it follows that the support of $\mu$ leaves invariant a linear subspace of $P^{m-1}$ contrary to our irreducibility hypothesis. Apply (1.6) to the cocycle $r(g, u)$ defined above. There will exist $u_1, u_2, \ldots, u_n$ for which (1.6) is valid such that the vectors $\tilde{u}_1, \tilde{u}_2, \ldots, \tilde{u}_n$ form a basis of $C^m$. We then conclude that

$$(1.8) \qquad \frac{1}{n}\log\|p(X_n)p(X_{n-1})\ldots p(X_1)\tilde{u}_j\| \to \alpha_{\mu,v}(r)$$

with probability 1. Since $\{\tilde{u}_j\}$ form a basis we find

$$(1.9) \qquad \limsup \frac{1}{r}\log\|p(X_n)p(X_{n-1})\ldots p(X_1)\| \leqslant \alpha_{\mu,v}(r)$$

For any $v \in C^m$. Apply Theorem 1.2 to any other $\mu$-stationary measure $v'$. We then obtain from (1.8) and (1.9) that

$$\alpha_{\mu,v'}(r) \leqslant \alpha_{\mu,v}(r).$$

Hence $\alpha_{\mu,v}(r)$ is independent of $v$ so that Theorem 1.3 applies. This yields Theorem 1.4 with $\beta_p(\mu) = \alpha_{\mu,v}(r)$. □

Note that Theorem 1.4 essentially implies Theorem 1.1 but for the slightly strengthened hypothesis. It also gives more since it implies that the various columns of the product matrix $p(X_n)p(X_{n-1})\ldots p(X_1)$ have the same rates of growth.

We also point out a fundamental difference between the situation in the foregoing theorems and the classical law of large numbers. Classically the limit

$$\lim \frac{X_1 + \cdots + X_n}{n} = E(X) = \int_{-\infty}^{\infty} x \, d\mu(x)$$

is expressed directly (and linearly) in terms of the measure $\mu$. For our case the limit of

$$(\log\|p(X_n)p(X_{n-1})\ldots p(X_1)\|)/n$$

requires an auxiliary measure $v$ for its evaluation. In particular there does not appear to be a simple connection between $\beta_p(\mu)$ for the convex combinations of measures $\mu'$ and $\mu''$.

We conclude this section with a comment regarding the finiteness hypothesis which occurs in each of our laws of large numbers. Let us say that a measurable function $L(g)$ is a 'length function' on $G$ if $L(g) \geqslant 0$ and $L(g_1 g_2) \leqslant L(g_1) + L(g_2)$.

DEFINITION 1.3. A measure $\mu$ on $G$ is of *finite length* if for every length function $L$ on $G$ we have

$$(1.10) \qquad \int L(g)\,\mathrm{d}\mu(g) < \infty.$$

An example of a length function occurs if $G$ is generated by a subset $K$. Setting $L(g)$ equal to the minimal number of terms $l$ in $g = g_1 g_2 \ldots g_l$ where $g_i \in K$ or $K^{-1}$. Then $L(g)$ is a length function. If $p$ is a representation of $G$ then $\log^+ \|p(g)\|$ is a length function. It is easily shown that if $G$ is generated by a compact subset, then if (1.10) is valid for the length function determined by this subset it is valid for all length functions so that $\mu$ is of finite length.

## 2. The Space of Strong Laws

Let $G$ be a group as above and consider the space $\mathcal{M}$ of all probability measures of finite length on $G$ which moreover have the property that their support generates $G$ as a closed group. Suppose $\psi$ is a function on $G$ such that for any $\mu \in \mathcal{M}$ and $\{X_n\}$ a sequence of independent identically distributed $G$-valued random variables with distribution $\mu$, we have

$$\lim\,(1/n)\psi(X_n X_{n-1} \ldots X_1) = \beta(\mu)$$

exists with probability 1. Denote the linear set of the functionals $\beta$ defined in this way by $V$. When $G = Z$ it is easy to see that $V$ is two-dimensional with $\beta(\mu) = \int (ax + b|x|)\,\mathrm{d}\mu(x)$. In general, $V$ will be infinite-dimensional. Nonetheless the functionals $\beta_p$ defined by rates of growth of all irreducible representations often span a finite-dimensional subspace.

THEOREM 2.1. *If $G$ is a connected Lie group, then the subspace of $V$ spanned by the functionals $\beta_p$ where*

$$\beta_p(\mu) = \lim\,(1/n) \log \|p(X_n X_{n-1} \ldots X_1\|)$$

*and $p$ ranges over all irreducible representations of $G$ is finite-dimensional.*

*Proof.* The crucial fact is that $G$ has a solvable subgroup $S$ with $G/S$ compact. Moreover $H(G, G/S)$ is finite-dimensional. To prove this last fact observe if $x_0 \in G/S$ is the point corresponding to the coset $S$, and $r \in L(G, G/S)$ then $r(s, x_0)$ defines a homomorphism of $S$ into $R$. Assume that this homomorphism is identically 0. Set $f(gx_0)=r(g, x_0)$ for $g \in G$, and verify that this is well-defined. then

$$f(g'gx_0)-f(gx_0)=r(g'g, x_0)-r(g, x_0)=r(g', gx_0)$$

so that $r \in B(G, G/S)$. It now follows that dim $H(G, G/S)$ is no larger than the dimension of the space of homomorphisms of $S$ into $R$. To prove the theorem we will show that to every irreducible representation $p$ there is some $r \in Z(G, G/S)$ with $\beta_p(\mu)=\alpha_{\mu,\nu}(r)$. Then any linear relation between the cocycles $r$ modulo $B(G, G/S)$ translates into a linear relation between the functionals $\beta_p$.

Let $p$ be an irreducible representation of $G$ in $GL(m, C)$. Since $S$ is solvable we can find a vector $v_0 \in C^m$ which is an eigenvector of $p(S)$. We define a cocycle in $Z(G, G/S)$ by setting

$$r(g, g'S)=\log \|p(gg')v_0\|/\|p(g')v_0\|, \tag{2.1}$$

which is meaningful since replacing $g'$ by $g's$, $s \in S$, only multiplies $v_0$ by a scalar. But now $\beta_p(\mu)=\alpha_{\mu,\nu}(r)$ for any $\mu$-stationary measure $\nu$ on $G/S$ since both expressions are obtained by evaluating the limit a.e. of

$$r(X_n X_{n-1} \ldots X_1, u)/n$$

for some $u \in G/S$. This completes the proof.

Now let $G$ be a semisimple Lie group and $\Gamma$ a lattice subgroup. If all the simple factors of $G$ have $R$-rank $>1$ then Margulis' superrigidity theorem [13] states that every unbounded representation of $\Gamma$ is the restriction of a representation of $G$. Then the foregoing argument establishes the finite-dimensionality of the subspace of $V(\Gamma)$ spanned by $\beta_p$, $p$ irreducible representation of $\Gamma$.

It is not difficult to show that this phenomenon does not take place for all discrete groups $\Gamma$, even finitely generated $\Gamma$ occurring as lattice subgroups of semi-simple Lie groups. To see this let us introduce the notion of *dominance* of representations. We say that $p_1$ dominate $p_2$ if there exists a constant $c$ such that for every $\mu \in \mathcal{M}$,

$$\beta_{p_2}(\mu) \leqslant c \max \{\beta_{p_1}(\mu), 0\}.$$

When the subspace of $V(\Gamma)$ spanned by $\beta_p$, $p$ irreducible, is finite-dimensional, it is easy to see that there is some (reducible) $p$ which dominates all the rest. On the other hand, for a free group it is easy to see that no representation dominates all the representations. It follows that for lattice subgroups of $SL(2, R)$ the foregoing phenomenon is no longer valid.

## 3. Qualitative Behavior at $\infty$ and Boundaries of a Group

Let $G$ be a locally compact, $\sigma$-compact group as before, $\{X_n\}$ a sequence of identically distributed independent $G$-valued random variables. We consider the random products of the $X_n$ which we now find it convenient to write $X_1 X_2 \dots X_n$ and the associated random walk $W_n(g) = gX_1 X_2 \dots X_n$ which initiates at the element $g \in G$. One can consider two extreme possibilities of behavior:

(i) $W_n(g) \to \infty$, and

(ii) for any neighborhood $V$ of $g$, $W_n(g)$ returns to $V(g)$ infinitely often.

A basic theorem in the general theory of random walks states that one or the other of these two possibilities occurs with probability one. Which takes place depends in the underlying measure $\mu$ and the random walk is called *transient* in the first case, *recurrent* in the second. For certain groups $G$ any measure $\mu$ whose support generates $G$ leads to a transient random walk. For example if $G = R^d$ or $Z^d$ with $d \geqslant 3$ then the random walk is necessarily transient. For $G = R$ or $Z$ the walk will be recurrent if $E(X_i) = 0$ whereas for $G = R^2$ or $Z^2$ more precise information is needed to determine recurrence on transience but both are possible. In the ensuing discussion we will always assume that $\mu$ is not supported by a proper closed subgroup of $G$. Two general results can be proved for the non-commutative case.

THEOREM 3.1. *If $G$ is non-unimodular then every random walk on $G$ is transient.*

THEOREM 3.2. *If $G$ is non-amenable then every random walk on $G$ is transient.*

For a proof of Theorem 3.1 we refer the reader to [2]. A proof of Theorem 3.2 is given in [6]. We indicate how Theorem 3.2 may be proved in certain cases using the laws of large numbers established in Section 1. Assume $G$ is a non-amenable connected Lie group. Every connected Lie group $G$ contains a solvable subgroup $H$ such that $G/H$ is compact. Now $G$ cannot leave a measure invariant on $G/H$; otherwise it may be shown that $G$ is itself amenable. Now let $\mu$ be a smooth measure on $G$ of finite length and let $\nu$ be a $\mu$-stationary measure on $G/H$. Since $\mu * \nu = \nu$, $\nu$ is itself a smooth measure and the cocycle

$$r(g, u) = \log\,(dg^{-1}\nu/d\nu)(u)$$

will be continuous. By Theorem 1.2 we will have for some $u$

$$(3.1) \qquad r(X_n X_{n-1} \dots X_1, u)/n \to \iint \log (dg^{-1}v/dv)(u)\, d\mu(g)\, dv(u)$$

with probability one. If the integral in (3.1) is shown to be different from zero it will follow that $X_n X_{n-1} \dots X_1$ is transient. It follows that $X_1^{-1} X_2^{-1} \dots X_n^{-1}$ is transient. Now by Jensen's inequality

$$\int \log (dg^{-1}v/dv)(u)\, dv(u) \leqslant \log \int (dg^{-1}v/dv)(u)\, d(u) = 0$$

with equality only if $(dg^{-1}v/dv)(u) \equiv 1$. But the latter would imply that $v$ is an invariant measure and this would contradict the non-amenability of $G$. We remark that Theorem 3.2 applies to all non-compact semi-simple Lie groups.

Now assuming that the random walk $W_n(g) = gX_1X_2 \dots X_n$ is transient we can ask for a more precise description of how it tends to $\infty$. To illustrate this assume that $G \subset GL(m, R)$ and that $\mu$ is a smooth measure of compact support. Then Theorem 8.4 of [4] implies the following behavior of $W_n = X_1 X_2 \dots X_n$. If we consider the column vectors of $W_n$ as vectors in $R^m$, these all tend a single line in $R^m$ as $n \to \infty$. This line, or point of $P^{m-1}$, is itself a random variable and can be thought of as the 'direction' in which $X_1 X_2 \dots X_n$ goes to $\infty$.

To make this more precise we define the notion of a *boundary space* of a random walk $W_n(g)$ on $G$. First note that $G$ may be regarded as a $G$-space by having $G$ operate on itself by multiplication from the left.

DEFINITION 3.1. A compact $G$-space $M$ is a *boundary space* for the random walk $W_n(g) = gX_1X_2 \dots X_n$ if

(a) $M$ can be attached to $G$ so that $G$ is an open subset of $G \cup M$ and $G \cup M$ is a $G$-space, with a metric topology.

(b) with probability one, $W_n(g)$ converges in $G \cup M$ to a point of $M$.

For example consider $P^{m-1}$ as a $G$-space for $G \subset GL(m, R)$ and attach $P^{m-1}$ to $G$ by taking as a neighborhood of a point $u \in P^{m-1}$ the set of matrices all of whose columns are within some angle of $u$. With appropriate conditions on $\mu$ we can say that $P^{m-1}$ is a boundary space of $G$ for the corresponding random walk.

The notion we have introduced of a boundary space of a random walk on $G$ is somewhat awkward because in order to describe the boundary space it is necessary to show how it is to be attached to the group $G$. We shall see however that there is an alternative characterization of a boundary space by which the space 'attaches itself' to the group in a natural way. The key to this is in the following proposition.

PROPOSITION 3.3. *Let $M$ be a boundary space for the random walk $W_n(g)=gX_1X_2\ldots X_n$ and define the $M$-valued random variable $z=\lim X_1X_2\ldots X_n$ where the convergence is in $G\cup M$. Let $\nu$ be the measure on $M$ which gives the distribution of the random variable $z$. Then*

(i) *$\nu$ is a $\mu$-stationary measure* (Definition 1.2).

(ii) *the sequence of random measures*

$$\nu_n=X_1X_2\ldots X_n\nu$$

*converges with probability one to the point measure $\delta_z$.*

We remark that the convergence of measures on $M$ is taken in the weak sense: $\nu_n\to\delta_z$ means that for every continuous function $f(x)$ on $M$,

$$\int f\,\mathrm{d}\nu_n\to\int f\,\mathrm{d}\delta_z=f(z).$$

*Proof.* If we write $z'=\lim X_2X_3\ldots X_{n+1}$ it is clear that $z$ and $z'$ have the same distribution $\nu$. On the other hand $z=X_1z'$ and $z'$ are independent. Hence we can write

$$\int f(x)\,\mathrm{d}\nu(x)=E(f(x))=E(f(X_1z'))=\iint f(gy)\,\mathrm{d}\mu(g)\,\mathrm{d}\nu(y)$$

so that $\nu=\mu*\nu$.

The second statement of the theorem requires more background, and depends on the theory of vector-valued martingales. We claim that for any $\mu$-stationary measure $\nu$ the sequence of measure-valued random variables $\{X_1X_2\ldots X_n\nu\}$ forms a martingale. In fact

$$\begin{aligned}E(X_1X_2\ldots X_{n+1}\nu|X_1,X_2,\ldots,X_n)\\ &=X_1X_2\ldots X_nE(X_{n+1}\nu|X_1,X_2\ldots X_n)\\ &=X_1X_2\ldots X_nE(X_{n+1}\nu)=X_1X_2\ldots X_n\mu*\nu\\ &=X_1X_2\ldots X_n\nu.\end{aligned}$$

As a result, for any $\mu$-stationary measure $\nu$, the sequence $X_1X_2\ldots X_n\nu$

will converge with probability one. (This vector-valued martingale convergence depends on the metrizability of $M$.) We want to show that when $\nu$ is the distribution of $z$ then the limit measure is $\delta_z$. For this we point out that $z$ is a random variable measurable with respect to the $\sigma$-field generated by $X_1, X_2, X_3, \ldots$ since $z = \lim X_1 X_2 \ldots X_n$. As a result for any function $f(x)$ on $M$ we have

$$\lim E(f(x)|X_1, X_2, \ldots, X_n) = f(z).$$

But if we write $z^{(k)} = \lim_{n\to\infty} X_k X_{k+1} \ldots X_{k+n}$ then

$$z = z^{(1)} = X_1 X_2 \ldots X_n z^{(n+1)}$$

and

$$E(f(z)|X_1 X_2, \ldots, X_n) = E(f(X_1 X_2 \ldots X_n z^{(n+1)})|X_1, X_2, \ldots, X_n)$$
$$= \int f(X_1 X_2 \ldots X_n x)\, d\nu(x) = \int f\, d\nu_n.$$

Here we have used the fact that $z^{(n+1)}$ has the same distribution as $z$ and that it is independent of $X_1, X_2, \ldots, X_n$. This completes the proof. □

Proposition 3.3 suggests the following definition.

DEFINITION 3.2. With $\mu$ a probability measure on $G$, a $(G, \mu)$-space will be a pair $(M, \nu)$ where $M$ is a $G$-space and $\nu$ is a $\mu$-stationary measure.

Let $(M, \nu)$ be a $(G, \mu)$-space. We "attach" $M$ to the group $G$ as follows. For each point $x \in M$ consider the neighborhoods of $\delta_x$ in the space of probability measures on $M$ with the weak topology. If $N$ is such a neighborhood we define the neighborhood $N^* \subset G \cup M$ by

$$N^* = \{y|\delta_y \in N\} \cup \{g|g\nu \in N\}$$

thus $g_n \to x$ will mean that

$$\int f\, dg_n \to f(x).$$

It could of course happen that no sequence $g_n \nu$ converges to a point measure in which case $G \cup M$ would be a discrete union.

Proposition 3.3 says that when $M$ is a boundary space and $\nu$ is the distribution of $z$, the foregoing construction provides a non-discrete topology on $G \cup M$ with an equivalent notion of convergence to the boundary of the random walk $W_n(g)$. We are thus led to

DEFINITION 3.3. A $(G, \mu)$-space $(M, \nu)$ is a *boundary space* if $M$ is a compact metric space and with probability 1, $X_1 X_2 \dots X_n \nu$ converges to a point measure on $M$.

Actually Proposition 3.3 states that every boundary space in the sense of Definition 3.1 gives rise to one in the sense of Definition 3.3. But the converse is clearly true, since the topology on $G \cup M$ when $(M, \nu)$ is a boundary space is such that $W_n(g) = gX_1 X_2 \dots X_n$ necessarily converges (a.e.) to a point of $M$. The advantage of the latter definition over the former is in that for a *boundary space* in the sense of Definition 3.3, the manner of attaching $M$ to $G$ is automatically determined by specifying the measure $\nu$ on $M$.

There is still another characterization of a boundary space $(M, \nu)$ which is useful, although it is less descriptive.

If $M$ is any $G$-space we can define transition probabilities for an $M$-valued random walk by setting

$$p(x, A) = \mu\{g | gx \in A\}.$$

With these transition probabilities a Markov process $z_0, z_1, z_2, \dots$ is defined on $M'$ as soon as the initial distribution of $z_0$ is given. If this is chosen to be a $\mu$-stationary measure then we will find that all the $z_n$ have the same distribution and, in fact, $\{z_n\}$ is a stationary Markov process. Moreover this process can be combined with the independent random variables $\{X_n\}$ to form a compound stationary process which is characterized in the following definition.

DEFINITION 3.4. *A $\mu$-process* $\{\{X_n, z_n\}; -\infty < n < \infty\}$ is a stationary $G \times M$ process satisfying:

(i) the distribution of $X_n$ is $\mu$ and the distribution of $z_n$ is $\nu$ where $\nu$ is a $\mu$-stationary measure,

(ii) $X_1, X_2, X_3, \dots X_n$ are jointly independent of $z_0, z_{-1}, z_{-2}, \dots,$

(iii) $z_{n+1} = X_{n+1} z_n$.

Every $\mu$-stationary measure determines an essentially unique $\mu$-process where the independent $X_n$ represent the transitions of the random walk $z_n$. We can now state

PROPOSITION 3.4. *A $(G, \mu)$-space $(M, \nu)$ is a boundary space if and* *only if in the corresponding $\mu$-process $\{X_n, z_n\}$ each $z_n$ is measurable with respect to the past $\{X_n, X_{n-1}, X_{n-2}, \ldots\}$.*

The proof is straightforward and can be found in [6]. Here are two consequences of Proposition 3.4. Let $(M, \nu)$ be a boundary space of $(G, \mu)$ and suppose $\varphi: M \to M'$ is an equivariant map into a $G$-space $M'$. The image $\varphi(\nu) = \nu'$ is again a $\mu$-stationary measure so that $(M', \nu')$ is again a $(G, \mu)$-space. It is clear from Definition 3.3 that if $\varphi$ is continuous then $(M', \nu')$ is again a boundary space. However from Proposition 3.4 it follows that even if $\varphi$ is just a *measurable* equivariant map, then $(M', \nu')$ is again a boundary space. This fact plays an important role in the sequel.

Proposition 3.4 also enables one to extend the notion of a boundary space to $G$-spaces that are not metrizable. In the non-metric case one cannot assert that lim $X_1 X_2 \ldots X_n$ exists with probability one so that Definition 3.3 becomes empty. On the other hand, the existence of a $\mu$-process satisfying the conditions of Proposition 3.4 is meaningful.

The main result of this section is the existence of a *universal boundary space.* To simplify matters we shall make the harmless assumption that if $(M, \nu)$ is a $(G, \mu)$-space no proper closed $G$-invariant subset of $M$ has $\nu$-measure equal to one.

THEOREM 3.5. *There exists a* universal boundary space $(M_0, \nu_0)$ *for a given pair $(G, \mu)$ such that any boundary space $(M, \nu)$ is a continuous equivariant image of $(M_0, \nu_0)$. Moreover $(M_0, \nu_0)$ is unique up to isomorphism.*

The construction of $(M_0, \nu_0)$ which is just a matter of piecing together the – possibly uncountable – family of boundary spaces $(M, \nu)$, is carried out in [6]. From the construction it is to be expected that $M_0$ is non-metrizable. A much deeper result which is also more useful is proved in [3].

THEOREM 3.6. *Let $G$ be a semisimple Lie group with finite center and let $\mu$ be an absolutely continuous measure on $G$. There is a boundary space $(M_1, \nu_1)$ such that $M_1$ is a compact* homogeneous *space of $G$, $\nu_1$ a smooth measure on $M_1$ which is universal in the sense that any other boundary space is a* measurable *equivariant image of $(M_1, \nu_1)$. Moreover $M_1$ is in every case one of a finite number of covering spaces of a uniquely determined homogeneous space $B(G)$.*

We mention that $B(G)$ is the quotient of $G$ by a minimal parabolic subgroup. We also remark that if $K$ is a maximal compact subgroup of $G$ then $K$ acts transitively on any covering space of $B(G)$. $B(G)$ is universal for the following property which is reminiscent of Definition 3.3:

(*) For every probability measure $\pi$ on $B(G)$ there exists a sequence $\{g_n\}$ in $G$ with $g_n\pi \to$ a point measure.

## 4. $\mu$-HARMONIC FUNCTIONS

There is another approach to the objects discussed in the last section, and this is by way of the notion of a $\mu$-harmonic function. Consider the probability measure $\mu$ on the group $G$ and suppose $\nu$ is a $\mu$-stationary measure on some compact $G$-space $M$. Let $\varphi$ be a bounded measurable function on $M$ and form

$$h(g) = \int_M \varphi(gx)\, d\nu(x). \tag{4.1}$$

The fact that $\nu$ is $\mu$-stationary means that for every function $f(x)$,

$$\int f(x)\, d\nu(x) = \iint f(g'y)\, d\mu(g')\, d\nu(y),$$

so, in particular,

$$h(g) = \iint \varphi(gg'y)\, d\nu(y)\, d\mu(g') = \int h(gg')\, d\mu(g').$$

DEFINITION 4.1. A $\mu$-harmonic function on $G$ is a function satisfying

$$h(g) = \int h(gg')\, d\mu(g') \tag{4.2}$$

for every $g \in G$.

It will be convenient to add the hypothesis that $h(g)$ is left-uniformly-continuous, i.e., that as $u_n \to$ identity $h(u_n g) \to h(g)$ uniformly.

Notice that when $\varphi(x)$ is continuous in (4.1) the resulting function $h(g)$ will be left-uniformly-continuous.

PROPOSITION 4.1. *Assume* $(M, \nu)$ *is a boundary space for* $(G, \mu)$. *Then*

(4.1) *sets up a* 1–1 *correspondence between continuous functions on M and a uniformly closed subspace of the bounded $\mu$-harmonic functions on* G. (The assumption of the last section is still in force, i.e., we assume no proper G-invariant closed subset of $M$ has full $\nu$ measure.)

*Proof.* The map $\varphi \to h$ is clearly norm-non-increasing and it will suffice to prove that it is norm preserving. Suppose first that $\varphi$ achieves its maximum on the support of $\nu$. Use the fact that $(M, \nu)$ is a boundary space for the random walk $W_n(g) = gX_1X_2 \dots X_n$ and that specifically, $X_1X_2 \dots \dots X_n\nu \to \delta_z$ where $\nu$ is the distribution of $z$. Now with positive probability $\varphi(z) > \max \varphi - \varepsilon$, and so with positive probability

$$\int \varphi(x)\,dX_1 \dots X_n\nu(x) > \max \varphi - \varepsilon. \tag{4.3}$$

But the integral in (4.3) is

$$\int \varphi(X_1X_2 \dots X_nx)\,d\nu(x) = h(X_1X_2 \dots X_n).$$

Hence $\sup h(g) \geqslant \max \varphi - \varepsilon$. This proves that in this case $\sup h(g) = \max \varphi$. If the maximum of $\varphi$ is off of the support of $\nu$ we can still find some $g \in G$ with $\max_{x \in \operatorname{supp}\nu} \varphi(gx) > \max \varphi - \varepsilon$ (by the minimality hypothesis on $M$). The foregoing argument shows that $\sup h(g) \geqslant \max \varphi - 2\varepsilon$ and again we have the desired conclusion.

As we shall see in the next section, the representation (4.1) is the analogue of the Poisson representation for classical harmonic functions in the unit disc in terms of their boundary values. This motivates

DEFINITION 4.2. A *Poisson space* of $(G, \mu)$ is a $(G, \mu)$ space $(M, \nu)$ for which the Poisson representation

$$h(g) = \int_M \varphi(gx)\,d\nu(x)$$

defines an isometry between continuous functions on $M$ and bounded $\mu$-harmonic functions on $G$.

THEOREM 4.2. *If M is a compact metric G-space, $\nu$ a $\mu$-stationary measure on M, then $(M, \nu)$ is a Poisson space of $(G, \mu)$ if and only if it is a boundary space.*

One direction of this is proven in Proposition 4.1. We leave the other direction to the reader. One more ingredient is still needed before the theory of the foregoing section can be invoked to give us the theory of $\mu$-harmonic functions.

PROPOSITION 4.3. *Every bounded $\mu$-harmonic function is represented on $G$ by* (4.1) *for an appropriate Poisson space* $(M, \nu)$, *and an appropriate boundary function $\varphi$.*

*Proof.* Here we shall invoke the assumption that $h(g)$ is left-uniformly-continuous. By this hypothesis the set of right translates $h(gg')$ is equicontinuous and its closure in the topology of uniform convergence on compact sets defines a compact space $M$. $M$ is a $G$-space for we may define a $G$-action on $M$ by $g_1 f(g) = f(gg_1)$. Now we claim that $M$ is a boundary space of $(G, \mu)$. Here it is most convenient to adopt Definition 3.1. Namely attach $M$ to $G$ by setting

$$g_n \to f \Leftrightarrow h(gg_n) \to f(g),$$

the convergence on the right being uniform on compact sets. Since $h(g)$ is $\mu$-harmonic $h(gX_1X_2 \ldots X_n)$ is a martingale and so it converges with probability 1 and the limit is clearly in $M$. Hence $M$ is a boundary space and the stationary measure $\nu$ on $M$ is the distribution of $\lim h(gX_1X_2 \ldots X_n)$ as a random function of $g$. From this and the fact that

$$h(g) = E(h(gX_1X_2 \ldots X_n))$$

we obtain

$$h(g) = \int f(g)\, \mathrm{d}\nu(f)$$

$$= \int_M gf(e)\, \mathrm{d}\nu(f) = \int_M \varphi(gf)\, \mathrm{d}\nu(f)$$

where $\varphi(f) = f(e)$ is a continuous function on $M$. This gives the desired Poisson representation. Note that since $(M, \nu)$ is an equivariant image of the universal boundary space we have, in particular

$$h(g) = \int_{M_0} \hat{h}(g\xi)\, \mathrm{d}\nu_0(\xi)$$

for a unique function $\hat{h}(\xi)$. □

Combining the foregoing results with those of the last section we obtain:

THEOREM 4.4. *For any probability measure* $\mu$ *on* $G$ *there is a universal Poisson space* $(M_0, \nu_0)$ *such that every bounded* $\mu$*-harmonic function is represented almost everywhere by* (4.5) *for a unique continuous boundary function on* $M_0$. *If* $G$ *is a semisimple Lie group with finite center and* $\mu$ *is absolutely continuous then there is a compact homogeneous space* $M_1$ *of* $G$ *and a smooth measure* $\nu_1$ *on* $M_1$ *such that every bounded* $\mu$*-harmonic function admits a Poisson representation in terms of a unique measurable function on* $M_1$.

The reader is referred to [15] for a rather different and instructive approach to this subject.

## 5. Some Examples

If a random walk $W_n(g) = gX_1X_2 \ldots X_n$ is recurrent, naturally its boundary behavior is trivial and the corresponding Poisson boundary $(M_0, \nu_0)$ is trivial; i.e., $M_0$ reduces to a point. There are some cases when the random walks are transient and nevertheless $(M_0, \nu_0)$ is trivial. Notably we have

THEOREM 5.1. *If* $G$ *is Abelian and* $\mu$ *is not supported on a proper closed subgroup of* $G$ *then its universal Poisson boundary is trivial.*

This is equivalent to the assertion that under the stated conditions on $G$ and $\mu$, there are no non-constant bounded harmonic functions on $G$. This is due to Choquet and Deny and their elegant proof goes as follows. Consider the subset of the unit ball of $L^\infty(G)$ consisting of $\mu$-harmonic functions. In the weak $*$-topology this forms a compact convex set and is spanned by extremals. However each translate of a $\mu$-harmonic function is again $\mu$-harmonic, so the formula

$$h(x) = \int h(x+t)\, d\mu(t) \tag{5.1}$$

expresses $h$ as a linear combination of other $\mu$-harmonic functions. If $h(x)$ is extremal we must have $h(x) = h(x+t)$ for almost all $t$ with respect to $\mu$. Since the support of $\mu$ generates $G$, $h$ must be constant.

Another example in which a complete description of the universal Poisson boundary may be given is the following. Let $G$ be a semisimple connected Lie group with finite center and let $K$ be a maximal compact subgroup. Assume that $\mu$ is an absolutely continuous measure on $G$ satisfying $k\mu=\mu$ for $k\in K$. This means that $\mathrm{d}\mu(g)=\lambda(g)\,\mathrm{d}g$ where $\lambda(kg)=\lambda(g)$. Consider the universal Poisson space $(M_1, \nu_1)$ described in Theorem 3.6, according to which $M_1$ is a homogeneous space of $G$ on which $K$ acts transitively. We have $\mu * \nu_1=\nu_1$ and since $k\mu=\mu$ when $k\in K$ we find that $\nu_1$ is also $K$-invariant. Since $K$ is transitive on $M_1$, $\nu_1$ must be the unique $K$-invariant measure on $M_1$, say, $\nu_1=m_{M_1}$. Since $(M_1, \nu_1)$ is a boundary space there exists some sequence of elements in $G$ with $g_n\nu_1\to$point measure. If this is true when $\nu_1$ is a smooth measure supported on all of $M_1$ it is easy to conclude that this property holds for all measures on $M_1$. But now $M_1$ is a covering space of $B(G)$ where $B(G)$ is universal for this property. From this we conclude that $M_1=B(G)$. Let us now write $\nu_1=m_{M_1}=m_B$. Thus we have fully identified the boundary spaces for $K$-invariant smooth measures $\mu$. This translates into a characterization of the $\mu$-harmonic functions on $G$.

To complete the characterization of bounded $\mu$-harmonic functions in this case we introduce another notion. If $f(g)$ is a function in $L^\infty(G)$, form the weak $*$-closure of the set of its right translates $\{f(gg')\}\subset L^\infty(G)$ together with their convex linear combinations. Call this convex set $V_f$. In the weak $*$-topology $V_f$ is a compact set. $G$ acts on $V_f$ by right translation: $R_{g_1}\varphi(g)=\varphi(gg_1)$. (When $f(g)$ is right-uniformly-continuous $V_f$ consists of continuous functions).

Now suppose $h(g)$ is a $\mu$-harmonic function on $G$. There exists a measurable boundary function $\psi(x)$ in $B(G)$ with

$$(5.2)\qquad h(g)=\int\limits_{B(G)}\psi(gx)\,\mathrm{d}m_B(x).$$

From (5.2) we can deduce a generalized mean-value property:

$$\int\limits_K h(gkg')\,\mathrm{d}k=\iint\limits_{KB(G)}\psi(gkg'x)\,\mathrm{d}m_B(x)\,\mathrm{d}k=\int\limits_{B(G)}\psi(gy)\,\mathrm{d}m'(y)$$

where $m'$ is defined by

$$\int\varphi(y)\,\mathrm{d}m'(y)=\iint\varphi(kg'x)\,\mathrm{d}m_B(x)\,\mathrm{d}k.$$

But this shows that $m'$ is $K$-invariant, so $m' = m_B$. This gives

$$\text{(5.3)} \qquad \int_K h(gkg')\,\mathrm{d}k = h(g).$$

From this it follows that $V_h$ is irreducible as a convex $G$-space. In other words no proper closed convex subset of $V_h$ is invariant under $G$. For from (5.3) we conclude that for any $f \in V_h$

$$\text{(5.4)} \qquad h(g) = \int_K f(gk)\,\mathrm{d}k$$

this being true for $f = R_{g'}h$ and these span $V_h$. Now (5.4) shows immediately that $V_h$ is irreducible.

Conversely, suppose that $h(g)$ is a bounded function on $G$ with the property that $V_h$ is irreducible and suppose furthermore that $h(gh) = h(g)$ for $k \in K$. We invoke an important property of the minimal parabolic subgroup $P \subset G$ for which we have $B(G) = G/P$. $P$ is an amenable group which means that whenever $P$ acts on a compact convex set by affine maps it has a fixed point. Thus $V_h$ contains a function $\tilde{\psi}(g)$ *with* $\tilde{\psi}(gu) = \tilde{\psi}(g)$ for $u \in P$. $\tilde{\psi}$ corresponds to a function $\psi$ on $B(G) = G/P$. By irreducibility of $V_h$, we can approximate $h(g)$ by some convex combination $h'(g) = \Sigma\alpha_i \tilde{\psi}(gg_i)$; since $h(g) = h(gk) = \int_K h(gk)\,\mathrm{d}k$ we can also approximate $h(g)$ by

$$h''(g) = \Sigma\alpha_i \int \tilde{\psi}(gkg_i)\,\mathrm{d}k = \int_B \tilde{\psi}(gx)\,\mathrm{d}m_B(x).$$

This shows that $h(g)$ satisfies (5.2). Again we conclude that it satisfies (5.3). Now a function satisfying (5.3) is in fact $\mu$-harmonic for any $\mu$ in the class we are describing.

These functions are related to classical harmonic function as follows. Every $\mu$-harmonic function satisfies

$$h(g) = \int h(gg')\lambda(g')\,\mathrm{d}g' = \int h(gg')\lambda(k^{-1}g')\,\mathrm{d}g'$$
$$= \int h(gkg')\lambda(g')\,\mathrm{d}g' = h(gk)$$

for $k \in K$. So $h(g)$ corresponds to a function on the Riemannian symmetric space $G/K$. We have the correspondence $f \to \bar{f}$ of functions $f(g)$ on $G$

satisfying $f(gk)=f(g)$ and functions on $G/K$ satisfying $\Delta h=0$ where $\Delta$ is a Laplace–Beltrami operator, i.e., an elliptic second-order differential operator on $G/K$. The corresponding function $h(g)$ will be annihilated by an invariant elliptic operator on $G$. If we solve the heat equation for this operator we find that $h(g)$ is invariant under convolutions by functions $\lambda(t, g)$ where $\lambda(t, kg)=\lambda(t, g)$ for $k \in K$. But this means that $h(g)$ is in the class of $\mu$-harmonic functions under consideration. Conversely, if $h(g)$ is $\mu$-harmonic and bounded we have seen that (5.3) is valid. But it follows readily from this that $\bar{h}(g)$ is annihilated by any Laplace–Beltrami operator on $G/K$. Putting this together we have

THEOREM 5.2. *If $h(g)$ is a bounded function on $G$ satisfying $h(gk)=h(g)$ for $k \in K$, the following conditions are equivalent*:
(i) *Setting $\bar{h}(gK)=h(g)$, $\bar{h}$ satisfies $\Delta\bar{h}=0$ for some Laplace–Beltrami operator on $G/K$.*
(ii) *$h(g)$ satisfies $h(g)=\int h(gg')\lambda(g')\,dg'$* for some *$\lambda(g)\geqslant 0$ with $\lambda(kg)=\lambda(g)$ for $k \in K$, $\int \lambda(g)\,dg=1$.*
(iii) *$h(g)=\int_{B(G)}\psi(gu)\ dm_B(u)$ for a bounded measurable function $\psi(u)$.*
(iv) *$h(g)=\int_K h(gkg')\,dk$ for all $g' \in G$*
(v) *The convex set $V_h$ is irreducible.*

We call these functions simply harmonic functions on $G$ or on $G/K$.

The foregoing theorem tells among other things where to look for boundary values of harmonic function in Riemannian symmetric spaces. We remark that there are examples of bounded symmetric domains in complex space whose universal Poisson space does not coincide with any part of the natural topological boundary of the domain as a subset of $C^n$. This fact explains the anomalous boundary behavior of harmonic functions in these domains.

The boundary $B(G)$ can be determined quite explicitly. When $G=GL(n, R)$, $B(G)$ is the ‘flag manifold’ whose points consist of all flags $V_1 \subset V_2 \subset \cdots \subset V_{n-1}$, where $V_r$ is an $r$-dimensional subspace of $R^n$. $B(SL(n, R))$ has as an equivariant image the projective space $P^{n-1}$ as well as the various Grassman spaces $G_{n,r}$. This, by the way, shows that in considering the boundary behavior of random products of matrices the direction of the column matrices do not give the full picture, since they correspond to the boundary space $P^{n-1}$.

With regard to the boundary spaces of matrix groups we mention without proof (see [6]):

THEOREM 5.3. *Let G be a closed non-compact irreducible subgroup of SL(n, R). If μ is a measure on G not supported by a proper closed subgroup, then some Grassman space* $G_{n,r}$, $1 \leqslant r \leqslant n-1$, *is a boundary space for* $(G, \mu)$.

We conclude this section with some applications to the theory of lattice subgroups of Lie groups. We begin with the following proposition.

PROPOSITION 5.4. *Let μ be a measure on the group G and suppose G′ is a subgroup of finite index in G. Consider the random walk* $W_n(g) = gX_1X_2 \ldots X_n$ *and let*

$$N = \inf \{n \mid W_n(e) \in G'\}.$$

*N is a random variable which is finite a.e. and set* $X' = X_N$, *and let μ′ be the distribution of X′. μ′ is a measure on G′ and*

(i) *the restrictions of bounded μ-harmonic functions in G to G′ are precisely the μ′-harmonic functions on G′;*

(ii) *the universal boundary space* $(M_0, \nu_0)$ *of* $(G, \mu)$ *is the universal boundary space of* $(G', \mu')$.

The proof of this is relatively straightforward. A partial analogy is given by

THEOREM 5.5. *Let G be a semisimple Lie group with finite center and Γ a lattice subgroup of G. There exists a measure μ on Γ such that the universal boundary space of* $(\Gamma, \mu)$ *in the measurable sense coincides with* $(B(G), m_B)$. *A bounded function on X is μ-harmonic if and only if it is the restriction of a harmonic function on G (with respect to some maximal compact subgroup K).*

This theorem plays a role in studying rigidity questions for the group Γ. Suppose we have an irreducible representation of Γ by an unbounded group of matrices in SL(n, R). According to Theorem 5.3, some $G_{n,r}$ is a boundary of $(\Gamma, \mu)$. Hence we will obtain an equivariant measurable map from $B(G)$ to $G_{n,r}$. Under various extra hypotheses Margulis is able to deduce from this that the representation extends to a representation of $G \supset \Gamma$. We refer to [13] for details.

*Hebrew University, Jerusalem*

## REFERENCES

[1] Azencott, R., *Espaces de Poisson des Groupes Localement Compacts*, Lecture Notes in Math. 148, Springer, 1970.

[2] Brunel, A., Crepel, P., Guivarc'h, Y., and Keane, M., 'Marches aleatores recurrentes sur les groupes localement compacts', *Comptes Rendus* **275** (1972).

[3] Furstenberg, H., 'A Poisson formula for semi-simple Lie groups', *Ann. of Math.* (2) **77** (1963), 335–386.

[4] Furstenberg, H., 'Non-commuting random products', *Trans. Amer. Math. Soc.* **108** (1963), 377–428.

[5] Furstenberg, H., 'Boundaries of Riemannian symmetric spaces', *Symmetric Spaces* Short courses presented at Washington University, New York, 1972.

[6] Furstenberg, H. 'Boundary theory and stochastic processes on homogeneous spaces' in *Harmonic Analysis on Homogeneous Spaces*, Symposia in Pure Math., A.M.S., Williamstown, Mass., 1973.

[7] Glasner, S., *Proximal Flows*, Lecture Notes in Math. 517, Springer, 1976.

[8] Grenander, U., *Probabilities on Algebraic Structures*, Wiley, New York, 1963.

[9] Guivarc'h, Y., 'Une loi des grands nombres pour les groupes de Lie', to appear.

[10] Guivarc'h, Y., Keane, M., and Roynette, B., *Marches Aleatoires sur les groupes de Lie*, Lecture Notes in Math. 624, Springer, 1977.

[11] Kingman, J. F. C., 'Subadditive ergodic theory', *Annals of Probability* **1**, 883–909.

[12] Raugi, A., 'Fonctions harmoniques sur les groupes localement compacts a base denombable', *Bull. Soc. Math de France* (1977)

[13] Tits, J., 'Travaux de Margulis sur les sous-groupes discrets des groupes de Lie', Seminaire Bourbaki, 1975/76, No. 482.

[14] Tutubalin, V. N., 'Some theorems of the type of the law of large numbers', *Theory of Probability*, 1969, pp. 313–319.

[15] Zimmer, R., 'Amenable group actions and an application to Poisson boundaries of random walks', *J. Func. Anal.* **27**, (1978) 350–372.

# SUBJECT INDEX

MATHEMATICAL PHYSICS AND APPLIED MATHEMATICS

*Editors:*

M. FLATO *(Université de Dijon, Dijon, France)*
R. RĄCZKA *(Institute of Nuclear Research, Warsaw, Poland)*

1. M. Flato, Z. Maric, A. Milojevic, D. Sternheimer, and J. P. Vignier (Eds.), *Quantum Mechanics, Determinism, Causality, and Particles.* 1976, x+252 pp.
2. K. Maurin and R. Rączka (Eds.), *Mathematical Physics and Physical Mathematics.* 1976, xviii+504 pp.
3. M. Cahen and M. Flato (Eds.), *Differential Geometry and Relativity.* 1976, xii+304 pp.
4. M. Flato, C. Fronsdal, and K. A. Milton (Eds.), *Selected Papers (1937–1976) of Julian Schwinger.* 1979, xxviii+413 pp.
5. J. A. Wolf, M. Cahen, and M. De Wilde (Eds.), *Harmonic Analysis and Representations of Semisimple Lie Groups.* 1980, viii+495 pp.